BIOLOGICAL SAMPLING IN THE DEEP SEA

BIOLOGICAL SAMPLING IN THE DEEP SEA

Edited by

Malcolm R. Clark, Mireille Consalvey and Ashley A. Rowden

National Institute of Water and Atmospheric Research, Wellington, New Zealand

WILEY Blackwell

Registered Office
John Wiley & Sons, Ltd, The Atrium, Southern Gate, Chichester, West Sussex, P019 8SQ, UK

Editorial Offices
9600 Garsington Road, Oxford, OX4 2DQ, UK
The Atrium, Southern Gate, Chichester, West Sussex, P019 8SQ, UK
111 River Street, Hoboken, NJ 07030-5774, USA

For details of our global editorial offices, for customer services and for information about how to apply for permission to reuse the copyright material in this book please see our website at www.wiley.com/wiley-blackwell.

Library of Congress Cataloging-in-Publication Data

Names: Clark, Malcolm R., 1956– editor. | Consalvey, Mireille, editor. | Rowden, A. A. (Ashley Alun), editor.
Title: Biological sampling in the deep sea / edited by Malcolm R. Clark, Mireille Consalvey, Ashley A. Rowden.
Description: Hoboken, New Jersey : Wiley-Blackwell, 2016. | Includes bibliographical references and index.
Identifiers: LCCN 2015026507 | ISBN 9780470656747 (hardback)
Subjects: LCSH: Abyssal zone. | Deep-sea biology. | Abyssal zone–Research–Equipment and supplies. | Deep-sea ecology. | Habitat (Ecology) | BISAC: NATURE / Animals / Marine Life.
Classification: LCC QH91 .B49 2016 | DDC 578.77/9–dc23 LC record available at http://lccn.loc.gov/2015026507

A catalogue record for this book is available from the British Library.

Wiley also publishes its books in a variety of electronic formats. Some content that appears in print may not be available in electronic books.

Cover image: The deep-sea research vessel *Tangaroa*, typical of many offshore vessels, deploying camera equipment during a survey of seamounts off New Zealand. [photo Malcolm Clark, NIWA].

Set in 9.5/11.5 Times by SPi Global, Pondicherry, India
Printed and bound in Singapore by Markono Print Media Pte Ltd

1 2016

Contents

Contributors

Sophie Arnaud-Haond, Institut Francais de Recherche pour l'Exploitation de la Mer, Brest, France

Neil W. Bagley, National Institute of Water and Atmospheric Research, Wellington, New Zealand

Maria Baker, National Oceanography Centre and University of Southampton, Southampton, United Kingdom

Elanor M. Bell, Australian Marine Mammal Centre, Australian Antarctic Division, Hobart, Australia

Mairi Best, Consultant on Ocean Observing and Research Infrastructures

Brian J. Bett, National Oceanography Centre, Southampton, United Kingdom

Jérôme Blandin, Institut Francais de Recherche pour l'Exploitation de la Mer, Brest, France

David A. Bowden, National Institute of Water and Atmospheric Research, Wellington, New Zealand

Nils Brenke, German Centre for Marine Biodiversity Research, Wilhelmshaven, Germany

Tin-Yam Chan, National Taiwan Ocean University, Keelung, Taiwan, Republic of China

Bernd Christiansen, Universität Hamburg, Hamburg, Germany

Malcolm R. Clark, National Institute of Water and Atmospheric Research, Wellington, New Zealand

Mireille Consalvey, Victoria University of Wellington, Wellington, New Zealand

Ian J. Doonan, National Institute of Water and Atmospheric Research, Wellington, New Zealand

Brian Harley, Centre for Environment, Fisheries and Aquaculture Science, Lowestoft, United Kingdom

Maia Hoeberechts, Ocean Networks Canada, University of Victoria, Victoria, Canada

Nancy Jacobsen Stout, Monterey Bay Aquarium Research Institute, Moss Landing, United States of America

Alan J. Jamieson, Oceanlab, University of Aberdeen, Aberdeen, United Kingdom

Daniel O.B. Jones, National Oceanography Centre, Southampton, United Kingdom

S. Kim Juniper, Ocean Networks Canada, University of Victoria, Victoria, Canada

Stefanie Kaiser, German Centre for Marine Biodiversity Research, Wilhelmshaven, Germany

Christopher Kelley, Hawaii Undersea Research Laboratory, University of Hawaii, Honolulu, United States of America

Terrence Kerby, Hawaii Undersea Research Laboratory, University of Hawaii, Honolulu, United States of America

Kostas Kiriakoulakis, John Moores University, Liverpool, United Kingdom

Geoffroy Lamarche, National Institute of Water and Atmospheric Research, Wellington, New Zealand

Peter Lamont, Scottish Association for Marine Science and Scottish Marine Institute, Oban, United Kingdom

Dhugal J. Lindsay, Japan Agency for Marine-Earth Science and Technology, Yokosuka, Japan

Marjolaine Matabos, Institut Francais de Recherche pour l'Exploitation de la Mer, Brest, France

Gui M. Menezes, University of the Azores, Horta, Portugal

Lenaick Menot, Institut Francais de Recherche pour l'Exploitation de la Mer, Brest, France

John S. Mitchell, National Institute of Water and Atmospheric Research, Wellington, New Zealand

Christian Mohn, Aarhus University, Roskilde, Denmark

Tina N. Molodtsova, P.P. Shirshov Institute of Oceanology, Moscow, Russia

Bhavani E. Narayanaswamy, Scottish Association for Marine Science and Scottish Marine Institute, Oban, United Kingdom

Timothy D. O'Hara, Museum Victoria, Melbourne, Australia

Alan R. Orpin, National Institute of Water and Atmospheric Research, Wellington, New Zealand

Arne Pallentin, National Institute of Water and Atmospheric Research, Wellington, New Zealand

Benoît Pirenne, Ocean Networks Canada, University of Victoria, Victoria, Canada

Eva Ramirez-Llodra, Norwegian Institute for Water Research, Oslo, Norway

Bertrand Richer de Forges, Muséum National d'Historie Naturelle, Paris, France

Katleen Robert, National Oceanography Centre and University of Southampton, Southampton, United Kingdom

Ashley A. Rowden, National Institute of Water and Atmospheric Research, Wellington, New Zealand

Henry A. Ruhl, National Oceanography Centre and University of Southampton, Southampton, United Kingdom

Pierre-Marie Sarradin, Institut Francais de Recherche pour l'Exploitation de la Mer, Brest, France

Jozée Sarrazin, Institut Francais de Recherche pour l'Exploitation de la Mer, Brest, France

Thomas A. Schlacher, The University of the Sunshine Coast, Maroochydore, Australia

Stefano Schiaparelli, University of Genoa, Genoa, Italy

Kareen E. Schnabel, National Institute of Water and Atmospheric Research, Wellington, New Zealand

Timothy M. Shank, Woods Hole Oceanographic Institution, Woods Hole, United States of America

Michael F. Sigler, Alaska Fisheries Science Center, National Oceanic and Atmospheric Administration, Juneau, United States of America

Samantha Smith, Blue Globe Solutions, Toronto, Canada

Karen I. Stocks, Scripps Institution of Oceanography, La Jolla, United States of America

Andrew D. Thaler, Duke University, Beaufort, United States of America

Derek P. Tittensor, World Conservation Monitoring Centre, Cambridge, United Kingdom

Paul A. Tyler, National Oceanography Centre and University of Southampton, Southampton, United Kingdom

Cindy Lee Van Dover, Duke University, Beaufort, United States of America

Sybille van den Hove, Autonomous University of Barcelona, Barcelona, Spain

Michael Vardaro, Rutgers University, New Brunswick, United States of America

Martin White, National University of Ireland, Galway, Ireland

Foreword

Prepare for an orgy of the senses. Such anticipation might surprise a buyer of a thick book titled *Biological Sampling in the Deep Sea*. But this book is inevitably a carousel of taste, smell, touch, sight and sound.

Life in the sea probably began about three billion years ago, and it probably began deep and moved up. To survive, the early bacteria had to have knowledge of their environment, which means they were endowed with sensors and computing machinery. The limits of sensors in quality and quantity, and the computing system behind them, essentially bound limits to knowledge. What strikes me is that while humans knew little about the deep sea until the twentieth century, other forms of life knew a lot. The job of humans during the past hundred years or so has been to catch up and surpass other life in knowledge of the oceans. To catch up, we need to use all the senses, as ocean life itself does.

The discoveries about diversity, distribution and abundance of marine life in the Census of Marine Life program, carried out by the 60 authors of the book and their colleagues, astonish with the sensory and computational brilliance of the organisms themselves. Now the book astonishes us too with the brilliance of humanity as a marine animal. First, we learn that nature had already overcome many limits to knowledge of the oceans. Many other forms of life could sense what humans could not. Next, we learn that humanity has largely caught up and now even zoomed ahead.

Our breakthroughs owe to sensors and their elaborators, bound to mechanisms and materials used to build them. What do humans now see, hear and touch deeply? With gravity measures made from satellites and multibeam sonars mounted on the hulls of vessels, we map the seafloor in fine detail. We invent new devices to carry sensors, fixed, drifting, towed and propelled. With passive acoustics, we locate some kinds of life and with active acoustics other kinds. With taste, we lure animals to our cameras. We sieve seawater for DNA and use short sequences to ascertain what species have operated recently nearby. We knit our deep-sea sensors into informationally connected monitoring arrays that allow us not only to spot organisms but also to encompass ecosystems. We use robotic arms, hooks and nets to capture precious samples to bring to the surface world, where we store specimens and build data libraries.

Thus, we have vastly expanded the knowable for humans beyond our own senses. That is the story of the 19 chapters of this sensuous book. Thanks to editors Malcolm Clark, Mireille Consalvey and Ashley Rowden for this astonishing extension of our limits to knowledge. The tools and techniques codified here promise many more decades of discoveries that allow us to appreciate, and hopefully protect, the ancient beauty and genius of the deep sea.

Jesse H. Ausubel
Director, Program for the Human Environment,
The Rockefeller University
Program manager, Alfred P. Sloan Foundation,
Census of Marine Life (2000–2011)

Preface

The ocean covers 71% of the surface of the earth and the deep sea (considered here to be at depths greater than the continental shelf at 200–300 m) comprises the vast majority of this area. Seafloor habitats in the deep sea are many and varied, and include extensive areas of lower continental slope incised with canyons, beyond which there are large numbers of seamounts among plateaux, ridges, rises, plains and trenches, as well as hydrothermal vents and cold seeps. Until the late nineteenth century the depths beyond the photic zone were thought to be largely devoid of life. However, this concept has been disproven as scientific expeditions have discovered, and continue to find, a specialist deep-sea fauna that is among the most diverse on the planet.

Scientific knowledge of the deep sea has increased rapidly, but there exist large gaps in sampling coverage. As researchers have collectively worked towards increasing spatial coverage, inconsistencies have become apparent in how different habitats and regions have been sampled. Ultimately the extent to which data can be integrated will constrain how much our understanding of deep-sea ecosystems can develop. Furthermore, the pace of human impacts and changes in the deep sea are outstripping knowledge growth. There is growing pressure on deep-sea researchers to obtain data to understand the effects of fisheries and other human impacts that is vital for effective management and conservation.

In the wider scientific literature and online there already exist a number of publications and 'grey literature' reports discussing deep-sea sampling. In particular there are several methodological books (e.g. Eleftheriou & McIntyre, 2005; Eleftheriou, 2013; Danavaro, 2010) and general deep-sea texts (e.g. Gage & Tyler, 1991; Tyler, 2003). However, this book represents the first comprehensive compilation of deep-sea sampling methodologies for a range of habitats, and by dedicating a book specifically to deep-sea methods we are recognizing that the field of deep-sea research now warrants its own volume. The book fills a niche in the scientific and management communities, reviewing the real-life applications of current (and in some instances developing) deep-sea sampling tools and techniques across a range of deep-sea habitats and under a variety of conditions. In creating this book we have been able to draw upon the experiences of those at the 'coal face' of deep-sea sampling, expanding on the existing methodological texts while encompassing a level of detail often omitted from journal publications.

Ultimately we hope that the book will promote international consistency in sampling approaches and data collection, advance the integration of information into global databases, and facilitate improved data analyses and consequently uptake of science results for management and conservation of the deep sea.

Origin and scope of the book

The concept of this book was initially discussed by a small group of researchers comprising a 'Standardization Working Group' of the Census of Marine Life field project on seamounts (CenSeam). However, it rapidly became apparent that we could not, and should not, sample seamounts in isolation and the scope of the book was widened to encompass all deep-sea habitats.

The majority of sampling methods covered in the book target the mega- and macrofaunal assemblages associated with the deep seafloor. However, any study of the deep sea cannot exclude the wider influence of the entire water column and we also consider pelagic and environmental sampling. The main focus of the book by its very nature is biological sampling, but deep-sea research is a truly multidisciplinary venture. However, to encompass oceanographic and geological sampling in any detail would call for additional volumes that were beyond the realistic scope of our efforts. Nevertheless we hope all biologists can engage with other disciplines and encourage ongoing collaboration to obtain a wider understanding of deep-sea ecosystems.

In gathering authors to contribute to this book we have tried to cover as wide a range habitats and sampling equipment as possible. However, the content reflects personal experience and inevitably there will be omissions, but hopefully the book manages to build a solid foundation for future discussions of how to develop and progress biological sampling in the deep sea.

We hope that this book will appeal to a range of readers. Early career researchers can benefit from the experiences shared by those who have been on the trawl deck for many years, wisdom that is often not recorded in scientific papers and is often learnt the hard way. We also hope that the book will provide a scholarly platform and a solid foundation for students, as well as a valuable resource for environmental and science managers and policy-makers in detailing not only what tools are available to address issues in deep-sea management and conservation, but also in explaining some of the inherent challenges. The content of this book is based on the collective experiences of many deep-sea scientists, but we anticipate that even the most seasoned researcher will find this book of use. By documenting even the seemingly simplest of steps we might better identify where differences in approach occur, thereby providing opportunities for discussion and standardization across the research community to facilitate more robust data comparisons and analyses. Finally, the role of every research ship's complement in making deep-sea research happen cannot be understated, and we hope that the book may find a place on board research vessels around the world as a reference to crews for the rationale behind some of the deployment strategies, and subsequent use of data.

The book cannot deal with every situation or gear type. Hence the chapters are cross-referenced and contain an extensive list of citations. The references included in the book are intended to give readers access to the main papers that can facilitate more detailed investigation of particular issues.

The book has been 5 years in the making, as the authors involved have tried to juggle the demands of their scientific work with the preparation of the various chapters. Research never stands still, and hence technological and sampling advances have occurred over that time period. We have tried to incorporate such changes wherever possible, and trust there are no major omissions that affect current 'best practice' advice or information.

Structure of *Biological Sampling in the Deep Sea*

The first part of the book sets the background scene with Chapters 1 and 2 considering general characteristics of deep-sea habitat and fauna, in the context of their influence on sampling methodologies. Chapter 3 moves towards the details of deep-sea sampling and discusses survey and sampling design. Recognizing that the motivations behind research surveys will vary, this book does not aim to be prescriptive, but rather to provide a toolbox of experience for users to utilize; with this in mind Chapter 3 draws heavily on case studies which bring to bear experiences from several research institutes.

Chapters 4–14 cover the major deep-sea sampling methods and equipment that are employed today. A 'hands on' description of the sampling operations includes information that is largely overlooked in the published literature. Each chapter encompasses a detailed description of the method, including information on gear specifications, modifications as well as notes on gear maintenance and handling; and also specific advice for application to different habitats and fauna. Treating sampling on the basis of gear type means there is some duplication and repetition, but this approach enables researchers to learn quickly about a particular gear type available to them, and keeps all relevant information together in concise sections. This structure emphasizes the need for consistency and comparability in the set-up and use of particular equipment, which is one of the priority objectives of the book. Chapter authors also detail what information must be captured for future cross-comparisons and analyses, as well as offering guidance on quality assurance. Sample sorting and processing is considered in the context of each method, and Chapter 15 includes a detailed discussion on the preservation and curation of biological samples.

However, important as ocean-going research and shipboard sampling are, ultimately they are only the beginning of the sequence of scientific research. Chapter 16 tackles what happens when ashore, and outlines best practices for developing and implementing deep-sea data management. Chapter 17 builds on this discussion to consider the peculiarities of deep-sea data analysis, with the authors drawing heavily on their own experiences and lessons learnt about data-poor situations and what to do and what not to do.

Scientific curiosity drove much of the early exploratory deep-sea research, but the current-day motivation to better understand this environment is now driven by an increasing need to manage and conserve deep-sea habitats and communities given a wide suite of anthropogenic threats. The knowledge base needed to effectively manage and conserve the deep sea is underpinned by good sampling design and practice, as well as effective transfer of information. Chapter 18 considers management approaches and outlines a process for developing biological sampling programmes in the deep sea that deliver both what scientists desire and managers or policy-makers need.

Ongoing technological advances drive continued exploration of the deep, and just as early researchers might never have anticipated the wealth of life that we now know exists, the deep sea is so poorly sampled we know there is a lot more to discover. Therefore the book ends with Chapter 19 considering how advances in technology and capability might drive further exploration, and what the future of deep-sea sampling might hold.

MALCOLM R. CLARK, MIREILLE CONSALVEY, AND ASHLEY A. ROWDEN

Acknowledgements

We thank the contributing chapter authors and many reviewers for their enthusiasm and efforts in bringing this book to fruition. CenSeam researchers are acknowledged for their role in developing the idea and concept, especially Alan Williams (CSIRO) for generating the original book proposal and helping launch the process. The contributions of other deep-sea Census of Marine Life projects are substantial, and we appreciate the willing involvement and commitment of ChEss, CeDAMAR, MAR-ECO and COMARGE. The book has taken 5 years to prepare and publish, and at times authors must have wondered if the idea and their efforts had slowly died. Their commitment to the process, and their acceptance of the delays and changes, are greatly appreciated.

Funding for much of this effort was from the Sloan Foundation, which was always supportive through the umbrella of the Census of Marine Life. The editors also received some funding for time from a NIWA research programme (Vulnerable Deep-Sea Communities, FRST contract CO1X0906).

We also express our thanks for the assistance of Wiley Blackwell's publishing staff, especially Nigel Balmforth, Carys Williams, Kelvin Matthews, Harriet Stewart-Jones, Lea Abot, and Sandeep Kumar at SPi Global. The undertaking took much longer than originally anticipated, and their patience and understanding was greatly appreciated.

References

Danavaro, R. (ed.) (2010). *Methods for the study of deep-sea sediments, their functioning and biodiversity*. Boca Raton, FL: CRC Press.

Eleftheriou, A. (ed.) (2013). *Methods for the study of marine benthos* (4th edn). Oxford: Wiley Blackwell.

Eleftheriou, A., McIntyre, A. (eds) (2005). *Methods for the study of marine benthos* (3rd edn). Oxford: Blackwell Science.

Gage, J.D., Tyler, P.A. (1991). *Deep-sea biology: a natural history of organisms at the deep-sea floor*. Cambridge, UK: Cambridge University Press.

Tyler, P.A. (ed.) (2003). *Ecosystems of the deep oceans*. Amsterdam: Elsevier.

Chapter 1
Deep-Sea Benthic Habitats

Paul A. Tyler[1], *Maria Baker*[1] *and Eva Ramirez-Llodra*[2, 3]*
[1] National Oceanography Centre and University of Southampton, Southampton, United Kingdom
[2] Institut de Ciències del Mar, Barcelona, Spain
[3] Norwegian Institute for Water Research, Oslo, Norway

Abstract

The oceans were observed to be deep during the great age of exploration in the early to mid-nineteenth century. Subsequent exploration demonstrated that the ocean was bisected by underwater mountain ranges and dotted with abyssal hills. With the advent of the echosounder and latterly multichannel swath bathymetry, we now know that the deep ocean has topography as diverse as found on land. In the last 30 years, with an increase in deep-sea scientific activity and the use of underwater vehicles, we have learned that the deep sea consists of a series of habitats and ecosystems interconnected by hydrography and topography. The more recent challenges have been how to sample and analyse these separate habitats and ecosystems. This chapter describes the different environments and briefly outlines the main methods of sampling for each habitat or ecosystem. More detailed aspects of these sampling methods are found in subsequent chapters.

Keywords sampling, continental slope, canyon, cold-water coral reef, cold seep, mud volcano, mid-ocean ridge, hydrothermal vent, abyssal plain, trench, seamount

1.1 Introduction

Since the great age of oceanic exploration sailors have recognized that the oceans get deeper as their ships move away from the coast. The abruptness of this increase in depth was known to vary throughout the global ocean, with extensive shallow shelves off some coastlines, such as northwestern Europe, and precipitous increases in depth close to shore in other areas, as seen in the southeastern Pacific. In an age of cast lines to determine depth, data on oceanic depths was limited and accumulated slowly; it was only when there was a technology imperative, such as the seabed survey by HMS *Cyclops* in the North Atlantic prior to the laying of the first transatlantic telegraph cable in 1857, that our knowledge increased. This survey demonstrated that, on leaving Europe, the cable would cross the continental shelf, sink across the continental slope and cover most of its distance across the Atlantic on abyssal plains at depths greater than 5000 m (Murray & Hjort, 1912). However, the great depth of the Atlantic appeared to be bisected by a linear structure we now call the Mid-Atlantic Ridge that forms an element in a submarine mountain chain that circles the globe. These 'primitive' methods of sounding continued well into the twentieth century, so our knowledge of the ocean seafloor was composed of spot depths on naval charts.

*This chapter was submitted in February 2012.

Biological Sampling in the Deep Sea, First Edition. Edited by Malcolm R. Clark, Mireille Consalvey and Ashley A. Rowden.

The advent of the echosounder in the 1930s gave a continuous record of the depth of the seabed traversed, and increased the resolution of depth measurements considerably. It was the compilation of all the data to date in the early 1960s that gave rise to the iconic seafloor figure produced by Bruce Heezen and Marie Tharpe, which integrated for the first time depth measurements into a single three-dimensional figure. And what a figure this turned out to be. This and subsequent maps of the world's seafloor revealed that the oceans were more complicated than simple deep basins covering ~70% of the surface of the Earth (Fig. 1.1). The most prominent features were apparently flat plains between 3 and 6 km depth that cover 50% of the surface of the Earth (Figs 1.1 and 1.2).

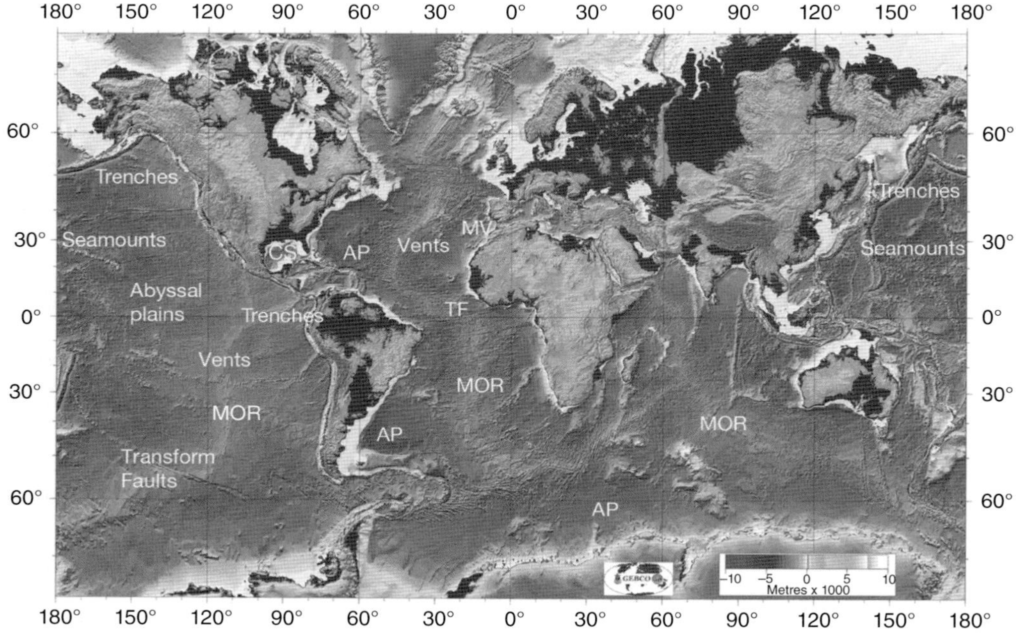

Fig. 1.1 Bathymetry of the global ocean showing a selection of the main features. MOR, mid-ocean ridge; AP, abyssal plain; TF, transform faults; MV, mud volcanoes.

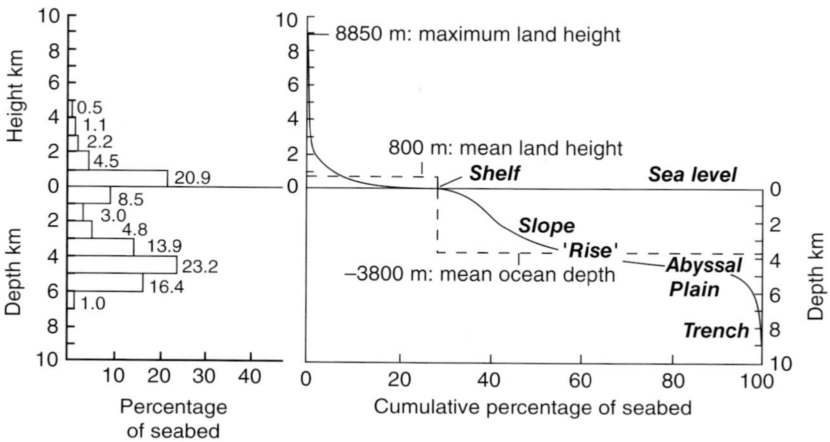

Fig. 1.2 Hypsographic curve of the percentage of the global ocean at 1000 m depth intervals.

In many places, these plains were not only separated by submarine mountain ranges but were pockmarked by individual mountains or groups of mountains now termed seamounts. Particularly obvious in the Pacific was a series of very deep linear basins referred to as trenches, which are now known to occupy only 1% of the ocean but extend to the greatest depths of just short of 11 000 m. As we zoom in on the seabed topography, we see a steep slope between the shallow continental shelf and ocean basin referred to as the continental slope, the sediment-draped base being called the continental 'rise' or 'fan', although there is some dispute about this terminology. From the 1990s, widespread use of precise satellite navigation and a new generation of multibeam echosounders producing 3D swath images of the sea floor (Fig. 1.3) revealed further details. The continental slope may be smooth or bisected by submarine canyons that, at first glance, appear as gashes normal to the submarine contours. The continental slope is also a major site for cold seeps, one of the forms of chemosynthetically driven ecosystems now discovered, particularly in sedimentary areas. A similarly functional ecosystem is found along the mid-ocean ridges (MORs) and other volcanic areas, such as back-arc basins and subduction zones. Here, hydrothermal vents, so-termed owing to the expulsion of hot fluid, rich in sulfide, methane and dissolved minerals, support abundant local communities. Related faunal communities are found on even more localized habitats, such as whale falls and other organic inputs such as wood and kelp (at a scale of metres) that form their own ephemeral ecosystem on the seabed.

It is the scale variation from thousands of kilometres on abyssal plains or along mid-ocean ridges to the small decimetre scale of vents and organic falls, together with the type of substratum, whether sedimentary, rock or biogenous, that drives the different sampling strategies employed to understand deep-sea ecosystems. It is only in the last 20 years or so that we have truly resolved the deep sea into its component ecosystems. Although available information is still limited and we do not know the exact extension of most of these ecosystems, deep-sea research has greatly developed in the last decades (Danovaro *et al.*, 2015; Ramirez-Llodra *et al.*, 2010a).

We will now describe the main features that will define sampling strategies for these different systems. Our approach is to examine the deep sea by moving out from the shore over the continental

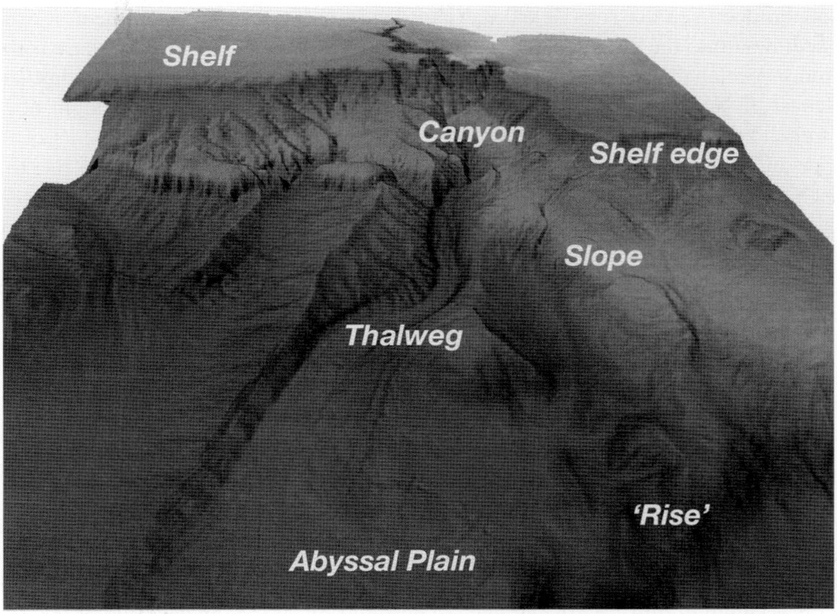

Fig. 1.3 Three-dimensional image of the Nazaré submarine canyon and nearby continental slope off the coast of Portugal showing the main features of the continental slope and a canyon. The view is to the east from open ocean.

shelf and margin, out over the abyssal plains to reach the MOR systems and eventually the trenches or subduction zones. Embedded in these global scale features are smaller scale habitats such as cold-water coral reefs, canyons, seamounts, cold seeps and hydrothermal vents. We will define each habitat, examine the main sampling methodologies and discuss their limitations (Table 1.1).

1.2 Ecosystem and habitat diversity in the deep sea

1.2.1 Moving into deep water

Although we now recognize different ecosystems and habitats within the deep sea, it is essential to acknowledge that they are all interlinked *inter alia* through hydrography in the water column, primary productivity arising from vertical flux of surface production or microbially mediated chemosynthesis and variation in the composition of the seabed from soft sediments to rocky outcrops.

As we move away from shore, the seabed sinks gently to a depth of 200 m. This is the continental shelf (Figs. 1.2 and 1.3) that provides most of the current ecosystem services from the ocean. In Antarctica the shelf occupy depths down to ~600 m as a result of the weight of the icecap on the surrounding continent. Throughout the world, the edge of the continental shelf is defined by an increase in the angle of the seabed (the shelf edge) (Fig. 1.3) and from this point down to depths of ~3000 m we find the continental slope (Figs 1.2 and 1.3). Together these are referred to as the continental margin and represent the most diverse collections of habitats in the ocean (Menot *et al.*, 2010), including soft sediment slopes, canyons, cold seeps, cold-water corals and brine pools.

Continental slopes vary throughout the global ocean, mainly determined by whether they are passive or active margins. Bordering Europe, the eastern United States, eastern South America, all Africa and Australia, margins are passive and extend for thousands of kilometres. In many cases the continental slope increases in depth at an angle of about 8° (equivalent to the relatively gentle slope on a hillside). Although rock outcrops may be present, most of the slope is covered with sediment (Fig. 1.4a) and at the base of these slopes there is a decrease in the angle, giving rise to the continental deep-sea fan (or rise). In certain conditions where there has been a catastrophic influx of sediment into the deep sea by turbidity currents (Talling *et al.*, 2007), there is the formation of a more specialized sedimentary feature called a turbidite. Along active margins, such as off the west coast of North and South America, the continental slope plunges straight down to the adjacent trench and has no obvious continental fan. The sedimentary environments of the continental slope and fan (especially on passive margins, although not turbidites) give rise to some of the highest biodiversity in the deep sea (Grassle & Maciolek, 1992). Since the nineteenth century, the traditional methods of sampling these areas have included trawls, sleds and box-corers, and these methods are still widely used today (Table 1.1), although they have been supplemented by the use of submersibles and remotely operated vehicles (ROVs) for small-scale work and autonomous underwater vehicles (AUVs) for wider scale surveys.

Not all margins follow this simple pattern and it is the heterogeneity of environments that leads to the apparent high species diversity found at these bathyal depths between 200 and 3000 m (Levin *et al.*, 2010). Where the continental slope is steep enough to prevent settlement of sediment and bedrock forming, the margin is exposed (Tyler & Zibrowius, 1992). Hard substrata also occur at higher latitudes in all oceans where drop stones from melting icebergs, both past and present, increase the local heterogeneity of the seabed. Such drop stones interact with the local hydrography, modifying the sedimentology of the immediately adjacent seabed (Fig. 1.4b). In such a situation, sessile attached megafauna may be found, which are not easily sampled other than by the use of submersibles or ROVs (Tyler & Zibrowius, 1992). The ripples and winnowed sediment observed in Figure 1.4 are evidence of current flow along the contours of the continental slope.

Table 1.1 A summary of the main deep-sea habitats, their definitions, and the types of sampling methods, as well as the main considerations for each method

Habitat	Definition	Sampling methods	Sampling considerations
Continental shelf	The seabed between low water spring tide and the shelf edge usually at ~200 m depth (deeper on the Antarctic continental shelf)	Trawls, sledges, grabs, corers, drop and towed cameras	Often well sampled or fished so data may be biased by past sampling/exploitation
Continental slope	From about 200 m depth where there is an increase in gradient down to ~3000 m depth)	Trawls, sledges, grabs, corers, submersible, ROV, epibenthic sledges, drop and towed cameras, landers	Generally straightforward to sample except in very steep and/or rocky areas
Submarine canyons	Irregular fissure-like incisions in the continental slope crossing the depth contours from 50 m to >4000 m	Submersible, ROV, drop cameras; in well-known canyons trawls, epibenthic sledges, towed cameras and corers have been used	Steep-sided and narrow and difficult for trawls and accurate coring
Cold-water coral reefs	Ahermatypic corals that form reefs down to 2000 m depth, the most common species being *Lophelia pertusa*	Submersible, ROV, towed cameras	Delicate structure vulnerable to disturbance or destruction by trawls, epibenthic sleds etc.
Cold seeps	Sedimentary chemosynthetic ecosystems fuelled by thermogenic or biogenic methane and/or hydrogen sulfide	Corers, drop and towed cameras, submersible, ROV	Small spatial extent
Oxygen-minimum zones	Zones along the continental slope where the water column oxygen-minimum zone impinges and results in low oxygen values close to the seabed	Corers, trawls epibenthic sledges, drop and towed cameras, ROV	Often 'soupy' consistency of surface sediment that makes trawling and coring difficult
Mud volcanoes	A form of cold seep looking like a 'cow pat' that often exudes fluid mud	Drop and towed cameras, corers, submersible, ROV	Small spatial extent
Methane hydrates	Methane held in a water matrix under particular temperature/pressure condition forming a solid	Submersible, ROV	Small spatial extent if exposed at the seabed at all
Pock marks	'Holes' in the sediment on various scales caused by burrowing fauna or geological movement	Submersible, ROV	Small spatial extent
Deep-sea fan	Sedimentary area where the incline of the continental slope decreases and ultimately merges with the abyssal plain, composed of alluvial sediment	Trawls, epibenthic sledges, corers, drop and towed cameras, ROV	The presence of turbidites results in sediment heterogeneity over a wide range of scales

(*continued*)

Table 1.1 (Continued)

Habitat	Definition	Sampling methods	Sampling considerations
Abyssal plains	The extensive 'flat' sedimentary seabed between 3000 and 6000 m deep	Trawls, epibenthic sledges, corers, ROV, AUV, drop and towed cameras, landers	Shows spatial heterogeneity on a variety of scales from kilometres to centimetres. Small, soft-bodied forms dominate
Hypersaline pools	High-salinity anoxic small-scale habitats found at abyssal depths	ROV	Small spatial extent, anoxic, very limited metazoan fauna
Seamounts and guyots	Seamounts are usually extinct volcanoes that rise hundreds or thousands of metres above the surrounding seafloor. Guyots are flat-topped seamounts	Submersible, ROV, drop and towed cameras, epibenthic sledges, corers	Steep sides, mostly hard and rugged substrate (apart from top of guyots where sediment accumulates). Can have delicate fauna vulnerable to disturbance from trawl, sled sampling
Mid-ocean ridge including transform faults	The 60 000 km ridge system, also called the 'spreading centre', often down the centre or off-centre in the main oceans. Transform faults are large 'slashes' in the MOR	Submersible, ROV, trawls	Mainly a rock fauna and often very sparse
Hydrothermal vents	Chemosynthetically driven ecosystems fuelled by hydrogen sulfide formed by water/rock interactions at specific sites along the MOR and back-arc volcanoes and basins	Submersible, ROV, AUV, drop and towed cameras	Small spatial extent, and some sites have a short 'life' span. High biomass. Some delicate communities that would be disturbed by destructive gears
Trenches	Found along active margins where one plate subducts below another plate to form the deepest parts of the ocean from 6 km to 11 km depth	Trawls, landers, ROV	Very deep and a long time to collect each sample
Organic food falls	Whale falls, wood falls and macroalgal and seagrass falls	Submersible, ROV, towed cameras and landers	Very small spatial extent and limited 'life' span

Continental margins can also be impacted by oxygen-minimum zones (OMZs) (Levin, 2003). These zones occur at bathyal depths, particularly in the eastern Pacific and the Arabian Sea, where sinking organic matter from high surface production is remineralized, with heterotrophic bacteria stripping the water column of oxygen. Where the OMZs impinge on the continental slope, the fauna at the seabed tends to have reduced diversity, although some species are adapted to hypoxic conditions (Creasey *et al.*, 1997). There is no special sampling strategy for OMZs, although sampling of

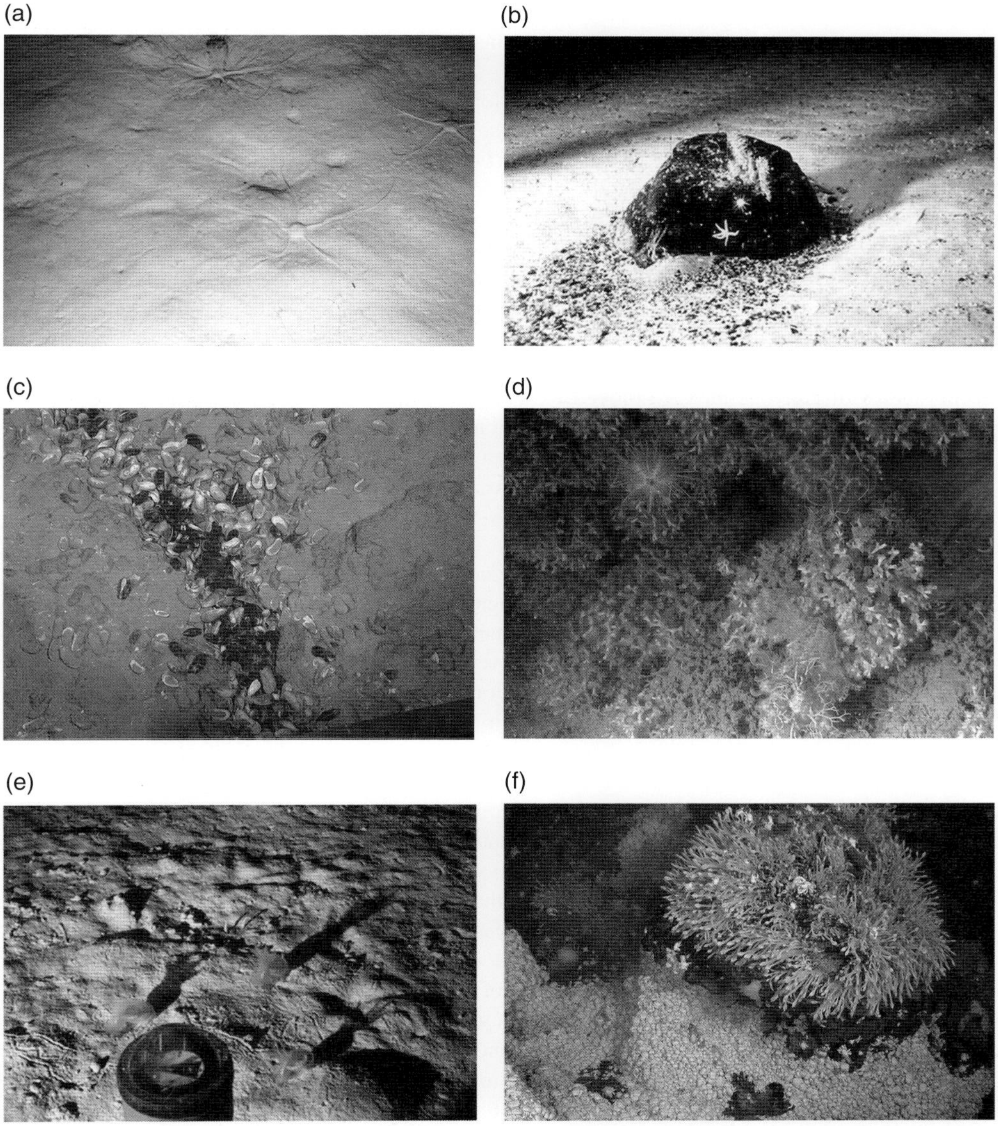

Fig. 1.4 (a) The continental slope seabed at 2200 m depth in the Northeast Atlantic showing ophiuroids on foraminifera ooze. (Photograph by P.A. Tyler.) (b) A drop stone in the continental slope at 2639 m in the Northeast Atlantic showing the winnowing of sediment downstream of the drop stone and current generated ripples in the background. (Photograph by P.A. Tyler.) (c) *Bathymodiolus mauritanicus* alive and dead on the Darwin mud volcano at 1100 m depth. (Photograph courtesy of NOC/NERC. Reproduced with permission.) (d) The frame-building cold-water coral *Lophelia pertusa* in the Whittard submarine canyon at 1300 m depth in the Northeast Atlantic. (Photograph courtesy of NOC/NERC. Reproduced with permission.) (e) The Porcupine Abyssal Plain at 4800 m in the Northeast Atlantic showing tracks, trails and pits in which phytodetritus can accumulate. The small holothurian is *Amperima rosea*. (Photograph courtesy of DSM Billett NOC/NERC. Reproduced with permission.) (f) Dense populations of the yeti crab *Kiwa tyleri* and the barnacle *Vulcanolepas* sp. at hydrothermal vents along the southern part of the East Scotia Ridge, Antarctica, 2500 m depth. (Photograph by ChEsSo consortium.) (Please see Plate 1 for colour representation of the figure.)

the water column and interstitial salinity from sedimentary cores become important. The use of perspex cores in a megacore allows visual examination of the cores to show the various layers of change down the upper sedimentary column.

The apparent 'smoothness' of the continental slope is bisected in many areas of the global ocean by submarine canyons. These arise close to the shelf edge or part way across the shelf and form a deep chasm normal to the contours of the slope (Fig. 1.3). Canyons are subjected to specific geochemical, hydrographic and sedimentological processes (Canals *et al.*, 2013) that enhance organic matter transport (Puig *et al.*, 2003) and shape faunal community composition and structure (Ramirez-Llodra *et al.*, 2010b; Schlacher *et al.*, 2010). The formation of submarine canyons has led to much debate, whilst the environmental conditions within canyons such as strong currents and turbidity flows have hindered their study. Deployed equipment, such as current meters, can be swept away and lowered corers may be difficult to position where intended (Canals *et al.*, 2006). Trawls are invariably lost, except in canyons where the topography is particularly well known. The most successful recent studies of canyons have employed ROVs (Huvenne *et al.*, 2011), but still the use of such powered vehicles present challenges even in moderate currents with suspended sediments in canyons (Tyler *et al.*, 2009). A second major problem for canyon analysis studies is the extreme heterogeneity (of depth, substratum type, seabed morphology, etc.) at different scales from thousands of kilometres down to metres where seabed type and depth may vary, whereas currents and suspended sediment load may vary on a tidal time scale.

Another feature of the continental margin are cold seeps (Sibuet & Olu-Le Roy, 2002; Levin, 2003). Cold seeps are environments where biogenic or thermogenic methane and biologically mediated hydrogen sulfide form the basis for primary production for the local food web (Tunnicliffe *et al.*, 2003). The name of the environment derives from the seepage of these compounds in the sediment and that there is no increase in temperature (Levin, 2003). One of the most common forms of cold seep is the mud volcano, of which classic examples occur in the Gulf of Mexico, Barents Sea (Haakon Mosby Mud Volcano), Mediterranean Sea and the Gulf of Cadiz (Fig. 1.4c) (Olu-Le Roy *et al.*, 2004; Vanreusel *et al.*, 2008). Mud volcanoes look like the classic 'cow pat' and can be 100–200 m diameter. Infrequently mud pours out of the mud volcano and forms a slide down the flank of the volcano. Typical macrofauna include siboglinid tubeworms and bathymodiolid mussels (Fig. 1.4c) as found at hydrothermal vents. In addition, some mud volcanoes support colonies of octocorals, presumably feeding on the local bacterial production (Levin, 2005). On passive margins are found 'pock marks' (Ondréas *et al.*, 2005; Olu-Le Roy *et al.*, 2007). Pock marks are conical dips in the seabed sediment that often have associated seeps because of gas loss. In the Gulf of Mexico, a highly specialized form of cold seep is the 'brine pool' (MacDonald *et al.*, 1990). This pool is some 30 m long and contains water at 120 ppt salinity. At the base of the pool is a salt dome and arising through the waters of the pool and the surrounding sediment is methane that fuels the methanogenic bacteria endosymbiotic with the mussel *Bathymodiolus childressi*. This mussel bed has a distinct fauna that contribute to the local diversity. Other small forms of cold seep are methane hydrates that appear to support an endemic species, the ice worm *Hesiocaeca methanicola* (MacDonald *et al.*, 2003). Another highly specialized type of cold seep are the Campeche asphalt volcanoes found in the southwest Gulf of Mexico (MacDonald *et al.*, 2004). These are unique habitats characterized by the episodic intrusions of semi-solid hydrocarbons that spread over and form structures with a significant vertical relief: the Chapapote knolls. Finally, one of the most recently discovered novel forms of seepage are the 'hydrothermal seeps'. These hybrid systems, where methane seepage and diffuse hydrothermal flow are found together, have been documented along the Costa Rica margin (Levin *et al.*, 2012).

Cold seeps, mud volcanoes, pock marks and other such features present their own sampling challenges. In most cases they are relatively small and, although sedimentary, they are not easily sampled by corer from surface ships. The most successful programmes for the analysis of cold seeps have involved the use of visual observation, faunal sampling and coring from submersibles and ROVs.

A final specialized environment of the continental margin is that of cold-water coral reefs (Freiwald, 2002). Most people are familiar with tropical shallow water reefs, the corals of which rely on symbiotic zooxanthellae for the deposition of calcium carbonate. Typically at bathyal depths in the ocean, there is a suite of corals that are azooxanthellate (having no zooxanthellae) but are successful in building reefs. The best known of these corals is *Lophelia pertusa* (Fig. 1.4d), but they also include the coral genera *Oculina*, *Madrepora* and *Solenosmilia*. These corals are 'frame builders', the frame being calcium deposited by the coral. The interstices of the frames provide a habitat for a wealth of invertebrate fauna and fish, making them an exceptionally high biodiverse environment. These reefs may be found throughout the global ocean but are particularly common in the North Atlantic (Zibrowius, 1985; Mortensen *et al.*, 2008; Davies *et al.*, 2008). They are especially vulnerable to deep-sea fishing where the corals are destroyed by the nets of heavy fishing gear that may clear an area of its corals and all the associated fauna (Hall-Spencer *et al.*, 2002). Cold-water corals present difficulties for sampling in both practical and legal terms (e.g. some deep coral reefs are protected from disturbance and only non-destructive scientific sampling is allowed). Today, the main sampling effort is by the use of ROVs, but cold-water corals may also be surveyed by side-scan sonar, towed cameras and the use of AUVs.

1.2.2 The most extensive benthic environment on Earth

As we reach the bottom of the continental slope, the seabed starts to level off and between 3000 and 6000 m depth we have the most extensive benthic environment on Earth: the abyssal plains. These plains cover 50% of the total Earth surface (Fig. 1.2) and, although they vary depth and contain both basins and hills on a scale of kilometres, their slope is imperceptible. All the oceans of the globe have abyssal plains (Fig. 1.1). The original perception was that they were flat and featureless and, when viewed by low-resolution multibeam swath bathymetry from a surface vessel, this may appear to be the case. The traditional method for sampling the abyssal plains has been the use of trawls and sleds for larger fauna, box-corers for macrofauna, and multicores for meiofauna. Photography gives an *in situ* view of the seabed and it was soon realized that, although flat, there was considerable heterogeneity within this mainly sedimentary environment (Fig. 1.4e). Simply, there are bumps and dips often at the sub-metre scale. The bumps may be mounds formed by burrowing organisms living in the sediment and close examination shows that this disturbance affects the local biodiversity. Tracks and trails also create very small-scale disturbance that can lead to heterogeneity on the sub-metre scale. Dips in the seabed often get filled with phytodetritus from surface production, which has sunk to the seabed and collected in these hollows. In the Pacific, manganese nodules are scattered on the abyssal plain and form their own unique 'province', with the manganese nodules supporting a fauna composed mainly of foraminifera (Mullineaux, 1987; Miljutina *et al.*, 2010). In the Mediterranean Sea, anoxic hypersaline pools have been found at abyssal depths (van der Wielen *et al.*, 2005). Such environments were thought to be the preserve of microorganisms only, but Danovaro *et al.* (2010) have shown that at least three new species of the phylum Loricifera are capable of living under these conditions.

This habitat variation on the extensive abyssal plains leads to high diversity but low biomass amongst small infauna, including polychaetes, nematodes and peracarid crustaceans (Vanreusel *et al.*, 1995; Lambshead *et al.*, 2001; Brandt *et al.*, 2007; Levin *et al.*, 2010) although biomass is low as a result of low food availability (Smith *et al.*, 2008). Thus the problem for sampling abyssal plains is that, by trawling over a couple of kilometres of seabed, the trawl will integrate the diversity of the seabed, whereas coring or sampling by submersible or ROV will differentiate the biodiversity on a micro-scale. Mosaicking from ROVs is developing as a tool to examine the sub-metre scale variation on abyssal plains.

Although extensive, the abyssal plains may be dotted, or even peppered as in the Pacific, by seamounts. There is a very specific definition that states seamounts are volcanoes, usually extinct, that rise hundreds or thousands of metres above the surrounding seafloor (Koslow, 2007; Pitcher *et al.*, 2007) (Fig. 1.5). They are formed at MORs and are carried across the floor of the ocean by tectonic

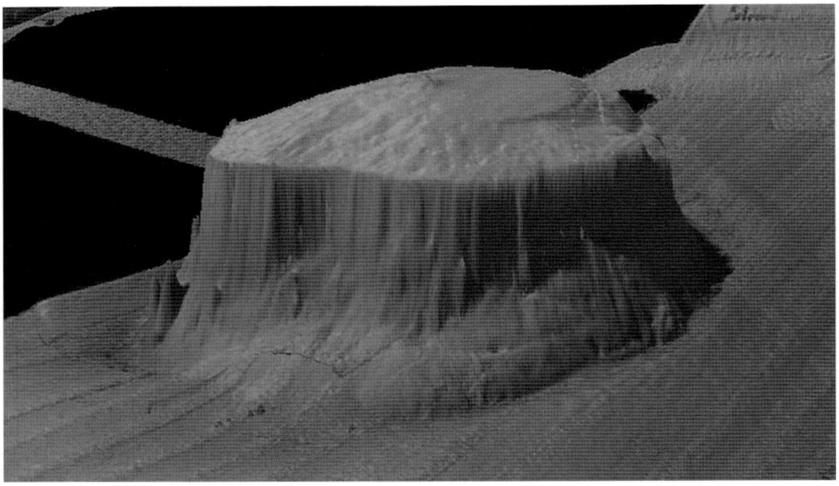

Fig. 1.5 Swath bathymetry of the Anton Dohrn seamount with its flat top, technically a guyot. (Image courtesy of Colin Jacobs, NOC.)

plate movement. This is seen particularly elegantly in the Emperor seamount chain formed by the Pacific hotspot that is now creating the Big Island of Hawaii. As these islands are moved to the northwest, they are eroded away to below sea level to form seamounts or guyots (where the top of the seamount is particularly flat, as in the Anton Dohrn seamount in the Northeast Atlantic (Fig. 1.5)). All seamounts are steep sided and their shape modifies the flow past them, with flow accelerating near the summit (Genin *et al.*, 1986). Because of the steep sides and the accelerated flow over seamounts, the associated fauna is often sessile and filter feeding and includes a high abundance of all types of coral (Clark *et al.*, 2010). Such an environment, as with cold-water coral reefs, attracts a high invertebrate and vertebrate diversity. This, together with the trapping of vertically migrating pelagic organisms, makes seamounts potentially intense fishing grounds and some seamounts in the global ocean have been devastated by deep-water fishing (Rogers, 1999; Althaus *et al.*, 2009).

An additional stress on seamounts is when the top of the seamount protrudes into the OMZ, which results in a natural reduction in biodiversity over the top of the seamount (Wishner *et al.*, 1990). As with many of these specialized environments in the deep sea, traditional sampling methods have limitations, and submersible technology is often the preferred way of studying seamounts, although cameras towed from surface vessels have been used successfully (e.g. Rowden *et al.*, 2010).

1.2.3 Mountain chains and hot fluid

The abyssal plains do not extend from continent to continent. As one moves across the ocean, there is a decrease in water depth, an increase in the ruggedness of the terrain and an increase in the proportion of rock at the seabed. This is the foothills of the MOR (Fig. 1.1), a 60000-km-long structure that runs along all oceans. The MOR is the site of formation of new ocean floor. As the ocean plates are dragged down into the subduction zones, new seafloor material is added at the MORs. Mid-ocean is a bit of a misnomer, as these 'spreading centres' are roughly mid-ocean in the Atlantic and Indian oceans but are asymmetrically offset to the east in the Pacific (Fig. 1.1). The correct term should be spreading centres, but MOR is in common use. We know very little about the environment of MORs except for hydrothermal vents. The cross-section of the MOR can vary from a deep summit axial graben of 1000 m (as seen in the Atlantic) to the very shallow axial valley seen along the East Pacific Rise (Van Dover, 2000). More distinctive features of the MORs are the transform faults (Fig. 1.1) that appear as giant slits at right angles to the main axis of the ridge. These transform faults allow deep circulation between the deep basins either side of the MOR and may prove to be important conduits of reproductive propagules between the different ocean basins. Although of great linear extension, the MOR has not been sampled extensively, although the MAR-ECO and ECOMAR projects have examined the Mid-Atlantic Ridge close to the Charlie–Gibbs Fracture Zone (Bergstad *et al.*, 2008). Remarkably, they were able to trawl areas of the ridge as well as using corers, submersibles and ROVs.

A distinctive feature of spreading centres are the hydrothermal vents. These were first discovered in 1977, when geophysicists could not explain the heat loss of the Earth's interior by conductive heat loss alone, citing the expected presence of convective heat loss. Along the Galapagos Rift, they found the predicted convective heat loss and serendipitously discovered what became known as the hydrothermal vent fauna (Corliss *et al.*, 1979). A hydrothermal vent occurs when seawater has percolated through the oceanic crust, reacted with the subsurface rock, been heated by magma and had its chemical composition changed by removing all oxygen and magnesium, reducing sulfate to sulfide and incorporating metals. The water released at the hydrothermal vent is rich in hydrogen sulfide, which reacts with seawater to precipitate as sulfide (Van Dover, 2000; Tolstoy *et al.*, 2008). The formation of the chimney from which hydrothermal fluid emanates varies in structure and may affect the distribution of local fauna. In the Northeast Pacific, 'flanges' control the escape of vent fluid, leading to the strongest thermal gradient on Earth, whilst in some Atlantic vents 'beehive diffusers' are found (Van Dover, 2000). The hydrogen sulfide provides an energy source for microbially mediated primary production by chemolithoautotrophic microorganisms found both free living, forming bacterial mats, and as endosymbionts in certain endemic vent species (Childress & Fisher, 1992). This metazoan fauna, although of relatively low biodiversity, produce a magnificently high biomass of fauna at the vent, surrounded by the very low biomass of the adjacent deep sea (Fig. 1.4f). Hydrothermal vents are also found at subduction zones as well as spreading centres.

The discovery of vents changed the basic paradigms of marine biology for ever. However, sampling and studying of vents could only ever have been carried out by the use of submersibles and more recently ROVs, as the scale at which the sampling takes place is sub-metre and even at centimetre scale.

1.2.4 Into hades

At the other end of oceanic plates, destruction takes places in a subduction zone where, usually, the heavier oceanic plate subducts beneath the lighter continental plate (Fig. 1.1). Because both plates are sinking, we find the deepest ocean on Earth, the trenches, between 6000 and 11 000 m, the best examples of which are found in the western Pacific (Fig. 1.1). Trenches occupy only 1% of the globe and by their tectonic morphology are both deep and very linear. Trenches were first sampled only in the early 1950s by the Danish *Galathea* and Russian *Vitjaz* expeditions that used trawls (Jamieson *et al.*, 2010). Piccard and Walsh dived into the Marianas Trench in 1960 in the bathyscaphe *Trieste*, but after only a brief glimpse of the deepest point on Earth, they dropped their ballast and rose back to the surface. The only other dive to the deepest point on Earth was made in 2012 by James Cameron, who piloted solo the DEEPSEA CHALLENGER submersible. So far, we know that trenches support high bacterial abundance and biomass (Danovaro *et al.*, 2003), as well as an important and diversified benthic fauna (Belyaev, 1989). Sibuet *et al.* (1988) have also reported the existence of a cold seep fauna in the upper parts of the subduction zone off Japan, and Fujikura *et al.* (1999) reported the deepest recorded seep at over 7000 m in the Japan Trench.

New technology of sampling such depths is now available and are permitting exciting investigations of the hadal environment. Hybrid remote vehicles are coming on line to be able to sample such environments routinely. Two deep-sea robotic vehicles, *Nereus* (USA) and *Kaiko* (Japan), have already reached the deepest parts of the Marianas Trench. To date, some of the most successful sampling has been with landers equipped with baited cameras and traps that have identified taxa not thought to occur at hadal depths (Jamieson *et al.*, 2009).

1.2.5 Special cases

The habitats and ecosystems described above are geographically determined by the tectonic activity or history of the great tectonic plates. There are, however, a number of 'minor' ecosystems that are not necessarily geographically determined. These include whale and wood falls as well as seagrass and algal clumps. Generally, these ecosystems are transient, although there is evidence that whale falls may have an impact on the seabed that lasts decades (Smith & Baco, 2003). The main effect of these mini ecosystems is to increase local biomass, although local species diversity immediately adjacent to the organic input may decrease and cause local chemosynthetic activity, particularly in the sediments (Bernardino *et al.*, 2010; Smith & Baco, 2003). There has also been the implication that these 'minor' ecosystems are important stepping stones for the dispersal of reproductive propagules throughout the deep sea (Distel *et al.*, 2000). Because of their small scale, these organic inputs can only be examined by the use of submersibles or ROVs.

1.3 Conclusions

Sampling of the deep seabed has changed considerably since the heroic age of deep-sea exploration. The pioneers used trawls and dredges which sampled relatively flat surfaces of sediment or rock. In the 1960s, a move to quantitative determination of the biodiversity of the deep sea led to the development of quantitative samplers, the USNEL box-corer being the workhorse for some 40 years, only recently being replaced by the megacorer. Such corers were limited to sedimentary seabeds and still caused disturbance as they landed. Technology has helped dampen this effect but not eliminated it. The scale of sampling ability was also reduced to sub-metre. Even by the early 1970s, environments with steep slopes and/or rocks, such as canyons and seamounts, were very rarely sampled.

Two notable innovations stimulated the next era of sampling in the deep sea. The development of the submersible *Alvin* and the movement of oil exploration into offshore waters led to the development of ROVs. With these two aids, sampling could be achieved at small scales and the sub-metre heterogeneity found at cold-water coral reefs, vents and cold seeps could be analysed. Many scientific sampling and analysis packages have been developed and deployed by submersibles and ROVs and with the fantastic improvement in position fixing through GPS, detailed photograph mosaicking of the seabed can occur. In addition, landers and long-term observatories have become an important tool in the armoury of the deep-sea ecologists. In the final analysis, however, there are still areas of the deep sea that are difficult to sample. Canyons with their steep sides, underhangs and strong currents remain a challenge, as do the ocean trenches. Deep-sea ecologists are also very aware of *their* sampling impact on the deep-sea bed and sampling strategy is now devised to cause the minimal amount of damage to the environment. Sampling technology will continue to develop at all scales within the deep sea, and include a focus on temporal variation. Already we are seeing the establishment of 'remote sensing' networks in the Northeast Pacific, and the use of AUVs, both of which will allow access from desk top to the deep sea without the need to board a ship!

Acknowledgements

We acknowledge the support of the Sloan Foundation through the Census of Marine Life Field Programme Chemosynthetic Ecosystems (ChEss). Three reviewers made a number of very useful comments.

References

Althaus, F., Williams, A., Schlacher, T.A., *et al.* (2009). Impacts of bottom trawling on deep-coral ecosystems of seamounts are long-lasting. *Marine Ecology Progress Series*, 397: 279–294.

Belyaev, G.M. (1989). *Deep-sea ocean trenches and their fauna*. Moscow: Nauka Publishing, 385pp. Translated to English by Scripps Institution of Oceanography, USA.

Bergstad, O.A., Falkenhaug, T., Astthorsson, O.S., *et al.* (2008). Towards improved understanding of the diversity and abundance patterns of the mid-ocean ridge macro-and megafauna. *Deep-Sea Research II*, 55: 1–5.

Bernardino, A.F., Smith, C.R., Baco, A., Altamira, I., Sumida, P.Y.G. (2010). Macrofaunal succession in sediments around kelp and wood falls in the deep NE Pacific and community overlap with other reducing habitats. *Deep-Sea Research I*, 57: 708–723.

Brandt, A., Gooday, A.J., Brandao, S.N., *et al.* (2007). First insights into the biodiversity and biogeography of the Southern Ocean deep sea. *Nature,* 447: 307–311.

Canals, M., Puig, P., Durieu de Madron, X., Heussner, S., Palanques, A., Fabres, J. (2006). Flushing submarine canyons. *Nature*, 444: 354–357.

Canals, M., Company, J.B., Martin, D., Sanchez-Vidal, A., Ramirez-Llodra, E. (2013). Integrated study of Mediterranean deep canyons: novel results and future challenges. *Progress in Oceanography*, 118: 1–27.

Childress, J.J., Fisher, C.R. (1992). The biology of hydrothermal vent animals: physiology, biochemistry, and autotrophic symbioses. *Oceanography and Marine Biology: An Annual Review*, 30: 337–441.

Clark, M.R., Rowden, A.A., Schlacher, T., *et al.* (2010) The ecology of seamounts: structure, function, and human impacts. *Annual Review of Marine Science*, 2: 253–278.

Corliss, J.B. Dymond, J., Gordon, L.I., *et al.* (1979). Submarine thermal springs on the Galapagos Rift. *Science*, 203: 1073–1083.

Creasey, S., Rogers, A.D., Tyler, P., Young, C., Gage, J. (1997). Population biology and genetics of the deep sea spider crab *Encephaloides armstrongi* Wood-Mason 1891. *Philosophical Transactions of the Royal Society B*, 352: 365–379.

Danovaro, R., Della Croce, N., Dell'Anno, A., Pusceddu, A. (2003). Depocenter of organic matter at 7800m depth in the SE Pacific Ocean. *Deep-Sea Research I*, 50: 1411–1420.

Danovaro, R., Dell'Anno, A., Pusceddu, A., Gambi, C., Heiner, I., Kristensen, R.M. (2010). The first metazoa living in permanently anoxic conditions. *BMC Biology*, 8: 1–10.

Danovaro, R., Snelgrove, P., Tyler P.A. (2015). Challenging the pargdigms of deep-sea ecology. *Trends in Ecology and Evolution*, 29: 465–475.

Davies, A.J., Wisshak, M., Orr, J.C., Roberts, J.M. (2008). Predicting suitable habitat for the cold-water coral *Lophelia pertusa* (Scleractinia). *Deep-Sea Research I,* 55: 1048–1062.

Distel, D.L., Baco, A.R., Chuang, E., Moril, W., Cavanaugh, C., Smith, C.R. (2000). Do mussels take wooden steps to deep-sea vents? *Nature,* 403: 725–726.

Freiwald, A. (2002). Reef-forming cold-water corals. In: Wefer, G., Hebbeln, D., Jorgensen, B.B., Schluter, M., van Weering, T.C.E. (eds), *Ocean margin systems*. Springer, Heidelberg, pp. 365–385.

Fujikura, K., Kojima, S., Tamaki, K., Maki, Y., Hunt, J., Okutani, T. (1999). The deepest chemosynthesis-based community yet discovered from the hadal zone, 7326 m deep, in the Japan Trench. *Marine Ecology Progress Series*, 190: 17–26.

Genin, A., Dayton, P.K., Lonsdale, P.F., Speiss, F.N. (1986). Corals on seamount peaks provide evidence of current acceleration over deep-sea topography. *Nature*, 322: 59–61.

Grassle, J.F., Maciolek N.J. (1992). Deep-sea species richness: regional and local diversity estimates from quantitative bottom samples. *American Naturalist*, 139: 313–341.

Hall-Spencer, J., Allain, V., Fossa, J.H. (2002). Trawling damage to Northeast Atlantic ancient coral reefs. *Proceedings of the Royal Society London B*, 269: 507–511.

Huvenne, V.A.I., Tyler, P.A., Masson, D.G., *et al.* (2011). A picture on a wall: Innovative mapping reveals cold-water coral refuge in submarine canyon. *PLoS ONE*, e28755, doi:10.1371/journal pone0028755

Jamieson, A.J., Fujii, T., Solan, M., Matsumoto, A.K., Bagley, P.M., Priede, I.G. (2009). First findings of decapod crustacea in the hadal zone. *Deep-Sea Research I,* 56: 641–647.

Jamieson, A.J., Fujii, T., Mayor, D.J., Solan, M., Priede, I.G. (2010). Hadal Trenches: the ecology of the deepest places on Earth. *Trends in Ecology & Evolution*, 25: 190–197.

Koslow, A. (2007). *The silent deep*. Chicago, IL: University of Chicago Press.

Lambshead, P.J.D., Tietjen, J., Glover, A., Ferrero, T., Thistle, D., Gooday, A.J. (2001). Impact of large-scale natural physical disturbance on the diversity of deep-sea North Atlantic nematodes. *Marine Ecology Progress Series*, 214: 121–126.

Lambshead, P.J.D., Brown, C.J., Ferrero, T.J., *et al.* (2002). Latitudinal diversity patterns of deep-sea marine nematodes and organic fluxes: a test from the central equatorial Pacific. *Marine Ecology Progress Series*, 236: 129–135.

Levin, L.A. (2003). Oxygen minimum zone benthos: adaptation and community response to hypoxia. *Oceanography and Marine Biology: An Annual Review,* 41: 1–45.

Levin, L.A. (2005). Ecology of cold seep sediments: interactions of fauna with flow, chemistry and microbes. *Oceanography and Marine Biology: An Annual Review*, 43: 1–46.

Levin, L.A., Sibuet, M., Gooday, A.J., Smith, C., Vanreusel, A. (2010). The roles of habitat heterogeneity in generating and maintaining biodiversity on continental margins: an introduction. *Marine Ecology*, 31: 1–5.

Levin, L.A., Orphan, V.J., Rouse, G.W., *et al.* (2012). A hydrothermal seep on the Costa Rica margin: middle ground in a continuum of reducing ecosystems. *Proceedings of the Royal Society London B*, 279: 2580–2588.

MacDonald, I. R., Reilly, J.F., Guinasso Jr, N.L., *et al.* (1990). Chemosynthetic mussels at a brine-filled pockmark in the northern Gulf of Mexico. *Science*, 248: 1069–1099.

MacDonald, I.R., Sager, W.W., Peccini, M.B. (2003). Gas hydrate and chemosynthetic biota in mounded bathymetry at mid-slope hydrocarbon seeps: Northern Gulf of Mexico. *Marine Geology*, 198: 133–158.

MacDonald, I.R., Bohrmann, G., Excobar, E., *et al.* (2004). Asphalt volcanism and chemosynthetic life in the Campeche Knolls, Gulf of Mexico. *Science*, 304: 999–1002.

Menot, L., Sibuet, M., Carney, R.S., *et al.* (2010). New perceptions of continental margin biodiversity. In: McIntyre, A. (ed.), *Life in world's oceans: diversity, distribution, and abundance*. Oxford: Blackwell, pp. 79–101.

Miljutina, M.A., Miljutin, D.M., Mahatman, R., Galeron, J. (2010). Deep-sea nematode assemblages of the Clarion-Clipperton Nodule Province (Tropical North-Eastern Pacific). *Marine Biodiversity*, 40: 1–15.

Mortensen, P.B., Buhl-Mortensen, L., Gebruk, A.V., Krylova, E.M. (2008). Occurrence of deep-water corals on the Mid-Atlantic Ridge based on MAR-ECO data. *Deep-Sea Research II*, 55: 142–152.

Mullineaux, L.S. (1987). Organisms living on manganese nodules and crusts: Distribution and abundance at three North Pacific sites. *Deep-Sea Research*, 34: 165–184.

Murray, J., Hjort, J. (1912). *The depths of the ocean*. London, Macmillan.

Olu-Le Roy, K., Sibuet, M., Fiala-Medioni, A., *et al.* (2004). Cold seep communities in the deep eastern Mediterranean Sea: composition, symbiosis and spatial distribution on mud volcanoes. *Deep-Sea Research I*, 51: 1915–1936.

Olu-Le Roy, K., Caprais, J.C., Fifis, A., *et al.* (2007). Cold seep assemblages on a giant pockmark off West Africa: spatial patterns and environmental control. *Marine Ecology*, 28: 115–130.

Ondréas, H., Olu, K., Fouquet, Y., *et al.* (2005). ROV study of a giant pockmark on the Gabon continental margin. *Geo-Marine Letters*, 25: 281.

Pitcher, T. J., Morato, T., Hart, P.J.B., Clark, M.R., Haggan, N., Santos, R.S. (2007). *Seamounts: ecology, fisheries and conservation*. Oxford: Blackwells.

Puig, P., Ogston, A.S., Mullenbach, B.L., Nittrouer, C.A., Sternberg, R.W. (2003). Shelf-to-canyon sediment transport processes on the Eel Continental Margin (Northern California). *Marine Geology*, 193: 129–149.

Ramirez-Llodra, E., Brandt, A., Danovaro, R., *et al.* (2010a). Deep, diverse and definitely different: unique attributes of the world's largest ecosystem. *Biogeosciences*, 7: 2851–2899.

Ramirez-Llodra, E., Company, J.B., Sardà, F., Rotllant, G. (2010b). Megabenthic diversity patterns and community structure of the Blanes submarine canyon and adjacent slope in the Northwestern Mediterranean: a human overprint? *Marine Ecology*, 31: 167–182.

Rogers, A.D. (1999). The biology of *Lophelia pertusa* (Linnaeus 1758) and other deep-water reef-forming corals and impacts from human activities. *Internationale Revue der gesamten Hydrobiologie*, 84: 315–406.

Rowden, A.A., Schlacher, T.A., Williams, A., *et al.* (2010). A test of the seamount oasis hypothesis: seamounts support higher epibenthic megafaunal biomass than adjacent slopes. *Marine Ecology*, 31 (suppl. 1): 95–106.

Schlacher, T.A., Williams, A., Althaus, F., Schlacher-Hoenlinger, M.A. (2010). High-resolution seabed imagery as a tool for biodiversity conservation planning on continental margins. *Marine Ecology*, 31: 200–222.

Sibuet, M., Olu-Le Roy, K. (2002). Cold seep communities on continental margins: structure and quantitative distribution relative to geological and fluid venting patterns. In: Wefer, G., Hebbeln, D., Jorgensen, B.B., Schluter, M., van Weering, T.C.E. (eds), *Ocean margin systems*. Berlin: Springer Verlag, pp. 235–251.

Sibuet, M., Juniper, S.K., Pautot, G. (1988). Cold-seep benthic communities in the Japan subduction zones: Geological control of community development. *Journal of Marine Research*, 46: 333–348.

Smith, C.R., Baco, A.R. (2003). Ecology of whale falls at the deep-sea floor. *Oceanography and Marine Biology: An Annual Review*, 41: 311–354.

Smith, C., De Leo, F.C., Bernardino, A.F., Sweetman, A.K., Martinez-Arbizu, P. (2008). Abyssal food limitation, ecosystem structure and climate change. *Trends in Ecology and Evolution*, 23: 518–528.

Talling, P.J., Wynn, R.B., Masson, D.G., *et al.* (2007). Onset of submarine debris flow deposition far from original giant landslide. *Nature*, 450: 541–544.

Tolstoy, M., Waldhauser, F., Bohnenstiehl, D.R., Weekly, R.T., Kim W.-Y. (2008). Siesmic identification of along axis hydrothermal flow on the East Pacific Rise. *Nature*, 451: 181–184.

Tunnicliffe, V., Juniper, K.S., Sibuet, M. (2003). Reducing environments of the deep-sea floor. In: Tyler, P.A. (ed.), *Ecosystems of the World*, Volume 28, *Ecosystems of the deep oceans*. London: Elsevier, pp. 81–110.

Tyler P.A., Zibrowius, H. (1992). Submersible observations of the invertebrate fauna on the continental slope southwest of Ireland (NE Atlantic Ocean). *Oceanologica Acta*, 15: 211–226.

Tyler, P.A., Amaro, T., Arzola, R., *et al.* (2009). Europe's Grand Canyon: Nazaré Submarine Canyon. *Oceanography*, 22: 46–57.

Van Dover, C.L. (2000). *The ecology of deep-sea hydrothermal vents*. Princeton, NJ: Princeton University Press.

Van der Wielen, P.W.J.J, Bolhuis, H., Borin, S., *et al.* (2005). The enigma of prokaryotic life in deep hypersaline anoxic basins. *Science,* 307: 121–123.

Vanreusel, A., Vincx, M., Bett, B.J., Rice, A.L. (1995). Nematode biomass spectra at two abyssal sites in the NE Atlantic with a contrasting food supply. *Internationale Revue der gesamten Hydrobiologie*, 80: 287–296.

Vanreusel, A., Andersen, A.C., Boetius, A., *et al.* (2008). Biodiversity of cold seep ecosystems along the European Margin. *Oceanography*, 22: 110–127.

Wishner, K., Levin, L., Gowing, M., Mullineaux, L. (1990). Involvement of the oxygen minimum in benthic zonation on a deep seamount. *Nature*, 346: 57–59.

Zibrowius, H. (1985). Scléractinaires bathyaux et abyssal de l'Atlantique nord-oriental. In: Laubier, L., Monniot, C. (eds), *Peuplements profonds du Golfe de Gascogne. Campagnes BIOGAS*. Brest: IFREMER, pp. 311–324.

Chapter 2
Deep-Sea Fauna

Stefano Schiaparelli[1]*, Ashley A. Rowden*[2] *and Malcolm R. Clark*[2]
[1] Department of Earth, Environment and Life Sciences and Italian National Antarctic Museum, Section of Genoa, University of Genoa, Genoa, Italy
[2] National Institute of Water and Atmospheric Research, Wellington, New Zealand

Abstract

After 150 years of exploration, we now know the deep sea represents the largest complex of ecosystems on our planet which embrace the greatest number of animal species and biomass of the living world. However, there is still a great deal more information that is needed to improve our knowledge of deep-sea faunal dynamics which demands more and better targeted sampling. In this chapter we outline the main biological characteristics of deep-sea organisms, describing aspects of animal life forms, behaviours and adaptations to the deep sea that could affect sampling techniques and survey design. Aspects of spatial and temporal distribution patterns of diversity and abundance are also described for both benthic and pelagic fauna, and examples given of various related sampling issues. The deep sea hosts a huge variety of animal types and faunal communities, with highly variable characteristics. Consideration of these characteristics provides a useful biological context for the sampling techniques that are covered in subsequent chapters of this book.

Keywords fauna, plankton, nekton, benthos, abundance, diversity, temporal and spatial patterns, adaptations, symbioses

2.1 Introduction

Progress toward knowledge of the deep ocean and of the living forms which inhabit it has been slow compared to the study of the more accessible shallow marine environments, hampered by obvious technical, logistic and economic constraints.

The first global biological survey of the deep sea, where the ocean floor was sampled in a systematic way, was undertaken from the HMS *Challenger* in the years 1872–1876 (Murray, 1895). During this expedition, hundreds of deep-sea stations were completed and a large number of species new to science were collected, resulting in more than 40 volumes of reports. About 80 years after the *Challenger*, a series of 'exploration records' in the deep sea were broken during the decade 1950–1960, which is often remembered as the apex of the 'heroic age of deep-sea exploration'. In these years, for example, the *Galathea 2* expedition (1950–1952) retrieved biological material from the record depth of 10 190 m and the bathyscaphe *Trieste* (1960) reached the bottom of the Marianas Trench at 10 902 m. However, most deep-sea species had never been observed alive, and their ecologies and behaviours remained largely unknown.

Biological Sampling in the Deep Sea, First Edition. Edited by Malcolm R. Clark, Mireille Consalvey and Ashley A. Rowden.

Fig. 2.1 A pictorial representation of the life in a deep sea canyon presented in the February 1953 issue of *Life* magazine. The original legend stated: 'Canyon's mouth at base of a continental slope, jellyfish, viperfish, sea devils, sea cucumbers, quill worms move through lightless depths a mile below last glimmer of sun. Cataract in background is a slump of seaslide of green mud and sand, similar to one that broke the Atlantic cables in 1929'. (© Real Estate of James Lewicki. Reproduced with permission.) (Please see Plate 2 for colour representation of the figure.)

An example of this situation can be found in the illustrations of deep-sea creatures presented in the popular science series 'The World We Live In', published by the American magazine *LIFE* in 1952–1954. Deep-sea organisms had to be painted, as no good photographs were available at the time. In the chapter entitled 'The miracle of the sea' (February 1953), the American illustrator James Lewicki depicted deep-sea ceratioid anglerfish and stomiids swimming together in schools (Fig. 2.1). Modern-day deep-sea biologists would immediately know the schooling of these anglerfish is very unlikely. These fish are a now well-known example of one of the most dramatic adaptations occurring in the deep sea to compensate for the extremely low abundance of individuals: sexual parasitism (Munk, 2000; Pietsch, 2004). In this group of fish, dwarfed males fuse their tissues with those of the larger females and spend all their lives as parasites, but only 'if they are fortunate enough to meet [a female]' while they are 'floating about in the darkness of the middle depths of the ocean', as stated by Regan (1926), the scientist who first described this adaptation.

Lewicki's representation of over 60 years ago was clearly the result of an artistic contingency, the painter having to do his best to give a pictorial representation of mostly unknown deep-sea creatures, and injecting into his artwork knowledge of the more familiar coastal fish, which usually do form schools. It was only about 20 years later, in the 1970s, that the use of manned submersibles and remotely operated vehicles (ROVs) allowed for the documentation and photographing of deep-sea species *in situ*. The use of such equipment not only allowed the observation of the behaviour of deep-sea organisms, but also enhanced the quantitative and experimental study of deep-sea ecosystems.

On the whole, after 150 years of exploration of the deep, our perception of this unfamiliar environment has changed dramatically (Koslow, 2007), moving from an initial belief that the deep sea was an homogeneous place almost devoid of life, to the awareness that bathyal (200–3000 m), abyssal (3000–6000 m) and also hadal (6000–11 000 m) depths, despite their distance from the productive sea surface, have diverse life which provide important ecosystem services (Smith *et al.*, 2008; Jamieson *et al.*, 2010; Ramirez-Llodra *et al.*, 2010). Today we are also aware that the deep sea is the largest ecosystem on our planet, embracing the greatest number of animal species, the greatest biomass and the greatest number of individual organisms of the living world (Ramirez-Llodra *et al.*, 2010). As such, there is an urgent need to understand its dynamics and evolution, as well as to protect both its pelagic (Robison, 2009) and benthic (Tyler, 2007) components. However, notwithstanding the rapid increase in deep-sea knowledge that has occurred in the last few decades, it is evident that we still have much to learn to address even the most basic questions and more sampling is needed (McClain & Mincks Hardy, 2010).

In this chapter, the main deep-sea faunal groups, their adaptations and their distribution patterns will be briefly reviewed with the aim of introducing the following book chapters, where the gears and the techniques used to study and sample deep-sea organisms are described. A complete review of deep-sea fauna is not attempted here, for there are already several excellent books, which specifically focus on the biology and ecology of deep-sea organisms (e.g. Gage & Tyler, 1991; Herring, 2002; Koslow, 2007). The information presented in this chapter will consider the characteristics of deep-sea fauna with respect to how they may influence the selection of appropriate sampling equipment or approaches. Beside the 'classic' (but always necessary) destructive gears such as dredges and trawls, a special focus is placed on the new generation of non-destructive techniques (e.g. the latest generation of ROVs, submersibles, towed and drop cameras), which allow important ecological and behavioural data to be obtained while avoiding an impact on highly vulnerable deep-sea habitats.

2.2 Life forms

Virtually all known phyla may be found in the deep ocean. However, some benthic groups are more readily encountered in the deep sea, such as the Hexactinellida (glass sponges), Tanaidomorpha, Asellota, Galatheoidea (squat lobsters), Pycnogonida (sea spiders), Priapulida (cactus worms), Echiura (spoon worms) and Enteropneusta (acorn worms) (Fig. 2.2a), Sipunculida (peanut worms), stalked crinoids (sea lilies) and the molluscan Aplacophora and Monoplacophora. In the water column, several groups of invertebrates (gelatinous cnidarians such as ctenophores, and appendicularians; copepods; decapods; chaetognaths; cephalopods) and vertebrates (mesopelagic fish such as lanternfish) characterize the deep sea ocean.

All size classes from the microbiota (that include viruses, bacteria and Archaea), through the meiofauna, macrofauna and megafauna are represented in the deep sea. The meiobenthos is dominated by nematodes, although if the protozoan fraction is also considered, foraminiferans can be equal, or even exceed, the nematodes. The dominant macrofaunal components, in term of densities, are the polychaetes, followed by crustaceans, bivalves and gastropods. Echinoderms and crustaceans dominate the megafauna.

Regardless of the taxa or size, animals with ossified carbonate skeletons (e.g. ophiuroids, asteroids and echinoids) become less represented with depth, due to the increase in carbonate solubility. At depth, these taxa are substituted by soft-bodied organisms, such as holothurians and organic-walled foraminifera, the latter even flourishing at the deepest point in the world's oceans (10 896 m) in the Challenger Deep (Todo *et al.*, 2005; Gooday *et al.*, 2008) and by organisms with silicate skeletons (e.g.

hexactinellid sponges). This means that different gear types, or modifications to existing gear, are needed in order to retain whole specimens. For example, trawls and dredges will damage fragile sponges and corals, and sampling with ROVs or submersibles will be necessary to collect these intact. Similarly, photographic methods will be needed to obtain accurate density estimates or individual animal counts.

2.3 Life habits

Deep-sea organisms can be ascribed to the classic categories of plankton, benthos and nekton, according to their life habits (see Glossary). The 'catchabilities' of deep-sea species do not differ from those of shallow-water species, being determined by the characteristics of deployed gear and the species itself, such as size, abundance, behaviour, mobility and location relative to the seafloor.

If a full census of the organisms living in a given deep-sea area is the target of the study, no single sampling gear will allow for the retrieval of 'the complete set' of species present, due to the sampling bias of any gear and differences in life habits of each species. It is therefore necessary to deploy a variety of gears, each one having a better performance in a given habitat and for specific groups of organisms with similar lifestyle. As a general rule, the more complex the topography of an area, the greater the number of potential habitats, which in turn involves deploying a greater variety of sampling gears.

For example, in the study of benthic infauna, box-corers, multiple corers and grabs will be the elective methods to obtain quantitative estimates of the meio- and macrofaunal species present (see Chapter 10). On the other hand, these tools will not sample larger benthic megafaunal species well, which would instead be better collected by deploying towed gears such as trawls and dredges (see Chapter 7), towed cameras or ROVs (see Chapters 12 and 13). In some cases, gear has been specifically designed to obtain certain taxa, such as the Brenke sled (Brenke, 2005) to sample organisms that inhabit the 'benthic boundary layer' just above the seafloor (see Chapter 9).

Benthopelagic organisms, which include fish, cephalopod and decapod species, are capable of strong swimming and considerable vertical and horizontal movements. Sampling their distribution is particularly difficult, as they occupy different benthic habitats, migrating from one to another and perform diel and seasonal migrations. The overall result is a highly heterogeneous distribution of species at the regional scale. In these situations towed cameras provide a good option, which allows for estimates of habitat occupancy, as well as the recognition of behavioural patterns (e.g. D'Onghia *et al.*, 2011).

Both the technical characteristics of sampling gear and an organism's life habits need to be taken into account when selecting gear that will maximize the collection of the target benthic, pelagic or nektonic group of organisms (e.g., benthic fauna see Table 2.1). However, there is always a 'bias' (for the species/sizes that are more easily sampled by it) and this needs to be properly considered during statistical treatment of the data (see Chapter 17).

2.4 Adaptations

2.4.1 Miniaturization and gigantism

Examples of both dwarfism and gigantism have been repeatedly documented in deep-sea species. Gigantism may be favoured by the fact that the energetic cost per unit of mass is lower in big invertebrates, while dwarfism would allow for a greater number of conspecifics to coexist, increasing the possibility of finding mates. Although these two opposite strategies are both present in deep-sea organisms, their adoption varies according to taxonomic group and the taxonomic level considered (Ramirez-Llodra *et al.*, 2010 and references therein). In general, several studies have indicated that the size structure of deep-sea communities tends to get smaller with increasing depth. This trend has been confirmed for microfauna to megafauna (Rex *et al.*, 2006). The general miniaturization of most

Table 2.1 Benthic sampling gear types and their associated target fauna, and the habitats and substrates on which they are typically deployed. [note that gears can also be used to obtain samples or information that can be used to characterise the seafloor substrates.]

Gear type	Target fauna	Habitats/substrate type
Trawls, dredges (fish, beam, Agassiz-type)	Fish, mega- and macroinvertebrates (especially sessile and epifauna)	Mainly large flat areas of soft sediment substrate in all habitats
	Fauna attached to or associated with large rocks (but damage likely)	**Qualitative to semi-quantitative**
ROVs, submersibles, towed cameras	Mega- and macro-invertebrates (mainly epifauna but some information about infauna from *lebensspuren*)	All habitats and substrate types
	Includes fauna attached to or associated with rocks	**Qualitative** to **quantitative**
Epibenthic sledges	Mega- and macroinvertebrates (mostly epifauna, but hyperbenthos or some infauna can be sampled depending on sledge type)	Suitability for habitat and substrate depends on sledge type (but some sledge types can sample all habitats/ substrate types)
	Can include fauna attached to or associated with small rocks	**Semi-quantitative**
Grabs	Macro- and meiofauna (epifauna and infauna)	Soft sediments only in all habitats
	Can include fauna attached to or associated with small rocks	**Semi-quantitative to quantitative**
Corers	Macro- and meiofauna (epifauna and infauna)	Soft sediments only in all habitats
	Can include fauna attached to or associated with small rocks	**Quantitative**

Fig. 2.2 (a) The Ross Sea (Antarctica) deep seafloor at 1072 m. Note the heavily reworked substrate with mounds, burrows and grooves. On the right there is a large enteropneust which is leaving behind a trail of mud, after this has been ingested and passed through the digestive system. This kind of organism is rarely collected intact with destructive methods. (© NIWA, IPY-CAML Voyage 2008. Reproduced with permission.) (b) The Ross Sea (Antarctica) at about 2200 m. A large carnivorous ascidian is facing the bottom current waiting for prey. This very fragile organism was documented only with a towed camera, no specimens were retained by a sled and a beam trawl deployed at the same site. (© NIWA, IPY-CAML Voyage 2008. Reproduced with permission.) (c) A luxuriant assemblage of gorgonians, soft corals, zoanthids and sponges flourishing on the flank of Scott Island (Antarctica). The presence of particular local conditions support the high biomasses of suspension feeders. (© NIWA, IPY-CAML Voyage 2008. Reproduced with permission.) (d) A 'cold-water coral' assemblage of *Solenosmilia variabilis* hosts featherstars and sponges, with orange roughy swimming close to the seafloor at 950 m depth on a small seamount off New Zealand. (© NIWA. Reproduced with permission.) (e) The stegocephalid amphipod *Parandania* illustrates two of the main adaptations occurring in many mesopelagic crustaceans: the lack of eyes and a 'mimetic' purple coloration (already partially faded). (scale bar 1 cm). (© NIWA. Reproduced with permission.) (f) New discoveries continue to surprise scientists, as in the case of the 'barreleye' fish *Macropinna microstoma*. The apparent function of the transparent shield is to protect the tubular eyes from contact with stinging tentacles of cnidarians, the prey of *Macropinna*. (© 2003 MBARI. Reproduced with permission.) (g) Sampling a whale bone carrying *Osedax frankpressi* by the manipulator arm of MBARI's remotely operated vehicle *Tiburon* at a depth of about 3000 m in the Monterey Canyon. (© 2003 MBARI. Reproduced with permission.) (h) Specimens of *Osedax* with the palps emerging out of a whale bone. (Photograph by Greg Rouse, Scripps Institute of Oceanography. Reproduced with permission.) (Please see Plate 3 for colour representation of the figure.)

deep-sea soft-bottom taxa implies that smaller meshes have to be used to retain a significant fraction of the species and, to this end, smaller mesh sizes are used to sample macrofauna in the deep sea than in shallow waters (e.g. 300 μm versus 500 μm or 1 mm) (e.g. Chapter 10).

However, within particular taxonomic groups gigantism can commonly occur. For example, it is found in scavenger amphipods and isopods, where maintaining a large size increases their capability to forage on carcasses. In these cases sampling is less difficult and such species can be obtained in quantities from baited traps (see Chapter 11). Gigantism is also known in some groups of deep-sea protists such as xenophyophores, gromiids, astrorhizacean, komokiacean and allogromiid foraminifera, which all include some very large species (up to 10 cm in size). These organisms can be best sampled using images (from towed cameras or ROVs), as they are often fragile and break up with direct sampling (e.g. sleds).

Given the extreme size range of deep-sea organisms, when sampling activities are planned, a variety of sampling gears may need to be used in order to maximize sampling efficiency of organisms present in a given area.

2.4.2 Mobility

Marine organisms living in the water column may be capable of considerable movements and migrations either due to passive drift or limited swimming movements (as in most planktonic taxa), or vigorous active swimming (species of nekton). While some benthic organisms are also capable of moving actively, most are typically sedentary or sessile (i.e. permanently fixed to the substrate).

Motility is typically found in predatory or scavenging pelagic and benthic taxa, which need to actively search for food. A sedentary or sessile habit characterizes suspension feeder species that rely on water-carried food resources. Many infaunal and epifaunal species are deposit feeders (i.e. rely on food particles deposited on the seafloor) and may actively move for short distances in order to reach and exploit a new area.

Several deep-sea species that live permanently in the water column or occasionally swim to reach another foraging ground have developed gelatinous body structures to enhance floatation and passive transport by bottom currents. These adaptations allow for a considerable reduction in the amount of energy necessary to move to another feeding ground. One of the best examples of this behaviour is that of the cirrate octopods *Cirroteuthis* and *Grimpoteuthis*, which have well-developed webs and semi-gelatinous bodies that allow neutral buoyancy and an 'umbrella-style drifting' to take advantage of near-bottom water currents. These move these octopods with a very low energy-expenditure (Villanueva *et al.*, 1997). Other invertebrates, such as the deep-sea holothuroids (e.g. elasipodids) also use the same drifting strategy, with currents determining the direction of horizontal displacement (Miller & Pawson, 1990). These species have a soft body and as a result are extremely fragile, meaning that classic sampling gears such as trawls generally cause considerable damage. ROV/submersible technology provides the opportunity to collect such delicate species, as well as study these species *in situ* (see Chapter 13).

Mobile benthic species are typically firmer bodied than pelagic taxa, and are often effectively sampled by trawl nets and sleds. Fish trawls can be towed at a speed of several knots across the seafloor, and cover a relatively large swept area, making them effective at sampling fish and larger invertebrates. Beam trawls can also be towed fast enough to catch benthic fish and mega-invertebrate species, and sleds are able to sample smaller sized mobile animals (see Chapters 7 and 9).

Several deep-sea sessile groups have elongated or stalked bodies that allow them to elevate from the substrate, taking advantage of bottom currents and intercepting water-carried food particles. Examples can be found in both soft and hard substrate habitats, including the pennatulaceans (octocorals), gigantic hydrozoans (e.g. *Branchiocerianthus*), both solitary (e.g. the carnivorous genus *Megalodicopia*, Fig. 2.2b) and compound ascidians, and all stalked crinoids. In these cases, sampling with destructive gears returns highly damaged and incomplete specimens and only manipulator arms and suction devices of ROVs and submersibles allow the collection of fully intact specimens.

2.4.3 *Coloration and bioluminescence*

Below about 200 m water depth light quickly disappears. Deep-sea organisms show different adaptations to the low amount of available light (or its absence) by, for example, developing 'mimetic' coloration and communication systems based on self-produced light.

In the mesopelagic and bathypelagic layers the fish are silver grey or deep black, while invertebrates show purple or brilliant red colorations (Fig. 2.2). All these colours make species invisible in darkness and less visible in the 'twilight zone'. Bright red organisms are being perceived as black due to the adsorption of the red wavelength in the first 5 m of water depth, whereas deep-sea demersal fish appear 'whitish' because they lack any colour.

Chemical reactions that allow bioluminescence mean that communication by the emission of light signals is possible in the dark. It is used for defence, offence and mate attraction (reviewed in Haddock *et al.*, 2010). Various arrangements and patterns of photophores and bioluminescent organs are present in many midwater vertebrates and invertebrates, especially fish and squid. In some mesopelagic fish (e.g. Sternoptychidae), bottom-oriented photophores are also used to produce a light of similar intensity to ambient light (counter-illumination), virtually eliminating the organism's silhouette and making it difficult for a predator looking from below to see potential prey.

Bioluminescence fades almost immediately in species caught and brought to the surface, and the integrity of light organs may also be compromised due to the pressure shock experienced by the organisms during sampling. The study of bioluminescence is typically performed *in situ*, by means of ROVs and landers, all other classic sampling gears causing considerable damage to these fragile organisms and structures.

2.4.4 *Vision*

A common adaptation in deep-sea fauna is the extreme reduction or absence of eyes, especially in abyssal species (Warrant & Locket, 2004). The 'loss' of eyes is a phenomenon that may have occurred several times in deep-sea benthic groups, as molecular data are now allowing us to appreciate (e.g. Raupach *et al.*, 2009), according to the complexity and timing of colonization events of the deep sea from the shallow seas (and vice versa).

With mesopelagic fish and invertebrates (e.g. cephalopods and amphipods), eyes can be very large, allowing the capture of all available light. One of the best examples is that of the opisthoproctid fish *Macropinna microstoma* (the 'barreleye' fish), described in 1939 but only recently observed alive off Monterey Bay thanks to ROV technology (Fig. 2.2f). In this species the complex tubular eyes are 'immersed' in a unique, transparent cephalic shield that protects the eyes from the stinging nematocysts of cnidarian prey. Thanks to its transparency the shield poses no limits to eye movements; the eyes can be turned inside this structure to be directed dorsally or rostrally. This feature has never been observed before in other *Macropinna* specimens collected by midwater trawls due to its extreme fragility (Robison & Reisenbichler, 2008).

Certain deep-sea species may be attracted by artificial luminescence, and hence light traps have been applied to standard sampling gears, and are responsible, for example, for catch increases of cephalopods in midwater trawls (Clarke & Pascoe, 1998) and of demersal fish in bottom trawls (Gordon *et al.*, 2002).

Conversely, bright white lights (i.e. with full visible spectrum), as typically used on ROVs, may present a potential impact by causing retinal damage to the eyes of some deep-sea organisms (Herring *et al.*, 1999). For this reason, experiments using a special autonomous deep-sea viewing platform (the 'Eye-in-the-Sea', EITS) were performed to test whether the use of red (685 nm) and far-red (695 nm) light could represent an unobtrusive alternative to the use of white light. This test showed a higher number of visits by fish to baited traps with red and far-red compared to white light, validating the efficacy of the technique for a variety of scientific uses in the deep sea (Widder *et al.*, 2005; Raymond & Widder, 2007).

2.4.5 Symbioses

In recent years several macroscopic 'symbiotic' interactions between different invertebrate species have been documented in the deep sea. These findings are as a result of the increasing use of ROV technology and towed camera systems, which enable direct and *in situ* observation of species.

The most widespread interaction seems to be epibiotic, where a smaller species lives permanently attached to the body of a larger sessile host, looking for shelter, stealing its food or just taking advantage of an elevated position to intercept food particles (e.g. Buhl-Mortensen *et al.*, 2010; Shank, 2010). Parasitic interactions occurring in the deep sea have been discovered from samples collected with destructive gears (e.g. trawls), but supplementary information about host preferences and parasite occurrence has been provided by examining images taken by ROVs and towed camera platforms. Examples include polynoid polychaetes (Schiaparelli *et al.*, 2010) and tanaids (Alvaro *et al.*, 2011), which live in association with several deep-sea holothuroids, and a recently described association of polar shrimps with sea anemones for defence (Schiaparelli *et al.*, 2015). ROV technology has also allowed the documentation of more subtle relationships, such as that of the liparid fish genus *Careproctus*, whose females lay their eggs in the gill chambers of king crabs (Somerton & Donaldson, 1998; Stein *et al.*, 2006).

2.5 Spatial distribution patterns

2.5.1 Large-scale benthic patterns

Ecosystems below 200 m lack any primary photosynthetic production and in general have to rely on allochthonous inputs of organic matter from shallower depths (Gage & Tyler, 1991; Smith *et al.*, 2008). This situation often results in an extremely food-limited environment where heterotrophic communities dominate. The variability of deep-sea bottom morphology (which is highest along continental margins, at mid-oceanic ridges and seamounts) and hydrological patterns redistribute the 'rain' of particulate matter from shallower depths and produce discontinuities in its occurrence and abundance. This high variability in the quality and quantity of available organic matter is, in turn, responsible for the observed differences in the spatial distribution of organisms (e.g. Johnson *et al.*, 2007). Except in chemosynthetic situations (e.g. hydrothermal vents, seeps), food limitation can therefore be considered one of the main drivers of spatial and temporal distribution (on an ecological time scale) and adaptation (on an evolutionary time scale) of benthic deep-sea organisms (e.g. Levin *et al.*, 2001; Rex & Etter, 2010). Describing and understanding these patterns requires particular sampling approaches.

2.5.1.1 Geographic patterns in species distribution

In the deep sea, many groups of organisms that inhabit the seafloor, especially nematodes, foraminifera, molluscs and echinoderms, have often been considered to be cosmopolitan, at least in the case of morphologically identified species. In recent years, this observation has been broadly supported by genetic data, which provides evidence for extensive horizontal gene flow and genetic homogeneity in vent, seamount and abyssal plains environments (Baker *et al.*, 2010; McClain & Mincks Hardy, 2010). This pattern can be explained by the longer duration of the pelagic larval development (determined by metabolic constraints), which allows greater dispersal distances for the propagules of deep-sea fauna compared to those of shallow-water fauna (Bradbury *et al.*, 2008). In general, predator species show narrower ranges than deposit feeders, while megafaunal benthic species tend to have wider depth ranges than macro-infauna (Carney, 2005).

However, precise assessment of range statistics is generally hampered by undersampling of highly diverse species pools and by taxonomic uncertainty (Carney, 2005). Moreover, molecular investigations have also proved the existence of complexes of cryptic species, which were previously grouped into single morphological species with a wider distributional range (Vrijenhoek, 2008), suggesting caution in assessing 'general trends'.

This variety of distributions is the natural result of past events which conditioned (on an evolutionary time scale) the biogeographical patterns of species, and the current drivers, which have influences at much finer spatial and temporal scales (on an ecological time scale). Species' ranges may therefore vary greatly, ranging from largely widespread to more restricted or even endemic (e.g. to a specific trench).

Given this variability, it is clear that a variety of sampling strategies and spatial scales have to be taken into account (e.g. Kingsford & Battershill, 1998, and see Chapter 3). According to the species/community of interest, the chosen sampling grid of stations may be very large (i.e. up to hundreds of kilometres between stations) if the species/community has a broad distributional range, or a much more dense sampling grid (kilometres to tens of kilometres between stations) if it is known to have a narrower distribution, or have an extreme variability even at a small spatial scale (metres to hundreds of metres between stations).

2.5.1.2 Bathymetric patterns in species distribution

Depth-related trends and range restriction in faunal distribution have been known since the beginning of deep-sea exploration. Major changes in composition are found with increasing depth due to the variation in physical factors (pressure, temperature, salinity, sediment type, currents topography) and resource availability (nutrients and organic matter) (e.g. Carney, 2005 and references therein).

A bathymetric pattern that can generally be observed is a shift in species composition at the transitions between the shelf and slope (300–500 m), at upper slope depths (1000 m), and between the lower slope and the abyssal plain (2000–3000 m) (Carney *et al.*, 1983). These 'transitional zones' become less marked with increasing latitude and depth, due to the broadening depth distribution of species (Carney, 2005). The main factor underlying this zonation pattern is the increasing distance between the productive surface layers and the deep seafloor – that is food availability. At the continental shelf edge, there are species pools which show a high degree of geographic differentiation and are characterized by large population sizes sustained by an intense benthic–pelagic coupling due to the short distance from the high levels of primary productivity of the surface water (Carney, 2005). At the bottom of the continental slope and below the permanent thermocline, species have smaller population sizes that are food limited due to the more distant link with productive surface layers (Carney, 2005). Between these two extremes there is an intermediate set of species with distributions that merge into the distribution of the other two biotas. On the abyssal seafloor (3000–6000 m), where the environment is relatively homogeneous compared to shallower depths, there are no abrupt depth changes in species composition. Here, although the species pool is less known, there are strong indications of a marked horizontal heterogeneity in species occurrences, with a patchiness probably due to different detritus influxes (i.e. pulsed vs. continuous but low influx) (Carney, 2005).

Hadal habitats have high levels of endemism (e.g. Wolff, 1970; Belyaev 1989), although sampling has always been difficult because of the extreme depths. In recent years there has been an increase in trench research with deployment of deep baited landers and fish traps (see Chapter 11), especially in the Pacific Ocean, revealing considerable variation in physical characteristics and faunal community composition within and between trenches (e.g. Jamieson *et al.*, 2010). Trench communities show distinct differences from adjacent abyssal plains, with depth being a key driver, with less overlap in species as depth increases (Blankenship-Williams & Levin, 2009; Jamieson *et al.*, 2011).

Genetic differentiation along bathymetric gradients has been found in several taxa, with changes occurring even within few kilometres. This genetic differentiation tends to decrease with depth, as environmental conditions become more homogeneous (Etter *et al.*, 2005) showing a degree of cosmopolitanism that increases with depth – up to 40% in bathyal zone and up to 60% in the abyssal zone (Ebbe *et al.*, 2010).

With the exception of abyssal plains, where bathymetry does not vary to a great extent, sampling designs generally need to include some form of depth stratification in order to take into account the faunal shifts related to depth. At transition zones, sampling frequency with depth may need to be increased in order to resolve the precise depth at which species composition changes within a region or a particular habitat. Sampling issues are described further in Chapter 3, and aspects related to analysis of data are covered in Chapter 17.

2.5.1.3 Abundance and biomass

As a general rule in the deep sea, community standing stock (abundance, measured as either population density or biomass) shows an exponential decline with depth in metazoan meiofauna, macrofauna and megafauna (Rowe, 1983; Rex *et al.*, 2006). This pattern has been explained by the progressive reduction of the flux of particulate organic matter to the seafloor with increasing depth (Johnson *et al.*, 2007). Bathymetric declines in standing stock have been revealed using traditional sampling gear. For example, sampling with trawls determined that demersal fish and cephalopod biomass and abundance declines with depth along coastal margins and from the summit of mid-oceanic ridges to the middle rises (e.g. Merrett & Haedrich, 1997). However, very few samples using traditional sampling methods, particularly towed gears such as trawls and sleds, have been recovered from below 6000 m. The increased use of baited traps at hadal depths has shown that communities dominated by carnivorous and scavenging amphipods can attain very high densities (Jamieson *et al.*, 2010), an observation which may begin to challenge the widely accepted pattern of an exponential bathymetric decline in abundance/biomass.

Abundance and biomass of species may show two contrasting patterns, ranging from (a) high abundances shown by few large taxa living in specific habitats, such as seamounts and vent systems, where they aggregate for the substrate and food, to (b) abyssal seafloor, where species often occur with relatively few individuals. In the first case, random sampling is not generally possible due to the limited spatial extent of the communities, but also because standard sampling gears such as corers and dredges may have high potential impact on vulnerable species and environments. In the second case, devices that sample large surface areas, or multiple smaller samples, are required to obtain robust estimates of abundance/biomass (see Chapter 3). In the abyss the fauna demonstrate another general characteristic of deep-sea communities, that of a low degree of dominance or high evenness. This feature, which has few equivalents in shallow-water ecosystems, can be explained as an optimized partition of species and niches for the exploitation of the limited food sources available in the deep sea (Ramirez-Llodra *et al.*, 2010). In the abyss this translates to 'rare is common', for nearly all species are collected as singletons or doubletons at 1–2 sites only (Ebbe *et al.*, 2010).

2.5.1.4 Diversity

The fact that deep-sea sediments host a high diversity of macrofauna became evident only in the 1960s, when the classic 1 mm mesh size of the sieves used in coastal surveys was substituted by smaller sizes (Hessler & Sanders, 1967). Subsequently it became standard practice for deep-sea studies of macrofauna to use a sieve with a 300 μm mesh size (see Chapter 10), revealing an unexpected richness of smaller macrofauna, especially infauna.

Among the most frequently recorded patterns of deep-sea diversity is a unimodal peak at 2000–3000 m. However, although this parabolic distribution is shared across a number of taxa, macrofaunal diversity patterns cannot be generalized for the deep sea. For example, the maximum depth of diversity varies according to location and the considered taxon (Ramirez-Llodra *et al.*, 2010 and references therein). Such differences cannot be explained by a single factor; physical changes in substrate, amount of organic matter input and several other determinants together influence the bathymetric distributional pattern of diversity at a specific site. In some areas the distributional curves are actually inverted from the predicted pattern and diversity minima for some taxa (e.g. polychaetes and foraminiferans) can be found at mid-slope depths where an oxygen-minimum zone (OMZ) is present (Menot *et al.*, 2010). These OMZs, although they have a depressed alpha-diversity, represent major contributions to regional diversity, hosting a suite of different species supported by characteristic processes (Levin, 2003).

With demersal fish fauna, the most important factors in determining richness and taxonomic diversity are depth and, to a lesser extent, latitude. Diversity decreases with increasing depth and shows a complex relationship with latitude, while richness peaks at depths around 1400 m (Campbell *et al.*, 2011). Although diversity assessments are available for most areas, several fish groups that comprise small-bodied species or a body form that is slender may regularly be undersampled and multiple gear types and mesh sizes are necessary to evaluate diversity patterns across their full taxonomic and size ranges (see Chapter 3).

2.5.2 *Large-scale pelagic patterns*

Sampling of pelagic metazoans has always been (and still is) challenging, because of technical constraints of sampling large areas, and the enormous volume of the deep pelagic environment (Ramirez-Llodra *et al.*, 2010). This situation is reflected in chronic underexploration of the deep pelagic ocean, which, at the moment, has to be considered the least studied environment in the deep sea (Webb *et al.*, 2010).

Through the water column, the major factors affecting an organism's distribution are related to depth and the associated changes in physical variables of light, temperature and pressure (Bucklin *et al.*, 2010a). The water column is also a highly dynamic environment. Here different circulation patterns affect nutrient resuspension and upwelling, as well as the transport and rain of fine particles, delineating markedly different water masses and layers (Carney *et al.*, 1977). This dynamism is reflected in the distribution of organisms, whose abundance, biomass and diversity may vary considerably, even at small temporal scales.

2.5.2.1 Geographic patterns of species distribution

At the global scale, the thermohaline 'conveyor belt', which rules the ocean circulation, not only redistributes heat (Gage & Tyler, 1991) but also plays a pivotal role in the distribution of pelagic species, which may disperse following these deep-sea corridors and colonize deep-sea basins from shallower environments (McClain & Mincks Hardy, 2010). Several Antarctic invertebrates appear to have spread from the Antarctic shelf into the world's deep-sea basins via this circulation pattern (e.g. Tyler *et al.*, 2000; Strugnell *et al.*, 2008). At regional scales, distributional patterns of pelagic communities are to a large extent driven by oceanographic features, such as gyres, eddies, upwelling and the transitional areas between water masses (e.g. UNESCO, 2009). These oceanographic features determine distinct assemblages of species, compatible with spatial patterns of change in ecosystem function and/or productivity (e.g. Longhurst, 1998). The boundaries of these assemblages are, however, sometimes 'fuzzy' and may be greatly affected by vertical and/or horizontal movements of water masses. From a sampling viewpoint this variability means that survey designs always need to take into account both temporal and spatial components together, as the boundary between water masses which host different assemblages may move through time.

The pelagic realm is relatively unaffected by interactions at its upper and lower limits (i.e. ocean's surface and bottom), with the exception of the ocean ridges which represent marked topological discontinuities that exert an influence on the structure of the overlying pelagic communities (Vecchione *et al.*, 2010). Changing patterns of communities means that sampling needs to cover the gradient of conditions over and along ridges, as well as a considerable distance off the ridge axis (e.g. Wenneck *et al.*, 2008).

2.5.2.2 Bathymetric patterns in species distribution

Changes in the physical characteristics of the pelagic environment have been used historically to define different layers, each one having a peculiar set of physical characteristics. These zones, although useful to describe the main physical features encountered, have to be considered as broad generalizations only, their boundaries (with the exception of very pronounced pycnoclines) representing transitional areas where conditions change progressively rather than sharp discontinuities. Species living in the water column maintain specific depth ranges with respect to these conditions. Bathymetric patterns can be investigated through direct sampling by using midwater trawls or plankton nets. However, fine-scale reconstruction of the water column structure is best achieved by using multi-opening net systems such as MOCNESS (Multiple Opening and Closing Net and Environmental Sensing System), which enable stratified oblique tows coupled with collection of real-time environmental data (see Chapters 4 and 6).

Indirect sampling with acoustic methods can also be effective at discriminating community patterns. Thanks to the differing acoustic properties of several groups of species living in the water column, multi-frequency acoustic data can separate groups of animals, and describe the general composition and distribution of pelagic communities, once the acoustic target strength of the species under has been assessed (e.g. Kloser *et al.*, 2002; O'Driscoll *et al.*, 2011). Such techniques are also important where the high densities of aggregating species limit the use of trawls because catch rates can be too high in a localized area (e.g. Clark, 1996).

It is important that sampling designs for pelagic fauna include a temporal component, for the bathymetric position of a species may vary during the day/night because many species undertake diurnal migrations in order to obtain food or escape predation (e.g. Herring, 2002).

2.5.2.3 Abundance and biomass

The abundance and biomass of pelagic organisms vary with differing eutrophic, mesotrophic or oligotrophic conditions in certain areas. Standing stock also varies with depth, shifting from the highest values near the surface, to intermediate ones in the mesopelagic and very low levels in the bathypelagic, and then increasing again in the benthopelagic (Marshall, 1979).

Each taxonomic group also shows specific trends and patterns. For example, the abundance of copepods is highest in shallow water, and then decreases exponentially with depth. Decapods also have their highest abundances in the first 1000 m of the water column. Gelatinous planktonic organisms (cnidarians, ctenophores and tunicates) also show peaks of biomass usually shallower than 1000 m. Among the major contributors to biomass at the global scale are the scyphomedusae belonging to the genera *Periphylla* and *Atolla*, which are grazed by several fish and amphipod species (e.g. *Parandania*, Fig. 2.2e). For the fish, general trends for biomass indicate an exponential decline in the deep-water column below 1000 m in open ocean ecosystems while, along mid-ocean ridges, fish biomass peaks between 1500 and 2500 m (Sutton *et al.*, 2008). System-wide models of upper trophic levels suggest that in some areas an important component of pelagic fish biomass is still unquantified. This 'missing biomass' is probably represented by myctophids and other mesopelagic micronekton that may abound in deep water but are not easy to sample, their specific composition and role remaining largely unknown (Johnson *et al.*, 2011). As mentioned earlier in this chapter, deep-towed acoustics with multiple frequencies may be an option in the future to quantify this part of the ecosystem (and see Chapter 6).

Abundance and biomass can also greatly vary temporally, as species undertake diurnal vertical or seasonal horizontal migrations to a specific location (e.g. breeding grounds). Discrete bottom features, such as seamounts, are known to trap migrating zooplankton during its dawn descent (the trapping of the 'deep scattering layer'), determining ephemeral higher concentrations of organisms that are consumed by the resident predators (Genin & Dower, 2007; Clark *et al.*, 2010). This requires repetitive temporal sampling above and off the seamount, with multiple gear types, to sample the plankton, as well as predatory fish.

Given the high likely variation in abundance and biomass of planktonic organisms, it is clear that sampling design and surveys need to consider both spatial and temporal sources of variability in order to obtain consistent and meaningful quantitative data about abundance and biomass.

2.5.2.4 Diversity

Both *in situ* observation and direct sampling have shown a marked variability of diversity patterns with depth, with differences in macroplanktonic assemblages occurring even in depth layers separated by only a few tens of metres (e.g. Vinogradov, 2005). As a general trend, the number of zooplanktonic species decreases with depth (Longhurst, 1995), and may show a peak in midwater layers (Bucklin *et al.*, 2010b).

Greater numbers of pelagic species at all depths can be found at lower latitudes compared to assemblages at higher latitudes (Angel, 2003; Vecchione *et al.*, 2010). Latitude and temperature are the two major determinants of the diversity of zooplanktonic taxa (e.g. Copepoda, Decapoda, Euphasiacea, Amphipoda). Copepod diversity patterns in particular are strongly correlated with annually averaged sea surface temperatures and, to a lesser extent, with salinity at the latitudinal basin scale (Rombouts *et al.*, 2009). Their diversity is negatively correlated with average sea surface chlorophyll (Rombouts *et al.*, 2009). The diversity patterns of taxa with hard exoskeletons have been largely determined using traditional sampling techniques such as trawls and nets (see Chapter 7).

The 'gelatinous' fraction of the zooplankton, encompassing cnidarians (hydromedusae, scyphomedusae, siphonophores), ctenophores and tunicates (salps, doliolids, pyrosomes, appendicularians) represents one of the most significant discoveries in recent years, achieved through the use of deep-diving midwater vehicles which allowed estimation of the abundance and diversity of these organisms (see Chapters 12 and 13), previously severely underestimated with sampling based on nets only. This group is now is considered a 'hotspot' for discoveries of species new to science (Bucklin *et al.*, 2010a, 2010b).

2.5.3 Small-scale benthic and pelagic patterns

Beside depth-related factors, habitat heterogeneity is one of the other main drivers of biodiversity, providing differentiated niches and other opportunities for many species to exist in the deep sea (Rex & Etter, 2010).

Among the most obvious examples of habitat heterogeneity in the deep sea is found where chemosynthetic environments (e.g. vents and seeps) or biogenic substrates (e.g. whale bones, coral reefs) occur surrounded by relatively bare sand or rock seafloor. In these cases, at the spatial scale of hundreds or tens of metres, faunal communities change from having high abundances and biomasses of few highly adapted macro-epifaunal species to the very low biomasses present on the neighbouring, food-limited substratum.

In the deep sea, the first kind of these biomass-rich 'oases' to be discovered was hydrothermal vent habitat in 1977. Since then more than 200 species have been described from these environments, many of them belonging to new genera (e.g. *Alviniconcha*, *Ifremeria* and *Lepetodrilus,* among molluscs) and new families (e.g. Kiwidae among decapods) (Baker *et al.*, 2010).

Whale falls represent another example of localized habitat for rich communities of microbes and invertebrates which progressively degrade flesh, blubber and even the more refractory components of the carcass, chitin and bones (Smith & Baco, 2003; Goffredi *et al.*, 2004). Whale carcasses, together with wood and kelp remains, are considered to be similar habitats to vents and seeps, due to the fact that the microbes found here have similar metabolic pathways to their vents and seeps counterparts and support similar communities of invertebrates (Smith & Baco, 2003). After the first whale fall was found in 1987 (Smith *et al.*, 1989), these habitats have been documented down to about 3000 m in several of the world's ocean basins, where they seem to represent, at least on evolutionary time scales, the 'stepping-stones' for the dispersal of deep-sea vent faunas (Smith *et al.*, 1989; Distel *et al.*, 2000).

Immediately after a whale carcass reaches the seafloor, specialized opportunistic species colonize the body following a precise succession of four stages; namely the 'mobile-scavenger' stage, the 'enrichment opportunists' stage, the 'sulfophilic' stage and the 'reef' stage (Smith & Baco, 2003). This succession of communities requires about 3 years to be completed, although the length of each stage may vary according to the bottom depth and the seawater temperature (Smith & Baco, 2003; Fujiwara *et al.*, 2007).

In these habitats, some of the most peculiar organisms are the whalebone-eating siboglinid worms of the genus *Osedax* (the 'zombie worm') (Fig. 2.2g,h), which represent a very important element in the ecological succession of whale falls, being able to bore into bones and hence to completely degrade whale carcasses (Rouse *et al.*, 2004; Glover *et al.*, 2005).

Like vents and seeps, the localized nature of whale falls means that they are most efficiently and effectively sampled using submersibles or ROVs. After these patchy habitats have been located, such vehicles can operate cameras, position push-cores with the manipulator arm or collect samples without compromising the integrity of the habitat. Several studies of vents, seeps and whale falls have included a time series of sampling events, and as such rely upon the vehicles being able to return to the point of observation with the aid of GPS as well as physical markers previously placed at the study site (see Chapter 13).

Cold-water or deep-sea corals (Fig. 2.2c,d) that form 'reef-like' structures represent another conspicuous patch type of habitat in the deep sea. While there are many coral taxa, the principal habitat-forming species are the scleractinians (Freiwald *et al.*, 2004). These corals are all azooxanthellate and occur at temperatures between 4 and 13°C (Freiwald, 2002), especially on raised geomorphological structures, such as seamounts, where they take advantage of the particular hydrological conditions than enhance the supply of suspended food particles. Their importance in deep-sea ecosystems is their biogenic three-dimensional (3D) structure, which provides habitat for a variety of fish and associated invertebrates (e.g. squat lobsters, sea stars, brittle stars, polychaete worms and crabs). These thrive on the structures mainly as epibionts (Shank, 2010), but also promote the biodiversity of the meiofauna living in the surrounding sediments (Bongiorni *et al.*, 2010).

Deep-sea corals also represent one of the most vulnerable habitats in the deep sea, especially to trawling (and fishing impact in general), due to their fragile composition and general slow growth. Deep-sea coral reefs may require decades to recover from impact, or may even not recover at all (e.g. Althaus *et al.*, 2009; Williams *et al.*, 2010), and destructive methods of sampling should to be avoided as much as possible. Towed cameras, ROVs, autonomous underwater vehicles (AUVs) or manned submersibles are ideal for sampling the fauna of these fragile habitats (see Chapters 12 and 13).

Habitat heterogeneity is also observable on the abyssal plain, which is usually considered homogeneous. For example, differences in infaunal communities (meio- and macrofauna) at scales of 1–5 km can be found between nodule fields compared to adjacent soft sediment areas (Miljutina *et al.*, 2009).

Beside between-habitat differences, deep-sea communities may also show differences within the same habitat, due to patchiness related to variation in sedimentary carbon content, physical disturbance or micro-habitat provision. For example, patchiness in macrofaunal species at bathyal depths can be observed at the scales of centimetres to hundreds of metres, which matches the distribution of organic matter in the sediment (e.g. McClain *et al.*, 2011). The physical disturbance associated with the creation of bioturbatory features on the seafloor, such as mounds and pits, has been shown to influence the small-scale distribution of infauna. Small-scale differences in nematode communities have also been observed surrounding small (1–2 cm in diameter) sponges occurring on the flat abyssal seafloor (e.g. Hasemann & Soltwedel, 2011). In order to document these small-scale patterns in the benthos it was necessary to exploit the precision sampling capability of ROVs.

2.6 Temporal patterns

The apparent constancy of the major abiotic characteristics of the deep sea led to the initial belief that this environment was one of the most stable on Earth. However, it was later shown that this is not the case and that the deep sea is highly dynamic (Tyler, 1988). Both long-term and short-term events have been recognized, as well as regular and episodic ones, which affect species distribution and fluctuations in their abundance.

2.6.1 Short-term patterns

In the case of deep-sea benthic organisms, the most important short-term and regular phenomenon is the seasonal deposition of phytodetritus, which brings high-quality food resources in a food-limited environment, triggering biological responses including seasonal reproduction (e.g. Young, 2003). This phenomenon is obviously more marked in areas with high seasonal primary productivity; here, changes on the deep seafloor occur within few weeks after the surface phytoplankton bloom, with macrofauna showing an immediate response to food availability and the microbiota a delayed one (Witte *et al.*, 2003).

Stochastic short-term events are represented by the fall of whale carcasses, fish and jellyfish remnants, which, at the very local scale, produce an additional organic input that profoundly influences the faunal community for at least 3 years. Benthic storms, which usually last only a matter of minutes to hours, can dramatically change substrate texture and precipitate community shifts which may persist for many years (Lambshead *et al.*, 2001).

These sorts of events, both short and long term, require repeat sampling at the same site. Fixed stations and time-lapse photographic methods have been important in documenting variability in environmental conditions and faunal responses (e.g. Billet *et al.*, 2010; see Chapters 3 and 14). The use of ROVs and manned submersibles has also been important in resampling sites with a positional precision that many towed gears cannot achieve.

Bioturbation can locally alter sediment characteristics, with a cascade of effects conditioning microbial metabolism, rates of organic-carbon degradation and the activity of the surrounding infauna. As such, it can influence spatial patterns, but these may vary on short times scales. For example, at bathyal depths, the sediment mounds produced by megafaunal species, such as echiuran worms as they feed, are periodically made and subsequently eroded as the organism relocates itself from one patch of food to another. The influence this sediment turnover has upon the local infauna will therefore not be constant (Smith *et al.*, 1986).

As already noted, the study of bioturbation and its effects on local infaunal communities is typically performed by manned submersibles and ROVs which are able to manoeuvre at the smallest spatial scale and collect the necessary replicated cores in proximity of the mound or the biogenic structure of interest, and return to the location of the same structure at a later date.

2.6.2 Long-term patterns

Deep-sea communities are also affected by long-term events, for example, decadal-scale shifts in primary production principally caused by climate-related oscillations. During major El Niño/La Niña transitions changes occur in phytoplankton composition and abundance in surface layers, which may translate to large-scale effects on benthic communities. Changes of two orders of magnitude in abundance of abyssal megafauna have been documented following an abrupt demographic increase of the holothuroid *Amperima rosea* (Billet *et al.*, 2010; see Chapter 3).

The study of long-term patterns in the deep sea requires the disentangling of seasonal or intra-annual changes from inter-annual and, if possible, decadal ones. Methods and techniques may be no different from those required for monitoring short-term variability, but need to occur over much longer time periods. This needs a major research and funding commitment over a time scale that exceeds immediate management time frames and short-term objectives. Hence it involves developing a different type of research approach which needs very careful evaluation of sampling gear and survey design. These issues are covered in different chapters in this book (e.g. Chapters 3 and 14).

2.7 Concluding remarks

Deep-sea biologists need to physically collect samples to complete taxonomic inventories and quantify diversity. However, they are also often operating in fragile habitats, where even scientific sampling activities may have an impact. Hence they are faced with the difficult task of collecting enough specimens or data to accomplish a study, but to do so without altering the communities they study. The development of towed cameras, submersibles and ROVs now enables non-destructive or selective sampling and observation which helps preserve the environment and the fauna. However, only the physical sampling of specimens enables correct taxonomic classifications, and more invasive sampling gear is needed for sampling the wide range of faunal characteristics described in this chapter, and especially given the spatial and temporal variability of the faunal communities.

Acknowledgements

We are indebted to Bruce Robison, Kim Fulton-Bennett and Nancy Jacobsen Stout (Monterey Bay Aquarium Research Institute, MBARI) for permission to use the images of *Macropinna microstoma* (Fig. 2.2f) and the ROV arm sampling a whale bone (Fig. 2.2g). Greg Rouse (Scripps Institution of Oceanography) kindly made available the *Osedax* picture of Fig. 2.2h. Three anonymous reviewers are acknowledged for their suggestions and comments which greatly improved a first draft of the chapter.

References

Althaus, F., Williams, A., Schlacher, T.A., *et al.* (2009). Impacts of bottom trawling on deep-coral ecosystems of seamounts are long-lasting. *Marine Ecology Progress Series*, 397: 279–294.

Alvaro, M.C., Blazewicz-Paszkoycz, M., Davey, N., Schiaparelli, S. (2011). Skin-digging tanaidaceans: the unusual parasitic behaviour of *Exspina typica* Lang, 1968 in Antarctic waters and worldwide deep basins. *Antarctic Science*, 23 (4): 343–348.

Angel, M.V. (2003). The pelagic environment of the open ocean. In: Tyler, P.A. (ed.), *Ecosystems of the deep ocean*. Elsevier, Amsterdam. pp. 39–79.

Baker, M.C., Ramirez-Llodra, E.Z., Tyler, P.A., *et al.* (2010). Biogeography, ecology, and vulnerability of chemosynthetic ecosystems in the deep sea. In: McIntyre, A.D. (ed.), *Life in the world's oceans*. Oxford: Blackwell, pp. 161–182.

Belyaev, G.M. (1989). *Deep sea ocean trenches and their fauna*. Moscow: Nauka, 255 pp (in Russian, translation courtesy of Scripps Institution of Oceanography Library).

Billet, D.S.M., Bett, B.J., Reid, W.D.K., Boorman, B., Priede, I.G. (2010). Long-term change in the abyssal NE Atlantic: the '*Amperima* event' revisited. *Deep-Sea Research II*, 57: 1406–1417.

Blankenship-Williams, L. E., Levin, L.A. (2009). Living deep: a synopsis of hadal trench ecology. *Marine Technology Society Journal*, 43(5): 137–143.

Bongiorni, L., Mea, M., Gambi, C., Pusceddu, A., Taviani, M., Danovaro, R. (2010). Deep-water scleractinian corals promote higher biodiversity in deep-sea meiofaunal assemblages along continental margins. *Biological Conservation*, 143: 1687–1700.

Bradbury, I.R., Laurel, B., Snelgrove, P.V.R., Bentzen, P., Campana, S.E. (2008). Global patterns in marine dispersal estimates: the influence of geography, taxonomic category and life history. *Proceedings of the Royal Society B Biological Sciences*, 275: 1803–1809.

Brenke, N. (2005). An epibenthic sledge for operations on marine soft bottom and bedrock. *Marine Technology Society Journal*, 39(2): 10–21.

Bucklin A., Ortman B.D., Jennings R.M., *et al.* (2010a). A 'Rosetta Stone' for metazoan zooplankton: DNA barcode analysis of species diversity of the Sargasso Sea (Northwest Atlantic Ocean). *Deep-Sea Research II*, 57: 2234–2247.

Bucklin, A., Nishida, S., Schnack-Schiel, S., *et al.* (2010b). A census of zooplankton of the global ocean. In: McIntyre, A.D. (ed.), *Life in the world's oceans*. Oxford: Blackwell, pp. 247–265.

Buhl-Mortensen, L., Vanreusel, A., Gooday, A.J., *et al.* (2010). Biological structures as a source of habitat heterogeneity and biodiversity on the deep ocean margins. *Marine Ecology*, 31: 21–50.

Campbell, N., Neat, F., Burns, F., Kunzlik, P. (2011). Species richness, taxonomic diversity, and taxonomic distinctness of the deep-water demersal fish community on the Northeast Atlantic continental slope (ICES Subdivision VIa). *ICES Journal of Marine Science*, 68: 365–376.

Carney, R.S. (2005). Zonation of deep biota on continental margins. *Oceanography and Marine Biology: An Annual Review*, 43: 211–278.

Carney, R.S., Haedrich, R.L., Rowe, G.T. (1983). Zonation of fauna in the deep sea. In: Rowe, G.T (ed.), *The sea*, Vol. 8: *Deep-sea biology*. New York: Wiley-Interscience, pp. 97–122.

Carney, R.S., Haedrich, R.L., Rowe, G.T. (1977). Zonation of fauna in the deep sea. In: Keegan, B.F., Ceidigh, P.O., Boaden, P. (eds). *Biology of benthic organisms*. New York: Pergamon Press, pp. 372–398.

Clark, M.R. (1996). Biomass estimation of orange roughy: a summary and evaluation of techniques for measuring stock size of a deepwater fish species in New Zealand. *Journal of Fish Biology*, 49 (suppl. A): 114–131.

Clark, M.R., Rowden, A.A., Schlacher, T., *et al.* (2010). The ecology of seamounts: structure, function, and human impacts. *Annual Review of Marine Science*, 2: 253–278.

Clarke, M.R., Pascoe, P.L. (1998). The influence of an electric light on the capture of oceanic cephalopods by a midwater trawl. *Journal of the Marine Biological Association of the United Kingdom*, 78: 561–575.

D'Onghia, G., Indennidate, A., Giove, A., *et al.* (2011). Distribution and behaviour of deep-sea benthopelagic fauna observed using towed cameras in Santa Maria di Leuca cold-water province. *Marine Ecology Progress Series*, 443: 95–110.

Distel, D.L., Baco, A.R., Chuang, E., Morrill, W., Cavanaugh, C., Smith, C.R. (2000). Do mussels take wooden steps to deep-sea vents? *Nature*, 403: 725–726.

Ebbe, B., Billett, D.S.M., Brandt, A., *et al.* (2010). Diversity of abyssal marine life. In: McIntyre, A.D. (ed.), *Life in the world's oceans*. Oxford: Blackwell, pp. 139–160.

Etter, R.J., Rex, M.A., Chase, M.R., Quattro, J.M. (2005). Population differentiation decreases with depth in deep-sea bivalves. *Evolution*, 59: 1479–1491.

Freiwald, A. (2002). Reef-forming cold-water corals. In: Wefer, G., Billett, D.S.M., Hebbeln, D., Jørgensen, B.B., Schlüter, M., Van Weering, T. (eds), *Ocean margin systems*. Berlin: Springer, pp. 365–385.

Freiwald, A., Fosså, J.H., Grehan, A. Koslow, T., Roberts, J.M. (2004). *Cold-water coral reefs*. Cambridge, UK: UNEP-WCMC, 84 pp.

Fujiwara, Y., Kawato, M., Yamamoto, T., *et al.* (2007). Three-year investigation into sperm whale-fall ecosystems in Japan. *Marine Ecology*, 28: 219–232.

Gage, J.D., Tyler, P.A. (1991). *Deep-sea biology. A natural history of organisms at the deep-sea floor*. Cambridge, UK: Cambridge University Press.

Genin, A., Dower, J.F. (2007). Seamount plankton dynamics. In: Pitcher, T.J., Morato, T., Hart, P.J.B., Clark M.R., Haggan N., Santos R.S. (eds), *Seamounts: ecology, fisheries, and conservation*. Oxford: Blackwell, pp. 85–100.

Glover, A.G., Källström, B., Smith, C.R., Dahlgren, T.G. (2005). World-wide whale worms? A new species of *Osedax* from the shallow north Atlantic. *Proceedings of the Royal Society B Biological Sciences*, 272: 2587–2592.

Goffredi, S.K., Paull, C.K., Fulton-Bennett, K., Hurtado, L.A., Vrijenhoek R.C. (2004). Unusual benthic fauna associated with a whale fall in Monterey Canyon, California. *Deep-Sea Research I*, 51: 1295–1306.

Gooday, A.J., Todo, Y., Uematsu, K., Kitazato, H. (2008). New organic-walled Foraminifera (Protista) from the ocean's deepest point, the Challenger Deep (western Pacific Ocean). *Zoological Journal of the Linnean Society*, 153: 399–423.

Gordon J.D.M., Bergstad O.A., Pascoe P.L. (2002). The influence of artificial light on the capture of deep-water demersal fish by bottom trawling. *Journal of the Marine Biological Association of the UK*, 82: 339–344.

Haddock, S.H.D., Moline, M.A., Case, J.F. (2010). Bioluminescence in the sea. *Annual Review of Marine Science*, 2: 443–493.

Hasemann C, Soltwedel T (2011). Small-scale heterogeneity in deep-sea nematode communities around biogenic structures. *PLoS ONE*, 6(12): e29152.

Herring P.J., Gaten E., Shelton P.M.J. (1999). Are vent shrimps blinded by science? *Nature*, 398: 116.

Herring, P.J. (2002). *The biology of the deep ocean*. Oxford: Oxford University Press.

Hessler, R., Sanders, M.L. (1967). Faunal diversity in the deep sea. *Deep Sea Research*, 14: 65–78.Jamieson, A.J., Fujii, T., Mayor, D.J., Solan, M., Priede, I.G. (2010). Hadal trenches: the ecology of the deepest places on Earth. *Trends in Ecology and Evolution*, 25(3): 190–197.

Jamieson, A.J., Kilgallen, N.M., Rowden, A.A., *et al.* (2011). Bait-attending fauna of the Kermadec Trench, SW Pacific Ocean: evidence for an ecotone across the abyssal–hadal transition zone. *Deep-Sea Research I*, 58: 49–62.

Johnson, C.L., Runge, J.A., Curtis, K.A., *et al.* (2011). Biodiversity and ecosystem function in the Gulf of Maine: pattern and role of zooplankton and pelagic nekton. *PLoS ONE*, 6(1), e16491.

Johnson, N.A., Campbell, J.W., Moore, T.S., *et al.* (2007). The relationship between the standing stock of deep-sea macrobenthos and surface production in the western North Atlantic. *Deep-Sea Research I*, 54: 1350–1360.

Kingsford, M., Battershill, C.N. (eds) (1998). *Studying marine environments: a handbook for ecologists*. Canterbury, New Zealand: Canterbury University Press, 335 pp.

Kloser, R.J., Ryan, T., Sakov, P., Williams, A., Koslow, J.A. (2002). Species identification in deep water using multiple acoustic frequencies. *Canadian Journal of Fisheries and Aquatic Sciences,* 59: 1065–1077.

Koslow, T. (2007). *The silent deep: the discovery, ecology, and conservation of the deep sea*. Chicago, IL: University of Chicago Press.

Lambshead, P.J.D., Tietjen, J., Glover, A., Ferrero, T., Thistle, D., Gooday, A.J. (2001). Impact of large-scale natural physical disturbance on the diversity of deep-sea North Atlantic nematodes. *Marine Ecology Progress Series*, 214: 121–126.

Levin, L.A. (2003). Oxygen minimum zone benthos: adaptation and community response to hypoxia. *Oceanography and Marine Biology: An Annual Review*, 41: 1–45.

Levin, L.A., Etter, R.J., Rex, M.A., *et al.* (2001). Environmental influences on regional deep-sea species diversity. *Annual Review of Ecology and Systematics*, 32: 51–93.

Longhurst, A. (1995). Seasonal cycles of pelagic production and consumption. *Progress in Oceanography*, 36: 77–167.

Longhurst, A. (1998). *Ecological geography of the sea*. London: Academic Press.

Marshall, N.B. (1979). *Deep sea biology: developments and perspectives*. New York: Garland, STPM Press, p. 566.

McClain, C.R., Mincks Hardy, S. (2010). The dynamics of biogeographic ranges in the deep sea. *Proceedings of the Royal Society B Biological Sciences*, 277: 3533–3546.

McClain, C.R., Nekola, J.C., Kuhnz, L., Barry, J.P. (2011). Local-scale faunal turnover on the deep Pacific seafloor. *Marine Ecology Progress Series*, 422: 193–200.

Menot, L., Sibuet, M., Carney, R.S., *et al.* (2010). New perceptions of continental margin biodiversity. In: McIntyre, A.D. (ed.), *Life in the world's oceans.* Oxford: Blackwell, pp. 79–101.

Merrett, N.R., Haedrich, R.L. (1997). *Deep-sea demersal fish and fisheries*. London: Chapman and Hall.

Miljutina, M.A., Miljutina, D.M., Mahatma, R., Galéron, J. (2009). Deep-sea nematode assemblages of the Clarion-Clapperton Nodule Province (Tropical North-Eastern Pacific). *Marine Biodiversity*, 40: 1–15.

Miller, J.E., Pawson, D.L. (1990). Swimming sea cucumbers (Echinodermata: Holothuroidea): a survey, with analysis of swimming behavior in four bathyal species. *Smithsonian Contributions to the Marine Sciences*, 35: 1–18.

Munk, O. (2000). Histology of the fusion area between the parasitic male and the female in the deep-sea anglerfish *Neoceratias spinifer* Pappenheim, 1914 (Teleostei, Ceratioidei). *Acta Zoologica*, 81: 315–324.

Murray, J. (1895). A summary of the scientific results obtained at the sounding, dredging and trawling stations of the HMS Challenger. *Challenger Reports, Summary Research*, 2: 797–1608.

O'Driscoll, R., Macaulay, G.J., Gauthier, S., Pinkerton, M., Hanchet, S. (2011). Distribution, abundance and acoustic properties of Antarctic silverfish (*Pleuragramma antarcticum*) in the Ross Sea. *Deep-Sea Research*, 58: 181–195.

Pietsch, T.W. (2004). Dimorphism, parasitism, and sex revisited: modes of reproduction among deep-sea ceratioid anglerfishes (Teleostei: Lophiiformes). *Ichthyological Research*, 52: 207–236.

Ramirez-Llodra, E., Brandt, A., Danovaro, R., *et al.* (2010). Deep, diverse and definitely different: unique attributes of the world's largest ecosystem. *Biogeosciences*, 7: 2851.

Raupach, M.J., Mayer, C., Malyutina, M., Waegele, J.W. (2009). Multiple origins of deep-sea Asellota (Crustacea: Isopoda) from shallow waters revealed by molecular data. *Proceedings of the Royal Society B Biological Sciences*, 276: 799–808.

Raymond, E.H., Widder, E.A. (2007). Behavioral responses of two deep-sea fish species to red, far-red, and white light. *Marine Ecology Progress Series*, 350: 291–298.

Regan, C.T. (1926). The pediculate fishes of the suborder Ceratioidea. *Dana Oceanographic Reports*, 2: 1–45.

Rex, M.A., Etter, R.J. (2010). *Deep-sea biodiversity: pattern and scale*. Boston, MA: Harvard University Press, 368 pp.

Rex, M.A., Etter, R.J., Morris, J.S., *et al.* (2006). Global bathymetric patterns of standing stock and body size in the deep-sea benthos. *Marine Ecology Progress Series*, 317: 1–8.

Robison, B.H. (2009). Conservation of deep pelagic biodiversity. *Conservation Biology*, 23(4): 847–858.

Robison, B.H., Reisenbichler, K.R. (2008). *Macropinna microstoma* and the paradox of its tubular eyes. *Copeia*, 4, 780–784.

Rombouts, I., Beaugrand, G., Ibañez, F., Gasparin, S., Chiba, S., Legendre, L. (2009). Global latitudinal variations in marine copepod diversity and environmental factors. *Proceedings of the Royal Society B Biological Sciences,* 276: 3053–3062.

Rouse, G.W., Goffredi, S.K., Vrijenhoek, R.C. (2004). *Osedax*: Bone-eating marine worms with dwarf males. *Science*, 305: 668–671.

Rowe, G.T. (1983). Biomass and production of the deep-sea macrobenthos. In: Rowe, G.T. (ed.), *The sea*, Vol. 8: *Deep-sea biology*. New York: Wiley Interscience, pp. 97–121.

Schiaparelli, S., Alvaro, M.C., Bohn, J., Albertelli, G. (2010). 'Hitchhiker' polynoid polychaetes in cold deep waters and their potential influence on benthic soft bottom food webs. *Antarctic Science*, 22(4): 399–407.

Schiaparelli, S., Ahyong, S.T., Bowden, D. (2015). Evidence of niche conservatism and host fidelity in the polar shrimp Lebbeus kiae n. sp. (Decapoda: Caridea: Thoridae) from the Ross Sea, Antarctica. *Hydrobiologia*, 761: 45–69, doi: 10.1007/s10750-015-2403-1.

Shank, T.M. (2010). Seamounts: deep-ocean laboratories of faunal connectivity, evolution, and endemism. *Oceanography*, 23(1): 108–122.

Smith, C.R., Jumarst, P.A., DeMastert, D.J (1986). *In situ* studies of megafaunal mounds indicate rapid sediment turnover and community response at the deep-sea floor. *Nature*, 323: 251–253.

Smith, C., Baco, A. (2003). The ecology of whale falls at the deep-sea floor. *Oceanography and Marine Biology: An Annual Review*, 41: 311–354.

Smith, C.R., De Leo, F.C., Bernardino, A.F., Sweetman, A.K., Martinez Arbizu, P. (2008). Abyssal food limitation, ecosystem structure and climate change. *Trends in Ecology and Evolution*, 23(9): 518–528.

Smith, C.R., Kukert, A., Wheatcroft, R.A., Jumars, P.A., Deming, J.W. (1989). Vent fauna on whale remains. *Nature*, 341: 27–28.

Somerton, D.A., Donaldson, W. (1998). Parasitism of the golden king crab, *Lithodes aequispinus*, by two species of snailfish, genus *Careproctus*. *Fishery Bulletin*, 96: 871–884.

Stein, D.L., Drazen, J.C., Schlining, K.L., Barry, J.P., Kuhnz, L. (2006). Snailfishes of the central California coast: video, photographic and morphological observations. *Journal of Fish Biology*, 69: 970–986.

Strugnell J.M., Rogers A.D., Prodöhl P.A., Collins M.A., Allcock A.L. (2008). The thermohaline expressway: the Southern Ocean as a centre of origin for deep-sea octopuses. *Cladistics*, 24: 853–860.

Sutton, T.T., Porteiro, F.M., Heino, M., *et al.* (2008). Vertical structure, biomass and topographic association of deep-pelagic fishes in relation to a mid-ocean ridge system. *Deep-Sea Research II*, 55(1–2): 161–184.

Todo, Y., Kitazato, H., Hashimoto, J., Gooday, A.J. (2005). Simple Foraminifera flourish at the 10,896 ocean's deepest point. *Science*, 307: 689.

Tyler P.A., Young C.M., Clarke A. (2000). Temperature and pressure tolerances of embryos and larvae of the Antarctic sea urchin *Sterechinus neumayeri* (Echinodermata: Echinoidea): potential for deep-sea invasion from high latitudes. *Marine Ecology Progress Series*, 192: 173–180.

Tyler, P. (2007). The deep sea: if we do not understand the biodiversity, can we assess the threat? In: Hester, R.E., Harrison, R.M. (eds). *Biodiversity under threat*. London: Royal Society of Chemistry Publishing, pp. 81–106.

Tyler, P.A. (1988). Seasonality in the deep-sea. *Oceanography and Marine Biology*, 26: 227–258.

UNESCO. (2009). *Global Open Oceans and Deep Seabed (GOODS) – Biogeographic Classification. IOC Technical Series, 84*. Paris: UNESCO-IOC.

Vecchione, M., Bergstad, O.A., Byrkjedal, I., *et al.* (2010). Biodiversity patterns and processes on the Mid-Atlantic Ridge. In: McIntyre, A.D. (ed.), *Life in the world's oceans*. Oxford: Blackwell, pp. 103–121.

Villanueva, R., Segonzac, M., Guerra, A. (1997). Locomotion modes of deep-sea cirrate octopods (Cephalopoda) based on observations from video recordings on the Mid-Atlantic Ridge. *Marine Biology*, 129: 113–122.

Vinogradov, G.M. (2005). Vertical distribution of macroplankton at the Charlie-Gibbs Fracture Zone (North Atlantic), as observed from the manned submersible 'Mir-1'. *Marine Biology*, 146: 325–331.

Vrijenhoek, R.C. (2008). Cryptic species, phenotypic plasticity, and complex life histories: assessing deep-sea diversity with molecular markers. *Deep-Sea Research II*, 56: 1713–1723.

Warrant, E.J., Locket, N.A. (2004). Vision in the deep sea. *Biological Review*, 79: 671–712.

Webb, T.J., Vanden Berghe, E., O'Dor, R. (2010). Biodiversity's big wet secret: the global distribution of marine biological records reveals chronic under-exploration of the deep pelagic ocean. *PLoS ONE*, 5(8): e10223.

Wenneck, T.deL., Falkenhaug, T., Bergstad, O.A. (2008). Strategies, methods, and technologies adopted on the R.V. *G.O. Sars* MAR-ECO expedition to the Mid-Atlantic Ridge in 2004. *Deep-Sea Research II*, 55: 6–28.

Widder, E.A., Robison, B.H., Reisenbichler, K.R., Haddock, S.H.D. (2005). Using red light for in situ observations of deep-sea fishes. *Deep Sea Research I*, 52(11): 2077–2085.

Williams, A., Schlacher, T.A., Rowden, A.A., *et al.* (2010). Seamount megabenthic assemblages fail to recover from trawling impacts. *Marine Ecology*, 31 (suppl. 1): 183–199.

Witte, U., Wenzhöfer, F., Sommer, S., *et al.* (2003). *In situ* experimental evidence of the fate of a phytodetritus pulse at the abyssal sea floor. *Nature*, 424: 763–765.

Wolff, T. (1970). Concept of hadal or ultra-abyssal fauna. *Deep-Sea Research*, 17: 983–1003.

Young, C.M. (2003). Reproduction, development and life history traits. In: Tyler, P.A. (ed.), *Ecosystems of the world, Vol. 28: Ecosystems of the deep oceans*. London: Elsevier, pp. 381–426.

Chapter 3
Survey and Sampling Design

Malcolm R. Clark[1], Thomas A. Schlacher[2], Gui M. Menezes[3], Tina N. Molodtsova[4] and Ian J. Doonan[1]

[1] National Institute of Water and Atmospheric Research, Wellington, New Zealand
[2] The University of the Sunshine Coast, Maroochydore, Australia
[3] University of the Azores, Horta, Portugal
[4] P.P. Shirshov Institute of Oceanology, Moscow, Russia

Abstract

The fundamental aspects of survey and sampling design in the deep sea are generally the same as in shallower shelf and coastal waters. However, limitations imposed by depth, topography and by the extra time required to survey large areas of the deep sea can be challenging. In this chapter we focus on some of the key elements that are required to plan a valid survey and sampling strategy. Particular attention needs to be directed at establishing clear objectives, and a strategy of fewer tasks carried out well is generally preferable to a larger number done too poorly to address or test hypotheses. Spatial sampling aspects of survey design need especially to consider replication, scale, dispersion and independence of sampling units, and to avoid confounding of multiple environmental factors. Several case studies are presented to illustrate some of these elements, and to help the reader become familiar with practical aspects of biodiversity surveys, abundance surveys, surveys directed at ecological function, surveys that take a wide 'ecosystem' view, and the benefits of long-term time series. This chapter is not meant to be a comprehensive account of survey design, but addresses some of the most important elements a researcher should be aware of.

Keywords survey design, sampling, biodiversity, abundance, replication, time series, ecosystem surveys

3.1 Introduction

There are a large number of texts dealing with issues of survey and sampling design. It is not the purpose of this chapter to recount in detail the principles and practice that have already been described for sampling shallow seas (e.g. Kingsford & Battershill, 1998). Rather, as regular biologists going to sea, with a reasonable background knowledge of statistics and the need to undertake robust analyses of survey data, we use this chapter to highlight some of the key issues that people planning and undertaking surveys of the deep sea need to be aware of.

There is little that is inherently different in the deep sea from shallower water survey design. The fundamental elements of good science are the same, starting with the crucial need to identify the problem clearly, and define specific objectives that a survey is to address. These two (seemingly)

Biological Sampling in the Deep Sea, First Edition. Edited by Malcolm R. Clark, Mireille Consalvey and Ashley A. Rowden.

simple tasks then lead onto evaluating what type of survey design is appropriate to achieve the objectives, given the limitations of available gear, time and resources at sea. However, in the deep sea sampling is inherently more challenging than in shallow water, and surveys are rarely as straightforward. The expense of vessel time generally means that there are multidisciplinary teams on board with potentially competing or conflicting objectives, and there are invariably time constraints, either imposed by the often rough weather offshore, the need to undertake many different tasks during a single voyage, or simply by the amount of time that deploying any sampling gear to the deep seafloor takes.

Each of the following chapters that describe various gear types refer to aspects of their use and selectivity, which link to how applicable such gear might be for meeting certain objectives, and Chapter 17 on data analysis details some of the issues associated with data-limited situations, which also feeds back into elements of survey design. Hence in this chapter we intend to describe some of the more basic issues that most scientists designing deep-sea surveys should take into account, and then illustrate some of these issues by describing several case studies that are based on the practical experience of the authors, or provide particularly useful examples of survey and sample design.

3.2 General survey design

3.2.1 Spatial design considerations

The seemingly straightforward task of identifying and summarizing the key features of spatial sampling designs in deep-sea ecology is challenging. Because ecological deep-sea science is an extraordinarily diverse discipline (Gage & Tyler, 1991; Levin *et al.*, 2010; Ramirez-Llodra *et al.*, 2010; Schlacher *et al.*, 2010a), covering an expansive field of themes and research questions, it is well-nigh impossible to give specific recommendations on spatial sampling design that would do justice to the diversity of the science. Thus, specific and detailed advice will not cover the breath of the science in all instances and may, at worst, impose unreasonable demands on investigators; hence, here we limit coverage to provide some general guidelines, which should have broad applicability in the deep sea, and illustrate these with examples from the literature.

3.2.1.1 The primacy of hypotheses

The primary criterion – one which must always override all other considerations – in designing a study, is that the spatial design must match the study objective(s), and especially address a particular hypothesis. This may seem a curious point to restate here, but it is surprisingly common to encounter 'poor' designs which fail to provide data that can be used confidently to test an hypothesis with adequate power and rigour (Underwood, 1997). Clearly defining and stating the objectives of a study (arguably best done in the form of a predictive hypothesis) and embedding them into a strong ecological, conservation or management context must occur before planning the spatial layout of the sampling and going to sea. Often this can be an iterative process in which ship time, gear type and sampling protocols limit the scope of the work and hence require recasting of the hypotheses. It is generally better to address a few objectives well, using the best design possible, rather than stretching resources too thinly over too many objectives.

Irrespective of the nature of a cruise and its specific objectives, investigators will generally face four, broad, interrelated issues at the planning stage:

- *Scale*: Over what distance should sampling units be placed?
- *Spatial arrangement*: How should sample units be dispersed and arranged?

- *Number*: How many replicate samples should be collected?
- *Environmental drivers*: How to account for habitat and environmental heterogeneity, including depth?

3.2.2 Scale

The spatial scale or geographic extent over which samples are to be collected is a fundamental design criterion in all deep-sea studies. Most ecological processes vary with spatial scale (Levin, 1992), the distribution of many assemblages have 'spatial structure' which varies with scale (Thrush, 1991; Legendre *et al.*, 1997), environmental drivers that structure communities operate at different scales (Levin *et al.*, 2001), and habitat heterogeneity can be scale-dependent (Williams *et al.*, 2010a). Spatial scale also influences the degree to which results from a study are likely to be more widely applicable, with generality usually increasing at greater spatial scales or where experiments are replicated in different geographic areas (Thrush *et al.*, 2000; Knapp *et al.*, 2004).

Deep-sea science spans a broad range from detailed small-scale studies (Smith *et al.*, 1986) to basin-wide or global investigations (Vanreusel *et al.*, 2010). As is the case with all other criteria, the objectives of a study are the prime criterion to determine the appropriate scale of sampling. For example, working in submarine canyons, McClain and Barry (2010) investigated the influence of habitat features and heterogeneity on benthic structures over a range of scales (1 m–1 km) that included detailed examinations of patterns over comparatively small distances. Similarly, seeps, pockmarks, vents and whale falls represent distinct hotspots of biological activity in the deep sea, and detailed mapping – including very fine scales – of fauna and environmental conditions is available for several of these systems (Cuvelier *et al.*, 2001; Smith & Baco, 2003; Roy *et al.*, 2007; Menot *et al.*, 2010; Bowden *et al.*, 2013, Boschen *et al.*, 2015).

At regional scales, the interest is often to identify environmental drivers and habitat attributes that influence biological community metrics (e.g. biomass, abundance, diversity). Many studies encompass replicate sites that are thought to represent a cross-section of regional environmental heterogeneity (Narayanaswamy *et al.*, 2010; O'Hara *et al.*, 2010; Rowden *et al.*, 2010; Schlacher *et al.*, 2010b; Williams *et al.*, 2010a; Schlacher *et al.*, 2014). This is a sensible approach, provided that the number of different sources of environmental heterogeneity incorporated into the design is not too large to prevent appropriate replication of sampling within each source.

At the largest scales (ocean basins to global) nearly all studies will have to draw on compiling data from multiple sources, or using models to predict occurrences of biota for areas not physically sampled (Rex *et al.*, 2006; Tittensor *et al.*, 2009; Clark *et al.*, 2010b; Vanreusel *et al.*, 2010; Davies & Guinotte 2011; Yesson *et al.*, 2012). Investigators often have little choice over spatial design and the main challenges are quality control of records, matching spatial scales of environmental predictors to the fauna, obtaining accurate taxonomy and accounting for possible confounding (these are discussed in Chapter 17).

3.2.3 Sample independence

Independence between samples is a critical requirement for all sampling designs. If samples are not independent, then it is difficult to interpret the outcomes of statistical tests correctly, as reported error rates and probabilities may be wrong. Assuming independence when it does not hold is a fatal flaw. One of the most prevalent aspects of non-independence in the ecological literature is called pseudoreplication, defined by Hurlbert (1984) as the use of inferential statistics to test for treatment effects with data from experiments where either treatments are not replicated (though samples may be) or replicates are not statistically independent.

Hurlbert's (1984) seminal paper on pseudoreplication must be an arrow in every investigator's design quiver. Although there are at least three different forms of pseudoreplication (simple, sacrificial and temporal), all have as their underlying cause the problems of collecting samples that are not independent, treating multiple samples as though they were independent assessments or treatments, or both (Hurlbert, 1984).

Pseudoreplication is particularly troublesome when testing putative environmental impacts, and when experiments comprise many treatments that are poorly replicated (Underwood, 1997, and see www.ma.utexas.edu/users/mks/statmistakes/pseudorep.html). Basic remedies are to (a) adequately disperse treatments and samples in space, (b) unambiguously distinguish between truly replicated 'treatments' and repeated measurements within individual treatments, and (c) use designs that specifically test for environmental impacts such as 'beyond-BACI' (Underwood, 1994).

In deep-sea benthic ecology, subsampling from box-corers and grabs is common and multi-corers are widely used (Bett *et al.*, 1994; see Chapter 17). Similarly, frame grabs from video or multiple still images from cameras towed over the seabed are customarily treated as 'samples' (Howell *et al.*, 2010; McClain *et al.*, 2010; see Chapter 12). It is debatable whether these units represent truly independent samples, and analysis methods must take this into account (see Chapter 17). A key argument is the degree to which environmental connectivity couples organisms between 'samples' (or subsamples) biologically to make them non-independent. In addition, mobility, size and the behaviour of organisms in relation to the physical size of the collecting gear are important when assessing the independence of samples. For example, in the same block of mud collected by a box-corer, abundance estimates of relatively large and highly mobile isopods derived from subcores may be closely linked, whereas lower connectivity could be argued for smaller nematodes. Unless there are solid reasons to assert independence of samples taken in close physical proximity, it is sensible to adjust sampling designs to improve independence of samples, achieve adequate dispersion and to avoid pseudoreplication.

There are no hard and fast rules about separation distance between sampling sites. This depends on the size of the sampling gear and the nature of the habitat. Fisheries surveys using large trawl nets typically have tows that are several kilometres apart (e.g. ICES, 2010 and see Chapter 7). McClain *et al.* (2010) used transects that were 100 m long, and separated by several hundred metres depth to investigate patterns of faunal distribution with depth on Davidson Seamount. In contrast, surveys on small conical seamount features will almost always, by necessity, have sampling sites that are close together. In this case it is possible to reduce spatial correlation by separating tows by a certain angle (e.g. 20°) which will generally sample different substrate, depth or oceanographic aspect (e.g. Clark & Rowden, 2009). An additional consideration is that if there is limited ability to separate samples and reduce the risk of non-independence, then efforts can be made at the analysis stage to account for it (e.g. applying a nested sample approach, such as a nested analysis of variance). Chapter 17 covers several such issues of sampling limitations.

3.2.4 Replication

Replication is extensively covered in all standard texts of univariate statistics and sampling design (Green, 1979; Sokal & Rohlf, 1995; Zar, 1996; Quinn & Keough, 2002; Gotelli & Ellison, 2004). In essence, replicate samples are collected to estimate variability (differences between samples both within and between treatment), achieve the required precision of the study and ensure hypotheses are adequately tested. Underwood and Chapman (2005) state very simply that 'spatial replication is a mandatory component of any benthic study'. The more variable the environment or the fauna, the more replicates will be needed to estimate statistics at a given precision which will affect the power of statistical tests. Power analysis (Underwood & Chapman, 2003; Maxwell *et al.*, 2008) is an important tool in designing studies, especially those where putative environmental impacts are investigated (Underwood, 1993, 1994). It requires, however, prior knowledge of variability, which is seldom available in deep-sea studies. The alternatives are to use estimates of variance from other studies or to approximate replication levels from well-designed and analysed studies in similar environmental settings.

Replicate samples must always be taken for any combination of space, time, habitat type or whatever other factor is to be tested (Underwood, 1997). In shallow-water settings there is no excuse for a lack of replication and deep-sea studies must similarly be adequately replicated when aiming to test hypotheses (McClain & Barry, 2010). There are, however, a few circumstances in the deep sea when replication can be justifiably limited; this may arise during synoptic mapping of large areas or of multiple seascape features using gear that provides samples which are physically pooled in space (e.g. towed trawls and dredges). For example, to derive preliminary conservation rankings of submarine canyons as part of a regional management plan off southeast Australia, megabenthos was collected with trawls that crossed depth contours in several canyons; here the objective was to maximize information on biodiversity (hence depth-integrated samples) and geographic coverage at the cost of high replication levels within a canyon (Schlacher *et al.*, 2007). In such instances (Williams *et al.*, 2010a), investigators must be explicit about the objectives (e.g. synoptic mapping, biodiversity surveys) and be aware that such studies can generate hypotheses but cannot yet test them. By contrast, to detect whether bottom trawling impacts the deep megabenthos of seamounts, replication of sampling within and between different 'fishing treatments' (closed vs. open to fishing) was necessary to rigorously test the putative effects of fishing (Althaus *et al.*, 2009; Clark & Rowden, 2009; Williams *et al.*, 2010b).

3.2.5 Environmental linkage and confoundment

Many deep-sea ecological studies seek to relate biological patterns to 'environmental drivers', postulated to structure assemblages through a variety of processes (e.g. effects on recruitment, reproduction, food supply, metabolism, dispersal). Some fundamental considerations at the spatial design stage include: (a) the spatial grain and scale of sampling the fauna should match that of environmental variables (if there is previous information on an area this may give some clues to this); ideally, measurements of environmental conditions co-occur with faunal collections over the same geographic area, (b) environmental variables that have a known and strong influence on the fauna are factored into the sampling design, including coverage of different habitat types if of interest, and (c) faunal and environmental factors are estimated at similar degrees of precision so that relationships can be analysed.

Designs that purport to test for specific effects but ignore the coupling between the environment and fauna can introduce significant sources of error. In situations where strong correlations between biological metrics and environmental conditions occur, where such relationships are predicted based on the literature, or where species are closely associated with particular habitat types, all spatial designs must guard against confounding the primary hypothesis test with effects stemming from environmental sources. For example, depth is known to be related to deep-sea diversity, community structure and trophic processes (Carney, 2005; McClain *et al.*, 2010; Priede *et al.*, 2010). Consequently, if the chief question is to compare any of these metrics between treatments (e.g. areas with different fishing intensity, carbon flux or upwelling), samples in each treatment should be collected from sites with similar environmental conditions, such as comparable depths. Failing to do so will confound the main test of interest (e.g. differences in biomass caused by fishing?) with depth-related effects (e.g. differences in biomass caused by depths?). In these instances, no logical conclusion can be reached as the effects of the hypothesized agent of change cannot be separated from those of the confounding factor. A similar logic applies to other drivers: any variable which can measurably influence the metric of interest needs to be explicitly controlled for.

The most effective remedy to guard against confoundment at the design stage is to locate samples in similar conditions for all variables which are suspected of introducing confoundment (depth will almost invariably be a confounding factor). Post-cruise solutions to confounding are possible, but potentially more problematic and hence should not be relied upon as a matter of course (see Chapter 17). Before statistical analysis, all data sets need to be checked for possible confoundment (e.g. comparison of depth statistics between treatments); if significant, samples can either be culled to balance treatments with respect to the confounding factor, or statistical techniques can be employed to lessen the influence of a confounding variable numerically. Howell *et al.* (2010) provide an example of a well-executed study where confounding by depth, sediment type and seabed topography was carefully controlled at the design stage, through some sample culling, and through statistical techniques.

3.3 Case studies

3.3.1 Biodiversity surveys of seamount habitat

Seamounts are important habitats for biodiversity in the deep sea, often representing the only shallow environments against a backdrop of the abyssal plain, and also being sites of hard substrate in place of the predominantly soft sediment of flat topography (e.g. Clark, 2009). They can also be important for fisheries and sites of potential seabed mining (Clark *et al.*, 2010c) and hence environmental and resource managers often need to know the abundance of species (e.g. deep-sea commercial fish species such as orange roughy (*Hoplostethus atlanticus*), alfonsino (*Beryx splendens*)) or the composition of benthic bycatch which is affected by impacts of bottom trawling or dredging. However, depth range and substrate heterogeneity make them a difficult habitat to survey in a robust quantitative fashion. New Zealand seamounts have often been surveyed for their demersal fish and benthic invertebrate fauna, and below we outline some of the key elements that have worked, and not worked, in these surveys.

Trawl surveys have been attempted on seamounts, where trawls were towed from the summit of the seamount (necessary to land and stabilize the gear) down the flanks to the base, in a random direction from the peak (with 20° separation), and at different times of the day to reduce correlation between catches from a small area (Clark, 1994). A random direction was possible on some seamounts, but not on others where the rough seafloor prevented trawling. Hence, first conducting a multibeam survey, or even single-beam echosounder survey along prospective paths, was important to establish whether the seafloor is suitable for the particular gear (e.g. is the bottom smooth and soft enough to use a beam trawl?) and minimize risk of gear damage. However, the performance of trawls down the flanks of seamounts can be variable depending on the nature of the seafloor, and towing speed varies with the steepness of the seamount, and how much wire is being payed out, the degree of bottom contact and headline height are also variable throughout a tow (see Chapter 7).

Catchability in trawls can vary by species and fish size, but trawl surveys can nevertheless yield good fish composition data to enable comparisons between seamounts as well as other deep-sea habitats (e.g. Tracey *et al.*, 2004). Given the damage to biogenic habitat on seamounts, trawling is not a preferred option for sampling new seamounts for biodiversity, but if fish are the main fauna of interest, there may be few alternatives.

New Zealand seamounts are routinely surveyed for benthic invertebrate fauna by using photographic techniques, in conjunction with small-scale direct sampling gear. Survey design is similar irrespective of whether towed cameras (see Chapter 12) or more sophisticated remotely operated vehicles (ROVs) are used (see Chapter 13). We have found straight-line transects to be an efficient method of mapping biodiversity on a conical seamount. At least four photographic transects across the entire site are arranged in a regular 'starburst' pattern centred on the seamount peak (Fig. 3.1). If a towed frame system is used, it is recommended to tow downslope, as upslope or horizontally along a contour the gear hits the seabed where the flanks of the seamount are steep or rough with large boulders. The combination of video and still camera deployment can enable both large-scale habitat mapping and megafaunal distribution to be determined, while the higher resolution still shots enable more detailed identification, as well as quantification of faunal densities (see Chapter 12). If the feature is more elongated (e.g. a ridge), then a parallel transect design will be more appropriate, with the spacing between transects being either constant (for mapping) or random (for more quantitative analysis).

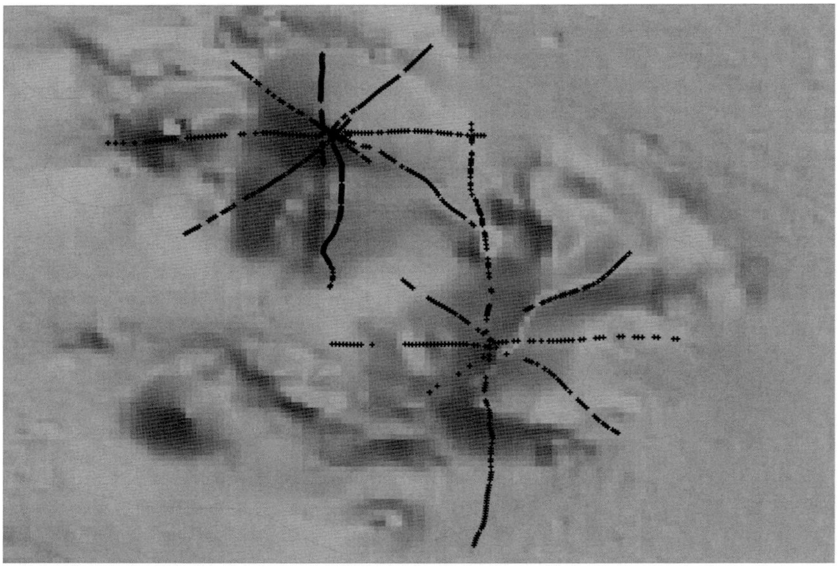

Fig. 3.1 The distribution of still camera stations taken on two small seamounts east of New Zealand during a biodiversity survey by NIWA. This illustrates the radial transect ('starburst') survey design. (Clark *et al.*, 2010a.)

Direct sampling for benthic invertebrate fauna has successfully been carried out using epibenthic sledges specifically designed for use on rough seafloor (see Chapter 9). There are several ways in which sampling stations can be allocated, depending on the primary objectives:

- Stations can be randomized over the entire seamount, by a combination of random direction from the seamount peak, and random depth down the slope. This is good for comparisons between seamounts that are at similar depths and have similar elevations (e.g. Clark & Rowden, 2009).
- The seamount can be stratified into depth zones, with stations allocated either to summit, flank and base strata (e.g. International Seabed Authority, 2007) or to specific depths which approximate such a division (e.g. Clark & Roberts, 2008). This approach enables within-seamount comparisons where faunal communities may differ given oceanographic or substrate variability between regions of the feature (Clark *et al.*, 2010b).
- Targeted station placement can be made according to topography. Often seamounts have areas of hard substrate which are more likely to host biogenic fauna (corals, sponges) and associated communities, or a seamount might have complex topography and be flat-topped with smaller peaks in areas on the flanks. If the objective is simply to describe the fauna on an individual seamount, then spreading stations around a variety of topography and substrate (soft as well as hard) will likely increase the number and variety of species captured (e.g. Bowden *et al.*, 2011). However, this sampling arrangement may end up reducing the degree of replication unless the number of stations is increased accordingly.
- Tows can be targeted solely on the basis of the photographic images, where new or unusual fauna are seen. This approach can be suitable if a seamount or area has already been well sampled, and the survey has the primary objective to monitor changes over time and where the main sampling tool is photographic.

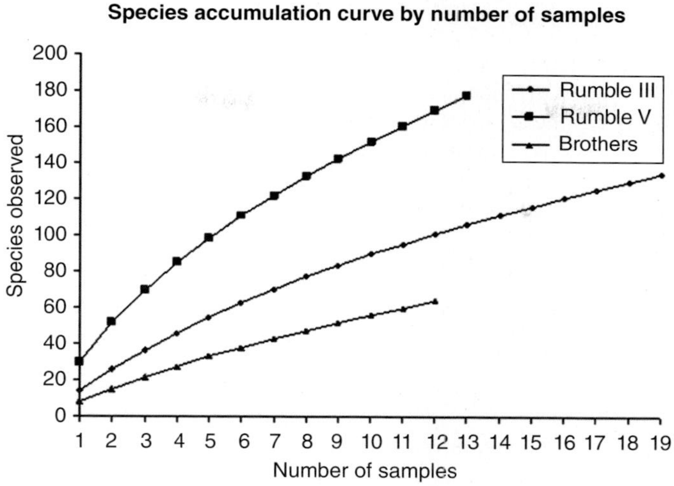

Fig. 3.2 Species accumulation curves for three seamounts off New Zealand. (NIWA, unpublished data.)

The number of direct samples on a seamount will generally be constrained by the amount of time available. It is highly likely that species accumulation curves for mega- and macrofauna will not asymptote, as this requires a high level of sampling effort (Clark *et al.*, 2012). The number and variety of animals caught typically continues to increase with the number of samples (including replicate samples) that are taken. Full biodiversity (i.e. number and range of species) is usually not well described, even for intensively sampled sites (e.g. seamounts, Fig. 3.2) and even when just the large (mega) fauna are considered. This makes estimates of 'rarity' and endemism very difficult. In the case of Rumble V seamount in Fig. 3.2, if trawling had stopped after 10 samples, 20% of the megafaunal species would have been unobserved. Even after 13 samples, at least five new species are recorded with each additional sample. Species accumulation curves for macrofauna can similarly continue to increase substantially, even after 30–40 core samples (e.g. McClain & Hardy, 2010). The most common sampling approach is to undertake the same number of tows or sampling deployments per feature or per unit area. That enables a relative comparison based on an equivalent level of sampling, or an estimated total species number based on the same sampling effort (if an accumulation curve approach is used). For megafaunal comparisons, at least four samples per seamount (e.g. Clark & Rowden, 2009) or five (International Seabed Authority, 2007; see Chapter 18) are recommended minima. This needs to be evaluated for each situation, based on characteristics of the location, faunal distribution, and survey objectives.

An issue with a starburst type of sampling approach, whether by sleds and cameras as described above, or acoustic equipment used for some deep-water fish species surveys (e.g. Branch, 2001) is that the summit of the seamount is sampled at a greater density than the flanks and base-the density of sampling decreases as the transects extend further down the flanks. A circle at the summit has a much smaller area than at the base, but sampling generally has a fixed area. Therefore a point near the summit has a higher probability of being sampled than one at the base. The application of strata for sled tows that divide the seamount into summit, flanks and base avoids this problem. The potential bias in this type of sampling has been resolved for acoustic surveys of orange roughy aggregations on seamounts by weighting data on the basis of position on the transect relative to the centre (equivalent to the seamount summit) by converting from Cartesian to polar coordinates, and also changing the method of variance estimation (Doonan *et al.*, 2003).

3.3.2 *Fish abundance surveys*

There is an extensive ecological literature on estimating the abundance of animals, driven in the marine environment largely by fisheries surveys to estimate the stock size of commercial species (e.g. Saville, 1977; Sissenwine *et al.*, 1983). Fisheries objectives have resulted in the development of several methods to estimate the abundance of deep-sea fish species, with the most common techniques being trawl surveys (e.g. papers in Doubleday & Rivard, 1981) and acoustic surveys (MacLennan & Simmonds, 1992), but also egg production methods (e.g. Hunter & Lo, 1983), and the use of baited landers (Priede & Merrett, 1996). Each method has strengths and weaknesses (see examples in Clark (1996) and chapters in this book). However, given that fisheries trawl surveys are still one of the most frequently used platforms to obtain biodiversity information, and that trawl survey design is often applied to other sampling gear, it is useful to review briefly some of the key elements of such design in the deep sea.

The type of survey design comes back to the objectives, in particular target species or faunal groups, and the degree of precision required in the results. In situations where there have been no previous surveys, and there is little information on which to base a survey design, then a systematic approach is often appropriate: either a regular grid-type pattern of sampling stations or strip transects with sampling at regular depths or intervals along the transect. These methods can give good information on the distribution of species, and an idea of their patchiness in occurrence or density. However, there are a range of difficulties in estimating overall abundance and variance.

A more common design is based on random sampling, which is either simple (assuming distribution of species is relatively homogeneous) or stratified (where the survey area is divided into subareas (strata) of more uniform density and sampling is randomized within each stratum). This latter approach is the standard for most fisheries or other resource surveys where prior knowledge of habitat or species distributions can enable effective stratification but retain the critical need for randomness to satisfy the mathematical assumptions of the commonly used statistical methods. However, it can be difficult to get the stratification right every time. The distribution of mobile fish and invertebrates can change from year to year. Stratification might still be correct, but varying densities between strata mean that any preconceived allocation of station numbers within each stratum may be wrong. A way of dealing with this is to adopt an adaptive strategy, whereby a 'two-phase survey' is carried out, where the results of Phase I are used to allocate the remaining stations in Phase II (Francis, 1984). This approach is routinely used for deep-sea surveys off New Zealand. It is important, however, to recognize that these surveys incur a small negative bias, but this is more than compensated for by the reduction in variance, especially from any extreme catches. The investigator should not be afraid of this bias since it is usually well below sampling error and given that most trawl surveys are relative rather than absolute, the bias does not count if it remains the same across the time series.

Once a decision is made to use a stratified random design, there are a number of questions that commonly arise. These are discussed in the sections below.

3.3.2.1 How many stations should be allocated for each stratum?

At least three are needed to estimate the variance of catch rates in each stratum. Where densities are expected to be higher, then more stations should be allocated. The advantage of a two-phase approach is that additional stations can be added where they are most needed to increase precision. Where a survey has previously been carried out, a rough approximation for station distribution by stratum is given by:

$$fi = \left(\frac{siAi}{T} \right)$$

where fi is the proportion of stations to be allocated to that stratum, si is the standard deviation of the stratum catch rate, Ai is the stratum area, and T is total $siAi$ summed for all strata. Since this procedure is approximate and distributions can change between years, an upper limit should also be put on *fi* so that most samples are not put into one stratum in the case where one stratum dominates the variance.

3.3.2.2 How far apart should each station be?

The separation of stations should be sufficient to minimize the correlation between samples, so that the assumption of independent samples is improved (as much as it can be). However, this depends upon the spatial scale of the faunal communities, stratum size, as well as steaming time between stations (there is a trade-off between more stations close together and fewer separated by large distances). A minimum distance of 3 nautical miles is used in New Zealand trawl surveys, although this is arbitrary and ensures tows do not cross over one another. In most cases it is a reasonable assumption that environmental differences, and hence biological composition or abundance, change as a linear function of distance. This is, however, not necessarily the case as patch dynamics and microhabitat variation are important drivers of species composition on one hand (e.g. McClain & Hardy, 2010), and on the other many deep-sea species have large dispersal distances (e.g. McClain & Hardy, 2010). The likelihood of interdependence between stations depends on a potentially complex relationship between geographic distance and biological 'connectedness'. McClain *et al.* (2011) derived distance–decay curves for macrofauna, and detected an increase in dissimilarity with distance, despite a high variability in the similarity of macrofauna between core samples.

3.3.2.3 What happens if the station is on rough ground that cannot be sampled?

If rough ground is encountered, then a search is made within a radius of 3 nautical miles of the position. If no suitable ground is found, then that station is discarded and the next random position is chosen. This means that some seafloor types will not be sampled, with consequences if composition or density of fish are substantially different on such ground and the fraction of such area is large, then a bias will occur. In this case it is necessary to rely on the survey being a relative index and assuming that the ratio of abundance for sampled ground to excluded ground is constant over time and across the abundance range.

3.3.2.4 How long should each sampling tow be?

This will depend upon target species, catch rate and time available. Highly mobile species can often swim or move in front of trawl gear for some time (15–20 minutes) and so the tow needs to be long enough to sample the faunal composition and abundance in a representative way. Differences in gear selectivity and catchability will, however, always be an issue to consider (see Chapter 7). In deep water, it can take a long time to shoot and haul back the gear, and so bottom time can be increased (up to 1 hour, about 3 nautical miles) without compromising the overall progress of the survey. However, the longer the tow, the more benthic impact occurs, which needs to be considered in fragile habitats.

When trawling on seamounts off Australia and New Zealand, where stony coral 'reefs' can occur, bottom time of trawls or sledges is kept low to minimize seafloor damage yet achieve adequate representation of catch and catch rate. Epibenthic sledges or dredges can fill with rocks and biogenic fauna such as corals within as small a tow distance as 0.4–0.5 km. Examination of previous data, prior camera tows or carrying out several trial tows are all ways to help determine an appropriate standard tow duration that balances sweeping enough area to characterize the general fauna in an area, yet not have problems of gear saturation or excessive damage to sensitive benthic fauna.

3.3.2.5 Should the same stations be repeated each year or new random stations generated?

This is a difficult question to answer. There are advantages in repeating the same fixed stations because comparisons between years are then based on exactly the same sites, and there are time savings by not having to search for new tow lines (a significant time cost in rough terrain). However, if fish or mobile invertebrates are moving around between surveys, randomizing each year can pick up any systematic movement that could be missed by fixed stations. Randomization is typically an important assumption of the statistical methods applied, and hence fixing stations can introduce a correlation between years. There is a trade-off between a bias and a reduced error rate in the abundance trend. A compromise solution is to fix Phase I stations, but randomize the location of Phase II tows. Appropriate stratification can help ensure that the survey picks up geographical variation in population changes. A similar problem arises with the order of stations. It can be most efficient for vessel time to tow in a certain sequence, often in one direction. However, if the fauna are moving in a set direction, there can be a bias one way or the other. It is preferable to vary the order in which stations are filled, so there is not a consistent pattern in time or space with the survey design.

3.3.2.6 How many surveys are needed to reliably estimate trends in abundance?

Trawling for demersal fish has unknown levels of escapement from the path of the gear, or herding by doors and sweeps. Hence surveys are usually regarded as giving relative estimates of fish abundance. Given that the precision obtainable with trawl surveys is often not high (coefficients of variation of 15–30%), it will be necessary to have 3–5 surveys to be able to determine reliably a trend in biomass against individual survey variability, and to compute a statistically significant difference between the surveys (at least the first and last), and calculate absolute abundance. Where sampling stations are based on a random design, the results of past surveys can be used to optimize the next survey. This can help avoid the chance of 'lost' surveys where results have such a high variance to be of limited value.

3.3.3 Approaches to design of 'ecosystem surveys'

Knowledge of the structure and dynamics of marine populations is essential to apply ecosystem-based management of marine resources. In order to describe and understand the functioning of deep-sea ecosystems we need to determine how physical and biogeochemical factors influence the distribution and structure of deep-sea communities. However, deep-sea surveys aiming to study the entire ecosystem are, by necessity, complex and expensive, and hence uncommon. Nevertheless, several recent projects have taken on the task of attempting such an integrated approach. The OASIS project (www1.uni-hamburg.de/OASIS/) studied seamounts in the Northeast Atlantic; MarEco (www.mar-eco.no) surveyed the Mid-Atlantic Ridge from Iceland to the Azores; EcoMar (www.mar-eco.no) focused on the Charlie–Gibbs Fracture Zone; the CONDOR project (www.horta.uac.pt/intradop/index.php/the-condor-seamount) is studying seamount function off the Azores; ECOMARG (www.ecomarg.net/index_en.html) aimed to study the benthic–demersal ecosystem in the Galician and Cantabrian Sea continental margins; and the most ambitious and extensive deep-sea ecosystem-based studies of all, HERMES (www.eu-hermes.net) and its successor HERMIONE (www.eu-hermione.net). These types of projects target all major ecosystem components, including plankton, micronekton, macronekton and benthos, many examine multiple habitats (e.g. slope, canyons, seeps, cold-water coral reefs) and aim to quantify the processes that link these components and habitats. The programmes are often large-scale and of short duration, such as MarEco and HERMES, but also sometimes focus on more restricted areas with repetitive surveys (e.g. OASIS on several seamounts, GoMA in the Gulf of Maine) (Fig. 3.3).

The ultimate output of these types of studies is often the production of an ecosystem model (Soetaert & Van Oevelen, 2009) that can be used for management of the natural resources of the study area, most often with a focus on fisheries (e.g. Armstrong *et al.*, 2009). Ecosystem-based studies have also been successful at demonstrating the impact of management decisions and practices, and the societal value of deep-sea ecosystems, to policymakers and public alike (Grehan *et al.*, 2009). However, the complexity of their undertaking can be a deterrent. Studies that are extensive in aim and area, such as HERMES and HERMIONE, have taken a multidisciplinary 'integrated project'

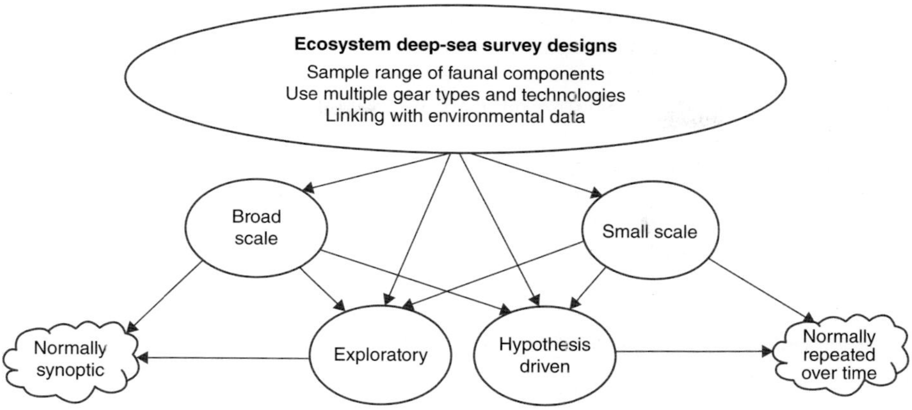

Fig. 3.3 Schematic of linkages between higher level aspects of ecosystem survey design.

approach (Weaver & Gunn, 2009; Weaver *et al.*, 2009), where a consortium of partners execute a number of semi-independent projects that are ultimately integrated to produce a whole ecosystem understanding. There is, therefore, a great deal of careful planning and project management required to successfully undertake an ecosystem-based study.

'Ecosystem-based' surveys will typically have more elements with a variable spatial and temporal scale than surveys based on a single habitat type or species group. Before beginning such a survey all accessible information concerning the selected study area should be compiled. This includes the general characterization of the topography, hydrology and oceanography of the study area and the compilation of biological data. The plan for an optimal sampling design will not only have to account for the spatial and temporal scale of the study, but will need to integrate sampling of the biological and physical environment according to available ship time, costs and equipment. Obtaining as much information as possible from a range of experts is essential for planning the field phase, and planning meetings should always involve representatives from the different scientific disciplines taking part in the survey.

Multidisciplinary projects and surveys of the deep sea normally involve the use of a multipurpose research vessel, or several vessels at one time. Vessel capabilities are critical to the success of ecosystem-type surveys. Ideally, a sophisticated multipurpose vessel will undertake a range of simultaneous biological and oceanographic sampling. However, there are relatively few vessels capable of carrying out an ecosystem study in a single survey, so most require several sampling legs – such as one to sample the pelagic component, another the benthos, another the physical environment, or alternatively a series of voyages each to address different questions about the whole ecosystem.

Most ecosystem survey approaches combine point sampling from the vessel with long-term data acquisition from moored instruments (e.g. echosounders on moorings, cameras on benthic landers, satellite imagery of ocean colour) (e.g. Wenneck *et al.*, 2008). Permanent sampling stations should be at key locations of the study area to measure temporal and spatial variability of key factors affecting the survey and broader project objectives (e.g. Wenneck *et al.*, 2008).

3.3.3.1 Multi-gear sampling times

Ecosystem surveys require collection and/or observation of organisms at sizes ranging from millimetres to metres and a variety of sampling methods must be used. Strategies and methods adopted during a single multidisciplinary survey must be selected in order to maximize data and sample collection in both pelagic and benthic habitats. Gear selectivity issues (see Chapter 2) mean that any ecosystem-type survey will use multiple gear types. During MarEco the limitations of bottom trawling only being possible on soft sediment was complemented by the deployment of an ROV, and carrying out longline surveys, in regions of rough bottom or sites too deep for trawling. Furthermore, in the water column, a combination of midwater trawling, ROV transects and the Multinet gave a more complete understanding of gelatinous fauna in the water column (Hosia *et al.*, 2008).

In deep waters, the time taken to deploy and retrieve sampling gear can be considerable. Some of the individual gear-type chapters in this book give specific details, but in Table 3.1 we present a general guide to the amount of time required for a variety of equipment that is commonly used on multipurpose surveys. This table can be used to estimate the overall time required for a site where multiple gear deployments are planned. For example, if a site is at 4000 m depth, and one CTD (conductivity–temperature–depth) cast, one longline set and one ROV transect is scheduled, from the table we estimate that about 17 hours is required for deployment and retrieval, with additional time for the longline soak (e.g. 3 hours), and ROV time along a transect (e.g. 4 hours), making a total time of 24 hours required at this site.

A further consideration for timing, however, is ensuring an allowance for mapping the terrain. Where the vessel has multibeam echosounder (MBES) capability, then several runs should be made

Table 3.1 Guide to deployment and retrieval times (hours) of various gear types commonly used in multipurpose surveys

	Vertical deployed gear			Fixed long gears	Towed gears			Multibeam mapping (average swath width, km)
Depth (m)	1 m/s	2 m/s	3 m/s	Longline	Bottom trawls	Pelagic trawls	ROVs	
200	0.11	0.06	0.04	0.56	0.56	0.44	0.22	1.00
400	0.22	0.11	0.007	1.11	1.11	0.89	0.44	2.00
600	0.33	0.17	0.11	1.67	1.67	1.33	0.67	3.00
800	0.44	0.22	0.15	2.22	2.22	1.78	0.89	4.00
1000	0.56	0.28	0.19	2.78	2.78	2.22	1.11	5.00
1200	0.67	0.33	0.22	3.33	3.33	2.67	1.33	6.00
1400	0.78	0.39	0.26	3.89	3.89	3.11	1.56	7.00
1600	0.89	0.44	0.30	4.44	4.44	3.56	1.78	8.00
1800	1.00	0.50	0.33	5.00	5.00	4.00	2.00	9.00
2000	1.11	0.56	0.37	5.56	5.56	4.44	2.22	10.00
2200	1.22	0.61	0.41	6.11	6.11	4.89	2.44	11.00
2400	1.33	0.67	0.44	6.67	6.67	5.33	2.67	12.00
2600	1.44	0.72	0.48	7.22	7.22	5.78	2.89	13.00
2800	1.56	0.78	0.52	7.78	7.78	6.22	3.11	14.00
3000	1.67	0.83	0.56	8.33	8.33	6.67	3.33	15.00
3200	1.78	0.89	0.59	8.89	8.89	7.11	3.56	16.00
3400	1.89	0.94	0.63	9.44	9.44	7.56	3.78	17.00
3600	2.00	1.00	0.67	10.00	10.00	8.00	4.00	18.00
3800	2.11	1.06	0.70	10.56	10.56	8.44	4.22	19.00
4000	2.22	1.11	0.74	11.11	11.11	8.89	4.44	20.00
4200	2.33	1.17	0.78	11.67	11.67	9.33	4.67	21.00
4400	2.44	1.22	0.81	12.22	12.22	9.78	4.89	22.00
4600	2.56	1.28	0.85	12.78	12.78	10.22	5.11	23.00
4800	2.67	1.33	0.89	13.33	13.33	10.67	5.33	24.00
5000	2.78	1.39	0.93	13.89	13.89	11.11	5.56	25.00

The depth range covered is 200–5000 m. Vertical gears include CTD/rosettes, box-corers, multi-corers, landers, drop cameras, etc., and three velocities are given which will depend on the nature of the equipment. For total station times, the tow or transect time needs to be added.

over the site first, to establish the topography, and also backscatter as it relates to sampling multiple (or single) substrate types. A CTD deployment, or a faster sound velocity probe cast, will need to be done to calibrate the settings of the MBES in that particular area (see Chapter 5).

3.3.3.2 Critical aspects to consider

The MarEco project was one of the most coordinated deep-sea ecosystem projects. It is well described in Wenneck *et al.* (2008), and results from analyses of MarEco data and samples are still being produced, and will be for some time. Below we provide a very brief evaluation of what we found to be critical to the success of planning and carrying out a complex ecosystem-type multidisciplinary survey.

- *Conduct thorough background research*: Collect existing information of the target ecosystem: compile all available bathymetry, geological studies, existing faunal lists, environmental data, bibliography of known reports and accessible information.
- *Ensure there is clear definition of objectives*: Survey aims, and the research questions to be addressed, must be unambiguous. Priorities will also need to be determined to reduce the chances of issues arising at sea when time is running out. Sort out the 'must do' versus the 'nice to do'.
- *Establish responsibilities*: Who does what? Decide on the main research components to be studied, and select a person to be responsible for the overall conduct and coordination of that science theme. Also identify links between the different research areas (e.g. oceanography and hydrographic data biological components) and ensure that good communication and coordination is in place. Early on, the rights and 'rules' regarding participation, data sharing, general hierarchical decision levels, etc., need to be made clear to avoid misconceptions, false expectations, etc. A small steering group is a good idea to oversee the multitude of communication 'issues' needed for the variety of tasks with an ecosystem survey.
- *Determine the range of sampling gear and type of technologies to be used*: The objectives of the survey will determine what is most appropriate. This will depend also upon the space and capabilities of the research vessel or science platform being used.
- *Determine the survey design and sampling regime*: These will depend on ship time, scheduling and budget, but once those parameters are known, then meeting the key scientific objectives will determine the most appropriate survey design. This is where the previous priority setting is important, as there is always a trade-off between doing fewer things very well, and more things less well (as discussed earlier in this chapter). A common approach is to have two tiers of survey, where there are 'major' sites where a full suite of gear is deployed, and 'minor' sites where there is less sampling (e.g. Bowden, 2011). The MarEco project approach was to combine continuous sampling along the ship's track, point observations with several samplers at predetermined sites and opportunistic sampling with less equipment to study particular phenomena or carry out exceptional tasks (Bergstad *et al.*, 2008). The evaluation of the preferred survey design will also incorporate back-up plans for gear failure, or poor weather, or simply finding a different situation to that which was expected. That is, have a Plan B for contingencies.
- *Conduct pre-survey trials*: If possible, undertake a short pilot survey, perhaps one-half to one day, where everything is tested before heading off into distant deep-sea survey grounds. Aspects include the sampling sequence, deployment and retrieval times of each gear operation, times to handle the samples, the onboard biological sampling scheme. In particular with any new, novel or modified equipment, it can be much easier to get working or repair close to shore, or in the period before the vessel is committed to heading offshore. If at all possible, ensure there are back-ups of critical gear. This is standard with bottom-contact equipment, where there is a known risk of damage or loss on rough ground, but the concept of redundancy also needs to apply to sophisticated electronic equipment, camera gear, etc.

- *Determine sampling and processing protocols*: There must be an agreed and standard strategy for processing samples, data sheets, onboard database design, data archiving capabilities, etc. (see Chapters 15 and 16). It can be a great help for biological objectives to have the capability onboard to undertake basic taxonomic sorting and pre-sorting of species.
- *Ensure post-voyage facilities on land are available*: This seems obvious, but many voyages that collect huge amounts of data or samples can face major issues when returning to port, where, for example, freezer space is not adequate, or databases are not in place to handle the volume and variety of electronic data.

3.3.4 Time-series sampling

Environmental conditions experienced by animals change over time, and this limits the utility of one-off surveys to provide an insight in how ecosystems function. Time series of observations are potentially able to provide a much greater understanding of ecological processes and how these may change in the future as oceanic conditions vary. While long-term stations have been in operation for many decades in shallow or pelagic environments (e.g. CalCOFI since 1949), and the German Bight since the early 1960s), the deep seafloor has only in recent years seen efforts to develop long-term time-series studies of benthic biodiversity and ecosystem change in the deep sea. Aspects of time series studies are related to seafloor observatories as described in Chapter 14; here we describe briefly some more traditional research carried out at deep-sea sites that are subject of long-term investigation.

One of the earliest time-series programmes was the DISCOL project (DISturbance and reCOLonization of a manganese nodule area in the South Pacific Ocean), which was a long-term and large-scale *in situ* experimental study simulating effects of commercial mining for nodules. The study area was located off western Central America at water depths of 4150 m. Launched in 1988 it measured the response to disturbance of ~11 km^2 of the seafloor and then monitored fixed stations with direct sampling and photographic observations that were carried on before disturbance (1989), and in 0.5, 3 and 7 years (1996) after disturbance (Thiel *et al.*, 2001). It was shown that even 7 years after impact the disturbance track was still visible, and recovery rates from the baseline situation varied between faunal groups (Bluhm, 2001). The slow growth rates of many deep-sea taxa mean that long-term monitoring is necessary to adequately assess the impacts and recovery potential. DISCOL also incorporated sediment, geochemical and oceanographic studies that improved interpretation of changes, and made results more applicable to other geographic regions. Key elements of the success of this project was that the survey was designed to address a specific objective(s) associated with the likely impact of a certain type of seabed mining, and it had from the outset a certainty of long-term support that enabled a multidisciplinary science approach, and for five surveys over 7 years to be carried out (Thiel *et al.*, 2001). Further surveys have been conducted in 2015 to extend the recovery observations.

The Porcupine Abyssal Plain, off the southwest coast of Ireland, is another region where long-term studies have taken place. The programme started in 1989 to study seasonal changes in the deep ocean, and in particular bentho-pelagic coupling between the surface and abyssal seafloor. A permanent site at 4800 m, known as the ‘BENGAL Station’ (Billett & Rice, 2001), was selected as far as possible from continental slope to the east and the Mid-Atlantic Ridge to the west. The site is now known as the Porcupine Abyssal Plain Sustained Observatory (PAP-SO). Various research projects had been done in the area, but the BENGAL project concentrated the multidisciplinary power of 14 research cruises between 1995 and 1998 (Billett & Rice, 2001), and a further 7 between 1999 and 2005 (Billett *et al.*, 2010). Sampling took place around the central BENGAL position (up to a 40 km radius) with multiple gear types, including several types of benthic lander, a variety of corers, photographic equipment, otter trawls, beam trawls, epibenthic sleds, a MOCNESS, sediment traps and CTDs (Billett & Rice, 2001, and papers within a special edition of *Progress in Oceanography*).

Studies have revealed significant variations in benthic faunal densities and community composition at seasonal and annual time scales. A striking change was discovered after 1996, when a small holothurian *Amperima rosea* increased in abundance from one or two individuals per hectare to more than 6000 (Bett *et al.*, 2001), before decreasing in 1999 and peaking again in 2002 (Billett *et al.*, 2010). Changes in *Amperima* abundance, as well as the wider faunal community, have been attributed to fluctuations in particulate organic carbon fluxes, and especially composition of pigments, such as carotenoids, which play a key role in reproductive processes of benthic invertebrates (Billett *et al.*, 2010; Lampitt *et al.*, 2010; Fitz-George-Balfour *et al.*, 2010). Resuspension of particulate matter by megafauna reworking and feeding in the sediment was thought to be an important aspect of this process (Ginger *et al.*, 2001). These conclusions were possible because of the many different sampling approaches included in the study design, and how the long-term monitoring enabled rare/infrequent events to be measured that could be missed by irregular surveys or those designed for single or specific objectives.

HAUSGARTEN is another long-term deep-sea set of stations ('observatory') located in the Fram Strait between the North Atlantic and central Arctic Ocean. The project was established in 1999 to measure and monitor the impact of large-scale environment changes. It includes 15 permanent sampling stations covering depths from 1000 to 5500 m along north–south and east–west transects (Soltwedel *et al.*, 2005). The central station where the transects cross is an experimental site. Regular autonomous sampling has been undertaken using moorings and landers, supplemented by data on faunal distributions from towed cameras. An ROV has been used at regular intervals for targeted sampling, servicing autonomous measuring instruments and undertaking video transects. Multidisciplinary research activities at HAUSGARTEN cover almost all aspects of marine ecosystems from pelagic zone to the benthic realm. The projects tend to focus on benthic processes such as the role of organic matter from phytodetritus sedimentation (e.g. Schewe & Soltwedel, 2003), but also cover water column studies and the interface with the seafloor. The long-term measurements have enabled good description of seasonal cycles, including considerable variation in particulate matter released during times of the year by ice melt in the Fram Sea (Hebbeln, 2000). This ice melt cycle links to differences in the flux of organic carbon between some sites.

The station design set along latitudinal and longitudinal transects has enabled a wide range of studies examining geographical, as well as temporal, variability (see Soltwedel *et al.*, 2005). For example, the deepest station at 5500 m recorded very high meiofaunal densities compared to shallower sites, and also to other global data, which was caused by high levels of organic matter in the sediments (Soltwedel *et al.*, 2003). It is possible that long-term increases in temperature (such as experienced during future climate change) may affect trends in flux, with consequences for deep-sea ecosystems (Soltwedel *et al.*, 2005).

There are numerous other examples of long-term scientific stations producing very strong science that cannot be adequately described here: the Sigsbee Abyssal Plain in the Gulf of Mexico (e.g. Escobar–Briones, 2003), Sagami Bay (e.g. Gooday, 2002), Abyssal Station M in the Northeast Atlantic (e.g. Smith *et al.*, 2006; Ruhl *et al.*, 2008) and chemosynthetic sites at the Håkon Mosby Mud volcano in the Barents Sea, and Lucky Strike vent field and Trans-Atlantic Geotraverse hydrothermal mound on the Mid-Atlantic Ridge (e.g. Copley *et al.*, 2007). The reader is referred to excellent accounts by Glover *et al.* (2010) and Larkin *et al.* (2010) which describe a number of these studies and give further references.

The success of such time-series stations is in part due to ensuring the initial sampling design is formulated to collect data to enable biological and environmental processes and drivers to be determined, and related to the observed changes in faunal communities over time. However, there are always trade-offs in such situations between measuring a few things well, and measuring many things in less detail. Survey design considerations are in most cases constrained by the logistics; costs of regular sampling over many years being the key factor. Permanent stations often cover a small number of sites or habitats, and as such can be limited in terms of some of the major issues discussed at the start of this chapter, such as scale and replication. These are two critical aspects that need to be considered with the compromises needed for this type of study:

- *Scale*: A small site is often easy and relatively fast to survey, thereby enabling more surveys to be done than if each survey was much more detailed. However, care is needed to ensure that the characteristics of the location are such that the results of the work can be applied more generally to other deep-sea environments. Other issues of scale relate to the sampling density and gear type used. It is desirable to sample as wide a range of animals as possible to understand the variability of responses, but if that is impractical, then reasons for targeting certain faunal groups need to be carefully considered, as often some of the 'unexpected' results are the most significant. However, the spatial scale of sampling will vary depending upon the faunal groups and their likely distributional variability.
- *Replication*: This is often sacrificed at permanent stations. The depth range is sometimes very limited, or where a number of depths are included the number of stations at each depth is small. Different habitat types (e.g. substrate composition, flat versus elevated topography) is the same issue. Wherever possible, replication of sampling stations within depth or habitat strata need to be factored into the survey design. The decision of fixed over random stations is another related aspect; fixed has the advantage of enabling direct comparison of a specific location over time, and avoiding potential variability when sampling over a wider area. However, enough sampling needs to occur over a wider area to ensure that the 'microscope' is not missing the bigger picture.

Time-series surveys have the primary objective of monitoring temporal changes, and cannot do everything. However, the principal objectives need to be very carefully considered at the outset, and the fundamental survey design choices made clear. They can then provide a solid foundation for other research to be added on to the core when and if specific research questions arise, or the results indicate that other studies are required.

3.4 Concluding remarks

Every survey is different in some way. We have not tried in this chapter to cover all situations, but to stress general principles of design that apply to most, and provide case studies to illustrate some applications. We stress again the need to think carefully about project objectives, and what the survey or sampling programme is intended to achieve. That leads to the next tier where general issues of spatial and temporal scale are determined, and beyond that again to the actual sampling detail. There never seems to be enough time on deep-sea surveys to do as much as the shipboard team would like, but as long as the big picture is retained, the appropriate hierarchy of design and sampling should fall into place.

References

Althaus, F., Williams, A., Schlacher, T.A., *et al.* (2009). Impacts of bottom trawling on deep-coral ecosystems of seamounts are long-lasting. *Marine Ecology-Progress Series*, 397: 279–294.

Armstrong, C.W. Grehan, A.J., Kahui, V., Mikkelsen, E., Reithe, S., Van Den Hove, S. (2009). Bioeconomic modeling and the management of cold-water coral resources. *Oceanography*, 22: 86–91.

Bett, B.J., Vanreusel, A., Vincx, M., *et al.* (1994). Sampler bias in the quantitative study of deep-sea meiobenthos. *Marine Ecology-Progress Series*, 104: 197–203.

Bett, B.J., Malzone, M.G., Narayanaswamy, B.E., Wigham, B.D. (2001). Temporal variability in phytodetritus and megabenthic activity at the seabed in the deep Northeast Atlantic. *Progress in Oceanography*, 50: 349–368.

Bergstad, O.A., Falkenhaug, T., Astthorsson, O.S., *et al.* (2008). Towards improved understanding of the diversity and abundance patterns of the mid-ocean ridge macro-and megafauna. *Deep Sea Research II*, 55: 1–5.

Billett, D.S.M., Rice, A.L. (2001). The BENGAL programme: introduction and overview. *Progress in Oceanography*, 50: 13–25.

Billett, D.S.M., Bett, B.J., Reid, W.D.K., Boorman, B., Priede, I.G. (2010). Long-term change in the abyssal NE Atlantic: The '*Amperima* Event' revisited. *Deep-Sea Research II*, 57: 1406–1417.

Bluhm, H. (2001). Re-establishment of an abyssal megabenthic community after experimental physical disturbance of the seafloor. *Deep-Sea Research II*, 48: 3841–3868.

Boschen, R.E., Rowden, A.A., Clark, M.R., Barton, S.J., Pallentin, A., Gardner, J.P.A. (2015). Megabenthic assemblage structure on three New Zealand seamounts: implications for seafloor massive sulfide mining. *Marine Ecology Progress Series*, 523: 1–14.

Bowden, D.A. (2011). Benthic invertebrate samples and data from the Ocean Survey 20/20 voyages to the Chatham Rise and Challenger Plateau, 2007. *New Zealand Aquatic Environment and Biodiversity Report*, No. 65. 46 p.

Bowden, D.A., Schiaparelli, S., Clark, M.R., Rickard, G.J. (2011). A lost world?Archaic crinoid-dominated assemblages on an Antarctic seamount. *Deep-Sea Research II*, 58: 119–127.

Bowden, D.A., Rowden, A.A., Thurber, A.R., Baco, A.R., Levin, L.A., Smith, C.R. (2013). Cold Seep epifaunal communities on the Hikurangi Margin, New Zealand: composition, succession, and vulnerability to human activities. *PLoS ONE*, 8: e76869.

Branch, T.A. (2001). A review of orange roughy *Hoplostethus atlanticus* fisheries, estimation methods, biology and stock structure. In: Payne, A.I.L., Pillar, S.C., Crawford, R.J.M. (eds), *A decade of Namibian Fisheries Science. South African Journal of Marine Science*, 23: 181–203.

Carney, R.S. (2005). Zonation of deep biota on continental margins. *Oceanography and Marine Biology – an Annual Review*, 43: 211–278.

Clark, M.R. (1994). Aspects of random trawl survey design for deepwater species on pinnacle and drop-off features. MAF Fisheries Greta Point Internal Report, 227: 23 p. (unpublished report, available NIWA, Wellington).

Clark, M.R. (1996). Biomass estimation of orange roughy: a summary and evaluation of techniques for measuring stock size of a deepwater fish species in New Zealand. *Journal of Fish Biology*, 49 (suppl. A): 114–131.

Clark, M.R. (2009). Seamounts: biology. In: Gillespie, R.G., Clague, D.A. (eds), *Encyclopaedia of islands.* Oakland, CA: University of California Press, pp. 818–821.

Clark, M.R., Roberts, C.D. (2008). Fish and invertebrate biodiversity on the Norfolk Ridge and Lord Howe Rise, Tasman Sea (NORFANZ voyage, 2003). *Aquatic Environment and Biodiversity Report*, No. 28, 131 p.

Clark, M.R., Rowden, A.A. (2009). Effect of deepwater trawling on the macro-invertebrate assemblages of seamounts on the Chatham Rise, New Zealand. *Deep-Sea Research Part I – Oceanographic Research Papers*, 56: 1540–1554.

Clark, M.R., Bowden, D.A., Baird, S.J., Stewart, R. (2010a). Effects of fishing on the benthic biodiversity of seamounts of the 'Graveyard' complex, northern Chatham Rise. *New Zealand Aquatic Environment and Biodiversity Report*, No. 46. 40 p.

Clark, M.R., Althaus, F., Williams, A., *et al.* (2010b). Are deep-sea demersal fish assemblages globally homogenous? Insights from seamounts. *Marine Ecology – an Evolutionary Perspective*, 31: 39–51.

Clark, M. R., Schlacher, T.A., Rowden, A.A., Stocks, K.I., Consalvey, M. (2012). Science priorities for seamounts: research links to conservation and management. *PLoS ONE*, 7(1).

Clark, M.R., Rowden, A.A., Schlacher, T., *et al.* (2010c). The ecology of seamounts: structure, function, and human impacts. *Annual Review of Marine Science*, 2: 253–278.

Copley, J.T.P., Jorgensen, P.B.K., Sohn, R.A. (2007). Assessment of decadal-scale ecological change at a deep Mid-Atlantic hydrothermal vent and reproductive time-series in the shrimp *Rimicaris exoculata. Journal of the Marine Biological Association of the UK*, 84: 859–867.

Cuvelier, D., Sarradin, P.-M., Sarrazin, J., *et al.* (2001). Hydrothermal faunal assemblages and habitat characterisation at the Eiffel Tower edifice (Lucky Strike, Mid-Atlantic Ridge). *Marine Ecology*, 32: 243–255.

Davies, A.J., Guinotte, J. (2011). Global habitat suitability for framework-forming coldwater corals. *PLoS ONE*, 6: 1–15. e18483.

Doonan, I.J., Bull, B., Coombs, R.F. (2003). Star acoustic surveys of localized fish aggregations. *ICES Journal of Marine Science*, 60: 132–146.

Doubleday, W.G., Rivard, D. (eds). (1981). Bottom trawl surveys: proceedings of a workshop held at Ottawa, November 12–14, 1980. *Canadian Special Publication of Fisheries and Aquatic Sciences*, 58.

Escobar-Briones, E. (2003). The effect of the 1997–1998 ENSO on the benthic macrofaunal biomass in the southern Gulf of Mexico: *Geofísica Internacional*, 42: 517–522.

Fitz-George-Balfour, T., Billett, D.S.M., Wolff, G.A., Thompson, A., Tyler, P.A. (2010). Phytopigments as biomarkers of selectivity in abyssal holothurians: interspecific differences in response to a changing food supply. *Deep-Sea Research II*, 57: 1418–1428.

Francis, R.I.C.C. (1984). An adaptive strategy for stratified random trawl surveys. *New Zealand Journal of Marine and Freshwater Research*, 18: 59–71.

Gage, J.D., Tyler, P.A. (1991). *Deep-sea biology*. Cambridge, UK: Cambridge University Press.

Ginger, M., Billett, D.S.M., Mackenzie, K., *et al.* (2001). Organic matter assimilation and selective feeding by holothurians in the deep sea: some observations and comments. *Progress in Oceanography*, 50: 407–422.

Glover, A.G., Gooday, A.J., Bailey, D.M., *et al.* (2010). Temporal change in deep-sea benthic ecosystems: a review of the evidence from recent time-series studies. *Advances in Marine Biology*, 58: 1–95.

Gooday, A.J. (2002). Biological responses to seasonally varying fluxes of organic matter to the ocean floor: a review. *Journal of Oceanography*, 58: 305–332.

Gotelli, N.J., Ellison, A.M. (2004). *A primer of ecological statistics*. Sunderland, MA: Sinauer.

Green, R.H. (1979). *Sampling design and statistical methods for environmental biologists*. New York: John Wiley & Sons.

Grehan, A.J., Van Den Hove, S., Armstrong, C.W., *et al.* (2009). HERMES: Promoting ecosystem-based management and the sustainable use and governance of deep-water resources. *Oceanography*, 22: 154–165.

Hebbeln, D. (2000). Flux of ice-rafted detritus from sea ice in the Fram Strait. *Deep-Sea Research II*, 47: 1773–1790.

Hosia, A., Stemmann, L., Youngbluth, M. (2008). Distribution of net-collected planktonic cnidarians along the northern Mid-Atlantic Ridge and their associations with the main water masses. *Deep Sea Research II*, 55: 106–118.

Howell, K.L., Mowles, S.L., Foggo, A. (2010). Mounting evidence: near-slope seamounts are faunally indistinct from an adjacent bank. *Marine Ecology*, 31 (suppl. 1): 52–62.

Hunter, J.R., Lo, N. C-H. (1983). Icthyoplankton methods for estimating fish biomass–introduction and terminology. *Bulletin of Marine Science*, 53: 723–727.

Hurlbert, S.H. (1984). Pseudoreplication and the design of ecological field experiments. *Ecological Monographs*, 54: 187–211.

Kingsford, M., Battershill, C. (eds) (1998). *Studying temperate marine environments: a handbook for ecologists*. Christchurch, New Zealand: Canterbury University Press, pp. 335.

Knapp, A.K., Smith, M.D., Collins, S.L., *et al.* (2004). Generality in ecology: testing North American grassland rules in South African savannas. *Frontiers in Ecology and the Environment*, 2: 483–491.

ICES (2010). *Manual for the international bottom trawl surveys in the western and southern areas. Revision III. Addendum II*. Copenhagen: International Council for the Exploration of the Sea.

International Seabed Authority (2007). Polymetallic sulphides and cobalt rich ferromanganese crust deposits: establishment of environmental baselines and an associated monitoring programme during exploration. *Proceedings of the International Seabed Authority's workshop*, Kingston, Jamaica, 6–10 September 2004.

Lampitt, R.S., Salter, I., de Cuevas, B.A., Hartman, S., Larkin, K.E., Pebody, C.A. (2010). Long-term variability of downward particle flux in the deep northeast Atlantic: causes and trends. *Deep-Sea Research II*, 57: 1346–1361.

Larkin, K.E., Ruhl, H.A., Bagley, P., *et al.* (2010). Benthic biology time-series in the deep sea: indicators of change. In: Hall, J., Harrison, D.E., Stammer, D. (eds), *Proceedings of OceanObs'09: Sustained Ocean Observations and Information for Society, Vol. 2*. ESA Special Publication WPP-306. Noordwijk, The Netherlands: European Space Agency. doi: 10.5270/OceanObs09

Legendre, P., Thrush, S.F., Cummings, V.J., *et al.* (1997). Spatial structure of bivalves in a sandflat: scale and generating processes. *Journal of Experimental Marine Biology and Ecology*, 216: 99–128.

Levin, L.A., Etter, R.J., Rex, M.A., *et al.* (2001). Environmental influences on regional deep-sea species diversity. *Annual Review of Ecology and Systematics*, 32: 51–93.

Levin, L.A., Sibuet, M., Gooday, A.J., Smith, C.R., Vanreusel, A. (2010). The roles of habitat heterogeneity in generating and maintaining biodiversity on continental margins: an introduction. *Marine Ecology*, 31: 1–5.

Levin, S.A. (1992). The problem of pattern and scale in ecology. *Ecology*, 73: 1943–1967.

MacLennan, D.N., Simmonds, J.E. (1992). *Fisheries acoustics*. London: Chapman and Hall.

Maxwell, S.E., Kelley, K., Rausch, J.R. (2008). Sample size planning for statistical power and accuracy in parameter estimation. *Annual Review of Psychology*, 59: 537–563.

McClain, C.R., Barry, J.P. (2010). Habitat heterogeneity, disturbance, and productivity work in concert to regulate biodiversity in deep submarine canyons. *Ecology*, 91: 964–976.

McClain, C. R., Hardy, S.M. (2010). The dynamics of biogeographic ranges in the deep sea. *Proceedings of the Royal Society B*, 277: 3533–3546.

McClain, C.R., Lundsten, L., Barry, J., DeVogelaere, A. (2010). Assemblage structure, but not diversity or density, change with depth on a northeast Pacific seamount. *Marine Ecology*, 31 (suppl. 1): 14–25.

Menot, L., Galeron, J., Olu, K., *et al.* (2010). Spatial heterogeneity of macrofaunal communities in and near a giant pockmark area in the deep Gulf of Guinea. *Marine Ecology*, 31: 78–93.

Narayanaswamy, B.E., Bett, B.J., Hughes, D.J. (2010). Deep-water macrofaunal diversity in the Faroe-Shetland region (NE Atlantic): a margin subject to an unusual thermal regime. *Marine Ecology*, 31: 237–246.

O'Hara, T.D., Consalvey, M., Lavrado, H.P., Stocks, K.I. (2010). Environmental predictors and turnover of biota along a seamount chain. *Marine Ecology*, 31 (suppl. 1): 84–94.

Priede, I.G., Merrett, N.R. (1996). Estimation of abundance of abyssal demersal fishes; a comparison of data from trawls and baited cameras. *Journal of Fish Biology*, 49: 207–216.

Priede, I.G., Godbold, J.A., King, N.J., Collins, M.A., Bailey, D.M., Gordon, J.D.M. (2010). Deep-sea demersal fish species richness in the Porcupine Seabight, NE Atlantic Ocean: global and regional patterns. *Marine Ecology*, 31: 247–260.

Quinn, G.P., Keough, M.J. (2002). *Experimental design and data analysis for biologists*. New York: Cambridge University Press.

Ramirez-Llodra, E., Brandt, A., Danovaro, R., *et al.* (2010). Deep, diverse and definitely different: unique attributes of the world's largest ecosystem. *Biogeosciences*, 7: 2851–2899.

Rex, M.A., Etter, R.J., Morris, J.S., *et al.* (2006). Global bathymetric patterns of standing stock and body size in the deep-sea benthos. *Marine Ecology Progress Series*, 317: 1–8.

Rowden, A.A., Schlacher, T.A., Williams, A., *et al.* (2010). A test of the seamount oasis hypothesis: seamounts support higher epibenthic megafaunal biomass than adjacent slopes. *Marine Ecology – an Evolutionary Perspective*, 31: 95–106.

Roy, K.O.L., Caprais, J.C., Fifis, A., *et al.* (2007). Cold-seep assemblages on a giant pockmark off West Africa: spatial patterns and environmental control. *Marine Ecology*, 28: 115–130.

Ruhl, H.A., Ellena, J.A., Smith, K.L. Jr. (2008). Connections between climate, food limitation, and carbon cycling in abyssal sediment communities, *Proceedings of the National Academy of Sciences USA*, 105: 17006–17011.

Saville, A. (ed.). (1977). Survey methods of appraising fishery resources. *FAO Fisheries Technical Paper*, 171.

Schewe, I., Soltwedel, T. (2003). Benthic response to ice-edge-induced particle flux in the Arctic Ocean. *Polar Biology*, 26: 610–620.

Schlacher, T.A., Schlacher-Hoenlinger, M.A., Williams, A., Althaus, F., Hooper, J.N.A., Kloser, R. (2007). Richness and distribution of sponge megabenthos in continental margin canyons off southeastern Australia. *Marine Ecology Progress Series*, 340: 73–88.

Schlacher, T.A., Rowden, A.A., Dower, J.F., Consalvey, M. (2010a). Seamount science scales undersea mountains: new research and outlook. *Marine Ecology – an Evolutionary Perspective*, 31: 1–13.

Schlacher, T.A., Williams, A., Althaus, F., Schlacher-Hoenlinger, M.A. (2010b). High-resolution seabed imagery as a tool for biodiversity conservation planning on continental margins. *Marine Ecology – an Evolutionary Perspective*, 31: 200–221.

Schlacher, T., Baco, A., Rowden, A.A., *et al.* (2014). Seamount benthos in a Cobalt-rich crust region of the Central Pacific: implications for conservation challenges posed by future seabed mining. *Diversity and Distributions*, 20(5): 491–502.

Simmonds, J.E., MacLennan, D.N. (2005). *Fisheries acoustics: theory and practice*. Oxford: Wiley-Blackwell.

Sissenwine, M.P., Azarovitz, T.R., Suomala, J.B. (1983). Determining the abundance of fish. In: MacDonald, A.G., Priede, I.G. (eds), *Experimental biology at sea*. London: Academic Press, pp. 51–101.

Smith, C.R., Jumars, P.A., DeMaster, D. (1986). In situ studies of megafaunal mounds indicate rapid sediment turnover and community response at the deep-sea floor. *Nature*, 323.

Smith, C.R., Baco, A.R. (2003). Ecology of whale falls at the deep-sea floor. *Oceanography and Marine Biology*, 41: 311–354.

Smith, K.L., Baldwin, R.J., Ruhl, H.A., Karru, M., Mitchell, B.G., Kaufmann, R.S. (2006). Climate effect on food supply to depths greater than 4,000 metres in the northeast Pacific. *Limnology and Oceanography*, 51: 166–176.

Soetaert, K., Van Oevelen, D. (2009). Modeling food web interactions in benthic deep-sea ecosystems: a practical guide. *Oceanography*, 22: 128–129.

Sokal, R.R., Rohlf, F.J. (1995). *Biometry: the principles and practice of statistics in biological research.* New York: W.H. Freeman.

Soltwedel, T., Miljutina, M., Mokievsky, V., Thistle, D., Vopel, K. (2003). The meiobenthos of the Molloy Deep (5600 m), Fram Strait, Arctic Ocean. *Vie et Milieu*, 53: 1–13.

Soltwedel, T., Bauerfeind, E., Bergmann, M., *et al.* (2005). HAUSGARTEN: multidisciplinary investigations at a deep-sea, long-term observatory in the Arctic Ocean. *Oceanography*, 18: 47–61.

Thiel, H., Schriever, G., Ahnert, A., Bluhm, H., Borowski, C., Vopel, K. (2001). The large-scale environmental impact experiment DISCOL – reflection and foresight. *Deep-Sea Research II*, 48: 3869–3882.

Thrush, S.F. (1991). Spatial patterns in soft-bottom communities. *Trends in Ecology & Evolution*, 6: 75–79.

Thrush S.F., Hewitt J.E., Cummings V.J., Green J.M., Funnell G.A., Wilkinson M.R. (2000). The generality of field experiments: interactions between local and broad-scale processes. *Ecology*, 81: 399–415.

Tittensor, D.P., Baco, A.R., Brewin, P.E., *et al.* (2009). Predicting global habitat suitability for stony corals on seamounts. *Journal of Biogeography*, 36: 1111–1128.

Tracey, D.M., Bull, B., Clark, M.R., MacKay, K.A. (2004). Fish species composition on seamounts and adjacent slope in New Zealand waters. *New Zealand Journal of Marine and Freshwater Research*, 38: 163–182.

Underwood, A.J. (1993). The mechanics of spatially replicated sampling programs to detect environmental impacts in a variable world. *Australian Journal of Ecology*, 18: 99–116.

Underwood, A.J. (1994). On beyond baci – sampling designs that might reliably detect environmental disturbances. *Ecological Applications*, 4: 3–15.

Underwood, A.J. (1997). *Experiments in ecology: their logical design and interpretation using analysis of variance*. Cambridge, UK: Cambridge University Press.

Underwood, A.J., Chapman, M.G. (2003). Power, precaution, Type II error and sampling design in assessment of environmental impacts. *Journal of Experimental Marine Biology and Ecology*, 296: 49–70.

Underwood, A.J., Chapman, M.G. (2005). Design and analysis in benthic surveys. In: Eleftheriou, A., McIntyre, A. (eds), *Methods for the study of marine benthos* (3rd edn). Oxford: Blackwell, pp. 1–42.

Vanreusel, A., Fonseca, G., Danovaro, R., *et al.* (2010). The contribution of deep-sea macrohabitat heterogeneity to global nematode diversity. *Marine Ecology*, 31: 6–20.

Weaver, P.P.E., Gunn, V. (2009). HERMES: Hotspot ecosystem research on the margins of European seas. *Oceanography*, 22: 12–15.

Weaver, P.P.E., Boetius, A., Danovaro, R., *et al.* (2009). The future of integrated deep-sea research in Europe: The HERMIONE Project. *Oceanography*, 22: 178–191.

Wenneck, T.deL., Falkenhaug, T., Bergstad, O.A. (2008). Strategies, methods, and technologies adopted on the R.V. *G.O. Sars* MAR-ECO expedition to the Mid-Atlantic Ridge in 2004. *Deep-Sea Research II*, 55: 6–28.

Williams, A., Althaus, F., Dunstan, P.K., *et al.* (2010a). Scales of habitat heterogeneity and megabenthos biodiversity on an extensive Australian continental margin (100–1100 m depths). *Marine Ecology*, 31 (suppl. 1): 222–236.

Williams, A., Schlacher, T.A., Rowden, A.A., *et al.* (2010b). Seamount megabenthic assemblages fail to recover from trawling impacts. *Marine Ecology*, 31 (suppl. 1): 183–199.

Yesson, C., Taylor, M.L., Tittensor, D.P., *et al.* (2012). Global habitat suitability of cold-water octocorals. *Journal of Biogeography*, 39: 1278–1292.

Zar, J.H. (1996). *Biostatistical analysis*. Upper Saddle River, NJ: Prentice Hall.

Chapter 4

Environmental Sampling

Martin White[1], Christian Mohn[2] and Kostas Kiriakoulakis[3]

[1] National University of Ireland, Galway, Ireland
[2] Aarhus University, Roskilde, Denmark
[3] John Moores University, Liverpool, United Kingdom

Abstract

This chapter provides an overview of selected instrumentation, methodologies and procedures for surveying important physical quantities in deep-sea environments. Due to the focus of the book, emphasis is put on ship-based sampling techniques that are considered a robust standard for use in environmental *in situ* sampling of deep-sea marine systems. Instruments considered are the conductivity, temperature and depth (CTD) profiler, acoustic Doppler current profiler (ADCP) and several types of instruments for determination of suspended particle concentration. In combination, these instruments can provide a detailed description of the dynamic processes occurring in the region of interest and subsequent influence on the habitat under study. Basic types, configurations and usage of the instruments are presented, together with a discussion of considerations for their use within an overall sampling strategy.

Keywords CTD, ADCP, sediment traps, particulate organic matter, sampling strategies

4.1 Introduction

To fully understand biological processes and subsequent faunal distribution and functioning in the deep sea, a significant knowledge and appreciation of environmental conditions is a basic requirement. Physical processes that require consideration are turbulent mixing of heat, salt and chemical properties, both tidal and residual circulation, and other wave- or wind-induced dynamics such as upwelling, Langmuir circulation or baroclinic wave motions. These processes provide the basic foundation for the oceanic water mass structure, chemical cycling of nutrients, concentrations of organic matter and essential gases and drive the subsequent ecosystem and habitat response and development.

Physical processes occur on a myriad of temporal and spatial scales, resulting in a complex dynamic environment to which the biological community will respond. The development of an optimal approach to sampling strategy, therefore, necessitates a detailed understanding of the environmental conditions present. Physical time/space scales vary from the very small (second and sub-millimetre) up to large basin-scale processes related to atmosphere-ocean forcing and climatic variability on temporal scales of years and covering spatial scales of thousands of kilometres (Fig. 4.1). Overlap between physical and biological scales is a direct consequence of the relationship between the physical forcing and biogeochemical responses. At small scales, biological processes may dominate the biological response directly, although molecular processes and turbulence will

Biological Sampling in the Deep Sea, First Edition. Edited by Malcolm R. Clark, Mireille Consalvey and Ashley A. Rowden.

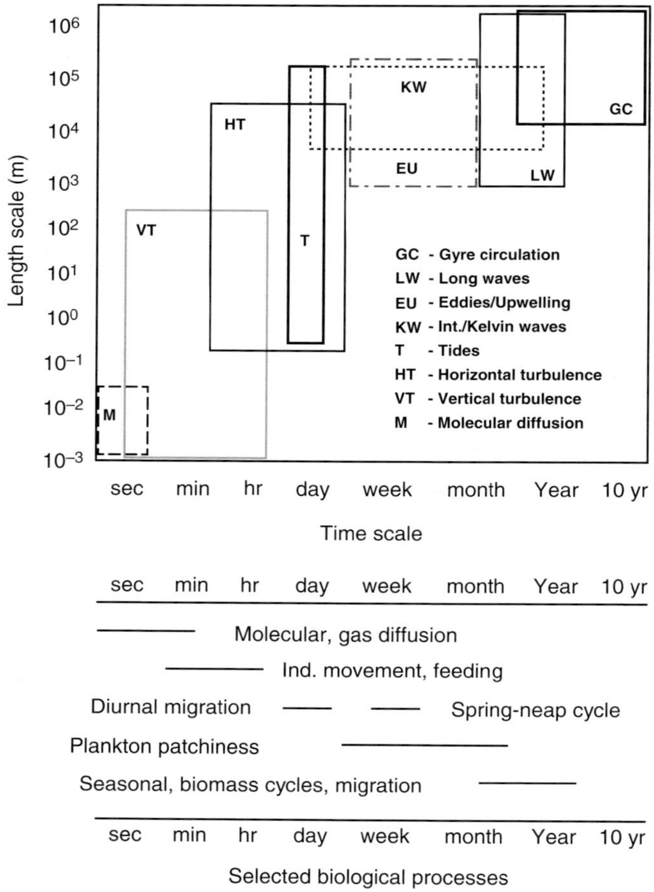

Fig. 4.1 Schematic illustrating the typical length and time scales of (a) relevant physical processes and (b) selected biological processes within the deep-sea environment of relevance to this book.

also directly affect gas diffusion, feeding and respiration within ecosystems (e.g. Anderson & Charters, 1982). As scales get larger, physical processes will become more influential. For example, the principal tidal periods (semi-diurnal and diurnal) will likely interact strongly with the natural diurnal plankton migration patterns that may alias planktonic advection (e.g. Goldner & Chapman, 1997). Scales of horizontal variability and stirring will play a significant role in plankton patchiness through features such as eddies (Martin, 2003) or from flow interaction with topography (Mohn & White, 2010). At the largest scales where interannual and climatic processes may be dominant, response of biological communities to climatic forcing can only be elucidated from large temporal or spatial scale monitoring of ocean structure and communities, such as through the ARGO float (Gould *et al.*, 2004) or the continuous plankton recorder (CPR) programmes (e.g. John & Reid, 2001; Richardson, 2008).

4.1.1 Sampling gear types

There are a large number of gear types available for sampling in the deep sea. However, the emphasis in this book is on ship-based sampling, and hence in this chapter we focus on common sampling equipment that is routinely used to provide standard measurements of *in situ* environmental conditions: namely the CTD, ADCP and sediment traps (Fig. 4.2). These will be discussed in more detail below.

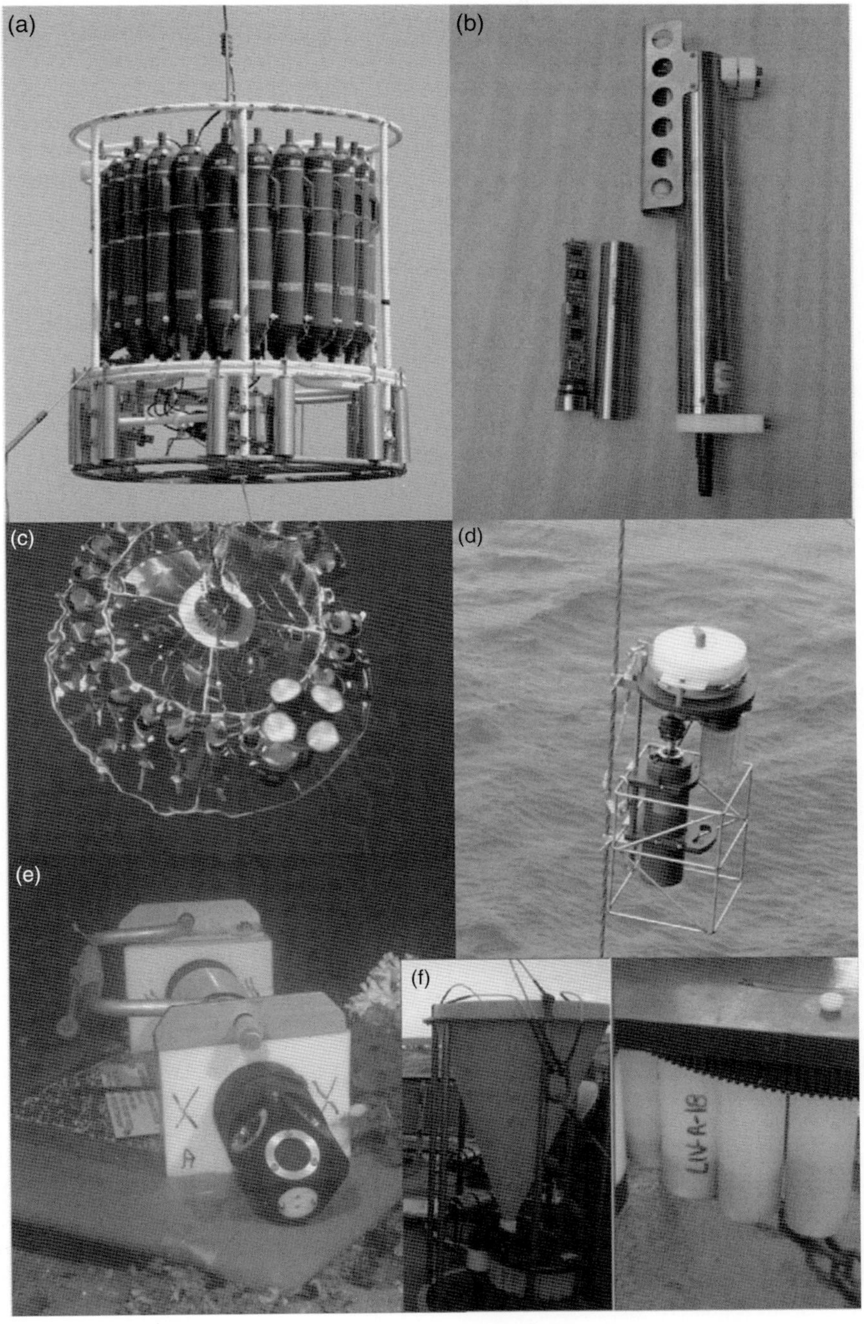

Fig. 4.2 Examples of the instruments discussed in this chapter. (a) Real-time CTD showing rosette of water bottles and electronic package at the bottom of the frame. (Photograph by Kostas Kiriakoulakis.) (b) Two self-recording instruments, an Sbe37 temperature–conductivity sensor and an RBR temperature recorder. (Photograph by Damien Guihen. Reproduced with permission.) (c) LADCP attached to CTD frame. (IOC Photo Library.) (d) SAPS instrument on winch line ready for deployment. (Photograph by Kostas Kiriakoulakis.) (e) ROV-deployed three-beam ADCP on base plate. (Photograph by Tomas Lundälv. Reproduced with permission.) (f) A sediment trap on deck with close-up of the carousel of sample containers after recovery. (Photograph by Kostas Kiriakoulakis.) (Please see Plate 4 for colour representation of the figure.)

4.2 Conductivity, temperature and depth

4.2.1 Principles, gear type and deployment

The CTD is the workhorse instrument for hydrographic observations when sampling the deep sea from research ships, autonomous platforms, or from towed platforms in a real-time or self-recording configuration. A CTD may be used from a winch on board a vessel in real time, when connected to a conducting cable, or as a self-recording instrument. CTD-type instruments can also be fixed on trawls, a mooring line or seabed lander systems (e.g. see Chapters 7 and 11), within a floating buoy (surface or subsurface), such as with ARGO floats (Gould *et al.*, 2004), towed as an undulating instrument (e.g. SEASOAR; Pollard, 1986), fixed on a biological or fisheries sampling instrument (e.g. Majewski *et al.*, 2009; Neat *et al.*, 2010) or as micro-instruments attached to marine animals (e.g. Myers *et al.*, 2006; Rikardsen *et al.*, 2007). In addition, expendable CTD/temperature sensors (XCTD, XBTs) are also commonly used (Mizuno & Watanabe, 1993) as well as specialized fast-response sensors, often free falling, used to quantify turbulence and energy dissipation (Dewey *et al.*, 1987).

CTD electronic systems have developed from the mechanical-era instruments of reversing thermometers and water sample bottles, triggered by weights attached to the winch line (Lawson & Larson, 2001). A basic CTD package comprises a temperature sensor (fast-response thermistor), conductivity cell and pressure sensor that can be connected to a conducting cable for real-time use or as a self-recording instrument (Lawson & Larson, 2001; Williams, 2009). Additional sensors may also be attached to the sensor recording package, such as those that measure fluorescence (to determine chlorophyll), optical transmission (turbidity) and oxygen (see Table 4.1). The basic sensor

Table 4.1 Basic, derived and supplementary water column properties that are typically measured by CTDs and additional attached sensors in whatever deployment fashion employed (real-time vessel, towed, moored etc.)

Parameter	Method	Biological significance
Basic measurements		
Conductivity (C)	Inductive resistance	General water column structure, water mass properties
Temperature (T)	Thermistor	
Pressure (P)	Piezo-electric sensor	
Derived parameters		
Depth	Use pressure and latitude (Saunders & Fofonoff, 1976)	Location in water column
Salinity (S)	Using C,T, P (Practical Salinity Scale – PSS 78)	Salt content and for calculating density
Density (ρ)	Using S, T, P (Eqn. of state)	Water column structure and dynamics (e.g. calculation of geostrophic currents)
Θ Sigma_θ (σ_θ)	Potential Temperature & Density (Eqn of state)	Comparison of water column structure over large vertical ranges where pressure effects on temperature need to be taken into account
Main additional measurements		
Oxygen	Optode	Oxygen levels, Anoxia, O_2 minimum layer
Fluorescence	Fluorescence	Chlorophyll levels (plankton abundance)
Turbidity	Light transmission	Suspended particulate concentrations
Water chemisitry & calibration	Rosette Bottles	Water chemistry, plankton, suspended particulates, salinity, oxygen

measurements (conductivity, temperature, pressure) enable secondary variables to be calculated (e.g. salinity, depth, density), while the other sensors may be connected to a system for a particular biological purpose. On a shipboard, vertically profiling CTD with real-time data transmission via conducting cable, the CTD package will have the electronic sensors located near the bottom of the containing frame. The CTD may also include a rosette of sampling bottles attached to the frame (Fig. 4.2a). These bottles, open at the start of any vertical profile, may be 'fired' from the desk command to close at chosen depths to allow water samples to be taken for both calibration of the electronics or to take biological/chemical measurements on board or back in the laboratory.

Real-time or self-recording CTD-type units may be employed in various configurations depending on the accuracy/resolution (space/time) requirements for any data that will be collected. For survey teams or vessels without a real-time cable/CTD system, there are small, self-recording CTD or similar instruments available from many manufacturers with a range of accuracies/resolution of the primary sensors together with additional sensors (e.g. Fig. 4.2b). These instruments can be invaluable for collection of appropriate supporting environmental data for biological surveys (e.g. Majewski *et al.*, 2009; Neat *et al.*, 2010). Care is required to locate these sensors in relation to the biological gear, particularly those that may disturb the seabed, so that the sensors are sampling the least-contaminated environment (e.g. placing units upstream of trawls or towed sledges so that the conductivity sensor is not affected by suspended sediment).

Data collection as described above will be determined by the use of the primary sampling equipment. The self-recording instruments will be able to map hydrographic properties over the spatial scale of the sample tows/transects at a high temporal resolution and can be useful, for example, in locating features such as fronts and eddies that may influence the biological distributions or sampling.

For real-time vertical profiles two CTD deployments are commonly used. For general applications, the CTD package is lowered at a constant speed to close to the seabed and back to the surface. For hydrographic observations, typical lowering speeds of between 0.5–1 m s^{-1} (30–60 m/min, Fig. 4.3a)

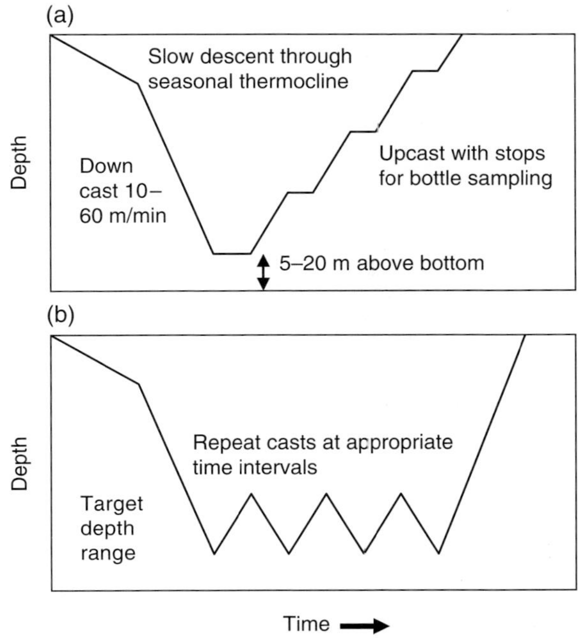

Fig. 4.3 Illustration of the depth profile of a CTD package to show typical deployment methods. (a) The CTD is lowered at constant speed to the seabed or target depth, before being raised and stopped at target depths for discrete water sampling. In (b) the CTD is lowered to the target depth and then repeatedly raised/lowered over a smaller depth range to increase temporal resolution in the target depth range.

are used, unless there is only interest in a particular depth range which may be at greater depths. Slower speeds might be used within the upper layer of ~100 m (seasonal thermocline) where vertical gradients of temperature/density are largest. The down profile is used for the water column structure measurements and the up profile for stopping the CTD to sample the water column using the rosette to obtain discrete water samples. The second common way to use the CTD is to 'yoyo' (sometimes termed 'jojo') the package with repeated descending and ascending profiles over a target depth range while the ship is fixed at a location or slowly drifting across a region/topographic feature of interest (Fig. 4.3b). This helps to improve the temporal resolution of the water column structure, for example to resolve tidal variability within the benthic boundary layer (e.g. White, 1994 and see later example in Fig. 4.5).

The rosette bottles on a CTD are a very useful sampling tool, although samples are limited to a finite number and volume depending on the size of the CTD frame (and hence bottle number/size) being employed. It is usual to close these bottles during the up-cast of a vertical profile, as the rosette is typically housed in the upper portion of a CTD frame. It can sometimes be hard to position the CTD bottles at the exact depth preferred by the scientist, and one should be realistic about the accuracy of the vertical location at which samples are taken. This is due to inherent characteristics of the CTD, particularly those involving response times of sensors which result in vertical offset in features for both up/down profiles (see next section), and not least because of the heaving/rolling of the CTD tethered to a ship at the surface of a restless ocean!

4.2.2 *Problems, pitfalls and processing*

Modern CTD systems generally record data at high temporal resolution, typically at 8–24 Hz, which would suggest a possible vertical resolution of 2–10 cm per sample at typical profiling speeds of 0.5–1 m s^{-1}, but is never achieved. This is due to a number of factors including the response times of the individual sensors to adjust to the surrounding environment, thermal inertia effects and the fact that the CTD package may be disturbed significantly by surface wave/swell action, as well as the result of drag on the profiling line due to the vertical distribution of currents. With appropriate data treatment a vertical resolution of measurements in the order of 1 m for general hydrography and 0.1 m for some fine-scale structure may be achieved if the sensor package is profiled slowly. There are a number of errors that are associated with such a complex instrument and an appreciation of this will help in appropriate data interpretation and identification of problems that may arise when sampling. Errors with CTD data arise for a number of reasons:

- There is a finite response time for each of the sensors, which varies with sensor, resulting in two particular characteristics. Water column features will often seem deeper/shallower than in reality for the down/up portion of the CTD cast, respectively. This may be particularly acute for oxygen as typical O_2 sensors have long response lags. Different lags in the response of the temperature (T) and conductivity (C) sensors may result in spiking in the resultant calculated salinity (S) as essentially the measurements of T and C come from different parcels of water. This may be important in regions of strong vertical gradients.
- There are also effects related to the movement of the CTD package through the water column, such as physical mixing and heating effects – both related to the contact of the CTD with the water and self-heating by the electronics. In addition, both surface wind/wave actions, as well as the currents within the water column, will disturb the descent of the CTD.

To reduce these errors, a number of steps can be completed in the processing of the data. The steps below may be used as a guideline as to a basic strategy for quality control, but are not hard and fast rules. Most CTD instruments will have manufacturer software that will accomplish these steps. It should also be noted that many of these steps will not be required for data collected at a sampling period of 1 s or greater (e.g. for the case of self-recording instruments attached to other sampling equipment).

(1) *Align temperature and conductivity sensors.* This accounts for different lag of the temperature and conductivity sensors on a real time CTD. Salinity may be calculated for a number of data samples with the conductivity sample offset by different measurement lags relative to the temperature to obtain the 'least spiky' salinity dataset.
(2) *Run a median filter through the data.* A median filter is used instead of a normal running mean to remove noise/spikes in the data, as this does not artificially change or smooth real vertical structure in any of the variables.
(3) *Perform a temperature calibration* if calibration data are available.
(4) *Calculate salinity and other secondary variables.* Use the filtered temperature, conductivity and pressure. Use an approved and tested routine for calculation of salinity (e.g. a UNESCO formula).
(5) *Filter salinity data.* Use a median filter through the salinity data with the same window size as previously used.
(6) *Calibrate salinity.* Perform a salinity calibration of salinity using bottle data.
(7) *Check for looping in pressure signal.* Data are discarded during the CTD up-cast or parts of the down-cast when the CTD is passing through a depth that has already been obtained. This is because the sensor package is located at the bottom of the CTD rosette system and therefore it is the down-cast that is passing through water that is relatively undisturbed by the CTD package.
(8) *Average the data* to the required vertical (profiling CTDs) or horizontal resolution (for towed and self-recording instruments).

A typical upper layer profile of T, C, S (Fig. 4.4a–c) and density (Fig. 4.4d) can be used to illustrate some of the issues raised above. The untreated profiles indicate a surface mixed layer and seasonal thermocline. Temperature and conductivity profiles are similar and also show smaller mixed layers (steps in the profile) of a few metres in vertical extent through the depth range of the seasonal thermocline. The salinity profile is spiky, however, in particular at the top (~35 m) and bottom (~60 m) of the seasonal thermocline when the mismatch in temperature and conductivity measurements is greatest in the raw data. The treated profiles smooth out the salinity errors but retain the water column structure, including the unusual decrease in salinity just below the mixed layer (the profile of density in Fig. 4.4d indicates this is not an artefact but the water column is stably stratified at that depth).

The profiles in Fig. 4.4e,f show the effect of 'looping' of the pressure records (this problem was eliminated in the profiles in Fig. 4.4a–d for clarity) from the same CTD profile. In the surface layer, surface wave action and ship movement significantly affects the CTD descent. The apparent periodicity of four cycles in about 800 samples (or ~32 s at 24 Hz) indicates a modulation at a period of 8 s, a typical deep ocean swell period. Modulation of the descent is apparent at greater depths (Fig. 4.4f) but is much reduced.

Data from a series of yoyo profiles show how a CTD can be manipulated to sample the water column adjacent to the seabed at relatively high temporal resolution (Fig. 4.5). Data are from a survey of the slope of the Porcupine Bank, Northeast Atlantic to assess the dynamics over a carbonate mound (see Mienis *et al.*, 2007). The CTD was lowered to close to the seabed (depth 700–750 m) from a depth of 400 m. Between each up/down profile the ship was edged slowly by a few hundred metres in a transect so a profile was made every ~20 minutes. The resultant temperature profiles show the variability in the benthic boundary layer (BBL) height (Fig. 4.5a). Use of a beam transmissiometer indicated a benthic nepheloid layer (BNL) or suspended particulate material associated with the BBL (Fig. 4.5b). The temperature profiles also indicated the likely presence of an overturning internal wave in progress, seen as the inversion of temperature in three consecutive vertical profiles between the depths of 650–700 m. If a single profile, perhaps made as part of a coarser resolution transect across the slope, had shown this temperature inversion, it may have been interpretated as a temporary mis-function (e.g. blocked sensors) as the associated

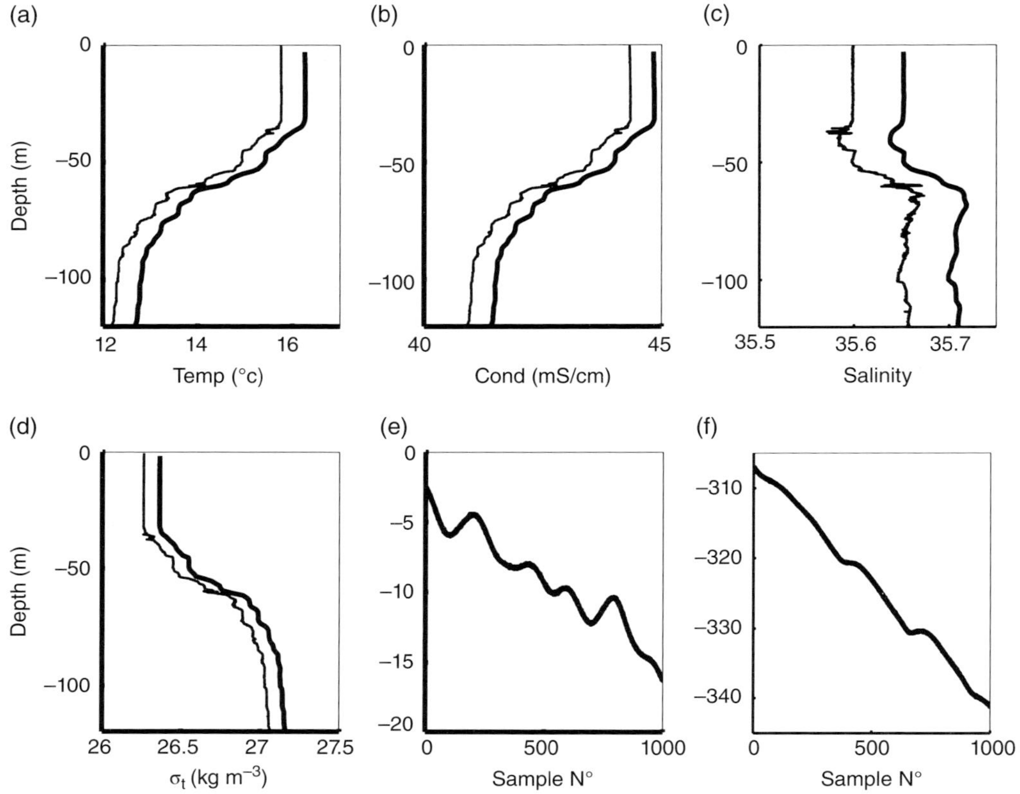

Fig. 4.4 An example vertical CTD profile showing the upper 120 m of (a) temperature, (b) conductivity, (c) salinity and (d) density from deep water in the Rockall Trough, Northeast Atlantic. The thin line shows the unaveraged, untreated data and the thick line treated data, offset in value for clarity. The variation of pressure with sample number for the same downward CTD profile is shown for portions of the cast in (e) and (f).

inversion in density would not be stable). Here the three consecutive profiles made within 40 minutes showing a consistent feature is persuasive, given the likely time scale for an overturn to occur and particularly as the region is known to have high internal wave energy (Dickson & McCave, 1986; Mienis *et al.*, 2007).

4.3 Acoustic Doppler current profilers

4.3.1 Principles and use

Acoustic Doppler current profilers (ADCPs) are sonar technology to measure speed and direction of ocean currents based on the Doppler effect. ADCPs transmit a sound pulse at a given frequency and take continuous records of the backscattered echoes from sound scatterers (e.g. biological particles, Fig. 4.6a). Any frequency shift between the source and return signals is a result of the motion

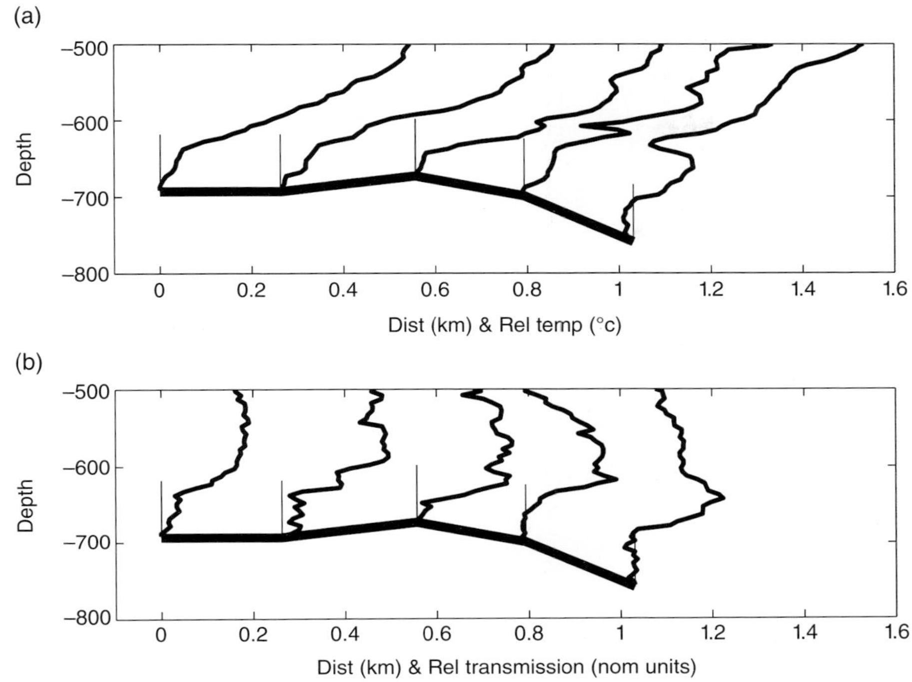

Fig. 4.5 Vertical profiles of near-seabed (a) temperature and (b) light transmission across a small carbonate mound on the northwest flanks of Porcupine Bank, Northeast Atlantic (see Fig. 4.9a for local bathymetry and station locations). Values of each property are relative to the value at the bottom, indicated by the short thin vertical line, and each profile is offset relative to the distance across the mound. (Data courtesy of Henko de Stigter, Royal Netherlands Institute for Sea Research.)

of the sound scatterers relative to the sound source (ADCP). The sound pulse is transmitted through a set of transducers with a predefined beam configuration and angle, with three, or often four, beams at an angle of 30°. Vertical profiles of current velocity are obtained through range-gating of the return signals along each beam (RD Instruments, 1996), separating the return signals into discrete depth bins based on the relationship of return time and sound velocity in seawater (Fig. 4.6a). A single pulse of sound is referred to as a ping, and usually an ensemble of pings within a certain measurement period is used to determine the current. In theory only three beams are required to resolve three-dimensional (3D) currents and the fourth beam adds redundancy and also a measure of data quality control. This is achieved through quantification of an 'error velocity', a statistical measure of data quality based on the difference between two vertical velocity profiles. Correlation between individual beams is also assesed. The beams spread further apart with distance from transducer, and this means that water is sampled by different beams at ever-increasing separation and hence increases the minimum length scale of coherent motions that will be adequately measured.

ADCPs are used for various purposes including monitoring river discharge rates (e.g. Muste *et al.*, 2004), collecting time series of ocean currents at individual mooring locations (e.g. van Haren, 2009) and long-term sampling of special areas of interest from moving vessels as part of repeated surveys or ship of opportunity programmes (Beal *et al.*, 2008; Rousset & Beal, 2010). ADCPs also provide other useful information, such as temperature at the transducer depth, strength of the backscattered signal (e.g. to determine accumulation and migration of biological particles) and beam statistics.

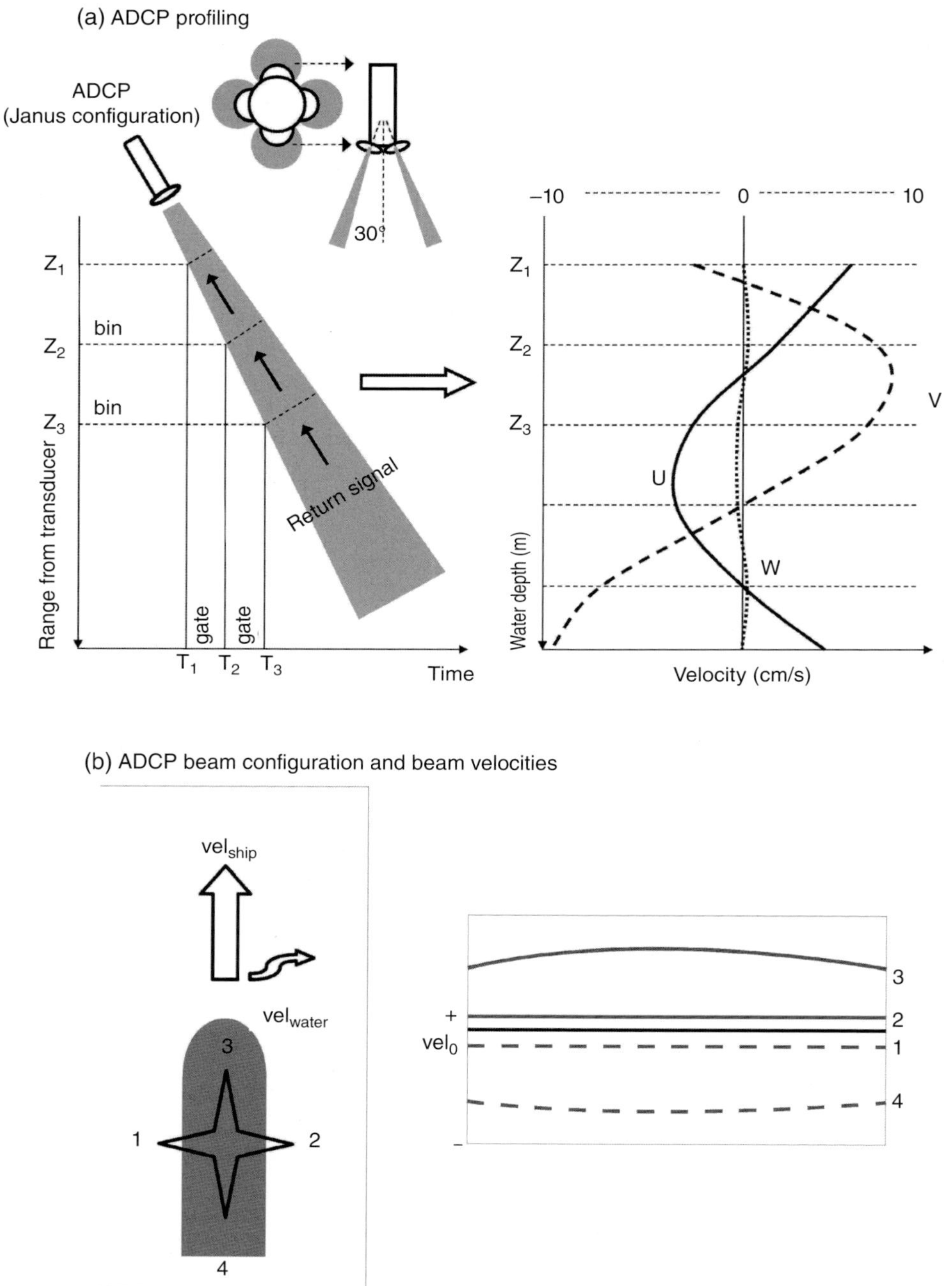

Fig. 4.6 (a) ADCP profiling. Depth bins are determined through range-gating of the backscattered signals. The return signals are matched to water depth by their individual return times and the sound velocity in water. The result is a vertical profile for all velocity components. (b) Schematic showing the default ADCP beam set-up (left) and corresponding raw beam velocities (right) that would be measured by the ADCP in this reference configuration.

Depth range and vertical resolution depend on the system's specification and set-up (see Table 4.2). The most common systems in oceanography are the following:

- Vessel-mounted four-beam ADCPs (VM-ADCP) are used for continuous on-station and underway sampling. VM-ADCP surveys may provide useful snapshots of the spatial structure of the circulation in areas of steady or quasi-steady flow conditions within the limits of the survey coverage.
- Lowered ADCPs (L-ADCP) are simultaneously operated with a CTD and mounted to a rosette sampler. L-ADCPs provide the large benefit of top to bottom profiling in abyssal waters deeper than the range of a VM-ADCP. Spatial coverage is restricted to the number of individual CTD stations per survey.
- Moored ADCPs (M-ADCP) are used to obtain long-term records of ocean currents at a single location. M-ADCPs may be mounted at the seafloor on a frame or lander (Fig. 4.1e), or on fixed moorings and surface platforms.

ADCP systems are available for a wide range of operating frequencies depending on the scales and processes of interest (see Table 4.2 for an overview). High operating frequencies provide more precise measurements and higher vertical resolution at the expense of profiling range, and are common on landers (e.g. like that shown in Fig. 4.1c). Low-frequency ADCPs can penetrate deeper into the water column at the expense of vertical resolution and give noisier data. The interplay of the three key parameters of profiling range, vertical resolution and random noise is commonly referred

Table 4.2 Table summarizing the typical acoustic properties of ADCPs used for oceanographic research, together with general environmental processes that can be quantified

Frequency	Typical range (m)	Vertical resolution (m)	Wavelength (cm)	Absorption coefficient (dB/m)	Examples of processes (platform independent)
2 MHz	2–4	0.1	0.075	1.2–1.6	Fine-scale turbulence, air–sea interaction
1.2 MHz	10–15	0.2	0.125	0.4–0.7	Benthic boundary layer dynamics, sediment dynamics
600 kHz	40–60	0.5	0.25	0.14–0.2	Small-scale currents, internal waves, dynamics at topography
300 kHz	80–120	1	0.5	0.06–0.08	Large/meso-scale near-surface and bottom currents, frontal dynamics, migratory behaviour of biological scatterers
75 kHz	400–600	4	2	0.02–0.03	Large-scale upper ocean currents, mixed layer, shelf–slope dynamics, behaviour of scatterering layers
38 kHz	1000+	8	4	0.007–0.01	Large-scale upper and intermediate ocean currents, biological scatterers

Values for typical range are based on both manufacturer's specifications and user experience.

to as the trade-off triangle (Spain, 2007), and an improvement of one in the instrument set-up will typically result in degradation of the other two.

4.3.2 *Problems, pitfalls and processing*

There are a number of factors affecting the accuracy, resolution and depth range of acoustic current measurements. The most important factors are sound absorption, operating frequency, signal strength of the transmitted sound pulse, lack or excess of scatterers, random noise and ship motion (in the case of VM-ADCPs). ADCP data post-processing is therefore a crucial task to transfer single-ping raw data into accurate, reliable and realistic estimates of ocean currents. This is especially important for VM-ADCPs where the ship's motion adds additional noise to the data. Major error sources are the possible misalignment of the transducer relative to the ship's keel, interference by physical objects, ship-induced turbulence and positioning errors in periods of acceleration of the ship. Ignoring these error sources may lead to incorrect estimates of current speeds and directions. One of the most comprehensive post-processing analytical suites for VM-ADCP data is the CODAS (Common Ocean Data Access System) toolbox developed at the University of Hawaii (http://currents.soest.hawaii.edu/docs/adcp_doc/index.html). Major post-processing steps for VM-ADCP data include the following (Firing & Hummon, 2010):

- Averaging of single-ping raw ADCP data profiles to obtain time-mean ensembles. Ensemble averaging reduces the impact of measurement errors and random noise. A typical averaging period is 5 minutes, but may be adjusted to ship speed and size of the sampling area (a 5 minute average at an underway ship speed of 10 knots gives a horizontal resolution of 1500 m).
- Quality control of ensemble velocity profiles to retain individual depth bins and/or profiles unaffected by physical disturbances (e.g. through bubbles, object interference, bottom reflection).
- Obtaining the scale factors for transducer amplitude and orientation relative to the ship heading by performing a water or bottom track calibration. The aim is to obtain accurate estimates of flow magnitude and direction.
- Calculating absolute current velocities by removing the best estimate of ship velocities per ensemble from the corresponding relative ADCP velocities.
- Optional de-tiding of absolute ADCP current velocities to separate the dominant barotropic/baroclinic tidal currents within a certain region from the residual (wind- and density-driven) flow.

Compensating for any misalignment of the ADCP transducer relative to the ship is important to avoid spurious currents. A standard reference frame can be defined by using the beam 3 heading when the ship is heading north (Fig. 4.6b). A potentially large misalignment can be identified by examining single-ping raw underway velocity records based on the assumption that the ship speed (8–10 knots, 4–5 m s^{-1}) is an order of magnitude greater than typical speeds of ocean currents (approx. 0.5 cm s^{-1}). If fully aligned with the ship's keel, data at beam 3 (forward beam) would show positive values and corresponding negative values at beam 4 (rear beam, Fig. 4.6b). These values represent the ship's transit speed slightly modified by the speed of ocean currents. In contrast, values at beams 1 and 2 would be very small or near zero representing ocean currents only. Any constant offset in the transducer alignment would cause substantially stronger current records at beams 1 and 2 with values greater than typical ocean currents. An example of this effect is shown for a survey in the Northeast Atlantic west of Ireland using an RDI Ocean Surveyor 75 kHz system (Fig. 4.7a,b). The strong difference between current records as a result of (i) uncorrected and (ii) corrected transducer misalignment illustrates the need for compensation prior (if known from previous cruises) or after the cruise.

4.3.3 *Correction for tidal flow*

ADCP data hold information on the total oceanic flow field composed of periodic (tidal and inertial) and sub-tidal (i.e. wind- and density-driven) constituents. For many applications the tidal variability is not desired and requires separation from the residual flow pattern. VM-ADCP data often represent instantaneous, non-synoptic measurements and provide only a snapshot of instantaneous tidal flow at

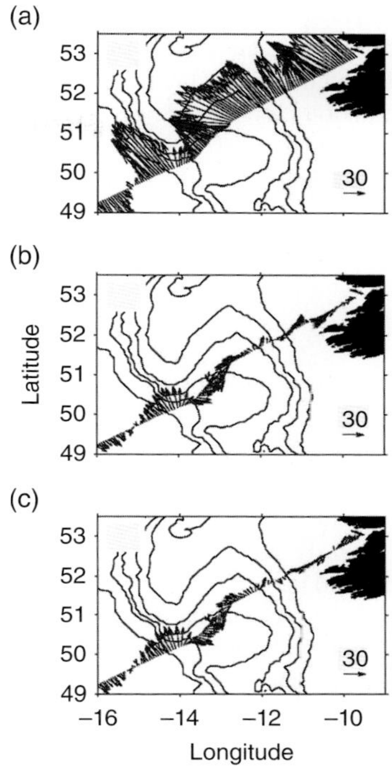

Fig. 4.7 Depth-averaged VM-ADCP currents (cm s^{-1}) from a 2007 R.V. *Celtic Explorer* survey in the Northeast Atlantic southwest of Ireland using a 75 kHz RDI Ocean Surveyor. (a) Data uncorrected for ADCP transducer misalignment. (b) Data corrected for transducer misalignment; there is a −53° transducer misalignment relative to the ship's keel. (c) De-tided residual currents based on tidal predictions from the OSU inverse tidal model (Egbert & Erofeeva, 2002).

any given velocity profile time and location. VM-ADCP data, therefore, pose a particular challenge for separating the spectrum of motions. De-tiding procedures include the use of tidal models, least squares tidal-fitting techniques and combinations of both. The classical method by Candela *et al.* (1992) applies least squares tidal fits and polynomial spatial functions to the VM-ADCP velocity components taking into account the spatial variability of the tidal constituents. Other approaches benefit from repeated transects as part of the survey strategy covering multiple tidal cycles (e.g. Valle-Levinson & Matsuno, 2003; Flagg *et al.*, 2006). Finally, a combination of different techniques can be used to account for spatio-temporal data gaps and local tidal dynamics (e.g. Carillo *et al.*, 2005). Comparative studies demonstrate that there is no general standard technique, but that the choice is rather dependent on the local characteristics of the physical regime (e.g. Foreman & Freeland, 1991; Isobe *et al.*, 2007).

An example of the effect of de-tiding VM-ADCP data is shown in Fig. 4.7c, using the same data set as shown in Fig. 4.7b. Tidal current predictions from the OSU inverse tidal model (Egbert & Erofeeva, 2002) were used to remove the tidal signal from the ADCP data. The resultant data set indicates that large-scale flow features (eddies) over abyssal depths away from the shelf edge and also at coastal boundaries are largely retained. On the shelf, however, the measured circulation is strongly modified due to the presence of tidal currents with changing tidal phase.

4.3.4 Biological observations using ADCPs

Use of ADCPs requires that there is something in the water column to scatter and reflect the transmitted acoustic signal. Although physical disturbance in the water column, such as from turbulence, can cause such reflections (e.g. Seim *et al.*, 1995; Holbrook *et al.*, 2003), scattering normally occurs from biological constituents or suspended inorganic material in the water column. The strength of the backscattered acoustic signal for a particular frequency ADCP, therefore, can provide very useful information on biological biomass and behaviour or suspended particulate matter (SPM) estimates. This includes estimation of zooplankton biomass (Heywood *et al.*, 1991), zooplankton distribution (Roe & Griffiths, 1993) and vertical migration (Cisewski *et al.*, 2010; van Haren, 2014).

The strength (echo intensity, EI) of the backscattered signal from transmitted acoustic pulse will be dependent on four things: (i) the transmitted power of each transducer, (ii) the strength of the backscatter (dependent on scatterer size/concentration/orientation), (iii) the spreading of the acoustic beam with distance travelled, and (iv) the absorption of the acoustic energy with distance travelled through the medium (RD Instruments, 1996; Deines, 1999), or in terms of sound decibel (dB) level:

$$\mathrm{EI}\left(\mathrm{dB}\right)=\mathrm{SL}\left(\mathrm{dB}\right)+\mathrm{Sv}\left(\mathrm{dB}\right)-20\log\left(R\right)-2\alpha R+\mathrm{constant}$$

where EI is the backscattered signal and SL the transmit power in decibels. The beam spreading is represented by the 20log(R) term, where R is the distance from the transducer and $2\alpha R$ is the absorption term and α is the frequency-dependent absorption coefficient in decibels (dB)/m (Table 4.2). The constant represents the fact that echo intensity strength is expressed in units of counts and not decibels. A calibration of each transducer is required to make a conversion (Deines, 1999). The wavelengths (λ) of the acoustic signal for common ADCP frequencies and typical absorption values (α) are given in Table 4.2 showing increased absorption with higher frequency. This wavelength determines the minimum size of particles that will influence the sound scattering (Sv) and this minimum value is about 0.25–0.5λ. Different frequencies may be used, therefore, for different observations – high frequencies (>1 MHz) for suspended sediment concentrations (e.g. Hoitink, & Hoekstra, 2005) or lower frequencies, like the ship hull-mounted 75/38 kHz ADCPs, for zooplankton and micronekton distributions (Heywood *et al.*, 1991). It is hard to correlate or estimate biomass exactly with backscatter strength for ADCPs, however, because of the scatter size dependency of the return amplitude and also the variation of backscatter strength with orientation of some scatterers (e.g. Zedel *et al.*, 2005).

Despite these limitations, useful information can be gained from the acoustic backscatter, even if treated with basic processing if no raw count–decibel calibration is available. Figure 4.8 shows a short length of acoustic backscatter from an upward-looking 300 kHz RDI Workhorse ADCP deployed on the seabed at a cold-water coral reef in Scandinavian waters (Lavaleye *et al.*, 2009). As well as the measured backscatter signal (Fig. 4.8a), also shown is a measure of backscatter anomaly (Fig. 4.8b). This is calculated by subtracting the mean backscatter amplitude for any depth bin from the time series at that depth bin. The measured signal shows the decrease in backscatter strength with distance due to absorption and beam spreading. Other variations in signal strength are less readily apparent although a diurnal zooplankton migration pattern is apparent (Fig. 4.8a). The time series of backscatter anomaly, while not fully quantitative, does however highlight the diurnal signal and its variability, including when the high backscatter signals of the migrating zooplankton reach the seabed (Fig. 4.8b), which is not as easily observed from the raw signal time series (Fig. 4.8a).

4.4 Particulate organic matter

4.4.1 Introduction and gear type

Particulate organic matter (POM; >0.4 μm) accounts for a small fraction (<10%, Libes, 1992) of the total organic matter in the ocean but it influences the transport of energy and material at depth and it is the main food source for many benthic and pelagic ecosystems. Therefore its accurate measurement and characterization is important, particularly in areas with complex topography and oceanographic regime, such as seamounts (e.g. Kiriakoulakis *et al.*, 2009; Vilas *et al.*, 2009). There are two

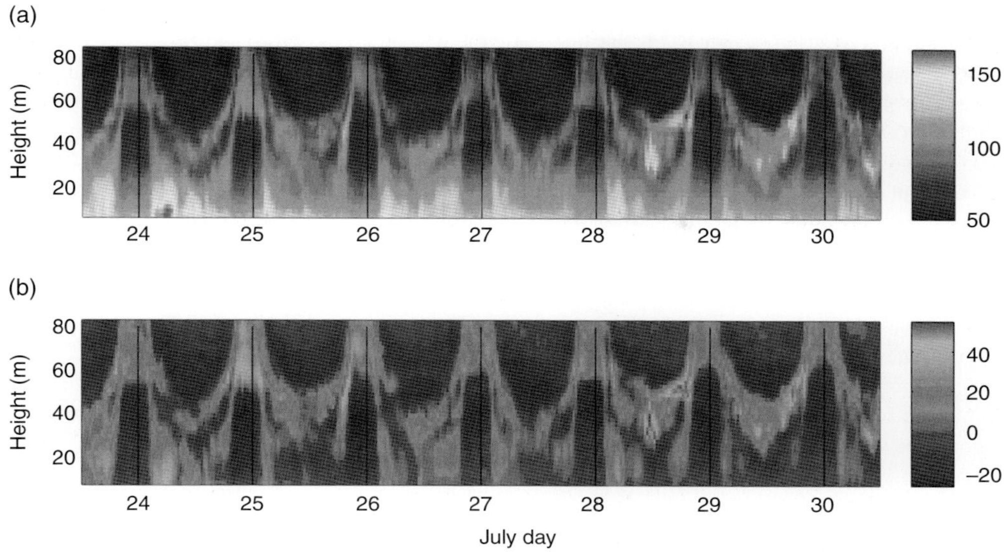

Fig. 4.8 Time series of the vertical variation of (a) acoustic backscatter (counts) and (b) acoustic anomaly (counts) after subtraction of the mean vertical profile of backscatter amplitude. Data are from an upward-looking 300 kHz ADCP at Tisler cold-water coral reef in a water depth of 120 m. A scale for backscatter in raw counts is given by the greyscale bar.

main ways of sampling POM: (i) sediment traps and (ii) various types of water samplers, most commonly Niskin bottles, bottles from CTD casts and/or large-volume *in situ* filtration systems (commonly referred to as SAPS or standalone pumping systems).

4.4.1.1 Sediment traps

Sediment traps are inverted conical funnels that capture falling particles and direct them to a receptacle at the bottom. They are generally used as a moored instrument for an extended period of time (days to months) and their collection mechanisms may consist of a series of sampling vessels that are cycled through (with a flexible sampling resolution – from days to weeks per vessel). Short-term deployments on landers or moorings could be considered for the duration of a survey campaign. A recent development is free drifting traps, often named neutrally buoyant sediment traps (NBSTs) or PELAGRA traps (Lampitt *et al.*, 2004), that move with the surrounding ocean currents (though they may remain at a fixed depth) in a quasi-Lagrangian manner. These may be used during any vessel survey of sufficient length. Recovery is less straightforward for these traps as they must surface at a predetermined time, and report their position (usually via satellite), which can be time consuming and labour intensive, particularly in rough weather.

4.4.1.2 Small- and large-volume water samplers

There are various types of water samplers that are used to determine the concentration of ocean particles, although fluxes cannot be calculated unless an independent estimate of current velocity is provided. These techniques, however, are perhaps more suitable in areas with complex topography and strong and complicated currents (e.g. seamounts) or where the capability to deploy and recover sediment traps does not exist.

Niskin bottles were developed early in the twentieth century and consist of open-ended tubes (typical capacity 1–5 L) equipped with spring-loaded cups that can close at required depths at both ends with the help of a 'messenger' (weight that travels down a wire). Niskin-type bottles are now often mounted on CTD frames (with varying capacity up to 20 L per bottle) and fired electronically. SAPS consist of a (usually) large filter holder (up to 293 mm diameter) that is linked to a water pump, operated by electronically controlled batteries that are housed in a titanium case (operational depth >5000 m). They can be deployed on the CTD wire (i.e. together with a CTD to save ship time) or independently and they have moderate requirements of ship time (1–8 h), depending on depth of deployment and location.

4.4.2 *Problems, pitfalls and processing*

Each sampling method for POM has advantages and disadvantages (Table 4.3) and it is thought that they sample different pools of POM (Wakeham *et al.*, 1984; Lee and Wakeham, 1989; Gardner *et al.*, 2003; Turnewitsch *et al.*, 2007; Buesseler *et al.*, 2007; Kiriakoulakis *et al.*, 2009; Vilas *et al.*, 2009; Liu *et al.*, 2009). Ultimately, the decision on the choice of the appropriate technique is made in relation to the scope of the study, logistics, costs, location and hydrodynamic regime of the area. The approaches will be considered separately below.

4.4.2.1 Sediment traps

Data from traps are often extrapolated over wide areas to calculate the export material over larger scales, although with limitations and uncertainties (Buesseler *et al.*, 2007). Tethered traps may be subject to hydrodynamic bias due to lateral advection and/or turbulence, and can lead to lower particle capture efficiency (undersampling) by up to a factor of 2 (Gardner 1980a, 1980b, 1985, 2000; Buesseler *et al.*, 2007). When traps are deployed for long periods (i.e. more than a few days) preservation of the collected material (by introducing chemical 'poisons'; see Buesseler *et al.*, 2007) is necessary in order to prevent sample decomposition and consumption by 'swimmers'. Swimmers are zooplankton or fish that enter the vessels and modify the sample by feeding and/or defecating on it (e.g. Michaels *et al.*, 1990). Swimmers have to be manually picked out after recovery, however the non-standardization of 'swimmer' removal techniques has been recognized to introduce errors (Buesseler *et al.*, 2000). Furthermore, many of the preservation chemicals (commonly formalin, mercuric chloride or a salt brine) can alter the biochemical composition (and fluxes) of organic particles by selectively degrading important constituents (e.g. polyunsaturated fatty acids) of the sample (Hedges *et al.*, 1993; Wakeham *et al.*, 1993). By contrast, Sampei *et al.* (2009) showed evidence that some 'swimmers' may fall into the trap

Table 4.3 Summary of methodologies to sample suspended particulate material in the water column with principal advantages/disadvantages of each

Sampling method	Main advantages	Main disadvantages
Sediment traps	Fluxes of particles (and their constituents) easy to calculate. Time series and flexible sampling resolution	Availability, hydrodynamic bias, 'swimmer' contamination, selective decomposition of certain biochemicals, may miss very large (rare) and small particles
Niskin bottles	Widely available, easy to operate and maintain	Practical depth limitation, small volumes, vertical resolution
Bottles from CTD rosettes	Widely available, precise sampling, high vertical resolution	Time-consuming filtering, vertical resolution, small volumes
In situ pumps	Large volumes, thus capturing a greater range of particles. Many analyses on the same filter	Not widely available, poor depth resolution due to costs and bulk of apparatus. Design and large filters may introduce sampling bias (e.g. DOM absorption, washing out, zooplankton bias)

naturally (due to death) and they should be included in the flux calculations and compositional characterization of POM. In practice this is difficult to evaluate, therefore most workers remove 'swimmers' assuming they are 'contamination'.

Upper ocean traps (<1000 m depth) are often shown to be affected by the adhesive properties of 'fresh' organic material that may adhere to the side walls of the sediment trap for an unknown length of time and fall at a later date, thus leading to errors in the time series calculation of fluxes and composition of material (Lampitt, personal communication). Sediment traps that are deployed in the deeper parts of the open ocean (i.e. below 1000 m) are less affected by most of the drawbacks mentioned above. This is because (i) there is generally less turbulence or horizontal movement in the deep open ocean, (ii) fewer zooplankton live at these depths, therefore the potential for 'swimmers' entering the vessels is smaller and (iii) the material seems to be less 'sticky' (Lampitt, personal communication), although the reason for this is unclear. Deep sediment traps (but not too close to the ocean floor – see below) are often considered more reliable and they have routinely been used for estimating large-scale particle fluxes in the open ocean (e.g. Lampitt *et al.*, 2001). An exception to this is traps that are deployed within the benthic boundary layer (i.e. within 100 m of the bottom), as benthic resuspended material that may enter the trap can lead to flux overestimates and compositional biases (e.g. Lampitt *et al.*, 2001; Kiriakoulakis *et al.*, 2001). This is even more pronounced in areas at continental margins that are often characterized by rough topography and high turbidity (e.g. submarine canyons; see de Stigter *et al.*, 2007).

The time resolution of multi-vessel sampling with carousels often depends on equipment specification and logistics and if not sufficient it may miss important short-lived events in the water column. For instance, the trap deployments in the BATS and HOT time-series programmes are only collecting the flux for about 10% of the total time (e.g. monthly 3-day deployments; Buesseler *et al.*, 2007). In addition, it has been shown that the traps may not sample a representative size range of particles, either missing large, rare particles that could 'carry' most of the flux or small suspended particles that may escape (Wakeham *et al.*, 1984; Lee & Wakeham, 1989).

4.4.2.2 Small- and large-volume samplers

Since small-water volumes are usually required for most chemical and biological analyses (typically from hundreds of millilitres to 1–2 L), water from the same bottle can be used for the determination of many parameters (e.g. particulate organic carbon (POC), chlorophyll, carbonate, pigments, microscopy). Material is usually collected onto filters (commonly glassfibre filters but various membrane filters can be used depending on types of analyses) after gentle vacuum filtration to avoid particle disintegration or loss through the pore sizes of the filter. However there are issues. Filtering is often a time-consuming process, mostly due to filter clogging, that requires considerable manpower and effort, particularly when different parameters are measured. Filter clogging depends on pore size, location and/or water depth, for example it may take several hours to filter 1 L of particle-rich, productive, surficial water onto GF/F (nominal pore size 0.7 μm) filters; this could lead to severe backlogs, leading to operator fatigue that can ultimately affect sampling resolution and strategy. In addition the small filter volumes pose problems with respect to weighing accuracy, replication and small-scale patchiness.

By contrast large-volume *in situ* filtration systems (SAPS) can usually filter hundreds or thousands of litres per hour, depending on particle loading which is controlled by the biogeochemical regime (eutrophic/oligotrophic) and water depth (surface/mid-water/close to bottom) (e.g. Kiriakoulakis *et al.*, 2004, 2009). Open ocean stations usually require pumping times of up to 2 hours due to scarcity of particles. Due to the lack of SAPS instruments that might be available for deployment, vertical resolution may not be as great as the CTD bottle casts, but one may see up to six units on the same wire at different depths. The volume filtered and the large area of the filter allows multiple analyses to be carried out on the same filter. They are suitable for suspended as well as large rarer sinking particulate material as they are thought to collect a greater range of particles than traps, although ambiguities exist regarding the nature and amount of particles they collect (e.g. Turnewitsch *et al.*, 2007; Kiriakoulakis *et al.*, 2009; Liu *et al.*, 2009). Disadvantages include costs (they are expensive) and handling (large and awkward to move, set up and recharge). Because they require about 2 hours to charge if used for more than 2 hours, they cannot be used for successive long deployments.

It should be noted that POC concentrations measured using CTD bottles and large *in situ* filtration systems have been shown to differ significantly, with the CTD yielding usually higher values than SAPS (up to 200 times; Miquel, 2005; Gardner *et al.*, 2003). Factors such as pressure effects during filtration, particle formation during handling, contamination, adsorption of dissolved organic carbon (DOC) onto filters, washing out of particles during ascent and bias against zooplankton sampling by the pumps have been suggested to be responsible for these differences (Turnewitsch *et al.*, 2007; Liu *et al.*, 2009). In particular, Turnewitsch *et al.* (2007) made recommendations regarding filter sizes and stacking, and advised against using very large filters. These appear to absorb high amounts of nitrogen-rich dissolved organic matter (DOM), so it is suggested to use two filters per deployment, the bottom serving as a blank (i.e. subtracting the values of the bottom one from the top filter).

4.5 Sampling strategies

There are many challenges with surveying the deep-sea environment. Ship-based surveys are essentially snapshots of the ocean experiencing a number of dynamic and biological processes acting over a myriad of space and timescales (Fig. 4.1). In addition, different degrees of measurement accuracy are required to adequately resolve different processses or environmental characteristics. Lateral and vertical property changes in the deep sea may be small. For example, temperatures may change by 1 or 2°C within a vertical range of ~50 m across the seasonal thermocline, whereas over a similar vertical range below the permanent thermocline, temperature changes over a 50 m interval are typically of the order of 0.01°C, 2 orders of magnitude smaller. Such small changes in the deep-sea environment may not be important over small spatial scales, but are crucial to resolve changes in the ocean 'climate' over basin or gyre scales.

The consideration of the dominant process scales and the methodologies to resolve them will be a significant factor in designing the hydrographic component of any survey (see Chapter 3 for a general account of survey design considerations). The ocean contains a wide spectrum of wave motions such as tides, internal waves and trapped topographic waves. For example, while tidal variability is generally small in the deep ocean, tidal motions may be amplified near topographic features such as seamounts. In theory, it is likely that a minimum of 5–6 observations per period/wavelength of variability will be required to resolve tidal and higher frequency motions. In the deep ocean, it may take 2 hours to complete a full CTD cast in 2000 m water depth and move to the next station 10 km away. It is likely, therefore, that tidal variability could bias some results in such a transect, or not be resolved unless a yoyo technique for a limited depth range close to the topographic feature is utilized to increase temporal resolution. CTD station grid design, therefore, requires careful consideration of station separation (spatial resolution and steaming time between stations based on the vessel's capability), time to complete the vertical profile (how the CTD is lowered) or whether repeat profiles at certain stations at half tide/wave period would be useful.

Two very different CTD sampling grids are shown in Fig. 4.9 to illustrate the resolution of scales: the CTD grid used to make observations over a carbonate mound, one portion of which is shown in Fig. 4.5, and the south Pacific spanning World Ocean Circulation Experiment (WOCE) hydrographic transect P06. The yoyo technique of CTD deployment employed for Fig. 4.9a allowed high temporal resolution measurements near the seabed approximately every 20 minutes to highlight tidal and short-term variability in the benthic boundary layer and associated suspended particulate material

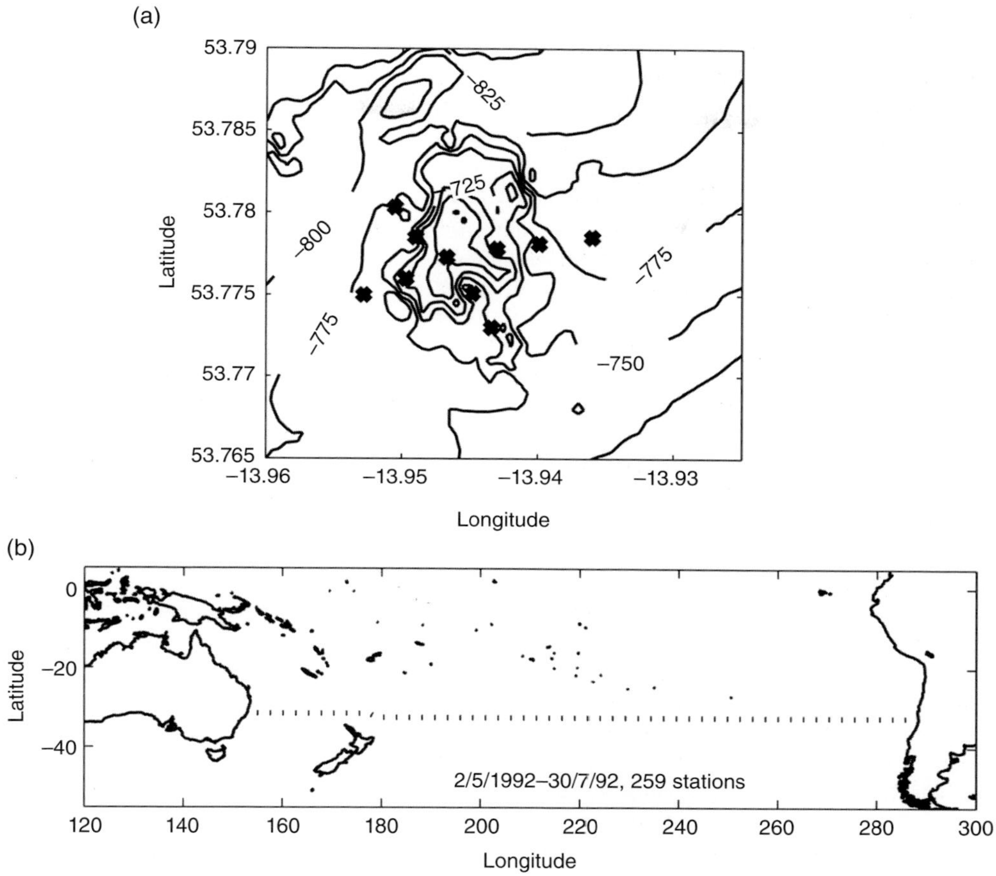

Fig. 4.9 Examples of a CTD station grid for measurements made (a) over a small carbonate mound, northwest Porcupine Bank (Northeast Atlantic), by 'yoyo' CTD from 400 m depth to bottom (the north–south transect data are shown in Fig. 4.5), taking 4 hours and (b) World Ocean Circulation Experiment (WOCE) line P06E,C,W occupied between May and July 1992 and comprising 259 full-depth CTD stations.

concentrations (Fig. 4.5). Conversely, the WOCE P06 section measured full-depth physical and chemical water column parameters and was conducted for the assessment of the long-term state of the ocean, driven by interannual or climatic forcing. In total, 259 stations were occupied during 70 days at sea occurring over a period of three months. Such a transect would not resolve the mesoscale (eddy) variability in the ocean, which would appear as a form of small-amplitude noise within the large-scale oceanic structure.

Appropriate resolution in the measurement interval, therefore, is vital to elucidate the process under investigation and to avoid problems whereby features of smaller time or space scale relative to the measurement interval may bias (or alias) the observation. One-off (essentially instantaneous) surveys, therefore, can be misleading and not be representative of a long-term steady or quasi-steady state of the ocean. This may lead to over-interpretation of data. Care is required, therefore, to establish observational strategies to resolve both spatially and temporally the physical processes of interest. Data analysis is not a focus of this chapter, but will follow from the acquisition of a good-quality data set. The authors can reccomend the book *Data Analysis Methods in Physical Oceanography* (Emery & Thomson, 2004).

4.6 Future outlook and summary

We have described the basic principles in regard to a relatively small range of ship-based instrumentation, and are only scratching at the surface of what can be undertaken to sample the deep-sea environment. New sampling methodologies are constantly being developed; in particular remote sensing techniques are becoming more powerful and useful. While satellite-based sensors generally only measure the surface layer of the ocean, they still can provide a basic idea of what features may lie near surface or below, for example temperature fronts (Miller, 2004, 2009). In addition, near real-time surface imagery of sea surface temperature or ocean colour can be invaluable to plan an upcoming, or adapt an on-going, survey (e.g. Smyth *et al.*, 2001). The increasing array of ARGO floats can provide high-quality regional hydrographic information (Gould *et al.*, 2004). The increasing reliability of autonomous gliders (Rudnick *et al.*, 2004) will likely provide exciting opportunities to collect vast amount of data all year round, even when weather conditions would halt vessel operations. Gliders might be used from a survey vessel to provide supporting hydrographic data, freeing ship time for other operations, or as autonomous recording instruments, deployed from coastal locations and providing large-scale and/or repeat transect data over long time periods.

From an *in situ* perspective, there is now increasing availability of fast-response sensors, including CTDs, for use for turbulence measurements. These might be considered for the assessment of biophysical interactions between turbulence, biogeochemical fluxes and small-scale biological processses. An example is the increased use of the 'eddy correlation' technique, developed in atmospheric research, to measure gas fluxes in the ocean (e.g. Lorrai *et al.*, 2010; Chipman *et al.*, 2012; Reimers *et al.*, 2012).

A good appreciation of the environmental background is a solid foundation to any deep-sea biological sampling programme. If appropriate sampling strategies (e.g. sampling frequency, duration and spatial resolution) are realized and utilized then a wealth of important environmental data can be used in conjunction with the biological sampling to be described in the following chapters. Careful attention is required, however, to avoid possible misinterpretation, or over-interpretation, of measurements through inappropriate sampling strategies or inadequate data processing. It should always be borne in mind that a survey is a snapshot of the vast spectrum of processes that occur in the oceanic environment.

Acknowledgements

The authors would like to thank the CenSeam scientific committee for the opportunity to contribute to the book and to two anonymous reviewers for very helpful comments on an earlier draft. We also thank Tomas Lundälv and Danien Guihen for the use of photographs.

References

Anderson, S., Charters, A. (1982). A fluid-dynamics study of sea-water flow through gelidium-nudifrons. *Limnology and Oceanography*, 27(3): 399–412.

Beal L.M., Hummon, J.M., Williams, E., Brown, O.B., Baringer, W., Kearns E.J. (2008). Five years of Florida Current structure and transport from the Royal Caribbean Cruise Ship Explorer of the Seas, *Journal of Geophysical Research*, 113, doi:10.1029/2007JC004154.

Buesseler, K.O., Steinberg, D.K. Michaels, A.F, *et al.* (2000). A comparison of the quantity and quality of material caught in a neutrally buoyant versus surface-tethered sediment trap. *Deep Sea Research*, 47: 277–294.

Buesseler, K.O., Sarin, M., Steinberg, D.K., *et al.* (2007). An assessment of the use of sediment traps for estimating upper ocean particle fluxes. *Journal of Marine Research*, 65(3): 345–416.

Candela, J., Beardsley, R.C., Limeburner, R. (1992). Separation of tidal and subtidal currents in ship-mounted acoustic Doppler current profilers. *Journal of Geophysical Research*, 97: 769–788.

Carillo, L., Souza, A.J., Hill, A.E., Brown, J., Fernand, L., Candela, J. (2005). Detiding ADCP data in a highly variable shelf sea: the Celtic Sea. *Journal of Atmosphere and Ocean Technology*, 22: 84–97.

Chipman, L., Huettel, M., Berg, P., *et al.* (2012). Oxygen optodes as fast sensors for eddy correlation measurements in aquatic systems. *Limnology and Oceanography: Methods*, 10(5): 304–316.

Cisewski, B., Strass, V.H., Rhein, M., Krägefsky, S. (2010). Seasonal variation of diel vertical migration of zooplankton from ADCP backscatter time series data in the Lazarev Sea, Antarctica. *Deep Sea Research*, 57: 78–94.

De Stigter, H.C., Boer, W., Mendes, P., *et al.* (2007). Recent sediment transport and deposition in the Nazaré Canyon, Portuguese continental margin. *Marine Geology*, 246: 144–164.

Deines, K.L. (1999). Backscatter estimation using broadband acoustic Doppler current profilers. In: *Proceedings of the IEEE Sixth Working Conference on Current Measurements*, San Diego, CA, 11–13 March 1999, pp. 249–253.

Dewey, R K., Crawford, W.R., Gargett, A.E., Oakey, N.S. (1987). A microstructure instrument for profiling oceanic turbulence in coastal bottom boundary layers. *Journal of Atmosphere and Ocean Technology*, 4: 288–297.

Dickson R.R, McCave, I.N. (1986). Nepheloid layers on the continental slope west of Porcupine Bank. *Deep Sea Research*, 33: 791–818.

Egbert, G.D., Erofeeva, S.Y. (2002). Efficient inverse modeling of barotropic ocean tides. *Journal of Atmosphere and Ocean Technology*, 19: 183–204.

Emery, W.J., Thomson, R.E. (2004). *Data analysis methods in physical oceanography* (2nd edn). New York: Elsevier, 634 pp.

Firing, E., Hummon, J.M. (2010). Shipboard ADCP measurements. *The GO-SHIP repeat hydrography manual: a collection of expert reports and guidelines*. IOCCP Report No. 14, ICPO Publication Series No. 134, version 1.

Flagg, C.N., Dunn, M., Wang, D-P., Rossby, H.T., Benway, R.L. (2006). A study of the currents of the outer shelf and upper slope from a decade of shipboard ADCP observations in the Middle Atlantic Bight. *Journal of Geophysical Research*, 111, doi:10.1029/2005JC003116.

Foreman, M.G.G., Freeland, H.J. (1991). A comparison of techniques for tide removal from ship-mounted acoustic Doppler measurements along the southwest coast of Vancouver Island. *Journal of Geophysical Research*, 96: 17007–17021.

Gardner, W.D. (1980a). Sediment trap dynamics and calibration, a laboratory evaluation. *Journal of Marine Research*, 38: 17–39.

Gardner, W.D. (1980b). Field assessment of sediment traps. *Journal of Marine Research*, 38: 41–52.

Gardner, W.D. (1985). The effect of tilt on sediment trap efficiency. *Deep Sea Research*, 32: 349–361.

Gardner, W.D. (2000). Sediment trap technology and surface sampling in surface waters. In: Hanson, R.B., Ducklow, H.W., Field, J.G. (eds), *The changing ocean carbon cycle, a midterm synthesis of the Joint Global Ocean Flux Study*. Cambridge, UK: Cambridge University Press, pp. 240–281.

Gardner, W.D., Richardson, M.J., Carlson, C.A., Hansell, D., Mishonov, A.V. (2003). Determining true particulate organic carbon: bottles, pumps and methodologies. *Deep Sea Research II*, 50: 655–674.

Goldner, D.R., Chapman, D.C. (1997). Flow and particle motion induced above a tall seamount by steady and tidal background currents. *Deep Sea Research*, 44: 719–744.

Gould, J., Roemmich, D., Wijffels, S. *et al.* (2004). Argo profiling floats bring new era of in situ ocean observations. *Eos Transactions of the AGU*, 85(19): 185–191. doi:10.1029/2004E0190002.

Hedges, J.I., Lee, C., Wakeham, S.G., Hernes, P.J., Peterson, M.L. (1993). Effect of poisons and preservatives on the fluxes and elemental composition of sediment trap materials. *Journal of Marine Research*, 51: 651–668.

Heywood, K.J., Scrope-Howe, S., Barton, E.D. (1991). Estimation of zooplankton abundance from shipborne ADCP backscatter. *Deep Sea Research*, 38: 677–691.

Hoitink, A.J.F., Hoekstra, P. (2005). Observations of suspended sediment from ADCP and OBS measurements in a mud-dominated environment. *Coastal Engineering*, 52(2): 103–118.

Holbrook, W.S., Paramo, P, Pearse, S., Schmitz, R.W. (2003). Thermohaline fine structure in an oceanographic front from seismic reflection profiling. *Science*, 301: 821–824.

Isobe, A., Kuramitsu, T., Nozaki, H., Chang, P.-H. (2007). Reliability of ADCP data de-tided with a numerical model on the East China sea shelf. *Journal of Oceanography*, 63: 135–141.

John, A., Reid, P.C. (2001). Continuous plankton recorders. In: Steele, H., Turekian, K.K., Thorpe, S.A. (eds), *Encyclopaedia of ocean sciences*. London: Academic Press, pp. 502–512.

Kiriakoulakis, K., Stutt, E., Rowland, S.J., Vangriesheim, A., Lampitt, R.S., Wolff, G.A. (2001). Controls on the organic chemical composition of settling particles in the Northeast Atlantic Ocean. *Progress in Oceanography*, 50: 65–87.

Kiriakoulakis, K., White, M., Bett, B.J., Wolff, G.A. (2004). Organic biogeochemistry of the Darwin Mounds, a deep-water coral ecosystem, of the NE Atlantic. *Deep Sea Research*, 51: 1937–1954.

Kiriakoulakis, K., Vilas, J.C., Blackbird, S.J., Arístegui, J., Wolff, G.A. (2009). Seamounts and organic matter – is there an effect? The case of Sedlo and Seine Seamounts. Part II. Composition of suspended particulate organic matter. *Deep Sea Research II*, 56: 2631–2645.

Lampitt, R.S., Bett, B.J., Kiriakoulakis, K., *et al.* (2001). Material supply to the abyssal seafloor in the Northeast Atlantic. *Progress in Oceanography*, 50(1–4): 27–63.

Lampitt, R.S., Sanders, R.J., Boorman, B., *et al.* (2004). Particulate export in the northeast Atlantic: an integrated attack using production rates, tracers and a novel drifting sediment trap. Paper presented at the ASLO/TOS Ocean Research Conference, Honolulu, HI, 87.

Lavaleye, M., Duineveld G., Lundälv T., *et al.* (2009). Cold-water corals on the Tisler Reef. Preliminary observations on the dynamic reef environment. *Oceanography*, 22: 76–84.

Lawson, K., Larson, N.G. (2001). CTD. In: Steele, H., Turekian, K.K., Thorpe, S.A. (eds), *Encyclopaedia of ocean sciences*. London: Academic Press, pp. 579–588.

Lee, C., Wakeham, S.G. (1989). Organic matter in seawater: biogeochemical processes. In: Riley, J.P. (ed.), *Chemical oceanography*, Vol. 9. London: Academic Press, pp. 1–51.

Libes, S.M. (1992). *An introduction to marine biogeochemistry*. New York: John Wiley & Sons.

Liu, Z., Cochran, J.K. Lee, C., Gasser, B. Miquel, B., Wakeham, S.G. (2009). Further investigations on why POC concentrations differ in samples collected by Niskin bottle and *in situ* pump. *Deep Sea Research II*, 56: 1558–1567.

Lorrai, C., McGinnis, D.F., Berg, P., Brand, A., Wüest, A. (2010). Application of oxygen eddy correlation in aquatic systems. *Journal of Atmosphere and Oceanic Technology*, 27: 1533–1546.

Majewski, A.R., Reist, J.D., Park, B.J., Lowdon, M.K. (2009). Fish catch data from offshore sites in the Mackenzie River Estuary and Beaufort Sea during the Open Water Season, August 2006, aboard the CCGS *Nahidik*. *Canadian Data Report of Fisheries and Aquatic Sciences*, 1218: 37 pp.

Martin, A.P. (2003). Phytoplankton patchiness: the role of lateral stirring and mixing. *Progress in Oceanography*, 57: 125–174.

Mienis, F., de Stigter, H.C., White, M., Duineveld, G., de Haas, H., van Weering, T.C.E. (2007). Hydrodynamic controls on cold-water coral growth and carbonate-mound development at the SW and SE Rockall Trough Margin, NE Atlantic Ocean. *Deep Sea Research*, 54: 1655–1674.

Michaels, A.F., Silver, M.W., Gowing, M.M., Knauer, G.A. (1990). Cryptic zooplankton 'swimmers' in upper ocean sediment traps. *Deep Sea Research*, 37: 1285–1296.

Miller, P.I. (2004). Multispectral front maps for automatic detection of ocean colour features from SeaWiFS. *International Journal of Remote Sensing*, 25(7–8): 1437–1442.

Miller, P.I. (2009). Composite front maps for improved visibility of dynamic sea-surface features on cloudy SeaWiFS and AVHRR data. *Journal of Marine Systems*, 78(3): 327–336.

Miquel J-C. (2005). Why do POC concentrations measured using Niskin bottle collections sometimes differ from those using in-situ pumps? *Deep Sea Research*, 52: 1324–1344.

Mizuno, K., Watanabe, T. (1993). Preliminary results of in-situ XCTD/CTD comparison test. *Journal of Oceanography*, 54: 373–380.

Mohn, C., White, M. (2010). Seamounts in a restless ocean: Response of passive tracers to sub-tidal flow variability, *Geophysical Research Letters*, 37, doi:10.1029/2010GL043871.

Muste, M., Yu, K., Spasojevic, M. (2004). Practical aspects of ADCP data use for quantification of mean river flow characteristics; Part I: moving-vessel measurements Flow *Measurement and Instrumentation*, 15: 1–16.

Myers, A.E., Lovelle, P., Hays, G.C. (2006). Tools for studying animal behaviour: validation of dive profiles relayed via the Argos satellite system. *Animal Behaviour*, 71: 989–993.

Neat, F., Kynoch, R., Drewery, J., Burns, F. (2010). Deepwater trawl survey manual. *Marine Scotland Science Report*, 03/10.

Pollard, R. (1986). Frontal surveys with a towed profiling conductivity /temperature/depth measurement package (SeaSoar). *Nature*, 323: 433–435.

RD Instruments (1996). *Acoustic Doppler current profiler: principles of operation – a practical primer*. USA: RD Instruments, 57 pp.

Reimers, C.E., Özkan-Haller, H., Berg, P., Devol, A., McCann-Grosvenor, K., Sanders, R.D. (2012). Benthic oxygen consumption rates during hypoxic conditions on the Oregon continental shelf: Evaluation of the eddy correlation method. *Journal of Geophysical Research: Oceans*, 117: 1978–2012. C02021.

Richardson, A.J. (2008). In hot water: zooplankton and climate change. *ICES Journal of Marine Science*, 65: 279–295.

Rikardsen, A.H., Diserud, O.H., Elliot, J.M., Dempson, J.B., Sturlaugsson, J., Jensen, A.J. (2007). The marine temperature and depth preferences of Arctic charr (*Salvelinus alpinus*) and sea trout (*Salmo trutta*), as recorded by data storage tags, *Fisheries Oceanography*, 16: 436–447.

Roe H.S.J., Griffiths, G. (1993). Biological information from an acoustical doppler current profiler. *Marine Biology*, 115(2): 339–346.

Rousset, C., Beal, L.M. (2010). Observations of the Florida and Yucatan Currents from a Caribbean Cruise Ship. *Journal of Physical Oceanography*, 40: 1575–1581.

Rudnick, D.L., Davis, R.E, Eriksen, C.C., Fratantoni, D.M., Perry, M.J. (2004). Underwater gliders for ocean research. *Marine Technology Society Journal*, 38(2): 73–84.

Sampei, M., Sasaki, H., Hattori, H., Forest, A., Fortiera, L. (2009). Significant contribution of passively sinking copepods to the downward export flux in Arctic waters. *Limnology and Oceanography*, 54: 1894–1900.

Saunders, P.M., Fofonoff, N.P. (1976). Conversion of pressure to depth in the ocean. *Deep Sea Research*, 23: 109–111.

Seim, H.E., Gregg, M.C., Miyamoto, R.T. (1995). Acoustic backscatter from turbulent microstructure. *Journal of Atmosphere and Oceanic Technology*, 12: 367–380.

Smyth, T.J., Miller, P.I, Groom, S.B., Lavender, S.J. (2001). Remote sensing of sea surface temperature and chlorophyll during Lagrangian experiments at the Iberian margin. *Progress in Oceanography*, 51: 269–281.

Spain, P. (2007). Optimizing your ADCP setup. Technical Note TN032, Teledyne RD Instruments, 6 pp. http://www.rdinstruments.com/tips/tips_archive/optimizesetup_1203.aspx

Turnewitsch, R., Springer, B.M., Kiriakoulakis, K., *et al.* (2007). Determination of particulate organic carbon (POC) in seawater: the relative methodological importance of artificial gains and losses in two glass-fiber-filter-based techniques. *Marine Chemistry*, 105: 208–228.

Valle-Levinson, A., Matsuno, T. (2003). Tidal and subtidal flow along a cross-shelf transect on the East China Sea. *Journal of Oceanography*, 59: 573–584.

van Haren, H. (2009). Using high sampling-rate ADCP for observing vigorous processes above sloping [deep] ocean bottoms. *Journal of Marine Systems*, 77: 418–427.

van Haren, H. (2014). Internal wave – zooplankton interactions in the Alboran Sea (W-Mediterranean). *Journal of Plankton Research*, 36: 1124–1134.

Vilas, J.C., Arístegui, J., Kiriakoulakis, K., *et al.* (2009). Seamounts and organic matter – is there an effect? The case of Sedlo and Seine Seamounts; Part 1. Distributions of dissolved and particulate organic matter. *Deep Sea Research II*, 56: 2618–2630.

Wakeham, S.G., Lee, C., Farrington, J.W., Gagosian, R.B. (1984). Biogeochemistry of particulate matter in the ocean: results from sediment trap experiments. *Deep Sea Research*, 31: 509–528.

Wakeham, S.G., Hedges, J.I., Lee, C., Pease T. (1993). Effect of poisons and preservatives on the composition of organic matter in a sediment trap experiment. *Journal of Marine Research*, 51: 669–696.

White, M. (1994). Tidal and sub-tidal variability in the sloping benthic boundary layer. *Journal of Geophysical Research*, 99: 7851–7864.

Williams, A.J. III (2009). CTD (conductivity, temperature and depth) profiler. In: Steele, J.H., Thorpe, S.A., Turekian, K.K. (eds), *Measurement technology, sensors and platforms*. London: Academic Press, pp. 23–33.

Zedel, L., Patro, R., Knutsen, T. (2005). Fish behaviour and orientation-dependant backscatter in acoustic Doppler profiler data. *ICES Journal of Marine Science*, 62: 1191–1201.

Chapter 5
Benthic Habitat Mapping

Geoffroy Lamarche, Alan R. Orpin, John S. Mitchell and Arne Pallentin
National Institute of Water and Atmospheric Research, Wellington, New Zealand

Abstract

Making deep-sea benthic habitat maps requires expensive and complex technologies to acquire, process and interpret remotely sensed data. Recent technologies that have advanced the field significantly include side-scan sonar, multibeam echosounders and autonomous underwater vehicles. Acquisition of bathymetric and backscatter data using these tools underpins modern habitat mapping, but requires carefully considered survey strategies and end-product development. Practical issues such as launch and recovery operations, navigation, accuracy and acquisition of additional data are highly dependent on the type of equipment used and can affect data quality. Processing of bathymetry and backscatter data can yield quantitative information about the seafloor, which in turn may be used as proxies to characterize the habitat. Data for proxies such as bathymetry, slope, rugosity or reflectivity can be partitioned into multiple segments with similar multivariate values (segmentation), which can be further grouped spatially to form a connected region of similar habitat (classification). Finally, the acquisition of ground-truthing data is critical to validate any benthic habitat map generated using such means. We illustrate the application of habitat mapping methodologies through examples selected primarily from studies conducted within the New Zealand region.

Keywords habitat, Multibeam echosounders, side-scan sonar, autonomous underwater vehicle, bathymetry, backscatter

5.1 Introduction

Humankind has long sought an association with the ocean and the quest to harness its resources stretches back millennia, driven by the desire to sustain growing populations. Along with our desire to exploit the ocean we have sought to understand it. However, with more than 70% of the Earth beneath water it comes as no surprise that much of the seafloor remains relatively unexplored and unknown. Although our knowledge of the seafloor environment has improved tremendously since the vision speculated by Jules Verne in *Twenty Thousand Leagues Under the Sea*, published in 1870, the shear variety of benthic habitats and the technological advances only recently achieved mean our understanding is still incomplete and patchy.

Improving our knowledge of the seafloor environment and benthic habitats has three primary aims (Harris, 2011): (1) to conduct scientific research aimed at generating knowledge of seafloor geology and benthic ecosystems; (2) to conduct non-living and living seabed resource assessments for economic and environmental management purposes; and (3) to support spatial

Biological Sampling in the Deep Sea, First Edition. Edited by Malcolm R. Clark, Mireille Consalvey and Ashley A. Rowden.

planning and management, including the design of Marine Protected Areas (MPAs). To attain these objectives, we increasingly require benthic habitat maps that are based on quantitative information collected by ship-based or remote technologies and produced using robust analytical procedures.

This chapter provides an overview of recent developments in seafloor habitat mapping and, where appropriate, a practical guide to methods and strategies. We cover aspects of acquisition of remotely sensed data of broad extent, using ship-based or remotely controlled equipment; key elements of survey design for habitat mapping; data processing, habitat categorization and map generation; and ground-truthing to validate or improve map products. We illustrate applications through examples selected mostly from our experience of studies conducted in New Zealand.

5.2 Habitat – what do we mean?

The term 'habitat' captures the combination of environmental and biological conditions that promote occupancy by a given group of seabed species. From a biological perspective, benthic habitat might more simply be defined as the place where a microorganism, plant or animal lives with associated physical descriptors of that place, such as seafloor morphology, geological substrate or oceanography. In that context, habitat mapping aims to represent different types of habitat, delineated spatially by distinct combinations of physical, chemical and biological conditions (Greene *et al.*, 1999; Kostylev *et al.*, 2001). As such, the term 'habitat map' is necessarily broad and encompassing, and might be limited to just one or two physical or biological components, or be a very complex amalgamation of many abiotic and biotic components. This contrasts with a biotope map, which always incorporates biological components and some aspect of the environment.

Mapping habitats is inherently problematic, as boundaries between mapped habitats assume that discontinuities in environmental gradients occur, and that distinct living assemblages can be paired with distinct environmental factors (Snelgrove & Butman, 1994). As such, boundary recognition across large areas of the seascape is fundamental to habitat mapping despite our understanding that obvious discontinuities are not always apparent in nature. There are three key aspects that need to be considered in deriving habitat maps, and which will influence the methodology and analyses in any particular situation (Fig. 5.1):

- *Scale*: The scale used should provide the optimal balance between habitat variability and similarity across expanses of the seafloor. The choice of scale can yield either simplicity (a reductionist approach) on the one hand or complexity (emphasis on variability and outlier data) on the other. The choice of scale should be sensitive to the interrelationships between the size of organisms and that of their physical environment.
- *Surrogate variables*: The term surrogate, also called proxy or indicators, refers to the biophysical variables that can be mapped with a quantifiable correspondence to the occurrence of substrate, benthic species or communities. Some abiotic variables are, in general, more easily and less expensively obtained than geological or biological observations. For example, acoustically derived data (e.g. water depth, backscatter) can be used as surrogates to predict certain physical properties of the seabed. However, the use of surrogates, though practical and scientifically defensible, means that the habitat map includes an abstraction from reality that should be kept in mind.
- *Data resolution*: The density of data acquired across the mapped area will impact upon the interpretation of the habitat. This is particularly true where ground-truth point data do not align closely with the level of spatial heterogeneity revealed by the continuous mapping of the physical environment. Also important to consider is the impact of temporally separated data

used to generate a habitat map. For example, biological data from historical samples of a sandbank used to make a habitat map may not accurately align with a modern acoustic survey if the bank position or structure has changed.
- *Validation*: Habitat maps provide a spatial depiction of data generated from models using statistical methodologies, and as such require testing. This validation is usually done through ground-truthing in order to attach a degree of confidence to the end map product.

Developments in remote-sensing equipment and other automated survey technologies over the last 20 years or so, in particular the advent of multibeam echosounders (MBES), remotely operated vehicles (ROVs) and autonomous underwater vehicles (AUVs), have had a profound impact on the way we are able to image, sample and map the seafloor. Using MBES, it is now cost-effective to image broad regional expanses of the seafloor at scales and levels of detail (resolution) that were simply not possible 20 years ago. The data generated provide baseline information from which thematic maps of the seabed environment, including maps of benthic habitat, can be derived and interpreted in conjunction with *in situ* ground-truthing from observational and point-sample data (e.g. grabs, cores, video transects, etc.). The concomitant development of Geographic Information System (GIS) has resulted in new and improved methodologies for developing maps of various forms. The use of both MBES and GIS can be resource costly (money and time). They also require levels of expertise to operate that are not always available. There are shortcomings with incorporating only limited environmental parameters (Kostylev, 2012; Kostylev & Hannah, 2007), yet point-sampling methods and visual interpolation are still used for the production of benthic habitat maps (e.g. Brown *et al.*, 2011).

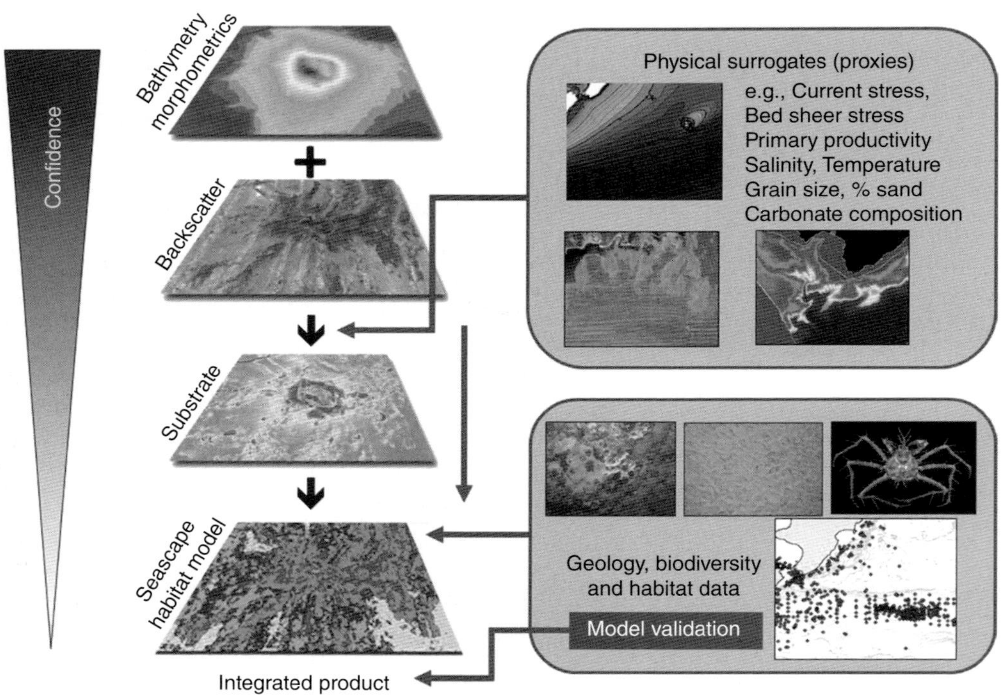

Fig. 5.1 Pathway to habitat mapping. Schematic portrayal of the data components that contribute to form a habitat map. The final – integrated – product has an identifiable, replicable and consistent relationship with marine biota. (Please see Plate 5 for colour representation of the figure.)

5.3 Acquisition of remote-sensed data

Currently, most deep-sea equipment used for benthic habitat mapping is mounted on, or deployed from, ocean-going research vessels. However, the technical characteristics and capabilities of research vessels vary tremendously, just as the performance of equipment can vary from vessel to vessel. In this section we first consider the acquisition of acoustic data from echosounders mounted on the hull of a research vessel or towed from the vessel, and second we outline the use of ROVs or AUVs to obtain similar and supplementary data that can be used for habitat mapping. The acquisition of image data, which can also be used for habitat mapping, using the same or similar equipment, is dealt with in Chapter 12.

5.3.1 Echosounders

There are three types of remote-sensing equipment commonly used for bathymetric and backscatter data acquisition, and which often form the basis for habitat mapping studies in the deep sea: single-beam echosounders (SBES), side-scan sonars (SSS) and MBES (Fig. 5.2 and Table 5.1).

5.3.1.1 Multibeam echosounders

MBES are the most commonly used technology for seafloor mapping, and the primary source of remotely sensed data applied for habitat mapping of the deep sea (Pandian *et al.*, 2009; Brown *et al.*, 2011; Harris & Baker, 2011). Due to the efficiencies afforded by their superior seafloor coverage, for broad-scale mapping of the deep sea, MBES has superseded all other types of conventional acoustic survey systems such as SBES and SSS.

The principle of MBES systems is a two-dimension extension of SBES (which will not be considered separately here), in that they transmit many beams (up to 1000 for some instruments) in a fan-like swath on each side of the ship's track, spanning up to 20 times the water depth for the most

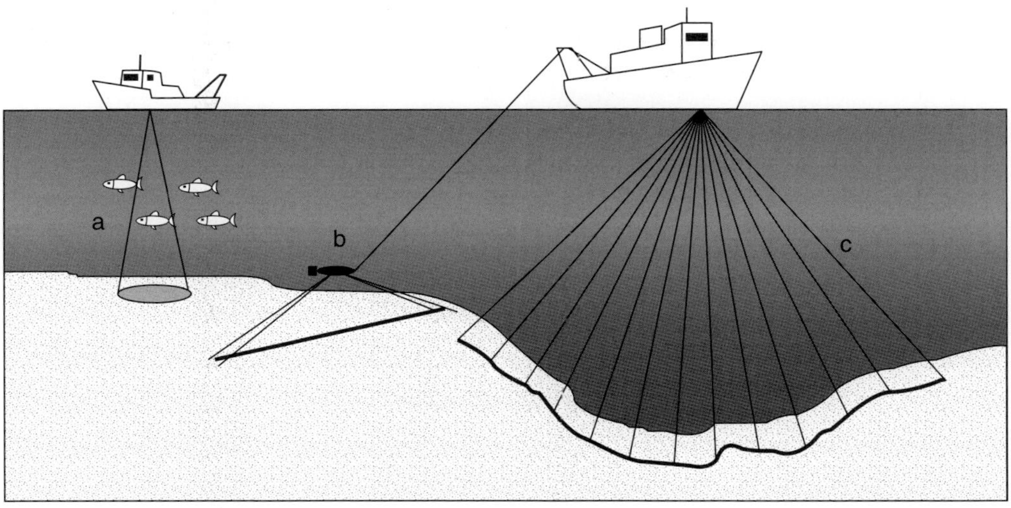

Fig. 5.2 The basic types of echosounders and their respective acoustic footprints. (a) Single-beam echosounder; (b) side-scan sonar; and (c) multibeam echosounder. The insonified patches of acoustically illuminated substrate – the footprint – on (and below) the seabed from the single-beam systems are indicated by the ellipse, whereas the broad- and narrow-swath footprints from the side-scan and multibeam sounders are schematically shown by the thick black lines. (From Lurton, 2010.)

Table 5.1 Advantages and disadvantages of the three echosounder systems most commonly used for mapping seafloor habitat

Single-beam echosounder (SBES)			
Typical characteristics			
Frequency range	12–700 kHz	Signal duration:	0.1 ms
Beam aperture	5–15°	Resolution	7.5 cm
Along-track aperture	0.2–1°	Across-track aperture	10–60°
Advantages		**Disadvantages**	
Least expensive		Limited to line coverage; interpolation usually required across lines	
Readily available on most vessels		Resource intensive if full coverage required	
Can be operated from smaller vessels		Limited angular response collected	
Simplicity of operation		Max. range depends on frequency, and is limited for high frequencies (e.g. 150 m (400 kHz), 400 m (100 kHz))	
Limited amount of processing			
Ability to conduct routine calibration			

Side-scan sonar (SSS)			
Typical characteristics			
Frequency range	100–500 kHz	Signal duration:	0.1–1 ms
Beam aperture	5–15°	Resolution	7.5–75 cm
Advantages		**Disadvantages**	
Increased seabed coverage over SBES		Poor spatial geo-referencing capability	
Faster than SBES (for similar coverage)		Low surveying speed	
Best vertical resolution		Requires specialist operators	
Good ability to image objects on seabed		Difficulty to maintain consistent elevation	
Less subject to vessel motion than SBES		Tow-body motion in rough seas	
Produce photo-like images of the seafloor		Backscatter calibration usually not possible	
Much less expensive than MBES		Shadows (but can be used to derive object height)	
Simple processing routines		No data at nadir	
Simplicity of operation		Not all systems produce bathymetry	
Can be operated from smaller vessels		Flat bottom assumption for correction	
Reduced effect of sound velocity in water			

Multibeam echosounder (MBES)			
Typical characteristics			
Frequency range	12 to >200 kHz	Number of beams	120; 288 (EM302) 800 (EM2040); higher for some systems Double for dual head systems
Total aperture	120–150°		
Advantages		**Disadvantages**	
Full horizontal coverage		Expensive initial purchase	
High accuracy (0.2–0.5% water depth)		Significant expense in terms of ship resources	
Quantitative backscatter response		Hull-mounted systems require demanding installation requirements	
Accurate high-frequency motion compensation		Complexity of operation	
Excellent geo-referencing		Significant computational requirements	
Less shadow effect than SSS		Complex processing requirements	
Fast survey speed (cost effective)		Requires specialist operators	
Ability to detect multiple angles of incidence			
Can include water column backscatter			
Operates up to 12 000 m water depth			

Table 5.2 Typical MBES operation capabilities (maximum depth in parentheses)

Frequency (kHz)	Domain	Typical depth (m)	Swath width	km²/day	Horizontal resolution (m)
12	Deep water	4000 (12 000)	20 km	10 000	100
30	Continental slope	1500 (8000)	6 km	2 000	25
100	Continental shelf	100 (500)	1 km	400	5
>200	Shallow water	10 (100)	100 m	40	<1

sophisticated systems (Fig. 5.2, Table 5.2). The primary advantage of MBES is their ability to insonify (acoustically illuminate) the seafloor with sufficient density of soundings to effectively provide full spatial coverage, thus eliminating the need for interpolation between transects as for SBES. Over the last 30 years MBES systems have progressively been refined, and precision today is commonly within ~1% of water depth.

Echosounders operate by the principle of sending sound waves to the seafloor and measuring the time it takes for the signals to return (Lurton, 2010). As the acoustic wave leaves the transmitting array and propagates into the water column, the amplitude of the signal decays as a direct function of the distance travelled. This decay is called the transmission loss (TL). Sound velocity in seawater depends on water salinity and temperature, and in turn, these parameters vary with depth, time and space. Hence, knowledge of the velocity structure of the water column is essential for the accuracy of MBES data. Typically, water masses on the continental slope and abyssal plain are more stable than shallower and more variable waters of the continental shelf, but even so, regular sound-velocity profiles (SVP) are needed for most surveys.

In addition, vessel motion also directly affects the geometry, quality and accuracy of the outgoing and incoming sound signals of the MBES. For MBES technology to operate appropriately in the open ocean, most modern vessels have the ability to accurately measure the vessel movement and position (attitude) in real time, including heading, position, velocity, heave, pitch and roll, using a combination of Global Positioning System (GPS) and motion sensors. Corrections for vessel position and motion are applied in real time to the incoming sound data, as part of the complex computations that are undertaken to generate the bathymetry.

5.3.1.2 Side-scan sonar

Side-scan sonar (SSS) systems most usually consist of two single-beam transducers mounted on each side of a tow fish, and angled downward toward the seabed. This fish is towed at a constant height close to the seafloor (usually 10–50 m above it) through a combination of vessel speed, tow-wire length and wire angle. Tow speeds are generally <5 knots across ground. SSS uses very-high-frequency sounders ranging from 100 to 500 kHz, and produces photo-like acoustic images of the seafloor that are very sensitive to any variability in the fine-scale seabed topography, such as sand waves, cables, wrecks, moorings and outcropping rocks (see ICES, 2007, for a full review of SSS).

The fundamental advantages of SSS include the ability to discern objects on the seabed, simpler processing routines than that for MBES, lower operational costs and their reasonably good handling in rough sea conditions (Table 5.1). SSS has been successfully used to delineate deep-sea habitats such as cold seeps that have hard and relatively high-relief carbonate structures associated with them (Jones *et al.*, 2010). However, the ability to tow the SSS at a constant height above the seabed can be a significant challenge in heavy seas, and the use of tow cables increases the uncertainty in the georeferencing of the side-scan images. Furthermore, while backscatter maps generated from MBES and SSS compare well qualitatively (Schimel *et al.*, 2009), unlike MBES, SSS backscatter data cannot yield quantitative information.

Today, some of the most sophisticated means of collecting data for seabed mapping include the use of echosounders mounted on AUVs and ROVs.

5.3.2 *Autonomous underwater vehicles*

AUVs are programmed to operate and collect data independently from the research vessel. They come in three broad classes: scientific research (universities, government and private research organizations); commercial survey (pipelines, cables, oil and gas surveys); and military deployment (mine countermeasures). Within these broad classes there are a wide range of size, depth range and payload variants. Scientific AUVs tend to be one-off designs and are continually being modified to meet the varying requirements of the research. Commercial and military AUVs are usually built to specific design criteria that allow for the production of multiple versions of the same vehicle designed for the task in hand; generally a commercial off-the-shelf design purchased as a full 'turn-key operation'. AUVs built for deep-water applications are extremely complex and require significant investment in infrastructure and personnel, both for operation and maintenance.

AUVs typically consist of an operating section which houses standard components required for communication and navigation; an energy section containing the vehicle's batteries and energy equipment; a payload section which carries survey specific sensors and equipment; the electric thrusters for propulsion; and lastly, the casing which is equipped with release, retrieval and safety mechanisms. As AUVs are based on a free-flooding design the electronics are housed in one or more pressure housings.

There are many types of AUVs in use, with costs reflecting complexity, size and performance (Wynn *et al.*, 2014). Currently for deep-sea applications there are a number of AUVs built by or for scientific institutions as well as commercial systems that are sometimes purchased, leased or chartered for habitat-mapping purposes. Scientific AUVs include Sentry and ABE (Woods Hole Oceanographic Institution (WHOI), United States of America), Urashima (Japan Agency for Marine-Earth Science and Technology (JAMSTEC), Japan), Autosub 6000 (National Oceanography Centre (NOC), United Kingdom) and MARUM SEAL (University of Bremen, Germany) (Fig. 5.3). Commercial systems are primarily from three manufacturers: Kongsberg with their Remus and Hugin AUVs; Bluefin Robotics; and International Submarine Engineering (ISE). Mid-price range systems, such as the Bluefin 9 and Remus 600, are generally up to 0.5 m in diameter and weigh approximately 250 kg, have a broad depth range, and are able to carry a complete payload in a single mission. More expensive AUVs have a greater depth range and can be quite large and heavy. For example, the Hugin 1000 and Bluefin 21 have depth ranges of 3000 m, while the Remus 6000 is able to operate to a depth of 6000 m. These AUVs have large diameters ranging from 0.5 to 1 m and weigh 650–2000 kg. Such AUVs usually operate at speeds of up to 1.5 m s^{-1}, which means that ocean currents can interfere with, or limit, their operational ability.

5.3.2.1 Payload

AUVs can be fitted with a wide range of sensors and equipment, including: compasses, depth sensors, geophysical instruments (MBES, SSS, seismic profiler, magnetometers), geochemical instruments (redox, methane sniffers), seafloor-imaging tools (high-definition digital still camera, video and spotlights) and oceanographic instruments (conductivity temperature depth (CTD), sound-velocity profilers, acoustic Doppler current profilers (ADCPs)). These instruments are all useful for collecting data for habitat mapping. Only the largest AUVs can carry a full suite of sensors in a single mission of many hours duration. The trend with modern AUVs is for a modular construction which allows for the addition of new sensor packages by adding pre-balanced modules. This approach speeds up mission changes and removes the need for extended calibration periods while the AUV is rebalanced.

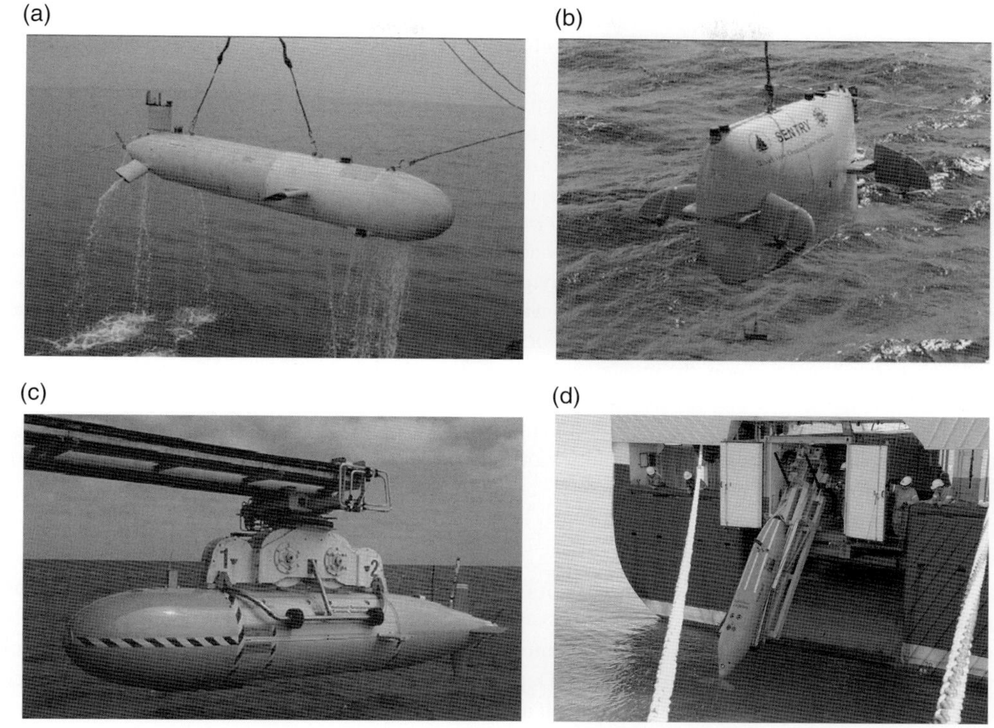

Fig. 5.3 Examples of deep-sea AUVs and launch and recovery systems. (a) The MARUM SEAL AUV of the University of Bremen. (Source: Gerrit Meinecke, MARUM, www.mpi-bremen.de.) (b) The Sentry AUV of Woods Hole Oceanographic Institution. (Photograph courtesy of John Mitchell. Reproduced with permission.) (c) The UK Natural Environment Council Autosub 6000 AUV. (From Wynn *et al.*, 2014). (d) The Hugin 3000. (Photograph courtesy of John Mitchell. Reproduced with permission.) (Please see Plate 6 for colour representation of the figure.)

Modern AUVs use rechargeable batteries to power the propulsion thrusters and onboard sensors. These are generally pressure-tolerant lithium polymer batteries, which in most AUVs give a mission time of around 24 hours. Some of the larger deep operating AUVs such as the Hugin 3000 use an aluminium/oxygen semi-fuel cell which allows for mission times of up to 60 hours, but this power source has the downside of higher maintenance and hazardous waste products.

5.3.2.2 Launch and recovery systems

The AUV deployment procedure varies depending on vehicle size and the launch and recovery system (LARS) (Fig. 5.3). Most deep-sea AUVs need to be either deployed from a stationary vessel using a crane or require the vessel to be steaming ahead at 1–2 knots while the AUV is launched down a stern ramp.

5.3.2.3 Navigation

A very important aspect of the AUV operation is navigation. Without an accurate knowledge of the vehicle's spatial position and route, the data collected are of limited use for habitat mapping. There are several ways to navigate an AUV: GPS; ultrashort baseline (USBL); long baseline (LBL); inertial; and a blended solution.

With GPS navigation, the AUV positions itself at the start of the survey mission, prior to diving and subsequently surfaces at regular intervals to recheck its position. This method is mostly useful with shallow AUVs.

USBL navigation uses an acoustic link from the surface vessel to calculate the position of the subsea vehicle relative to the known GPS position of the surface vessel, and transmits this updated positional information to the AUV. For LBL navigation a set of transponders are deployed on the seafloor at known positions and the AUV triangulates its position autonomously from these transponders by using acoustic signals (Jalving *et al.*, 2004).

Inertial navigation systems (INS) use a very accurate gyroscope and a Doppler velocity log (DVL) so the AUV can estimate its position while navigating through the water column. The DVL provides information on the speed and direction of the water flowing past the AUV and once the AUV comes close to the seafloor the DVL provides an accurate measure of its ground speed.

The blended solution is now the most common option, especially with deep-diving AUVs. In order to address the drift of the inertial solution, data from the USBL system transmitted from the surface can be merged into the computed position. This approach serves to correct for any drift that has occurred and improves the absolute accuracy of the system compared to the relative accuracy provided by an inertial system.

5.3.2.4 Obtaining data

An AUV's mission (i.e. its track and the data it is to collect) is computerized and preloaded. Mission planning and vehicle control require specific software. Once launched, the AUV can operate totally independently of the support vessel in a full autonomous mode but in most operations the position of the AUV is monitored from the support ship. Depending on the survey accuracy required, the surface vessel can continue with other operations such as sampling or camera work, while undertaking periodic checks of the AUV location. Where accuracy is paramount the surface vessel will mirror the AUV mission, keeping in constant contact using the USBL link to update the AUV's navigation, in effect using a virtual tether. This link can also be used to update the mission plan or, if required, abort it and bring the AUV to the surface.

A typical mission for a deep AUV requires the surface vessel to survey the area using its hull-mounted MBES to produce a digital elevation model of the area of interest. This model is then used for mission planning and is also loaded into the AUV to enable it to use these data to aid its navigation and plan for obstacle avoidance. When collecting the standard seafloor information required for habitat mapping, such as bathymetry and backscatter, the AUV will be tasked to 'fly' 30–50 m off the bottom. Once this mapping component of the AUV mission is completed successfully, the high-resolution bathymetry can be used to generate a follow-up mission plan and the AUV redeployed to fly closer to the bottom (around 8 m) to collect seafloor imagery.

5.3.3 Gliders

Gliders are a form of AUV, and are small (<2 m length), light (<100 kg) and relatively cost-effective (e.g., Davies *et al.*, 2003; Schofield *et al.*, 2007). Gliders have no propulsion and move by the relocation of internal ballast and alteration of the trim of their wings. By this method they are able to glide through the water column, moving forward at modest velocities and preset angles, and return to the surface by inflation of oil-filled bladders that make very small adjustments in buoyancy of <0.5% of its total mass. The continual repetition of this cycle moves the gliders across the ocean, resulting in their characteristic saw-tooth dive path. Their internal battery capacity offer possibilities for long-range autonomous deployments of tens of days to many months, down to depths of 1500 m, at advance speeds of tens of centimetres per second. They can also adaptively sample based on data collected in real time and can undertake missions in a wide range of weather conditions as the majority of their deployment is spent underwater.

Despite their relative affordability and logistical advantages, for the time being glider technology is limited to collection of water column data and they lack the ability to carry heavy and large payloads of additional geophysical equipment, including equipment to image the seafloor. However, it is probable that in the future they will be able to obtain data useful for benthic habitat mapping.

5.3.4 Remotely operated vehicles

Unlike AUVs, ROVs are tethered to a surface vessel, which supplies power and control via an umbilical. ROVs are deployed from a relatively stationary mother ship that holds station, often achieved through the use of a dynamic positioning system. Precise handling is needed to target seabed features and areas of interest. ROVs are covered in some detail in Chapter 13, because they are primarily used to obtain biological information either from seafloor imagery or from direct sample collection. However, it is important to note here that echosounders are sometimes mounted on ROVs, and because of their operational proximity to the seabed can be used to collect bathymetry and backscatter data for the production of fine-scale habitat maps. For example, ROVs have been used to acquire data for the production of habitat maps that show the location of hydrothermal vent communities on seamounts or other seafloor features (German *et al.*, 2008; Kumagai *et al.*, 2010).

5.4 Key elements of survey design for habitat mapping

A habitat-mapping survey, whatever gear type is used, should follow a systematic and purposeful approach to data acquisition. A number of documents have been written to help with designing surveys for habitat mapping, such as ICES (2007) and Boyd *et al.* (2006). Below we summarize the key elements commonly used in surveys by illustrating how to combine the acquisition of MBES and ground-truthing data.

5.4.1 Pre-survey actions

All information available over the survey region should be retrieved and collated prior to planning the acquisition of MBES and ground-truth data. General knowledge of the water depth and seafloor morphology will aid significantly with navigation planning for the survey lines and estimating survey time. Knowledge of point samples already taken will allow for some appreciation of expected substrate type, and therefore what type of sampling gear for geological and biological samples will be most suitable. An MBES survey may be required to a specific standard, which will also influence survey planning and the calculation of expected MBES survey time.

The calculation of expected survey time (T in hours) is proportional to the survey area (A in km^2), vessel speed (V in kph) and the expected swath width (W in km). The latter is determined from the survey area average depth and the swath aperture.

$$T = A/\left(D \times W \times V\right)$$

where D is the number of hours planned per surveying per day. If the area is complex with a variable depth range then the survey should be divided into several areas of similar water depth. The survey plan can be refined depending on the type of survey. For example, hydrographic surveys must adopt strict coverage and overlap criteria (100–200%) whereas science surveys are usually more concerned with maximizingg multibeam coverage and can be conducted with swath overlaps as low as 10–20% (Fig. 5.4).

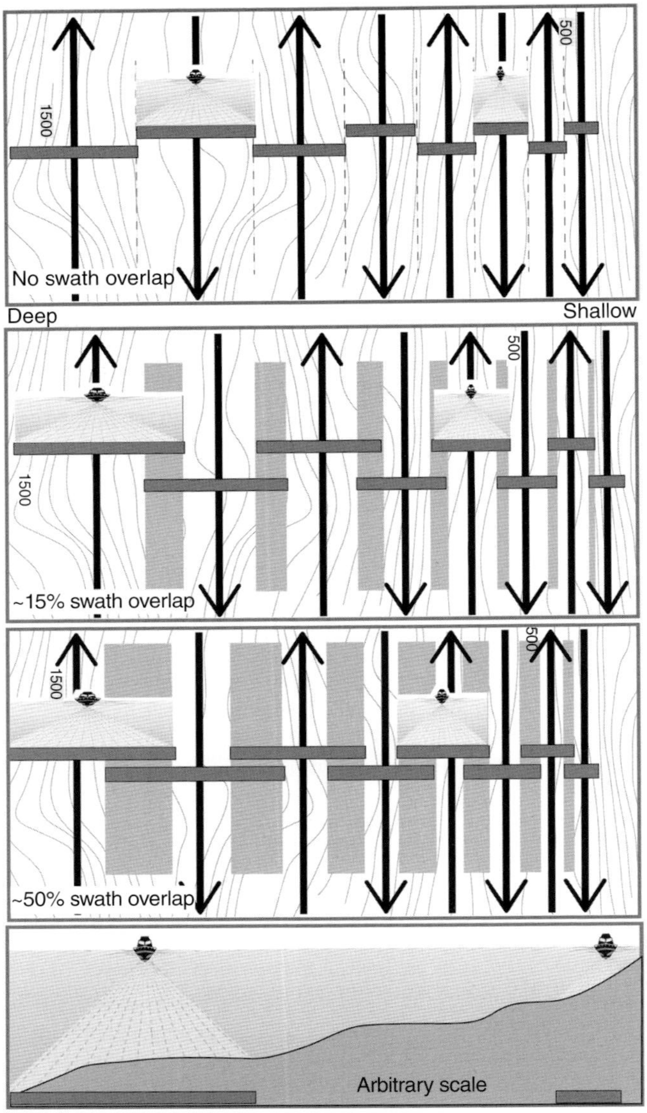

Fig. 5.4 A schematic showing varying survey strategies to manage multibeam echosounder swath overlap. There is a continuum between no overlap (top left), which optimizes time and area surveyed, to large overlap (bottom right) which optimizes resampling (mid-grey area) and minimizes uncertainty. The dark grey bands represent the swath width. The bottom frame shows the variations in asymmetry and width of the swath with water depth and morphology.

Pre-survey vessel mobilization and equipment installation needs to be carefully orchestrated, as it can be time consuming. It consists of two steps, subsequent to installation of all equipment on board:

(1) Ensure all locations of the critical components, such as GPS antennae, transducer position and depth, and inertial motion unit (IMU), are accurately referenced relative to the vessel's reference point. These measurements give the multibeam software values needed to calculate the position of the beam on the seafloor.

(2) Undertake a patch test or offset calibration. This test is used to measure any latency, pitch, roll or heading biases in the positioning and attitude systems. Once these two steps are completed the data collected should be accurate and aligned.

The sound velocity in the water should be determined immediately prior to commencing a survey, as discussed earlier. A specific sound-velocity probe can be deployed, or it can be calculated using data from a CTD profile.

5.4.2 *Surveying*

Once the survey commences, the critical issues are managing survey speed and optimizingg beam angle and overlap. If either of these are pushed too high then the quality of the resulting bathymetric data will suffer. Too fast a survey speed will result in noisy data and a reduction in swath width. Too wide a beam angle will result in the outer beams starting to 'flap' (another symptom of poor signal quality and noise), giving poor bathymetric data in the swath overlap areas. If the overlap is too low then all these factors combine, resulting in data holes (areas where there are no soundings) and quality issues between lines (Fig. 5.4). In most circumstances it is recommended to run survey routes parallel to contours (along slope) which will maintain relatively consistent swath widths along-track and allow control of the desired amount of swath overlap. It is always a compromise best managed by a skilled operator in the field. Getting the balance of factors correct will optimize the quality of the final data set, which once processed can be used to define habitat areas.

In a perfect survey environment, survey lines should be parallel to each other and to the seafloor contour. Modern inertial navigation systems of an MBES will settle within minutes after a turn but still require a lead-in time after a sharp turn onto line. Alterations in course during a survey line should be kept to a minimum and have a slow rate of turn. Too fast a turn will give the data a turn bias as the horizontal accelerations impact the IMU. A drastic turning angle also exposes the backscatter to anisotropic effects.

The last item that can have a significant impact on a survey is weather and the resulting sea state. The survey plan should be adaptable, so if the wind and sea direction is causing poor data (usually bubbles being drawn down along the ship's hull and under the transducers; so-called 'bubble sheeting') the transit line direction can be altered to minimize the issue. Transiting with the weather (downwind and parallel to the swell) can be effective, whereas transiting into the weather generally results in poor data recovery. Transiting across the weather can achieve variable results depending on vessel size, hull design and the wave/swell height. Air bubbles and turbidity caused by rough sea state also have a major negative qualitative and quantitative impact on backscatter data (see Lurton and Lamarche, 2015 for a full review of seafloor backscatter data acquisition). The seafloor backscatter intensity is angle dependent, and so overlap between adjacent swathes are of great importance. This redundancy may provide further information for downstream users interested in examining any angular response of the seafloor.

5.5 Data processing, categorization and map generation

Provided that a link exists between a measured remotely sensed parameter and the existence of a biological entity such as a species or community, such data can be used to generate a benthic habitat map. A large number of habitat surrogates are available for making habitat maps, but here we confine ourselves to examples based on bathymetric and seafloor backscatter data obtained from MBES surveys.

5.5.1 *Data processing*

5.5.1.1 *Multibeam echosounder bathymetry data*

MBES data processing follows three essential stages:

(1) The raw recorded data are processed using dedicated software (e.g. CARIS HIPS, MB-System, QINSy, IVS Fledermaus, Hypack). Commonly this stage involves separating navigation,

attitude and sounding elements of the raw data (compiled in the so-called 'datagram') into individual files to improve accessibility and data handling. At this stage data errors can be corrected, but in most modern systems there is no need for any detailed changes.

(2) The sounding data are merged with additional sensor data such as SVP, real-time heave, post-processed navigation and tide data. These are associated either by time (navigation, attitude and heave) or by location (SVP, tide) to the MBES soundings. The result is a corrected depth–position data set. These data still contain erroneous soundings that are commonly caused by external noise, oceanographic anomalies, and benthic or water-column flora and fauna. At this stage the MBES data should be examined line-by-line. Different software packages offer a variety of editing tools, commonly displaying the soundings in across-track and along-track directions, but also in three-dimensional view, which allows the user to compare soundings from multiple overlapping transects. Usually, the soundings that are deemed incorrect are flagged rather than deleted, so that they are not used to construct a bathymetric grid.

(3) The last stage is the gridding process that generates a regular digital terrain model (DTM) of interpolated depths from the cleaned and processed soundings at a user-selected resolution. The grid can be generated using the mean of all soundings in a cell or be shoal-biased, which is the norm for hydrographic surveys. As the chain of stages to manually clean and process MBES data is very labour intensive, blanket statistical filters are used to reduce the number of soundings, for example by excluding spikes. The average processing time can be as much as three times the data-acquisition time.

In an effort to further improve efficiency and the effectiveness of data cleaning a CUBE (combined uncertainty bathymetry estimate) model has become a standard data-processing tool over the last decade. In an additional computational step the TPU (total propagated uncertainty) – a measureless quality indicator – is calculated for each sounding, which combines information about positional accuracy (e.g. speed, acceleration, beam angle) into one value. The TPU is used to compute the statistically most likely bathymetric surface. The resulting surface and the CUBE solutions are examined in a three-dimensional viewer with associated two-dimensional slices (Fig. 5.5). The final, cleaned bathymetry surface, be it from manual cleaning or CUBE processing, can be exported to common formats for further analyses in other geospatial software.

5.5.1.2 Backscatter data

Backscatter echo is the component of the transmitted sound signal that is reflected back from the seafloor, or water column, and recorded by the sonar receiving arrays (Fig. 5.6). MBES systems have routinely collected the backscatter signal concomitantly to obtaining bathymetry data, but the backscatter information was for a long time considered merely a by-product, and disregarded as too complex to process or too ambiguous to interpret. However, there is now a greater appreciation of the information that can be derived from seafloor and water column backscatter data and how to best process them (see Lurton and Lamarche, 2015 for a full review of seafloor backscatter related issues). The terminology associated with backscatter data has been used inconsistently in the literature, and backscatter signal and image products have been interchangeably called retrodiffusion (now obsolete), imagery or reflectivity.

Backscatter data relate to seafloor micro-topography (or roughness) and volume heterogeneity, which relates to sediment grain-size and sediment composition (Jackson & Briggs, 1992). Hence, the backscatter signal has been used as a qualitative indication of the nature of the substrate (Hughes Clarke *et al.*, 1996). Despite inherent complexities, the potential of using backscatter data to provide quantitative information on substrate and micro-topography has long been recognized (e.g. de Moustier, 1986), and because the substrate is the physical habitat for benthic fauna, backscatter data are well suited for use in spatial studies of habitat biodiversity (Cochrane & Lafferty, 2002). However,

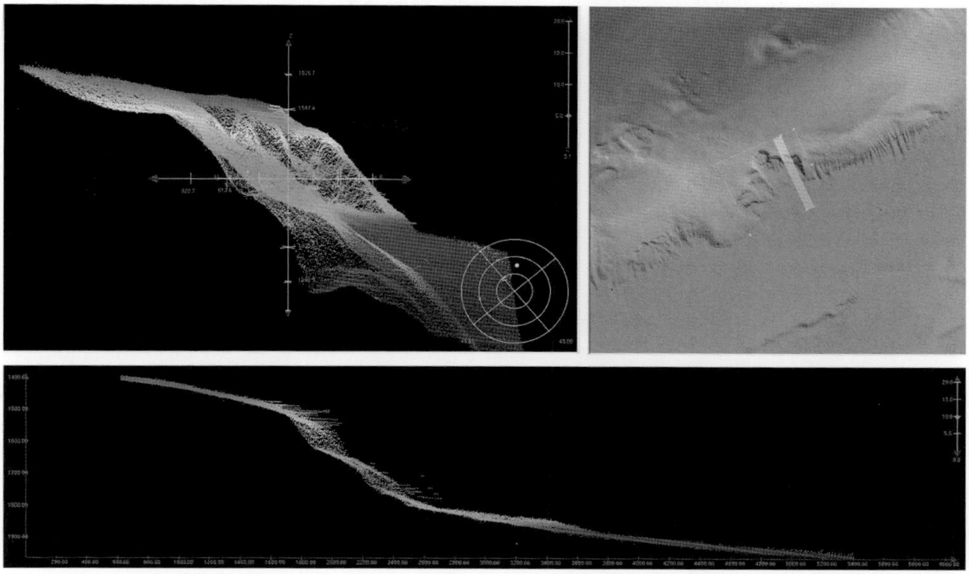

Fig. 5.5 Example of area-based CUBE model editing. An area of seafloor data (right panel) has been selected, and projected as a two-dimensional across-track stack of soundings (bottom panel) and in three-dimensions as fully editable rendered surface (left panel). The individual soundings are coloured according to depth, and excluded soundings are in grey. (Please see Plate 7 for colour representation of the figure.)

regional-scale quantitative applications of backscatter data for habitat mapping are recent and technically challenging (e.g. Brown & Blondel, 2009; Lamarche *et al.*, 2011a; Lucieer & Lamarche, 2011).

Backscatter data are inherently noisy and require complex processing and post-processing procedures. Processing refers to the acquisition stage of the survey, during which time the quantitative backscatter signal must be retrieved and decoded from the echo level at the receiving array (Lurton, 2010; Lurton and Lamarche, 2015). The processing routine includes system-dependent corrections that are usually the responsibility of the hardware manufacturer, and are undertaken on-board during the survey.

Post-processing, or processing post data acquisition, refers to the decoding of the raw backscatter measurements generated at the processing stage to provide an interpretable image, and quantitative parameters relating to the geological properties of the seafloor (Fig. 5.6). Backscatter imagery refers to the qualitative data and is usually only referenced in grey levels. In contrast, backscatter strength (BS) is quantitative and expressed in scalar acoustic units of decibels (dB). Post-processing aims to remove artefacts generated by the recording equipment, seafloor topography and water column properties (Hellequin *et al.*, 2003; Le Gonidec *et al.*, 2003), and includes: (i) removing the effects of energy loss from sound propagation in the water column; (ii) compensating for the transmitting and receiving array sensitivities and directivities; (iii) normalizing the signal with regards to the area and volume insonified; and (iv) correcting for the angular dependence of the signal. Usually, this creates a backscatter mosaic, which is a product that is geo-referenced and transferable on a map.

One of the most valuable steps for end-users is the attenuation of the strong signal associated with the specular reflection (sometimes shortened to just 'the specular'), which can obliterate the backscatter signal at the vessel nadir. This attenuation is performed by statistical compensation of the mean BS angular function (Augustin & Lurton, 2005), which consists of subtracting a bias value of the BS calculated as a function of the incidence angle to the calibrated image.

Once processed, backscatter images emphasize topographic and geological features not necessarily recognized with conventional surveying or readily obtained or conveniently presented from the

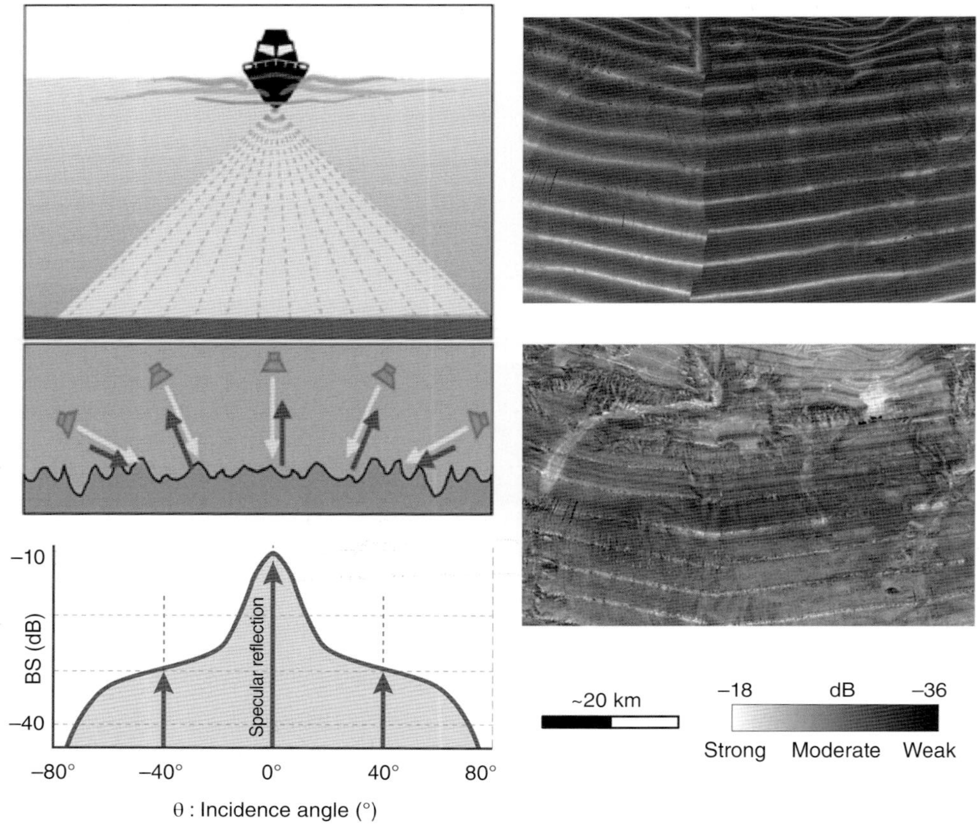

Fig. 5.6 Left column: Multibeam echosounder in operation viewed front-on; the backscatter (dark grey arrows) intensity depends directly on the incidence angle of the sound wave to the seafloor, as shown by the backscatter strength (BS) angular profile (bottom left). Top right: the strong signal (narrow white tracks) at the nadir occur vertically beneath the vessel and often obliterate the useful signal in data. Bottom right: Same image after processing using the SonarScope® software of IFREMER, France. Images from North Island, New Zealand.

bathymetry alone (Dartnell & Gardner, 2004; Goff *et al.*, 2000; Lamarche *et al.*, 2011a). The backscatter imagery can yield fine detail, in particular sediment composition (Fig. 5.7), but also information on the origin, nature and structure of bedforms. The combined use of backscatter and micro-topographic processing enhances the interpretation of fine-scale geological and topographical features visible in multibeam data, such as post-glacial scouring, erosional landforms and carbonated platform concretions. For example, strong reflectivity on canyon floors may suggest gravel deposits, indicating active sediment delivery of coarse-grained bedload, strong up- or down-canyon bottom currents that have winnowed away fine-grained sediment, or intra-canyon debris flows (Harris & Baker, 2011; Lamarche *et al.*, 2011b). These features may provide different habitat for benthic fauna.

Often, the built-in corrections applied to raw MBES data by the manufacturer's proprietary software are not accurate enough and create artefacts precluding any quantitative interpretation of the backscatter (Lurton and Lamarche, 2015). Thus, specialist software needs to be used in order to obtain quantitative data for use in habitat mapping. Furthermore, backscatter signal calibration is usually not undertaken in routine acquisition operation. This consists of computing corrections

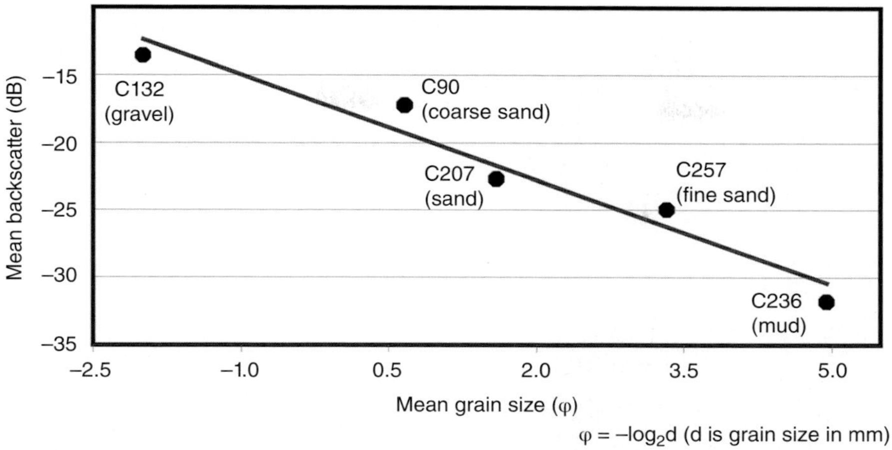

Fig. 5.7 Relationship of backscatter intensity versus mean grain size obtained using a laser granulometer. (From Lamarche *et al.*, 2011a.)

specifically adapted to the sounder. The calibration should ideally be computed at the start of the survey, on an established reference seafloor surface which is homogeneous in terms of sediment type and water depth (Lurton and Lamarche, 2015). If no such reference surface is available, self-calibration based on fitting a relevant model of array directivity on data recorded over homogeneous seafloor areas can be applied (Augustin & Lurton, 2005). This method is similar to that applied to satellite-borne radar data.

Specialist software packages for data processing backscatter data include commercial products such as CARIS HIPS and IVS Fledermaus, a scientific product such as SonarScope from IFREMER (France), or an open-source product such as MB Systems.

MBES systems now usually have additional functionality to record the backscatter signal generated within the full water column. Backscatter data recovered from the water column can indicate the presence of 'flares' from gas or fluid expulsion at the seabed which can be indicative of the presence of habitat for cold-seep communities (e.g. Bowden *et al.*, 2013).

5.5.2 *Data categorization and map generation*

5.5.2.1 Seabed morphology

Bathymetry data obtained from MBES surveys provide a means to easily generate parameters that help characterize seafloor morphology. Morphometric parameters include the various derivatives of the bathymetric data: slope, curvature, aspect, rugosity and bathymetric position index (BPI), sometimes called topographic position index (TPI) (Fig. 5.8). BPI compares an individual grid cell's characteristics to its neighbouring cells. At a broad scale, BPI can differentiate between crests (positive values greater than the mean in the positive direction), depressions (negative values greater than the mean in the negative direction), flats (near zero BPI, slope <5°) and slopes (near zero BPI, slope >5°). Each of the five parameters potentially gives important information for the delineation and characterization of seafloor morphology (Wilson *et al.*, 2007). However, errors and artefacts associated with bathymetry data, the intrinsic accuracy of the technology used, as well as uncertainties related to interpolation and gridding processes required to compute DTMs compound and directly impact the morphometric parameters and need to be addressed (Dolan & Lucieer, 2014).

GIS tools for marine habitat classification are now commonly available and can be used to make benthic habitat maps from bathymetric-derived morphometric parameters (Iampietro *et al.*, 2005;

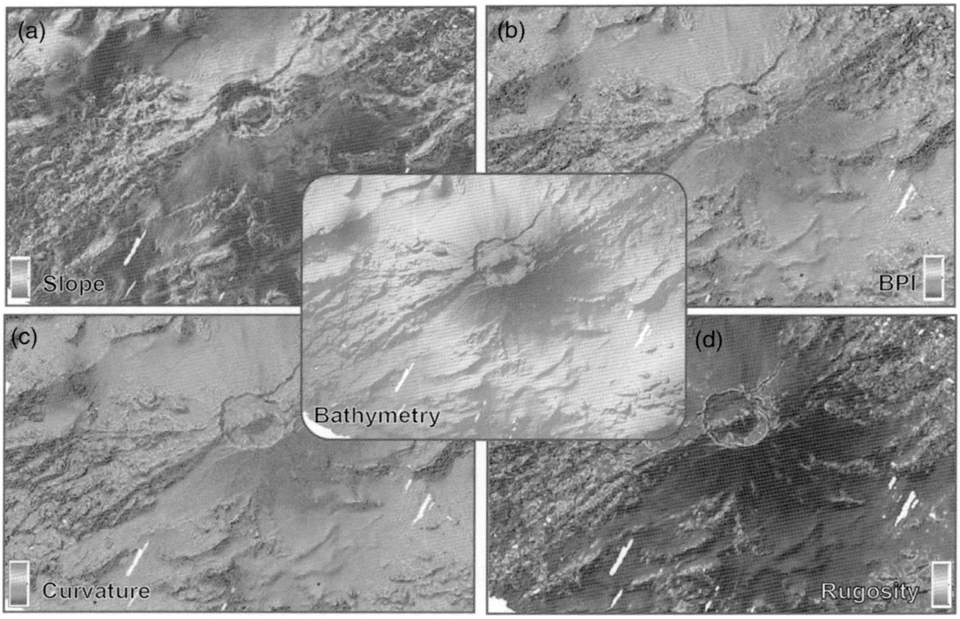

Fig. 5.8 Morphometric parameters that can be derived from bathymetry data: (a) slope, (b) BPI, (c) curvature (mean, plan or profile) and (d) rugosity. (Adapted from Barton *et al.*, 2014.) (Please see Plate 8 for colour representation of the figure.)

Lundblad *et al.*, 2006). For example, the automatic software Benthic Terrain Modeler (BTM), which is a sub-routine available for ArcGIS, uses the BPI metric. The BPI grids are classified into geomorphic or habitat classes by means of a classification dictionary. Boundaries between structural classes are stipulated according to 1 standard deviation differences in slope and depth. The result is a morphometric map showing the undersea landscape from flat areas to ridges.

The application of BTM to bathymetric data from a cluster of small seamounts off New Zealand (Fig. 5.9) resulted in five geomorphic classes that were considered to be biologically relevant (i.e. habitats) (Clark *et al.*, 2010):

- ridges – hard, suitable for attaching organisms such as corals, sponges, crinoids;
- steep slopes – generally hard but potentially less water flow for filter feeders;
- gradual slopes – potentially more mixed substrate types, less continuous hard cover, and likely to have patchier faunal distribution;
- gullies – a depression in the bathymetry, likely to have more soft sediment, with fewer corals and sponges;
- flats – generally surrounding the base of the seamounts and likely covered with soft muddy sediment.

5.5.2.2 Seafloor substratum

Habitat mapping has benefited from the immense increase in computational capabilities that now enable extremely large data sets to be processed and analysed. Of particular interest are the processes of image segmentation and classification which can be used to derive benthic habitat maps from backscatter data and other MBES-derived data (e.g. Boyd *et al.*, 2006; Lucieer & Lamarche, 2011; Lucieer *et al.*, 2013).

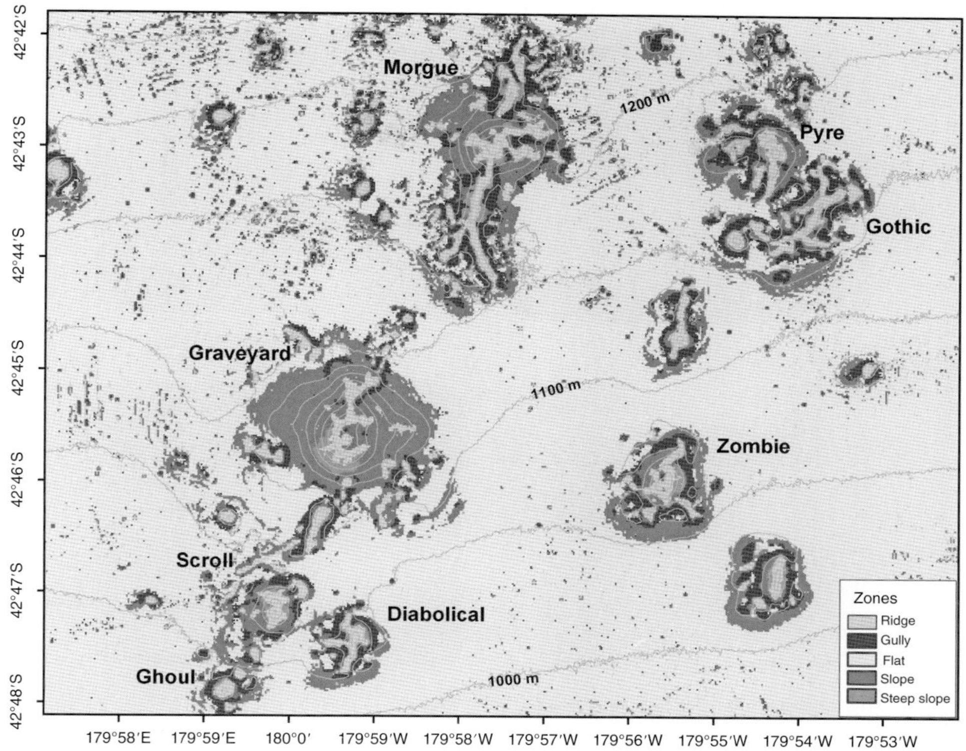

Fig. 5.9 Benthic seabed morphology map for a cluster of small seamount features at around 1000 m depth off New Zealand, derived from multibeam data using BTM. (Clark *et al.*, 2010.)

Segmentation is the process of partitioning a data set in clusters or regions of contiguous pixels (individual grid cells), which are similar with respect to a range of selected parameters. Pixels in the same segments have similar multivariate (spectral) values or belong to a similar pattern and form a connected region. The resulting segmented image typically includes a large number of segments, with each segment representing a mixture of the parameters used.

Segmentation can be pixel-based, where each pixel is compared with adjacent ones, or object-based, where the shape of the segment is used for classification (Fig. 5.10). Spatial homogeneity of pixels plays an important role in object-based segmentation, as objects are formed because of their spatial correlation, not just because of their thematic similarity.

Multivariate segmentation is a processing technique where additional ancillary information is used as a proxy for specific classification purposes. Seafloor backscatter and other physical data can be used to make habitat maps of substratum type. In the geomorphologically diverse region of Cook Strait, New Zealand, Lucieer and Lamarche (2011) used an object-based method of segmentation to generate a four-class classification map, with uncertainty in class attribution quantified by a confusion index layer. Here, backscatter served as a proxy for seafloor sediment grain size and heterogeneity. The resulting classification map had some correlation with the published sediment chart for the region, whereas comparisons with structural and morphological maps indicated that the object-based classification might also recognize seascape features such as sediment-wave fields and scarps, independently of the water depth. However, the method used by Lucieer and Lamarche (2011) acknowledges caution is needed with the application of imprecise and/or uncertain data through the integration of different approaches.

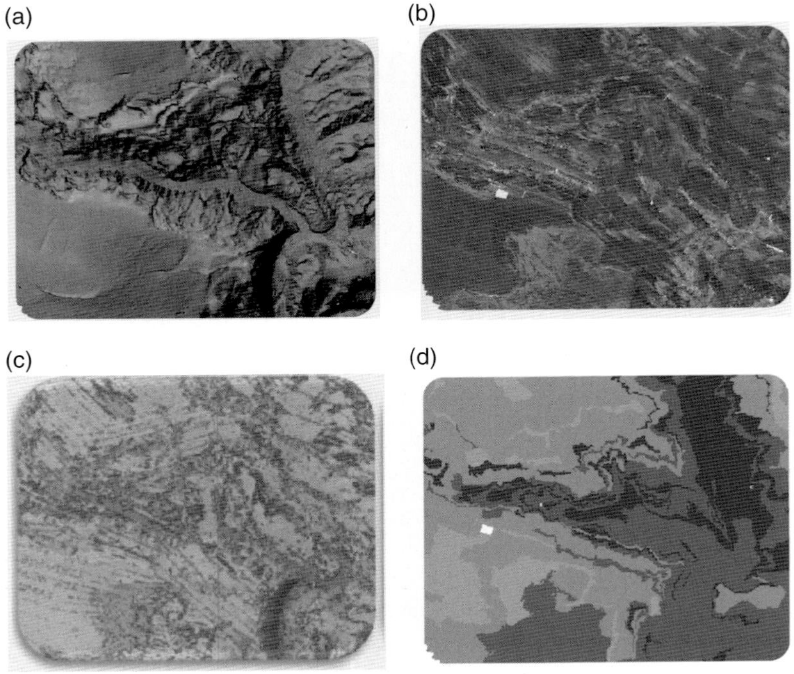

Fig. 5.10 (a) Bathymetry and (b) backscatter data used to generate (c) pixel-based segmentation and (d) object-based segmentation of an area of the Cook Strait, New Zealand. (Lucieer & Lamarche, 2011. Reproduced with permission of Elsevier.) (Please see Plate 9 for colour representation of the figure.)

Classification, which follows segmentation, is the process where segments are grouped into homogeneous subsets of objects. There are two overarching types of segmentation and classification: unsupervised and supervised (e.g. ICES, 2007; Lucieer & Lucieer, 2009).

- *Unsupervised classification* – also called objective classification – is where the elements of the maps are assembled first, and subsequently assigned a meaningful class (following a method to 'assemble first, predict later'). Here the remotely sensed data and *in situ* data are organized before they are combined (Brown *et al.*, 2011). The classes are unknown but are differentiated by the algorithm partitioning the data set into homogeneous subgroups. The process is repeatable and independent of user selection.
- *Supervised classification*, also called subjective classification, is when the classes are known or defined prior to segmentation, so that the objects to classify must be sorted out into these classes ('predict first, assemble later'). The classification algorithm is then trained on a subset of ground-truthed samples of the seabed, where the classes are set a priori by the user ('supervisor') according to their knowledge and objectives, such as classification of substrate types or other forms of habitat. Supervised classifications are highly dependent on the training data set (size, representativeness and information content) defining the existing classes (Stephens & Diesing, 2014).

Diesing *et al.* (2014) compared a number of approaches that used MBES and substrate data (including manual interpretation, geostatistics, object-based analysis and machine learning), and found that most map outputs were similar. They concluded, however, that more studies were required on the factors affecting classification performance, as well as the performance of different approaches. Combination of various methods, hybrid approaches, and multi-method ensembles may all help.

5.6 Acquisition of ground-truth data

Habitat maps integrate statistical measures of the interrelationships between sets of remotely sensed data. They can therefore be considered as an hypothesis that must be tested (Kostylev & Hannah, 2007). Category or class validation most importantly includes ground-truthing against targeted physical or biological samples (Anderson *et al.*, 2008). Although the samples should be collected specifically for that purpose, this is not often the case (Boyd *et al.*, 2006; ICES, 2007), but they must be independent of data used for the classification process. Because it is often not possible to run two consecutive surveys in the same area, for classification and for ground-truthing, a common way is to randomly partition the sample data set, 70% for classification and 30% for validation. However, the ground-truthing after the production of a habitat map using the maximum amount of available data is the preferred method.

Geological and biological seabed attributes can be ground-truthed through a range of methods, but is generally achieved through physical sampling of the seabed or assessed visually through camera or video techniques. Each approach has its advantages and disadvantages. Point sampling provides ground-truth data over a smaller spatial extent, but the data are accurate with respect to aspects like sediment grain size, and species identification. Observational data from photographic methods are generally over a larger area but the data are less certain – broad classes only of substrate, and typically faunal identification will not be at species level. Such decisions will be driven by the objectives of the research and scale of application.

The ground-truthing can enable the habitat maps to be improved. This may involve some adjustment of boundaries, perhaps even a major revision. The key issue, however, is there will be a measure of confidence associated with the mapping products.

Tools and methods that can be used to collect ground-truth data are detailed in other chapters in this book: seafloor photography and videos (see Chapters 12 and 13), sleds (see Chapter 9), trawls (see Chapter 7) and dredges and corers (see Chapter 10).

5.7 Synthesis

This chapter presents some routine protocols and methodologies adopted to acquire remotely sensed data with the aim to develop map products that describe elements of the benthic habitat. There is a wide range of high-end technologies available for collecting and processing remotely sensed data, with multibeam echosounders the most widely used equipment used to collect data to make benthic habitat maps. Basic survey design and equipment characteristics are used herein as examples, but the technology is constantly being advanced. Here, we have provided just a few examples of the sorts of habitat maps that can be produced using remotely-sensed data from our personal experience, and the reader is strongly encouraged to examine examples of other habitat mapping studies cited in this chapter, as well as review texts (e.g. Hughes Clarke *et al.*, 1996; Kenny *et al.*, 2003; Pickrill & Kostylev, 2007; Anderson *et al.*, 2008; Brown *et al.*, 2011; Diesing *et al.*, 2014; Calvert *et al.*, 2014; Diesing and Stephens, 2015).

It is important to remember that the utility and representativeness of habitat maps is, in part, balanced against the degree of quality control applied during the survey, and the acquisition of ground-truth data to test the veracity of the map product based on the remotely sensed data. The future of benthic habitat mapping lies in the development of quantitative procedures to complement qualitative approaches, with the ultimate aim to routinely and objectively characterize seafloor habitat. This goal will necessarily better utilize quantitative information obtained from different analytical approaches, recognizing the strengths and weaknesses of the various techniques. Whatever techniques are ultimately employed, benthic habitat mapping is now a standard tool to aid effective ocean management and conservation of the marine environment.

Acknowledgements

This chapter benefited from comments by two anonymous reviewers. Comments and support were also provided by editors Malcolm Clark and Ashley Rowden.

References

Anderson, J.T., Van Holliday, D., Kloser, R., Reid, D.G., Simard, Y. (2008). Acoustic seabed classification: current practice and future directions. *ICES Journal of Marine Science*, 65: 1004–1011.

Augustin, J.M., Lurton, X. (2005). Image amplitude calibration and processing for seafloor mapping sonars. In: *Oceans 2005 – Europe,* 20–23 June 2005, Vol. 1, pp. 698–701, doi:10.1109/OCEANSE.2005.1511799

Barton, S., Wysoczanski, R., Lamarche, G., Pallentin, A. (2014). Predictive substrate mapping in the active volcanic environment of the Kermadec arc. Paper presented at Geohab 2014, Lorne, Australia, 5–9 May 2014.

Bowden, D.A., Rowden, A.A., Thurber, A.R., Baco, A.R., Levin, L.A., Smith, C.R. (2013). Cold seep epifaunal communities on the Hikurangi Margin, New Zealand: composition, succession, and vulnerability to human activities. *PLoS ONE*, 8: e76869.

Boyd, S., Coggan, R., Birchenough, S., *et al.* (2006). The role of seabed mapping techniques in environmental monitoring and management. *Science Series Technical Report, Cefas*, 127: 170.

Brown, C.J., Blondel, P. (2009). Developments in the application of multibeam sonar backscatter for seafloor habitat mapping. *Applied Acoustics*, 70: 1242–1247.

Brown, C.J., Smith, S.J., Lawton, P., Anderson, J.T. (2011). Benthic habitat mapping: A review of progress towards improved understanding of the spatial ecology of the seafloor using acoustic techniques. *Estuarine, Coastal and Shelf Science*, 92: 502–520.

Calvert, J., Strong, J.A., Service, M., McGonigle, C., Quinn, R. (2014). An evaluation of supervised and unsupervised classification techniques for marine benthic habitat mapping using multibeam echosounder data. *ICES Journal of Marine Science*, doi: 10.1093/icesjms/fsu223

Clark, M.R., Bowden, D.A., Baird, S.J., Stewart, R. (2010). Effects of fishing on the benthic biodiversity of seamounts of the 'Graveyard' complex, northern Chatham Rise. *New Zealand Aquatic Environment and Biodiversity Report*, No. 46. 40 p.

Cochrane, G.R., Lafferty, K.D. (2002). Use of acoustic classification of sidescan sonar data for mapping benthic habitat in the Northern Channel Islands. *California. Continental Shelf Research*, 22: 683–690.

Dartnell, P., Gardner, J.V. (2004). Predicting seafloor facies from multibeam bathymetry and backscatter data. *Photogrammetric Engineering and Remote Sensing*, 70: 1081–1091.

Davis, R.E., Eriksen, C.E., Jones, C.P. (2003). Autonomous buoyancy-driven underwater vehicle gliders. In: Griffiths, G. (ed). *Technology and applications of autonomous underwater vehicles*. New York: Taylor & Francis, pp. 37–62.

de Moustier, C. (1986). Beyond bathymetry: mapping acoustic backscattering from the deep seafloor with Sea Beam. *Journal of the Acoustical Society of America*, 79: 316–331.

Diesing, M., Green, S.L., Stephens, D., Lark, R.M., Stewart, H.A., Dove, D. (2014). Mapping seabed sediments: comparison of manual, geostatistical, object-based image analysis, and machine learning approaches. *Continental Shelf Research*, 84: 107–119.

Diesing, M., Stephens, D. (2015). A multi-model ensemble approach to seabed mapping. *Journal of Sea Research*, 100: 62–69.

Dolan, M.F.J., Lucieer, V.L. (2014). Variation and uncertainty in bathymetric slope calculations using geographic information systems. *Marine Geodesy*, 37: 187–219.

German, C.R., Yoerger, D.R., Jakuba, M., Shank, T.M., Langmuir, C.H., Nakamura, K.-I., (2008). Hydrothermal exploration with the autonomous benthic explorer. *Deep Sea Research Part I: Oceanographic Research Papers*, 55: 203–219.

Goff, J.A., Olson, H.C., Duncan, C.S. (2000). Correlation of side-scan backscatter intensity with grain-size distribution of shelf sediments, New Jersey margin. *Geo-Marine Letters*, 20: 43–49.

Greene, H.G., Yoklavich, M.M., Starr, R.M., *et al.* (1999). A classification scheme for deep seafloor habitats. *Oceanologica Acta*, 22: 663–678.

Harris, P.T. (2011). Surrogacy. In: Harris, P.T., Baker, E.K. (eds), *Seafloor geomorphology as benthic habitat*: GeoHab *Atlas of seafloor geomorphic features and benthic habitats*. Amsterdam: Elsevier, pp. 93–108.

Harris, P.T., Baker, E.K. (eds) (2011). *Seafloor geomorphology as benthic habitat: GeoHab Atlas of seafloor geomorphic features and benthic habitats*. Amsterdam: Elsevier.

Hellequin, L., Boucher, J.M., Lurton, X. (2003). Processing of high-frequency multibeam echo sounder data for seafloor characterization. *IEEE Journal of Oceanic Engineering*, 28: 78–89.

Hughes Clarke, J.E., Mayer, L.A., Wells, D.E. (1996). Shallow-water imaging multibeam sonars: A new tool for investigating seafloor processes in the coastal zone and on the continental shelf. *Marine Geophysical Researches*, 18: 607–629.

Iampietro, P.J., Kvitek, R.G., Morris, E. (2005). Recent advances in automated genus-specific marine habitat mapping enabled by high-resolution multibeam bathymetry. *Marine Technology Society Journal*, 39: 83–93.

ICES (International Council for the Exploration of the Sea). (2007). *Acoustic seabed classification of marine physical and biological landscapes*. ICES Cooperative Research Report No. 286, pp. 183.

Jackson, D.R., Briggs, K.B. (1992). High-frequency bottom backscattering: Roughness versus sediment volume scattering. *Journal of the Acoustical Society of America*, 92: 962–977.

Jalving, B., Mandt, M., Hagen, O.K., Pøhner, F. (2004). *Terrain referenced navigation of AUVs and submarines using multibeam echo sounders*. Norwegian Defence Research Establishment.

Jones, A.T., Greinert, J., Bowden, D.A., *et al.* (2010). Acoustic and visual characterization of methane-rich seabed seeps at Omakere Ridge on the Hikurangi Margin, New Zealand. *Marine Geology*, 272: 154–169.

Kenny, A.J., Cato, I., Desprez, M., Fader, G., Schüttenhelm, R.T.E., Side, J. (2003). An overview of seabed-mapping technologies in the context of marine habitat classification. *ICES Journal of Marine Science*, 60: 411–418.

Kostylev, V.E. (2012). Benthic habitat mapping from seabed acoustic surveys: do implicit assumptions hold? *International Association of Sedimentologists*, 44: 405–416.

Kostylev, V.E., Hannah, C.G. (2007). Process-driven characterization and mapping of seabed habitat. In: Todd, B.J., Greene, H.G. (eds), *Mapping the seafloor for habitat characterization*. Geological Association of Canada, Special Paper 47, pp. 171–184.

Kostylev, V.E., Todd, B.J., Fader, G.B.J., Courtney, R.C., Cameron, G.D.M., Pickrill, R.A. (2001). Benthic habitat mapping on the Scotian Shelf based on multibeam bathymetry, surficial geology and sea floor photographs. *Marine Ecology Progress Series*, 219: 121–137.

Kumagai, H., Tsukioka, S., Yamamoto, H., *et al.* (2010). Hydrothermal plumes imaged by high-resolution side-scan sonar on a cruising AUV, Urashima. *Geochemistry, Geophysics, Geosystems*, 11: Q12013.

Lamarche, G., Lurton, X., Verdier, A.-L., Augustin, J.-M. (2011a). Quantitative characterization of seafloor substrate and bedforms using advanced processing of multibeam backscatter. Application to the Cook Strait, New Zealand. *Continental Shelf Research*, 31: S93–S109.

Lamarche, G., Rowden, A.A., Mountjoy, J.J., Lucieer, V.L., Verdier, A.-L. (2011b). The Cook Strait canyon, New Zealand: a large bedrock canyon system in a tectonically active environment. In: Harris, P.T., Baker, E.K. (eds), *Seafloor Geomorphology as Benthic Habitat*: GeoHab *Atlas of seafloor geomorphic features and benthic habitats*. Amsterdam: Elsevier.

Le Gonidec, Y., Lamarche, G., Wright, I.C. (2003). Inhomogeneous Substrate Analysis Using EM300 Backscatter Imagery. *Marine Geophysical Researches*, 24: 311–327.

Lucieer, V., Lamarche, G. (2011). Unsupervised fuzzy classification and object-based image analysis of multibeam data to map deep water substrates, CookStrait, New Zealand. *Continental Shelf Research*, 31: 1236–1247.

Lucieer, V.L., Lucieer, A. (2009). Fuzzy clustering for seafloor classification. *Marine Geology*, 264: 230–241.

Lucieer, V.L., Hill, N., Barrett, N., Nichol, S.L. (2013). Do marine substrates 'look' and 'sound' the same? Supervised classification of multibeam acoustic data using autonomous underwater vehicle images. *Estuarine, Coastal and Shelf Science*, 117: 94–106.

Lundblad, E.R., Wright, D.J., Miller, J., *et al.* (2006). A benthic terrain classification scheme for American Samoa. *Marine Geodesy*, 29: 89–111.

Lurton, X. (2010). *An introduction to underwater acoustics. Principles and applications* (2nd edn). London: Springer Praxis Books & Praxis Publishing.

Lurton, X., Lamarche, G. (eds) (2015). Backscatter measurements by seafloor-mapping sonars. Guidelines and Recommendations. GeoHab report, 200p. http://geohab.org/wp-content/uploads/2014/05/BSWG-REPORT-MAY2015.pdf

Pandian, P.K., Ruscoe, J.P., Shields, M., *et al.* (2009). Seabed habitat mapping techniques: an overview of the performance of various systems. *Mediterranean Marine Science*, 10: 29–43.

Pickrill, R.A., Kostylev, V.E. (2007). Habitat mapping and national seafloor mapping strategies in Canada. In: Todd, B.J., Greene, H.G. (eds), *Mapping the seafloor for habitat characterization*. Geological Association of Canada, Special Paper, pp. 483–495.

Schimel, A., Healy, T.R., McComb, P., Immenga, D. (2009). Comparison of a self-processed EM3000 multibeam echosounder dataset with a QTC View habitat mapping and a sidescan sonar imagery, Tamaki Strait, New Zealand. *Journal of Coastal Research*, 26: 714–725.

Schofield, O., Kohut, J., Aragon, D., et al. (2007). Slocum Gliders: Robust and Ready. Journal of Field Robotics, 24 (6): 473–485.

Snelgrove, P.V.P., Butman, C.A. (1994). Animal–sediment relationships revisited: cause versus effect. *Oceanography and Marine Biology: An Annual Review*, 32: 111–117.

Stephens, D., Diesing, M. (2014). A comparison of supervised classification methods for the prediction of substrate type using multibeam acoustic and legacy grain-size data. *PLoS ONE*, 9(4): e93950, doi:10.1371/journal.pone.0093950

Wilson, M.F.J., O'Connell, B., Brown, C., Guinan, J.C., Grehan, A.J. (2007). Multiscale terrain analysis of multibeam bathymetry data for habitat mapping on the continental slope. *Marine Geodesy*, 30: 3–35.

Wynn, R.B., Huvenne, V.A.I., Le Bas, T.P., *et al.* (2014). Autonomous Underwater Vehicles (AUVs): their past, present and future contributions to the advancement of marine geoscience. *Marine Geology*, 352: 451–468.

Chapter 6
Deep-Sea Zooplankton Sampling

Bernd Christiansen
Universität Hamburg, Institut für Hydrobiologie und Fischereiwissenschaft, Hamburg, Germany

Abstract

This chapter focuses on the technical aspects that are important for sampling zooplankton at bathyal and abyssal depths. One of the biggest challenges in deep-sea zooplankton sampling is their low abundance, requiring large sampling volumes, which can usually be achieved only with net samplers. Multiple opening and closing nets are the standard tools for sampling deep-sea zooplankton, and examples of common gear types are described, along with some alternative methods for assessing zooplankton. Information is provided about ship requirements and gear operations in the deep sea, including the sampling of zooplankton close to the seafloor and collecting live specimens for experiments.

Keywords zooplankton, instruments, bathypelagic zone, benthopelagic plankton, gear operation

6.1 Introduction

Zooplankton sampling in the deep sea is effectively no different from sampling in the near-surface water layers and in general the same methods are employed. In 1887, deep-sea zooplankton was sampled during one of the first extensive oceanographic expeditions, Viktor Hensen's 'Plankton-Expedition' on the German research vessel *National*. Hensen used a fine-meshed vertical net with an opening and closing mechanism down to a depth of 3500 m, giving the first results on the vertical distribution of zooplankton in the deep sea. Below 2000 m Hensen obtained only 'Copepoden und gewisse Radiolarien' (copepods and certain radiolarians), whereas above this depth other taxonomic groups such as chaetognaths, ostracods, amphipods, decapods, salps and siphonophores were in the catch (Brandt, 1890). He also observed that the density of zooplankton in the deep sea was much lower than that in the upper water column.

The following decades saw a rapid development of plankton nets for different purposes, most of them being used only in shallow waters. A comprehensive overview of the development of zooplankton nets and advanced plankton assessment instruments is given in Wiebe and Benfield (2003).

Many aspects of zooplankton sampling have been described elsewhere. However, apart from numerous publications about specific sampling methods and gear types, there are only a few works dealing with the topic on a more general level. An early attempt was made by Tranter (1968), with a monograph on zooplankton sampling methods, including an introduction to the problems of zooplankton sampling and a description of several common sampler types, many of which are still in use. The *ICES Zooplankton Methodology Manual* (Harris *et al.*, 2000) is a comprehensive book about state-of-the-art zooplankton methods. It includes descriptions of some of the most important

Biological Sampling in the Deep Sea, First Edition. Edited by Malcolm R. Clark, Mireille Consalvey and Ashley A. Rowden.

zooplankton assessment methods and instruments, and instructions for their operation, as well as information on sampling design and specific methods of sample processing and analysis. An overview of nearly all types of plankton nets and of acoustical and optical sensors for zooplankton can be found in Wiebe and Benfield (2003), who also provide an outlook for future developments in the assessment of marine zooplankton.

This chapter will address some aspects which are important for sampling zooplankton mainly in the deeper bathyal and abyssal (below 1000 m), and in the near-bottom water layer of the deep sea. The focus will be on physically collecting micro-, meso-and macrozooplankton, but alternative methods for assessing zooplankton, such as *in situ* observations, will also be addressed. Examples of zooplankton samplers used in deep-sea studies are presented, along with information on sampling operations and personal experiences from the Deep-sea Zooplankton Group of the Institut für Hydrobiologie und Fischereiwissenschaft (IHF), University of Hamburg. Sample processing and their subsequent analysis will not be covered here extensively. The interested reader will find detailed descriptions in Harris *et al.* (2000).

6.2 General considerations in deep-sea zooplankton sampling

In this section, a short overview of the more general issues involved with deep-sea zooplankton sampling will be given.

6.2.1 Problems

Zooplankton sampling in the deep sea generally has to cope with the same problems as in the upper water layers (reviewed by Tranter, 1968). Zooplankton cover a wide size range, from a few micrometres in nanozooplankton to several decimetres and even metres in megazooplankton. Furthermore, size is often inversely correlated with concentration (i.e. smaller zooplankton are usually more abundant than larger zooplankton). This means that it is not possible to collect a representative sample of the whole plankton community in a given area with only one device. Thus different types of gear, or a combination of samplers, have to be used if the study covers a broad size spectrum of zooplankton.

Although in general zooplankton drift passively with the ocean currents, many groups have considerable locomotive capabilities and may detect and avoid approaching sampling gear. As there is no ambient light below 1000 m (except for bioluminescence), visual detection does not play a role in avoidance behaviour. However, as outlined in Clutter and Anraku (1968), the friction of the towing wire, bridle and net frame in the water, as well as the bow wave, particularly in nets with low filtration efficiencies, generate increased water flow and pressure gradients well ahead of the gear. Such disturbances can be detected by several groups, including fish larvae and invertebrates. This problem can be reduced to a certain degree through constructive measures; in particular, by maximizing the filtration efficiency of the sampler and keeping the mouth opening free from obstacles such as bridles. Clogging of the nets can be a major problem in near-surface tows, reducing the filtration efficiency during the tow considerably, but usually this does not occur in deep-sea tows. Finally, many zooplankton groups are extremely fragile, making it difficult to collect some taxa intact (e.g. gelatinous forms).

6.2.2 Main methods

Three principal methods of collecting marine zooplankton have been developed since the beginning of plankton research:

- water bottles (with subsequent filtration, sedimentation or centrifugation);
- plankton pumps;
- nets.

Water bottles have the advantage of providing discrete samples with a precise volume, but they are limited in their sample volume. They are mainly used for phytoplankton and microzooplankton, which occur in high densities. Plankton pumps can be used for discrete and integrated samples, and their sample volume can easily be adjusted within certain limits. Pumps are often used for specific purposes, like continuous sampling. Plankton nets are still the most important zooplankton samplers and are very flexible with respect to sample size. However, determining the exact volume filtered is not trivial, and sampling discrete depth strata requires a considerable technological effort. Plankton nets always integrate along a certain distance.

Other methods of assessing, but not physically collecting, zooplankton include visual observations from submersibles, divers, video and photographic records, and direct optical and acoustic measurements.

6.2.3 Zooplankton concentration and sampling volume

One of the biggest challenges in deep-sea zooplankton sampling is the low concentration of organisms at bathyal and abyssal depths. For example, the typical mesozooplankton abundance in the abyssal realm is in the range of 0.1–1 individuals m^{-3} for all taxa combined (e.g. Christiansen *et al.*, 1999), meaning that individual taxa may occur in extremely low abundance. Consequently, the sample volume must be very large, in the range of several hundreds to tens of thousands of cubic metres, even for small zooplankton. This generally excludes two of the three basic types of zooplankton samplers mentioned above. Water bottles, often used for sampling microzooplankton and sometimes small mesozooplankton in neritic areas, can only sample up to a few hundred litres, even if combined on a rosette sampler. Pumps are available as shipboard units or as underwater systems placed directly *in situ*. They require either a hose to a filtration unit on the ship, which is only applicable to maximum depths of about 200 m, or an *in situ* filtration device. While shipboard pumps may have the capacity to sample more than 1 m^3 min^{-1}, the capacity of *in situ* pumps is much lower, and their use is thus restricted to fairly high plankton concentrations, or very long pumping times are required.

Thus most zooplankton collecting devices in deep-sea applications are nets, at least for quantitative studies. Large-sample volumes can be achieved either by long tows or by large nets. Since ship time is very expensive, and long tows affect the condition of the collected animals, nets as large as possible should be employed. Closing or opening/closing devices are mandatory for deep-sea plankton nets, and multiple net systems will dramatically increase the efficiency of ship time usage for many applications.

6.2.4 Scientific objectives

Numerous scientific questions arise in deep-sea zooplankton studies. While the first studies often focused on taxonomic and biogeographic descriptions (which are still important topics in deep-sea planktology), the spectrum of scientific objectives has since broadened to include distributions in time and space, physiology, behaviour and food web analyses, to name a few. For methodological aspects, it is important to discriminate between qualitative and quantitative studies, which have different demands on the complexity of the sampling gear and on the sampling design. Furthermore, it is important to consider the size of the target groups, which will determine the mesh size and sample size required.

6.2.4.1 Qualitative sampling

Qualitative studies may include the collection of material for taxonomic purposes, or for biological analyses such as histology, biochemistry, genetics and physiology. Simple nets will often be sufficient, but closing or opening/closing devices are necessary to sample zooplankton from defined depth strata. In many of these studies the condition of the organisms is most important, so the tow duration should be kept to a minimum, and towing speed should be slow in order to decrease the filtration pressure. Because these two preconditions are somewhat contradictory, some balance between the two is necessary. Slow towing speeds (<0.5 m s^{-1}) can be achieved only with vertical nets, and unless very large volumes or very narrow depth layers are to be sampled, they are preferable for qualitative deep-sea zooplankton studies.

6.2.4.2 Quantitative sampling

Most studies in the deep sea will require quantitative sampling (i.e. the sampling method has to allow for an estimate of sample volume, minimize avoidance by the target groups and consider gear selectivity). The condition of the collected animals is also important, but often a compromise is necessary between minimizing damage, minimizing ship time and sampling a water volume that sufficiently represents the target taxa. Both vertical and horizontal/oblique tows can be used in quantitative deep-sea zooplankton sampling. Vertical tows integrate vertically; their sample volume is directly dependent on, and limited by, the depth interval sampled. Oblique tows can be regarded as extended vertical hauls. They integrate both vertically and horizontally, and their sample volume can be adjusted by varying the hauling speed. Horizontal tows integrate horizontally; theoretically, their sample volume is unlimited, but practically it is restricted by the maximum possible tow duration.

Consequently, horizontal tows should be chosen if narrow depth layers and large volumes have to be sampled; for example, close to the seafloor or in narrow-banded sound-scattering layers. Conversely, if small-scale horizontal differences are important, or if the sampling space is confined (e.g. at topographic features such as seamounts or canyons), vertical hauls are appropriate. If the sample volume has to be as large as possible and small-scale horizontal variations are less important (e.g. over the abyssal plain), oblique tows should be employed. Table 6.1 gives a comparison of sample volume and tow duration for vertical and oblique tows, using the example of a 0.25 m^2 MultiNet® with five nets, and depth strata of 500 m to a maximum depth of 2500 m. At the same wire speed, the oblique tow yields a fivefold larger sample volume compared to the vertical haul, but takes less than twice the sampling time.

6.2.5 Sampling design

The sampling design depends on the specific goals of a study, and cannot be discussed here comprehensively. It should ideally follow the same rules in deep-sea zooplankton studies as for shallow-water sampling. Some basic advice for designing a sampling strategy is given in Skjoldal *et al.* (2000). For quantitative analyses, a sufficient number of replicate samples is necessary in order to allow for statistical analysis (e.g. test for difference in mean abundance between two or more locations). However, the optimal number of hauls is often unrealizable, since ship time for open ocean cruises is limited and expensive, yet requirements in deep-sea studies are extremely high. For example, covering the water column over abyssal depths with a towed sampling system like the MOCNESS (see below), takes about 10 hours per haul. To compare two stations, with four replicates each, about three full days of station time are required, not considering the steaming time between hauls. Therefore, often some compromise has to be found between the desired number of stations and the statistical robustness of the data.

Table 6.1 Comparison of tow duration and sample volume in vertical and oblique hauls (0.25 m^2 MultiNet®, five depth strata of 500 m each, maximum depth 2500 m, assumed 1.75:1 wire length to depth ratio in oblique hauls)

	Vertical haul	Oblique haul	Oblique haul
Ship speed (knots)	0	2	2
Wire speed (m s^{-1})	0.5	0.5	0.3
Max. length of wire (m)	2500	4375	4375
Tow duration (only hauling) (min)	83	146	243
Sample volume per depth layer (m^3)	125	630	917

Plate 1 (a) The continental slope seabed at 2200 m depth in the Northeast Atlantic showing ophiuroids on foraminifera ooze. (Photograph by P.A. Tyler.) (b) A drop stone in the continental slope at 2639 m in the Northeast Atlantic showing the winnowing of sediment downstream of the drop stone and current generated ripples in the background. (Photograph by P.A. Tyler.) (c) *Bathymodiolus mauritanicus* alive and dead on the Darwin mud volcano at 1100 m depth. (Photograph courtesy of NOC/NERC. Reproduced with permission.) (d) The frame-building cold-water coral *Lophelia pertusa* in the Whittard submarine canyon at 1300 m depth in the Northeast Atlantic. (Photograph courtesy of NOC/NERC. Reproduced with permission.) (e) The Porcupine Abyssal Plain at 4800 m in the Northeast Atlantic showing tracks, trails and pits in which phytodetritus can accumulate. The small holothurian is *Amperima rosea*. (Photograph courtesy of DSM Billett NOC/NERC. Reproduced with permission.) (f) Dense populations of the yeti crab *Kiwa tyleri* and the barnacle *Vulcanolepas* sp. at hydrothermal vents along the southern part of the East Scotia Ridge, Antarctica, 2500 m depth. (Photograph by ChEsSo consortium.)

Biological Sampling in the Deep Sea, First Edition. Edited by Malcolm R. Clark,
Mireille Consalvey and Ashley A. Rowden.

Plate 2 A pictorial representation of the life in a deep water canyon presented in the February 1953 issue of *Life* magazine. The original legend stated: 'Canyon's mouth at base of a continental slope, jellyfish, viper-fish, sea devils, sea cucumbers, quill worms move through lightless depths a mile below last glimmer of sun. Cataract in background is a slump of seaslide of green mud and sand, similar to one that broke the Atlantic cables in 1929'. (© Real Estate of James Lewicki. Reproduced with permission.)

Plate 3 (a) The Ross Sea (Antarctica) deep seafloor at 1072 m. Note the heavily reworked substrate with mounds, burrows and grooves. On the right there is a large enteropneust which is leaving behind a trail of mud, after this has been ingested and passed through the digestive system. This kind of organism is rarely collected intact with destructive methods. (© NIWA, IPY-CAML Voyage 2008. Reproduced with permission.) (b) The Ross Sea (Antarctica) at about 2200 m. A large carnivorous ascidian is facing the bottom current waiting for prey. This very fragile organism was documented only with a towed camera, no specimens were retained by a sled and a beam trawl deployed at the same site. (© NIWA, IPY-CAML Voyage 2008. Reproduced with permission.) (c) A luxuriant assemblage of gorgonians, soft corals, zoanthids and sponges flourishing on the flank of Scott Island (Antarctica). The presence of particular local conditions support the high biomasses of suspension feeders. (© NIWA, IPY-CAML Voyage 2008. Reproduced with permission.) (d) A 'cold-water coral' assemblage of *Solenosmilia variabilis* hosts featherstars and sponges, with orange roughy swimming close to the seafloor at 950 m depth on a small seamount off New Zealand. (© NIWA. Reproduced with permission.) (e) The stegocephalid amphipod *Parandania* illustrates two of the main adaptations occurring in many mesopelagic crustaceans: the lack of eyes and a 'mimetic' purple coloration (already partially faded). This amphipod group is known to feed on deep-water medusae of the genus *Atolla*, which are among the most important contributors to cnidarian biomass in the deep water column (scale bar 1 cm). (© NIWA. Reproduced with permission.) (f) New discoveries continue to surprise scientists, as in the case of the 'barreleye' fish *Macropinna microstoma*. The apparent function of the transparent shield is to protect the tubular eyes from contact with stinging tentacles of cnidarians, the prey of *Macropinna.* (© 2003 MBARI. Reproduced with permission.) (g) Sampling a whale bone carrying *Osedax frankpressi* by the manipulator arm of MBARI's remotely operated vehicle *Tiburon* at a depth of about 3000 m in the Monterey Canyon. (© 2003 MBARI. Reproduced with permission.) (h) Specimens of *Osedax* with the palps emerging out of a whale bone. (Photograph by Greg Rouse, Scripps Institute of Oceanography. Reproduced with permission.)

Plate 4 Examples of the instruments discussed in this chapter. (a) Real-time CTD showing rosette of water bottles and electronic package at the bottom of the frame. (Photograph by Kostas Kiriakoulakis.) (b) Two self-recording instruments, an Sbe37 temperature–conductivity sensor and an RBR temperature recorder. (Photograph by Damien Guihen. Reproduced with permission.) (c) LADCP attached to CTD frame. (IOC Photo Library.) (d) SAPS instrument on winch line ready for deployment. (Photograph by Kostas Kiriakoulakis.) (e) ROV-deployed three-beam ADCP on base plate. (Photograph by Tomas Lundälv. Reproduced with permission.) (f) A sediment trap on deck with close-up of the carousel of sample containers after recovery. (Photograph by Kostas Kiriakoulakis.)

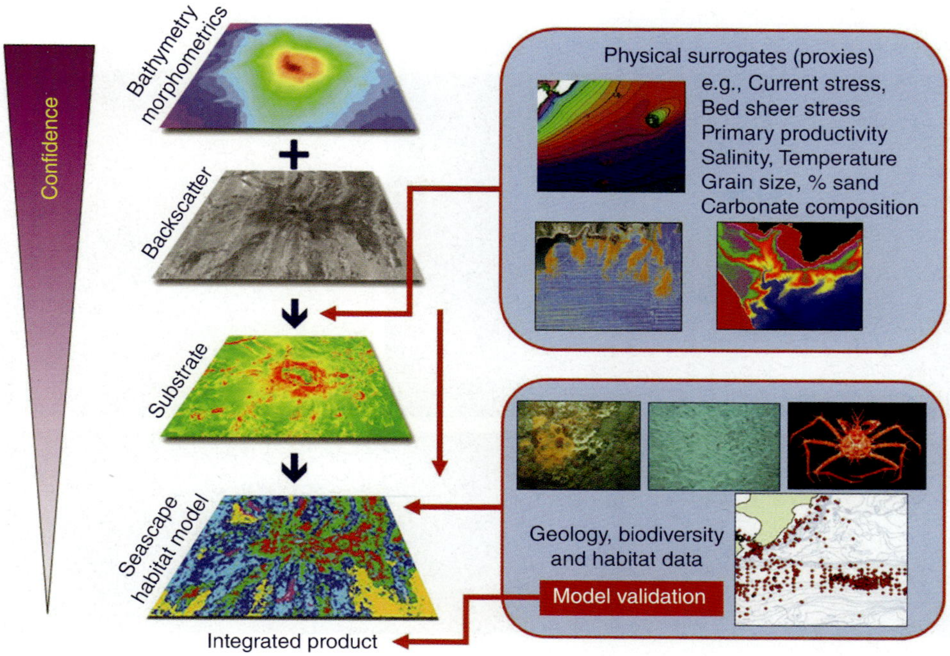

Plate 5 Pathway to habitat mapping. Schematic portrayal of the data components that contribute to form a habitat map. The final – integrated – product has an identifiable, replicable and consistent relationship with marine biota.

Plate 6 Examples of deep-sea AUVs and launch and recovery systems. (a) The MARUM SEAL AUV of the University of Bremen. (Source: Gerrit Meinecke, MARUM, www.mpi-bremen.de.) (b) The Sentry AUV of Woods Hole Oceanographic Institution. (Photograph courtesy of John Mitchell. Reproduced with permission.) (c) The UK Natural Environment Council Autosub 6000 AUV. (From Wynn *et al.*, 2014). (d) The Hugin 3000. (Photograph courtesy of John Mitchell. Reproduced with permission.)

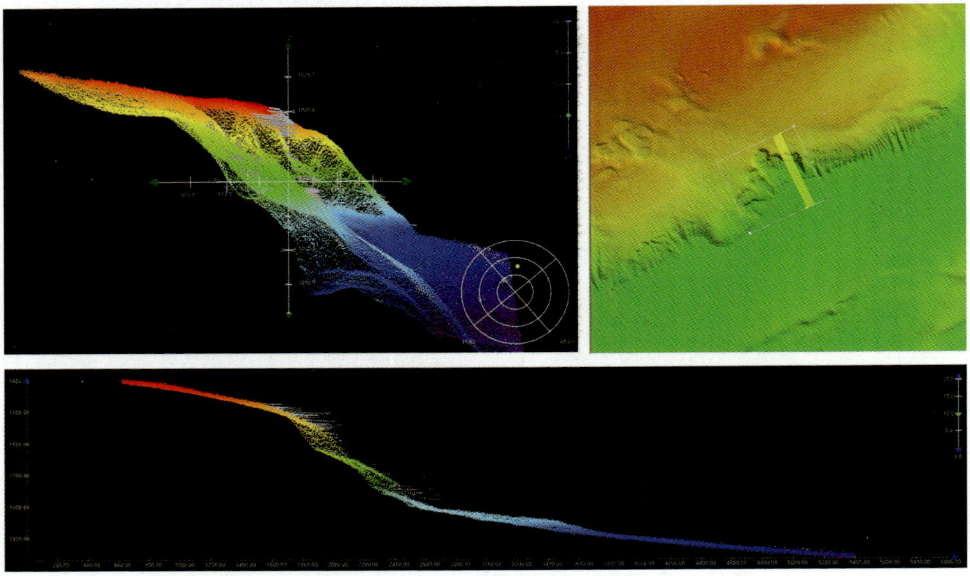

Plate 7 Example of area-based CUBE model editing. An area of seafloor data (right panel) has been selected, and projected as a two-dimensional across-track stack of soundings (bottom panel) and in three-dimensions as fully editable rendered surface (left panel). The individual soundings are coloured according to depth, and excluded soundings are in grey.

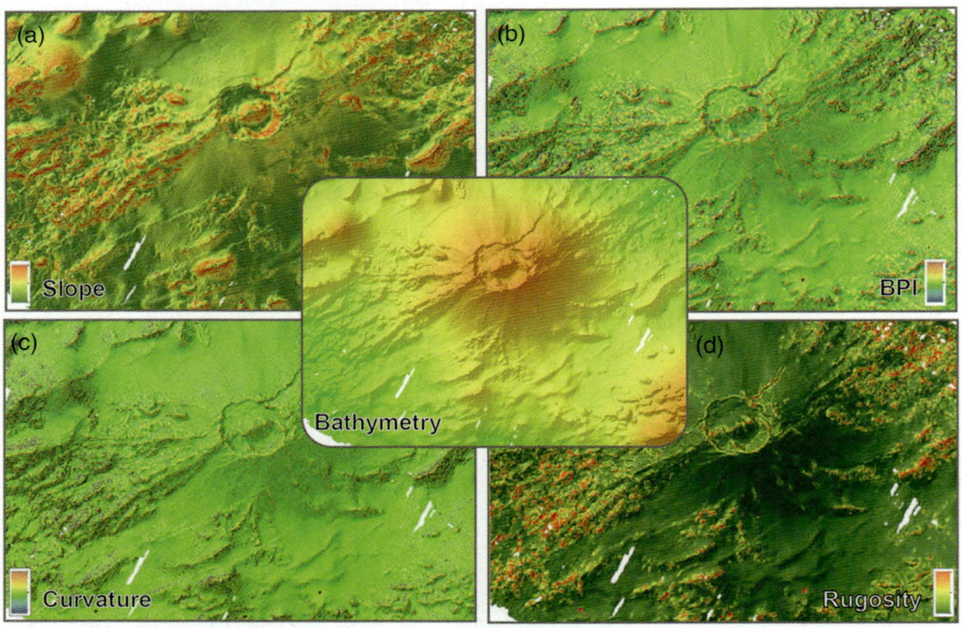

Plate 8 Morphometric parameters that can be derived from bathymetry data: (a) slope, (b) BPI, (c) curvature (mean, plan or profile) and (d) rugosity. (Adapted from Barton *et al.*, 2014.)

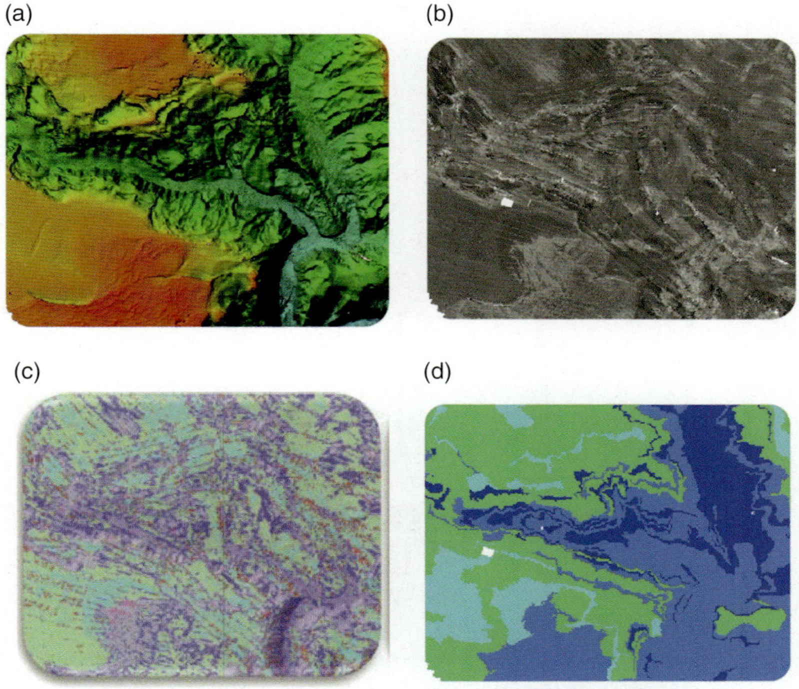

Plate 9 (a) Bathymetry and (b) backscatter data used to generate (c) pixel-based segmentation and (d) object-based segmentation of an area of the Cook Strait, New Zealand. (Lucieer & Lamarche, 2011. Reproduced with permission of Elsevier.)

Plate 10 The 1 m^2-double-MOCNESS. (Photograph by B. Christiansen.)

Plate 11 The 0.25 m^2 MultiNet® in vertical configuration. (Photograph by B. Christiansen.)

Plate 12 Agassiz trawl developed by the Netherlands Institute of Sea Research, being used off New Zealand. (Photograph courtesy of NIWA.)

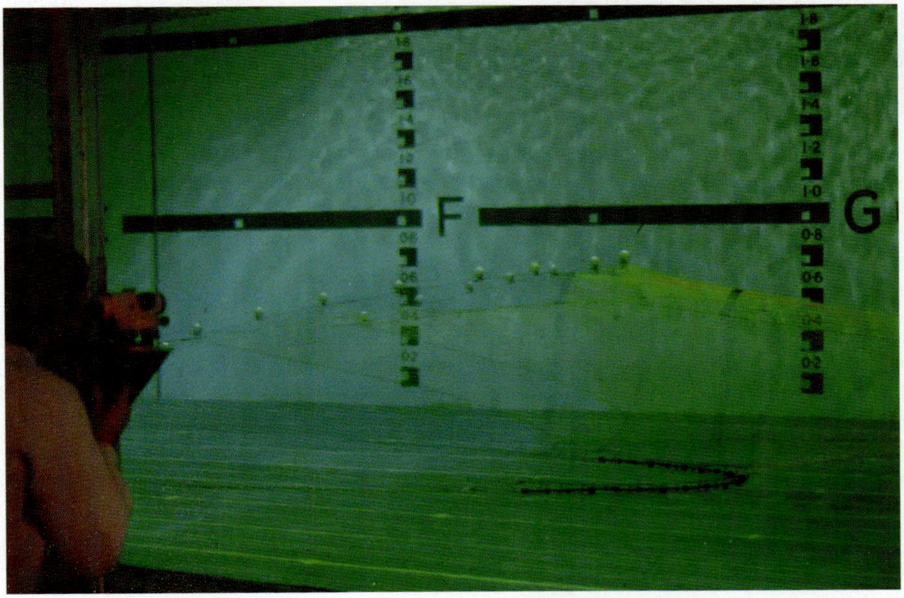

Plate 13 A scale model of a rough-bottom trawl being tested in a flume tank. Note the long upper wings but small lower wings on the leading edge of the net. (Photograph courtesy of NIWA.)

Plate 14 Aspects of operations during longline surveys in the Azores and Alaska, also showing some of the components used. (a) Small stone sinkers used on the Azores longline survey. (b) Main line hauling by a crew member. (c) Small metal sinkers used on the Alaska longline survey. (d) Retrieval and storage of the buoy and safety lines in the Azores longline survey. (e) Retrieval of the main line by a crew member using a hauler on the Alaska longline survey. (f) Deployment of longline off the stern of the vessel on the Azores longline survey. Main line units are joined, with a small float or a small stone. (g) The interior deck of a longliner showing part of the autoline system. (Photograph by G Jan Erik Dyb, MAR-ECO. Reproduced with permission.) (h) A sablefish caught by the longline used in Alaska. (i, j) Examples of small buoys used in the 'stone–buoy' longline in Azores surveys (red buoys used to 700 m and grey buoys used to 2000 m depth). (k) A small tub for storing the main line units used in the Alaska longline surveys. (a, b, d, f, i, j Azores longline surveys, University of Azores © ImagDOP. c, e, h, k Alaska longline surveys NOAA-NMFS Alaska Fisheries Science Center.)

Plate 15 The Triple-D (deep digging dredge) dredge is designed to sample burrowing fauna quantitatively. (Bergman, 2009. Reproduced with permission.)

Plate 16 (a) van Veen Grab. (Photograph by D Stevens. © NIWA, New Zealand. Reproduced with permission.) (b) The USNEL Mk II box-corer. (Photograph by B Bett. © National Oceanography Centre, UK. Reproduced with permission.) (c) The GOMEX box-corer ready for deployment. (Photograph by G Rowe. © Texas A&M University, USA. Reproduced with permission.) (d) The NIOZ box-corer being deployed. (Photograph courtesy of M Smit. © Royal Netherlands Institute for Sea Research, the Netherlands. Reproduced with permission.)

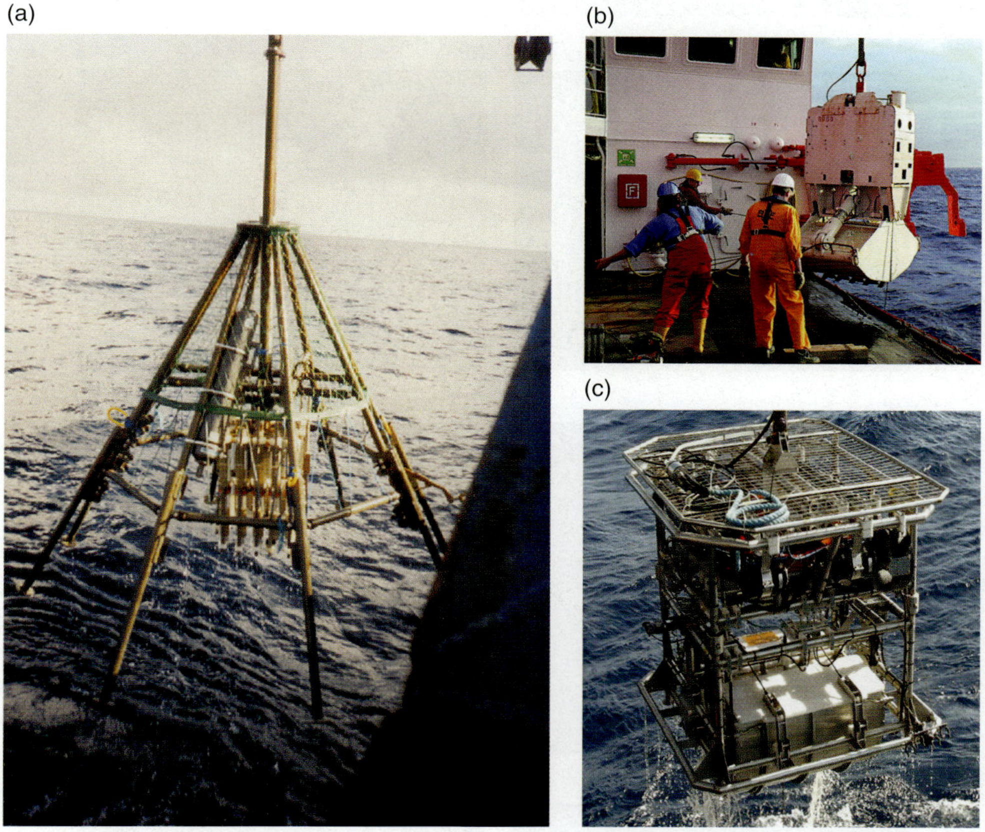

Plate 17 (a) The SMBA multiple corer being deployed. (Photograph by B Bett. © National Oceanography Centre, UK. Reproduced with permission.) (b) The IFM-GEOMAR TV-Grab ready for deployment deployed (Photograph by D Bowden. NIWA, New Zealand. Reproduced with permission.) (c) HyBIS, Hydraulic Benthic Interactive Sampler – a modern video grab. (Photograph by B Murton. National Oceanography Centre, UK. Reproduced with permission.)

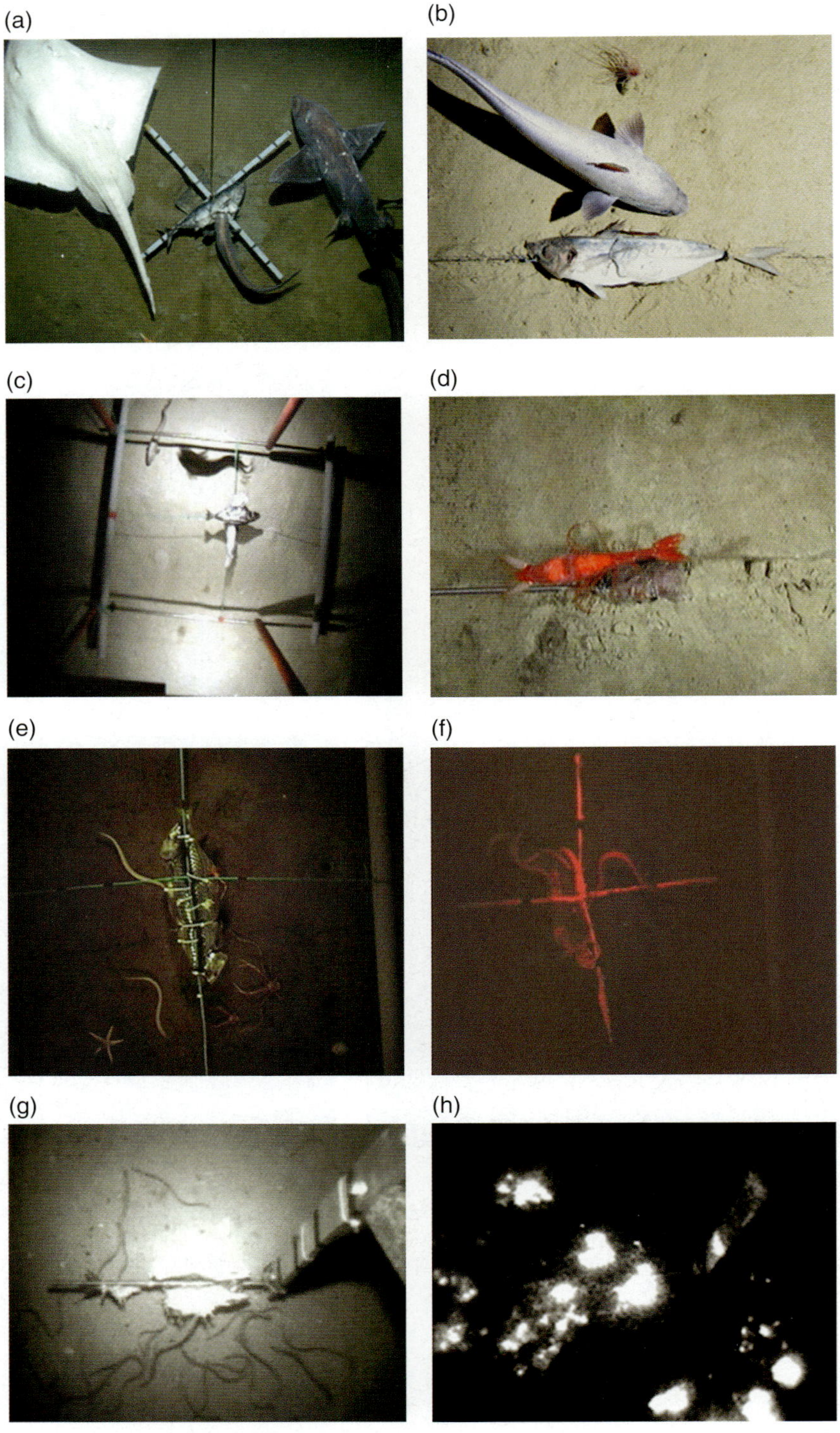

Plate 18 Examples of imaging types used in baited cameras. (a) Digital stills at 2 m above bottom (field of view (FOV) = 2.1 × 1.6 m; 3.4 m^2). (b) Digital stills at 1 m above bottom (FOV = 0.7 × 0.5 m; 0.3 m^2). (c) Frame grab from video at 2.8 m above bottom (FOV = 1.8 × 2.6 m; 4.7 m^2). (d) Frame grab from video at 1 m above bottom (FOV = 0.7 × 0.5; 0.4 m^2). (e) Normal colour CCD video camera with white light and (f) under red light (note the degradation in quality after switching to red light). (g, h) Ultra low-light Intensified Silicon Intensifier Target (ISIT) camera with white light on (g) and no illumination (h). When the lights are off the organisms cannot be seen but bioluminescent activity can. (Oceanlab, University of Aberdeen, UK. Reproduced with permission.)

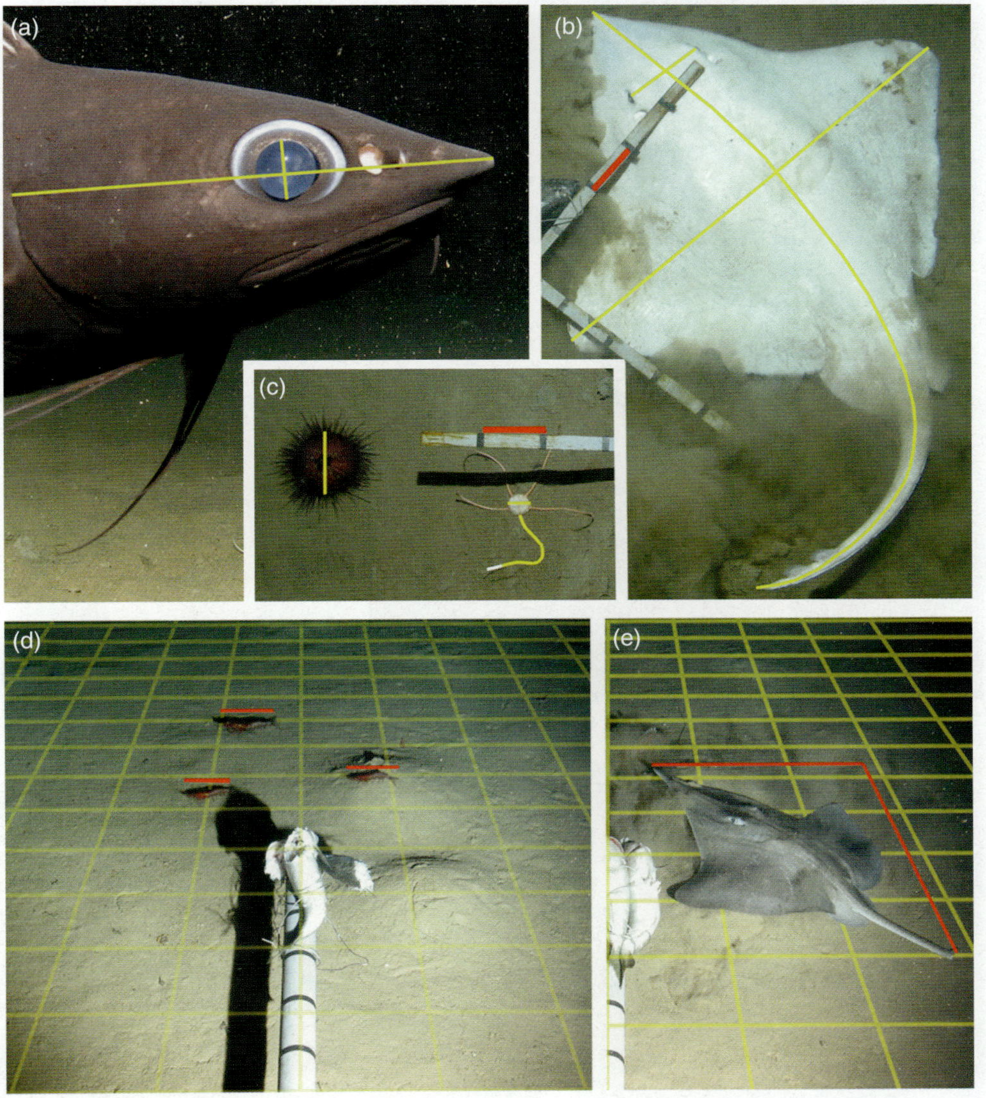

Plate 19 Digital measurements of deep-sea fauna from baited camera images. (a) Relative (%) measurements of eye diameter to head length from horizontal images. (b) Absolute body measurements from vertical photography. (c) Measurements of epifauna such as echinoid body diameter and ophuiroid arm length and disc diameter (the red line indicates the calibration measurement from the *in situ* scale bar). (d, e) Measurements from oblique images using a perspective grid; direct body lengths of natant decapods (d) and body lengths of a deep-water ray using trigonometry (e). (Oceanlab, University of Aberdeen, UK. Reproduced with permission.)

Plate 20 Towed camera platforms. (a) Deep Towed Imaging System (DTIS), National Institute of Water and Atmospheric Research, New Zealand. (Photograph courtesy of NIWA, New Zealand. Reproduced with permission.) (b) Deep towed camera system, Commonwealth Scientific and Industrial Research Organisation (CSIRO), Australia. (Photograph courtesy of CSIRO, Australia. Reproduced with permission.) (c) TowCam, Woods Hole Oceanographic Institute (WHOI), USA. (Photograph courtesy of WHOI, USA. Reproduced with permission.) (d) CamPod, Institute of Marine Research, Norway. (Photograph courtesy of the Institute of Marine Research (IMR), Norway.) (e) Benthic Optical Acoustic Grab (BOAGS), CSIRO, Australia. (Photograph courtesy of CSIRO, Australia. Reproduced with permission.) (f) Seabed high-resolution imaging platform (SHRIMP), National Oceanography Centre, UK. See Table 12.1 for technical specifications.

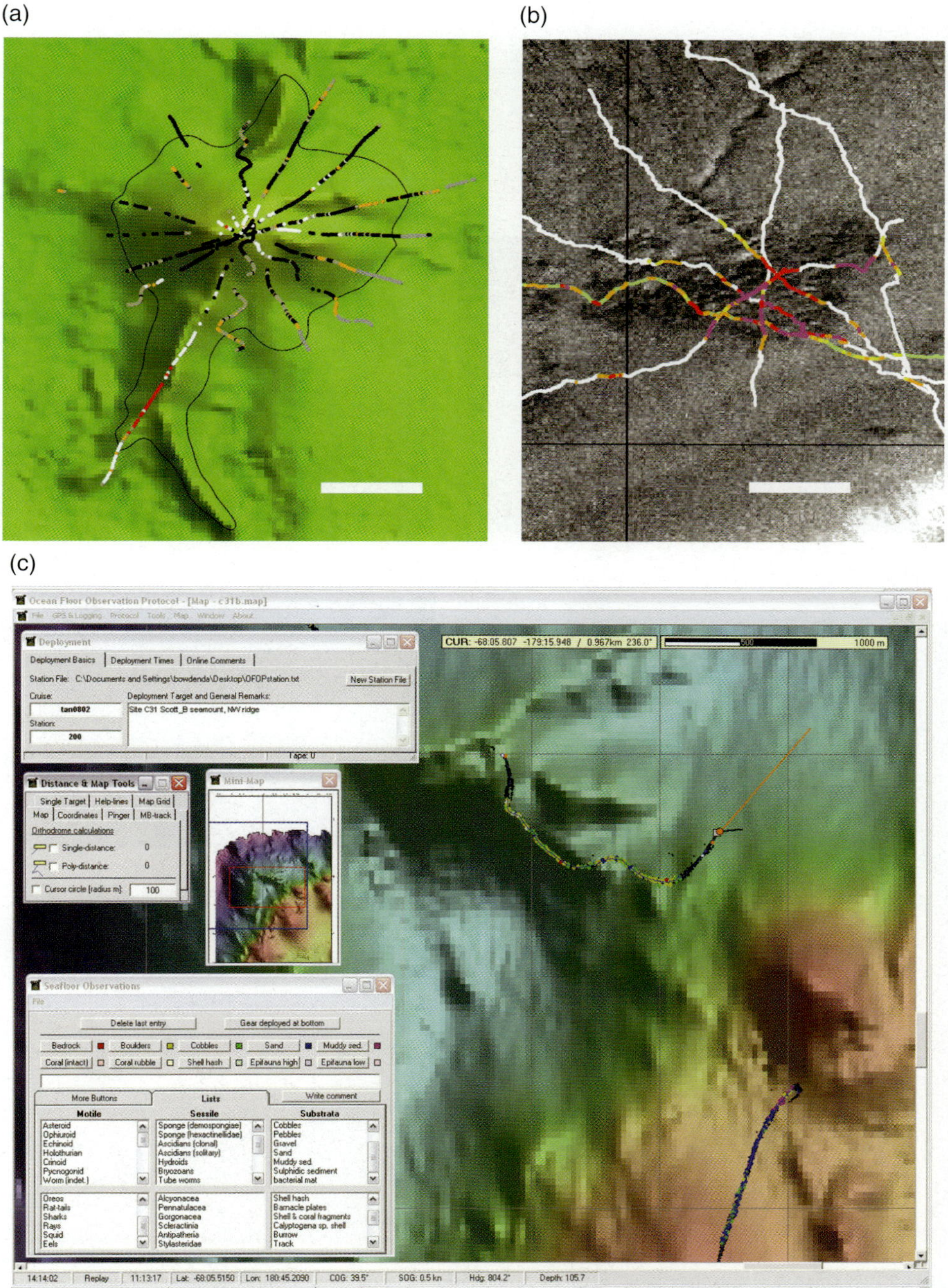

Plate 21 Towed camera survey strategy. (a) Multiple still image transects running downslope from summit to base of Morgue seamount, New Zealand, planned to map substrates and biota (Williams *et al.*, 2010). Points represent individual still photograph and are colour coded by substrate categories derived from analysis of images (scale bar, 500 m). (b) Intersecting video transects planned to locate and define the extent of North Tower cold seep, New Zealand (Baco *et al.*, 2010; Klaucke *et al.*, 2010). Tracks are displayed against high-resolution side-scan sonar imagery (the seep is the dark area at centre) and are colour-coded by substrate type (scale bar; 100 m). (c) Exploratory video transects on Scott B seamount, Antarctica (Bowden *et al.*, 2011). Screen shows tracking of the towed camera in real time (transect at upper right) against multibeam sonar bathymetry, while observations of fauna and substrata are recorded using Ocean Floor Observation Protocol software. Observations from an earlier transect are displayed simultaneously at lower right.

Plate 22 Examples of submersibles currently being used for biological sampling. (a) *Pisces V.* (Courtesy of NOAA-HURL Archives.) (b) *MIR 2.* (L. Murphy, Wikipedia Commons.) (c) *Alvin.* (Tom Kleindinst, Woods Hole Oceanographic Institution.) (d) *Nautile.* (IFREMER). (e) *Shinkai 6500.* (JAMSTEC. Reproduced with permission.) (f) *Jialong.* (Liu Feng, China National Deep Sea Center. Reproduced with permission.)

Plate 23 Examples of ROVs currently being used for biological sampling. (a) *Doc Ricketts*. (Todd Walsh © 2009 MBARI. Reproduced with permission.) (b) *Jason II.* (Photograph by Margo Edwards. Reproduced with permission.) (c) *Hercules.* (d) *Victor 6000.* (IFREMER.). (e) *ABISMO.* (JAMSTEC.) (f) *Deep Discoverer.* (Photograph by Art Howard. Reproduced with permission of NOAA Office of Ocean Exploration and Research.)

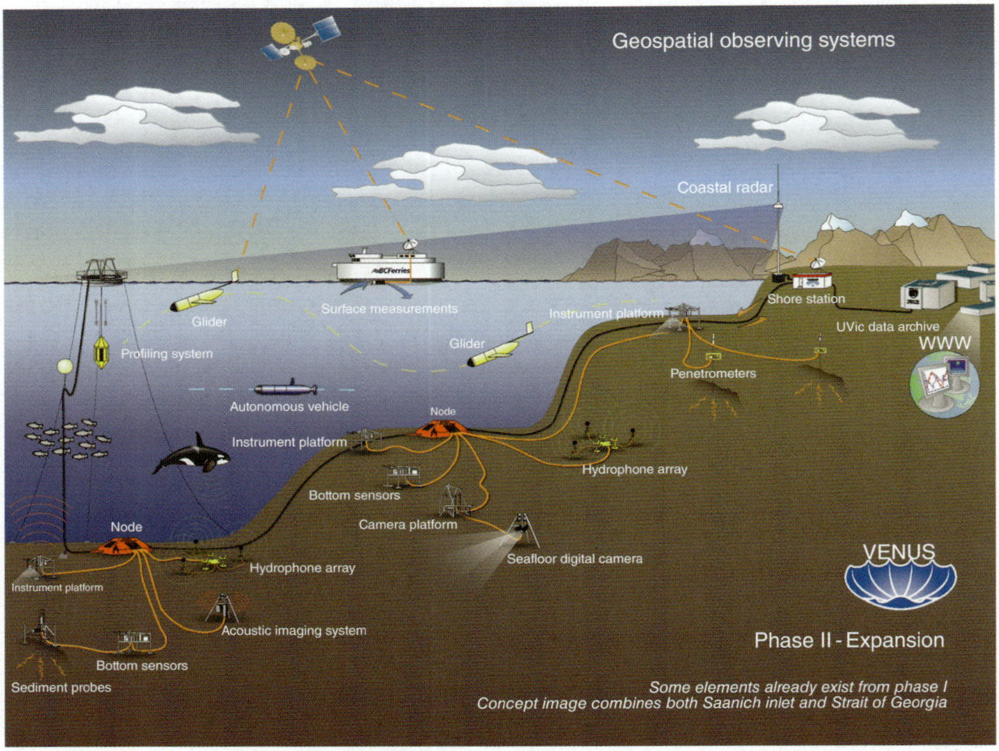

Plate 24 Sketch of a cabled observatory structure based on the example of VENUS (University of Victoria, BC, Canada). (Ocean Networks Canada. Reproduced with permission.)

(a)
(b)
(c)
TAN0802/109
ORH 109
19/02/08
(d)
(e)
056
TRAWL
(f)
(g)
(h)
(i)
(j)
(k)
ORH 106

Plate 25 Overview of the principal operations dedicated to sorting on deck and examples of catches from different gears. (a) Epibenthic sledge sample from a rocky bottom. The catch is naturally clear from fine sediment and the organisms are easily visible among the rocks. (b) Beam trawl catch with a high load of sediment. In this case the specimens sampled are not visible, being mainly within the mass of sediment. These need to be isolated, on a sorting table, after the fine sediment fraction has been washed away by spraying with seawater through a garden hose spray gun. (c) A very clean trawl catch. Here species have not been damaged by rocks and can be immediately processed for photography, tissue sampling, etc. (d) Large gorgonians (*Thouarella* spp.) up to 35 cm in length have been entangled at the mouth of the net. This kind of organism might not reach the cod-end of the net. In this case, the abundance of the octocorals makes their presence very obvious, however, when densities are lower they can be overlooked when entangled at the mouth of a long trawl net. (e) A large catch (about 150 kg) of mainly invertebrates. Large robust invertebrates (e.g. asteroids) are mixed with small, delicate invertebrates (e.g. crinoids) and small fishes. This is a more difficult situation to be handled, as each group of organisms has a different tolerance to air exposure and, due to the amount of material collected, the sorting can take a long time to be finished. Note the white board with the number of the station, which is a 10 × 10 cm square that can be used to give a rough indication of the size of the sampled organisms. (f) Another difficult case of a diverse catch in which there are species and associations which require different treatments. The large amount of wood requires special handling to isolate the associated fauna, which is a time-consuming process. (g) Sometimes a catch may be sorted directly on the deck. In this case, the deployed gear was a dredge with a mesh size of 2 cm: sieving is not required and sorting can take place with the naked eye. (h) During quantitative voyages (e.g. fish surveys), the catch has to be split into bins and weighed before species are isolated from the sediment. (i) In some cases, the large amount of collected biological material may require the involvement of most personnel. Sorting activities can be slow, meaning an accumulation of bins with unsorted biological material. Labelling bins is critical to avoid mixing catches. (j) Sorting of species in a sheltered lab. Note the number of transparent jars (in the white tray on the left corner of the sorting table) filled with seawater to keep specimens alive before photography and fixation. Running seawater is provided through the green tubes equipped with garden hose spray guns. (k) The sorting table after the subdivision of the catch into OTUs. Specimens are now ready to be placed in jars together with a label and fixed.

(a)
(b)
(c)
(d)
(e)
(f)
Corallimorphus profundus Moseley, 1877
CORALLIMORPHARIA
TAN0802 Stn 053
Frame 0194
(438-445 m)
TAN0802 Stn 031 Lot 411
NIWA 35728
Frame a
1 cm
TAN0802 Stn 031
ID 411
Frame e
TAN0802 Stn 031
ID 411
Frame c
Credits
Invertebrates pictures: Stefano Schiaparelli IPY-CAML (MNA Genoa, Italy), © LINZ
DITS images: NIWA - Dave Bowden, © LINZ
Classification and plate compilation: Stefano Schiaparelli IPY-CAML (MNA Genoa, Italy)
(g)
(h)
(i)

Plate 26 Examples of different photographic set-ups and of images taken on board. (a) Photographic set-up for large species in air. An example of a picture taken with this solution is given in (h). (b) To photograph small species in seawater, placed in containers with a black background, it is possible to use two small orientable flashes triggered by a wireless infrared commander. Examples of pictures taken by using this solution are in (d), (e) and (i). (c) Photography of fishes requires specific preparation (see Box 15.7 for details). (d) A small coralliomorpharian species still adhering to the pebble on which it has been sampled. It has been photographed in water, in a tray lined with adhesive black background. Note the extended tentacles and its almost 'live appearance'. (e) An asteroid (*Porania* sp.) photographed with the same technique. The scale (1 cm) is a piece of Resistall paper kept in place by inverted forceps. (f) An example of a 'working field guide' page. In the lower part of the plate, there are images taken in the lab oratory as shown in (d). (g) A lobster (*Puerulus angulatus*) of about 15 cm body length photographed in air using a black velvet background. Note that the image is a direct result from the camera without any digital editing. (h) A large fish photographed in air. Also included are the photographic colour scale (which will allow a perfect colour balance in post processing) and the specimen label. (i) An example of a small Antarctic mollusc (*Neactaeonina*) photographed in water. In this picture the scale bar, originally captured as in picture (e), was replaced in post production by a digitally added one.

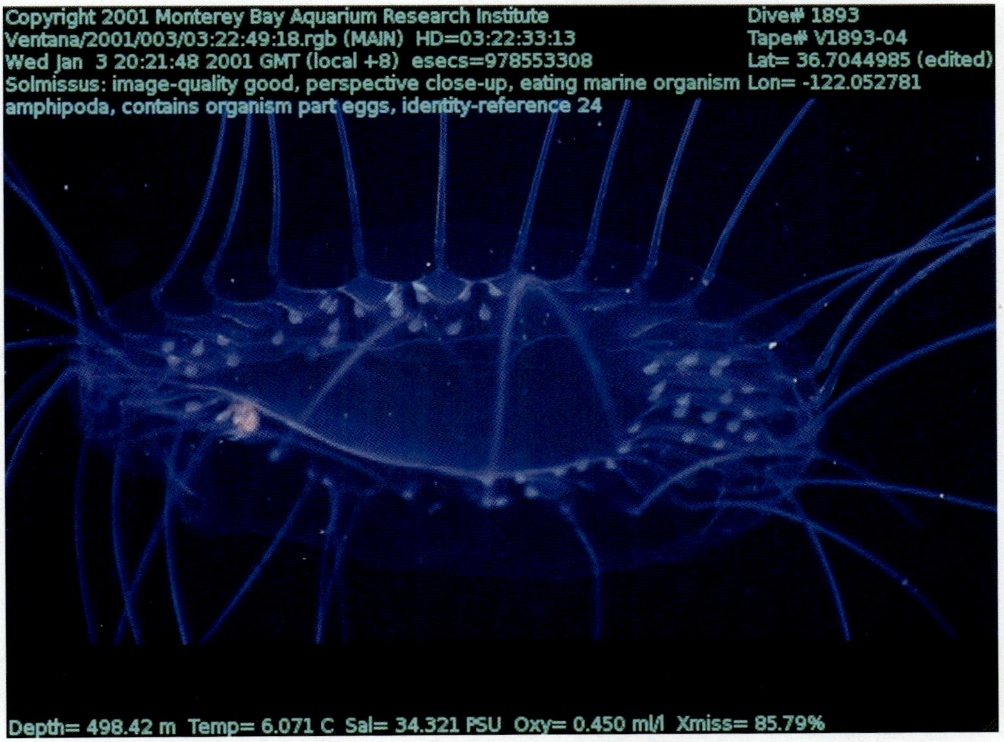

Plate 27 An example of a screen grab from a video overlaid with metadata, including the date, latitude, longitude and environmental conditions. Embedding these metadata within the image ensures they will not be lost as the images are transferred. (© MBARI. Reproduced with permission.)

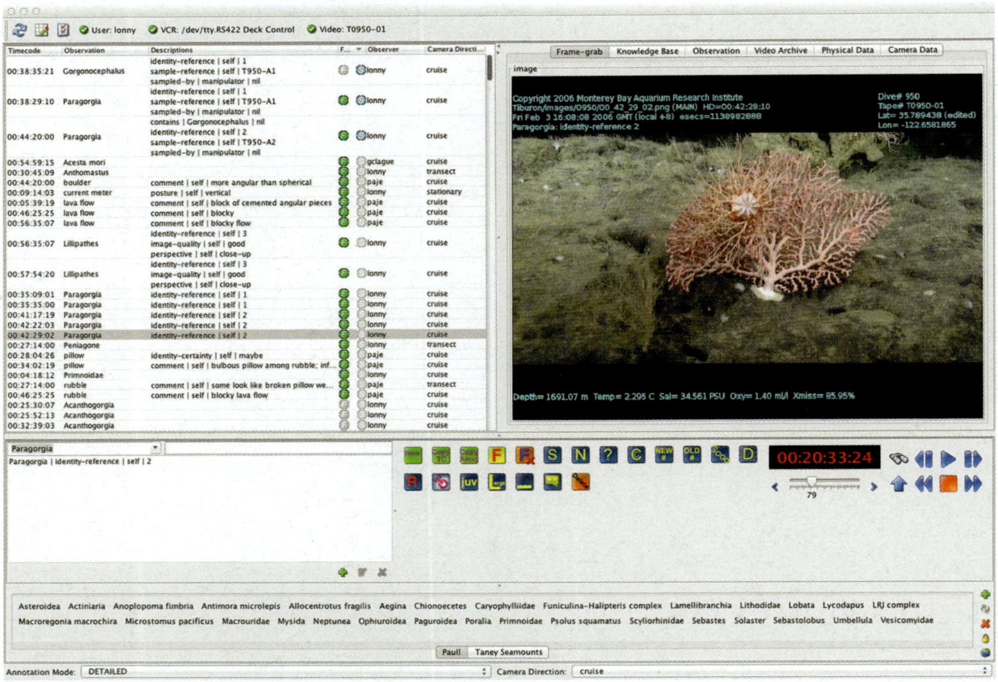

Plate 28 A screen from the Video Annotation Reference System (VARS), showing how an image can be annotated with terms describing the image's content, in this case the taxonomic name of the organism (*Paragorgia*) and its assigned identity reference number. Annotating with controlled terms allows large volumes of images to be efficiently searched for content of interest. (© MBARI. Reproduced with permission.)

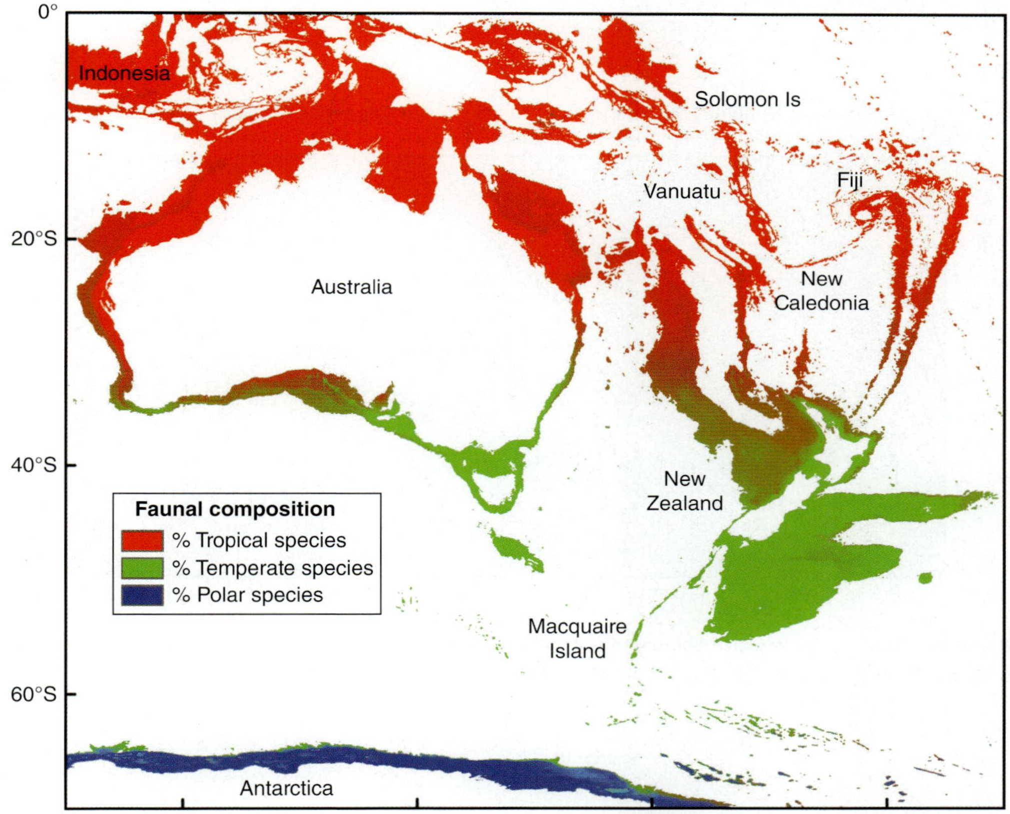

Plate 29 Map showing the overlapping distribution of tropical, temperate and polar brittle stars from the equator to Antarctica. Both shallow water (less than 250 m deep) and deeper (250–2000 m) faunas exhibit a strong latitudinal pattern. (Adapted from O'Hara *et al.*, 2011. Reproduced with permission of Elsevier.)

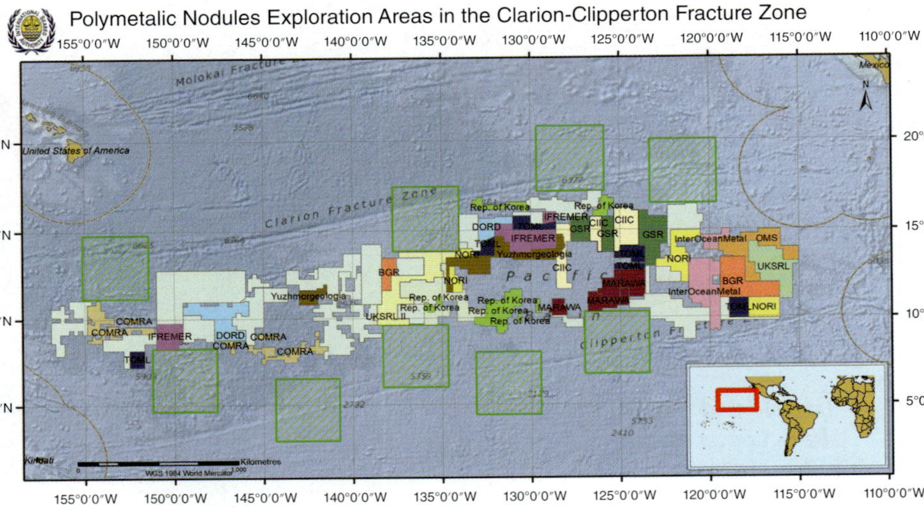

Plate 30 Areas of Potential Environmental Interest in the manganese nodule province of the Clarion–Clipperton Fracture Zone. (Courtesy of the International Seabed Authority.)

(a)

(b)

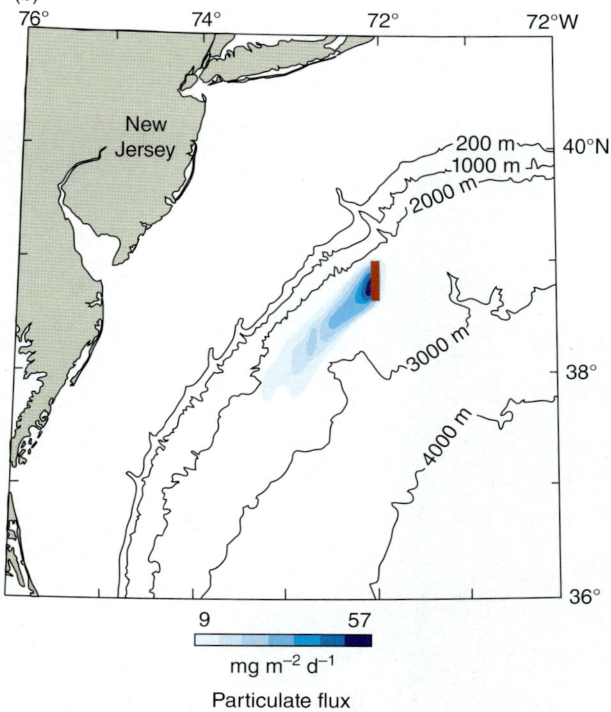

Plates 31 and 32 Sewage sludge from metropolitan New York and New Jersey in the late 1980s was barged to *Deepwater Dumpsite 106* and dumped into the water column (Plate 31), where sewage-derived organic carbon subsequently dispersed (Plate 32). Environmental impact assessments were initiated only after dumping was sanctioned and had commenced. (Plate 31 Courtesy of the US Environmental Protection Agency. Plate 32 Adapted from Fry & Butman, 1991.)

(a) (b)

Plate 33 Impact of trawling activity by deep-water fisheries. (a) Reef-like formations of stony coral on a small, unfished seamount off New Zealand (depth 950 m). (b) Coral rubble and gouges in the seafloor from trawl doors on a heavily fished seamount off New Zealand (depth 900 m). The speed at which environmental impact can occur from human activities emphasizes the need to integrate science and management from the outset of the activity. (Photographs courtesy of NIWA.)

Plate 34 Artist's impression of a future coordinated and integrated deep-sea sampling programme. (Erika Mackay, NIWA. Reproduced with permission.)

No general advice on the required number of replicates (per treatment, e.g. depth, location) can be given here, since this depends on the type and objectives of the study (see Chapter 3).

6.3 Examples of zooplankton samplers used in deep-sea studies

Many different types of zooplankton samplers have been used in deep-sea studies, and a comprehensive description is beyond the scope of this chapter. Here only short descriptions of some instruments that have been employed successfully in the deep sea are given.

6.3.1 Simple closing nets

In general, it is possible to sample deep-sea zooplankton with simple closing nets, but these are rarely used in deep-sea studies because they yield only a single sample per haul and are thus inefficient in utilizing ship time. However, they are relatively cheap, and they are a useful option for specific purposes, such as gently collecting animals to be used in morphological, histological or physiological studies.

6.3.1.1 WP2

One of the most commonly used simple closing nets is the WP2, developed as a standard plankton net for small mesozooplankton by an expert working group ("Working Party 2 - WP2") set up jointly by the International Council for the Exploration of the Sea (ICES), Scientific Committee on Ocean Research (SCOR) and United Nations Educational, Scientific and Cultural Organization (UNESCO) (Fraser, 1968). The net has a circular opening with a diameter of 56 cm, which is equivalent to an area of 0.25 m^2. The net itself is 261 cm long with a cylindrical and a conical section, made originally with 200 μm mesh. The filtration efficiency is given as 94%. The net can be equipped with a simple, messenger-operated Nansen closing system which uses a throttle line. The WP2 is a vertical net that is lowered open (cod-end first) to the desired maximum depth, then hauled to the minimum depth where the closing mechanism is triggered.

Messenger-operated closing devices are very simple and cheap, but they are inefficient at greater water depths because the messenger takes some time to trigger the release mechanism. For example, a messenger travelling down a wire to 1000 m takes about 20 minutes. During this time the net remains open with the danger of flushing, particularly if the ship's movements result in vertical accelerations of the net. So, although the WP2 is regarded as a quantitative plankton net, in the deep sea it is suitable only for qualitative sampling, particularly for retrieving undamaged organisms. A pressure trigger or an electrical/acoustical release mechanism could avoid the problems, but if quantitative samples are needed, it is usually more efficient to take advantage of modern multiple net samplers.

6.3.2 Multiple opening and closing net systems

Multiple opening and closing net systems are regarded as the standard tools for collecting deep-sea zooplankton. By yielding several samples per haul they make efficient use of ship time and can provide quasi-synoptic data series (e.g. several depth strata). Basically, these net systems consist of several (up to 20) single nets, which can be opened and closed sequentially on command from the ship, or automatically at pre-programmed depth intervals. The command can be transmitted acoustically or electrically. Acoustic telemetry has the advantage that no expensive conducting cable is required – a simple towing wire is sufficient. However, the bandwidth of acoustic telemetry is limited and at great depths the transmission may be problematic, in particular with towed systems, where the distance between the ship and the instrument may be up to 10 km. In these cases, to achieve maximum signal strength, the transducers on the ship and the gear must point toward each other. The quality of the signal depends not only on distance, but also on several other factors, such as noise from the ship's propeller, interference from other acoustic sources (e.g. echosounders), attenuation and reflection at discontinuity layers.

Modern ocean-going research vessels are usually equipped with conducting cables, and so most current systems make use of electrical telemetry. This allows for a more reliable data transmission between ship and instrument, making advanced electronic control systems and additional sensors with real-time transmission possible. Consequently, modern multiple opening and closing net systems provide real-time data on instrument depth, flowmeter rotations and net change, and are often equipped with additional sensors that yield estimates of bottom distance, temperature, conductivity, fluorescence and other parameters.

A wide variety of multiple opening and closing net systems have been developed in the last five decades; for an overview see Wiebe and Benfield (2003). Here two of the most widely used commercially available systems applied in deep-sea research will be presented in more detail.

6.3.2.1 MOCNESS

The MOCNESS (multiple opening and closing net and environmental sensing system, see Wiebe *et al.*, 1985) has a rectangular mouth opening with a Tucker-type net change system (Figs 6.1 and 6.2). It is suitable for horizontal and oblique tows. The bridles are attached to the top beam of the frame, keeping the net opening free from obstacles and thereby minimizing zooplankton avoidance.

Fig. 6.1 The 1 m²-double-MOCNESS. (Photograph by B. Christiansen.) (Please see Plate 10 for colour representation of the figure.)

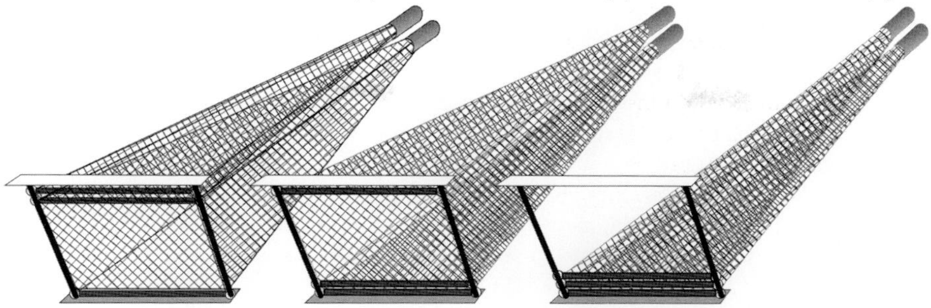

Fig. 6.2 Net change in MOCNESS-type multiple opening/closing nets. For simplification, only two nets are drawn. (a) First bar down, first net open; (b) first and second bar down, first net closed, second net open; (c) all bars down, second net closed.

A lead weight is fastened to the lower beam of the frame. This set-up results in a net opening which is not perpendicular to the water flow, but has a certain angle, ideally 45° at 2 knots, but may vary depending on actual towing speed and current conditions. The nets are tailored to match the oblique frame. Since the effective inflow area and thus the volume filtered vary with the angle of the net opening, this is measured by an inclinometer which is integrated in the underwater electronics unit. A mechanical flowmeter measuring the water flow is mounted on the upper frame outside the net opening, and hence is not affected by reduced flows if the net starts to fill, but this usually is not relevant, at least in deep-sea applications. A reed switch, which is actuated by the falling net bars during the closing and opening of the nets, signals net changes.

The MOCNESS is routinely equipped with conductivity and temperature probes and a pressure sensor, but the integration of further sensors into the system is possible. The system uses a LAN connection via conducting cable for transferring data and commands between the underwater unit on the instrument and a shipboard PC. The data are processed by PC software, and key information (e.g. instrument depth, horizontal and vertical velocity, volume filtered) is displayed in real time.

The MOCNESS comes in several sizes from 0.25 to 20 m^2, offering high flexibility concerning target size groups and required sample volume. All MOCNESS sizes use the same release system and electronics. They carry 5–20 nets; standard mesh sizes are 64 µm for the 0.25 m^2 system, 335 µm for the 1 m^2 version, and 3 mm for the larger systems. However, other mesh sizes are possible. For example, in deep-sea studies by the IHF the 1 m^2 MOCNESS was used with both 100 µm and 335 µm nets and the 10 m^2 MOCNESS with 1.5 mm nets, and Wiebe *et al.* (2010) successfully sampled deep-sea zooplankton with a 10 m^2 MOCNESS equipped with 335 µm nets.

6.3.2.2 MultiNet®

The MultiNet® (Fig. 6.3) is based on a modified Bé multiple plankton sampler (Weikert & John, 1981). In contrast to the MOCNESS, it can be used for both vertical and horizontal/oblique tows. The MultiNet® consists of a steel frame, with 5–9 nets attached. The square net openings are unfolded (opened) and folded (closed) successively by means of levers, which are activated by a battery-powered motor (Fig. 6.4). Net change is triggered from a deck command unit via conducting cable, but a programmable, self-contained release unit is also available. Data – in the standard version pressure and battery voltage – are transmitted via the conducting cable and displayed on the deck unit. Optional additional sensors include flowmeters, usually one mounted inside the net opening and one outside, and temperature and conductivity probes.

Fig. 6.3 The 0.25 m^2 MultiNet® in vertical configuration. (Photograph by B. Christiansen.) (Please see Plate 11 for colour representation of the figure.)

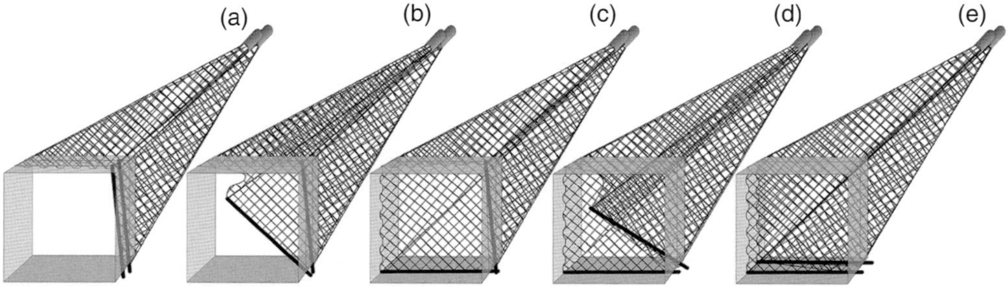

Fig. 6.4 Net change in MultiNet®. For simplification, only two nets are drawn. (a) Both levers vertical, both nets closed; (b) first lever turns, first net unfolds; (c) first lever horizontal, first net open; (d) second lever turns, first net folds, second net unfolds; (e) second lever horizontal, first net closed, second net open.

For horizontal/oblique tows, the MultiNet® is equipped with a V-fin depressor attached to the bottom of the net frame, and a two-legged bridle fastened on top of the frame, keeping the net opening free (apart from the optional flowmeter). For vertical hauls, a four-legged bridle is used, and the cod-end buckets are attached to a weighted frame which is supported by four auxiliary lines from the net frame (Fig. 6.5).

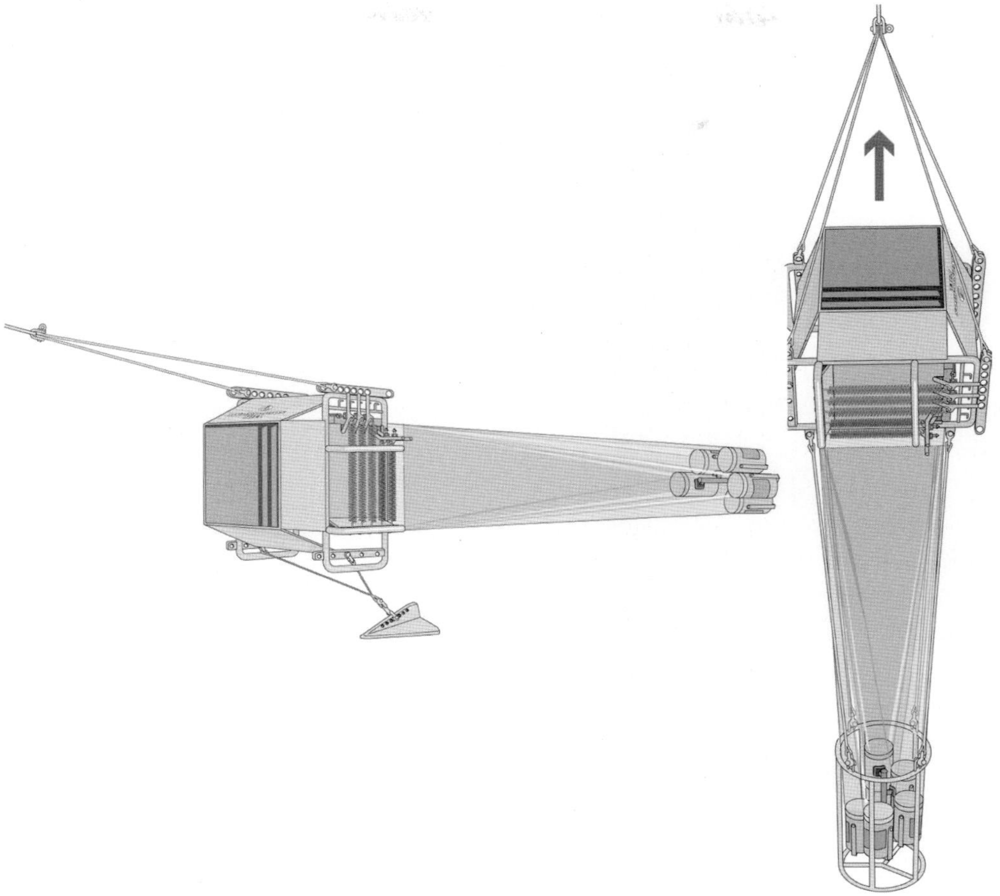

Fig. 6.5 MultiNet® in horizontal and vertical configurations. (Graphics by HydroBios Apparatebau. Reproduced with permission.)

The MultiNet® comes in three sizes: 0.125 m^2 (35.5 × 35.5 cm net opening), 0.25 m^2 (50 × 50 cm) and 0.5 m^2 (71 × 71 cm). The two smaller versions carry five nets each, the large one nine nets. Different mesh sizes are available for all types.

6.3.3 Large midwater trawls with multiple opening/closing cod-ends

The 10 and 20 m^2 MOCNESS are among the largest midwater trawls with rigid frames. These sizes require a large deck space (Fig. 6.6) and their handling on board is very difficult and can be dangerous in rough weather. For sampling even larger volumes at discrete depths, large midwater trawls have been equipped with multiple opening/closing cod-ends (e.g. Krygier & Pearcy, 1981; Wenneck *et al.*, 2008). Such cod-ends have also been used to convert standard, smaller plankton nets into multiple samplers (Wiebe & Benfield, 2003), but they have the major disadvantage that organisms are often retained in the nets for some time before being flushed into the cod-ends, which results in a 'blurring' of distribution patterns. Hence their use has been widely abandoned in favour of multiple opening and closing nets. However, for collecting large and very rare deep-sea zooplankton and micronekton in defined depth strata, multiple cod-ends in combination with large trawls can be the only option to achieve a sufficient sample volume.

Fig. 6.6 The 10 m^2 MOCNESS on the working deck of the RV *Meteor*. (Photograph by B. Christiansen.)

6.3.4 Alternative zooplankton samplers

Since the 1960s there has been a rapid development of instruments which do not physically collect organisms but assess zooplankton (and other particles) using their acoustical, optical or electrical properties (Wiebe & Benfield, 2003). For a detailed description of the principles and techniques of acoustical and optical methods see Foote and Stanton (2000) and Foote (2000). Many of these systems are not suitable for assessing deep-sea zooplankton, especially on a quantitative level, as their sample volumes are too small. However, some systems provide valuable insight into the deep-water pelagic community, particularly with respect to groups such as gelatinous zooplankton which are usually undersampled with conventional gear. Direct observations with submersibles and video systems have shown, for example, that gelatinous animals constitute a major component of the zooplankton community throughout the water column (e.g. Youngbluth *et al.*, 2008). Three examples used for direct assessments of zooplankton are presented here.

6.3.4.1 Underwater video profiler

The underwater video profiler (UVP) (Gorsky *et al.*, 2000) is one example of an optical zooplankton recorder successfully employed in deep water. Stemmann *et al.* (2008) used the UVP to assess macrozooplankton along the Mid-Atlantic Ridge down to a depth of 1000 m. The instrument proved to be particularly useful for the quantification of fragile, gelatinous plankton, which are poorly assessed using conventional methods. The UVP consists of a frame which is lowered vertically. It is equipped with two stroboscopes and two cameras positioned perpendicular to the light beams, one with a narrow angle and one with a wide-angle lens. The viewing fields of the cameras correspond to an illuminated volume of 1.25 L and 10.5 L, respectively. The sample frequency is 12 Hz, resulting in about 12 000 images for each camera per 1000 m cast and a total sample volume of about 120 m^3 for the wide-angle camera (Stemmann *et al.*, 2008). Images are screened automatically for larger objects, which are then identified visually.

6.3.4.2 *Submersibles: direct observations*

Direct visual observations from submersibles have proven to be a valuable tool for assessing larger, gelatinous zooplankton and their behaviour. Although a visual quantification of pelagic organisms in the three-dimensional space is difficult, systems have been developed which allow for enumeration of specimens. Vinogradov (2005), for example, used the Russian submersibles Mir-1 and Mir-2 to study the vertical distribution of large zooplankton (>1–2 cm) at the Charlie–Gibbs Fracture Zone from 170 m to 4361 m. Organisms were counted when they passed a 3 m^2 frame during the descent of the submersible, thus allowing for quantitative estimates of zooplankton abundance in different depth layers.

6.3.4.3 *Remotely operated vehicles: video surveys*

Similar to submersibles, remotely operated vehicles (ROVs) can be used as platforms for the observation and enumeration of zooplankton. One example of applying video transects to analyse the composition and distribution of zooplankton in deep water is given in Robison *et al.* (2010). Using the Monterey Bay Aquarium Research Institute (MBARI)-based ROV *Tiburon*, they made oblique transects between 1000 and 3500 m depth. In their set-up, a video camera was mounted on the ROV, pointing in the travel direction. The camera lens was adjusted to give a depth of field from 1 to 4 m in front of the ROV; illumination was provided by four lights. The minimum target size that could be identified reliably with this system was about 3 mm; using a high-definition video camera, this could be reduced to about 2 mm. The field of view at a focal distance of 1.5 m was 2.43 m^2; this area was used in combination with data from an acoustic current meter to calculate the sample volume of a transect. The video images were transmitted via the tether of the ROV to the research vessel and recorded there. The video images were analysed ashore by identifying and counting individual specimens as they passed through the camera's field of view.

The mean average speed through the water during the transects was 0.42 m s^{-1}, resulting in a total mean volume observed per transect of 9000 m^3. This is close to volumes that are achieved with towed nets.

6.4 Sampling operations

The following sections will provide practical information on the operation of zooplankton samplers in the deep sea, with a focus on net samplers. The basic operation procedures of other tools that are used as platforms for assessing deep-sea zooplankton, such as submersibles, ROVs, autonomous underwater vehicles (AUVs) are covered in chapters later in the book.

6.4.1 Ship requirements

One of the most important ship requirements for deep-sea zooplankton sampling is appropriate winches with sufficient wire and/or cable lengths. The winch controls should allow for a precisely adjustable wire speed in the range of 0.1–1 m s^{-1}. The exact length of wire should be indicated and an indication of wire tension is also useful. A simple wire is sufficient for simple nets and for more advanced net systems with acoustic control or self-contained releases. Most sophisticated sampling systems for deep-sea zooplankton will require a conducting cable for the transmission of commands and data, and possibly electrical power, or even a fibre optic cable if a high data rate and bandwidth are necessary (e.g. for real-time video transmission).

The cable length needed depends on the fishing depth, but also on the mode of operation, type of sampler, towing speed and currents. For vertical tows a length of little more than maximum instrument depth is sufficient, but for towed gear the scope (ratio of cable length to depth) ranges from about 1.5 to nearly 2.0, so cable lengths of up to twice the fishing depth are necessary. This means, for sampling the typical abyssal plain the winch has to carry more than 8000 m of cable. The required cable diameter and thus the breaking strength depend largely on the size and weight of the gear. Wiebe *et al.* (1985) made recommendations for cable diameters to be used with the various MOCNESS types, ranging from 7.4 mm for the 1 m^2 single version to 17.3 mm for the 20 m^2 MOCNESS. At the IHF either an 11 mm coax cable (breaking strength 89 kN) or an 18 mm coax cable (breaking strength 205 kN) is used for a 1 m^2 double MOCNESS, and the same 18 mm cable for the 10 m^2 MOCNESS. In general, a larger diameter is preferable because of a greater safety margin and a smaller depth:warp scope. However, the deployment of a lightweight plankton net with a heavy towing wire may be problematic because there is considerable slack in the wire when only a few hundred metres have been paid out, requiring very slow winch speeds until wire tension is sufficient.

Large nets with rigid frames require an A-frame with sufficient height and width. For example, the frame of the 10 m^2 MOCNESS is 3.4 m wide and 4.7 m high. Considering the length of the bridle and the dimensions of the connecting shackle and wire fitting, a clearance height of at least 7 m below the block is necessary, which is about the maximum currently available on large research vessels. For the 20 m^2 MOCNESS, a special launching and recovery system needs to be fitted to the ship.

Good manoeuvrability of the ship, particularly at low speeds, is necessary for towed samplers, many of which require a towing speed of no more than 2 knots through water. Manoeuvrability is also an important criterion for vertical gears in order to maintain position and to keep the wire vertical. Dynamic positioning (DP) is an extremely useful feature of modern research ships, allowing for automatic maintenance of the exact position or moving at slow speeds in any direction. For example, the German research vessel *Maria S. Merian* with pod propulsion, a 360° bow pump jet and DP can keep the exact position with an accuracy of a few metres even under adverse conditions. DP is also very important for the safe operation of ROVs (see Chapter 13). Constancy and easy adjustment of the ship speed is a further point which is particularly important if sampling close to the bottom with towed gear, since any variations in ship speed immediately result in changes of the fishing depth.

6.4.2 Weather and sea conditions

Weather and sea conditions are very important for sampling, even in the deep sea where there is also a substantial response of gear to vessel motion. There is no fixed rule as to the maximum wind force or wave height at which sampling is possible. This depends on several factors such as the size and type of ship, type and operation mode of gear, and direction of swell with respect to wind direction.

Usually the ship is positioned with the head to the sea, and vertical nets are preferably hauled over the side close to midships in order to minimize vertical accelerations due to the ship's pitch. If wind and sea come from different directions, rolling may be a problem, for the catch and for the instrument. In unfavourable conditions, vertical speeds of more than ±1 m s^{-1} can superimpose the wire speed (Maaß, 2007), so during hauling, in particular at low speeds, part of the catch may be flushed when gear velocity through the water becomes negative. If the instrument is too light, the wire/cable may come loose and eventually kink or even jump off the leaders and break.

The effects of sea and swell are similar for towed instruments. Although towed nets may be operated over the side as well, the larger nets are often towed over the stern, where the vertical movements can be much larger than midships, leading to considerable vertical accelerations of wire and gear, and variations in wire tension. Steaming against high swell also affects the constancy of the ship's speed and thus the fishing depth, making operations close to the seabed difficult.

6.4.3 *Gear operation*

The specific gear operation depends on the type of sampler. Here, I describe some of the basic operational procedures of the MultiNet® and the MOCNESS. These can be applied to most samplers, either vertical or towed.

6.4.3.1 Preparation of sampling gear

A thorough pre-deployment check of the gear is mandatory, and is even more important in deep-sea studies where one single haul may take several hours and often cannot be repeated if the instrument fails. The following checks should be made before each haul:

- Examine the net mesh for damage, especially in the case of delicate, fine mesh, and of the mesh on the drainage windows of the cod-end buckets.
- Ensure that all bolts and nuts are tightened.
- Ensure that cod-end buckets are secured. With towed systems, there is considerable drag on the cod-ends. In many nets rigid cod-end buckets are attached to a ring, which is fastened to the end of the net with a hose clamp. This hose clamp has to be fastened very tightly and checked regularly. In the MOCNESS, the bucket is attached to the ring with three clamps which open easily. They can be secured with tie wraps, but a good alternative is the use of rubber rings (5–10 cm wide) which fit tightly over the bucket and clamps and additionally prevent them from damaging the neighbouring nets.
- Inspect all electrical connections, and clean and lubricate them, if necessary.
- Power up the electronics system and check whether all components are working, particularly the release trigger, flowmeter and net response.
- Check the battery voltage and ensure that battery capacity is sufficient for the total tow duration (possibly several hours).
- Log a possible offset of the depth gauge.

6.4.3.2 Vertical tows

The operation of vertical nets is straightforward. It is important for the ship to maintain position, especially in areas with uneven topography, for example on the slopes of seamounts or in canyons, where water depth may change rapidly in a short distance. We once collected a rock with the MultiNet® when the ship drifted toward a rather steep slope (surprisingly without any damage to the gear!). Sometimes it is difficult to keep the wire vertical in strong currents, and it may be necessary to move the ship slightly with the current.

Instrument depth is usually measured with a pressure gauge, but can also be estimated without a sensor, if the wire length can be measured accurately, and the wire angle is close to vertical. Small differences from the vertical (<10°) have only a small effect on the ratio of wire length to sampling depth. In vertical tows, the sample volume can be calculated directly from the area of the net opening and the depth interval sampled, if the filtration efficiency of the sampler is known. In order to account for an unknown, or varying, filtration efficiency, it is preferable to use a calibrated flowmeter which is mounted in the net opening.

Cod-ends in vertical nets should be weighted to prevent them from surging up during lowering and entangling the bridle and towing wire, and to keep enough tension on the wire in strong currents and rough seas. In order to minimize strain on the net material, the weighted cod-end has to be connected directly to the net frame with auxiliary lines, which are adjusted so that they take the full weight of the cod-end. In many net systems the weight is already integrated into the cod-end, or in the case of the MultiNet® the frame supporting the cod-end buckets (Fig. 6.5).

Simple closing nets are lowered cod-end first in open condition to the maximum depth, then hauled to the minimum depth where the net is closed. The MultiNet® is lowered in closed condition to the maximum depth, then the first net is opened, and the system is hauled to the surface at constant speed. Some users stop the winch at the desired depths for net change, but we do not recommend this procedure, since animals may be flushed out of the net during the stop, in particular if rolling of the ship results in vertical acceleration of the gear. Instead, the trigger should be released a few seconds before reaching the depth for net change, in order to account for the time lag of 3–4 seconds between firing and actual net change.

The wire speed depends on the type of net, and on mesh size and porosity of the net material. For the MultiNet® with standard 300 μm nets, a wire speed of up to 1 m/s is recommended by the manufacturer. We usually employ a wire speed of 0.5 m/s, both for lowering and hauling, which in our experience is a good compromise between tow duration and stress on the animals collected. This will result in a total tow duration (both ways) of about 1 hour per 1000 m depth. For nets with smaller mesh sizes and lower filtration efficiency, it may be necessary to decrease wire speed further in order to reduce strain on the nets. For example, the IHF planktologists used to haul the MultiNet® equipped with 55 μm nets at 0.2 m/s, but had good results also with a speed of 0.5 m/s in (oligotrophic) oceanic waters.

6.4.3.3 Horizontal and oblique tows

Since the instrument is well behind the ship when towed systems are used, and the ship moves a considerable distance during the haul, additional care has to be taken with ship positioning.

During an average MOCNESS haul to abyssal depths the ship will travel a total distance of about 20 nautical miles, so it is important to ensure that there is enough free space ahead of the ship, because during the tow only small changes in course are possible. Shifts in wind direction must be considered, as it may require an adjustment of the ship's course. Although usually deep-sea studies are far away from the coast, political or legal boundaries may limit the operating area for the ship.

A further factor to consider is the bottom topography, ensuring that water depth is sufficient during the whole tow. The planned towing track should be checked on accurate charts. If these are not available, it is good practice to make at least one echosounder transect along the track before the haul. During the haul, the location of the gear should be checked against the bottom profile.

Modern ultrashort baseline (USBL) positioning systems such as POSIDONIA 6000 (Peyronnet *et al.*, 1998) facilitate exact monitoring of instrument location with respect to the ship's position, and by integrating Global Positioning System (GPS) data, calculation of the absolute position. These systems are widely used in ROVs, but are also applicable in towed instruments. The maximum range is about 8000 m, and the performance of the system depends particularly on the ambient noise level and on the angle between instrument and ship, which may constrain its use in towed nets at great depths.

In most cases the instrument position is estimated from wire length and depth. Assuming the wire is a straight line, the distance of the gear behind the ship (s) can easily be calculated after Pythagoras:

$$s = \sqrt{\left(l^2 - d^2\right)}$$

where l is length of wire and d is instrument depth.

In reality the wire is not straight – it forms a curve – but this has very little effect on the length and can be ignored. Often the gear is not exactly behind the ship, but is towed at an angle to the side, which is very difficult to estimate.

In many instances it may be necessary to start fishing at a defined location, for example, because a certain topographic feature has to be sampled (or avoided). In order to 'hit' this location, the gear has to be deployed some distance before the starting position. The deployment position for a given starting point and depth of a fishing track depends on towing direction (usually into the wind), ship speed, wire payout speed, fishing depth and scope (ratio of wire length to instrument depth), and can be determined using the following procedure:

(1) Mark the intended fishing track on the chart (considering towing direction) and extend it in both directions.
(2) Calculate the distance of gear behind the ship at the start of the fishing track (considering assumed wire length and depth) and mark the corresponding ship's position on the track.
(3) Calculate the time required for lowering the device to the fishing depth (considering wire length and wire speed) and multiply by ship's speed.
(4) Go back this distance on the trackline from the marked ship's position and mark the deployment position.

For example (see Fig. 6.7): the targeted fishing depth is $d = 4000$ m. Assuming a scope of 1.75, the predicted wire length is $l = 7000$ m, resulting in a distance of $s = 5745$ m between the ship and the gear at the start of the fishing track. Paying out 7000 m of wire at a speed of 0.5 m s^{-1} takes 233 minutes. During this time, the ship travels a distance of 14 420 m at 2 knots. So the gear has to be deployed 8675 m (i.e. 14 420 – 5745) or 4.7 nautical miles before the position where it reaches the targeted depth and starts fishing.

Instrument depth is usually measured with a pressure sensor. Although it is possible to calculate the depth from wire length and wire angle, this is very inaccurate. The scope depends on several factors, including towing speed, but also on the type of net, currents, and the diameter and weight of the towing cable, and may vary between 1.5 and 2.

Assuming a known filtration efficiency, the volume filtered can provisionally be estimated from the area of the net opening and the sampling distance:

$$V = A \times d \times F$$

where V is volume filtered, A is the area of the mouth opening, d is the sampling distance and F is filtration efficiency. The sampling distance d in oblique tows can be calculated from the distance

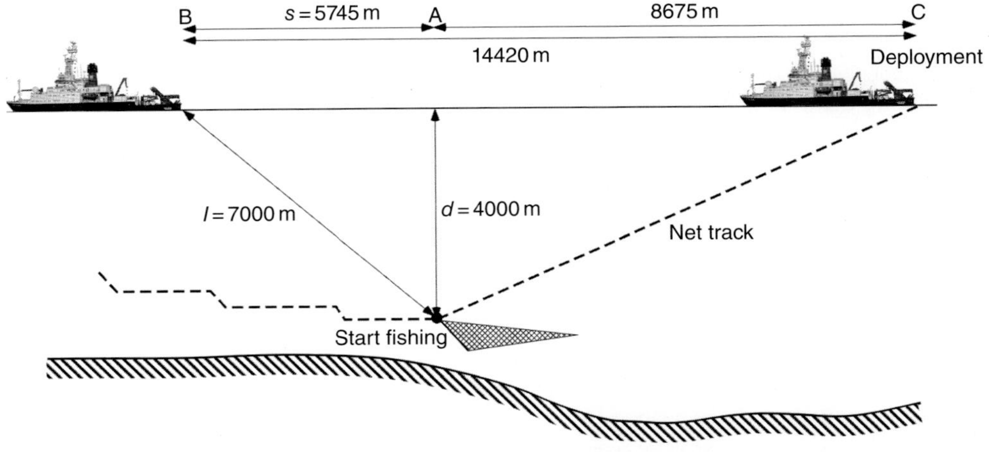

Fig. 6.7 Fishing position and deployment position of the vessel and gear in oblique/horizontal tows.

travelled by the ship, the length of wire hauled and the vertical sampling interval. However, if possible, the much more accurate measurement of the actual water flow through the net using a calibrated flowmeter is preferable.

Except for high-speed samplers, the towing speed in horizontal and oblique plankton hauls is usually about 2 knots (±0.5 knots) through the water. Since current velocity and direction can vary with depth, it may be necessary to control the actual instrument speed through water with a flowmeter. Wire speed during lowering is 0.5–0.8 m s^{-1}, allowing for sufficient water flow through the net for stabilization. Wire speed during hauling in oblique tows depends on the desired sampling volume per depth stratum and may range from 0.1 to 0.5 m s^{-1}.

During deployment and recovery of the net, water turbulence should be avoided as much as possible. This is not a problem if the gear is deployed over the side of the ship, but when launching over the stern, the propeller action can cause a destabilization and turning of the gear, and strain on the nets. Thus, the main propeller should be switched off and the speed necessary for stabilizing the sampler maintained by the bow thruster, until the net reaches a depth of about 20 m. With pod propulsion it is possible to turn the pods outward at a certain angle, leaving a rather quiescent zone behind the stern of the ship.

The MOCNESS is lowered with the first net open for stabilization. At depth, this net is closed, and the second and following nets are opened and closed successively according to the sampling design. The MultiNet® is lowered with all nets closed. It is stabilized through the hydrodynamic properties of the frame and by the depressor.

6.4.4 *Sampling zooplankton close to the seafloor*

The few observations of the near-bottom water layer in the deep sea have shown that the zooplankton community changes considerably close to the seafloor (e.g. Wishner, 1980a, 1980b), in particular within 10 m of the bottom (Christiansen *et al.*, 2010). Several methods and strategies have been applied to collect plankton close to the seafloor, including moored pumps (Mullineaux *et al.*, 2005) and nets (Dahms & Qian, 2004), slurp guns mounted on ROVs or submersibles (e.g. Smith Jr, 1982), traps, and (multiple) opening and closing nets, either as standalone systems (e.g. Christiansen *et al.*, 1999, 2010) or mounted on epibenthic sledges (e.g. Christiansen *et al.*, 2010), submersibles, deep-tow instruments and ROVs (e.g. Wishner, 1980b, Wishner & Meise-Munns, 1984). Many of these methods will not allow for an estimate of sample volume, or are limited to rather small sample volumes. For quantitative studies of the benthopelagic zooplankton community in the deep sea (with their characteristic very low abundances) usually only towed nets are feasible, together with direct assessments using submersibles and ROVs. Future developments could include camera systems mounted on terrain-following AUVs.

The exact position of the gear, knowledge of the bottom profile and additional monitoring of water depth during the tow, as described above, are crucial when sampling within a few metres of the seafloor. Depth variations along the fishing track should be gradual, as the near-bottom water layer above abrupt topography is almost impossible to sample with conventional (towed) gear.

In the following sections, the operation of pelagic nets towed close to the bottom and of zooplankton nets in combination with epibenthic sledges will be described, along with some alternatives. These may also be usable for sampling above rough terrain. Nets mounted on submersibles, ROVs and deep-tow instruments will not be covered here specifically, because the general operations of these instruments are treated in other chapters of this book.

6.4.4.1 Sampling with towed pelagic nets

Provided that distance to the seabed can be reliably controlled, any towed pelagic net can fish close to the bottom. On flat abyssal plains, without any depth variations, a calibrated pressure sensor in combination with the ship's echosounder may be sufficient to maintain a defined bottom distance (minimum safe bottom distance is about 20 m; Christiansen *et al.*, 1999) but usually it is preferable to measure directly the distance of the instrument to the bottom.

Pinger sounder systems use a sound source (pinger) on the instrument, which generates a sound signal at defined time intervals. This signal is received twice in a transducer on the ship, once as a direct signal, and again as a bottom-reflected signal. By analysing the runtime difference of both signals, the bottom distance is calculated and displayed graphically. This system works well in vertical gear and has been widely used in benthic research for controlling bottom contact in grabs and corers, but is less reliable in deep towed instruments with long distances and wide transmission angles.

Altimeters are small echosounder-like instruments which are mounted directly on the gear. They transmit a sound signal, which is reflected from the bottom and received by the same unit. Bottom distance is then calculated from the runtime of the signal and sound velocity in water. The data are transmitted either acoustically or via conducting cable to a deck display unit.

Since the ship speed usually varies to a certain degree, which affects the fishing depth, maintaining towed gear at a certain distance from the seafloor is difficult and needs continuous adjustments of wire length. Christiansen *et al.* (2010) used a Mesotech Simrad altimeter with a resolution of 0.1 m on a 1 m^2 double MOCNESS for sampling the near-bottom water layer on the Porcupine Abyssal Plain at a depth of 4850 m (Fig. 6.8). It was possible to maintain the targeted bottom distance of 15 m with a precision of ±5 m under moderate sea conditions, considering the variations in ship speed and the time lag between the detection of a changing distance, the adjustment of wire length and the actual response of the instrument. This means that the minimum safe bottom distance which can normally be achieved with a towed pelagic plankton net is about 10 m.

Instruments with hydrodynamic depressors are usually less sensitive to speed variations than those equipped with a dead weight and can probably be operated with a higher precision than described above.

Fig. 6.8 MOCNESS equipped with altimeter (arrow) for near-bottom sampling. (Photograph by Bernd Christiansen.)

6.4.4.2 *Sampling with bottom-touching gear*

An option to collect plankton very close to the seafloor is to mount a plankton net on a sledge-like frame which is towed along the bottom (Fig. 6.9, see also Chapter 9). This type of gear can be used to sample at a fixed distance from the seafloor, usually in the range of a few decimetres to about 1.5 m above bottom. Many epibenthic sledges make use of opening/closing devices which are mechanically activated through levers at bottom contact (e.g. Aldred *et al.*, 1976; Brandt & Barthel, 1995). Christiansen *et al.* (2010) equipped an epibenthic sledge with a MOCNESS-type multiple opening/closing net system which was remotely controlled from the ship (Fig. 6.10).

To avoid contamination of the catch with benthic organisms, the mouth opening of the sampler has to be kept well away from the mud cloud produced during bottom sampling by the sledge. One way of avoiding this problem is to place the opening of the so-called hyperbenthic or suprabenthic nets well ahead of the front edge of the skids, and during operation keep the bridle and the towing wire off the bottom. Monitoring the behaviour of a deep-sea epibenthic sledge, Christiansen and Nuppenau (1997) showed that the gear is extremely sensitive to the length of wire paid out. Too much wire results in sediment being stirred up by the bridle and towing wire; if too little wire is used, the gear immediately lifts off the bottom. Consequently, even small variations in ship speed, which are generally unavoidable, result in an immediate slackening and tightening of the towing wire, requiring a continuous adjustment of the wire length.

Hence, in order to obtain a clean, uncontaminated catch, the precise control of epibenthic sledge-based zooplankton samplers is extremely important, but very difficult to achieve. The optimum would be visual control with real-time video telemetry as described by Christiansen and Nuppenau (1997), monitoring not only bottom contact, but also the position of the bridle and the wire, and allowing for immediate adjustments. However, this set-up requires at least a high-quality coax towing cable, or a fibre optic cable, which are not often used for towed bottom-touching gear. The IHF zooplankton group also used a modified MOCNESS underwater unit with altimeter and inclinometer for approaching the bottom and monitoring bottom contact (Christiansen *et al.*, 2010). At lowering, the angle of the sledge was about 45°. When touching the bottom, the sledge tilted forward, and the wire length was adjusted so that the sledge was maintained at an angle of about 5°, which kept the bridle off the bottom. This system requires a conducting cable, but it is less demanding in terms of data rate than real-time video, and it should be possible in principle to modify it for acoustic telemetry. While Aldred *et al.* (1976) used

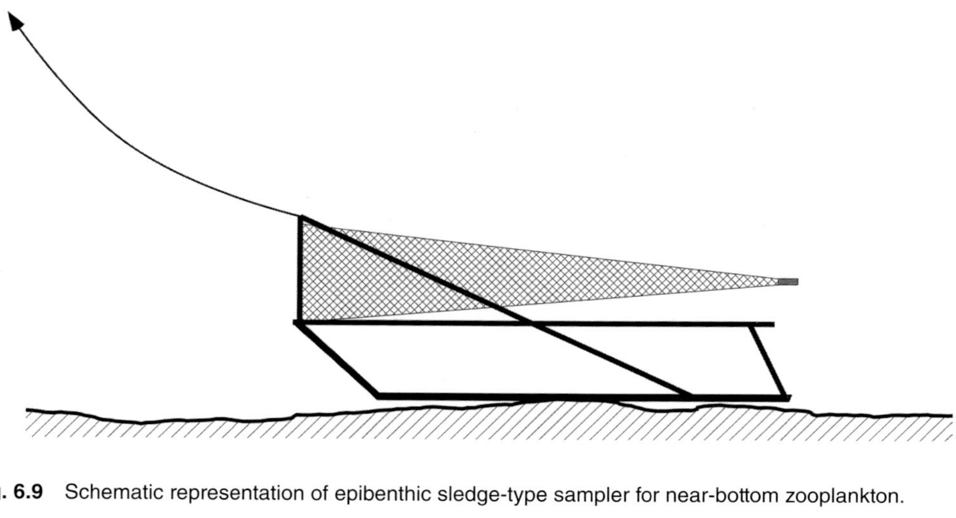

Fig. 6.9 Schematic representation of epibenthic sledge-type sampler for near-bottom zooplankton.

Fig. 6.10 Epibenthic sledge equipped with multiple opening and closing plankton net system (IHF-Fototrawl). (Photograph by Bernd Christiansen.)

a simple pinger system on an epibenthic sledge, which monitored only bottom contact and net opening, advanced deep-water wireless sensors (with acoustic transmission) are now available. These were originally developed for fish trawl surveillance during fishing, but are probably suited, with small modifications, to monitoring the behaviour of scientific gear like epibenthic sledges (Wenneck *et al.*, 2008).

6.4.4.3 Alternative near-bottom samplers

Moored pumps can be an option in areas which are not accessible to towed nets. As mentioned above, the capacity of *in situ* pumps is rather low, yielding a much smaller sample volume per time unit than nets. In order to obtain sufficient numbers of animals, it is necessary to pump for long time periods. However, this can affect the temporal and spatial resolution of the samples. Mullineaux *et al.* (2005) successfully used moored *in situ* pumps with a maximum capacity of 50 L min^{-1} to study larval distributions at hydrothermal vents. The pumps were arranged at 1, 20 and 175 m above the seafloor. Pumping for up to 24 hours per sample, the resulting sample volumes were 12 to >40 m^3, and up to 450 individuals per sample were obtained.

Acoustics are a widely used tool in fisheries ecology, and considerable effort has been made to apply this technology to plankton work. A comprehensive overview of the principles and techniques of acoustical zooplankton assessments, and examples for some instruments, can be found in Foote and Stanton (2000). Since the small sizes of the target groups in zooplankton research require high sound frequencies, and range is negatively correlated with frequency, hull-mounted acoustic transducers cannot be used in deep waters. The instruments have to be mounted either on submersibles, ROVs, AUVs or towed instruments, or can be moored on or above the seafloor. Sherman and Smith Jr. (2009) deployed a moored, upward-looking acoustic split-beam array (frequency 150 kHz) for assessing the benthic boundary layer community, with range bins of 10 cm, a maximum range of 200 m and the capability to resolve targets >1 cm. Since formal identification of acoustic targets usually is not possible, the acoustic array in this system is combined with a slave camera, which is positioned and activated when an acoustic target is detected.

One advantage of moored acoustic (and optical) systems is that they can operate autonomously for long time periods, and they can be integrated in cabled underwater networks allowing for the real-time transmission of data to a land base (e.g. Sherman and Smith Jr, 2009, and see Chapter 14).

6.4.5 Collecting live specimens

Most zooplankton taxa are very fragile. Collecting them live is difficult even in shallow water and has to be performed with great care (see Sameoto *et al.*, 2000, for some basics). Sampling specimens in deep water and bringing them live to the surface is a much greater challenge due to long tow durations, temperature changes and pressure differences. Several factors are important for keeping zooplankton alive during the haul:

- *Avoid physical damage*: Zooplankton can easily be damaged through forced contact with the sampler and the strands of the net, or through turbulence in the cod-end. Long tows increase this risk. The following measures are useful to minimize damage to the organisms: use soft net material with small mesh size, reduce filtration pressure by a slow towing speed and a sampler design with good filtration efficiency, use cod-end buckets without drainage or with closing systems, and reduce tow duration.
- *Avoid oxygen shortage*: This can occur if the cod-end volume is small relative to the plankton concentration, so a large cod-end volume and a short sampling time can reduce this effect. In deep-sea sampling this is usually not a major problem because abundances are low.
- *Keep temperature constant*: In most oceans, the temperature in deep water layers is considerably lower than at the surface, and many deep-sea animals are stenotherm, dying rapidly in warmer water. The effects of temperature differences between sampling depth and overlying waters can be reduced by using closed cod-ends with a large water volume and keeping the time in the warm water layers short, or by using special thermally protecting cod-ends (e.g. Childress *et al.*, 1978),
- *Keep pressure constant*: Decompression from several thousand metres can exert a considerable stress on the organisms. The physiological effects of pressure on aquatic animals have been studied in detail since the 1960s. Surprisingly, many zooplankton can be recovered live from depths of more than 2500 m, provided they are kept at *in situ* temperature and not physically damaged (Childress *et al.*, 1978). Collecting and keeping zooplankton at *in situ* pressure requires a large technological effort. Details are beyond the scope of this book, but we recommend Macdonald (1972) for an example of a plankton sampler maintaining constant temperature and pressure.

Many of the above problems can be avoided by performing *in situ* experiments, and the increasing availability of ROVs, lander systems and underwater laboratories will allow for further progress in this field. The basic approach is to collect zooplankton in an incubation chamber, which can be closed and kept at depth for the duration of the experiment. Early deep-sea *in situ* experiments with zooplankton were made, for example, by Wishner and Meise-Munns (1984) and Wishner and Gowing (1987). They used a multiple opening and closing net system mounted on a submersible to collect zooplankton, which were retained in cod-end incubation chambers equipped with closing doors and an injection mechanism allowing for the addition of labelled substrate and other substances. After the tow, the doors were closed and the zooplankton incubated for 0.5–3 hours. Adding radiolabelled amino acids allowed for the measurement of grazing, filtering and ingestion rates. To terminate the experiment, a killing agent was injected, the sample was brought to the surface and analysed in the laboratory.

Mixed macrozooplankton and individual larger zooplankton have been successfully sampled and incubated *in situ* with suction pumps from submersibles. For example, in the early 1980s, Smith Jr. (1982) used a 'slurp gun respirometer' for measuring respiration rates in deep-sea zooplankton. The animals were drawn into an incubation chamber with a centrifugal pump, then the doors of the chamber were closed, the chamber was attached to a free-vehicle mooring and left *in situ* for the incubation time of about one day, before it was recovered. Oxygen concentration was measured continuously with an oxygen probe. The same set-up was also used for measuring excretion rates.

6.4.6 *Catch handling and sample processing*

After hauling, nets should be rinsed thoroughly in order to flush all plankton into the cod-end buckets. This is best done when the nets are still hanging overboard, with one person unfolding the nets and a second person rinsing them with seawater from a hose. Longer nets which cannot hang freely should be washed down in sections from the mouth opening to the cod-end, one person always holding up the front part of the respective section. Particular attention should be paid to the canvas where the cod-end bucket is attached to the net.

After removing the cod-ends, the catch is usually poured gently through a sieve with appropriate mesh size. Remaining plankton in the cod-end are flushed into the sieve with filtered seawater; it is important that the drainage holes are rinsed very carefully.

The sample is then either flushed into sample bottles for fixation, or processed further. For a proper fixation of the catch in formaldehyde, the sample volume should be no more than a third of the bottle volume, in ethanol even less. If the samples cannot be fixed or processed immediately, they should be stored in a coldroom.

Rinsing the nets with warmer surface water exerts a mechanical and thermal stress on the organisms, and so collections of live zooplankton should be flushed immediately into a larger volume of filtered, cooled seawater and then kept at ambient temperature, with the nets being cleaned afterwards.

Further processing of the samples depends on the objectives of the study and is basically the same as in shallow-water zooplankton. See Harris *et al.* (2000) for a wide variety of laboratory methods used in plankton research.

6.5 Environmental impact of sampling operations

The environmental impacts of pelagic zooplankton samplers are negligible, and in most cases do not require any mitigation by scientists planning or carrying out the sampling operations. Sampling zooplankton close to the bottom, either with towed nets or samplers mounted on vehicles like submersibles and ROVs, may cause some damage to the benthic communities if the bottom is touched unintentionally; however, such impact is limited to very small areas. Similarly, moored instruments such as pumps and acoustic arrays can cause a local disturbance, but it is minor. In contrast, using epibenthic sledges may have a considerable impact on the benthos, and care is needed. Sledge tows, however, are often short because of the potentially high catch rates of benthic animals, or because of concerns to minimize bottom impact (see Chapter 9).

Acknowledgements

My thanks go to Horst Weikert and Werner Beckmann for establishing the study of deep-sea zooplankton at the Universität Hamburg, and to Rolf Koppelmann for technical advice and his contributions to improving the sampling systems.

References

Aldred, R.G., Thurston, M.H., Rice, A.L., Morley, D.R. (1976). An acoustically monitored opening and closing epibenthic sledge. *Deep-Sea Research*, 23: 167–174.

Brandt, A., Barthel, D. (1995). An improved supra-and epibenthic sledge for catching Peracarida (Crustacea, Malacostraca). *Ophelia*, 43: 15–23.

Brandt, K. (1890). Ueber die biologischen Untersuchungen der Plankton-Expedition. *Naturwissenschaftliche Rundschau*, 5: 112–114.

Childress, J.J., Barnes, A.T., Quetin, L.B., Robison, B.H. (1978). Thermally protecting cod ends for the recovery of living deep-sea animals. *Deep-Sea Research*, 25: 419–422.

Christiansen, B., Nuppenau, V. (1997). The IHF Fototrawl: experiences with a television-controlled, deep-sea epibenthic sledge. *Deep-Sea Research*, 44: 533–540.

Christiansen, B., Drüke, B., Koppelmann, R., Weikert, H. (1999). The near-bottom zooplankton at the abyssal BIOTRANS-site, northeast Atlantic: composition, abundance and variability. *Journal of Plankton Research*, 21: 1847–1863.

Christiansen, B., Bühring, S.I., Pfannkuche, O., Weikert, H. (2010). The near-bottom plankton community at the Porcupine Abyssal Plain, NE-Atlantic: structure and vertical distribution. *Marine Biology Research*, 6: 113–124.

Clutter, R.I., Anraku, M. (1968). Avoidance of samplers. In: Tranter, D.J. (ed.), *Zooplankton sampling. Part 1: Reviews on zooplankton sampling methods*. Monographs on Oceanographic Methodology. Paris: UNESCO, pp. 57–76.

Dahms, H.-U., Qian, P.-Y. (2004). Drift-pump and drift-net – two devices for the collection of bottom-near drifting biota. *Journal of Experimental Marine Biology and Ecology*, 301: 29–37.

Foote, K.G. (2000). Optical methods. In: Harris, R., Wiebe, P.H., Lenz, J., Skjoldal, H.-R., Huntley, M. (eds), *ICES zooplankton methodology manual*. London: Academic Press, pp. 259–293.

Foote, K.G., Stanton, T.K. (2000). Acoustical methods. In: Harris, R., Wiebe, P.H., Lenz, J., Skjoldal, H.-R., Huntley, M. (eds), *ICES zooplankton methodology manual*. London: Academic Press, pp. 223–258.

Fraser, J.H. (ed.) (1968). *Zooplankton sampling. Part 2: Standardization of zooplankton sampling methods at sea*. Monographs on Oceanographic Methodology. Paris: UNESCO.

Gorsky, G., Picheral, M., Stemmann, L. (2000). Use of the underwater video profiler for the study of aggregate dynamics in the north Mediterranean. *Estuarine, Coastal and Shelf Science*, 50: 121–128.

Harris, R., Wiebe, P.H., Lenz, J., Skjoldal, H.-R.,Huntley, M. (eds) (2000). *ICES zooplankton methodology manual*. London: Academic Press.

Krygier, E.E., Pearcy, W.G. (1981). Vertical distribution and biology of pelagic decapod crustaceans off Oregon. *Journal of Crustacean Biology*, 1: 70–95.

Maaß, B. (2007). Das neue eisrandfahige Forschungsschiff 'Maria S. Merian'. Eine Bestandsaufnahme der Planung, der Bauausfuhrung und des Betriebes und die hieraus resultierenden Erkenntnisse für weitere Neubauprojekte. Diplomarbeit, Fachhochschule Oldenburg/Ostfriesland/Wilhelmshaven.

Macdonald, A.G. (1972). An apparatus for the recovery and study of deep sea plankton at constant temperature and pressure. In: Brauer, R.W. (ed.), *Barobiology and the experimental biology of the deep sea*. North Carolina Sea Grant Program, University of North Carolina, Chapel Hill, NC, pp. 394–412.

Mullineaux, L.S., Mills, S.W., Sweetman, A.K., Beaudreau, A.H., Metaxas, A., Hunt, H.L. (2005). Vertical, lateral and temporal structure in larval distributions at hydrothermal vents. *Marine Ecology Progress Series*, 293: 1–16.

Peyronnet, J.-P., Person, R., Rybicki, F. (1998). POSIDONIA 6000: a new long range highly accurate ultra short base line positioning system. *OCEANS '98*. Nice, IEEE Oceanic Engineering Society.

Robison, B.H., Sherlock, R.E., Reisenbichler, K.R. (2010). The bathypelagic community of Monterey Canyon. *Deep Sea Research Part II: Topical Studies in Oceanography*, 57: 1551–1556.

Sameoto, D., Wiebe, P., Runge, J., *et al.* (2000). Collecting zooplankton. In: Harris, R., Wiebe, P.H., Lenz, J., Skjoldal, H.-R., Huntley, M. (eds), *ICES zooplankton methodology manual*. London: Academic Press, pp. 55–81.

Sherman, A.D., Smith Jr, K.L. (2009). Deep-sea benthic boundary layer communities and food supply: a long-term monitoring strategy. *Deep Sea Research II*, 56: 1754–1762.

Skjoldal, H.-R., Wiebe, P.H., Foote, K.G. (2000). Sampling and experimental design. In: Harris, R., Wiebe, P.H., Lenz, J., Skjoldal, H.-R., Huntley, M. (eds), *ICES zooplankton methodology manual*. London: Academic Press, pp. 33–53.

Smith Jr, K.L. (1982). Zooplankton of a bathyal benthic boundary layer: in situ rates of oxygen consumption and ammonium excretion. *Limnology and Oceanography*, 27: 461–471.

Stemmann, L., Hosia, A., Youngbluth, M.J., Soiland, H., Picheral, M., Gorsky, G. (2008). Vertical distribution (0–1000 m) of macrozooplankton, estimated using the underwater video profiler, in different hydrographic regimes along the northern portion of the Mid-Atlantic Ridge. *Deep Sea Research II*, 55: 94–105.

Tranter, D.J. (1968). *Zooplankton sampling. Part 1: Reviews on zooplankton sampling methods*. Monographs on Oceanographic Methodology. Paris: UNESCO.

Vinogradov, G.M. (2005). Vertical distribution of macroplankton at the Charlie-Gibbs Fracture Zone (North Atlantic), as observed from the manned submersible Mir-1. *Marine Biology*, 146: 325–331.

Weikert, H., John, H.-C. (1981). Experiences with a modified Bé multiple opening-closing plankton net. *Journal of Plankton Research*, 3: 167–176.

Wenneck, T.D.L., Falkenhaug, T.; Bergstad, O.A. (2008). Strategies, methods, and technologies adopted on the R.V. G.O. Sars MAR-ECO expedition to the Mid-Atlantic Ridge in 2004. *Deep Sea Research II*, 55: 6–28.

Wiebe, P.H., Benfield, M.C. (2003). From the Hensen net toward four-dimensional biological oceanography. *Progress in Oceanography*, 56: 7–136.

Wiebe, P.H., Morton, A.W., Bradley, A.M., *et al.* (1985). New developments in the MOCNESS, an apparatus for sampling zooplankton and micronekton. *Marine Biology*, 87: 313–323.

Wiebe, P.H., Bucklin, A., Madin, L., *et al.* (2010). Deep-sea sampling on CMarZ cruises in the Atlantic Ocean – an introduction. *Deep Sea Research II*, 57: 2157–2166.

Wishner, K. (1980a). Aspects of the community ecology of deep-sea, benthopelagic plankton, with special attention to gymnopleid copepods. *Marine Biology*, 60: 179–187.

Wishner, K.F. (1980b). The biomass of the deep-sea benthopelagic plankton. *Deep-Sea Research*, 27: 203–216.

Wishner, K.F.,Gowing, M.M. (1987). In situ filtering and ingestion rates of deep-sea benthic boundary-layer zooplankton in the Santa Catalina Basin. *Marine Biology*, 94: 357–366.

Wishner, K.F., Meise-Munns, C.J. (1984). In situ grazing rates of deep-sea benthic boundary-layer zooplankton. *Marine Biology*, 84: 65–74.

Youngbluth, M., Sornes, T., Hosia, A., Stemmann, L. (2008). Vertical distribution and relative abundance of gelatinous zooplankton, in situ observations near the Mid-Atlantic Ridge. *Deep Sea Research II*, 55: 119–125.

Chapter 7
Trawls

Malcolm R. Clark[1]*, Neil W. Bagley*[1] *and Brian Harley*[2]

[1] National Institute of Water and Atmospheric Research, Wellington, New Zealand

[2] Centre for Environment, Fisheries and Aquaculture Science, Lowestoft, United Kingdom

Abstract

Trawls are among the most commonly used types of sampling gear for surveying deep-sea faunal communities. In this chapter we outline the main types of trawl (otter, shrimp, beam, Agassiz, midwater), and describe aspects of their use given the type of seafloor and target faunal groups. Details are given of trawl deployment and operation, on ways to modify trawls for use on rough seafloor, and advice is provided on how to evaluate performance of trawl tows. Standardization of trawl gear, and the need for full description of its construction, set-up and operation, are stressed. Information is included on sample sorting and processing, as well as interpretation of trawl-derived data.

Keywords otter trawl, beam trawl, Agassiz trawl, shrimp trawl, standardization, trawling procedures, trawl performance

7.1 Introduction

Trawling is the action of towing nets across the surface of, or above, the seafloor. It is an old technique that dates back to Roman times when small bagnets with a wooden frame to hold the mouth open were used to catch oysters and flatfish. Over time this developed into the beam trawl, which was the main gear type worked from sailing vessels for several centuries. In the late nineteenth century with the development of steam-powered fishing vessels, bottom trawls were designed that did not rely on a rigid beam to hold the mouth open, but instead the net was spread and kept open horizontally by 'otter' boards. Today, trawling is one of the commonest commercial fishing methods, with both beam trawl and otter trawl gear used to catch a wide variety and large quantity of fish species throughout the world's oceans. There are many different designs and modes of operation of trawls depending on their purpose, and these are well described in commercial fishing texts such as Sainsbury (1996) and Gabriel *et al.* (2005).

Trawls are a standard component of the arsenal of sampling gear available to scientists, and are often a preferred gear method where the objectives are to sample large-bodied invertebrate epifauna or mobile fishes. Trawl surveys are often the main or sole data source for assessments of fishery stocks (e.g. Mesnil *et al.*, 2009), but are also often used for wider community composition studies (e.g. Clark *et al.*, 2010). They are able to cover relatively large areas of seafloor, and so can be used to describe broad spatial scales of species composition and relative abundance in the deep sea, where species densities can be low. However, trawls are generally regarded as only semi-quantitative, and are typically

Biological Sampling in the Deep Sea, First Edition. Edited by Malcolm R. Clark, Mireille Consalvey and Ashley A. Rowden.

applied in longer term monitoring of changes in faunal abundance over time, and measuring relative abundance rather than producing absolute density estimates (e.g. Clark, 1996; Billet *et al.* 2001).

In this chapter, we describe the general types of trawls commonly used in fisheries and biodiversity research in the deep sea, and focus on aspects of their operation that are particularly relevant to interpretation of the catch data for studies of deep-sea biodiversity.

7.2 General description of gear types

7.2.1 Bottom trawls

Bottom trawl is a generic term for trawl nets that are towed directly along the seabed. The bottom trawl net is basically a large conical bag made of netting, wide at the open mouth and tapering to the closed cod-end (the final section of the net) where the catch is retained. The bottom leading edge of the trawl net has a groundrope. This may be constructed from a combination of chain or wire, with rubber and/or metal discs, rollers or bobbins depending on the nature of the seafloor to be sampled and species targeted. The groundrope provides some protection to the trawl net and allows it to move over the seafloor without snagging or digging in. The top of the net is held open by floats which help give the trawl its vertical opening. Trawl nets can be spread between two vessels (pair trawling) but more commonly in research surveys just a single vessel is involved. There are four main types of bottom trawl: otter and shrimp trawls where the nets are usually spread and held open by trawl doors; and beam and Agassiz-type trawls where the mouth of the net is attached to a rigid frame. The two types differ in their fishing characteristics, and in the type of fauna they sample best.

7.2.1.1 Otter trawls

These trawls utilize trawl doors (termed 'otter boards') to spread the mouth of the net (Fig. 7.1). The angle of the doors, and hence the spreading power of them, is set by the position of the towing brackets to which the warps from the vessel are connected and wires connecting the trawl doors to the trawl net. Bottom contact and the spreading effect of the doors ('shear') can be reduced or increased depending on the target species and type of seafloor. Trawl doors need careful selection to match the vessel and net characteristics. Specialist advice is required on size, weight, shape, composition and structure of the doors.

Behind the doors are two backstrops which then connect to wires leading back towards the net. These generally consist of sweeps (single wires) and then bridles (2 or more wires) which are attached to the top and bottom of the net. This rigging can vary between nets and bottom type, but is usually based on what works well on commercial trawlers. However, sweep lengths are often shorter as research surveys are not intended to catch as much fish as a commercial operation, and to reduce environmental impact. The sweeps and bottom bridles kick up a plume of sand/mud which can help herd fish into the net. The headline of the net is held open in the vertical plane by buoyant floats which allow the net to open against the weight of the groundrope. Groundropes constructed of heavy bobbins or rubber rollers can give the overall trawl gear a weight of several tonnes (see Section 7.4). Detailed descriptions of an otter trawl and set-up are given in fishing gear books (e.g. Sainsbury, 1996; Gabriel *et al.*, 2005).

Trawls suitable for sampling flat and smooth seafloor are widely available, and most research institutes or vessels have such gear. However, deep-sea nets suitable for fishing rough seafloor, or features like seamounts, have different features, and a plan for a trawl type that is widely used in research and commercial fisheries off New Zealand and Australia is included in Appendix 7.1. This type of net has much-reduced lower wings and heavy ground gear. Sweeps and bridles are generally short (50 m in length). This net is broadly similar in characteristics to the Jackson 460 rockhopper whitefish trawl used in Scottish and Irish deep-sea trawl surveys (e.g. ICES, 2008).

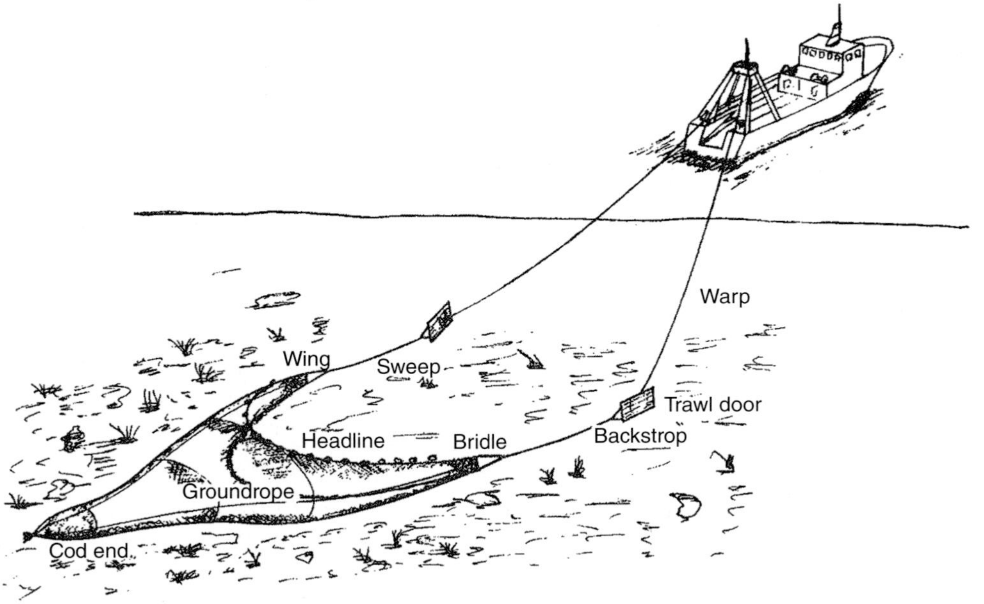

Fig. 7.1 Schematic of a standard bottom trawl being towed on the seafloor, with otter boards (trawl doors) spreading the net. (Adapted from Sainsbury, 1996. Reproduced with permission of John Wiley & Sons Ltd.)

Otter trawls are usually fished off two warps, as this enables maximum spreading efficiency from the doors, and reduces the likelihood of tangled gear and crossed doors. However, it is also possible to use a single warp with small otter trawls such as the Marinovitch semi-balloon otter trawl (OTSB) (Merret & Marshall, 1980; Gordon *et al.*, 1996). This net and gear has been successfully used to depths of 4000 m. The single warp was connected to a swivel with 50 m long bridles to the trawl doors. Headline buoyancy was provided by three glass floats.

Twin trawls, where two nets are towed by a single vessel at the same time, are often used in commercial bottom trawling (e.g. hoki (*Macruronus novaezelandiae*) fisheries off New Zealand) but are infrequently applied in research other than in specific experimental work such as determining fish selectivity with different mesh size.

7.2.1.2 Shrimp trawls

Shrimp trawls are similar to otter trawls, in that they usually rely on trawl doors to spread the net mouth. They are often towed by a single warp, but are sometimes deployed off outriggers extending either side of the vessel with each net having its own set of trawl doors. The trawls usually consist of two or three nets joined together with a skid (like a short ski) in between. These nets are smaller than a single net with the same mouth opening, and can be towed by smaller vessels. They generally have fine-mesh, light ground gear, and are used on relatively smooth bottom. They are not commercially used in most deep-sea environments beyond about 400–500 m.

7.2.1.3 Beam trawls

Beam trawls utilize a transverse beam which holds the mouth of the net open, with metal runners at each end. The net is typically a simple two-panel tapered bag, the top of which is lashed to the beam, and the bottom is like a standard otter trawl with a small groundrope that trails well behind the beam so that animals disturbed by the groundrope cannot escape over the top (Fig. 7.2). A tickler

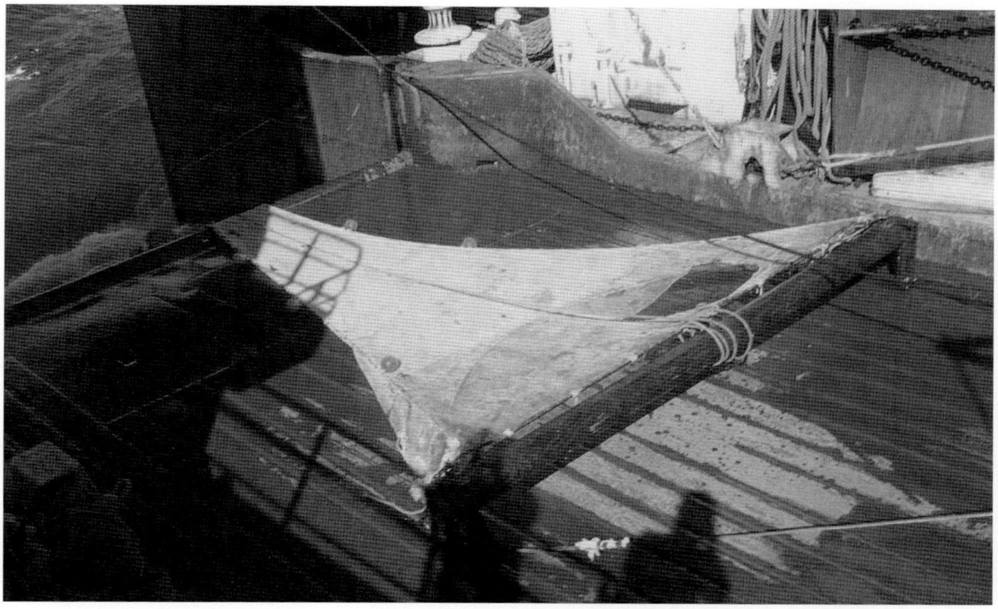

Fig. 7.2 Beam trawl as used by NIWA and IRD Noumea. The beam is 4 m wide, short enough to be hauled up the stern ramp of the trawler. (Photograph courtesy of NIWA.)

chain or chain mat is attached in front of the groundrope to dig out shallow animals and increase efficiency of the catch for small invertebrates.

Modern commercial beam trawlers can deploy two trawls either side of the vessel, and they can be as wide as 10–12 m. Beam trawls of that size can be difficult to handle on deck, and most research vessels will probably have beam widths of no more than 5–6 m. Small beam trawls with a 2 m width have been commonly used in studies of epibenthic fauna (e.g. Jennings *et al.*, 1999; Rees, 2009); trawls commonly used by IFREMER (France) have a width of 6 m; and those used by the National Institute of Water and Atmospheric Research (NIWA) (New Zealand), Research Development Institute (IRD) (New Caledonia) and the Centre for Environment, Fisheries and Aquaculture Science (CEFAS) (UK) are 4 m wide. Even this width needs careful control, especially when being deployed down the ramp of a stern trawler where guiding ropes are used to keep the beam square and up the right way until it is pulled up to the trawl block on the stern of the vessel for shooting. A short length of chain attached to the cod-end can help prevent the net being swept forward through the bridles by surface swells when shooting away. A design of beam trawl commonly used in the UK by CEFAS for flatfish surveys (ICES, 2009c) has a metal beam, and uses a similar net to commercial beam trawlers although has a cod-end liner to catch juvenile fish. A chain mat of 7/8 inch chain extends between the hardened steel shoes and the groundrope, which helps trawl performance when used on harder ground, and to reduce the amount of debris caught. Details of the design of this beam trawl are given in Appendix 7.2. With some beam trawls, the beam is made of wood, as this means it can break if the gear becomes snagged on the seafloor, and aids getting the gear free while still retaining some catch.

7.2.1.4 Agassiz-type trawls

Agassiz trawls (and the similar Sigsbee and Blake trawls) have been a standard scientific sampling tool since the early 1900s (e.g. Gage & Tyler, 1991). The Agassiz trawl is similar to a beam trawl in concept, although has a rigid and continuous frame at the front with oval skids that can generally be towed either way up. The trawl can therefore continue to sample in areas where one-sided gear may snag or be tipped

Fig. 7.3 Agassiz trawl developed by the Netherlands Institute of Sea Research, being used off New Zealand. (Photograph courtesy of NIWA.) (Please see Plate 12 for colour representation of the figure.)

over. The size of the frame, and nets, can be highly variable, ranging from small, very light trawls of 1 m width and 0.3 m height (e.g. used by NIWA) up to heavy trawls of 3.5 m width and 1 m vertical opening (e.g. Netherlands Institute for Sea Research) (Fig. 7.3). These can get quite sophisticated, with tripping mechanisms that open the mouth of the net only when the gear touches the bottom.

7.2.2 Midwater trawls

Midwater trawls are used to capture pelagic fish and cephalopod species that occur in the water column between the surface and the seafloor. The net is spread wide by trawl doors, but in contrast with bottom gear these doors are designed to 'fly' in midwater. The headline is held open by floats or kites, with long bridles and heavy weights attached to the wing-ends to pull the bottom of the net down. Groundropes are generally much lighter than on bottom trawls, although sufficiently strong for the gear to be towed lightly touching the bottom where the seafloor is smooth.

There are many designs of midwater trawls, which have got larger during the late twentieth century as commercial fisheries for schooling species such as herring (*Clupea harrengus*), mackerel (*Scomber* spp., *Trachurus* spp.), pollack (*Pollachius* spp., *Theragra chalcogramma*) and redfish (*Sebastes* spp.) developed in the North Atlantic and North Pacific Oceans. These are well described in Gabriel *et al.* (2005).

Large midwater trawls tend to be very long and are often difficult to handle on the limited deck spaces of research vessels unless they have good-sized net rollers. Commercial midwater trawls can be adapted for research to take fine-mesh cod-ends, but extension pieces before the cod-end may be needed to ensure correct water flow and to avoid pressure waves forming ahead of the cod-end. Smaller sized nets, such as Isaacs–Kidd midwater trawls, myctophid trawls or fixed-frame rectangular midwater trawls (RMTs), may be used in situations where sampling with fine-meshed nets in midwater is required, and various institutes have specialized midwater trawls designed for specific purposes (for a general description, refer to Gabriel *et al.*, 2005).

7.2.3 Electronics

There is a wide selection of electronic equipment that can be fitted to towed gears to help researchers measure gear characteristics, understand gear performance and record accurate information on trawl tows. Routine measurement of gear parameters is an integral part of research survey design, and has been recognized for a long time (e.g. Carrothers, 1981). The most common sensors include the headline transducer (also known as 'netsonde', Fig. 7.4) which gives depth of net, water temperature, headline height and information on bottom contact; distance sensors which are usually placed on the doors, and sometimes wing-ends; and catch sensors which indicate the degree of filling of the cod-end. Sensors can output real-time data so any problems with gear performance can be quickly identified. This type of equipment is expensive and some protection around the sensors may be needed depending on the gear type used.

Bottom-contact sensors are common and use changes in their angle when the gear is off the bottom (so more vertical) compared with on the bottom (more horizontal). Data are downloaded at the end of a deployment and can determine distance towed or bottom-contact time.

Self-tensioning winches are now standard on deep-sea commercial and most research trawling vessels, and these can be set up to adjust warp tension automatically to keep them the same length, or use sensors on the trawl that monitor trawl speed through the water and cross-currents to adjust warp lengths to ensure the net is fishing square.

(a) (b)

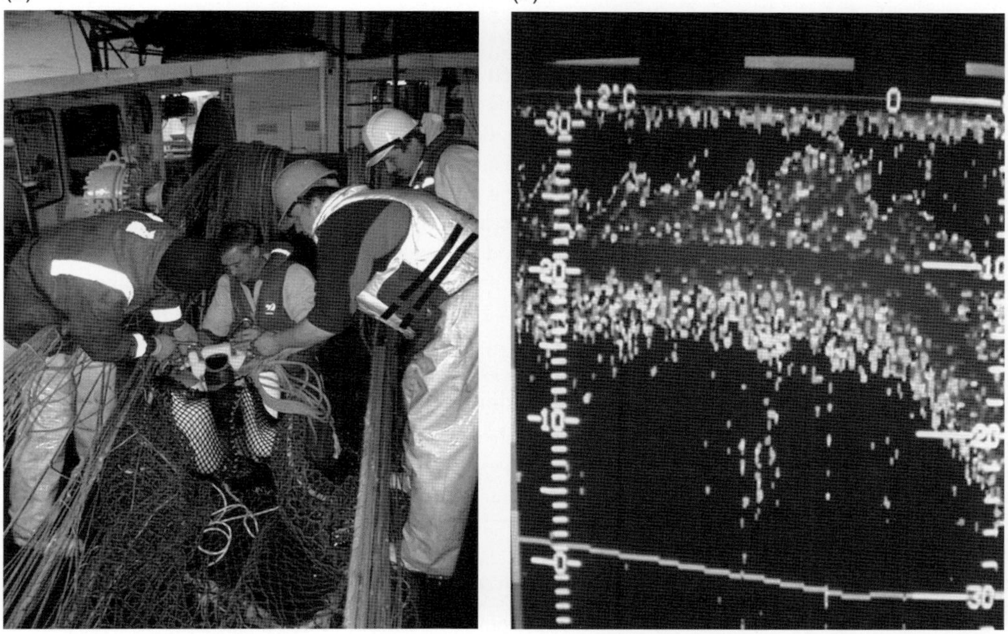

Fig. 7.4 (a) Net monitor mounted on the headline of an otter trawl and (b) the read-out seen on the bridge as the net lifts off the bottom (in this case with a catch of 40 t of orange roughy). In (b) the top line is the headline, with the first line below it being the groundrope of the trawl, and the second line below it being the seafloor. When the two lines converge, the net part of the trawl has landed on the seafloor and is fishing. In this case the speckled signal between the headline and bottom is fish (orange roughy). Indications after 2 minutes of fishing was that the catch was large, and hauling began, seen as the separation of the groundrope (at 10 m) and the seafloor. (Photographs courtesy of NIWA.)

7.3 Sampling operations

7.3.1 *Trawl selection*

The choice of gear will depend upon the objectives of the survey (e.g. the target species or taxa of interest to the researchers, depth of sampling, quantitative catch rates versus qualitative catch), capability of the vessel, working conditions and the availability of suitable gear (Table 7.1). No single gear type will sample all components of a deep-sea epifaunal community, and often multiple gears will be used to sample different faunal groups (see Chapter 2). This will involve a combination of trawls with sledges and grabs/corers (see Chapters 9 and 10).

Otter trawls are most effective at catching mobile demersal fish and large-bodied macro- and mega-invertebrates. The headline height above the seafloor can be several metres, and so they also catch animals which swim just above the bottom. If deployed from two warps, they are limited in depth by the amount of wire able to be carried on the trawl winches (commercially this is about 2000 m depth, based on 4000 m of 24–28 mm diameter wire).

Shrimp nets are smaller than large fish trawls, and can have finer mesh, making the gear suitable for catching small-bodied fish and smaller invertebrates. They can be deployed off smaller gauge wire (e.g. 16 mm) than heavier otter trawls, and hence the depth limit may be greater if more wire can be carried.

Beam trawls are used primarily in shrimp and flatfish fisheries, and hence are highly effective at sampling benthic invertebrates and small or slow-moving fish that are hard-down on the bottom. The gear is towed from a single warp and hence can be fished much deeper than twin-warp otter and shrimp trawls. Even relatively lightweight beam trawls have successfully been deployed at depths over 3000 m using 16 mm diameter wire.

Agassiz trawls are usually used only for research purposes, and are generally light and have fine-mesh nets. They are used to sample small epibenthic invertebrates, but do not catch mobile fish. They can be fished deep off a single wire of 14–16 mm, depending on the size and weight of the trawl.

The choice of trawl gear will be determined to an extent by seafloor topography. Trawl nets can be heavily constructed, and set up with ground gear and trawl doors to fish over rough seafloor (see Section 7.4). However, beam and Agassiz trawls are generally designed for smooth slope areas, and nets get ripped on an uneven bottom. A good knowledge of bathymetry is necessary before deciding on the type of trawl to be used. Unknown areas should be traversed before trawling, with the ships echosounder data used to determine how rough the seafloor may be, and hence what gear may be suitable.

Research vessels carrying out scientific surveys in the deep sea generally use trawl designs and trawl gear similar to those used by commercial fishing vessels. The size of the gear may be reduced,

Table 7.1 Summary of trawl gear suitable for general types of seafloor and faunal type. Yes has three levels of suitability (low–med–high)

	Seafloor type		Faunal type		
Gear type	Rough/hard	Smooth/soft	Infauna	Macro–megafauna	Fish
Light beam	No	Yes (high)	Yes (low)	Yes (high)	Yes (low)
Heavy chain – mat beam	Yes (low)	Yes (high)	Yes (low)	Yes (high)	Yes (med)
Agassiz trawl	No	Yes (high)	Yes (med)	Yes (high)	Yes (low)
Shrimp trawl	No	Yes (high)	Yes (med)	Yes (med)	Yes (low)
Otter trawl – full wing, small groundrope	No	Yes (high)	No	Yes (med)	Yes (high)
Otter trawl – rough-bottom heavy groundrope	Yes (high)	Yes (med)	No	Yes (low)	Yes (high)

or towing times shortened, as research vessels are not usually able to (nor need to for their purposes) handle large commercial-size catches. The trawls still need to be matched to the size and power of research vessels required to work offshore.

There are many designs of trawl nets for various commercial fishing applications. This means it is easy for a standard design to be used for research, perhaps with minor modifications or smaller mesh sizes. The use of commercial gear has advantages in availability of spares and easy maintenance, as well as greater acceptance of results used for fisheries stock assessment where issues of different gear selectivity and efficiency from commercially proven operations are reduced. However, research objectives need to be considered to ensure the trawl is appropriate for a range of scientific tasks.

Research or commercial vessels will be set up in a certain way to handle different types of trawl gear. Otter trawls in particular are matched to vessel size and power, and being larger in size will catch more than smaller beam or Agassiz trawls. The surface area and weight of trawl doors, and the diameter of trawl warp wire will also depend upon the vessel characteristics and net size. It is impractical, and beyond the scope of this chapter, to detail the deck gear necessary for deep bottom trawling, as it varies so much between vessels of different size designed for different purposes. However, in the sections below we highlight some of the aspects of trawling operations that may differ from shallow-water trawling, and are important for successful scientific sampling, irrespective of vessel type and set-up.

7.3.2 Standardization

It is essential for many research purposes that sampling gear, and its operation, are standardized. Details of gear plans and rigging set-up should be carefully and thoroughly documented. Deployment procedures need to be created and specified for researchers and ship's officers and crew. Standardizing the sampling procedures will ensure that data collected can be used for comparisons both within and between surveys. This must be a condition of any research survey if objectives include abundance estimation, or if the survey is (or likely to be) part of a time series.

The key components include the following:

- The sampling gear:
  - full net plans, including mesh types and sizes;
  - headline flotation plan, size and number of floats, position of floats;
  - groundrope composition, length, details of all components.
- Rigging plans:
  - sweep and bridle size and lengths;
  - layback of the headline (if any).
- Deployment procedures:
  - warp to depth ratios for amount of trawl wire;
  - standard electronics to be used, and acceptable values of certain measurements;
  - required towing speed.

The production of a formal technical report should be encouraged as a standard procedure for any time series of surveys. A good example of this is Neat *et al.* (2010) which describes the Scottish deep-water survey design, gear and sampling protocols in detail. Many institutes have 'internal' reports (e.g. Hurst *et al.*, 1992; McMillan, 1996), yet these are not readily accessible to other researchers. European research agencies through the International Council for the Exploration of the Sea (ICES) have put a lot of effort into standardizing national surveys where possible, with working groups producing a number of reports which include information and advice on improving consistency between trawl surveys (e.g. ICES, 2008, 2009a, 2010). A good account of trawl gear construction and maintenance was produced by the Study Group on Survey Trawl Standardization

(ICES, 2009b). A similar working group was established for beam trawl surveys (ICES, 2009c). The key to standardization is comprehensive documentation of the equipment and how it is set up, and clear description of how the trawls are operated during the survey.

The need for detailed net and gear notes has been emphasized recently with two surveys that have continued time series or undertaken repeat stations which require an identical set of sampling gear. In New Zealand an orange roughy (*Hoplostethus atlanticus*) trawl survey last carried out in the period 1992–94 was repeated in 2009. Although the net still used by the vessel was the same, it took considerable effort to find out exactly how the ground gear was originally configured, as the voyage reports had not detailed the specifics. Similarly off the UK, trawling with the OTSB was repeated in 2009–10 in the Porcupine Trough area, which had last been surveyed in the 1980s. It is easy for even small changes in the rigging of a net, or the configuration of the ground gear to alter over the years, and as scientific, technical and vessel staff change, the information can easily be lost.

Good maintenance, repair and replacement of parts is easily overlooked as gear is put away between surveys, yet is an important component of ensuring gear comparability and consistency of performance. Most research trawl gear and nets use commercially available components. Over time, these can change (e.g. mesh size regulations making a certain size obsolete: bobbins being replaced by rubber rollers and rockhopper discs) and so research institutes will need to ensure they have on hand adequate spares and supplies to be able to maintain the equipment in its original form. Commercial gear that is replaced by upgraded equipment is often cheap to purchase, and staff responsible for trawl gear should consider this if components in future may be hard to obtain.

7.3.3 *Trawl deployment*

Trawl nets, by their very nature of being towed along the seafloor, are prone to damage – getting ripped on underwater obstructions or getting stuck ('fast') on uneven ground. Hence it is important to survey the proposed trawl path using an echosounder before shooting the gear. This needs to be done at a slow enough speed to judge how rough the seafloor is. Generally 6 knots is a good compromise between steaming fast at 8–10 knots where the bottom on the echosounder looks relatively smooth, and steaming at towing speed which takes more time. Weather conditions and the ship's course will need to be taken into account to ensure a good-quality picture on the echosounder. If the research vessel has multibeam capability, then it is good practice to swath-map the feature or wider region first, so that sampling can be directed to optimize survey design (e.g. to avoid areas of foul ground, to tow across high and low backscatter areas to maximize diversity) and reduce the risks of towing on unknown substrate if the gear is substantially off to one side of the vessel (e.g. Clark & Roberts, 2008).

The design of a trawl survey may be based on random locations, or position tows where the seabed is rough. It is important to have experienced officers and deck crew who can provide trawling expertise to both determine if a tow line is workable, and to trawl it effectively. The conduct of a trawling survey should always be cooperative between scientists and the skipper, to balance the survey design and objectives with the safety of the gear and vessel.

7.3.3.1 Shooting the trawl gear

The vessel speed when shooting gear differs with gear type. With otter trawls the speed initially needs to be reasonably fast (4–6 knots) to maintain hydrodynamic shearing motion/stability of the doors so they maintain their spreading effect on the net. If vessel speed is too low, there is a risk of the doors becoming unstable and getting tangled ('crossed doors'). A check on the angle of the warps as they leave the ship can indicate good spreading of the doors (i.e. wires angle outwards). It is also good practice to pause shooting after a few minutes and check the netsonde readings to check the gear is settling properly. The vessel can slow down once the gear is deep, but it needs to speed up to 3 knots as the trawl approaches the bottom so that the gear lands with a good spread in place and is working efficiently from the time of touchdown.

With beam and Agassiz trawls, shooting speed needs to be reasonably slow depending on the size and weight of the gear. With heavier trawls (such as the 4 m steel beam trawl used by CEFAS) it is deployed and towed at similar speeds to an otter trawl. But with lighter beam trawls (e.g. as used by IFREMER, NIWA) they are shot away at 2–3 knots to keep the net streaming astern, and reduce the chances of swell or surges from the vessel (if it is pitching in rough weather) washing the cod-end up and over the beam or net frame, thereby tangling. A careful watch is needed at the stern of the vessel during the shooting operation to check this does not happen. Once the gear is sinking (with 50 m of wire out), the vessel should slow to 1–1.5 knots. The rate at which the wire is payed out also needs to be relatively slow (60–120 m min^{-1}) to make sure that the sinking rate of the trawl is faster than the wire to avoid tangling. Larger sized trawl warp of 20–28 mm diameter is heavy and sinks rapidly.

When fishing beam or Agassiz trawls in very deep water, the vessel can almost be stopped to aid speed of sinking of the gear. When enough wire has been payed out, the vessel can slowly speed up to towing speed. This approach does not enable good estimation of tow length, but such fishing is often not being carried out for quantitative comparison of catch rates, and hence ensuring the gear is on the bottom is a benefit. This cannot be done with otter trawls, as the doors will fall on their sides or cross over if the vessel speed drops too much.

7.3.3.2 Warp length

The amount of wire payed out relative to depth (warp length:depth ratio) decreases with depth, mainly due to the weight of the wire. When fishing otter trawls in water of 500 m or less it is common to use higher warp to depth ratios than in deeper water. Generally, between 500 m and 800 m depth ratios are about 2.0:1 to 2.2:1, and deeper than 800 m ratios are between 1.6:1 and 1.8:1. The smaller ratios are important to reduce the drag of the warps through the water or along the seafloor. Wire capacity of the winches may also be a limiting factor with the amount of wire used, as is the benefits of reduced time taken to shoot and haul. Deep-sea fisheries have often been based on aggregating species (e.g. orange roughy, alfonsino, oreos, roughnosed grenadier) and the optimal wire angles and door spread are less critical than in shallow water.

Reduced warp length to depth ratios apply also to beam and Agassiz trawls. The weight of the warps mean that they do not run straight between the ship and gear, but have a convex catenary. With the light gear deployed on larger diameter warps, some of the warp is going to be lying along the bottom in front of the trawl, and paying out more wire simply means that more will be in contact with the seafloor, and risk damage. It is advisable to use a length (50–100 m) of sacrificial wire (old trawl warp or sweep wire is ideal for this) to reduce wear and damage on the trawl warp. Below 500 m ratios of 1.5:1 or less are adequate.

7.3.3.3 Trawling speed

Most otter trawls used in deep-sea research that use or are based on commercial-style nets are designed to be towed at around 3–3.5 knots. Smaller and lighter beam and Agassiz trawls need to be towed more slowly to ensure consistent bottom contact, typically at 1–2 knots. Towing speed will have an effect on the type of fauna that will be caught, and the slower speed of the beam and Agassiz-type trawls mean they are less effective at catching fast-swimming fishes.

Trawling on rough seafloor, such as frequently found on seamounts, often requires the towing speed to be reduced in order to maintain good bottom contact. Otter trawls are almost always towed downslope on seamounts, unless the slope is gradual and the terrain reasonably smooth. This avoids the doors digging in or the groundrope snagging on protruding boulders. Seamounts often have lava flow structures, and these are layered in a downslope direction, and so towing upslope against the layering is a recipe for disaster! However, towing downslope means that the vessel speed may need to be reduced to 2.0–2.5 knots to keep the gear in contact with the bottom on the seafloor, as well as having to pay out more warp as depth increases down the flank. Changes in vessel speed take some time to transfer to changes in gear performance, and so active use of the trawl winches (constantly adjusting amount of warp out) is important to control the amount of bottom contact of the trawl gear. The headline netsonde can indicate when the headline is lifting, which could indicate more wire needs to be payed out or speed reduced to keep the gear stable. The netsonde is also important to monitor when gear is 'touching down' to ensure the gear is settled properly on the seabed and not just in light contact or flying slightly above the bottom. The type of trawl in Appendix 7.1 will have a headline height of 9–10 m when lightly touching, but is only fishing properly at between 6 and 7 m opening.

Towing speed is usually recorded as speed across the seafloor, because this is generally much easier to measure than speed through the water. The latter may be a better measurement of catch rates of animals like fish as water flow through the net is the most important aspect. However, it is difficult (if not impossible) to try to adjust every tow independently for current corrections, and hence careful monitoring of the vessel speed through the GPS is the most practical approach to take.

7.3.3.4 Trawl duration

Trawl duration during research sampling is highly variable, depending upon the nature of the survey, the type of seafloor and catch rates of various fauna.

Where surveys have an objective to estimate relative abundance of fauna, it is important that tows with the same gear are kept to the same duration throughout the survey. That avoids a bias in catchability of mobile species (see Section 7.6), which can vary with length of tow. Given the long time taken to deploy otter trawls to deep seafloor, and the generally low density of fish and most invertebrates at depths beyond 1500 m, we suggest a 60 minute tow duration will provide a good sample, it will have covered about 3 nautical miles of distance, will have caught any fish species that may have an initial burst of speed in response to the noise/sediment plume/physical presence of the trawl, and may also have covered a variety of bottom substrate types that will yield a wider selection of benthic fauna than shorter tows covering less distance.

However, on seamounts, tow duration of otter trawls will be shorter, and it may not be possible be standardize them because of variation in the size and depth range of different seamounts from summit to base. Although the species composition of fish and invertebrate fauna may not differ substantially from adjacent slopes, the relative abundance of certain species can be much higher on seamounts (e.g. Tracey *et al.*, 2004; Rowden *et al.*, 2010) and hence tows should be restricted to the seamount feature, and separate tows done on the slope as separate strata (see Chapter 3 on survey design).

Beam and Agassiz trawls fish hard down on the seafloor, more so than otter trawls which have ground gear that keeps the front of the net slightly elevated above the seafloor (see Section 7.4). The smaller mesh sizes often used in the beam and Agassiz trawls can clog rapidly if the gear is towed on soft sediment (e.g. mud) rather than firm gravel or sand. The nets are also more prone to damage by occasional cobbles or boulders. Hence shorter tows of about 30 minutes (covering 0.5–1.0 nautical miles) duration may be more appropriate. This is the duration of tows in the English beam trawl surveys for plaice (*Pleuronectes platessa*) and sole (*Solea solea*) in the English Channel and Irish Sea (ICES, 2009c).

Tow duration in almost all cases should be taken as the effective fishing time. This can be judged with otter trawls from a netsonde ('net monitor') which is mounted on the headline of the trawl. This is an acoustic transponder that is directed both up and down. The former can be useful in midwater trawling, but the bottom directed beam gives accurate information on the position of the trawl relative to the seabed. Figure 7.4b shows the read-out on the vessel of a trawl being hauled clear of the seafloor. The start of the tow is recorded when the headline closes up and stabilizes at 6–7 m (which is the normal towing opening and less than 10 m which is the opening when in midwater). The tow is finished when the separation between groundrope and headline is definite. The start of midwater trawls is recorded when the trawl reaches the target depth/layer being sampled.

Netsonde data can be used to indicate catch rates of fish and guide how much catch there may be. This is important when trawling aggregations of fish as large catches can be difficult for research vessels to handle, and wasteful if processing equipment is not available (it is present on some research vessels, but not most). Catch sensors placed along the cod-end and lengthener can help, as these display progressive filling of the net, and hauling can occur when it is apparent the net is getting full.

There are a number of commercial manufacturers that make and supply these netsonde units (e.g. Scanmar, Furuno). They can be rated to considerable depths, although the acoustic signal is often inconsistent beyond depths of about 1500 m (when the trawl is about 1.5 nautical miles behind the vessel). Factors such as weather, the use of paravanes (where the transponder is towed off the side of the ship and is several metres under the surface and clear of air bubbles and hull interference) instead of hull-mounted transducers (necessary for good signal reception at depths greater than 400–500 m), paravane receiver angles (most can be adjusted fore-and-aft) and internal net monitor settings can also affect the signal and readings in deeper water.

The finish of the tow is not always clear-cut, as it depends upon the vessel speed relative to the hauling speed of the winches. Generally the vessel speed is reduced slightly upon hauling, and the shorter wire ratios in deeper shots means otter trawl gear lifts off the bottom rapidly. Recording practices vary between research institutes. Some record the end of tow when the netsonde shows the groundrope lifting clear of the seafloor, as although the vessel speed slows, the hauling of the warps means the trawl is still moving forward over the seafloor. Others record from the moment hauling starts. It is important to be consistent, but in general if the vessel is still moving forward as hauling begins, the time the gear lifts off the bottom is the preferred option. Bottom-contact sensors can also give information on tow duration, which is used to adjust tow duration retrospectively, especially if netsondes are not used, or fail.

Netsondes are not normally used with beam and Agassiz trawls. The smaller size of the gear and lighter netting to attach a large transducer to, and the increased risk of damage and gear loss on rough ground, makes the risk to the expensive netsonde an issue for many research institutes. However, most modern research vessels have warp tension meters built into the winches. In this case, warp tension can be the main indicator of when the trawl is on the bottom. It is impossible to give absolute tension values, as this depends upon wire size, bottom type, weight of the gear and depth. However, there is generally an appreciable increase in the tension on the warps when bottom contact is made, and spikes in the loading when the gear digs in or bounces over rough or hard substrate. This requires officers or scientists to carefully monitor warp tension. Watching the warps is also a good test, as when on the bottom the warps can jerk and tremble, which does not happen when the gear is in midwater.

Independent depth sensors are small, and can be attached to parts of the beam trawl or Agassiz trawl where damage is less likely to occur. These can give read-outs in real time or data can be downloaded at the end of the tow to determine the distance trawled. The lack of change in depth read-out can indicate when the gear is on the seafloor. Depth sensors normally have a unidirectional signal, and so while the unit can be angled accordingly on a beam trawl, with the Agassiz trawl which can land either way up sometimes the sensor will be upside down and no signal will get to the vessel.

Vessel power is normally reduced almost completely when hauling beam trawls, as there are no trawl doors to worry about keeping spread and upright. Hence vessel speed is down, and hauling often moves the vessel backwards. Although the trawls may continue to move forward to a limited extent, warp tension is no longer a guide to bottom contact, and for practical purposes finish of the tow is usually recorded as the time when hauling begins.

7.3.3.5 Tow direction

Tow direction will be governed primarily by the survey design. This may require towing in certain directions according to the depth contours or substrate type. However, irrespective of survey design, the wind and sea conditions will ultimately dictate what is practical and safe. Towing into the wind and swell is recommended for most trawl operations, for two principal reasons. First, vessel and gear speed can be controlled by the amount of power on and wire out – the vessel can generally be speeded up, but not slowed down if being pushed along by a strong wind or swell. Second, if the gear becomes fast, vessel speed can be reduced immediately, wire slowly hauled back in, and the vessel can drift back over the gear to improve the chances of it coming free (i.e. changing the angle of pull on the gear). If this is done hauling back into the swell, things on deck can get very wet and dangerous with seas breaking over the stern or surging up the ramp of a stern trawler with a danger of damage to, or loss of, gear or even affecting safety of the vessel.

The direction of tow may also vary depending on the general topography of the seafloor. In areas of gradual continental slope, towing along the depth contour is generally favoured to reduce the effects of changes in fauna with depth. Similarly in areas of deep abyssal plain, the bottom is usually reasonably flat. However, seamounts present a different challenge for trawling, depending on the shape and size of the feature:

- If the seamount is large with a broad flat top then the summit can be fished in any direction, although the flanks should be towed downslope.
- If the seamount has a small summit area, then it becomes very important that the tow starts on the summit and proceeds downslope. The flatter area of the summit gives the gear a chance to settle before heading downhill, where rough bottom and the need to pay out more warp can make it difficult to get the gear stable – known as 'chasing the hill down'.

Towing upslope will increase the risk of the gear coming fast, and suffering extensive and expensive damage. Towing downslope is standard commercial fishing practice with most nations that have major deep-sea seamount fisheries (e.g. Clark, 1994). Typically Agassiz trawls and beam trawls are not used for seamount sampling because they are damaged too easily.

The final factor that influences tow direction is rough seafloor. The initial bathymetric mapping will have given an idea of where there are areas of seafloor that will be too rough or too steep to sample successfully. Some gear modification may help (see Section 7.4) but generally sampling of the rough areas will need to be left to gear that is more able to sample and not be damaged (e.g. epibenthic sled, dredge).

7.3.3.6 Trawl location

It is common practice at many research institutes to record vessel position for all sampling activities. This is understandable, as the vessel position is generally known with a high level of accuracy with GPS or differential GPS. It is therefore a consistent measure of location. However, towed gears in deeper water, trawls in particular, can be a considerable distance from the vessel, and at different depths from that under the vessel. This is especially true when fishing small seamounts or knolls, where the vessel is over flat bottom when the trawl is fishing on the seamount. When the tow position is plotted based on the vessel, it does not appear to be a tow on the seamount at all. Hence, wherever possible on steeply sloping terrain, attempts should be made to estimate and record the gear position.

An estimate of the distance behind the vessel can be obtained from trigonometry. Assuming there is no catenary in the warps (even though this is very unlikely), then calculations assuming a right-angled triangle are used (where depth is the short side, the hypotenuse is the amount of warp out – including sweeps and bridles). An example of these distances is given in Table 7.2.

Another calculation that is useful is the approximate distance behind the vessel by time, as when reading off echosounders there are usually time divisions but not distance. Examples of time behind are given for several towing speeds in Table 7.2.

Where depth is known, this can help indicate where the trawl is. For example, when trawling on a seamount, the vessel will steam over the seamount while shooting, and record the depth profile that

Table 7.2 Calculations of the distance and time (minutes) of the gear behind the vessel, given combinations of depth and amount of wire out (assuming a warp length to depth ratio of 1.8:1)

Depth (m)	500	600	700	800	900	1000	1100	1200	1300	1400	1500	2000	2500	3000
Wire out (m)	900	1080	1260	1440	1620	1800	1980	2160	2340	2520	2700	3600	4500	5400
Distance (m)	748	898	1048	1197	1347	1497	1646	1796	1946	2095	2245	2993	3742	4490
Distance (n.mile)	0.40	0.48	0.57	0.65	0.73	0.81	0.89	0.97	1.05	1.13	1.21	1.62	2.02	2.42
Time (at I knot)	24	29	34	39	44	49	53	58	63	68	73	97	121	145
Time (at 2 knots)	12	15	17	19	22	24	27	29	32	34	36	48	60	73
Time (at 3 knots)	8	10	11	13	15	16	18	19	21	23	24	32	40	49

the gear should follow if it is directly behind the vessel. Hence the gear position on the seamount can be estimated from the depth rather than distance or time behind.

Scientific hydroacoustic positioning systems have developed in recent years. These consist of a transponder beacon which can be attached to sampling gear. This gives angle and distance information relative to the vessel. It is commonly used on research equipment such as remote operated vehicles and drop camera frames. However, towed trawls have a much higher risk of damage, and such equipment is not fitted to trawls as standard. Some commercial software applications use winch- and trawl-mounted sensor information on wire out, bottom depth and ship's speed to automatically estimate the net position.

7.3.3.7 Trawl depth

The modern commercial netsonde, as well as giving information on bottom contact, normally has a pressure sensor as well, which gives a depth read-out on the net monitor screen. Alternatively depth is given by a separate depth sensor. This can be important information to record both for estimating gear position, and for knowing as accurately as possible the depths where species are caught.

Depth readings from net sensors are not usually calibrated at depth, and so it is advisable to check the sensor read-outs against the depth indicated by the vessel's echosounder. This can be done by a tow on flat bottom, which is often carried out routinely at the start of a survey to ensure gear is operating properly. The netsonde read-out needs to be corrected (deeper) by the headline height to give the bottom depth.

It is common for multiple echosounders on research vessel to differ in their depth read-outs. This is sometimes due to varying operating frequencies and beam widths. The settings should also be checked for whether vessel draught is included, and particular parameters entered for speed of sound in water.

7.4 Dealing with rough seafloor

In areas of steep slope at the edge of the continental shelf, and particularly on seamounts, the seafloor can be hard and rugged, with exposed bedrock (little sediment accumulation on top), boulders and cobbles. These can easily snag trawl nets, resulting in the loss of the catch, damage to the trawl, and in some cases loss of the trawl. There are several ways to reduce this likelihood, both with the gear itself, and the mode of operation.

Rough-bottom trawls typically have large ground gear, comprising bobbins and rubber rollers which help the trawl to roll, or bounce, over areas of rough bottom (Fig. 7.5). In comparison with smooth bottom ground gear (top two panels of Fig. 7.5) the fishing line of the rough-bottom trawl is raised clear of the seabed. The 'rockhopper' ground gear (a larger version of that shown in the third panel) can have rubber discs and rollers of large diameter for extremely rugged seafloor, while the bobbin gear (bottom panel) is generally limited to a maximum diameter of 600 mm. Note that these four examples are not exhaustive, and there are many ways of rigging bobbins and rollers, as well as the means of attaching the fishing line.

The same net can be used with different sets of ground gear. A decision on what type of ground gear would be most suitable for a given survey should be made in advance. Although it is possible to change ground gear at sea, depending on gear size and deck space, it may take a lot of time.

Cut-away lower wings are a feature of rough-bottom trawls used in commercial fisheries on seamounts for species such as orange roughy (Fig. 7.6). Damage to the lower wings is common on rough bottom where the leading sections of the net are vulnerable to being ripped. The absence of lower wings compared with nets where the ground gear is lighter and the net has full wings does not appear to decrease species composition (NIWA, unpublished data) but reduces catch rates (Smith *et al.*, 2008).

Wrapped wire looped chain

Rubber 'cookies' stamped from tyres, threaded onto chain

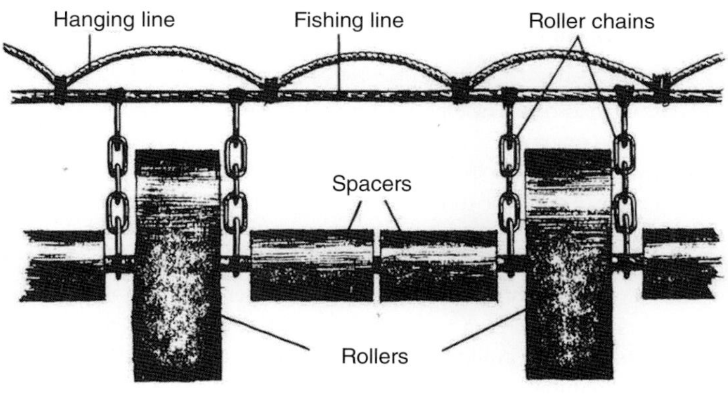

Heavy moulded rubber rollers separated by smaller spacers

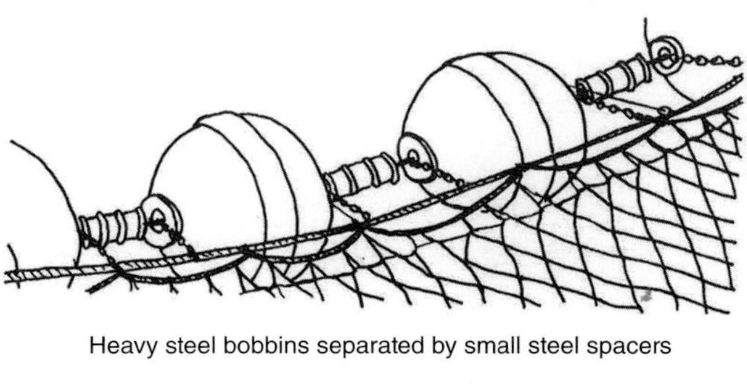

Heavy steel bobbins separated by small steel spacers

Fig. 7.5 Ground gear on bottom trawls, ranging from light and small roller gear (top panels) for smooth and soft bottom, to roller and bobbin gear for hard and rough seafloor (bottom panels). (Sainsbury, 1996. Reproduced with permission of John Wiley & Sons Ltd.)

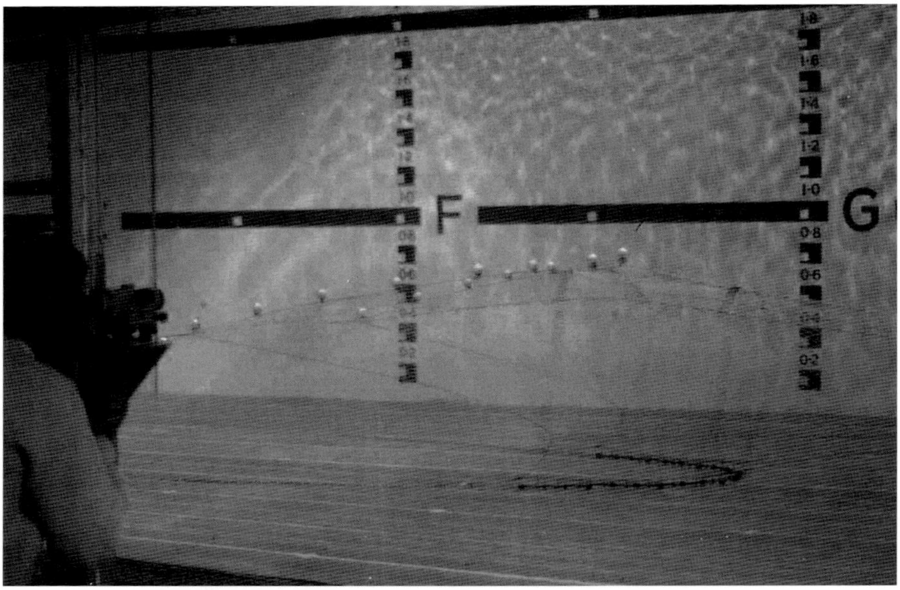

Fig. 7.6 A scale model of a rough-bottom trawl being tested in a flume tank. Note the long upper wings but small lower wings on the leading edge of the net. (Photograph courtesy of NIWA.) (Please see Plate 13 for colour representation of the figure.)

Stability of trawl doors can also be a factor affecting performance on rough seafloor or rapidly changing topography such as found on seamounts. Today there are many types and designs of trawl doors available for deep-sea fishing. Several types of doors are able to be used with both midwater trawls as well as with bottom trawls. The ratio of weight in the top and bottom sections of the door can be altered. A more even distribution of weight (less weight on the bottom of the trawl doors than normal) reduces the shearing action of doors to some extent, as the bottom contact is less, but also means the doors respond more rapidly to changes in the amount of warp. This distribution can give better control of the trawl on seamounts.

The ground gear, doors and net design are not easily altered from the gear inventory that a research institute or research vessel may have available for shallower shelf surveys. Hence the above comments are relevant mainly to new programmes going into deep-sea environments. The rigging of the net, however, can have an effect. Layback can be increased, whereby the top bridle is made longer than the lower bridle, which allows the headline to drop back, can reduce some pressure on the ground gear, and also increase the vertical opening. However, if the survey is part of a time series, then changes should not be made as these could affect comparability of surveys over time, even if the change would improve performance.

Net flotation is another important element. The buoyancy of floats varies with their depth rating. Deep floats (e.g. rated to 2000 m) have less lift than shallower floats (e.g. rated to 1500 m). More 2000 m floats are required for a certain net opening, which increases the resistance of the trawl through the water. Fewer 1500 m floats have been found by NIWA to give better gear performance trawling on seamounts, with greater headline height, and reduced instances of coming fast (Clark & Tracey, 1994).

7.5 Evaluation of trawl gear performance

It is important to assess each trawl shot and determine the success of a tow for later analysis. Trawl shots need to be classified at the time as acceptable (valid) or foul (rejected) due to poor performance or damage.

Net-mounted sensors provide real-time feedback on trawl gear performance (e.g. headline height, door spread, wing spread, net speed through the water, net symmetry, bottom contact and door tilt angles). These sensors are limited in depth capacity, and are often only rated to 2000 m. Acceptable parameters measured by such electronic equipment should be specified for a given set of trawl gear and checks made on the gear set-up if values fall outside an acceptable range. Such values should be determined from gear trials carried out before any serious survey work begins, and values derived from tows where the operation was felt to be successful (e.g. Hurst *et al.*, 1992; Neat *et al.*, 2010). There will always be variation in net parameters during a trawl tow, especially when towing over rough seafloor. However, average headline height, door spread, wingspread and towing speed should be able to be maintained close to 'average' values. A flowchart from the *ICES International Bottom Trawl Survey Manual* (ICES, 2010) shows the process of some of the checks during a trawl (Appendix 7.3).

In addition, other bridge equipment can provide some information of the success of a tow. Warp tension is often the first indication of gear fouling on the seabed, and a reduction of vessel speed can indicate gear clogging or fouling.

On retrieval there are visual observations that can aid in determining a successful trawl. Most of this is based on good bottom contact along the length of the gear, typically indicated by metal parts polishing with bottom contact:

- *Trawl warps* should show no signs of polish ahead of the doors (e.g. polish may indicate a trawl door has fallen over).
- *Trawl doors* should have a shiny surface along the shoe (foot) of the door indicating good bottom contact. No mud or polish on the upper parts of the door or towing brackets indicating the door may have fallen over during the tow.
- *Sweeps and bridles* should show polish along most of the sweep length. Depending on the back-strop set-up the first section of sweep immediately behind the door may not show any signs of polish. The bottom bridle should show signs of polish. Often there is chain joining the sweeps to the bridles and these should again show good signs of polish.

The metal parts of the trawl gear tend to discolour with rust in only a few hours, and so new polish signs should be visible for each trawl. Spray paint can be used to mark areas to assess bottom contact if needed.

The trawl ground gear can also provide an indication of bottom-contact and gear performance. Steel bobbins should show even polish along the length of the groundrope. As the ground gear forms a catenary the leading edge of the bobbins on each side should polish up more that the trailing edges. The bobbin in the centre of the groundrope should have even polish on both sides as an indication the net is not being towed in a skewed configuration (Fig. 7.7a). If there are no steel components to the ground gear small lengths of chain (tell-tails) can be attached at intervals along the groundrope (Fig. 7.7b).

Damage to the trawl gear and nets is usually evident when the trawl is hauled aboard. Consideration to how much damage is acceptable before a shot is deemed unacceptable needs to be included as part of the standardization procedures. Net damage is generally easy to identify, and a trawl should be classified as poor if there is sufficient ripping to have influenced either the catch composition or catch rate. Belly sections, lengtheners and lower wings are often ripped during trawling on rough ground, and these need to be checked for each tow and repaired before the gear is shot away again.

7.6 Sample sorting and processing

Trawl nets, unlike many other gear types, can mesh fauna throughout the length of the net (i.e. animals get caught in the netting rather than being swept along the net into the cod-end). It is therefore important to ensure the net is carefully cleared of all meshed fish and invertebrates after each trawl.

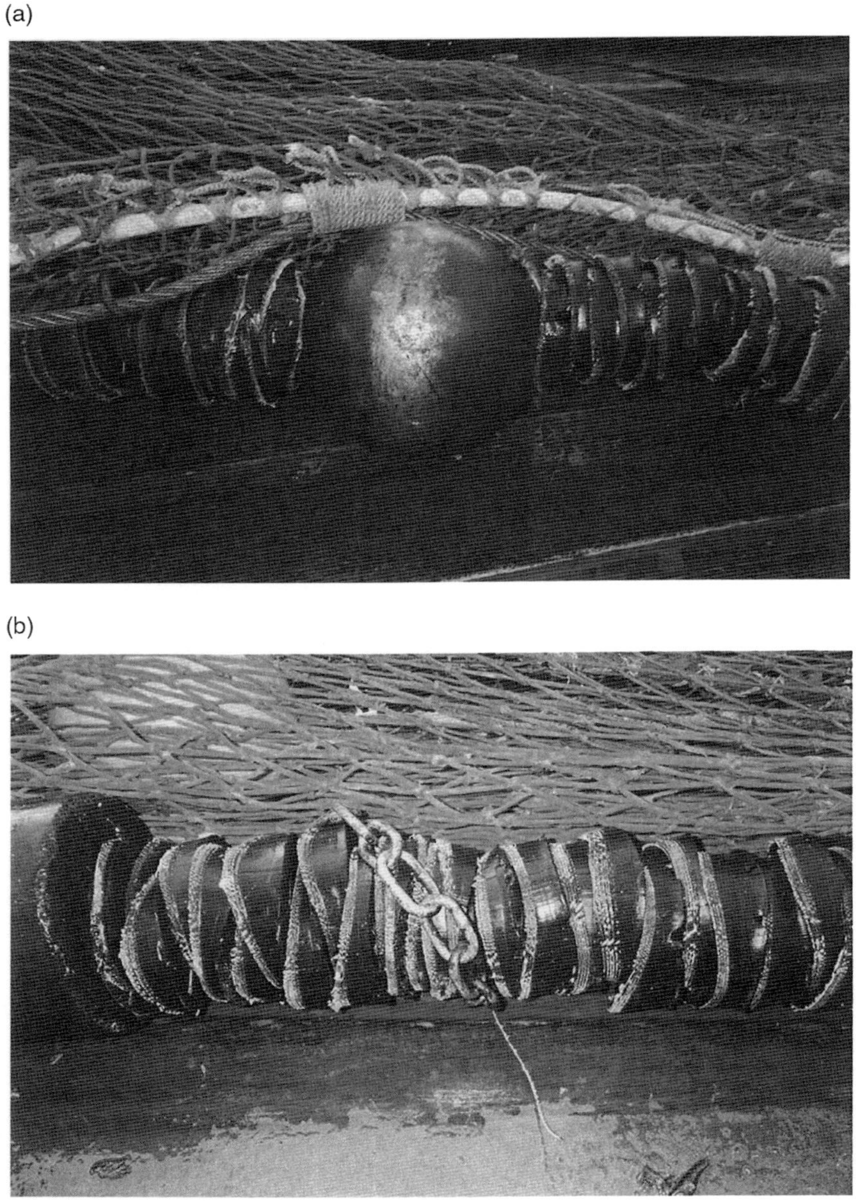

Fig. 7.7 Checks on gear performance. (a) Centre bobbin showing even polish which indicates towing is straight. (b) Part of a light chain tell-tail showing good polish after a trawl shot indicating bottom contact. (Photograph courtesy of NIWA.)

Otter trawls are often long, and several fleets (hauling a section of net) are needed to retrieve the cod-end onto the deck. The net drum should be stopped at regular intervals to enable meshed animals to be pulled out of the wings, belly and lengthener sections of the net. Checks should also be made after the cod-end is emptied.

Trawl nets can often catch large amounts of fish (when sampling an aggregation), or catch over a large distance. This situation can sometimes mean the back of the cod-end does not represent the species mix or size composition of the entire catch. For example with orange roughy, it has been found that different parts of the cod-end can have different sex ratios, and hence different size composition (Clark & Tracey, 1987; Pankhurst, 1988). Multiple samples from sections of the net should be taken in such a situation. A common practice at some institutes when trawls catch large quantities of fish is for a sample of 200 fish to be taken for every 5 t of catch. This partitions the sample throughout the net and when combined gives a more representative biological sample. Hence, when there is a need to subsample, we recommend taking several smaller samples from different parts of the net rather than a larger sample from one part.

Details of biological sampling and measurement methods should also be consistent between surveys. Most survey manuals should include a section on fish measurements so the same measurement is applied between species on different surveys. This is especially the case with species where there can be ambiguity of caudal fin measurements (e.g. if the caudal fins are ragged then standard or fork length should be used) or if tail breakage is common (such as with macrourid fish where pre-anal fin length can be used).

7.7 Interpretation of data

Each gear type has different selection capabilities and characteristics. Hence the interpretation of catch composition and relative abundance requires an understanding of what type and size of fauna are well sampled by the gear as deployed during a particular survey. There is an extensive literature on trawl gear efficiency and selectivity, and below we highlight some of the key factors to be aware of, rather than attempt to summarize a large number of reports and papers.

7.7.1 Trawl efficiency

Beam and Agassiz trawls are fixed width and therefore quantitative if there is good contact with the seafloor and a regular towing speed can be maintained. Nevertheless the efficiency of such gear has been measured in shallow water at 20–60% (e.g. review by Eleftheriou & Moore, 2005) but there are no studies known from deep-sea habitats where experimental measurement of efficiency is much more difficult, and where the lack of natural light may lead to differences from shallow water in the behaviour of animals in front of the trawl.

The performance of otter and shrimp trawls is not as consistent as beam or Agassiz trawls, and hence they are not regarded as fully quantitative (e.g. Gage & Bett, 2005) but sufficiently so for time series of surveys (see papers in Doubleday & Rivard, 1981, which cover aspects as relevant today as then). The spreading effect of the trawl doors will vary with amount of wire out, the setting of the warp-door attachment, weight of the doors, the substrate type and speed through the water. Door spread can show considerable variation during a tow and consequently the spread and opening height of the net. Often door spreads increase with depth, when towing into the prevailing weather (e.g. Saila & Mowbray, 1972; Hurst *et al.*, 1992). Towing up- or downslope can also affect gear performance.

7.7.2 Trawl selectivity

The design and set-up of trawls can affect the species, and size of animals caught. Species selectivity is affected by body size and shape (e.g. Tosunoglu *et al.*, 2003), and fish behaviour in response to an approaching trawl can result in success if the fish dives, or low catches if the fish swims upwards (e.g. Foster *et al.*, 1981). The heavy bobbin or rockhopper ground gear used on rough-bottom trawls will not be efficient at catching flatfish or other small-bodied species that are distributed, or dive hard down on the bottom. That situation is one where beam trawls are effective because of their smaller groundrope, and tickler chains which force animals out of the upper layer of the substrate. Large and small nets of the same design appear to have equal fish species selectivity (Dahm *et al.*, 2002), and so as long as the net and trawl gear are matched to the vessel size and power, smaller nets than used by commercial trawlers should be suitable for research vessels.

Size of the ground gear also has an effect. Selectivity experiments off Scotland showed marked improvement in species representation when rockhopper ground gear was changed from 21 to 16 inch rockhoppers (Neat *et al.*, 2010). Escapement of fish under large ground gear is a well-known effect (e.g. Walsh, 1992).

Mesh size and shape (e.g. square versus diamond mesh) have a clear effect on the size of fish and invertebrates able to be retained (e.g. Sainsbury, 1996). Small species can pass through a large mesh size in the cod-end, although some individuals are likely to be caught in the meshes even if their abundance is poorly represented. Francis *et al.* (2002) found little difference in community composition from different mesh sizes, and suggested that small-bodied fish can often be retained in large-mesh nets through the action of larger fish blocking the meshes, and mesh size can close when being towed.

Towing speed is an important factor in whether a trawl can catch large and faster swimming species (Dahm *et al.*, 2002). Gordon *et al.* (1996) found that catches differed significantly between a large Granton-type otter trawl fished off twin warps and a single warp OTSB. The latter trawl is towed more slowly, and was found to be a poor at catching large, mobile species, but highly efficient at catching slender-bodied eels (Gordon *et al.*, 1996). Trawling speed also affects the size composition of the catch, as swimming speeds of fish, and the time to fatigue, varies with species and size (e.g. reviewed by Foster *et al.*, 1981). Fast-swimming fish may be able to swim in front of a trawl for a short time, but eventually will tire and fall back into the net.

Standardization of trawl gear and its operation is critical with time series of surveys.

7.8 Environmental impact considerations

Bottom trawls are a major cause of damage to seafloor habitat, because of their weight and large areas covered during extensive trawling operations on fishing grounds. Various technical modifications can reduce bycatch, including changes in mesh size, the use of separator panels, sorting grids, various types of mesh and escape openings (Van Marlen, 2000; Kennelly & Broadhurst, 2002). Such changes have proven effective in reducing bycatch in many fisheries.

However, reducing the impact on benthic animals, sessile invertebrates in particular, is more difficult. Corals and sponges are fragile structures, and there is now little debate over the damage that bottom trawls can cause (Clark *et al.*, 2015). If the shape and nature of the seamount allows, then use of midwater gear close to the bottom may be effective in catching some species such as alfonsino (*Beryx splendens*) or armourhead (*Pseudopentaceros wheeleri*). The ground gear on midwater trawls is often light, but with nets consistently used close to the bottom, the groundrope can be made much heavier, and have a sacrificial link in the middle so it can part if the gear hits hard bottom rather than become fast and risk ripping the entire net or losing the entire set of gear.

If gear type cannot be changed, then the groundrope could be lightened, trawl doors balanced (weight distribution less concentrated on the lower part of the door) so they do not dig in as much, sweeps shortened so the trawl covers less area, and bobbins placed on the sweeps to ‘elevate’ them (Valdemarsen *et al.*, 2007; Rose *et al.*, 2010). The use of catch sensors that indicate filling rates of the net when towing into aggregations can also help to reduce tow times, as hauling can begin as soon as the catch is large enough, rather than towing for longer. Design of the ground gear should also be investigated. In some deep-sea fisheries (e.g. orange roughy (*Hoplostethus atlanticus*)) there has tended to be a shift from bobbin rigs to rockhopper gear, which is generally lighter, being designed to drag and bounce, rather than roll. Its impacts may differ, but may not necessarily be reduced. The likely application of such methods will vary with target species and habitat, and operationally may entail a trade-off between reduced bottom impact and catch rate of the fish species,

Gear modifications may be possible to reduce bottom contact, but it is important that the overall objectives of the survey are kept upper-most in mind, as well as other uses of the survey data. Gear changes should not compromise the catch if catch rates or species composition are important. For example, bycatch of some species can be reduced by altering the type of ground gear, and increasing mesh sizes will decrease the catch of small species or small-sized fish. However, research is increasingly directed towards describing entire communities and hence if trawling is a necessary method then bottom impact is unavoidable. Nevertheless the spatial scale of bottom trawling (tow duration, number of trawls) should be matched to that needed to achieve objectives.

Acknowledgements

We are grateful to Francis Neat (Fisheries Research Services, Marine Laboratory, Aberdeen) and Matthew Dunn (Victoria University of Wellington, Wellington) for their advice and helpful comments on a draft of this manuscript.

Appendix 7.1 Net, ground gear and rigging plans for a typical rough-bottom trawl used both commercially and for research on seamounts in the southern hemisphere (Reproduced with permission of NIWA)

Net plan

Mesh size	12 inch	12 inch	12 inch	9 inch	6 inch	4 inch
Twine size	5 mm olivene	5 mm olivene	5 mm olivene	5 mm olivene	5 mm oliv.	5 mm olivene
No. of meshes	50	20	30	50	50.5	50
Double twine	Wing tips 8 meshes deep		9.5 meshes in belly head			

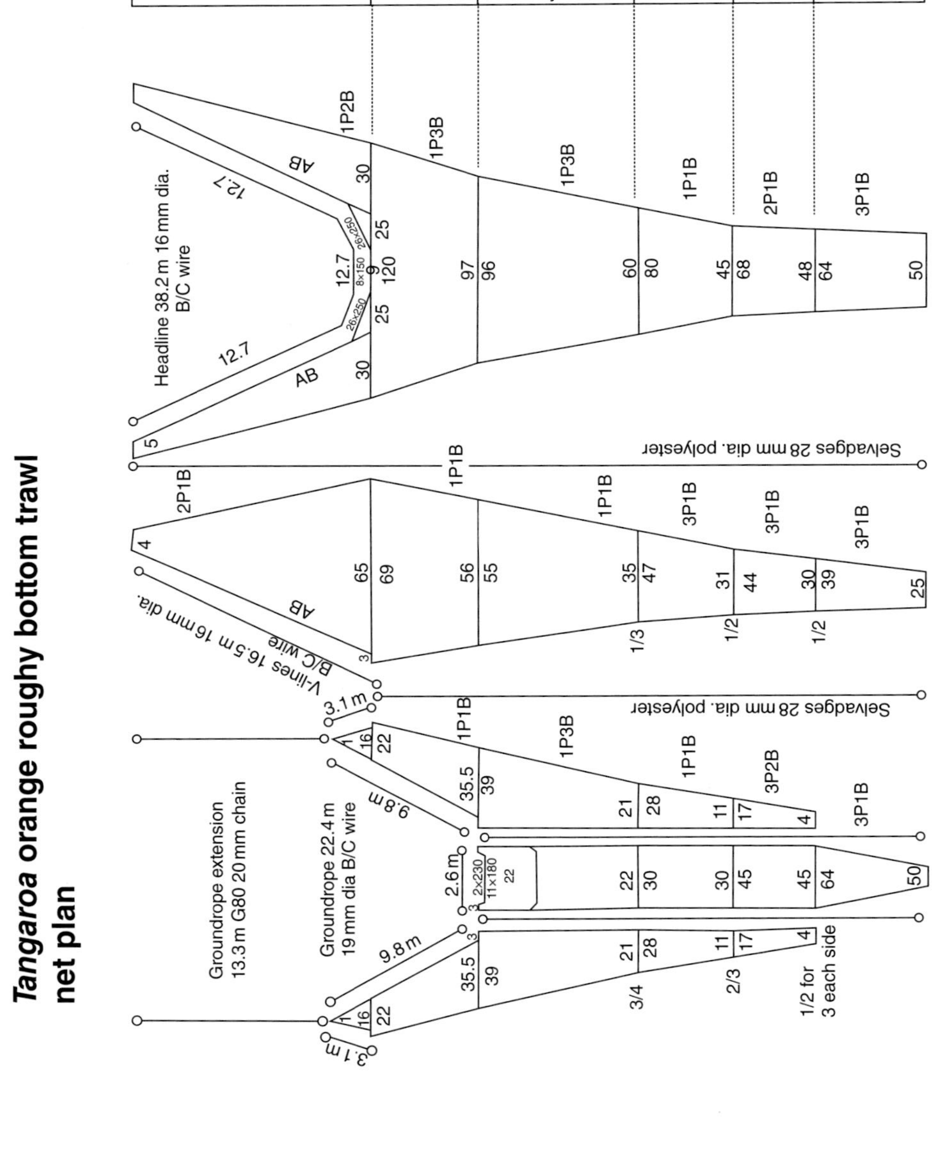

Headline float plan

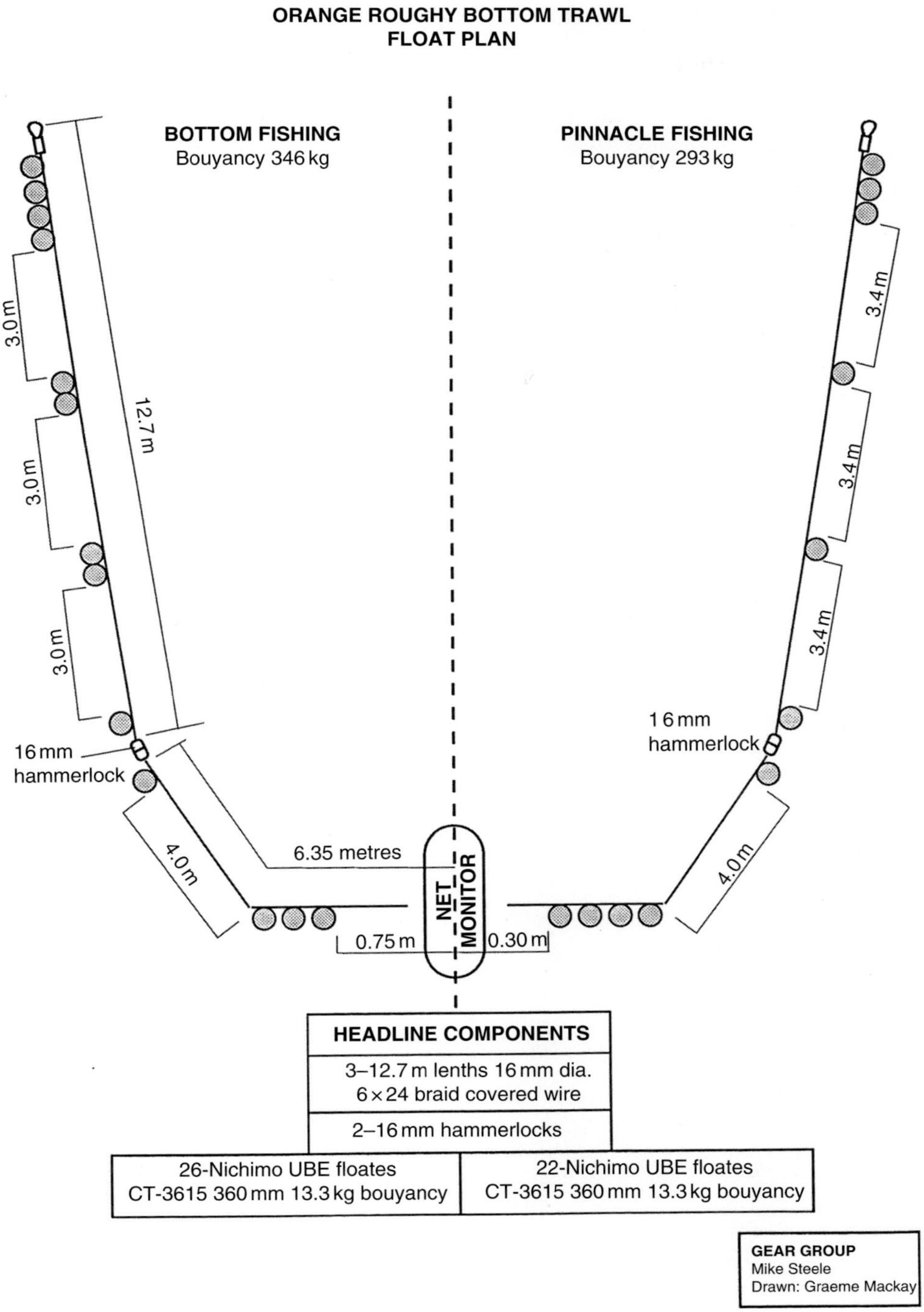

Ground gear plan

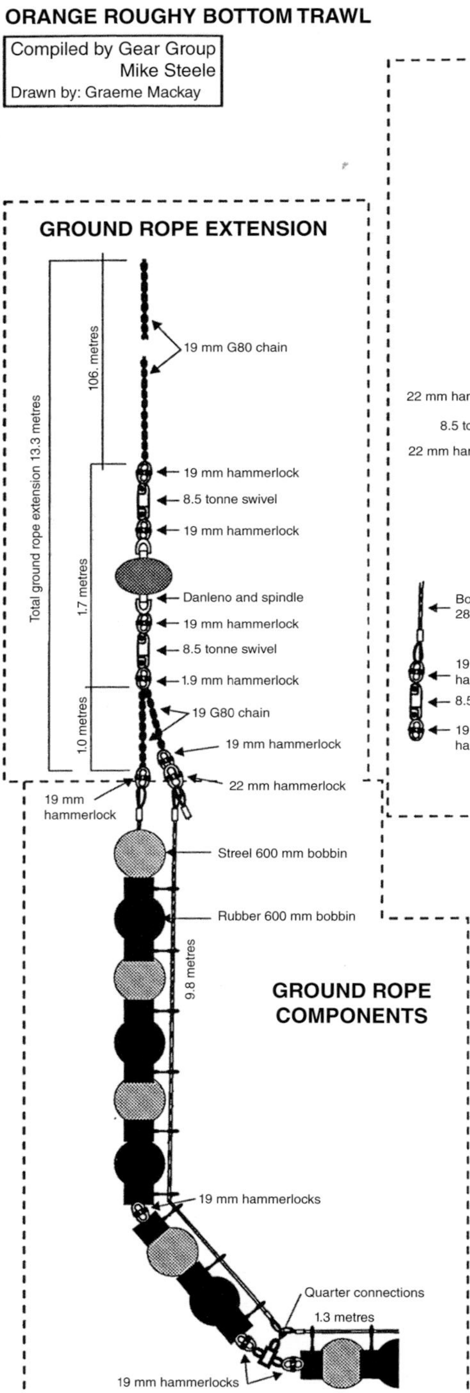

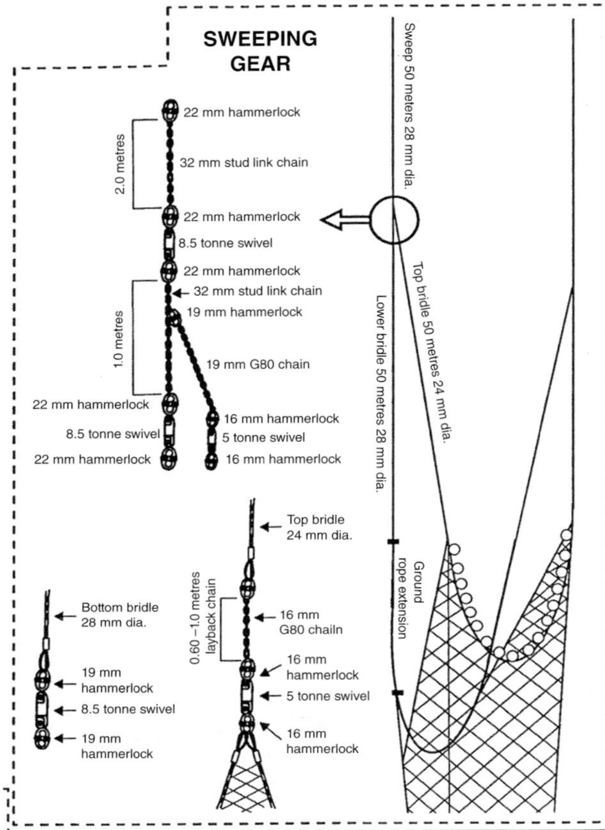

GROUND ROPE COMPONENTS
10 - 600 mm steel bobbins
9 - 600 mm rubber bobbins
22 - Lancasters
- Rubber spacers
2 - Quarter connections
6 - 19 mm hammerlocks
1 - 22 metre length of 16 mm G80 chain
1 - 3 section fishing line 18 mm 6 × 19
GROUND ROPE EXTENSION
4 - 1 metre lengths G80 19 mm chain
4 - 8.5 tonne swivels
12 - 19 mm hammerlocks
2 - 22 mm hammerlocks
2 - Danleno and spindle
2 - 10.6 metre lengths 19 mm G80 chain
SWEEPING GEAR
2 - 50 metre 24 mm 6 × 19 PPC wire rope
4 - 50 metre 28 mm 6 × 19 PPC wire rope
6 - 19 mm hammerlocks
4 - 5 tonne swivels
6 - 8.5 tonne swivels
2 - 3 metre length 32 mm stud link chain
2-1 metre 19 mm G80 chain
10 - 16 mm hammerlocks
10 - 22 mm hammerlocks
2 - 0.60–1.0 meter 16 mm G80 chain long link

Trawl door rigging plan

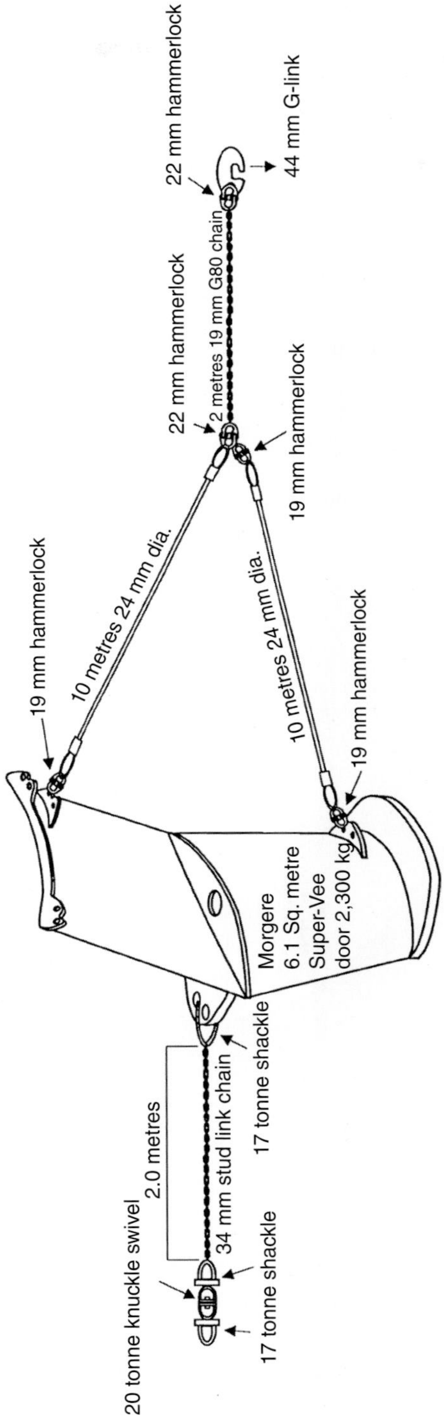

Appendix 7.2 Details of a beam trawl design used by CEFAS in European waters (CEFAS. Reproduced with permission)

Ground gear

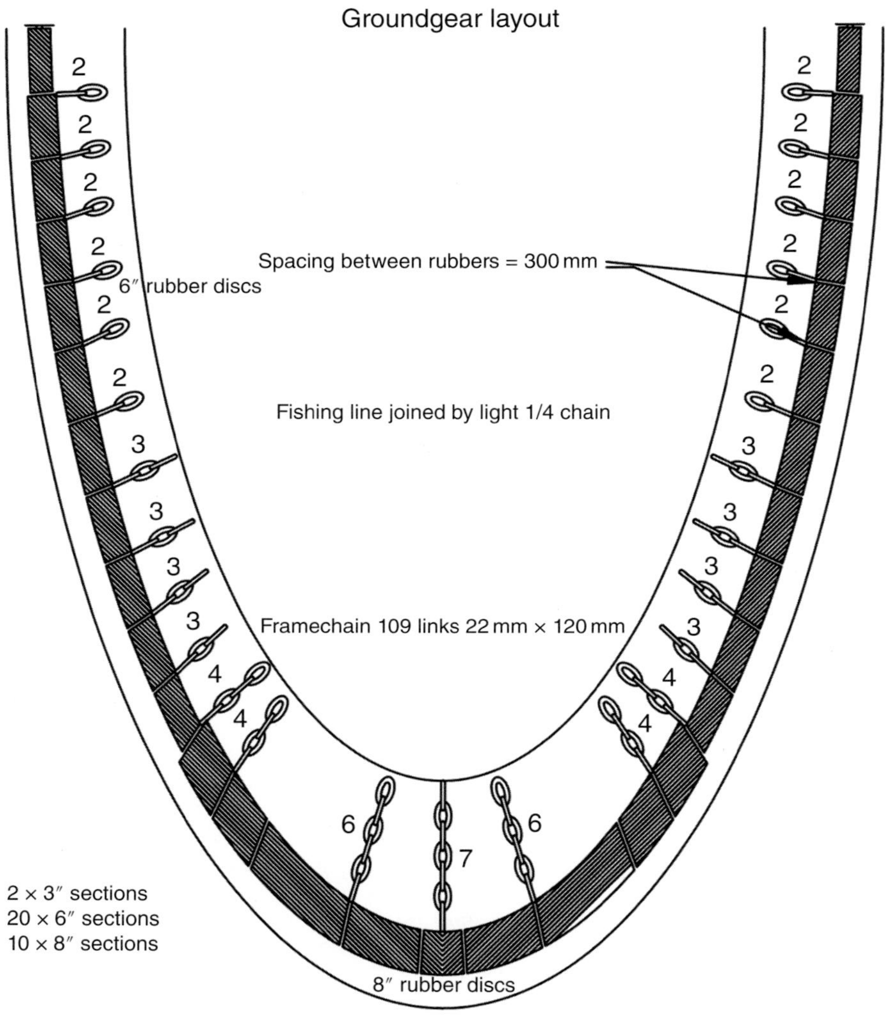

Trawl door rigging plan

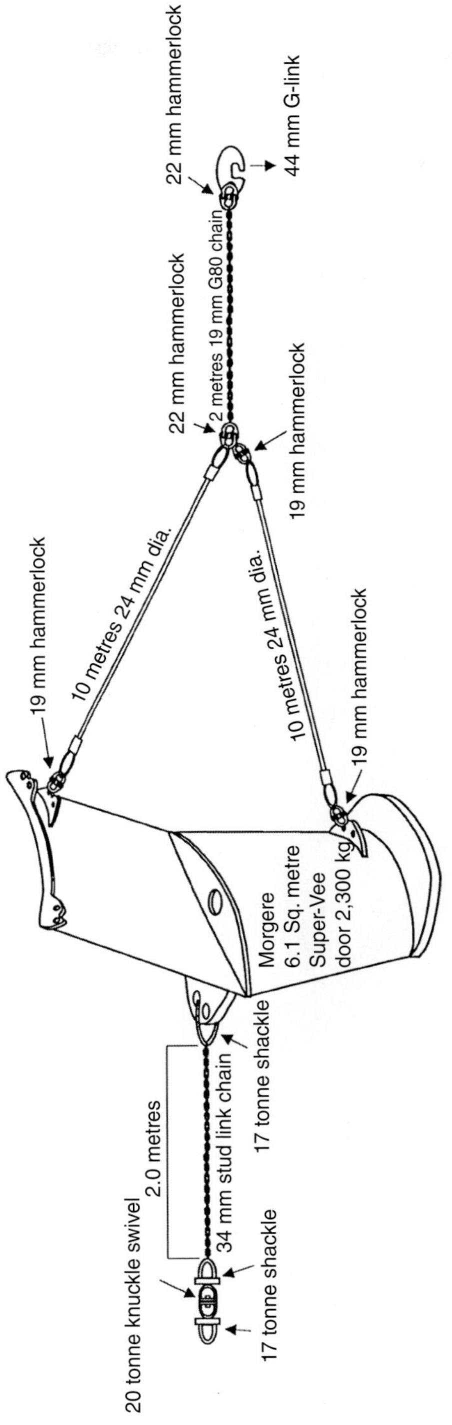

Appendix 7.2 Details of a beam trawl design used by CEFAS in European waters (CEFAS. Reproduced with permission)

Ground gear

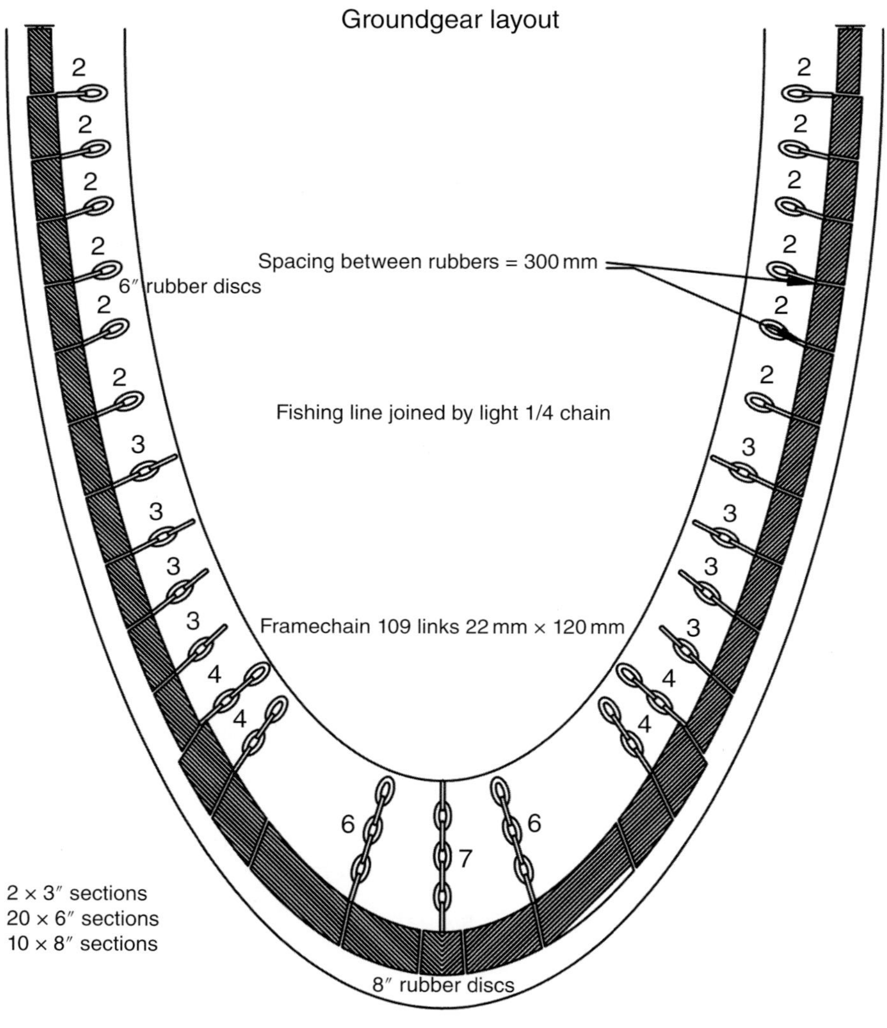

Chain mat

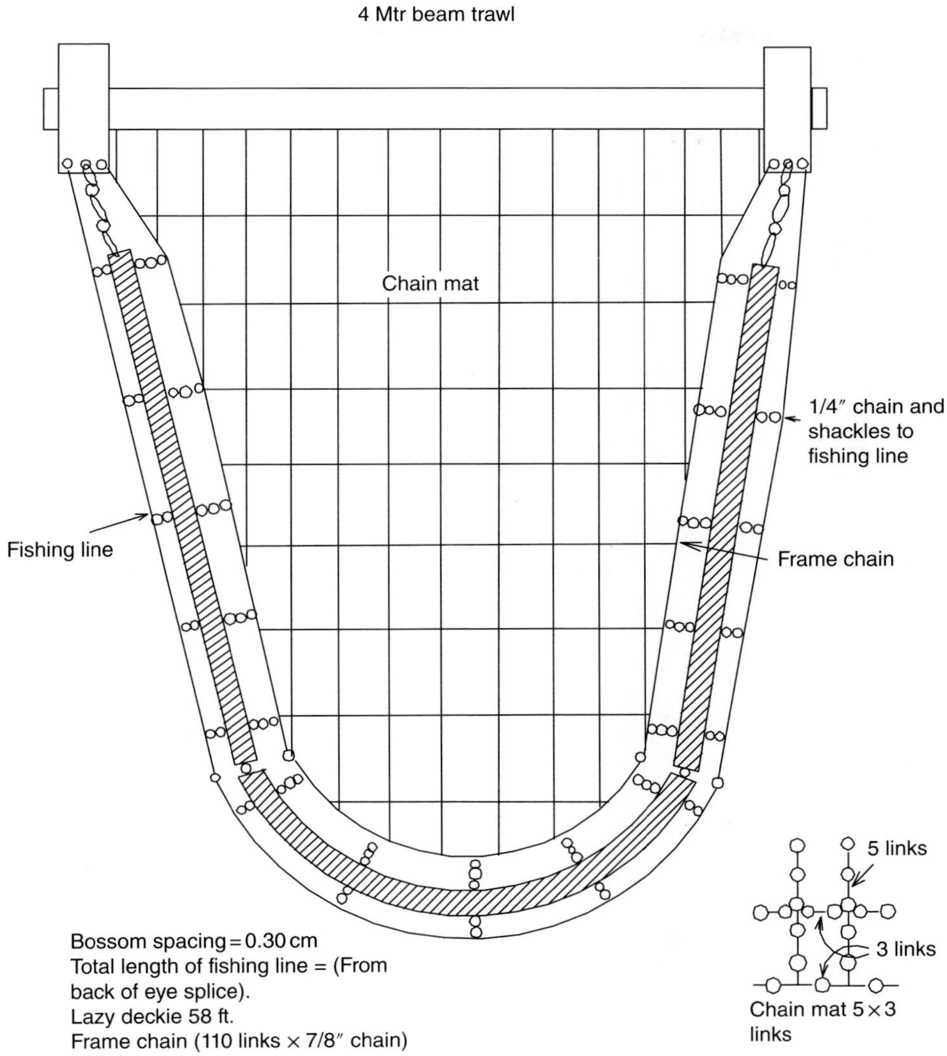
4 Mtr beam trawl
Chain mat
1/4″ chain and shackles to fishing line
Fishing line
Frame chain
5 links
3 links
Chain mat 5 × 3 links
Bossom spacing = 0.30 cm
Total length of fishing line = (From back of eye splice).
Lazy deckie 58 ft.
Frame chain (110 links × 7/8″ chain)

Net plan

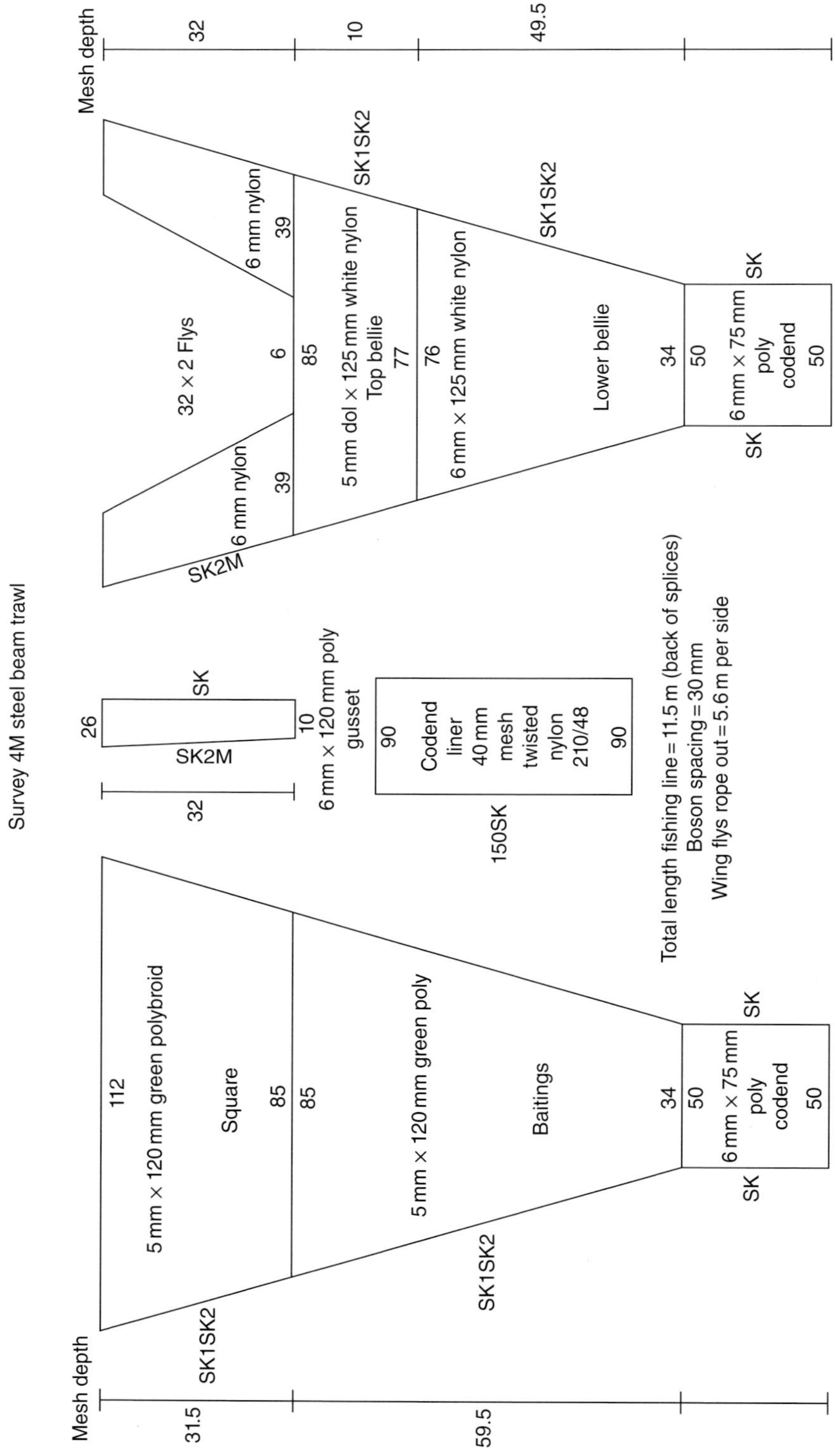

Survey 4M steel beam trawl
Mesh depth
32
10
49.5
32 × 2 Flys
6 mm nylon
39
6
39
6 mm nylon
SK2M
SK1SK2
85
5 mm dol × 125 mm white nylon
Top bellie
77
76
6 mm × 125 mm white nylon
SK1SK2
Lower bellie
34
50
6 mm × 75 mm
poly
codend
50
SK
SK
26
SK
SK2M
10
32
6 mm × 120 mm poly
gusset
90
Codend
liner
40 mm
mesh
twisted
nylon
210/48
90
150SK
Total length fishing line = 11.5 m (back of splices)
Boson spacing = 30 mm
Wing flys rope out = 5.6 m per side
112
5 mm × 120 mm green polybroid
Square
85
85
5 mm × 120 mm green poly
Baitings
SK1SK2
SK1SK2
34
50
6 mm × 75 mm
poly
codend
50
SK
SK
Mesh depth
31.5
59.5

Towing bridles

4 M Beam trawl towing points

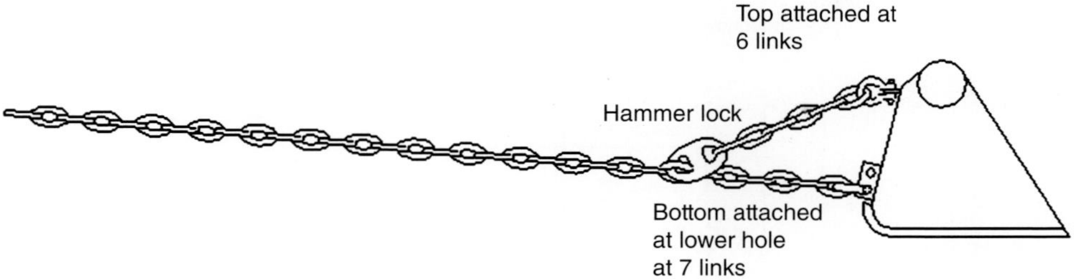

Appendix 7.3 Flow diagram of Scanmar sensor use from the *International Bottom Trawl Survey Manual* (Reproduced with permission. ICES, 2010)

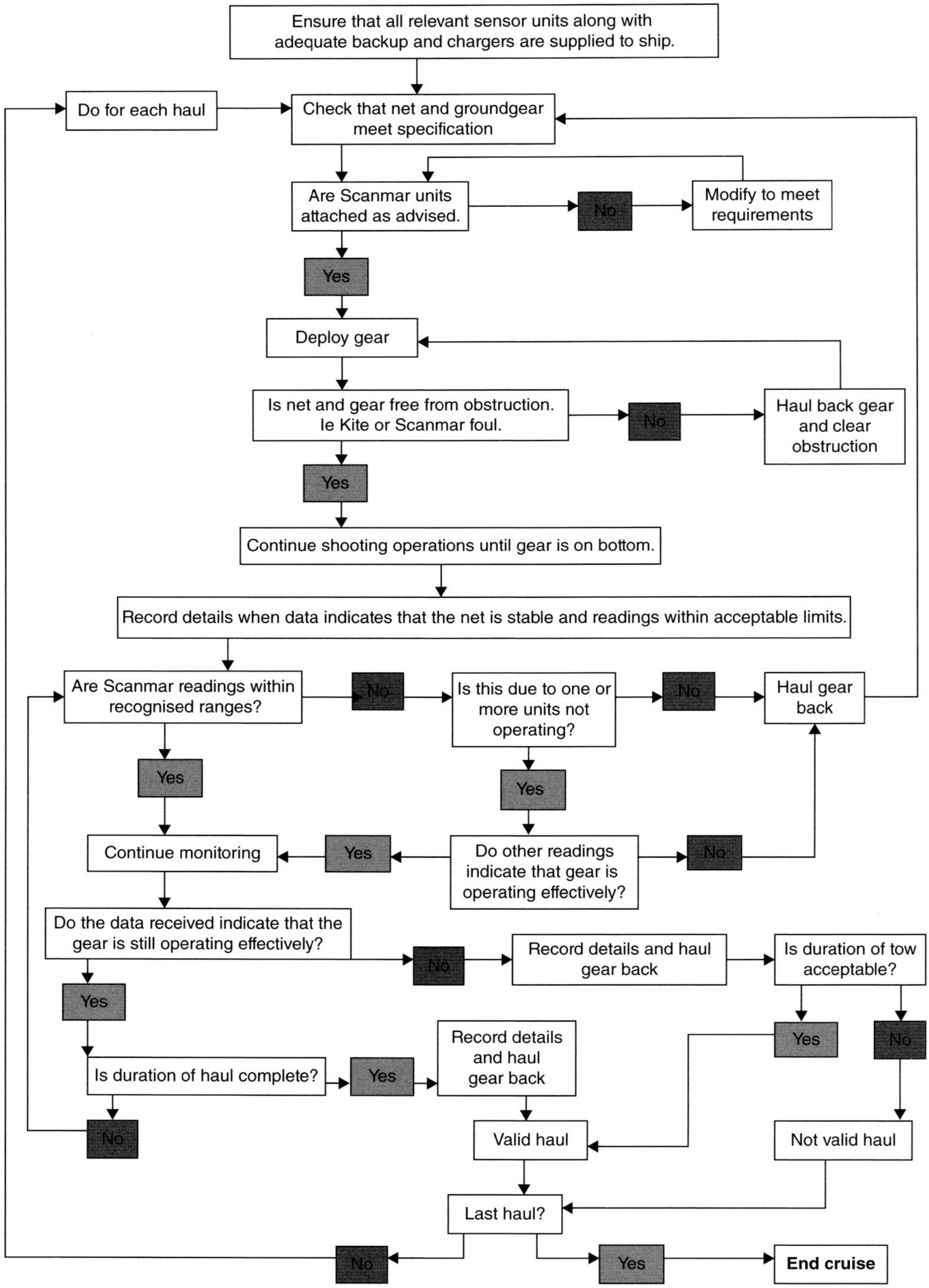

References

Billet, D.S.M., Bett, B.J., Rice, A.L., *et al.* (2001). Long-term changes in the megabenthos of the Porcupine Abyssal Plain (NE Atlantic). *Progress in Oceanography*, 50: 325–348.

Carrothers, P.J.G. (1981). Catch variability due to variations in groundfish otter trawl behaviour and possibilities to reduce it through instrumented fishing gear studies and improved fishing procedures. In: Doubleday, W.G., Rivard, D. (eds), Bottom trawl surveys: proceedings of a workshop held at Ottawa, November 12–14, 1980. *Canadian Special Publication of Fisheries and Aquatic Sciences*, 58: 247–257.

Clark, M.R. (1994). Aspects of random trawl survey design for deepwater species on pinnacle and drop-off features. MAF Fisheries Greta Point internal report 227: 23 pp. (draft report held at NIWA library, Wellington).

Clark, M.R. (1996). Biomass estimation of orange roughy: a summary and evaluation of techniques for measuring stock size of a deepwater fish species in New Zealand. *Journal of Fish Biology*, 49 (suppl. A): 114–131.

Clark, M.R., Roberts, C.D. (2008). Fish and invertebrate biodiversity on the Norfolk Ridge and Lord Howe Rise, Tasman Sea (NORFANZ voyage, 2003). *Aquatic Environment and Biodiversity Report*, No. 28. 131 pp.

Clark, M.R., Tracey, D.M. (1987). Report of an orange roughy trawl survey on Challenger Plateau, June–July 1987. Fisheries Research Centre internal report 86: 86 pp. (unpublished report held by NIWA library, Wellington).

Clark, M.R. Tracey, D.M. (1994). Trawl survey of orange roughy, black oreo, and smooth oreo in southern New Zealand waters, August–September 1992 (TAN9208). *NZ Fisheries Data Report*, 40: 37 pp.

Clark, M.R., Althaus, F., Williams, A., *et al.* (2010). Are deep-sea fish assemblages globally homogenous? Insights from seamounts. *Marine Ecology*, 31 (suppl. 1): 39–51.

Clark, M.R., Althaus, F., Schlacher, T., Williams, A., Bowden, D., Rowden, A.A. (2015). The impacts of deep-sea fisheries on benthic communities: a review. *ICES Journal of Marine Science*. doi: 10.1093/icesjms/fsv123

Dahm, E., Wienbeck, H., West, C.W., Valdemarsen, J.W., O'Neill, F.G. (2002). On the influence of towing speed and gear size on the selective properties of trawls. *Fisheries Research*, 55: 103–119.

Doubleday, W.G., Rivard, D. (eds) (1981). Bottom trawl surveys: proceedings of a workshop held at Ottawa, November 12–14, 1980. *Canadian Special Publication of Fisheries and Aquatic Sciences*, 58, 273 pp.

Eleftheriou, A., Moore, D.C. (2005). Macrofauna techniques. In: Eleftheriou, A., McIntyre, A. (eds), *Methods for the study of marine benthos*. Oxford: Blackwell Science, 418 pp.

Foster, J.J., Campbell, C.M., Sabin, G.C.W. (1981). The fish catching process relevant to trawls. In: Doubleday, W.G., Rivard, D. (eds), Bottom trawl surveys: proceedings of a workshop held at Ottawa, November 12–14, 1980. *Canadian Special Publication of Fisheries and Aquatic Sciences*, 58: 229–246.

Francis M.P., Hurst R.J., McArdle B.H., Bagley N.W., Anderson O.F. (2002). New Zealand demersal fish assemblages. *Environmental Biology of Fishes*, 65: 215–234.

Gabriel, O., Lange, K., Dahm, E., Wendt, T. (2005). *Von Brandt's fish catching methods of the world* (4th edn). Oxford: Blackwell, 53 pp.

Gage, J.D., Bett, B.J. (2005). Deep-sea benthic sampling. In: Eleftheriou, A., McIntyre, A. (eds), *Methods for the study of marine benthos*. Oxford: Blackwell, 418 pp.

Gage, J.D., Tyler, P.A. (1991). *Deep-sea biology: a natural history of organisms at the deep-sea floor*. Cambridge, UK: Cambridge University Press.

Gordon, J.D.M., Merrett, N.R., Bergstad, O.A., Swan, S.C. (1996). A comparison of the the deep-water demersal fish assemblages of the Rockall Trough and Porcupine Seabight, eastern North Atlantic: continental slope to rise. *Journal of Fish Biology*, 49 (suppl. A): 217–238.

Hurst, R.J., Bagley, N., Chatterton, T., Hanchet, S., Schofield, K., Vignaux, M. (1992). Standardisation of hoki/middle depth time series trawl surveys. MAF Fisheries Greta Point Internal Report No. 194. 89 pp. (unpublished report held at NIWA library, Wellington).

ICES (2008). Report of the Planning Group on the North-east Atlantic Continental Slope Survey (PGNEACS). ICES PGNEACS Report 2008. *ICES CM 2008/LRC:02*. 34 pp.

ICES (2009a). Report of the Planning Group on the North-east Atlantic Continental Slope Survey (PGNEACS). ICES PGNEACS Report 2009. *ICES CM 2009/LRC:03*. 55 pp.

ICES (2009b). Report of the Study Group on Survey Trawl Standardisation (SGSTS). ICES SGSTS Report 2009. *ICES CM 2009/FTC:09*. 127 pp.

ICES (2009c). *Manual for the offshore beam trawl surveys*. Revision 1.2, June 2009. Working Group on Beam Trawl Surveys. 30 pp. http://datras.ices.dk/Documents/Manuals/WGBEAM_Manual.pdf

ICES (2010). *Manual for the international bottom trawl surveys in the western and southern areas*. Revision III. Addendum II: IBTS manual on the western and southern areas 2010.

Jennings, S., Lancaster, J., Woolmer, A., Cotter, J. (1999). Distribution, diversity and abundance of epibenthic fauna in the North Sea. *Journal of the Marine Biological Association of the UK*, 79: 385–399.

Kennelly, S., Broadhurst, M. (2002). Bycatch begone: changes in the philosophy of fishing technology. *Fish and Fisheries*, 3: 340–355.

McMillan, P.J. (comp) (1996). Trawl survey design and data analysis procedures for deepwater fisheries research. MAF Fisheries Greta Point Internal Report No. 253. 26 pp. (unpublished report held at NIWA library, Wellington).

Merret, N.R., Marshall, N.B. (1980). Observations on the ecology of deep-sea bottom-living fishes collected off northwest Africa (08°–27°N). *Progress in Oceanography*, 9: 185–244.

Mesnil, B., Cotter, J., Fryer, R.J., Needle, C.L., Trenkel, V.M. (2009). A review of fishery-independent assessment models, and initial evaluation based on simulated data. *Aquatic Living Resources*, 22: 207–216.

Neat, F., Kynoch, R., Drewery, J., Burns, F. (2010). *Deepwater trawl survey manual. Marine Scotland Science Report* 03/10. 46 pp.

Pankhurst, N.W. (1988): Spawning dynamics of orange roughy, *Hoplostethus atlanticus*, in mid-slope waters of New Zealand. *Environmental Biology of Fishes*, 21: 101–116.

Rees, H.L. (ed.) (2009). Guidelines for the study of the epibenthos of subtidal environments. *ICES Techniques in Marine Environmental Sciences*, No. 42. 90 pp.

Rose, C.S., Gauvin, J.R., Hammond, C.F. (2010). Effective herding of flatfish by cables with minimal seafloor contact. *Fishery Bulletin*, 108: 136–144.

Rowden, A.A., Schlacher, T.A., Williams, A., *et al.* (2010). A test of the seamount oasis hypothesis: seamounts support higher epibenthic megafaunal biomass than adjacent slopes. *Marine Ecology*, 31 (suppl. 1): 95–106.

Saila, S.B., Mowbray, W.H. (1972). Trawl studies based on bottom current measurements. *Marine Fisheries Review*, 34(11–12): 32–47.

Sainsbury, J.C. (1996). *Commercial fishing methods: an introduction to vessels and gears* (3rd edn). Oxford: Fishing News Books, 359 pp.

Smith, M.H., Hart, A.C., McMillan, P.J., Macaulay, G. (2008). Acoustic estimates of orange roughy abundance from the northwest Chatham Rise, June–July 2005: results from the wide-area and hill surveys. *New Zealand Fisheries Assessment Report*, 2008/13, 42 pp.

Tosunoglu, Z., Ozbilgin, Y.D., Ozbilgin, H. (2003). Body shape and trawl cod end selectivity for nine commercial fish species. *Journal of the Marine Biological Association of the UK*, 83: 1309–1313.

Tracey, D.M., Bull, B., Clark, M.R., MacKay, K.A. (2004). Fish species composition on seamounts and adjacent slope in New Zealand waters. *New Zealand Journal of Marine and Freshwater Research*, 38(1): 163–182.

Valdemarsen, J.W., Jorgensen, T., Engas, A. (2007). Options to mitigate bottom habitat impact of dragged gears. *FAO Fisheries Technical Paper*, 29.

Van Marlen, B. (2000). Technical modifications to reduce the bycatches and impacts of bottom fishing gears. In: Kaiser, M.J., De Groot, S.J. (eds), *Effects of fishing on non-target species and habitats*. Oxford: Blackwell Scientific, pp. 198–216.

Walsh, S.J. (1992). Size-dependent selection at the footgear of a groundfish survey trawl. *North American Journal of Fisheries Management*, 12: 625–633.

Chapter 8
Longlines

Gui M. Menezes[1] *and Michael F. Sigler*[2]
[1] University of the Azores, Horta, Portugal
[2] Alaska Fisheries Science Center, National Oceanic and Atmospheric Administration, Juneau, United States of America

Abstract

Longlines are common and efficient baited gears used to sample marine fishes including deep-sea species, and are effective for sampling untrawlable seafloor and deep water. In this chapter we describe the configuration and use of several longline types: principally demersal and semi-pelagic longlines. A recommended standard sampling approach is a stratified random survey with representative samples collected from each stratum. Standard longline surveys have proved useful for monitoring fish abundance, supporting fisheries stock assessments, as well as for studying deep-sea fish assemblages. Most fish species caught with longlines are scavengers and predators and more than 45 different fish families have been collected with this method.

Keywords longlines, fish surveys, seamounts, demersal fish, deep-sea fish, sampling strategies, hooks, CPUE, diversity

8.1 General introduction

A longline is a long main line to which secondary branch (shorter) lines with baited hooks are attached. The longline is a common and efficient gear to sample marine fishes. Longlines and other line methods, including handlines, are passive or stationary capture techniques which are based on attracting fish to baited hooks. Fish are attracted by the smell and taste of baits and also use vision in clear waters to locate baits. The configuration of baited gears therefore strongly depends on understanding fish behaviour in relation to habitat and prey. In this chapter we focus on demersal and semi-pelagic longlines, which are the most suitable types for sampling deep-sea fish assemblages.

The handline with one or a few hooks is one of the oldest fishing methods and dates back to the Stone Age (Björdal & Løkkeborg, 1996). Hooks were made of stone, bone or shell; the lines were made of degradable materials like plant fibre and have rarely been found. The longline developed more recently and appeared once industrial production provided easily available and affordable hooks (Björdal & Løkkeborg, 1996). Longlines are now used in commercial fisheries worldwide (Bergstad & Hareide, 1996; Gordon *et al.*, 2003; Clark *et al.*, 2007; Bensch *et al.*, 2009), and despite variations in design, material and mode of operation, they are generally simple and easy to operate.

Longlines are effective in rough areas not accessible by bottom trawl, deep-sea areas and the steep topography on seamounts and continental slopes. They have frequently been used in ecological studies and exploratory deep-sea surveys in the North Atlantic by Norway, Ireland, Spain and Iceland

Biological Sampling in the Deep Sea, First Edition. Edited by Malcolm R. Clark, Mireille Consalvey and Ashley A. Rowden.

(ICES, 2002, 2005). Longlines have been used in most studies of deep-sea fish assemblages of the Atlantic islands and neighbouring seamounts (Uiblein *et al.*, 1996, 1998; Menezes, 2003; Menezes *et al.*, 2006). Norway organized deep-water exploratory surveys to Hatton Bank in 1991, 1998, 1999 and 2000. In 1993 and 1995 Ireland organized deep-sea fishery surveys on Rockall Trough, areas along the slope of the Porcupine Bank, and seamounts on the Mid-Atlantic Ridge (Connolly & Kelly, 1995). In 2004 and 2005 Spain conducted experimental fishery surveys of the Mid-Atlantic Ridge and Hatton Bank and adjacent waters (Rockall Bank and Div. VIa) (ICES, 2005).

Longlines can sample a variety of fish. For example, over 190 different species were recorded using the same longline configuration on surveys in the Azores, Madeira and Cape Verde islands and neighbouring seamounts (Menezes *et al.*, 2006; Menezes *et al.*, 2015). Longlines also can sample very deep waters. Depths to 2000 m were sampled in the Azores region (Menezes *et al.*, 2009) and on the Mid-Atlantic Ridge (Hareide & Garnes, 2001), to 1600 m in East Greenland (Fossen *et al.*, 2003), and to 4200 m in the Charlie–Gibbs Fracture Zone (Fossen *et al.*, 2008) (Table 8.1).

An advantage of longlines is that some fish species can be caught alive for mark and recapture studies. In addition, longlines impact the seafloor less than bottom trawls (Pham *et al.*, 2014). A disadvantage is that not all fish species are vulnerable to longlines; predator and scavenger fish species are the main species caught. Some abundant deep-sea species such as the commercially important orange roughy (*Hoplostethus atlanticus*) are not caught, thus limiting measurements of fish diversity.

Fish abundance can be monitored using standard longline surveys, and estimates have been used in stock assessments: for example in the North Pacific Ocean targeting sablefish (*Anoplopoma fimbria*) (Sigler & Fujioka, 1988; Sigler & Zenger, 1989) and Pacific halibut (*Hippoglossus stenolepis*) (Trumble, 1998; White *et al.*, 2010), Georges Bank targeting cod and haddock (Johnston & Hunt, 1999), Northwest Greenland targeting Greenland halibut (*Reinhardtius hippoglossoides*) (Simonsen & Boje, 1999; Simonsen *et al.*, 2000) and in the Azores targeting several demersal fish species (Menezes *et al.*, 2006; Menezes & Giacomello, 2013). However, longline surveys are uncommon, in part because longline catchability usually is unknown or difficult to model (Sigler, 2000). The area sampled by the longline gear is unknown and thus absolute abundances cannot be obtained directly from longline catches (Gunderson, 1993). Catch rates depend on factors that may vary locally and with time (Fernö & Huse, 1983; Løkkeborg *et al.*, 1989; Løkkeborg, 1990, 1994; Engås & Løkkeborg, 1994; Björdal & Løkkeborg, 1996; Løkkeborg & Fernö, 1998; Sigler, 2000). Nevertheless, time series measurements can be valuable for a number of reasons, and in longline surveys, the longline gear and operational procedures are standardized to the extent practicable to reduce these effects and provide reasonably accurate indices of abundance.

8.2 Gear description, specifications and modifications

8.2.1 Gear components

Basic components of a longline are the *main line* (to which the snoods with hooks are attached at regular intervals), *snood* (also called branch line or gangion), *hook* and *bait*. The length and number of main line units determines the length of the longline set, which may range from a few hundred to several thousand metres. Main line units are normally stored in baskets or small containers (Fig. 8.1) which may contain one or more units. Main line units are linked manually to each other when the gear is deployed and disconnected as it is being hauled. The main line is normally retrieved with a line hauler on the deck of the vessel. The gear is repaired and baited on land for shorter surveys, whereas for longer surveys the gear is repaired and baited aboard the vessel at sea. Hooks may be removed from snoods after each haul to avoid tangles, facilitate line repairs and slow hook rusting, or removed prior to storage between surveys. The number of crew required depends on the longline configuration being used and the amount of gear that is prepared and baited aboard the vessel each day. For

Table 8.1 Examples of deep-sea longline surveys and fisheries, with details of gear configuration

Title	Longline type	Main line	Snoods	Hooks	Baits	Depths (m)	References
Azores demersal and deep-water survey	Semi-pelagic 'buoy–stone'	Monofilament (1.80 mm) main line and multifilament (12 mm) safety line	Monofilament (0.8 mm)	Size 9, 6, 8	Salted chopped sardine, scomber, squid	Shallow water to 1200	Menezes *et al.*, 2006
Alaska sablefish survey	Demersal	Multifilament nylon (9.5 mm)	38 cm length no. 60 thread braided nylon at 2 m intervals	Circle hook Mustad size 13/0	Chopped squid	100–1000	Sigler & Zenger, 1989
Pacific halibut survey	Demersal	Multifilament	5.5 m intervals	Circle hook size 16/0	Chopped chum salmon	46–503	White *et al.*, 2010
Deep-water shark study	Demersal (Mustad Autoline)	Multifilament (9 mm and 11.5 mm)	40–70 cm length at 1.4 m intervals	Mustad size 13/O EZ and 7/O EZ	Squid and mackerel	500–2000	Clarke *et al.*, 2005
Mid-Atlantic Ridge study	Demersal (Mustad Autoline)	Multifilament (11.5 mm)	0.5 m	EZ hooks Mustad 12	No information	500–2000	Hareide & Garnes, 2001
Flemish Pass	Demersal (Mustad Autoline)	Multifilament (11.5 mm)	No information	EZ hooks Mustad 12	Squid	708–3028	Murua & Cárdenas, 2005
East Greenland	Demersal (Mustad Autoline)	Mustad line (9 mm)	1.3 m intervals	EZ hooks Mustad 12/0 and 3 circular hook types	Squid and granadier	360–1080	Woll *et al.*, 2001
East Greenland	Demersal (Mustad Autoline)	Multifilament (11.5 mm)	No information	EZ hooks Mustad 12/0	Mackerel and squid	300–1600	Fossen *et al.*, 2003
Mid-Atlantic Ridge study (MAR-ECO project)	Demersal (Mustad Autoline)	Multifilament (11.5 mm)	0.5 m	Circle 14/0 and EZ hooks (Mustad 13 and 13W)	Squid and mackerel	430–4300	Fossen *et al.*, 2008
Patagonian toothfish fishery in Falkland Islands	Demersal (Mustad Autoline)	Multifilament (11.5 mm)	1.2–1.4 m intervals	EZ hooks Mustad size 12	Squid and sardines	650–2000	Laptikhovskya & Brickle, 2005

Fig. 8.1 Aspects of operations during longline surveys in the Azores and Alaska, also showing some of the components used. (a) Small stone sinkers used on the Azores longline survey. (b) Main line hauling by a crew member. (c) Small metal sinkers used on the Alaska longline survey. (d) Retrieval and storage of the buoy and safety lines in the Azores longline survey. (e) Retrieval of the main line by a crew member using a hauler on the Alaska longline survey. (f) Deployment of longline off the stern of the vessel on the Azores longline survey. Main line units are joined, with a small float or a small stone. (g) The interior deck of a longliner showing part of the autoline system. (Photograph by G Jan Erik Dyb, MAR-ECO. Reproduced with permission.) (h) A sablefish caught by the longline used in Alaska. (i, j) Examples of small buoys used in the 'stone–buoy' longline in Azores surveys (red buoys used to 700 m and grey buoys used to 2000 m depth). (k) A small tub for storing the main line units used in the Alaska longline surveys. (a, b, d, f, i, j Azores longline surveys, University of Azores © ImagDOP. c, e, h, k Alaska longline surveys NOAA-NMFS Alaska Fisheries Science Center.) (Please see Plate 14 for colour representation of the figure.)

example, two experienced crew members can prepare, bait, deploy and retrieve about 1500 hooks per day, while six experienced crew members are needed to fish about 7000 hooks per day.

Longlines can also be operated using mechanized baiting methods. A metal rail storage system holds the hooks, snoods and main line. A magazine of about 1000 hooks is the typical main line unit size. As the vessel moves through the water, the main line and hooks are pulled from the storage rail through the automatic baiting machine and into the water. The baiting machine is capable of baiting 3 hooks per second with a boat speed of about 6–7 knots. Normally three crew members manage the setting and baiting operation. In addition to the line hauler that brings the line aboard, a secondary line hauler pulls the line from the hauling room to the hook room. Hooks are pulled through a hook cleaner to remove old baits. Machinery associated with the secondary line hauler also automatically stows hooks on the storage rack with the main line hanging below (Fig. 8.1). Two or three crew members repair the gear (e.g., replace lost hooks) as it comes off the secondary hauler. Up to 40 000 hooks per day can be deployed, two to three times that possible with manual baiting (Valdemarsen,

2001). Mechanized longlines are used on larger vessels in deep-sea fisheries such as for toothfish (*Dissostichus eleginoides* and *D. mawsoni*) off Argentina and in the Ross Sea, and high-volume fisheries such as Pacific cod (*Gadus macrocephalus*) in the Bering Sea. They have also been used in deep-sea fish surveys using commercial fishing vessels (Table 8.1).

Main lines are made of polyester, polyamide (nylon) and polypropylene, which, respectively, have greater breaking strength, greater stretch and lower density. Multifilament main line is used in, for example, sablefish and Pacific halibut surveys. The most common diameters are from 4 to 11 mm (Björdal & Løkkeborg, 1996) (Table 8.1). Monofilament main line, commonly used in smaller vessels, is useful for rough, steep bottom, such as that found on seamounts, because the material is less resistant to chafing and thus easier to break when the gear is stuck on the bottom. Multifilament main line can have a lead core (e.g. 50 g m^{-1}) to increase sinking rate and is commonly used in mechanized systems.

Snoods are usually made of the same materials as the main line. Monofilament snoods normally range from 0.3 to 1 mm in thickness and from 0.5 to several metres in length depending on the target species. Multifilament snoods are commonly made of two-strand rope ranging from 2 to 4 mm diameter with lengths from 0.3 to several metres (Björdal & Løkkeborg, 1996). Snoods sometimes are made from steel wire or chain to avoid cutting by the sharp teeth of species such as sharks or black scabbardfish (*Aphanopus carbo*). Snoods can be attached to the main line using swivels or a simple knot. Swivels reduce tangles and snood line breaks when fish struggle and twist after capture.

Hooks normally are made from steel. The basic parts are the shank, bend, point, barb and an eye or plate for attaching the snood. Hook size decreases with increasing ordinary numbers (1, 2, 3), and increases with increasing '/0' numbers (following the Mustad terminology) (Björdal & Løkkeborg, 1996) (see Appendix 8.1). The J- and C-shaped hooks are the two main hook types. In the Azores islands, the J-hook (no. 9), is effective for catching fish from 15 cm to 1.5 m in length and occasionally larger. Medium (13/0) and large (16/0) C-hooks (often called 'circle' hooks) are used in the North Pacific Ocean and are effective for catching fish sizes 25–35 cm and larger. C-hooks typically catch more fish than J-hooks and are easier to remove from fish because the fish usually is hooked in the mouth (e.g. operculum) and not the throat or stomach. This characteristic is useful for returning live fish to the sea, for example in mark–recapture studies.

Hook spacing varies from less than 1 m up to 7 m. The spacing is wider for target species that are less dense; for example, hook spacing in the North Pacific halibut fishery was wider in the 1970s (Skud & Hamley, 1978) when abundance was low. Of course, once hook spacing for a survey is chosen, the gear should be standardized for that spacing.

Baits include herring, mackerel, anchovy, sardine, sprat, pilchard, capelin and squid. These can be used whole or cut in pieces. Baits may be brined or salted to increase firmness, slow deterioration and increase durability. When longline gear is baited at sea, the bait should be kept frozen until needed and then partially thawed for chopping and baiting. The baited gear must be kept cool (e.g., in refrigerated rooms onboard) and fresh-smelling (not sour) until deployment. In research surveys, standardizing bait type, size and quality, deployment hours, soak time and longline materials are important to avoid variation in species catchability and selectivity.

8.2.2 Longline configuration and design

The three main types of longline are demersal, semi-pelagic and pelagic, which vary in configuration, material, mode of operation and fishing strategy (Fig. 8.2). Longlines can also be classified as drift and fixed. Drift longlines are allowed to drift with the current and are mainly used to target swordfish and tunas. Fixed longlines are the most common configuration for catching demersal and deep-sea fish species and hence are the focus of this chapter. With fixed longlines the fishing gear is fixed with anchors at each end of the gear with buoy lines attached to marker buoys at the surface.

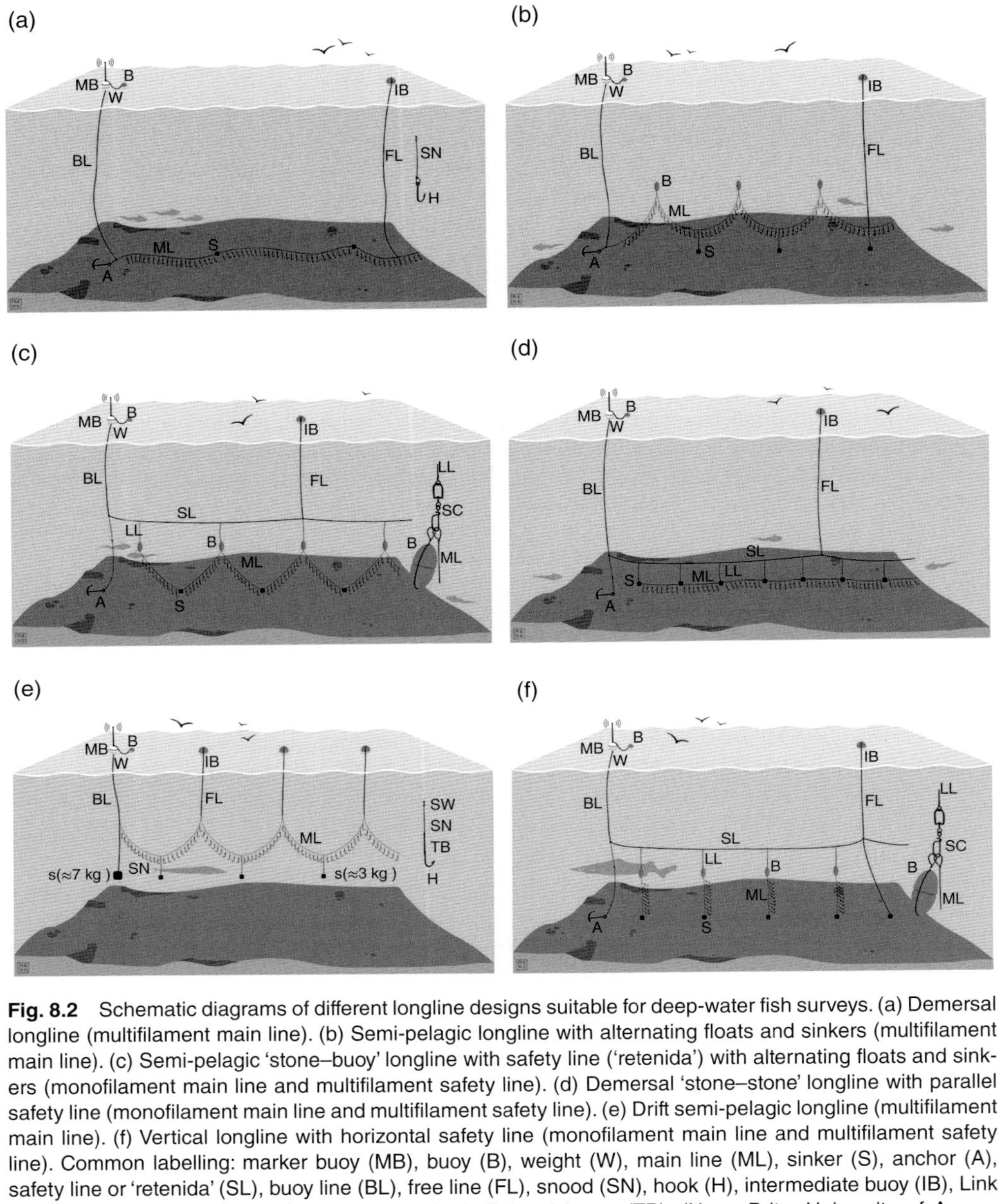

Fig. 8.2 Schematic diagrams of different longline designs suitable for deep-water fish surveys. (a) Demersal longline (multifilament main line). (b) Semi-pelagic longline with alternating floats and sinkers (multifilament main line). (c) Semi-pelagic 'stone–buoy' longline with safety line ('retenida') with alternating floats and sinkers (monofilament main line and multifilament safety line). (d) Demersal 'stone–stone' longline with parallel safety line (monofilament main line and multifilament safety line). (e) Drift semi-pelagic longline (multifilament main line). (f) Vertical longline with horizontal safety line (monofilament main line and multifilament safety line). Common labelling: marker buoy (MB), buoy (B), weight (W), main line (ML), sinker (S), anchor (A), safety line or 'retenida' (SL), buoy line (BL), free line (FL), snood (SN), hook (H), intermediate buoy (IB), Link line (LL), Snap swivel or clip (SC), swivel (SW), Twisted brass (TB). (Nuno Brito, University of Azores (cImagDOP). Reproduced with permission.)

Demersal longlines are commonly deployed to between 100 and 800 m depth in North Atlantic fisheries. In the toothfish fishery, demersal longlines are routinely used down to 1500 m, and in research surveys have been used even deeper (Fossen *et al.*, 2008). The only constraints on deep-sea sampling are line storage space and line hauler power. Demersal longlines are configured as a main line with hooks attached, and main line units attached to each other with or without small sinkers (1–3 kg), along the main line. The sinkers can be small stones or lead sinkers. This longline typically is only linked to the buoy lines at the two ends and all of the gear stays on the bottom. Intermediate buoys can be attached along the line in case the line breaks, although this may require a second line hauler to retrieve the intermediate buoy lines.

Some demersal species typically swim several metres off the bottom, and for these species, semi-pelagic longlines are more effective. In this configuration small floats alternate with small sinkers along the line, so that gear extends off the bottom and covers the depth distribution of the target species. This configuration is effective for catching benthopelagic fishes and is used in the hake, pollock, cod and haddock fisheries (Björdal & Løkkeborg, 1996). It is also used in the black scabbardfish fishery near the Madeira archipelago, where the gear is set at 550–1100 m and allowed to drift over bottom up to 1500 m depth (Merrett & Haedrich, 1997). Floats are made of plastic and come in different shapes, sizes and pressure resistance. Small oval floats are used in Azorean longline surveys and, depending on manufacturer and rating, can be used to about 800–1200 m depth. Other small floats resist crushing down to 2000–3000 m.

Vertical longlines catch species over a wide range of distances from the bottom and are appropriate for very rough seafloor because of reduced gear contact. In this case, main line units are not attached to each other but instead are attached to buoys or a horizontal safety line; their spacing equals at least their own length. These main line units have small floats and sinkers at each end; the number and spacing of hooks, and snood length, varies.

The stone–buoy longline is a type of semi-pelagic gear which adds a safety line oriented parallel to the main line. Snaps with swivels spaced along the safety line attach to tying-ropes on the main line units. The main line is monofilament (0.80 mm nylon) and the safety line and the tying-ropes are nylon multifilament with diameters of 10 mm and 4 mm, respectively (Fig. 8.1). Small floats attached at each tying-rope alternate with sinkers at each main line linkage. In an alternate configuration, small sinkers (1–2 kg) are attached at each tying-rope. In the first configuration, part of the main line stays slightly off the bottom and in the second configuration, all of the gear stays at the bottom (stone–buoy longline and stone–stone longline, respectively). In both cases the longline is hauled by the safety line, and if the main line breaks it is retrieved at the next linkage to the safety line. The main line with the snoods and hooks is hauled manually by one person or by a second hauler standing near a person operating the longline hauler where the safety line is retrieved. When hauling the gear, the snaps are detached from the main line, coiled and stored in a big container with the coiled safety line, ready for the next deployment (Fig. 8.1). The safety line is attached at both ends to the buoy lines (nylon multifilament 12 mm) and the anchor is normally attached to the first tying-rope of the safety line. This configuration, also known as the Spanish double-line system, has proven effective and reliable for Azorean seamount and island slope surveys with rough bottom and strong currents. Handling double-line semi-pelagic gear is more labour intensive than single-line demersal longline gear.

8.2.3 Other gear modifications

Trials are an important part of longline survey planning to enable development of the best deck layout for operating longlines and catch sampling. Small data-loggers that measure depth and time are useful for understanding gear performance and have proven easy to operate with longlines. They measure the sinking time of the gear, and once on bottom, the duration of the sampling period (i.e. on-bottom time).

Longline experiments are recommended to determine the catching efficiency of the research longline. Hook time recorders have been used (Somerton & Kikkawa, 1995; Sigler, 2000) to measure time to capture and have proven useful for understanding the effects of soak time and competition for hooks on abundance estimates. Soak time and hook spacing experiments can also be conducted (Sigler, 2000). Different baits alternated along the main line hooks have been used to test bait types and sizes. Selectivity can be studied by using different hook and bait sizes or different hook shapes in the same longline sets. Camera systems can help identify gear impacts on bottom habitats, longline 'behaviour' when deployed or during hauling, to estimate swept area of demersal longlines (Welsford & Kilpatrick, 2008), as well as to study fish behaviour (Løkkeborg *et al.*, 1989). Observations from manned submersibles are also occasionally made (e.g. Rodgveller *et al.*, 2011).

8.3 Sampling operations

8.3.1 Training and safety

Once the best longline configuration that meets survey goals, vessel characteristics and research funding has been decided, it is important to assign science and vessel crew members specific tasks. Well-trained, experienced individuals that coordinate with their team are essential for the success of longline surveys and the efficient use of ship time. Training and standardized operations also improve safety. There is a potential risk of injuries when deploying gear because line and hooks go overboard quickly and entanglements may occur. During retrieval the hauler may stop and the longline become extremely taut if the gear tangles on the seafloor. When this happens the line may suddenly slip from the hauler at high speed. Hence, like most gear-handling operations, the vessel and scientific crews must be alert during longline operations.

8.3.2 Deployment and retrieval

Before gear deployment the sampling area is surveyed with an echosounder to determine the bottom depths and topography. This information is important for positioning the longline and ensuring that the desired depths and habitats are sampled. Gear is set starting at the shallow end so that if the gear drifts, the anchor will reach the seafloor quickly and prevent the gear from drifting into deep water. Each longline should be deployed at similar light levels (e.g. the same time each day) because diurnal animal behaviour may affect catch rates, especially at lesser depths. Adding line without hooks is useful when the necessary sampling effort (e.g. number of hooks) for a habitat or depth strata has been reached – this saves bait and sampling time.

The gear is deployed from the vessel while underway. A buoy marking one end of the longline is set overboard first and once the buoy is in the water, the buoy line will be drawn overboard as the vessel moves forward. Subsequently an anchor and then the line with hooks are deployed. Finally the anchor and buoy at the far end of the gear are set once all of the line with hooks is deployed. Some line with no hooks (100–200 m long) is placed near the anchor to reduce the chance that hooks will tangle with the anchor. For longline baited by hand, the small containers or baskets that store the main line units are set up on a special deck table during the deployment and linked to each other (Fig. 8.1). Floats and small sinkers may also be attached to the main line. Two or more crew members are necessary for this operation, depending on the longline configuration. For longlines baited by machine, the longline is deployed from a magazine located immediately forward of the baiting machine.

Small sinkers may be deployed periodically along with the line with hooks to ensure that demersal longline gear remains on the seafloor. The distance between these intermediate sinkers depends on bottom roughness, with more sinkers deployed on rougher seafloor. For example, 3 kg sinkers are attached every 100 m on a survey of the upper continental slope of Alaska. Intermediate sinkers also help keep the gear in place in areas with strong currents. On smooth and flat seafloor such as continental shelves or low current areas (<0.5 knots), intermediate sinkers may not be necessary. For semi-pelagic longline gear, intermediate subsurface buoys are added to the line alternating with the intermediate sinkers to reduce tangling with the bottom in rocky areas. The gear may be marked (e.g. numbered floats) to guide data-recording and link catch data to location and depth information.

The longline is set in the direction of the current, reducing the risk of tangles which are more likely if the gear is set across or against the current direction. A reasonable setting speed is 6 knots. Over rough bottom, setting the gear faster tightens the line so that some of the gear lies off bottom by spanning gullies between high points of the seafloor. An indication that the line is too tight is sections of line retrieved with untouched bait that coincides with the location of a gully on the vessel's echosounder (i.e. the line was draped across the gully edges but did not reach the gully floor); then the deployment speed needs to be reduced.

The longline is also recovered in the direction of the current. The vessel will drift with the current and lift the longline. If the longline is recovered against the direction of the current, the line will be under more tension and if the gear is tangled on the bottom, will be more likely to break. However, when the weather is bad, recovering the gear into the wind and waves and against the current is necessary to keep the ship speed slow yet under good control. Longline recovery is time consuming. About 1200 hooks can be hauled per hour for both Alaska (demersal) and Azores (semi-pelagic) longline surveys. Buoy lines and other lines without hooks are hauled at about 1 m s^{-1} by hydraulic haulers. In bad weather or strong currents, hauling speed is slower and tangles and delays more common.

8.3.3 Essential ancillary information

Time, position and water depth are periodically recorded during longline deployment and retrieval. These records provide a map of the locations and depths sampled. If currents and gear drift are minimal, then these deployment and retrieval maps will match and provide a record of the location and depths sampled. If the gear drifts significantly, then the retrieval records are a better measure of the location and depths sampled but some verification is necessary. In deep areas where the seafloor slopes steeply, the recorded depth can overestimate the sampling depth because the longline gear is several hundred metres off bottom when brought aboard the vessel. Data-loggers deployed periodically along the line can help determine whether this bias is a practical concern, by comparing the logger measurements to the water depths recorded during retrieval.

As the longline is retrieved, the status of each hook could be recorded. Hook categories are fish species, empty hook, bait and ineffective (broken, bent, lost or tangled). These data can be recorded with an electronic data-logger and uploaded to a database following longline retrieval. Effort is computed as the total number of hooks minus the number of ineffective hooks. Typically catch per unit effort (CPUE) is analysed by species for each main line unit on a per hook basis.

8.3.4 Application to different habitats and fauna

Longline gear is suitable for sampling most types of habitat, with the exception of seeps and hydrothermal vents. Longline gear is therefore more versatile than trawl and other mobile gears that cannot access high-relief habitat.

The fauna targeted by the survey is determined by the gear configuration. Semi-pelagic gear floats off the bottom and may attract some fish species not susceptible to the demersal gear that lies on the seafloor. Hook size and shape, wire diameter and bait type and size also affect the fauna targeted. For example, a hook with a thinner wire diameter (1 mm) can capture small fish (<2 kg). If larger species are targeted (>3–5 kg), then a hook with a thicker wire diameter (2 mm) is used to increase fish retention and reduce hook breakage. In general, we recommend using small J-hooks for fish <2 kg, intermediate-size C-hooks for 3–10 kg fish and large C-hooks for >10 kg fish.

Depending on the survey objectives there are several possible sampling designs. Apart from the biological and ecological goals of the study, survey plans must take into account the costs, available ship time and the amount of gear available (Table 8.2). A standard sampling approach is to conduct a stratified random survey with representative samples collected from each stratum (Box 8.1). Stratum boundaries are often defined by depth intervals (e.g. 101–200 m depth stratum) and longline gear can be set either entirely within one depth stratum or across multiple depth strata. On flat

Table 8.2 Major sampling strategies for deep-sea longline surveys with some general considerations about design and analysis.

Main sampling strategies	Abundance surveys	Biodiversity surveys	Spatial distribution	Flat or homogeneous bottoms	Sloped bottom and seamounts	Tagging studies	Trophic studies	Growth, genetics, reproduction studies	Fish habitat	Multispecies studies	Single species studies	General considerations
Unequal number of hooks per tow or sampling unit	+	++	++	0	++	0	0	0	+	0	0	Catch data can be stratified by depth zones (strata) or longline units defined a priori Standardization of abundance and diversity data is an issue for comparison purposes Different soak times for parts of the gear must be considered if an issue (see text) Number of species caught must be standardized by sampling effort Species richness must be standardized by sampling effort More complex standardization models may be needed
Equal number of hooks per tow or sampling unit	++	++	+	++	+	0	0	0	++	0	0	More hauls and ship time are needed to cover vertical/depth gradients More than one tow per day unless time of the day is not an issue Number of species caught may be directly comparable among stations or sampling units Species richness and all traditional diversity and evenness indices can be applied Simplest quantitative analysis
Small number of longer hauls	+	++	++	+	++	–	–	+	++	++	+	Less time consuming when the goal is to cover several depth strata Useful when targeting a large number of species Independence of sampling units within the same tow may be an issue
Large number of shorter hauls	+	++	+	++	+	+	+	++	+	+	++	Useful when targeting a small number of species Appropriate for surveys of flat bottom or homogeneous habitat More time consuming when the goal is to cover several depth strata Time of day may be an issue in shallower waters Patchiness of fish distribution may be an issue

Codes: 0, not relevant to the chosen strategy; +, appropriate; ++, highly appropriate; and –, inappropriate.

Box 8.1 Example survey strategy for measuring deep-sea fish diversity on a seamount using longlines

Seamounts are active or extinct undersea volcanoes. Their summits and flanks are often rough and steep, which can preclude sampling using trawls. Longlines are an alternative sampling tool and have been used successfully in many studies.

Depth is the main ecological factor structuring marine fish assemblages; therefore an ideal seamount sampling strategy will orient along the depth gradient.

- In a 'long-rig' set strategy, the longline is set from the summit to the deepest water possible, covering several depth strata in one set. Seamount geomorphology and currents may vary with location, affecting the fish assemblage, and so we recommend multiple sets oriented in different directions. The resultant data are stratified by predefined depth strata (e.g. 50 m intervals). Position and depth of the main line units are recorded during longline deployment and the catch and other biological data are recorded by depth strata during longline retrieval.
- In a 'short-chain' strategy, deployments are made independently in each depth stratum, thus requiring more sets and ship time to cover the same depth range. Normally one standard gear configuration is deployed during surveys. However two gear configurations (e.g. two hook sizes) can be deployed in the same longline set to increase the number of species targeted. In either case, following standard sampling methodologies allows for comparisons among seamounts.

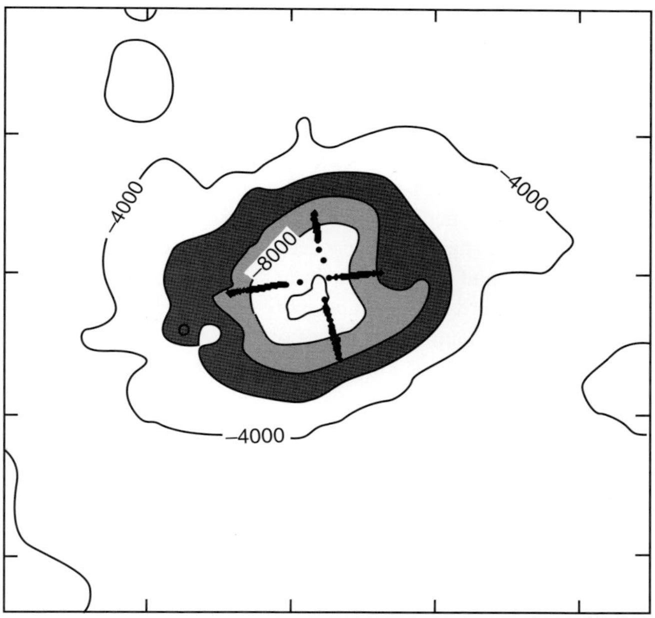

Hypothetical seamount longline sample data using a 'long-rig' strategy along depth gradients:

Depth strata (m)	Number of mainline units	Fishing effort (number of hooks)	Number of species	Number of specimens	Weight of specimens (kg)	CPUE (number/ hook)
351–400	3	90	10	40	120	0.444
401–450	5	150	7	35	50	0.233
…						
751–800	6	180	9	46	60	0.256
801–850	7	210	4	70	90	0.333
…						

bottom, laying the longline gear within one depth stratum is straightforward. On steep bottom, the longline gear is laid parallel to the depth contours to remain within one depth stratum. This approach can be problematic if the bottom is steep or irregular or the depth contour is hard to follow. An alternate approach for steep bottom is to lay the longline gear across multiple depth strata and keep track of the depth of each section of longline gear. This approach also minimizes the number of longline sets necessary to sample multiple depth strata and thus the gear handling time (e.g. less buoy line to deploy and retrieve). In general, we recommend one set per depth stratum on flat bottom. For bottom slope of more than a few degrees, we recommend setting gear across multiple depth strata. If the slope is shallow, then orient the gear perpendicular to the depth contours. If the slope is steep, then orient the gear diagonally across the depth contours. This angle is determined by the minimum amount of gear desired in each depth stratum (e.g. 10 main line units in each depth stratum). Table 8.3 summarizes some recent examples of deep-sea surveys using longlines, which can be referred to for more detail of survey design.

The minimum amount of gear desired in each stratum can be determined based on pilot samples to evaluate the variability of catch. Pilot data can be randomly sampled with replacement to determine the trade-off between sample effort, variance reduction and the ability to detect statistically significant differences (Monte-Carlo simulation). Standard survey data can be substituted if no pilot data are available for this analysis. From our experience, about 500 hooks are necessary as a minimum sample effort for each depth stratum. If catches of the target species are low, then more sample effort may be needed.

Fish assemblages can be characterized by depth-related patterns of size and species composition. For example, on North Atlantic seamounts, teleosts dominate at all depths, and animal sizes generally increase with depth; deep-water sharks are commonly caught in intermediate depth waters (700–2000 m), but sharks or elasmobranches are less abundant below ~3000 m (Priede *et al.*, 2006). Based on these general patterns, modified gear may be useful in some areas or at particular depths. For example, on North Atlantic seamounts, a deep-sea rig using slightly larger hooks and twisted wires or small chains in branch lines at intermediate depths (e.g. 1000–2000 m or deeper) broadens faunal sampling to sharp-toothed fish and sharks. In these deep-sea rigs, the main line and branch lines have slightly larger diameters (e.g. 2.2 and 1.8 mm), the branch lines are longer (1.8 m) and spaced farther apart, and the floats, when used, are stronger and more pressure resistant. This means that, in the same longline haul, it is possible to combine longline configurations (preferably alternating) to obtain a better representation of fish assemblages. Longline sets combining standard and deep-sea gear have been successfully made in several longline surveys of island slopes and seamounts in the Azores and Madeira regions (Table 8.3).

8.3.5 Handling and gear maintenance

Maintenance of sampling gear is essential for ensuring that the sampling gear operates consistently and that catch variability is attributable only to differences in fish abundance and not gear performance. The critical features of longline gear for standardization are hook shape, snood length, hook spacing and bait size:

- Hooks are normally used for several deployments, and damaged hooks must be replaced.
- Snoods are tied to a uniform length.
- If the main line breaks, then the damaged line is cut out and sufficient replacement line is spliced to maintain hook spacing.
- Lines are properly coiled for storage. Hooks usually are detached from the snoods when stored. It is normal to wash them in freshwater, allow them to dry and store them in a dry place or immerse in oil to avoid rust. High-quality stainless steel snaps are normally resistant to rust and need little maintenance.

Table 8.3 Examples of demersal longline surveys in the Atlantic Ocean including location, depth range covered and number of species caught

Longline survey	General survey characteristics	Depth range (m)	Latitudinal range	No. species caught	Reference
Mid-Atlantic Ridge (between 1996 and 1997)	339 hauls, 100 m depth strata, 2 longline designs	500–2000	61°–40°N	56 species, 27 families	Hareide & Garnes, 2001
Mid-Atlantic Ridge (MAR-ECO, 2004)	56 hauls, 2 longline designs	450–4300	54°–42°N	40 species, 17 families	Fossen *et al.*, 2008
Azores islands and several seamounts (1995–2010)	406 hauls, 50 m depth strata, 2 longline designs	25–1250	37°–39°N	143 species, 72 families	Menezes, 2003; Menezes *et al.*, 2006; personal observations
Seamounts: Great Meteor, Irving, Plato; Atlantis, Cavala, Voador, Monte Alto and Gigante, Formigas (2008)	28 hauls, 50 m depth strata, 3 longline designs	150–2000	30°–40°N	30 species, 35 families	Personal observations
Sedlo seamount (Azores, 2003 and 2004)	8 hauls, 50 m depth strata, 2 longline designs	700–2100	40°N	22 species, 12 families	Menezes *et al.*, 2009
Madeira islands and Seine seamount (1995–1997)	40 hauls, 50 m depth strata, 1 longline design	25–1200	33°–33°N	99 species, 48 families	Menezes, 2003
Seine seamount (Madeira archipelago, 2003)	8 hauls, 50 m depth strata, 2 longline designs	180–2100	33.5°N	41 species; 23 families	Menezes *et al.*, 2009
Cape Verde islands and two seamounts (2000)	20 hauls, 50 m depth strata, 1 longline design	25–1200	17°–15°N	98 species; 49 families	Menezes, 2003
Great Meteor seamount (1998)	2 hauls, 2 longline designs	600–1100	29.5°N	10 species, 7 families	Uiblein *et al.*, 1999
Lanzarote and Fuerteventura, Canary Islands (1994, 1995, 1997)	216 hauls, 1 longline design	18–1217	29°–28°N	73 species, 38 families	Uiblein *et al.*, 1996, 1998
Algarve (southern Portugal)	Fishery	200–700	37°N		Erzini *et al.*, 2001
West and North of Ireland (1993–2000)	22 hauls, 200 m depth strata, targeting deep-sea sharks	500–2000	50°–59°N	19 species, 12 families	Clarke & Borges, 2005

(*continued*)

Table 8.3 (Continued)

Longline survey	General survey characteristics	Depth range (m)	Latitudinal range	No. species caught	Reference
Seamounts: Great Meteor, Irving, Plato; Atlantis, Cruiser, Hyeres, Conception, Endeavour, Paps (2000)	220 hauls	300–900	28°–35°N 10°–5°N		Iglesias & Muñoz, 2001
Sierra Leone Rise (2000)	No details	200–2000	10°–5°N		Iglesias & Muñoz, 2001
Brazil – between Salvador and Cabo São Tomé, including seamounts (1996–1998)	206 hauls	19–1000	12°–22°S	125 species, 39 families	Martins *et al.*, 2005
Brazil between Chuí and Cabo de São Tomé (1996–1997)	206 hauls	100–510	34°–22°S	63 species	Haimovic *et al.*, 2004

8.3.6 *Quality assurance*

Longline gear is often retrieved with some broken, bent or tangled hooks. If the number of ineffective hooks is small, subtracting this number from the total number of hooks is a reasonable method for correcting effort for ineffective hooks. At some point, however, large numbers of hooks may be broken or tangled and the number of ineffective hooks is unclear. For example, the gear may be retrieved with a large tangle of hooks but with a few fish caught on some of the hooks within the tangle. In this case, hooks trapped within the tangle may be ineffective but the number of ineffective hooks is not obvious. A recommended approach is to exclude main line units with >10% ineffective hooks; otherwise correct main line units for the number of ineffective hooks. From our experience, few main line units are excluded from analysis (<1–2%).

Other sampling problems may affect data quality. Local currents can cause gear drift from the desired position during deployment. Gear tangles can occur sporadically, sometimes preventing proper coverage of depth strata. These can be common problems in areas of steep slopes. In shallow waters, gear can float near the surface and catch more pelagic species, which are sometimes also caught in the water column during line deployment or retrieval. These data are excluded from analysis of demersal fish assemblages. Bad weather can limit scientific data quality even when longline operations can continue. For example, fish lost during retrieval ('drop offs') increase when weather is poor and may bias catch rates. During sablefish longline surveys from a 47-m-long vessel in Alaska, sampling stops when wave height is greater than 3.0–3.7 m and wind speed is greater than about 30 knots.

8.3.7 *Sample sorting and processing*

Longline retrieval, catch sorting and biological sampling occur simultaneously because longline retrieval typically lasts several hours. A random subsample of fish is usually collected for biological studies of reproduction, growth or genetics. Longline surveys may be less useful for trophic studies because most fish caught have empty stomachs. Decreasing the hauling speed may improve this, but only for surveys dedicated to this purpose due to the longer time needed for gear retrieval. Fish caught with longlines typically are landed with minimal damage to skin and fins. Their good condition increases their suitability for taxonomic studies, reference collections, and high-quality photographs of specimens exhibiting their natural colours. In hot weather, we advise keeping the fish sample cold with ice.

8.4 Measurements, metrics and data considerations

There are three main reasons for conducting fish surveys with longlines:

- to obtain abundance indices of species of commercial importance (the most common reason);
- to describe the dominant fish species, diversity, assemblage composition and habitat associations;
- to collect specimens for tag–recapture, biological and ecological studies.

Other possible reasons are gear selectivity analysis, analysis of the impacts of the fishing gear on habitats and VMEs (vulnerable marine ecosystems) and studies on bycatch and less common species.

8.4.1 Evaluating sampling effectiveness

Understanding the longline capture process is important for evaluating sampling effectiveness. Three key parameters of the capture process are the rate at which fish encounter the gear, the probability that fish successfully locate an available bait and the probability that fish are hooked (Sigler, 2000). Encounter rate is affected by bait type. The encounter rate decreases over time as odour is released from the bait and as fish are caught. Locating baits depends in part on the number of available baits. This decreases as fish are captured, but fish appear to readily locate baits, even when few remain (Sigler, 2000). Hooking probability depends on the hook type and bait size. Longline catch rates can provide an accurate index of fish abundance if the on-bottom time is long enough to cover the period when most fish encounter the gear and the initial bait density is high enough that baits remain available throughout the soak (Sigler, 2000).

Optimizing longline sampling depends on balancing soak time and hook spacing. The minimum soak time can be determined by a soak time experiment that examines catch and species composition over time. Because the accumulated catch typically reaches an asymptote, the minimum soak time can be determined as the time when the catch approaches a certain fraction (e.g. 90%) of the asymptote. Hook spacing should be adjusted so that some baits remain when the gear is retrieved. If few baits remain (e.g. <10%) when the longline gear is retrieved, then the hook spacing should be decreased.

Longlines have a catch limit determined by the number of hooks they employ. In the fisheries literature, 'gear saturation' (Beverton & Holt, 1957) is described as the effect of hooked fish reducing the fishing power of a unit of fishing gear as the catch increases. Reduction of fishing power is due to the decrease of the number of available hooks, and is also affected by the loss or decreasing attractiveness of bait (Haimovic & Ávila-da-Silva, 2007). Mathematical models relating longline catch rate to fish abundance have been developed that enable catch rates to be corrected for gear saturation (Murphy, 1960; Rothschild, 1967; Somerton & Kikkawa, 1995). These models share two important assumptions: (i) encounters to the gear are random, independent, and occur at a constant rate through time and (ii) the probability of a fish locating a bait decreases linearly with the number of baits remaining. For example, Murphy (1960) expressed the rate of change of the number of hooked fish as a linear function of the number of baits remaining.

A more general mathematical model that includes the previous models as special cases was used to test the second assumption (Sigler, 2000). An improved model fit implied that for sablefish, the probability that a fish locates an available bait is not proportional to the number of baits available; instead, this probability remains constant, at least initially, for a wide range of available baits, likely because sablefish are good bait predators. Atlantic cod also are good bait predators; cod swim upstream directly to a baited line and then swim along the line for 100 m or more after locating the baited line (Løkkeborg & Fernö, 1998). A fish with this searching behaviour would probably locate a bait, unless large sections of line had hooked fish. Searching the line also makes biological sense as a natural food-seeking behaviour, implying that most species readily locate baited hooks, even when few remain. If so, gear saturation of longlines is unimportant as long as hook spacing is narrow enough that some baits remain at retrieval (as discussed in the previous two paragraphs).

Other factors that potentially affect sampling effectiveness include current speed (Løkkeborg *et al.*, 1989) and hunger state (Løkkeborg *et al.*, 1995). The influence of factors such as time of day on hunger state can be standardized in sampling designs. Other influences such as temperature and currents cannot be standardized but can be measured to test and account for effects. Finally some factors, such as hunger state, cannot be standardized or easily measured and thus add to the variability of abundance measurements. However, although all factors cannot be considered, to ensure consistency in longline-based research and monitoring, it is essential to control bait quality and size, hook size, deployment time, soak time and line type.

Longline catch rates can be adjusted for soak time:

$$\mathrm{CPUE}_{ik} = \frac{N_{ik}}{h_k * T_k} \times 1000$$

where N is the number (or weight) of fish caught of species i in strata k, h is the number of hooks, T is the soak time and N is expressed per thousand (1000) hooks. This adjustment is useful for short soak times when catch is still increasing linearly as a function of soak time. This adjustment will bias catch rates if applied to longer soak times when catch has reached an asymptote as a function of soak time. The preferred method is to allow the gear to soak long enough that the catching process has reached an asymptote so that no soak time adjustment is necessary.

8.4.2 *Data analysis*

A common index of fish abundance is the area-weighted catch rate. An area-weighted catch rate by region or depth strata can be computed from the summation of the average relative densities of each species (i) by region or depth strata multiplied by the corresponding area (A) of each region or depth stratum (k)

$$\sum A_{ik} \overline{\mathrm{CPUE}_{ik}}$$

The advantage of this index is that values can be used to obtain an overall (area-weighted catch rate) value. The bootstrap method has been used to test the statistical significance of differences in area-weighted catch rates, for example among years (Sigler & Fujioka, 1988).

Extensive literature is available on statistical methods to test for differences in fisheries catch rates, species diversity and multivariate analysis. In this section, we focus on aspects of these methods that are specific to analysis of longline survey data. Generalized linear models (GLMs) or general additive models (GAMs) (e.g. McCullagh & Nelder, 1989; Guisan *et al.*, 2002; Maunder & Punt, 2004; Zuur *et al.*, 2007) can be applied to test for differences in longline catch rates due to effects of independent variables such as depth. More complex models (e.g. generalized linear mixed models (GLMMs) (Maunder & Punt, 2004)) may be applied when the data have *a priori* problems of heterogeneity and non-independence. Common response variables are catch in number or weight, catch rate (CPUE) or diversity with explanatory variables such as year, area, month, depth and vessel. When using catch rather than catch rate as the response variable, the Poisson or negative binomial distributions must be used and fishing effort included as an explanatory variable (Maunder & Punt, 2004).

Several diversity indices are available to characterize community structure based on longline data. As in the case of the catch rate analysis, there are important considerations for the use and choice of diversity indices if fishing effort (the number of hooks) varies from tow to tow or area to area. Greater sampling effort increases the probability of catching more fish and more species, thus affecting estimates of the diversity indices. Using the species richness (*S*) diversity index (the number of species per sampling unit) when the number of hooks differs between sampling units is incorrect. Also a standardized species richness index (e.g. the average number of species caught per hook) should be used with care because this scalling may distort the numbers (Gotelli & Colwell, 2013). Species accumulation curves are useful to understand the level of sampling (e.g. number of tows or number of hooks) and the number of species expected according to the number of samples in a certain area. Other useful diversity indices are less sensitive to sampling effort, including rarefaction indices, Hill's index, dominance curves, and taxonomic-related diversity indices (e.g. Ludwig & Reynolds, 1988; Clarke & Warwick, 2001).

Multivariate analysis can be applied to longline data in the same way as other sources of species occurrence and abundance data to determine structure and distribution patterns of demersal and deep-sea fish assemblages (e.g. Legendre & Legendre, 2000; Zuur *et al.*, 2007). Results based on such data indicate that faunal change can be observed with latitudinal, longitudinal and depth gradients from the shelf to abyssal seafloor (Merrett & Haedrich, 1997); food supply, depth and water mass appear to be the main factors structuring demersal fish assemblages on continental slopes and rises, abyss, seamounts and oceanic islands (Bergstad *et al.*, 2008). Longline catch data can be standardized for effort, and square root or log-transformed before multivariate analysis (Clarke & Warwick, 2001). Species are usually selected for analysis according to a predefined percentage of occurrences to avoid including many zeros and excessive influence of rare species in the analysis.

8.5 Comparisons with other methods that sample fishes

8.5.1 Species and size composition

Trawl surveys normally catch more species than longline surveys (Connolly & Kelly, 1996; Hareide & Garnes, 2001; Clarke *et al.*, 2005; Bergstad *et al.*, 2008; Fossen *et al.*, 2008) but typically undersample the larger predators and scavengers (Jørgensen, 1995). Longlines, however, are able to catch many different species, and about 45 different demersal and deep-sea fish families have been collected with this method (Table 8.4). Some abundant species caught by trawls such as *Hoplostethus atlanticus* and *Coryphaenoides rupestris* are seldom caught with longlines (Hareide & Garnes, 2001) whereas several studies show that deep-sea longlines catch many more elasmobranches than teleosts and larger specimens of teleosts (Connolly & Kelly, 1996; Hareide & Garnes, 2001; Clarke *et al.*, 2005; Menezes *et al.*, 2006, 2009; Fossen *et al.*, 2008), which are usually top predators. For example Fossen *et al.* (2008) recorded a mean individual fish weight 15 times larger (2400 g) than the mean individual weight of (160 g) recorded in the catches from the bottom trawl onboard RV *G.O. Sars* during the MAR-ECO cruise in the Mid-Atlantic Ridge.

Table 8.4 Examples of families of deep demersal ichthyofauna caught on longline surveys

ORDER/families		
ANGUILLIFORMES	SYNGNATHIFORMES	POLIMIXIIFORMES
Congridae	Macrorhamphosidae	Polymixiidae
Muraenidae	HEXANCHIFORMES	RAJIFORMES
Synaphobranchidae	Hexanchidae	Dasyatidae
AULOPIFORMES	LOPHIIFORMES	Rajidae
Aulopodidae	Lophiidae	SCORPAENIFORMES
Chlorophthalmidae	OPHIDIIFORMES	Scorpaenidae
Paralepididae	Ophidiidae	Sebastidae
Synodontidae	OSMERIFORMES	Setarchidae
BERYCIFORMES	Alepocephalidae	Triglidae
Berycidae	PERCIFORMES	Anoplopomatidae
CARCHARINIFORMES	Chiasmodontidae	SQUALIFORMES
Pseudotriakidae	Epigonidae	Centrophoridae
Scyliorhinidae	Gempylidae	Dalatiidae
Triakidae	Polyprionidae	Etmopteridae
CHIMAERIFORMES	Serranidae	Somniosidae
Chimaeridae	Sparidae	ZEIFORMES
GADIFORMES	Trichiuridae	Zeidae
Gadidae	Nototheniidae	Oreosomatidae
Lotidae	Anarhichadidae	
Macrouridae	PLEURONECTIFORMES	
Moridae	Bothidae	
Phycidae	Scophthalmidae	

Scavenger fishes appear to predominate in deeper assemblages (Collins *et al.*, 2005), and longlines are effective in detecting this pattern (Menezes *et al.*, 2009). During the MAR-ECO surveys, 30 fewer species were caught by longline (Fossen *et al.*, 2008) compared with trawl (Bergstad *et al.*, 2008); even fewer species (about half) were caught by baited landers (Fossen *et al.*, 2008). Baited landers attracted mostly teleosts, while longlines caught more sharks (King *et al.*, 2006).

Longlines generally catch larger fish compared to other fishing gears (Løkkeborg & Björdal, 1992; Fossen *et al.* 2003). Gillnets are the most size-selective (catching the narrowest size range), followed by longlines, while trawls are the least size-selective (Björdal & Løkkeborg, 1996). Sampling bias may occur if only one gear is used, and consequently affect biological analyses (e.g. representative sex ratios, size composition) as shown by Huse *et al.* (1999) for Greenland halibut. Size selectivity is also important when fish segregate spatially by life stage (juvenile, adult), or if larger fish swim faster and find gear earlier than small fish. Larger fish sometimes also chase smaller fish from baits (Björdal & Løkkeborg, 1996). Bait size itself is an important influence on size selectivity, and hence choice of bait size will depend on the survey objectives and fish species of interest.

8.5.2 *Sex ratios*

A disproportionate number of females are often caught by longlines and other gears (Wirtz & Morato, 2001). The sex ratio sometimes is depth-dependent (e.g. trawl catch of *Coryphaenoides armatus*, Merrett & Haedrich, 1997). Males and females may be spatially segregated (e.g. deep-water sharks) and one of the sexes may be more common in the surveyed area. Longline selectivity, which often favours larger fish, may not reflect the sex ratio for sexually dimorphic species. In addition, larger fish are often female, and be more capable of escape from towed gears than smaller fish of different sex. This has been observed by Clarke *et al.* (2005) for the deep-water shark *Deania calcea*, where larger females were more common in longline than trawl catches. In contrast, smaller, more active males of *Coryphaenoides armatus* may avoid trawls but be attracted to bait (Merrett & Haedrich, 1997). Reproductive behaviour may affect fish availability as well. For example, gravid female sharks (*Centroscymnus coelolepis*) are found in shallower waters (Clarke *et al.*, 2005). Reproductive seasons may also affect longline catch rates, as spawning fish may not be feeding (and hence are not sampled), or may feed heavily following spawning. For example, high catch rates occur on spawning aggregations of *Beryx splendens* and *Epigonus telescopus* on Sedlo seamount (Menezes *et al.*, 2009), and higher catch rates of *Pagellus bogaraveo* are annually obtained during their spawning season in the Azores (authors unpublished data).

8.5.3 Lost gears and impacts on seabed

The loss of the longline gear may occur due to strong currents or rough or steep bottom. If the line breaks during hauling, the vessel travels to the other buoy line and resumes hauling the gear. If the line breaks again, then a drag is used to attempt recovery of the gear. Floating underwater line sections between main line units may facilitate successful retrieval of lines using a drag chain (Fenaughty, 2008). In some cases, the longline may tangle with gear previously lost in rough areas. With time these places may accumulate lost gear and should be avoided. However, ghost fishing by longlines is less of a problem than it is for gillnet or trap gear. Longline gear eventually stops fishing because the baits deteriorate or are eaten by scavengers and the hooks rust. Both monofilament and multifilament longlines may remain on the seafloor for many years, sometimes becoming covered with sediments or colonized by marine organisms.

Longline impacts on sessile fauna like corals and sponges have been observed (Fossa *et al.*, 2002; Mortensen *et al.*, 2008). These impacts depend on gear configuration, faunal morphology (e.g. branched or unbranched corals), bottom types and current. The longline may move laterally during hauling (Welsford & Kilpatrick, 2008), implying a wider longline footprint, and hence impact, than previously assumed. Longline floats likely reduce any damage by reducing bottom contact. Nevertheless, overall impact associated with demersal longlines is typically less than other fishing gears like trawls, deep-water gillnets and traps (Koslow *et al.*, 2001; Hall-Spencer *et al.*, 2002; Fossa *et al.*, 2002; Chuenpagdee *et al.*, 2003; Pham *et al.*, 2014).

8.5.4 Interactions with seabirds and marine mammals

Bird bycatch may occur but can be minimized by good fishing practises. Birds are susceptible to hooking when the gear is deployed by trying to steal baits at the surface. If the gear is weighted, the longline will generally sink quickly unless caught in prop wash and held near the surface for a time. ‘Tori lines’ are streamer lines attached to the stern of the vessel above the location where the longline is deployed. The streamer lines discourage seabirds by interrupting their path to the baits. This deterrent is widely used in global longline fisheries (Melvin *et al.*, 2004). Nighttime deployments also appear to be effective at decreasing seabird bycatch because most bird species are less active at night.

Interactions between cetaceans and longline vessels have been reported (Moreno *et al.*, 2008). These include the sablefish (*Anoplopoma fimbria*) longline fishery in the Gulf of Alaska (Mitchell *et al.*, 2002; Sigler *et al.*, 2008), the Patagonian toothfish (*Dissostichus eleginoides*) longline fishery (Ashford *et al.*, 1996), longline fisheries near South Georgia and the Falkland/Malvinas Islands (Nolan *et al.*, 2000) and near the Prince Edward Islands, South Africa (Tilney & Purves, 1999). Entanglement with gear and removal of bait are the main interactions (Moreno *et al.*, 2008). Little can be done to deter marine mammals, but their effects on sampling can be reduced by avoiding commercial fishing grounds (where a higher level of fishing activity, and bait and fish catch could attract mammals), and by surveying in seasons when marine mammals are not present, or at lower densities.

Acknowledgements

We appreciate the comments of Chris Lunsford (Alaska Fisheries Science Center, NOAA) and three anonymous reviewers, and the suggestions and comments of the editors.

Appendix 8.1 Characteristics of some longline component materials

Characteristics of nylon monofilament line

			Breaking strength (kg)	
Diameter (mm)	Length per kg (m)	Mass (g/1000 m)	Dry line unknotted	Wet line knotted
0.1	909 000	11	0.65	0.4
0.2	27 700	44	2.3	1.4
0.3	11 100	90	4.7	2.7
0.5	4170	240	12	6.5
0.7	2080	480	24	12.5
1.0	1090	920	42	22
1.3	650	1540	65	35
1.6	430	2330	98	52
1.9	300	3290	132	72

Adapted from FAO (1972).

Characteristics of synthetic fibre multifilament lines

	Polyamide (nylon)		Polyester		Polyethylene		Polypropylene	
Diameter (mm)	Mass (kg/100 m)	Dry-line breaking strength (kg)	Mass (kg/100 m)	Dry-line breaking strength (kg)	Mass (kg/100 m)	Dry-line breaking strength (kg)	Mass (kg/100m)	Dry-line breaking strength (kg)
4	1.1	320	1.4	295				
6	2.4	750	3	565	1.7	400	1.7	550
8	4.2	1350	5.1	1020	3	685	3	960
10	6.5	2080	8.1	1590	4.7	1010	4.5	1425
12	9.4	3000	11.6	2270	6.7	1450	6.5	2030
14	12.8	4100	15.7	3180	9.1	1950	9	2790
16	16.6	5300	20.5	4060	12	2520	11.5	3500

	Polyamide (nylon)		Polyester		Polyethylene		Polypropylene	
18	21	6700	26	5080	15	3020	14.8	4450
20	26	8300	32	6350	18.6	3720	18	5370
22	31.5	10000	38.4	7620	22.5	4500	22	6500
24	37.5	12000	46	9140	27	5250	26	7600
Floatability	Not float		Not float		Will float		Will float	

Adapted from FAO (1972).

Dimensions of common longline hooks

Ordinary hooks			Hammered hooks		
Reference number	Opening (mm)	Diameter (mm)	Reference number	Opening (mm)	Diameter(mm)
12	9.5	1	2	10	1
11	10	1	1	11	1
10	11	1	1/0	12	1
9	12.5	1.5	2/0	13	1.5
8	14	1.5	3/0	14.5	1.5
7	15	2	4/0	16.5	2
6	16	2	5/0	20	2.5
5	18	2.5	6/0	27	3
4	20	3	8/0	29	3.5
3	23	3	10/0	31	4
2	26.5	3.5	12/0	39	5
1	31	4	14/0	50	6
1/0	35	4.5			

Adapted from FAO (1972).

References

Ashford, J.R., Rubilar, P.S., Martin, A.R. (1996). Interactions between cetaceans and longline fishery operations around South Georgia. *Marine Mammal Science*, 12: 452–457.

Bensch, A., Gianni, M., Gréboval, D., Sanders, J.S., Hjort, A. (2009). Worldwide review of bottom fisheries in the high seas. FAO Fisheries and Aquaculture Technical Paper. No. 522. FAO, Rome, 145 pp.

Bergstad, O.A., Hareide, N.-R. (1996). Ling, blue ling and tusk of the north-east Atlantic. *Fisken og Havet*, 15: 126 pp.

Bergstad, O.A., Menezes, G., Høines, Å. (2008). Demersal fish on a mid-ocean ridge: distribution patterns and structuring factors. *Deep-Sea Research II*, 55: 185–202.

Beverton, R.J. H., Holt, S.J. (1957). *On the dynamics of exploited fish populations*. Fishery Investigations Series II, Vol. XIX. London: Ministry of Agriculture, Fisheries and Food.

Björdal, A., Løkkeborg, S. (1996). *Longlining*. Oxford: Fishing News Books, 156 pp.

Chuenpagdee, R., Morgan, L.E., Maxwell, S.M., Norse, E.A., Pauly, D. (2003). Shifting gears: assessing collateral impacts of fishing methods in the U.S. waters. *Frontiers of Ecology and Environment*, 10: 517–524.

Clark, M.R., Vinnichenko, V.I., Gordon, J.D.M., Beck-Bulat, G.Z., Kukharev, N.N., Kakora, A.F. (2007). Large-scale distant-water trawl fisheries on seamounts. In: Pitcher, T.J., Morato, T., Hart, P.J.B., Clark, M.R., Haggan, N.,Santos, R.S. (eds), *Seamounts: ecology, conservation and management*. Fish and Aquatic Resources Series. Oxford: Blackwell, pp. 361–399.

Clarke, K.R., Warwick, R.M. (2001). *Changes in marine communities: an approach to statistical analysis and interpretation* (2nd edn). Plymouth: PRIMER-E.

Clarke, M.W., Borges, L., Officer, R.A. (2005). Comparisons of trawl and longline catches of deepwater elasmobranchs West and North of Ireland. *Journal of Northwest Atlantic Fisheries Science*, 35: 429–442.

Collins, M.A., Bailey, D.M., Ruxton, G.D., Priede, I.G. (2005). Trends in body size across an environmental gradient: a differential response I scavenging and non-scavenging demersal deep-sea fish. *Proceedings of the Royal Society B*, 272: 201–2057.

Connolly, P.L., Kelly, C.J. (1995). Deep-water trawl and longline surveys in 1995. Marine Institute Fishery Leaflet 173.

Connolly, P.L., Kelly, C.J. (1996). Catch and discards from experimental trawl and longline fishing in the Rockall Trough. *Journal of Fish Biology*, 49 (suppl. A): 132–144.

Engås, A., Løkkeborg, S. (1994). Abundance estimation using bottom gillnet and longline – the role of fish behaviour. In: Fernö, A., Olsen, S. (eds), *Marine fish behaviour in capture and abundance estimation*. London: Fishing News Books, pp. 134–165.

Erzini, K., Gonçalves, J.M.S., Bentes, L., Lino, P., Ribeiro, J., Stergiou, K.I. (2001). The hake deep-water semipelagic 'pedra-bola' longline fishery in the Algarve southern Portugal. *Fisheries Research*, 51: 327–336.

FAO (Food and Agriculture Organization of the United Nations) (1972). *Catalogue of fishing gear designs*. London: Fishing News Books, 155 pp.

Fenaughty, J.M. (2008). The autoline system – an updated descriptive review of the method with recommendations to clarify CCAMLR conservation measures regulating longline fisheries within the convention area. CCAMLR Document WG-FSA-08/60.

Fernö, A., Huse, I. (1983). The effect of experience on the behavior of cod (*Gadus morhua* L.) towards a baited hook. *Fisheries Research*, 2: 19–28.

Fossa, J.H., Mortensen, P.B., Furevik, D.M. (2002). The deep-water coral *Lophelia pertusa* in Norwegian waters: distribution and fishery impacts. *Hydrobiologia*, 471: 1–12.

Fossen, I, Jørgensen, O., Gundersen, A.C. (2003). Roughhead grenadiers (*Macrourus berglax*) in the waters off East Greenland: distribution and biology. *Journal Northwest Atlantic Fishery Science*, 31: 285–298.

Fossen, I., Cotton, C.F., Bergstad, O.A., Dyb, J.E. (2008). Species composition and distribution patterns of fishes captured by longlines on the Mid-Atlantic Ridge. *Deep-Sea Research II*, 55: 213–217.

Gordon, J.D.M., Bergstad, O.A., Figueiredo, I., Menezes, G.M. (2003). Deep-water Fisheries in the Northeast Atlantic: I. Description and current trends. *Journal of Northwest Atlantic Fishery Science*, 31: 137–150.

Gotelli, N.J., Colwell, R.K. (2013). Estimating species richness. In: Magurran, A.E., McGill, B.J. (eds), *Biological diversity: frontiers in measurement and assessment*. Oxford: Oxford University Press, 39–54 pp.

Gunderson, D.R. (1993). *Surveys of fisheries resources*. New York: John Wiley & Sons, 248 pp.

Guisan, A., Edwards Jr., T.C., Hastie,T. (2002). Generalized linear and generalized additive models in studies of species distributions: setting the scene. *Ecological Modeling*, 157: 89–100.

Hall-Spencer, J., Allain, V., Fossa, J.H. (2002). Trawling damage to Northeast Atlantic ancient coral reefs. *Proceedings of Royal Society London B*, 269: 507–511.

Haimovic, M., Ávila-da-Silva, A.O. (2007). A comparison of methods for estimating relative abundances in bottom longline fishery surveys. *Fisheries Research*, 85: 226–231.

Haimovic, M., Ávila-da-Silva, A.O., Wongtschwski, R. (2004). *Prospecção pesqueira de espécies demersais com espinhel-de-fundo na Zona Econômica Exclusiva da região Sudoeste-Sul do Barsil*. São Paulo: Instituto Oceanográfico – USP, 112 pp.

Hareide, N-R., Garnes, G. (2001). The distribution and catch rates of deep water fish along the Mid-Atlantic Ridge from 43 to 61 N. *Fisheries Research*, 51: 297–310.

Huse, I., Gundersen, A.C., Nedreaas, K.H. (1999). Relative selectivity of Greenland halibut (*Reinhardtius hippoglossoides*, Walbaum) by trawl, longlines and gillnets. *Fisheries Research*, 44: 75–93.

ICES. (2002). Report of the Working Group on Biology and Assessment of Deep-Sea Fisheries Resources (WGDEEP), *ICES CM 2002/ACFM:16*. 256 pp.

ICES. (2005). Report of the Working Group on Biology and Assessment of Deep-Sea Fisheries Resources (WGDEEP). *ICES CM 2006/ACFM:07*. 202 pp.

Iglesias, S., Muñoz, P.D. (2001). Spanish investigations on deep-sea resources in northern Mid-Atlantic Ridge and adjacent areas. The Census of marine life Pilot Project. Patterns and Processes of the Ecosystem of the Northern Mid-Atlantic. Working Document, Bergen Workshop, 12–13, 7 pp.

Johnston, T.L., Hunt, J.J. (1999). Preliminary results of a groundfish longline survey on Georges Bank. DFO Canadian Stock Assessment Secretariat Research Document, 99/78, 27 pp.

Jørgensen, O. (1995). A comparison of deep water trawl and longline research fishery in the Davis strait. In: Hopper, A.G. (ed.), *Proceedings of NATO Advanced Research Workshop on deep water fisheries of the north Atlantic slope*. Amsterdam: Kluwer Academic, p. 420.

King, N.J., Bagley, P.M., Priede, I.G. (2006). Depth zonation and latitudinal distribution of deep sea scavenging demersal fishes of the Mid-Atlantic Ridge, 421–531N. *Marine Ecology Progress Series*, 319: 263–274.

Koslow, J.A., Gowlett-Holmes, K., Lowry, J.K., O'Hara, T., Poore, G.C.B., Williams, A. (2001). Seamount benthic macrofauna of southern Tasmania: community structure and impacts of trawling. *Marine Ecology Progress Series*, 213: 111–125.

Laptikhovsky, V., Brickle P. (2005). The Patagonian toothfish fishery in Falkland Islands' waters. *Fisheries Research*, 74: 11–23.

Legendre, P., Legendre, L. (2000). *Numerical ecology* (2nd edn). Amsterdam: Elsevier Science, p. 853.

Løkkeborg, S. (1990). Reduced catch of under-sized cod (*Gadus morhua*) in longlining by using artificial bait. *Canadian Journal of Fisheries and Aquatic Sciences*, 47: 1112–1115.

Løkkeborg, S. (1994). Fish behaviour and longlining. In: Fernö, A., Olsen, S. (eds). *Marine fish behaviour in capture and abundance estimation*. London: Fishing News Books, pp. 9–27.

Løkkeborg, S., Björdal, Å. (1992). Species and size selectivity in longline fishing – a review. *Fisheries Research*, 13: 311–322.

Løkkeborg, S., Fernö, A. (1998). Diel activity pattern and food search behaviour in cod, *Gadus morhua*. *Environmental Biology of Fishes*, 54: 345–353.

Løkkeborg, S., Björdal, Å., Fernö, A. (1989). Responses of cod (*Gadus morhua*) and haddock (*Melanogrammus aeglefinus*) to baited hooks in the natural environment. *Canadian Journal of Fisheries and Aquatic Sciences*, 46: 1478–1483.

Løkkeborg, S., Olla, B.L., Pearson, W.H., Davis, M.W. (1995). Behavioural responses in sablefish, *Anoplopoma fimbria*, to bait odor. *Journal of Fish Biology*, 46: 142–155.

Ludwig, J.A., Reynolds, J.F. (1988). *Statistical ecology*. New York: John Wiley & Sons, 337 pp.

Martins, A.S., Olavo, G., Costa, P.A.S. (2005). Recursos demersais capturados com espinhel de fundo no talude superior da região entre Salvador (BA) e o Cabo de São Tomé (RJ). In: Costa, P.A.S., Martins, A.S., Olavo, G. (eds), *Pesca e potenciais de exploração de recursos vivos na região central da Zona Econômica Exclusiva brasileira*. Rio de Janeiro: Museu Nacional, pp. 109–128 (Série livros no. 13).

Maunder, M.N., Punt, A.E. (2004). Standardizing catch and effort data: a review of recent approaches. *Fisheries Research*, 70: 141–159.

McCullagh, P., Nelder, J.A. (1989). *Generalized linear models* (2nd edn). New York: Chapman & Hall, 511 pp.

Melvin, E.F., Sullivan, B., Robertson, G., Wienecke, B. (2004). A review of the effectiveness of streamer lines as a seabird by-catch mitigation technique in longline fisheries and CCAMLR streamer line requirements. *CCAMLR Science*, 11: 189–201.

Menezes, G.M. (2003). Demersal fish assemblages in the Atlantic Archipelagos of the Azores, Madeira, and Cape Verde. PhD thesis, Universidade dos Açores, Horta.

Menezes, G.M., Sigler, M.F., Silva, H.M., Pinho, M.R. (2006). Structure and zonation of demersal fish assemblages off the Azores Archipelago (mid-Atlantic). *Marine Ecology Progress Series*, 324: 241–260.

Menezes, G., Rosa, A., Melo, O., Pinho, M. (2009). Demersal fish assemblage of the Seine and Sedlo seamounts (Northeast Atlantic). *Deep-Sea Research II B*, 2683–2704.

Menezes, G.M., Giacomello E. (2013). Spatial and temporal variability of demersal fishes at Condor seamount (Northeast Atlantic). *Deep Sea Research Part II: Topical Studies in Oceanography,* 98, Part A: 101–113.

Menezes, G,M., Tariche, O., Pinho, M.R., Sigler, M- F., Silva, H.M. (2015). Structure and zonation of demersal and deep-water fish assemblages off the Cape Verde archipelago (Northeast-Atlantic) as sampled by baited longlines. *Deep-Sea Research I*, 102: 118–134.

Merrett, N.R., Haedrich, R.L. (1997). *Deep-sea demersal fish and fisheries*. London: Chapman & Hall, 282 pp.

Mitchell, E.A., Mesnick, S.L., Allen, A.C. (2002). Sperm whale depredation in the demersal longline fishery for sablefish, *Anaplopoma fimbria*, in the Gulf of Alaska: research needs and approaches to mitigation. Working Paper for the Workshop on Cetacean Interactions with Commercial Longline Fisheries in the South Pacific region: Approaches to Mitigation, Apia, Samoa, November 2002.

Moreno, C.A., Castro, R., Mújica, L.J., Reyes, P. (2008). Significant conservation benefits obtained from the use of a new fishing gear in the Chilean Patagonian toothfish fishery. *CCAMLR Science*, 15: 79–91.

Mortensen, P.B., Buhl-Mortensen, L., Gebruk, A.V., Krylova, E.M. (2008). Occurrence of deep-water corals on the Mid-Atlantic Ridge based on MAR-ECO data. *Deep-Sea Research II*, 55: 142–152.

Murphy, G.I. (1960). Estimating abundance from longline catches. *Journal of the Fisheries Research Board of Canada*, 17: 33–40.

Murua, H., Cárdenas, E. (2005). Depth distribution of deepwater species in Flemish Pass. *Journal of Northwest Atlantic Fishery Science*, 37: 1–12.

Nolan, C.P., Liddle, G.M., Elliot, J. (2000). Interactions between killer whales (*Orcinus orca*) and sperm whales (*Physeter macrocephalus*) with a longline fishing vessel. *Marine Mammal Science*, 16: 658–664.

Pham, C.K., Diogo, H., Menezes, G., *et al.* (2014). Deep-water longline fishing has reduced impact on Vulnerable Marine Ecosystems. *Sci Rep* 4. doi: 04837 10.1038/srep04837

Priede, I.G., Froese, R., Bailey, D.M, *et al.* (2006). The absence of sharks from abyssal regions of the world's oceans. *Proceedings of the Royal Society London B*, 273: 1435–1441. doi:10.1098/rspb.2005.3461

Rodgveller, C.J., Sigler, M.F., Hanselman, D.H., Ito, D.H. (2011). Sampling efficiency of longlines for shortraker and rougheye rockfish using observations from a manned submersible. *Marine Coastal Fisheries*, 3(1):1–9. doi:10.1080/19425120.2011.558447

Rothschild, B.J. (1967). Competition for gear in a multiple-species fishery. *ICES Journal of Marine Science*, 31: 102–110.

Sigler M.F. (2000). Abundance estimation and capture of sablefish (*Anoplopoma fimbria*) by longline gear. *Canadian Journal of Fisheries and Aquatic Sciences*, 57: 1270–1283.

Sigler, M.F., Fujioka, J.T. (1988). Evaluation of variability in sablefish, *Anoplopoma fimbria*, abundance indices in the Gulf of Alaska using the bootstrap method. *Fishery Bulletin, U.S.*, 86: 445–452.

Sigler, M.F., Zenger Jr., H.H. (1989). Assessment of Gulf of Alaska sablefish and other groundfish based on the domestic longline survey, 1987. U.S. Dep. Commer., NOAA Tech Memo. *NMFS F/NWC-169*, 54 pp.

Sigler, M.F. Lunsford, C.R., Straley, J.M., Liddle, J.B. (2008). Sperm whale depredation of sablefish longline gear in the northeast Pacific Ocean. *Marine Mammal Science*, 24:16–27.

Simonsen, C.S., Boje, J. (1999). An assessment of the inshore Greenland halibut stock component in NAFO Division 1A. NAFO SCR Doc. 99/48.

Simonsen, C.S., Boje, J., Kingsley, M.C.S. (2000). A review using longlining to survey fish populations with special emphasis on an inshore longline survey for Greenland halibut (*Reinhardtius hippoglossoides*) in West Greenland, NAFO Division 1A. NAFO SCR Doc. 00/29.

Skud, B.E., Hamley, J.M. (1978). Factors affecting longline catch and effort. International Pacific Halibut Commission Scientific Report, Number 64. Seattle, Washington, USA.

Somerton, D.A., Kikkawa, B.S. (1995). A stock survey technique using the time to capture individual fish on longlines. *Canadian Journal of Fisheries and Aquatic Sciences*, 52: 260–267.

Tilney, R., Purves, M.G. (1999). The status of integrated fisheries monitoring in South Africa. In: Nolan, C.P. (ed.), *Proceedings of the International Conference on Integrated Fisheries Monitoring, Sydney, Australia, 1–5 February 1999*. Rome: FAO, 378 pp.

Trumble, R.J. (1998). Northeast Pacific flatfish management. *Journal of Sea Research*, 39: 167–181.

Uiblein, F., Bordes, F., Castillo, R. (1996). Diversity, abundance and depth distribution of demersal deep-water fishes off Lanzarote and Fuerteventura, Canary Islands. *Journal of Fish Biology*, 49: 75–90.

Uiblein, F., Bordes, F., Castillo, R., Ramos, A. (1998). Spatial distribution of shelf-and slope-dwelling fishes collected by bottom longline off Lanzarote and Fuerteventura, Canary Islands. *Marine Ecology*, 19: 53–66.

Uiblein, F., Geldmacher, A., Koster, F., Nellen, W., Kraus, G. (1999). Species composition and depth distribution of fish species collected in the area of the Great Meteor Seamount, eastern central Atlantic, during cruise M42/3, with seventeen new records. *Informes técnicos del Instituto Canario de Ciencias Marinas, Telde (Gran Canaria)*, 5: 47–85.

Valdemarsen, J.W. (2001). Technological trends in capture fisheries. *Ocean and Coastal Management*, 44: 635–651.

Welsford, D., Kilpatrick, R. (2008). Estimating the swept area of demersal longlines based on in-situ video footage. CCAMLR Document WG-FSA-08/58. Hobart, Australia.

White, E., Soderlund, E., Dykstra, C.L., Geernaert, T.O., Ranta, A.M. (2010). Standardized stock assessment survey. International Pacific Halibut Commission. www.iphc.washington.edu/publications/rara/2010/2010.377.2010Standardizedstockassessmentsurvey.pdf

Wirtz, P., Morato, T. (2001). Unequal sex ratios in longline catches. *Journal of the Marine Biological Association of the UK*, 81: 187–188.

Woll, A.K., Boje, J., Holst, R.,Gundersen, A.C. (2001). Catch rates and hook and bait selectivity in longline fishery for Greenland halibut (*Reinhardtius hippoglossoides*, Walbaum), at East Greenland. *Fisheries Research*, 51: 237–246.

Zuur, A.F., Ieno, E.N., Smith, G.M. (2007). *Analysing ecological data*. Berlin: Springer, 672 p.

Chapter 9
Epibenthic Sledges

Stefanie Kaiser and Nils Brenke
German Centre for Marine Biodiversity Research, Wilhelmshaven, Germany

Abstract

Epibenthic sledges and dredges have been widely used for the investigation of deep-sea macro- and megafaunal communities, providing strong insights into benthic biodiversity and distribution patterns. Across institutes and nations a great variety of gear types have been designed, reflecting the diversity of deep-sea habitats and overall aims of study. This chapter gives an introduction to the most widely used deep-sea sledge and dredge types and assesses various advantages and disadvantages, such as sampling efficiency and operational limitations. Practical advice for deployment and sample processing is provided, as well as recommendations for the interpretation of data obtained.

Keywords macrobenthos, megabenthos, sledges, dredges, seamounts, sampling, standardization, efficiency

9.1 Introduction

Most of what we know about macrobenthic biodiversity and distribution in the deep sea is based on analyses of material collected by trawls and sledges (e.g. Thistle, 2003). Yet, the investigation of the seabed by towed gears is technically extremely challenging due to the great variety of seafloor topography and benthic habitats. Depending on the aim of the survey, the design of the sampling gear differs greatly in terms of, for example, size, weight and mesh size. Brenke (2005) gave an historical overview of different sampling equipment – from early dredges to modern sledges – and differentiated three broadly different groups: dredges, bottom-plankton nets and sledges (sometimes called sleds).

- *Dredges* have been used for the study of deep benthos since the start of deep-sea explorations in the nineteenth century (e.g. during the expeditions of HMS *Lightning* (1868), HMS *Porcupine* (1869–70) and HMS *Challenger* (1872–76)). These early dredges, such as the 'naturalist dredge' (or Ball's dredge, Fig. 9.1) and 'oyster dredge', were derived from bottom trawls used in fisheries and provided some first insights into zonation and diversity patterns with increasing depth. Due to their robustness, dredges can be deployed under rough conditions and thus are often used for geological rock collections.

Biological Sampling in the Deep Sea, First Edition. Edited by Malcolm R. Clark, Mireille Consalvey and Ashley A. Rowden.

Fig. 9.1 The naturalist's dredge – one of the first dredges to be deployed in the deep sea, and used during the HMS *Challenger* expedition (1972–76). (Modified from Thomson & Murray, 1885, by Monika Hänel, Biocentre Grindel & Zoological Museum, University of Hamburg, Germany.)

- Finely woven *bottom-plankton nets* have been employed to sample floating and swimming organisms just above the seabed (e.g. Beardsley, 1973). Some of these are equipped with an opening and closing mechanism in order to constrain sampling to a distinct horizon above the seafloor.
- Unlike bottom-plankton nets and dredges, *sledge dredges and benthic sledges* have runners, on which they are towed along the seafloor, and are predominantly used for catching small macrobenthic soft-bottom fauna. Most sledge dredges lack a closing device. Benthic sledges combine the robustness of the latter with the control (mechanical or acoustical opening and closing mechanism)

of plankton nets, thus avoiding contamination by planktonic organisms on the long way through the water column (summary given in Brenke, 2005).

The transition from one gear type to another is seamless, however, and it is difficult to find a satisfying classification. Some sampling devices, for example, were formerly described as dredges, but should be placed among sledge-type sampling devices (e.g. Triple D dredge; Bergman & Van Santbrink, 1994). A 'sled' (e.g. WHOI epibenthic sled, Hessler & Sanders, 1967) describes a type of sledge that is characterized by smaller mesh size and different sampling characteristics than a sledge (Thistle, 2003). Both terms have often been used interchangeably across the literature (e.g. Gage, 1975), despite showing differences in sampling performance. In practice it is not important to discuss the specific terminology, but to compare sampling efficiency and handling. Therefore, sampling devices introduced here are named following their first description.

There is now a great variety of sledges and dredges available to maximize efficiency on different habitat and bottom types as well as capturing different types of fauna (Hessler & Sanders, 1967; Rothlisberg & Pearcy, 1976; Brunel *et al.*, 1978; Rice *et al.*, 1982; Brandt & Barthel, 1995; Blomqvist & Lundgren, 1996; Sneli, 1998; Brenke, 2005). Differences are often most obvious in robustness, mesh size, the presence of an opening/closing mechanism, as well as the number of nets; sledge-type samplers can consist of only one net or hold two to several nets on top of each other in order to assess the vertical distribution of fauna above the seafloor (e.g. Rothlisberg & Pearcy, 1976; Brunel *et al.*, 1978; Sorbe, 1983; Brandt & Barthel, 1995). Finally, nets can be fitted to a metal frame, rubber mats or canvas strips to protect them from mechanical damage and to make them particularly suitable for technically demanding deep-sea work (Brenke, 2005).

The purpose of this chapter is to give an introduction to different sledge types, their specific sampling characteristics and evaluate the advantages and disadvantages of each type of gear. The main focus will be on sledges, but dredges predominantly used for deep-sea work will also be covered. Sampling and sorting procedures will also be described. Finally the quality of abundance and diversity data will be discussed and recommendations given on aspects of analysing and comparing data.

9.2 Description of dredge and sledge types, specifications and modifications

9.2.1 Dredges and anchor dredges

Various dredges have been constructed to sample macro- and megabenthos from a range of bottom types (hard substrate to muddy seafloor; e.g. Fowler & Allen, 1928; Beauchamp, 1932; Picard, 1965; Clarke, 1972). Despite most dredges being only qualitative samplers, they have been proven to be a useful tool for preliminary assessment of sediment structure and faunal patterns. However, there have been attempts to design dredges for semi-quantitative and quantitative sampling of the deep-sea benthos, including a modified anchor dredge (Sanders *et al.*, 1965; Fig. 9.2a), an anchor box dredge (Carey & Hancock, 1965; Fig. 9.2b) and the 'Triple D' dredge (Bergman & Van Santbrink, 1994; Fig. 9.3).

9.2.1.1 Anchor dredge and anchor box dredge

Forster's (1953) intention in developing the prototype of an anchor dredge was to build a device which could operate on both packed sandy bottom and muddy ground. The Forster anchor dredge was specifically designed to collect infauna. Unlike dredges, anchor dredges are usually not towed, but designed to 'bite' into the sediment. The basic construction consists of a steel plate, which is

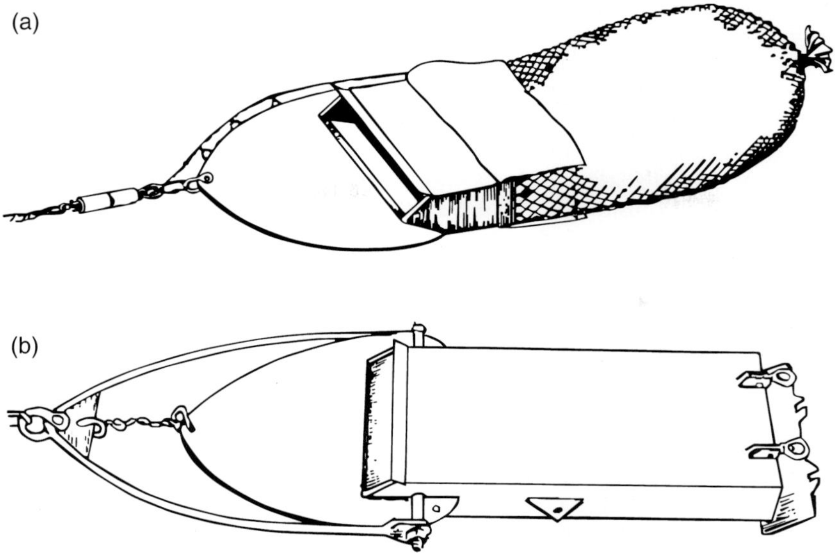

Fig. 9.2 (a) Anchor dredge and (b) anchor box dredge used for deep-sea sampling. (Adapted from Carey & Hancock, 1965, and Sanders *et al.*, 1965, by Birgit Baumgart, Biocentre Grindel & Zoological Museum, University of Hamburg, Germany.)

Fig. 9.3 The Triple-D (deep digging dredge) dredge is designed to sample burrowing fauna quantitatively. (Bergman, 2009. Reproduced with permission.) (Please see Plate 15 for colour representation of the figure.)

attached to a metal frame at an acute angle to the seafloor in order to dig deeply into the sediment. The warp is fitted to a towing arm to avoid disturbance of the surface and associated biota. Although it is argued to obtain semi-quantitative samples (Eleftheriou & Moore, 2005), Forster (1953) considered the anchor dredge to sample infauna only qualitatively.

Sanders *et al.* (1965) constructed a modified version of the Forster anchor dredge (Fig. 9.2a), which has been successfully deployed on both the shelf and in the deep sea. In contrast to the Forster anchor dredge, the Sanders *et al.* (1965) version can be operated double-sided and may sample quantitatively, as the penetrating depths of the dredge is consistent (but see section on efficiency of the gear below). A horizontal metal plate (= planing surface) is attached to a metal frame and divides the opening into two equal halves. While being dragged across the seafloor, the metal plate is supposed to skim over the surface, so that the anchor dredge can only sink to a certain depth (approx. 11 cm). Collection of sediment continues until the net is filled and sediment clogs the opening.

The anchor box dredge as described by Carey and Hancock (1965, Fig. 9.2b) is derived from the Sanders *et al.* anchor dredge and the McIntyre box dredge (McIntyre, 1964). Important changes from the former versions include a metal box protecting the net and a mechanical opening and closing mechanism. Gage (1975) compared sampling characteristics of the anchor dredge with the anchor box dredge, and found that both sample in a similar way. Although the anchor box dredge samples a smaller area (1.3 m^2 penetrating to 10 cm depths) than the anchor dredge, it is easier to handle and is more robust (see Gage, 1975).

9.2.1.2 Triple D dredge

The Triple D ('deep digging dredge') was designed to quantitatively sample larger infauna (to 10 cm depths) especially at low densities (Bergman & Van Santbrink, 1994; Bergman, 2009; Fig. 9.3). A steel cage is fitted into a metal frame attached to runners and is followed up by a net of 1.4 cm mesh size. The cage has two openings, one on the front to collect epifauna and one on the bottom side to sample infauna. Adjustable blades are attached to the underside opening altering the penetration depth and width. In order to assess the trawling distance an odometer is fitted to the runner. The Triple D has recently been updated (Bergman, 2009); the major difference is that the odometer (counter) wheels are now operated by pressure cylinders. Opening and closing of the pressure valves is programmed via an electronic switch board (G. Duineveld, Royal Netherlands Institute for Sea Research (NIOZ), personal communication).

The Triple D seems to be efficient in sampling burrowing animals, as it covers larger areas than a box core. However, the dredge is prone to damage of the nets and blades and can only be deployed on sandy and muddy bottoms without rocks and boulders.

9.2.2 Epibenthic sleds and sledges

9.2.2.1 WHOI epibenthic sled and modifications

Hessler and Sanders (1967) built the prototype of the WHOI epibenthic sled (Fig. 9.4) at the Woods Hole Oceanographic Institution (WHOI), and it became one of the first epibenthic sledges to be deployed in the deep sea. Data obtained by the sled refuted the idea of an impoverished deep sea, which had been the predominant view since early *Challenger* collections (Hessler & Sanders, 1967; Sanders & Hessler, 1969). The WHOI epibenthic sled combines characteristics of former sampling equipment such as the Riedl-Trawl and Okelmann's detritus sledge (Okelmann, 1964; Hessler & Sanders, 1967) and was designed to sample larger macro-benthic to smaller mega-benthic organisms on and above the deep seafloor.

In simple terms, the sled consists of a plankton-type (000 monofilament nylon) net surrounded by a flat metal frame (1.4 m wide and 30 cm high) and equipped with runners. The runners are arranged symmetrically on the upper and lower side of the sled and hence the sled can be operated double-sided.

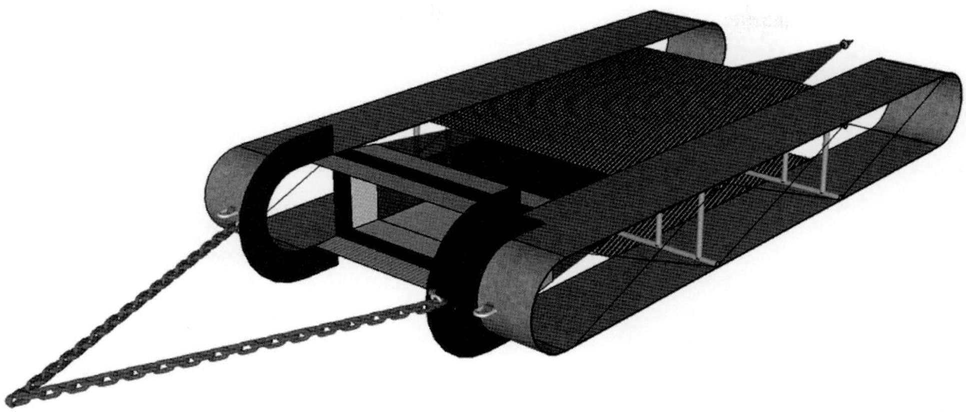

Fig. 9.4 The WHOI epibenthic sled developed at the Woods Hole Oceanographic Institution. (The figure is based on the original description by Hessler & Sanders (1967).)

Canvas strips attached to the rear end of the metal frame help to prevent chafing of the nets. The mouth area is 81 cm wide by 30 cm high. Adjustable cutting edges are located on top and bottom of the mouth area, and can be raised or lowered to increase or decrease the penetration depth. The main part of the net is protected by the frame and has a mesh size of 1 mm; the main net is followed by a 1.3 m extension (0.3–0.5 mm mesh) tapering towards the cod-end to avoid washing of the sample during retrieval.

A number of samplers have evolved from the WHOI sled (e.g. IOS sledge, Aldred *et al.*, 1976; Rice *et al.*, 1982, Fig. 9.5; Sneli sledge, Sneli, 1998, Fig. 9.6; benthic sled sensu Blomqvist & Lundgren, 1996). The acoustically monitored IOS epibenthic sledge (Fig. 9.5) was designed by members of the biology group at the UK Institute of Oceanographic Sciences (IOS, now National Oceanographic Centre, Southampton (NOC); Aldred *et al.*, 1976; Rice *et al.*, 1982). It had upgraded technical features: a mechanical opening and closing device to prevent planktonic contamination, a camera system to take images of the seafloor prior to sampling as well as an underwater acoustic beacon ('pinger') to assess when the sledge reaches the bottom and to facilitate calculation of the towing distance.

Most of the above sampling devices have been successfully deployed in the deep sea, but some are not robust enough to sample rough seafloor habitats, such as seamounts. Seamounts represent a challenging deep-sea environment characterized by, for example, strong currents and rugged topography (e.g. Clark *et al.*, 2010), which makes benthic sampling there technically demanding. The probability that gear gets stuck (and subsequently lost) can be very high, and nets are prone to damage. Lewis (1999, 2009) constructed a sledge, the CSIRO (Commonwealth Scientific and Industrial Research Organization) seamount epibenthic sampler (SEBS = Sherman sledge; Fig. 9.7), which addressed some of these requirements. Primarily, this sledge is much more robust and heavier (now 1200 kg) than the WHOI sled and is equipped with a novel weak-link recovery system. Due to the rather large mesh size of 25 mm, the Sherman sledge is biased towards sampling of larger megabenthos (e.g. hydroids, corals, fish and cephalopods), but also collects smaller (macrobenthic) groups such as isopods and polychaetes (Lewis, 1999). A 1 mm mesh sampler can be bolted inside the mouth area to collect smaller organisms (Lewis, 2009).

The Sherman sledge is large and heavy, and difficult to handle on small vessels at sea. Some of the key elements have been adopted by other research institutes (e.g. National Institute of Water & Atmospheric Research (NIWA), New Zealand) in building much lighter sledges that still incorporate the break-away system, but also feature the ability to tow either way up, and have the net completely

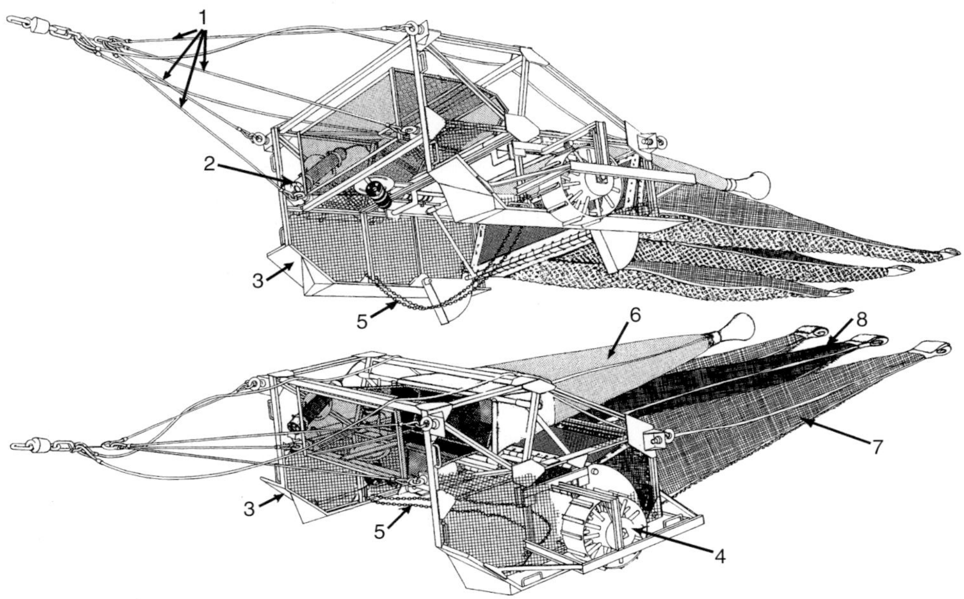

Fig. 9.5 A modified version of the IOS epibenthic sledge; three epibenthic nets as well as an upper suprabenthic net are fitted to the rear end of the frame: (1) towing bridle, (2) still camera; (3) runners; (4) odometer; (5) tickler chain; (6) suprabenthic net; (7) coarse mesh net; (8) fine mesh net. (Adapted from Gage & Bett, 2005. Reproduced with permission of B. Bett, National Oceanographic Centre, Southampton.)

Fig. 9.6 The Sneli sledge has been widely used across the North Atlantic and Arctic oceans. The wire running along the side is a security cable; if the sledge gets stuck, the cable will turn the sledge around, so it can be retrieved backwards. (Photograph by J.A. Sneli, Norwegian University of Science and Technology, Trondheim, Norway. Reproduced with permission.)

Fig. 9.7 An improved version of the CSIRO Sherman sledge designed for seamount sampling. (Photograph by M. Lewis, CSIRO. Reproduced with permission.)

enclosed within the sledge frame (Fig. 9.8) (Clark & Stewart, in review). Designed by scientists at NIWA, the sled is similar in general design to the CSIRO Sherman sledge but it is much smaller (155 cm long, 50 cm high and 130 cm wide) and lighter (400 kg fully rigged) (Table 9.1). The mesh size used for much of NIWA's deep-sea research is 25 mm, covered by an anti-chafing net of 100 mm. The sled can operate on both sides, and so it does not matter if it gets tipped over. It is rigged with the same break-away system of bridles and chains as the Sherman sledge. Even though the NIWA sledge is relatively light, it has proved very successful on seamounts and rough seafloor off New Zealand (Clark & Stewart, in review), where it has been used as a standard gear type to sample macro- and mega-invertebrate fauna over a wide range of seafloor topography and substrate composition (e.g. Clark & Rowden 2009).

A further gear successfully deployed on seamounts (i.e. Great Meteor and Hyères seamounts in the North Atlantic, authors personal observation) is the Brenke sledge (Brenke, 2005), described in detail in the following section.

9.2.2.2 Rothlisberg and Pearcy epibenthic sledge and modifications (Brenke sledge)

To add another faunal dimension, a number of sledges have been designed to not only sample in- and epifauna, but also organisms above the seabed (so-called supra- or hyperbenthos, Mees & Jones, 1997). These samplers usually consist of one or two (but sometimes more) horizontally divided, finely woven nets, sampling the fauna up to 1.5 m above the seafloor (e.g. Brunel *et al.*, 1978). Some of these sledges have a quite complicated design and are rather heavy, and consequently not easy to handle at sea, particularly in adverse weather conditions (e.g. Brandt & Barthel, 1995). Hence, a more lightly built epibenthic sampler developed by Rothlisberg and Pearcy (1976, Fig. 9.9)

Table 9.1 Pre-sampling considerations: comparison of important technical features (weight, opening–closing device, mesh size), selectivity (target fauna, bottom type) and sampling efficiency (quantitative vs. qualitative) in different sledge and dredge types

Gear	Target fauna	Bottom type	Sampling efficiency	Dimensions $l \times w \times h$ (cm)	Weight (kg)	Opening–closing device	Net mesh size (mm)	Cod-end mesh size (mm)	Reference
Anchor dredge	Macrobenthic infauna	Soft bottom	(semi-) quantitative	n.a.	250	–	0.42	0.42	Sanders *et al.*, 1965
Anchor box dredge	Macrobenthic infauna	Soft bottom	(semi-) quantitative	240 × 53 × 21	341	+	n.a.	n.a.	Carey & Hancock, 1965
Triple D dredge	Megabenthic in- and epifauna	Soft bottom without rocks and boulders	(semi-) quantitative	200 × 150 × 150	600	–	7–14	7–14	Bergman & Van Santbrink, 1994
WHOI sled	Macro- to smaller megabenthic epifauna	Soft bottom	(semi-) quantitative	230 × 140 × 30	~160	+/–[a]	1	0.3	Hessler & Sanders, 1967; Dahl *et al.*, 1976
IOS sledge	Larger megabenthic epifauna	Soft bottom	qualitative	193 × 229 × 61	n.a.	+	4.5	1	Aldred *et al.*, 1976; Rice *et al.*, 1982
Sneli sledge	Macrobenthic in-and epifauna	Soft bottom-rough	qualitative	200 × 80 × 20	80	–	n.a.	n.a.	Sneli, 1998
Sherman sledge	Macro- to larger megabenthic epifauna	Rugged seafloor (seamounts)	qualitative	242 × 160 × 80	1200	–	25	25	Lewis, 1999, 2009
NIWA sled	Macro- to larger megabenthic epifauna	Rugged seafloor (seamounts)	qualitative	155 × 130 × 50	400	–	25	25	Clark & Rowden, 2009; Clark & Stewart, in review
R-P sledge	Macrobenthic epifauna	Soft bottom	qualitative	150 × 120 × 60	150	+	0.571	0.571	Rothlisberg & Pearcy, 1977
Brenke sledge	Macrobenthic epifauna	Soft bottom-rough	qualitative	280 × 130 × 116	444	+	0.5	0.3	Brenke, 2005

Weight: measured empty, in air; opening–closing device: +, present, –, absent; dimensions of the frame only (without net): width (w) × length (l) × height (h).

[a] The original version by Hessler & Sanders (1967) was constructed without a closing mechanism.

Fig. 9.8 The NIWA sled on board RV *Tangaroa*. The cylinders on the side of the frame are for sediment samples. (Photograph courtesy of the NIWA, New Zealand.)

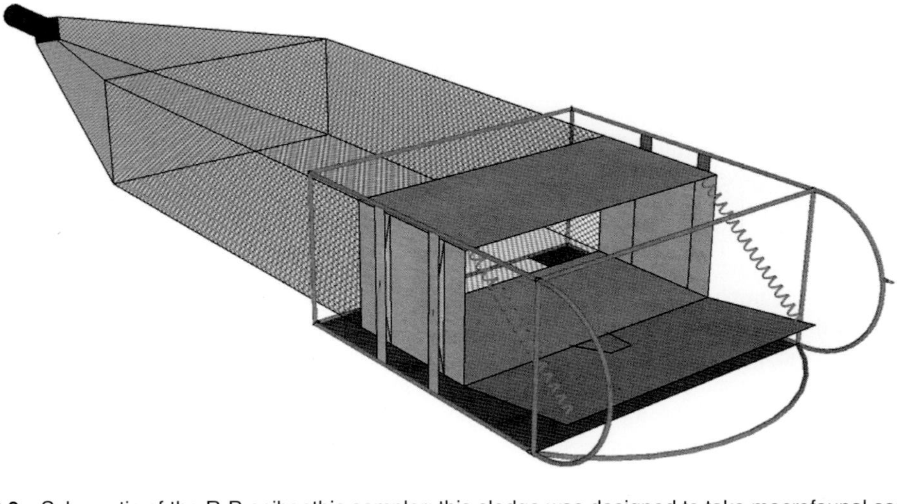

Fig. 9.9 Schematic of the R-P epibenthic sampler: this sledge was designed to take macrofaunal samples on and above the sediment–water interface. (The figure is based on the original description by Rothlisberg & Pearcy, 1976.)

and modifications of it have been widely used across the North Atlantic and the Arctic oceans (e.g. Buhl-Jensen, 1986; Brattegard & Fosså, 1991; Brandt & Schnack, 1999).

Rothlisberg and Pearcy's (1976) epibenthic sampler (R-P sledge) was designed for a study of the vertical distribution of larval stages in planktonic shrimps on, and slightly above, the seafloor. The general construction consists of a sampling box equipped with a simple closing device and fitted into a steel frame on runners. The opening and closing mechanism is characterized by a hinged door, which is attached to the bottom of the box. A projecting triangular shoe opens the door as soon as the sledge hits the seafloor. In the water column the door is closed due to springs inside the box. A canvas anti-chafing mat helps to protect the finely meshed (571 μm) nylon net, which is attached to the rear frame.

When first developed, the R-P sledge was mostly deployed in shallower waters. Since then, several modifications have been made in order to sample benthos at greater depths (e.g. Buhl-Jensen, 1986; Brattegard & Fosså, 1991). These include, for example, a reinforced rubber mat and weights at the rear of the mat to avoid tangling of net and mat. A further version was developed by Brandt and Barthel (1995), who added an upper supranet on top of the already existing epinet in order to sample fauna from 25–60 cm and 77–105 cm above the bottom; here, mesh sizes ranged from 300 to 500 μm.

Following on from Brandt and Barthel's (1995) version, Brenke (2005) constructed the next generation of R-P-type sledges. These have been widely used around the world, from hard substrate to soft bottom, from seamounts to Antarctic sponge communities and from shelf to the deep sea (e.g. Brenke, 2005; Kaiser *et al.*, 2011; authors' personal knowledge). The Brenke sledge (Fig. 9.10) is characterized by a large and strong metal frame entirely surrounding the epi- and the supranet and a modified opening and closing system (orientated in a 45° angle). Due to small mesh sizes (300–500 μm) the sledge can sample meiofaunal- and macrofaunal-sized epibenthic organisms. The Brenke sledge can optionally be divided into three subunits (front, centre and rear) to facilitate cargo transport. On smaller vessels, the Brenke sledge can be deployed in a reduced version (frontal and centre unit only), in which configuration rubber mats must be fitted to the rear of the frame to protect the nets, as in the original version of the R-P sledge. Furthermore, a cooling box for deployment in warm tropical and subtropical waters has been designed, which surrounds the cod-ends, to preserve the samples for subsequent genetic analyses (Fig. 9.10). This box is intended to keep the samples at an almost constant temperature from the cold deep to the (warm) surface waters in order to inhibit or slow down lyses of DNA during the long recovery process.

9.2.3 Further equipment and future development

Additional equipment has been developed in order to monitor and/or maximize performance, to minimize environmental impact of the gear, to improve quality of the data obtained and to gain additional data on density and distribution patterns of the fauna.

Deep-sea sledges equipped with camera systems, for example, have been used since the early 1970s (e.g. Thiel, 1970) and since then have been continuously improved and refined (e.g. Christiansen & Nuppenau, 1997). As sledges (like other gear types) only sample a fraction of the total fauna, photographic images can be useful to evaluate sampling efficiency of the sledge and to provide further qualitative information on community structure and habitat characteristics. Furthermore, the use of CTDs, RCM (recording current meter), O_2 sensors and other sensors can help to measure environmental parameters along the sampled track. Recently, a version of the Brenke sledge equipped with a still camera and video system, as well as CTD, RCM, DCS (Doppler current sensor) and O_2 sensor has been successfully deployed both in the Atlantic and Pacific Oceans (Brandt *et al.*, 2013).

Fig. 9.10 The Brenke sledge, shown being lowered from the aft deck of RV *Meteor* during the DIVA-3 expedition. Numbers indicate (1) weak link; (2) security wire; (3) crow's foot; (4) doors, closed in the water column; (5) lever unit; (6) cooling box surrounding the cod-ends of the epi- and supranet. (Photograph by S. Kaiser.)

To improve estimates of the trawling distance, an odometer can be fitted to the sledge (e.g. Koulouri *et al.*, 2003). Additional sensors, such as acoustic positioning systems, can provide telemetry data on the position of the sledge relative to the bottom and vessel. Finally nets can be added either to optimize sampling efficiency, to sample different size fractions or different types of fauna (e.g. Gage & Bett, 2005).

9.3 Sampling operations: how to choose and use a sledge

9.3.1 Pre-sampling considerations

All sampling approaches are selective, favouring certain faunal types, taxonomic groups and size classes. Hence, a wide range of bottom-plankton nets as well as sledge- and dredge-type equipment have been developed to meet the basic requirements of a survey, depending on the aims and spatial scale of the study, ecology and size spectrum of the target organisms as well as quality of data required. For example, many sledges and dredges may not be appropriate for the assessment of faunal densities and total species richness, but are suitable to obtain large numbers of specimens for taxonomic, phylogenetic or phylogeographic analyses.

Benthic surveys usually aim at limiting collection of fauna to a defined area in, on or just above the seafloor. An important feature of deep-sea sledges is therefore a sufficient opening/closing device to avoid capture of non-targeted planktonic organisms. Yet, some samplers (e.g. Sneli sledge, see Table 9.1) lack such a mechanism and hence may sample a high proportion of pelagic taxa such as copepods and chaetognaths, depending on the mesh size.

Sampling devices, such as the R-P sampler and Brenke sledge, have rather small mesh sizes (≤500 μm) and thus adequately sample meio- and macrofaunal groups. Others (e.g. Sherman, NIWA and IOS sledges) target the larger megafauna (>2000 μm), while the WHOI sled and the Triple D dredge are somewhere in between (Table 9.1). The anchor dredge and anchor box dredge were specifically designed to sample burrowing fauna (Forster, 1953; Carey & Hancock, 1965; Sanders *et al.*, 1965), whereas most sledges are more efficient in sampling fauna on and slightly above the sediment–water interface.

The choice of the most appropriate sampling device may also depend on logistic and technical limitations depending on, for example, the size of the ship, warp length, maximum load of the winch, availability of gear, and budget (number and duration of deep-sea tows). Many deep-sea sledges are quite large and heavy (e.g. WHOI sled and Sherman sledge; Table 9.1) and are probably best operated from medium-sized to large ships, whereas most dredges can be deployed from small vessels (Eleftheriou & Moore, 2005). The Forster anchor dredge, for example, operates most efficiently from small trawlers, as larger ships may drag the anchor dredge out of the sediment (Eleftheriou & Moore, 2005). The size of the Brenke sledge, on the other hand, can be adjusted to the ship's dimension by detaching the rear part of its frame (Brenke, 2005). There are two sizes of the NIWA sled, being simple and relatively cheap to build, appropriate for different-sized vessels.

Bottom type is a very important determinant of gear selection, in that some samplers are complicated and fragile and hence require careful handling at sea. The Triple D dredge, for instance, can only be deployed on soft bottom and under sea conditions of moderate swell and wave height, whereas others are quite robust, and can be operated on a variety of bottom surfaces (e.g. many dredges, Sherman, NIWA and to a lesser extent the Brenke sledge). Rough seafloor and weather conditions can also have profound impacts on the performance and deployment of a sledge, which will be discussed in more detail below.

9.3.2 Deployment

In common with almost all gear types described in this book, prior to deployment of sledges, it is necessary to map bathymetry and roughness of the seafloor, ideally using multibeam echosounder data especially along steep slopes or unknown, rugged ground (e.g. Aldred *et al.*, 1976; and see Chapter 5). As deep-sea hauls can take a long time (up to 8 hours), sea conditions should be also monitored carefully (wind speed, swell and wave height in particular). Christiansen and Nuppenau (1997) found that ship's movement directly influences the performance of the sledge, even with several kilometres warp length, because pitching of the ship may lead to poor bottom contact of the sledge and thus inefficient sampling. Rough weather conditions can, furthermore, lead to damage of the gear during deployment or retrieval, and loss of samples (e.g. Hessler & Sanders, 1967; Sneli, 1998).

An easy way to assess the performance of the sledge on the seafloor is to mark runners, sides and top of the sledge prior to deployment (simply with a waterproof pencil or tape). If the deployment was successful, runners should be polished.

A range of deployment procedures are possible for the various dredge and sledge types, most obviously differing in towing speed, duration and warp-to-water-depth ratio. Some of these will be described below.

9.3.2.1 *Towing bridles*

Arrangement (and consequent towing angle), length and diameter of the towing bridles are important determinants of the *working load limit* and *breaking load* of the weak-link system. In the Sherman sledge the recovery system consists of a set of towing bridles (10–12 mm), which are attached to the rear of the sledge. The angle of attack is determined by the length of the weak-link chains, which are fixed to the forward towing lugs. This gives the Sherman sledge a four-point break-away system (Lewis, 1999, 2009). In the Brenke sledge, a crow's foot links the swivel and weak link with the frontal unit of the sledge, while a security wire connects the swivel with the rear unit (Fig. 9.10). The bridle arrangement is gear- and ship-specific though (for example with regard to deck equipment, winch, etc.) and also depends on the experience of the crew and scientists, so it is not possible to provide universal guidance.

9.3.2.2 *Deployment and retrieval of the gear*

Deployment and recovery procedures can vary greatly depending on the gear type, ship (e.g. size, presence of deployment/recovery slip), as well as sea and weather conditions. Thus, there are several ways to deploy or retrieve gear. In the following sections the most common procedures are presented.

All gear types described here are operated from the stern of the ship stabilized by a rope, which is fitted to the frame. Usually a trawl wire of 13–18 mm in diameter is used for deep-sea deployments (e.g. Carey & Hancock, 1965; Christiansen & Nuppenau, 1997), while the Sherman sledge is deployed using a 24 mm wire (Lewis, 2009).

Anchor dredge and anchor box dredge There are some operational differences in the deployment of the Forster and the Sanders anchor dredge, for example in the warp length and performance of the gear (Forster, 1953; Sanders *et al.*, 1965). To avoid any confusion the following account will focus on the protocol presented by Sanders *et al.* (1965).

When the anchor dredge enters the water the ship drifts or steams at very slow speed. As soon as enough wire is paid out (about 1.3–1.5 times water depth), the anchor dredge is towed very slowly across the seabed until it reaches full 'biting depths'. For the deployment of the anchor box dredge (*sensu* Carey & Hancock, 1965) the ship has to hold position. When the warp length equals water depth, the ship speed increases to 1 knot (1 knot = 0.51 m s^{-1}), and more wire is paid out at 0.8–1.25 m s^{-1}. After the haul, the dredge is recovered at approx. 0.6–0.8 m s^{-1}.

Towed dredges and sledges During lowering of the gear, ship and winch speed need to be balanced, as the sledge should stay behind the ship and sink faster than the wire. If the winch speed is too high the sledge may become entangled with the wire, while a high ship speed may prevent sinking of the gear.

When the gear is lowered into the water, the ship usually steams at 1–1.4 knots over the ground, while warp is payed out at slow speed (0.25–0.5 m s^{-1}) until there is sufficient tension on the cable. Depending on the speed of the vessel and gear type, the winch speed is then increased to 0.5–0.7 m s^{-1} (Aldred *et al.*, 1976; Brenke, 2005). As soon as the wire length equals water depth, the ship speed is adjusted to 1 knot and the winch speed to 0.5 m s^{-1}, while more wire is payed out to an optimal warp–depth ratio (see below). Due to its greater weight, a ship speed of

2.5 knots is recommended for the deployment of the Sherman sledge (M. Lewis, personal communication).

At the end of the tow, the ship stops. As long as the gear is on the seabed it is hauled at slow speed (~0.5 $m\,s^{-1}$ maximum, for the IOS sledge 0.2 $m\,s^{-1}$ is recommended, see Aldred *et al.*, 1976). If the gear gets stuck, there are some procedures that can be followed:

(1) The winch and ship should be stopped immediately and wire payed out slowly to reduce wire tension. The gear is then hauled in at very slow speed (max. 0.1–0.25 $m\,s^{-1}$) until the equipment breaks free (this may be repeated). As soon as the gear clears the seafloor, the winch speed can be increased to 0.7–1 $m\,s^{-1}$).
(2) If the sledge is still stuck, the ship can steam back in an arc, so that it is positioned at 90° to the gear. During steaming the winch tension needs to be kept constant so that the tow wire is not damaged and the cable is kept clear of the seafloor to prevent entanglement. The gear is then hauled as described above. If the sledge is equipped with a weak-link system (e.g. Sneli, 1998; Brenke, 2005; Lewis, 2009) it should ensure safe recovery of the gear even if the weak link breaks.

9.3.2.3 Warp length

An optimal warp–depth ratio (scope) is crucial to guarantee good performance of the gear. A reduced scope (i.e. insufficient wire payed out) may lead to greater tension on the wire; as a consequence the gear may skip across the seafloor with only the rear section touching, and thus catch very little. A high scope, on the other hand, can cause entanglement of the gear, disturbance of the bottom by the wire (Christiansen & Nuppenau, 1997) and can also lead to unnecessarily long, time-consuming hauls. Therefore the warp length needs to be carefully adjusted, and depends also on the sea conditions and bottom topography (with regard to factors such as uneven ground and slope).

The scope for most sledge deployments is at least 1.5–1.8:1 (e.g. Hessler & Sanders, 1967; Brenke, 2005), and so a minimum of 4500 m wire needs to be payed out to sample the abyss (>3000 m water depth). For the Forster anchor dredge a warp length of five times water depth is recommended to prevent the anchor dredge from being pulled out of the sediment (Forster, 1953). The scope also depends on the water depth; with increasing depth the weight of the wire increases substantially relative to the total load. Hence, the scope should decrease with increasing depth (Carey & Hancock, 1965; Gage & Bett, 2005, and see Chapter 7). For example, in 1000 m depth the warp-to-water-depth ratio should be 1.8–2:1 (i.e. 1800–2000 m warp length), while in 2000 m depth the scope should not exceed 1.5:1 (Carey & Hancock, 1965).

9.3.2.4 Towing direction, speed and duration

Towed gears are usually deployed against the wind and thus surface current direction. This allows better speed control of the gear (see also Chapter 7) and prevents the sledge being pushed under the ship during recovery.

Sledges are preferably operated on even ground or only with slight changes in water depth. Sampling of steep slopes, seamounts or troughs can be quite challenging, as flat plateaus sufficiently large for the whole length of the tow are sometimes hard to detect. On seamounts, it is common practice to tow downslope (e.g. Clark & Roberts, 2008). This is in contrast to geological sampling with dredges, where the gear is towed upslope to improve the extent of 'digging in'. This can cause sledges to become stuck, and risk gear damage or loss. If the break-away system has worked, the sledge is recovered by pulling out backwards, but this means that the net is often everted, with loss of catch. Towing slowly downslope reduces gear damage or unsatisfactory tows, and appears to sample effectively (M. Clark, NIWA, personal communication).

Towing speed varies greatly between gear types; for the WHOI sled, NIWA sled, IOS and Brenke sledge a trawling speed of 1–1.5 knots over ground is recommended (e.g. Gage, 1975; Aldred *et al.*,

1976; Brenke, 2005; Clark & Stewart, in review), while the Sherman sledge and Triple D dredge are towed at much higher speed, between 2–2.5 and 3.7 knots, respectively (Bergman & Van Santbrink, 1994; Lewis, 2009). Tow duration also varies, between 10 minutes (e.g. Brenke sledge) and 1 hour (WHOI sled; Hessler & Sanders, 1967) in deep-sea hauls. Yet, in order to decrease impact of trawling on benthic fauna towing time should be reduced to a minimum. On the shelf, Gage (1975) reduced the towing duration of the WHOI sled to 5 minutes and still recovered a sufficient sample. It is also common practice to keep tow times short in habitats where catch rates of sessile fauna (such as corals and sponges) can be high. NIWA sled tows on seamounts around New Zealand are generally only 10–15 minutes, covering a distance of 300–400 m (Clark & Rowden, 2009).

Trawling time is sometimes standardized to allow comparison of data between surveys using complementary sampling equipment (e.g. Brenke, 2005; Brandt *et al.*, 2007). However, it is not clear whether trawling time is equivalent to sampling time, as in deep-sea tows the sledge can remain on the seafloor for a long time until enough warp is retrieved and it is hauled clear (depending on water depths and winch speed). It is very likely that throughout this period sediment is continuously transported into the net. This was demonstrated, for example, during the DIVA ('Latitudinal gradients of deep-sea biodiversity in the Atlantic Ocean') 3 expedition, which sampled along the rise and plateau of the Great Meteor Bank using a Brenke sledge. Here, towing time was usually standardized to 10 minutes; however, during this cruise a sledge was retrieved after a 1 minute tow yielding a sufficient sample size (authors' personal observation).

9.3.2.5 Monitoring the position of the gear

The observation of the gear's position relative to the ship and ground is important information to ensure safe performance of the gear and accurate calculation of the towing distance and hence sampled area.

There are various methods to estimate the towing distance, some of which were reviewed by Brenke (2005). Important variables for the assessment of the towed area are, for example, the GPS position of the ship as well as the time span between start and end of the tow (i.e. the time when the sledge clears the seafloor).

Different techniques have been applied to monitor the position of the gear in the water column and on the seabed, including telemetry systems ('pinger'; Aldred *et al.*, 1976) and observation of tension on the winch wire. The pinger is attached to the gear; it continuously sends acoustic signals ('pings'), which are recorded by sensors on the ship. Pinger systems have performed adequately attached on deep-sea trawls (e.g. during DIVA cruises) and on NIWA sled tows on seamounts off New Zealand (Clark & Stewart, in review), but it has been argued that they are not suitable for towed gears and that specialized acoustic telemetry systems should be used instead (Gage & Bett, 2005). There are also restrictions on their use: in Antarctic waters, the use of pingers is not allowed as they may attract whales. On seamount sledges in New Zealand, a depth sensor can be attached, which gives information on the depth of the gear, and this can be compared with seamount bathymetry to estimate its position on the feature (M. Clark, NIWA, personal communication).

9.3.3 Differences in bottom type

Bottom structure can strongly affect the behaviour of a sledge on the seafloor; that is, rugged ground can lead to poor or intermittent bottom contact, damage, overturning, blocking or even loss of the gear. Therefore the addition of weights and a robust construction makes some sledges particularly suitable for the operation on rough bottom such as the Sherman, NIWA and, to a lesser extent, the Sneli and Brenke sledges. Many sledges are equipped with a weak-link system to ensure recovery of the gear in case it gets trapped (e.g. Hessler & Sanders, 1967; Sneli, 1998; Lewis, 1999, 2009; Brenke, 2005). Extra weights can help to balance and stabilize the gear on the seabed and in the water column (Sneli, 1998; Brenke, 2005), although depending on the bottom type and current strength, the weight needs to be carefully adjusted. On soft bottom, an overweight sledge may sink too deeply into the sediment and not run smoothly, while a light sledge may have poor bottom contact.

Nets are particularly prone to damage and thus are usually mounted inside metal frames or have canvas mats on the outside to protect them from abrasion (see above). A modification of the Brenke sled was made for sampling the Antarctic benthos: a grid was fitted to the mouth opening of the upper and lower net to avoid intrusion of larger drop stones and hexactinellid sponges (N. Brenke, personal communication).

Sediment and rocks can also block the spring-shoe mechanism as used in the R-P and Brenke sledges; as a consequence, doors remain open in the water column and the sledge samples planktonic fauna on its retrieval. On very soft bottom, the doors may not open at all, as there is not enough resistance for the shoe to be lifted.

9.4 Sample sorting and processing

Single towed deep-sea deployments are typically several hundred to thousand metres long so are not 'point' samples. Sledges and trawls are effective though in collecting larger numbers of specimens in a vast but low-density environment such as the abyssal plains compared with corers, which sample precise, but relatively small areas. However, there are also some limitations, as data obtained are at most semi-quantitative. Thus estimates of diversity and abundance levels are approximate, as is any 'standardization' of data and sampling protocol (see below). Long trawling times may increase the chance of capturing rare fauna, but it also increases the chance of damaging very fragile specimens. Operational considerations, such as reducing towing speed and gentle sieving of the catch on board, may help to minimize mechanical impact though and provide good specimens for taxonomic and biogeographic studies.

9.4.1 Sieving and preservation

A number of methods have been described to separate macrofaunal organisms from the sediment fraction most efficiently and at the same time keep mechanical damage as low as possible (reviewed in Eleftheriou & Moore, 2005; see Chapters 10 and 15). A protocol developed by Sanders *et al.* (1965) for sample processing has been adopted by several authors (e.g. Hessler & Sanders, 1967; Gage, 1975). First, an estimation of the volume of the sample is made, particularly in 'quantitative' and semi-quantitative gear types such as the anchor dredge and anchor box dredge (cf. Carey & Hancock, 1965; Sanders *et al.*, 1965). The sample is then carefully elutriated with seawater; the largest specimens are picked out first and preserved separately. The elutriated fraction and heavier bulk are poured through a 420 μm sieve and subsequently fixed in formalin for 2–3 days. In samples collected by the Brenke sledge, supra- and epinet are usually elutriated separately and then carefully sieved through a 500 and 300 μm mesh. In order to protect fragile specimens from damage, strong water jets should be avoided wherever possible when either washing down nets or sieving.

Sledge samples are normally fixed in buffered 5% formol or 70% ethanol (see Chapter 15). Protocols for molecular genetic analyses involve keeping samples at a relatively constant temperature (between 0 and 4°C) at all times; samples are elutriated with cold seawater, sieved and fixed with 96% non-denatured pre-cooled ethanol. Samples should be then stored at −20°C for several days (at least 48 hours) prior to the start of the sorting process. During this time the samples should be moved continuously (every few hours) to avoid freezing and to ensure optimal preservation.

9.4.2 Sorting and subsampling

Sorting of macrobenthic material by hand can be time-consuming; processing a single sledge sample may take days up to several weeks. A number of methods have been developed to speed up the sorting process, including flotation, staining, elutriation and centrifugation techniques (e.g. Barmuta,

1984; Cromar & Williams, 1991). Most of these methods should not be applied, however, if specimens are to be used for genetic analyses.

Subsampling is another method to decrease time and cost of processing benthic samples (Barbour & Gerittsen, 1996). Gage (1982) has described a subsampling method for non-quantitative deep-sea samples retrieved by towed gears. In the Brenke sledge it has been suggested that the epi- and supranet do not sample substantially different faunas (N. Brenke, personal communication; Knox *et al.*, 2012), and hence the typically much cleaner supranet samples could be used as a substitute for the epinet to make sample processing much quicker.

However, sorting effort and value of the data obtained need to be carefully balanced. In some cases, such as for general environmental monitoring, it might be sensible to apply rapid sorting procedures to produce results more quickly. However, on the other hand with many species being 'rare' in the deep sea, sorting the entire sample should always be considered, particularly if high numbers of specimens are needed for phylogeographic or systematic studies.

It is useful to record samples directly on board (including individual numbers, station name, depth, geographic coordinates, collecting date and time, etc.) and to give each vial an individual numerical code. If possible, photographs may be taken of selected specimens. See Chapter 15 for more details of onboard sample handing and processing.

9.5 Interpretation of data

Factors such as ship speed, sediment structure, sea conditions and reduced scope (among others) can limit sampling efficiency of towed gears by reducing the bottom contact of the gear (see previous sections). Strong bottom currents may prevent light sledge gear reaching the seafloor, while poor weather conditions and the ship surging can cause washing of samples (e.g. Hessler & Sanders, 1967; Sneli, 1998). Furthermore, the bow wave effect, as known from multi- and box-corer sampling (Chapter 10), can also reduce the fraction of fauna sampled (e.g. Lampitt *et al.*, 1986). Hence, sledge data are unlikely to be fully quantitative, but care is also needed in assuming that the data are comparable in a relative sense.

Gage *et al.* (2002) found that mesh size can strongly affect quantification of biomass and diversity levels in the deep sea; a relatively coarse mesh size leads to undersampling of the smaller size classes, particularly juvenile stages. Finer meshes usually increase the number of individuals and species sampled, but also the time spent on sorting and identifications (Schlacher & Wooldridge, 1996). It has been suggested that for estimations of biomass a sieve size of 500 μm–1 mm is enough, while for assessment of species numbers a mesh size ≤300 μm should be used (Gage *et al.*, 2002).

All sampling devices are biased, in that they only sample a fraction of total fauna present. Even 'quantitative' gears such as a box-corer have certain limitations (e.g. Shirayama & Fukushima, 1995). Hence, a comparison between different gear types is necessary in order to evaluate the degree to which sledges and dredges provide samples that are representative of a faunal assemblage (in terms of faunal abundance, richness and composition). The reader is referred to Chapter 2 which describes some of the biological implications of different sampling gears.

9.5.1 Efficiency of the gear

Anchor dredges and anchor box dredges have been suggested to obtain quantitative samples (Sanders *et al.*, 1965). Hessler and Jumars (1974) argue, however, that the anchor dredge may not be at the same depth for the entire time of the tow and samples may only be semi-quantitative.

Dickinson and Carey (1975) compared sampling efficiency of the Carey–Hancock anchor box dredge with the Smith–McIntyre grab (Smith & McIntyre, 1954); they demonstrated there were no significant differences in relative abundances between gear types, but estimates of faunal densities were much higher in the grab samples. Gage (1975) investigated differences in diversity and faunal composition between the anchor box dredge, WHOI epibenthic sled and Van Veen grab. Differences were slight, with diversity being lower in anchor box dredge samples, possibly due to undersampling the more motile fauna.

It is clear that most sledge- and dredge-type samplers provide qualitative samples as it is hard to assess whether the gear is on the ground at all times, if nets are consistently filled with sediment (or rocks and other faunal catch), and how quickly this occurred. Camera systems have been used to monitor and improve the efficiency of towed gears (e.g. Rice *et al.*, 1979; Christiansen & Nuppenau, 1997). Rice *et al.* (1979) found large discrepancies between the fauna seen in photographs and the fauna sampled by the IOS sledge. The sledge underestimated faunal densities, particularly of larger epifauna. Other towed gears are also biased in favouring certain components of the benthic fauna. The WHOI sled and Brenke sledge, for example, have been shown to efficiently sample fauna associated with the sediment–water interface, whereas burrowing organisms are underrepresented (Gage, 1975; Brenke, 2005).

Even if the sledge has good bottom contact for the entire length of the tow, there are some further limitations. Based on scuba divers' observations, Gage (1975) found that sampling performance and efficiency of the WHOI epibenthic sled was affected by the position of the cutting blades; set in a horizontal position (i.e. lowered to maximum extent), the sled performed like an anchor dredge – the mouth blocked as soon the sled was filled with sediment and therefore samples obtained were very small. Raising the blades a little bit provided a larger sample, but probably changed sampling performance because of, for example, resuspension of sediment. In R-P samplers, the front shield may cause turbulence, which varies depending on the bottom type, and may increase the amount of sediment sampled by the sledge (Brenke, 2005 and discussion therein). Furthermore, strong bottom currents against the towing directions may increase transport of sediment into the net, which leads to an overestimation of abundance (Brenke, 2005).

9.5.2 Standardization and interpretation of data

The standardization of gear, sampling protocol and data recording is important (and stressed throughout this book) to allow comparison of abundance and diversity levels between surveys. Trawling speed, for example, has been standardized to 1 knot and tows to a certain area swept (1000 m^2 in the R-P sampler) in order to obtain comparable data sets (e.g. Brattegard & Fosså, 1991; Brandt & Schnack, 1999; Brenke, 2005). However, biodiversity data from sledges and dredges must be treated with care due to the previously discussed limitations. Performance and efficiency can differ greatly between sledge hauls (e.g. in WHOI sled, IOS sledge; see Gage *et al.*, 1980; Rice *et al.*, 1979), implying a potentially high and gear-specific systematic error. For example, Rice *et al.* (1979) suggested that the sledge may not sample consistently and thus between-tow comparability may be problematic. Brenke (2005) estimated that the Brenke sledge operates with a minimum variation of 25% between hauls and therefore abundance levels only represent relative values. Brattegard and Fosså (1991), however, found that diversity and abundance levels differed little between R-P sledge replicates, suggesting that one sample may be sufficient to assess suprabenthic diversity.

Gear performance needs to be evaluated for each tow, and tows may be rejected if there were obvious problems, such as entanglement, overturning of the sledge, problems with the opening–closing mechanism or torn nets. However, in most cases samples can still be used for taxonomic and biogeographic analyses.

9.6 Concluding remarks

Across many expeditions, institutes and nations, a vast amount of information on benthic biodiversity and large-scale distribution patterns in the deep sea has been collected by means of epibenthic sledges (e.g. Sanders & Hessler, 1969; Rex *et al.*, 1993) and also more recently (e.g. during ANDEEP (Antarctic benthic deep-sea biodiversity: colonization history and recent community patterns), BIOICE (Benthic Invertebrates of Icelandic waters), DIVA, NORFANZ (Norfolk Ridge and Lord Howe Rise), Louisville Ridge and Ocean Survey 20/20 cruises (e.g. Svavarsson, 1997; Clark & Roberts 2008; Brenke, 2005; Brandt *et al.*, 2007; Rowden *et al.*, 2010; Kaiser *et al.*, 2011; Knox *et al.*, 2012; Lörz *et al.*, 2013; Clark *et al.*, 2015). Although sledges and dredges only sample a small fraction of the existing fauna, they have been a very useful tool because they usually yield much larger numbers of species and specimens compared to 'quantitative' samplers, such as box corers (e.g. Brandt & Schnack, 1999). Sanders and Hessler's (1969) discovery of unexpectedly high biodiversity in the deep sea, for example, was based on epibenthic sled samples. More recently, sampling the Southern Ocean deep sea during the ANDEEP I-III expeditions brought up an unprecedented number (>600) of isopod species in only 40 sledge samples (Brandt *et al.*, 2007). The quality of data obtained, however, makes assessment and interpretation of biodiversity and distribution patterns complicated. We are not yet at a stage where we are able to build truly quantitative sledges.

As covered in other chapters (see Chapters 2 and 15 for example), there is not a single gear type to sufficiently (and solely) describe biodiversity and distribution patterns in the deep sea, as all have issues of faunal selectivity, sample size and performance. Thus, a combination of gear types is required to assess biodiversity on a local and regional scale, considering the low faunal densities and large numbers of 'rare' species in the deep sea. However, recognizing the limitations of gear type, sledges can provide valuable qualitative insights into the taxonomy, ecology and biogeography of deep-sea organisms, which can be used as a baseline for future studies to monitor environmental change – also with regard to prospected deep-sea exploitation (e.g. Gage, 2001; Kaiser & Barnes, 2008).

Acknowledgements

The authors are grateful to Brian Bett (NOC), Magda Bergman (NIOZ), Gerard Duineveld (NIOZ), Mark Lewis (CSIRO), Jon-Arne Sneli and Malcolm Clark (NIWA) for providing pictures and useful information on handling and technical details of each gear type. We would like to thank Birgit Baumgart (ZIM) and Monika Hänel (ZIM), who kindly made the drawings of the anchor dredge and naturalist's dredge respectively. David Bowden (NIWA) and David K.A. Barnes (British Antarctic Survey) are thanked for language corrections and helpful discussions. We are grateful to Malcolm Clark, Mark Lewis and Rhian Waller (WHOI) for valuable comments on an earlier draft of this manuscript. The first author acknowledges funding provided by the 'Innovationsfond' (University of Hamburg).

References

Aldred, R.G., Thurston, M.H., Rice, A.L., Morley, D.R. (1976). An acoustically monitored opening and closing epibenthic sledge. *Deep-Sea Research*, 23: 167–174.

Barbour, M.T., Gerittsen, J. (1996). Subsampling of benthic samples: a defense of the fixed-count method. *Journal of the North American Benthological Society*, 15(3): 386–391.

Barmuta, L.A. (1984). A method for separating benthic arthropods from detritus. *Hydrobiologia*, 112: 105–107.

Beardsley, A.J. (1973). Design and evaluation of a sampler for measuring the near-bottom vertical distribution of pink shrimp, *Pandalus jordani*. *Fishery Bulletin US*, 71: 243–253.

Beauchamp, R.S.A. (1932). A new dredge. *Internationale Revue der gesamten Hydrobiologie und Hydrographie*, 27(4–6): 467–469.

Bergman, M.J.N. (2009). Deep digging dredge (Triple-D). In: Rees, H.L. (ed.), *Guidelines for the study of the epibenthos of subtidal environments*. ICES techniques in Marine Environmental Sciences No. 42: 13.

Bergman, M.J.N., Van Santbrink, J.W. (1994). A new benthos dredge ('Triple-D') for quantitative sampling of infauna species of low abundance. *Netherlands Journal of Sea Research*, 33(1): 129–133.

Blomqvist, S., Lundgren, L. (1996). A benthic sled for sampling soft bottom. *Helgoländer Meeresuntersuchungen*, 50: 453–456.

Brandt, A., Barthel, D. (1995). An improved supra- and epibenthic sledge for catching Peracarida (Crustacea, Malacostraca). *Ophelia*, 43: 15–23.

Brandt, A., Schnack, K. (1999). Macrofaunal abundance at 79°N off East Greenland: opposing data from epibenthic-sledge and box-corer samples. *Polar Biology*, 22: 75–81.

Brandt, A., Gooday, A.J., Brix, S.B., *et al.* (2007). The Southern Ocean deep sea: first insights into biodiversity and biogeography. *Nature*, 447: 307–311.

Brandt, A., Elsner, N., Brenke, N., *et al.* (2013). Epifauna of the Sea of Japan collected via a new epibenthic sledge equipped with camera and environmental sensor systems. *Deep-Sea Research II*, 86–87: 43–55.

Brattegard, T., Fosså, J.H. (1991). Replicability of an epibenthic sampler. *Journal of the Marine Biological Association of the UK*, 71: 153–166.

Brenke, N. (2005). An epibenthic sledge for operations on marine soft bottom and bedrock. *Marine Technology Society Journal*, 39(2): 10–21.

Brunel, P., Besner, M., Messier, D., Poirier, L., Granger, D., Weinstein, M. (1978). Le traîneau suprabenthique MACER-GIROQ: appareil amélioré pour l'échantillonnage quantitatif étagé de la petite faune nageuse an voisinage du fond. (The MACER-GIROQ Suprabenthic Sled: an improved device for quantitative two-level sampling of the small swimming fauna near the bottom). *Internationale Revue der gesamten Hydrobiologie und Hydrographie*, 63(6): 815–829.

Buhl-Jensen, L. (1986). The benthic amphipod fauna of the west-Norwegian continental shelf compared with the fauna of five adjacent fjords. *Sarsia*, 71: 193–208.

Carey, A.J., Hancock, D.R. (1965). An anchor-box dredge for deep-sea sampling. *Deep Sea Research and Oceanographic Abstracts*, 12(6): 983–984.

Christiansen, B., Nuppenau, V. (1997). The IHF Fototrawl: experiences with a television-controlled, deep-sea epibenthic sledge. *Deep-Sea Research I*, 44(3): 533–540.

Clark, M.R., Roberts, C.D. (2008). Fish and invertebrate biodiversity on the Norfolk Ridge and Lord Howe Rise, Tasman Sea (NORFANZ voyage, 2003). Aquatic Environment and Biodiversity Report No. 28, 131 pp. http://fs.fish.govt.nz/Page.aspx?pk=113&dk=22301

Clark, M.R., Rowden, A.A. (2009). Effect of deepwater trawling on the macro-invertebrate assemblages of seamounts on the Chatham Rise, New Zealand. *Deep-Sea Research I*, 56: 1540–1554.

Clark, M.R., Stewart, R. (in review). The NIWA seamount sled: an effective epibenthic sledge for sampling epifauna on seamounts and rough seafloor. *Deep-Sea Research I*.

Clark, M.R., Rowden, A.A., Schlacher, T., *et al.* (2010). The ecology of seamounts: structure, function, and human impacts. *Annual Review of Marine Science*, 2: 253–278.

Clark, M.R., Anderson, O., Bowden, D., *et al.* (2015). Vulnerable Marine Ecosystems of the Louisville Seamount Chain: voyage report of a survey to evaluate the efficacy of preliminary habitat suitability models. *New Zealand Aquatic Environment and Biodiversity Report*, No. 149, 86 p.

Clarke, A.H. (1972). The Arctic dredge, a benthic biological sampler for mixed boulder and mud substrate. *Journal of the Fisheries Research board of Canada*, 29: 1503–1505.

Cromar, G.L., Williams, D.D. (1991). Centrifugal flotation as an aid to separating invertebrates from detritus in benthic samples. *Hydrobiologia*, 209: 67–70.

Dahl, E., Laubier, L., Sibuet, M., Stromberg, J.-O. (1976). Some quantitative results on benthic communities of the deep Norwegian Sea. *Astarte*, 9: 279–297.

Dickinson, J.J., Carey Jr., A.G. (1975). A comparison of two benthic infauna samplers. *Limnology and Oceanography*, 20(5): 900–902.

Eleftheriou, A., Moore, D.C. (2005). Macrofauna techniques. In: Eleftheriou, A., McIntyre, A.D. (eds), *Methods for the study of marine benthos*. Oxford: Blackwell, pp. 160–228.

Forster, G.R. (1953). A new dredge for collecting burrowing animals. *Journal of the Marine Biological Association of the UK*, 32: 193–198.

Fowler, G.H., Allen, E.J. (1928). *The science of the sea*. Oxford.

Gage, J.D. (1975). A comparison of the deep-sea epibenthic sledge and anchor-box dredge samplers with the Van Veen grab and hand coring by diver. *Deep-Sea Research*, 22: 693–702.

Gage, J.D. (1982). An aerated sedimentation column for subsampling large benthic samples from deep-sea sediments. *Deep-Sea Research Part A*, 29(5): 627–630.

Gage, J.D. (2001). Deep-sea benthic community and environmental impact assessment at the Atlantic Frontier. *Continental Shelf Research*, 21: 957–986.

Gage, J.D., Bett, B.J. (2005). Deep-sea benthic sampling. In: Eleftheriou, A., McIntyre, A.D. (eds), *Methods for the study of marine benthos*. Oxford: Blackwell, pp. 273–325.

Gage, J.D., Lightfoot, R.H., Pearson, M., Tyler, P.A. (1980). An introduction to a sample time series of abyssal macrobenthos: methods and principal sources of variability. *Oceanologica Acta*, 3: 169–176.

Gage, J.D., Hughes, D.J., Gonzalez Vecino, J.L. (2002). Sieve size influence in estimating biomass, abundance and diversity in samples of deep-sea macrobenthos. *Marine Ecology Progress Series*, 225: 97–107.

Hessler, R.R., Jumars, P.A. (1974). Abyssal community analysis from replicate box cores in the central North Pacific. *Deep-Sea Research*, 21: 185–209.

Hessler, R.R., Sanders, H.L. (1967). Faunal diversity in the deep-sea. *Deep-Sea Research*, 14: 65–78.

Kaiser, S., Barnes, D.K.A. (2008). Southern Ocean deep-sea biodiversity, sampling and predicting responses to climate change. *Climate Research*, 37: 165–179.

Kaiser, S., Griffiths, H.J., Barnes, D.K.A., Brandão, S.N., Brandt, A., O'Brien, P.E. (2011). Is there a distinct continental slope faunal in the Antarctic? *Deep-Sea Research II*, 58: 91–104.

Knox, M.A., Hogg, I.D., Pilditch, C.A., Lörz, A.N. (2012). Abundance and diversity of epibenthic amphipods (Crustacea) from contrasting bathyal habitats. *Deep-Sea Research I*, 62: 1–9.

Koulouri, P.T., Dounas, C.G., Eleftheriou, A.S. (2003). A new apparatus for the direct measurement of otter trawling effects on the epibenthic and hyperbenthic macrofauna. *Journal of the Marine Biological Association of the UK*, 83: 1363–1368.

Lampitt, R.S., Billett, D.S.M., Rice, A.L. (1986). Biomass of the invertebrate megabenthos from 500 to 4100 m in the northeast Atlantic Ocean. *Marine Biology*, 93: 69–81.

Lewis, M. (1999). CSIRO-SEBS (Seamount, Epibenthic sampler), a new epibenthic sled for sampling seamounts and other rough terrain. *Deep-Sea Research I*, 46: 1101–1107.

Lewis, M. (2009). Sherman the epibenthic sled for rough terrain. CSIRO Marine and Atmospheric Research paper; 029. Canberra, ACT: CSIRO Marine and Atmospheric Research, 15 pp. www.cmar.csiro.au/publications/cmarseries/index.htm

Lörz, A.N., Kaiser, S., Bowden, D.A. (2013). Macrofaunal crustaceans in the benthic boundary layer from the shelf break to abyssal depths in the Ross Sea (Antarctica). *Polar Biology*, 36: 445–451. doi:10.1007/s00300-012-1269-1

McIntyre, R.J. (1964). A Box Dredge for quantitative sampling of benthic organisms. *Limnology and Oceanography*, 9(3): 460–461.

Mees, J., Jones, M.B. (1997). The hyperbenthos. *Oceanography and Marine Biology: an Annual Review*, 35: 221–255.

Okelmann, K.W. (1964). An improved detritus sledge for collecting meiobenthos. *Ophelia*, 1–2: 217–222.

Picard, J. (1965). Recherches qualitatives sur les biosciences marines des substrats dragables de la région marseillaise. *Recueil des Travaux de la Station Marine d'Endoume*. Bull 36(52): 1–160.

Rex, M.A., Stuart, C.T., Hessler, R.R., Allen, J.A., Sanders, H.L., Wilson, G.D.F. (1993). Global-scale latitudinal patterns of species diversity in the deep-sea benthos. *Nature*, 365: 636–639.

Rice, A.L., Aldred, R.G., Billett, D.S.M., Thurston, M.H. (1979). The combined use of an epibenthic sledge and a deep-sea camera to give quantitative relevance to macro-benthos samples. *Ambio Special Report, 6, The Deep Sea: Ecology and Exploitation*: 59–63.

Rice, A.L., Aldred, R.G., Darlington, E., Wild, R.A. (1982). The quantitative estimation of the deep-sea megabenthos: a new approach to an old problem. *Oceanologica Acta*, 5: 63–72.

Rothlisberg, P.C., Pearcy, W.G. (1976). An epibenthic sampler used to study the ontogeny of vertical migrations of *Pandalus jordani* (Decapoda, Caridea). *Fishery Bulletin US*, 74: 994–997.

Rowden, A.A., Schlacher, T.A., Williams, A., *et al.* (2010). A test of the seamount oasis hypothesis: seamounts support higher epibenthic megafaunal biomass than adjacent slopes. *Marine Ecology*, 31 (suppl. 1): 95–106.

Sanders, H.L., Hessler, R.R. (1969). Ecology of the deep-sea benthos. *Science*, 163: 1419–1424.

Sanders, H.L., Hessler, R.R., Hampson, G.R. (1965). An introduction to the study of deep-sea benthic faunal assemblages along the Gay Head-Bermuda transect. *Deep-Sea Research*, 12: 845–867.

Schlacher, T.A., Wooldridge, T.H. (1996). How sieve mesh size affects sample estimates of estuarine benthic macrofauna. *Journal of Experimental Marine Biology and Ecology*, 201: 159–171.

Shirayama, Y., Fukushima, T. (1995). Comparisons of deep-sea sediments and overlying water collected using multiple corer and box corer. *Journal of Oceanography*, 51: 75–82.

Smith, W., McIntyre, A.D. (1954). A spring-loaded bottom-sampler. *Journal of the Marine Biological Association of the UK*, 3: 257–264.

Sneli, J.-A. (1998). A simple benthic sledge for shallow and deep-sea sampling. *Sarsia*, 83: 69–72.

Sorbe, J.C. (1983). Description d'un traîneau destiné à l'échantillonnage quantitatif étagé de la faune supra-benthique néritique [Description of a sledge used for quantitative two-level sampling of neritic suprabenthic fauna]. *Annales de l'Institut océanographique*, 59(2): 117–126.

Svavarsson, J. (1997). Diversity of isopods (Crustacea): new data from the Arctic and Atlantic Oceans. *Biodiversity and Conservation*, 6: 1571–1579.

Thiel, H. (1970). Ein Fotoschlitten für biologische und geologische Kartierungen des Meeresbodens. *Marine Biology*, 7: 223–229.

Thistle, D. (2003). The deep-sea floor: an overview. In: Tyler, P.A. (ed.), *Ecosystems of the World 28*. Amsterdam: Elsevier Science, pp. 5–37.

Thomson, C.W., Murray, J. (1885). *Scientific results of the voyage of the HMS Challenger during the years 1873–76*. Narrative – Vol. 1, Chapter II, 74.

Chapter 10
Corers and Grabs

Bhavani E. Narayanaswamy[1], *Brian J. Bett*[2], *Peter A. Lamont*[1], *Ashley A. Rowden*[3], *Elanor M. Bell*[1]* *and Lenaick Menot*[4]

[1] Scottish Association for Marine Science, Scottish Marine Institute, Oban, Argyll, United Kingdom
[2] National Oceanography Centre, European Way, Southampton, United Kingdom
[3] National Institute of Water and Atmospheric Research, Wellington, New Zealand
[4] Institut Français de Recherche pour l'Exploitation de la Mer, Brest, Plouzané, France

Abstract

A range of grabs and corers have been used for sampling deep-sea sediments. Grabs are comparatively cheap and simple compared to corers. With the advent of video-guided sampling, more sophisticated grab systems are being developed. Grabs, with their scooping action, return rather disturbed samples that should be regarded as semi-quantitative in nature. Corers collect higher quality samples more suited to quantitative projects. Box-corers and multiple corers are most commonly used by deep-sea ecologists, and are employed for the collection of macro-, meio- and micro-benthos, as well as a wide range of environmental samples. Here we review the range of sediment samplers used in deep-sea studies, providing some general recommendations and guidelines for their operation. Methods for processing samples taken by box-corers and multiple corers are also detailed.

Keywords box-corer, multi-corer, megacorer, grabs, sampling techniques, sampling protocols, macrofauna, meiofauna, microbes

10.1 Introduction

Historically, the first attempts at sampling the deep-sea benthos were undertaken using a variety of trawling nets that had large mesh sizes. This resulted in the majority of individual specimens passing through the nets and the assumption that benthic abundance, and hence diversity, in the deep sea was low (Gage & Tyler, 1991). Furthermore, mean individual size of invertebrate taxa decreases with depth so that the early use of coarse nets led to lower estimates of diversity (Gage & Tyler, 1991). Hessler & Sanders (1967) designed an epibenthic sled with a fine mesh to collect relatively large samples of the fauna that were sparsely distributed in the sediments and across the sediment–water interface (see Chapter 9). The drawback of the deep-sea epibenthic sled system is that it is essentially a qualitative, or at best, a semi-quantitative sampler. In order to quantify faunal richness and abundance, the size of area or volume sampled must be accurately known and the sampler must efficiently collect specimens from that area or volume.

*Current affiliation: Australian Marine Mammal Centre, Australian Antarctic Division, Tasmania, Australia.

Biological Sampling in the Deep Sea, First Edition. Edited by Malcolm R. Clark, Mireille Consalvey and Ashley A. Rowden.

The first benthic grab was designed for use in shallow waters (Petersen & Boysen Jensen, 1911), but beginning in the 1950s larger and heavier variants were used to sample the deep sea. These included a modified Petersen grab (0.2 m^2) used on the *Galathea* expedition (Spärck, 1956), and the Okean grab (150 kg, 0.25 m^2) (Lisitsin & Udintsev, 1955) used extensively by the Soviets in their worldwide sampling of the deep seafloor over a number of decades. Grabs are simple devices, so they can achieve high sample recovery rates. However, although the area of seafloor surface sampled can be determined, the scooping action of the grab means that the depth of the sample is uneven. This limitation to the quantitative nature of the grab prompted the development of coring devices for taking benthic samples. Early corers that influenced later designs include the Reineck box-corer (Reineck, 1963) and the Craib tube-corer (Craib, 1965). However, both of these corers only sampled a relatively small area of seabed, were relatively lightweight and as such were not particularly suited to recovering samples from the deep sea.

More recently, large versions of van Veen and Day grabs (0.2–0.25 m^2) have also been used to collect samples (McIntyre *et al.*, 1984) (Fig. 10.1a). Hessler and Jumars (1974) in conjunction with the United States Navy Electronic Laboratory designed the USNEL box-corer, which was essentially a larger and heavier version of the Reineck box-corer. USNEL-type box-corers (Fig. 10.1b) have become popular and widespread over time. Other varieties of box-corer that have been designed for sediment sampling in the deep sea include, for example, the Gulf of Mexico (GOMEX) (Boland & Rowe, 1991) and Nederlands Instituut voor Onderzoek der Zee (NIOZ) box-corers (see Gage & Bett, 2005) (Fig. 10.1c,d). The comparatively large area sampled (0.06–0.25 m^2) allows for more specimens to be collected in a quantitative fashion, allowing for a range of ecological parameters to be determined. However, the potential drawback of using a box-corer is that they may create a bow wave, resulting in disturbance of surficial sediments and the potential loss of associated fauna (Bett *et al.*, 1994).

Multiple corers (see Barnett *et al.*, 1984) (Fig. 10.2a), based on the principles of the Craib tube-corer, were built to collect samples that were virtually free from the effects of a bow wave. These corers are hydraulically damped, allowing for the collection of fauna at the sediment–water interface, including flocculent material such as phytodetritus (Billett *et al.*, 1983) that can be easily resuspended and blown aside by box-corer bow waves.

Corers that are not directly connected to a surface ship are also used to collect samples in the deep sea. So-called free-corers, or boomerang-corers, descend independently to the seafloor to take a small sample (6.6 cm diameter) before returning to the surface following the contact-release of ballast weight. They were designed and used in the 1960s to depths of 6400 m (Bowen & Sachs, 1964). However, these devices have not been widely used since in the deep sea. More recently, various 'mini-corers' have been designed for use by the manipulator arms of submersibles and remotely operated vehicles (ROVs). Push corers are the simplest and most widely used form, having a simple flap valve at the top to aid recovery of the sediment column and protect it during return to the surface.

A range of other devices have been developed for the collection of larger samples, including the modified Birge–Ekman box-corer, as described by Rowe and Clifford (1973) and the 'pot sampler' of Van Dover (2002), designed to operate in hydrothermal vent mussel beds (see Chapter 13 for details regarding ROV sampling).

Methods for processing samples taken by grabs and corers for macrofauna, meiofauna and microbes have not changed a great deal since at least the 1970s. However, there has been an ever increasing need to standardize processing techniques as more institutes around the world undertake deep-sea surveys, and there is a growing need for comparable data to support global data analyses.

Below, we describe the main grab and corer gear types, their operation and sample data. We note that useful complementary information can also be obtained from various chapters in benthic sampling textbooks by Eleftheriou and McIntyre (2005), Danovaro (2010) and Eleftheriou (2013).

Fig. 10.1 (a) van Veen Grab. (Photograph by D Stevens. © NIWA, New Zealand. Reproduced with permission.) (b) The USNEL Mk II box-corer. (Photograph by B Bett. © National Oceanography Centre, UK. Reproduced with permission.) (c) The GOMEX box-corer ready for deployment. (Photograph by G Rowe. © Texas A&M University, USA. Reproduced with permission.) (d) The NIOZ box-corer being deployed. (Photograph courtesy of M Smit. © Royal Netherlands Institute for Sea Research, the Netherlands. Reproduced with permission.) (Please see Plate 16 for colour representation of the figure.)

10.2 Description of gear types

10.2.1 Grabs

Grabs typically consist of two quarter-circle buckets or scoops that pivot around a central hinge; although there are versions that comprise single or multiple scoops. For most types of grabs the buckets are held open during descent to the seabed and close on contact with the bottom after the tension on the lowering wire is released. The buckets are brought together to enclose a sediment

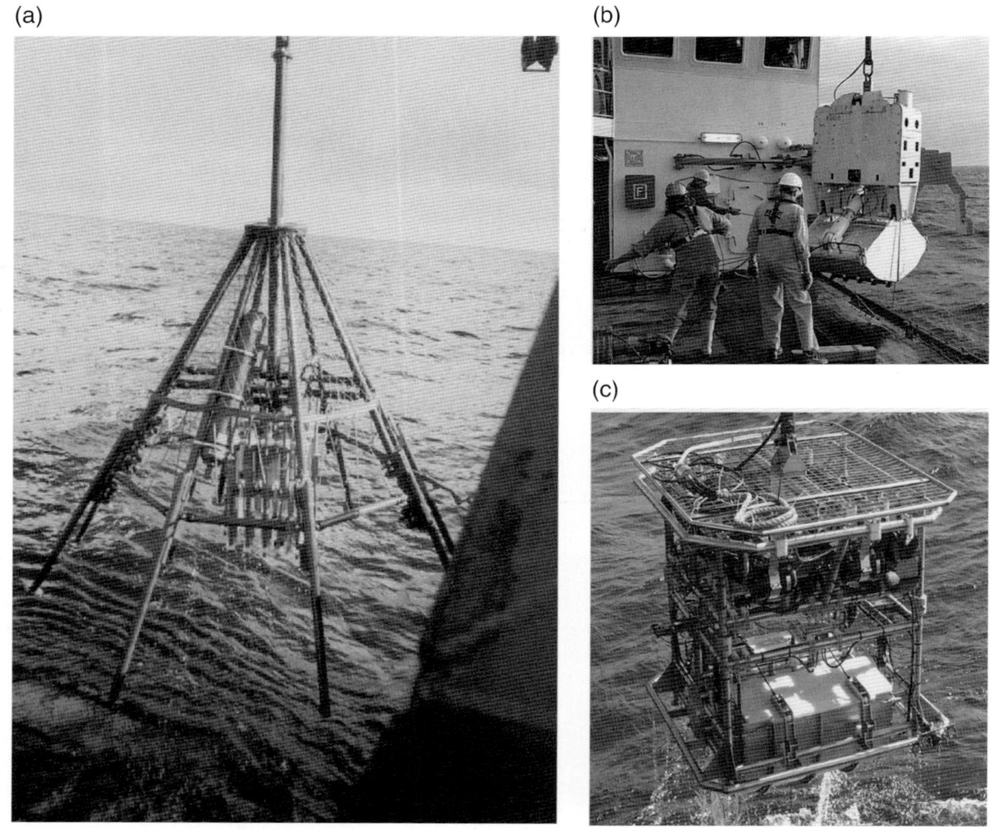

Fig. 10.2 (a) The SMBA multiple corer being deployed. (Photograph by B Bett. © National Oceanography Centre, UK. Reproduced with permission.) (b) The IFM-GEOMAR TV-Grab ready for deployment deployed (Photograph by D Bowden. NIWA, New Zealand. Reproduced with permission.) (c) HyBIS, Hydraulic Benthic Interactive Sampler – a modern video grab. (Photograph by B Murton. National Oceanography Centre, UK. Reproduced with permission.) (Please see Plate 17 for colour representation of the figure.)

sample when the wire is raised towards the surface vessel. A variety of mechanisms have been designed for assisting the opening and closing procedure, to produce a number of different grab types. Larger variants of shallow water grabs are used for sampling the deep-sea benthos. These include the previously mentioned Petersen, Okean, Day, and van Veen grabs, as well as Campbell grabs (410 kg, 0.55 m^2) (Hartmann, 1955). Grabs are not particularly well suited for quantitatively sampling coarse sediments, and, even in soft sediments, maximum penetration depths are typically less than 20 cm. In addition, the 'bite' profile of a grab is uneven (Riddle, 1989). The top of the grab buckets usually have ports covered with wire mesh or a one-way opening flap that attempt to reduce the bow wave that precedes the device as it is lowered to the seafloor. Sample integrity can be compromised by 'washout' through these ports or the gap between poorly sealed closed buckets. This means that vertical structure in a grab sample from the deep sea is often severely affected, or even lost.

Several researchers have employed video-guided grab systems, using a live video feed to the surface vessel to guide sample collection. These vary from simple lightweight systems, such as that described by Mortensen *et al.* (2000), to larger versions such as the Russian GTVD-2/Preussag grab (Sheremet & Efimova, 1996) and the IFM-GEOMAR TV-Grab (2000 kg, 1.8 m^2) (Fig. 10.2b) often used on the German research vessel *Sonne*. More sophisticated systems are now available, such as the National Oceanography Centre (Southampton) HyBIS (hydraulic benthic interactive sampler) (Fig. 10.2c) (Murton, 2010). HyBIS has an hydraulically activated 0.3 m^3 double-jaw grab that can be repeatedly closed and opened at depth, while the instrument module above the grab carries a number of cameras, arranged both vertically and obliquely, and a set of ROV-type thrusters that permit fine-scale positioning of the grab (rather than relying on the surface vessel position).

Large, hydraulic, video-guided grabs are frequently used for geological sampling in the deep sea, and they can be useful tools to recover associated biological material from patchy habitat such as sulphide-rich deposits. However, they are expensive to operate and are not considered the ideal choice for quantitative sampling of benthic organisms in the deep sea. Although grabs can be used to make collections of deep-sea fauna in locations or sediments that are not suitable for operating corers, box-corers and multi-corers should be used whenever possible for quantitative research. As such, no additional operational detail will be provided here for grabs.

10.2.2 Box-corers

Large box-corers generally comprise: (i) a removable square/round sample box, attached below (ii) a weighted central column carrying the box-closure mechanisms that may move through (iii) a gimballed support frame. These devices are generally large and heavy, and best deployed from the side gantry of a research vessel. Deployment from stern gantries is possible, but likely to significantly reduce success rates as a result of the greater heave (vertical motion) at the stern compared with an amidships position. Technical specifications of selected examples of box-corers are given in Table 10.1.

10.2.2.1 USNEL Mk II box-corer

The USNEL Mk II deep-sea box-corer (Hessler & Jumars, 1974) has a detachable, square, open-ended steel sample box with a removable spade closure for the bottom of the box (Fig. 10.1b). Vents at the top of the box are held open during the descent of the corer in an attempt to reduce the effect of a bow wave. Contact of the corer with the seabed triggers mechanical actions to close the top vents and allow the spade closure system to activate. The standard USNEL Mk II box-corer collects samples with a core length of about 40–50 cm with an area of 0.25 m^2 in 'typical' deep-sea sediments.

Table 10.1 Specifications of selected deep-sea corers

Corer type	Cross-section shape	Dimensions (*W*, *B*, *H*, cm)[a]	Area (cm^2)[b]	Typical laden weight (kg)[c]
USNEL box-corer	Square	50 × 50 × 50	2500	1000
GOMEX box-corer	Square (small)	20 × 30 × 50	600	250
	Square (large)	40 × 40 × 50	1600	650
NIOZ box-corer	Cylindrical	Ø 30 × 55	707	1100
	Square	30 × 20 × 55	600	
	Cylindrical	Ø 50 × 55	1964	1600
	Square	50 × 50 × 55	2500	
Multi-corer	Cylindrical	Ø 5.65 × 65	25.1 × 12	900
Megacorer	Cylindrical	Ø 10 × 60	78.5 × 12	900

[a] For multiple corers, the size of one sampling unit (tube) is given; for box-corers, *W* = width, *B* = breadth, *H* = height.
[b] For multiple corers, the area of one sampling unit × number of units.
[c] Weight of corer with standard ballast and sample.
Ø, diameter.

10.2.2.2 NIOZ box-corer

The basic concept and operation of the NIOZ box-corer (Fig. 10.1d) is very similar to the USNEL Mk II. However, its major design variations include: (i) a counter-balance arm that operates in opposition to the spade bottom-closure arm; (ii) a valve-like top closure; and (iii) optional use of square or round cross-section 'boxes'. The balance arm was added to counteract the turning force of the spade closure action, and is thought to reduce sediment column distortion. The valve-like top closure is designed to improve bottom water retention. The NIOZ box-corer is available with both round and square 'boxes' in two sizes, giving samples of an area 0.06–0.25 m^2 with a maximum core length of 55 cm.

10.2.2.3 GOMEX box-corer

The comparatively small GOMEX box-corer (Fig. 10.1c) (Boland & Rowe, 1991) was developed from a corer originally described by Jonasson and Olausson (1966). The GOMEX corer has a fixed sample box, a double-jaw bottom closure and a flap lid top closure. Contact of the corer with the seabed triggers mechanical actions to close the lid and transfers the ballast weight to close the jaws. The corer is available in different sizes offering sample areas of 0.06 and 0.16 m^2 with a potential core length of about 40 cm. It should be noted, however, that current commercially available versions of the GOMEX box-corer may differ to that described by Boland and Rowe (1991).

10.2.2.4 Multiple box-corers

A multiple box-corer has been designed and used in a limited number of studies (Gerdes, 1990). This device has not been widely adopted by deep-sea scientists, most probably due to the difficulties inherent in operating such a large and complicated device as well as issues of pseudoreplication (see also section on Multiple corers and Chapter 3).

10.2.3 Multiple corers

Multiple corers essentially comprise: (i) a coring head that carries a number of core tubes (ranging from 4 to 12), together with their individual top- and bottom-closure mechanisms, and which is connected to (ii) an hydraulic damper system, that in turn is connected to (iii) a seafloor landing frame. These devices can be almost as large and heavy as box-corers, and are generally best deployed from the side gantry of a research vessel for the same reasons as outlined for box-corers. Technical specifications of selected examples of multiple corers are given in Table 10.1.

10.2.3.1 The multi-corer

The Scottish Marine Biological Association (SMBA) multi-corer (Barnett *et al.*, 1984) is in essence a scaled-up multiple version of the Craib tube-corer (Craib, 1965). The frame carries 12 cores, each with an approximate internal diameter of 5 cm, and with its own top- and bottom-closure mechanisms (Fig. 10.2a). When the external frame lands on the seafloor, the coring head begins to descend slowly, against the resistance of the hydraulic damper. When the coring head encounters sufficient resistance as a result of sediment penetration a mechanical trigger activates the release of the top- and bottom-closure mechanisms. The bottom closures initially fall to the sediment surface, only swinging fully into place as the coring head is drawn back out of the sediment as the ship's winch slowly takes up tension on the wire. Several variants of the original SMBA pattern multiple corer have been developed, including: (i) a frame design modified to reduce deck footprint and height; and (ii) a core head design modified to carry fewer, larger core tubes (e.g. 4 × 25 cm diameter).

10.2.3.2 *Megacorer*

The Bowers and Connelly megacorer (see Barnett *et al*., 1984) is conceptually identical to the SMBA multiple corer, but has been extensively re-engineered to produce a much more compact device. In its standard form it carries a maximum of 12 cores each with an internal diameter of 10 cm, and each core is housed in an independent unit together with its top and bottom closures. The use of independent core units introduces useful flexibility, offering the option to: (i) reduce the total number of cores deployed, thereby increasing the sediment penetration force if required; and (ii) switch or mix between 10 cm and 5 cm diameter cores to meet particular sampling needs. In its standard arrangement (12 × 10 cm cores) it will return about 40 cm long cores in 'typical' deep-sea sediments.

10.2.4 Choice of a corer

The key attribute of any sampling device is that it should return a representative sample of the target benthic assemblage (see Table 10.2 for the pros and cons of each coring system). In very simple terms, for a representative biological sample, a core (and its subsequent processing) must return all of the specimens present in the sediment volume sampled by the device. This completeness is not the case for USNEL-type box-corers, and by analogy, likely not to be the case for the NIOZ box-corer. Tests of meiobenthos data indicate significant under-sampling by USNEL-type box-corers relative to SMBA pattern multi-corers (e.g. Bett *et al*., 1994). Similarly, tests of macrobenthos data indicate significant under-sampling by USNEL-type box-corers relative to megacorers (e.g. Gage & Bett, 2005; Narayanaswamy & Bett, 2011). This apparent loss of specimens is thought to be the result of the bow-wave effect. The simple open design of the GOMEX box-corer suggests that it may not suffer to such a large degree. Boland and Rowe (1991) present data that imply a greater number of individuals were collected using a GOMEX corer compared to an USNEL-type corer at the same depths and in the same locality.

Under-sampling will obviously underestimate the true standing stock (abundance, biomass) of the benthos. Loss of specimens from a sample is likely to occur in a non-random fashion: for example, superficially distributed and small-bodied specimens are more likely to be lost than larger infaunal

Table 10.2 Evaluation of the merits of various corers

Corer type	Pros	Cons	Habitat suitability
USNEL Mk II box-corer	Large sample area. Long history of use	Large/heavy. Bow-wave problem	Efficient in a wide range of soft sediments. Can also be used in relatively coarse sediments and in most deep-sea environments. However, difficult to sample among steep/rocky outcrops
NIOZ box-corer	Large sample area	Large/heavy. Bow-wave problem	As above
GOMEX box-corer	Relatively small/light. Little bow-wave effect	Small sample size	As above
SMBA Multi-corer	Little bow-wave effect. Long history of use	Small sample size	Efficient in a wide range of soft sediments. Can be used in many deep-sea environments. Care must be taken when sampling in rocky terrain
Megacorer	Little bow-wave effect. Relatively compact. Flexibility of independent core units	Small sample size	As above

organisms. As a consequence, species diversity and species composition assessments from such samples are like to be biased (i.e. be non-representative). Although it might be reasonable to apply a correction factor to standing stock estimates (e.g. Narayanaswamy *et al.*, 2005), it is very unlikely that any useful correction could be made to biased species diversity and species composition data. Therefore great caution should be applied when attempting to compare or compile data from sampling coring devices with different sampling performances.

Multiple corers are not a universal panacea for deep-sea sampling problems, but they should be regarded as the first choice instrument wherever they can be practically employed. This recommendation applies equally to biological (micro-, meio- and macro-benthos) and environmental sampling of the sedimentary environment. However, the availability of multiple non-independent samples can introduce the issue of pseudoreplication if more than one core from a single deployment is used to determine mean measures for any parameter (e.g. abundance, diversity, etc.) (see Section 10.5 and Chapter 3).

10.2.5 Additional sensors

The most commonly used additional item of equipment is the 'pinger' (or near-bottom echosounder). This sensor is typically attached to the wire about 50 m above the corer and enables the operator to determine the distance of the corer from the seabed as it approaches the bottom (see Gage & Bett, 2005, for the theory of operation). The pinger's function can also be fulfilled by an ultrashort baseline (USBL) transponder capable of acoustically telemetering depth data to the surface vessel, having the additional value of providing more precise navigational data for the coring site. However, present USBL technology is limited to a depth of about 6000 m.

A variety of other equipment may also be directly attached to the corer. For example, self-contained conductivity, temperature, depth (CTD) instruments and water bottles may be employed to measure the bottom water conditions. Other common additions include still and video cameras together with their associated lighting. Such equipment is variously operated: (i) continuously from launch, (ii) on a timer delay, (iii) by linkage to the corer's trigger mechanism, or (iv) by independent bottom-contact trigger (e.g. a weight suspended below the corer). Images of the coring operation and the sediment surface allow an appreciation of the effectiveness of the sampling, and provide additional information on the benthic assemblage.

10.3 Sampling operations

10.3.1 Essential information

As in all scientific operations, it is extremely important to keep comprehensive notes on all aspects of sampling operations (Table 10.3), and to ensure that all relevant metadata can be uniquely linked to the resultant samples. Most regular sea-going institutions have developed their own in-house log sheets to help ensure that standardized records are kept. Custom databases are also available (e.g. 'Biocean', Fabri *et al.*, 2006).

10.3.2 Generalized protocols for box-corer operation

Even among USNEL Mk II corers there can be significant variations in their detailed mechanical arrangements as supplied by different institutions/manufacturers. It is therefore very important that all involved with the sampling operation are familiar with the detailed mechanical arrangement of the specific corer in use. This is particularly critical for safety, including: (i) use of 'safety pins' that prevent triggering and the descent of the coring head, (ii) pre-triggering precautions to reduce risk of triggering during launch and/or deployment; and (iii) method or mechanism to lock the spade arm(s)

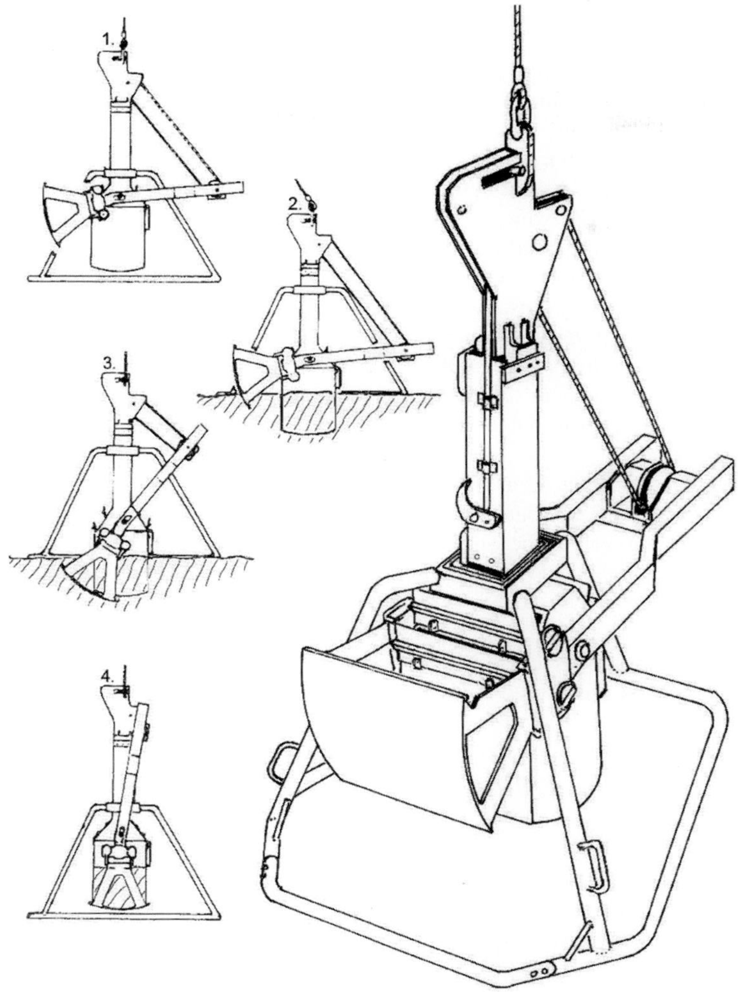

Fig. 10.3 A schematic showing the operation and collection of sediment using the USNEL Mk II box-corer (sequences 1–4). (Photograph by B Bett. © National Oceanography Centre, UK. Reproduced with permission.)

of box-corers in the fired position during the sample removal phase. Below we provide generalized notes on the operation of a large box-corer such as the USNEL-type (see also Fig. 10.3).

10.3.2.1 Deployment of the box-corer

(1) The corer should have its safety pins in place and the spade arm(s) lashed or locked in the up/fired position.
(2) The core box is fitted; note the box is likely to have a specific orientation (consider the location of fixing points, any opening panel, any draining panels, etc.).
(3) The spade is fitted; note the spade is likely to have a specific orientation (i.e. the leading, sharpened edge must be facing downwards once the spade arm is swung to the horizontal).
(4) The spade arm(s) is freed and carefully swung to the horizontal while feeding the 2.5 m activating wire warp(s) through the upper trigger mechanism.

(5) The termination of the activating warp(s) is carefully located in the upper trigger mechanism and the securing bolt drawn through it.
(6) The securing bolt is locked in place by cocking the trigger lever.
(7) Water venting mechanisms should be opened and locked in place ('bow-wave effect' reduction).
(8) Pre-trigger precautions are applied at this time (these may include wiring the trigger lever, or its linkage, to the central column to increase the resistance to firing; lashing the spade to the support frame with light twine to reduce swinging action).
(9) With the agreement of the Officer of the Watch the corer can now be lifted from the deck and swung just outboard.
(10) Once outboard the safety pins are removed.
(11) The winch pays out wire slowly; pay-out speed is increased gradually to a maximum of about 60 m min^{-1}. If the wire is payed out too fast there is a risk that the wire may 'overtake' the descending corer and 'snag'. Midwater tripping can also be an issue if the wire is payed out too fast, and can occur in rough weather with the slackening and tightening of the wire in a significant swell.
(12) When the corer is about 50 m above the seabed (determined by wire out, 'pinger' or other telemetry system), the pay-out speed is reduced to 10 m min^{-1} for the remainder of the descent. At 50 m above the seabed, the wire pay-out can also be momentarily ceased to allow the corer to straighten up on the end of the wire prior to the final descent at the slower pay-out rate. In higher sea states or swells a final pay-out rate of up to 20 m min^{-1} is sometimes used to reduce the likelihood of triggering during a glancing touchdown on the seabed. However, this may exacerbate the 'bow-wave effect'.
(13) On bottom contact (determined by wire tension), around a maximum of 10 m of additional wire may be payed out to reduce the effect of the corer being pulled over by the drift of the ship or out by vertical ship heave. Also an additional 2.5 m of wire incorporated in the spade activating arm will have been released.

10.3.2.2 Collection of box-corer sample

(1) As the corer touches the seabed, the spring-loaded cocking bar, which has been locked into position, is released by the linkage mechanism as the weighted central column of the corer passes through the gimbals, thus allowing the heavy box to sink into the sediment.
(2) The top of the box, open during the descent of the corer, also closes once the corer is on the seabed.
(3) The corer is allowed to settle for a few seconds (10–20 seconds).
(4) The spade arm activating wire warp is hauled in causing the spade to close under the box once heaving commences. The sediment core is then effectively sealed within the corer.
(5) As the corer is pulled out of the sediment, there is a marked increase in wire tension, and as the corer is freed from the sediment the tension notably decreases. These changes in wire tension are useful indicators that the corer has arrived on/lifted from the seafloor.

10.3.2.3 Retrieval of the box-corer sample

(1) Wire is then recovered at a slow rate (5 m min^{-1}) with low tension on the winch until the weight has been taken up on the wire allowing time for the spade to close under the box. Any haul rate will be additional to the ship heave rate.
(2) When the corer is clear of the bottom (determined by wire out or wire tension monitor), the wire hauling rate is increased gradually to a maximum of 60 m min^{-1}.
(3) On approach to the surface, hauling rate is reduced.
(4) With the agreement of the Officer of the Watch, the corer can now be lifted from the water.

(5) Resist any temptation to insert the safety pins at this stage – it will likely result in loss of sample. (*Should the corer return untriggered exercise extreme caution – the safety pins should be re-inserted with the corer still outboard and certainly before the corer comes in contact with the deck.*)
(6) Note that the corer should be landed in an appropriate orientation, with the side of the box that can be removed facing a sufficiently large area of open deck.
(7) The spade arm(s) must be firmly secured in the fired (up) position at this time (a serious accident may occur at step 12 below if this is not done).
(8) Details of core box removal vary between models, but may include: removal of a section of the support frame (and lazy spade) to permit trolley access, using a trolley to lift the spade and box (where a simple pallet truck is used wooden wedges may also be required), disconnecting the spade from its arm, disconnecting the box from corer column and/or opening the water venting mechanism.
(9) The corer is then hoisted slowly on the main wire to give clearance (e.g. 5 cm) between the corer column and the disconnected box and spade.
(10) The core box is then gently pulled clear from the corer.
(11) Insert safety pins at this point.
(12) The corer can now safely be lowered back to the deck and slack wire payed out.

10.3.3 Multiple corer operation

The SMBA pattern multiple corer (as described by Barnett *et al.*, 1984) has been produced by several manufacturers, introducing a number of modifications to the original design, particularly with respect to the style of the support frame and the number/diameter of core tubes carried. Similarly, the Bowers & Connelly megacorer represents a wholesale re-engineering of the original concept. It is therefore very important that *all involved with the sampling operation are familiar with the detailed mechanical arrangement of the specific corer in use*. This familiarity is critical for the 'locking-up' of the coring head, such that it is unable to descend through the support frame. Below we provide generalized notes on the operation of a multiple corer (see also Fig. 10.4).

10.3.3.1 General multiple corer deployment

(1) The head lock-up mechanism must be securely in place prior to any preparatory work.
(2) Clean core tubes are fitted and secured with locking collars or similar.
(3) The individual top and bottom closures for the cores are now cocked to the armed position. The order and manner in which this is done varies substantially between systems. For example with the SMBA multi-corer, tops must be done first; while with megacorers, bottoms must be done first.
(4) The corer can now be lifted from the deck and swung outboard.
(5) The head lock-up mechanism can now be released.
(6) The winch pays out wire slowly; pay-out speed is increased gradually to a maximum of about 60 m min^{-1} (or lower speed with lighter corers/wires, or in higher sea states).
(7) When the corer is about 50 m above the seabed (determined by wire out, 'pinger' or other telemetry system), the pay-out speed is reduced to 10 m min^{-1} for the remainder of the descent. At 50 m above the seabed, the wire pay-out can also be momentarily ceased to allow the corer to straighten up on the end of the wire prior to the final descent at the slower pay-out rate.

10.3.3.2 Collection of multiple-corer sample

(1) On bottom contact (determined by wire tension monitor), wire pay-out continues for a further 5–10 m.
(2) The corer is left on the seabed for 1–2 minutes to enable the hydraulically damped descent of the coring head to take place.

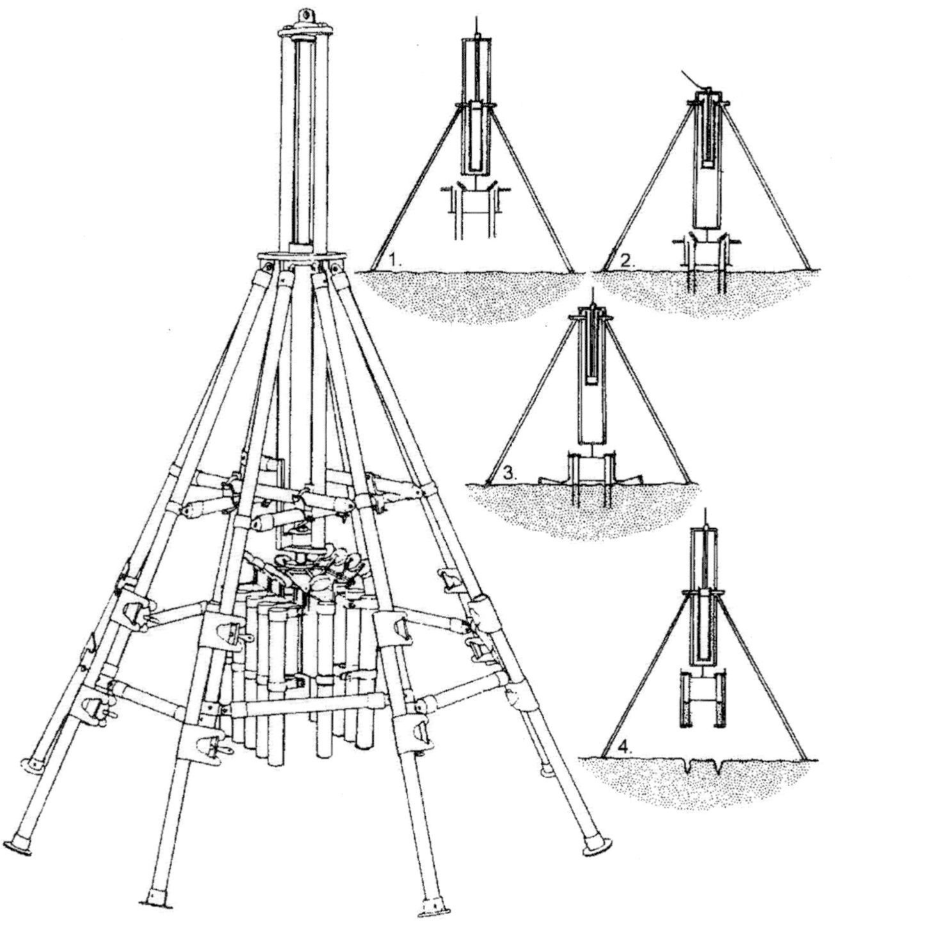

Fig. 10.4 A schematic showing the operation and collection of sediment using the SMBA multiple corer (sequences 1–4). (B Bett. © National Oceanography Centre, UK. Reproduced with permission.)

10.3.3.3 Retrieval of the multiple-corer sample

(1) Wire is then recovered at a slow rate (10 m min^{-1}) with low tension on the winch until the weight has been taken up on the wire. This reduces sample loss that can occur when the corer is pulled out quickly and under tension from the sediment.
(2) When the corer is clear of the bottom (determined by wire out, wire tension monitor), the wire hauling rate is increased gradually to a maximum of 60 m min^{-1} (or lower speed with lighter corers/wires, or in higher sea states).
(3) On approach to the surface, hauling rate is reduced.
(4) The corer can now be lifted from the water.
(5) Where possible the head lock-up mechanism should be engaged with the corer still outboard; however, many designs require the corer to be swung aboard and allowed to just touch deck (i.e. with their weight still taken on the wire) before the head lock-up mechanism can be engaged.
(6) With the head lock-up mechanism securely engaged the corer is fully landed on deck and slack wire payed out.

(7) Prior to any attempt to remove samples, the performance of each coring unit should be assessed and recorded (e.g. top and bottom closures fired, core recovered, its length, clarity of supernatant water, noting any obvious core surface and profile features).
(8) Depending on the corer design, cores may be removed while still attached to the instrument or the entire tube closing mechanism can be taken from the corer and hung on a rack for processing. The detailed methodology of core removal will depend on the specific instrument, but in general the following approach should work well with two people. The first operator applies downward pressure to the top cap, to ensure a good seal and limit any core slippage at the next stage. With an appropriate bung in hand, the second operator swings away the bottom closure and quickly positions the bung in its place. The top closure is then opened and the second operator pushes the bung firmly into the base of the core. The core locking collar is then removed; the core is eased from its holder and carefully passed to the first operator.
(9) The core should be transferred to an appropriate rack and some form of top cap or bung put in place.

10.4 Sample processing

10.4.1 Assessing core suitability

When to accept or discard cores collected can be quite subjective. Researchers from Scottish Association of Marine Sciences (SAMS) and Institut Français de Recherche pour l'Exploitation de la Mer (IFREMER) have used the following categories and criteria for acceptance/rejection of cores (Table 10.3) – primarily collected by the box-corer.

10.4.2 Sub-coring the box-core

Large box-core samples are frequently subsampled; this can be achieved either *in situ* or on deck after recovery of the corer. The best known of the *in situ* techniques is generally referred to as the 'vegematic'. This comprises a 5 × 5 array of 10 × 10 cm square sections that are fitted within the main core box. On recovery of the corer the individual sub-core rectangular sections can be removed and processed separately for different samples. However, when using this technique, Jumars (1975) established that the outer 16 sub-cores of the array returned fewer macrofaunal specimens compared to the inner 9 sub-cores and attributed this finding to the bow-wave effect. The presence of the sub-core array most likely exacerbates the bow-wave effect by further restricting water flow through the box during its descent.

Post-recovery subsampling of the box-core can be achieved by various means, including vegematic inserts, simple push cores, syringe-based piston cores and box quadrats. The overlying surface water should only be siphoned off once the vegematic inserts or sub-cores are placed into the box after

Table 10.3 Criteria used for the acceptance/rejection of cores based on quality

Category		Criteria
Accept	1	Full-length core, clear overlying water, undisturbed sediment surface with delicate fauna/tubes present
	2	Medium-length core, relatively clear overlying water, generally undisturbed sediment surface with some features such as tubes present
Reject	3	Core shallow or with cloudy overlying water, sediment surface obviously disturbed, no delicate organisms present
	4	Core shallow or with sediment surface obviously disturbed, no overlying water, no delicate organisms present

recovery, and the material in the water collected on a sieve with the pre-determined mesh size for the target fauna. Once the sub-core is removed from the box-corer, the sediment is extracted by placing the corer on an extruder which forces the sediment up into a metal/plastic collar which has been pre-cut or marked to the depths required. The sediment is sliced using a thin metal or plastic slicing plate and then placed in a sample vessel with a preservative or filtered seawater prior to sieving. Delicate surface fauna are removed prior to slicing and stored in separate jars with formalin. Sediment from each core and each fraction should be kept separate.

10.4.3 Subsectioning multi-cores

Prior to each multi-core being removed from the coring unit, the length of the core should be measured, and a record made of any features that are easy to see on the surface. The overlying surface water from the individual core should be siphoned off and the material in the water collected on a sieve with the pre-determined mesh size for the target fauna. The core is then placed on an extruder and gently extruded from below. As the sediment reaches the top of the core any visible delicate fauna can be removed and added to the overlying surface water material. This is then generally preserved in formalin. The core is further extruded into pre-cut polycarbonate collars marked with the required depth, the sediment is sliced and put in a sample vessel and, depending on the depth horizon, usually preserved in formalin (particularly for the upper depth horizons prior to sieving) or filtered seawater (used for lower depth horizons prior to sieving). Sediment from each core and each fraction should be kept separate.

10.4.4 Whole box-core processing

Once the box-core is back on deck, the overlying surface water is gently siphoned off and the material in the water collected on a sieve with the pre-determined mesh size for the target fauna. Then the front of the box can be removed allowing for initial general observations to be made followed by extraction of the material. General notes are made of any features that are of interest, such as tubes, burrow openings, etc. on the surface of the core, and the overall surface condition (e.g. does it slope in any direction), and finally the depth of box penetration into the sediment is measured. Any delicate surface fauna (e.g. xenophyophores) and any tubes are removed and placed in a separate sample vessel prior to further processing.

10.4.5 Macrofaunal protocols

Typically, cores processed for macrofauna are sliced into depth horizons. The vertical intervals of the slices taken from cores depend on the level of analysis required, the time available on the ship and the number of operators who can process the samples. Ideally, the sediment is sliced into 5 or 6 depth strata: 0–1 cm, 1–3 cm, 3–5 cm, 5–10 cm, 10–15 cm and >15 cm.

Whole box-cores are sliced by eye using builder's trowels once the depth horizons are marked by scoring across the exposed sediment profile after the front side is removed (note this approach is not possible with the circular NIOZ box). The precision of the desired horizontal sectioning is easier to achieve when using multi- or megacore samples since appropriately marked sections of core tubing can be placed over the sample cores and the desired sample extruded into the graduated tube. A variety of pistons mounted in different ways are used to push the sediment out of the core tube. Multi- and megacore fractions are small enough for the entire sample to be placed in formalin (or preservative of choice) and processed some time later after fixation. For sections from whole cores, the upper two sediment fractions may be placed immediately in fixative/preservative with the lower fractions being sieved first in filtered seawater. This allows time for tissues of fauna in the top fractions to be fixed before washing, which reduces damage to delicate animals.

10.4.6 *Meiofaunal protocols*

Collecting meiofauna from a box-core generally requires the use of a plastic core or syringe sub-core, which can have internal diameters of 3.6, 4.5 or 5.0 cm. The sub-core is pushed into the sediment and a stopper (either a rubber bung or film) is placed over the top so that no air is present, allowing for removal of the core. Once the sub-core is removed a bung is placed at the bottom of the core to prevent the sediment from falling out. The sub-core is required to be kept vertical to prevent the sediment layers from mixing and thus altering the distribution of the meiofauna. The objectives of the study will influence the number of depth horizons that the sub-core is sliced into. Multi- and megacores taken for meiofauna are also typically sectioned. If the vertical distribution of the meiofauna is a priority, then sectioning at 0–1 cm, 1–3 cm, 3–5 cm, 5–10 cm, >10 cm should be sufficient. The sediment core is extracted by placing the core on an extruder which gently forces the sediment up into a plastic collar which has been pre-cut or marked to the sections required. A thin metal or plastic plate is then used to section the sediment into the different depth horizons and each sediment horizon is placed into a separate sample vessel with the addition of preservative or formalin. Shake the sample vessel gently to ensure that all the sediment and fauna comes into contact with the preservative or formalin. If section slicing is not able to be undertaken immediately, the entire core can be frozen at −20°C to prevent the movement of the meiofauna through the core; this is particularly important if analysis of the vertical distribution of meiofauna is to be undertaken.

10.4.7 *Microbial protocols*

The methodology used for collection of microbial fauna depends on what groups are of interest. The main microbial organisms commonly sampled are the viruses, Archaea and bacteria, algae, protozoa and their resting stages. Collecting microbial samples from a box-core can be achieved in several ways: a volumetric spoon, cut-off sterile syringe, or small, sterile plastic sub-core can be used to sub-core the desired volume of sediment. Laboratory gloves should always be worn by the person carrying out the sampling. If the aim of the research is to determine microbial community distribution and diversity across a sediment depth profile, a sub-core should be sliced as described above, and quantitative samples taken from the desired depth horizons using either a cut-off sterile syringe or a volumetric spoon.

Collecting microbial samples from multi- and megacores requires the cores to be extruded as described above. Subsamples can either be taken from the upper part of the core using a cut-off, sterile syringe, once the surface of the sediment has been extruded to a point where the sediment surface is level with the coring unit, or preferably from slices of different depth horizons. When slicing cores from the multiple corer, the polycarbonate collars and slicers should be cleaned with seawater and then rinsed in 80–100% ethanol prior to slicing, as well as between slices of each depth strata. If the vertical distribution of the microbial community is a priority then selecting 5–6 strata should be sufficient (e.g. 0–1 cm, 2–3 cm, 3–4 cm, 5–6 cm, 7–8 cm, 9–10 cm). If comparisons with meio- or macrofaunal distributions are to be performed, we recommend selection of depth strata that correspond with these other sampling protocols. The simplest way to collect the subsamples is to sheer the slices off directly into a sterile ziplock bag/sampling bags with a wide opening. Alternatively, a sterile wide-necked container can be used. For most post-sampling analytical procedures, it is necessary to know the weight and/or volume of the sediment subsample collected, and therefore it is extremely useful to have pre-weighed, labelled and documented the collection containers prior to filling them with sediment. It is also important to have some spare, pre-weighed containers in reserve in case extras are required.

The subsamples should then be preserved and homogenized. Rapid freezing in liquid nitrogen is necessary when virus and bacteria samples cannot be processed within 24 hours. However, it is important to note that when defrosting microbial samples for which cell integrity is important, it should be done slowly to avoid cell lysis, such as overnight in a refrigerator.

10.4.8 *Sampling for molecular analyses*

Sampling for molecular analysis requires slightly different protocols. Care should be taken to ensure that sub-corers and sample containers are sterile (either manually pre-sterilized or purchased as such), that laboratory gloves are worn by the person carrying out the sampling, and as few transfer steps as possible are employed. When slicing cores from a multi-corer, the polycarbonate collars and slicers should be cleaned with seawater and then rinsed in 80–100% ethanol prior to beginning to slice, as well as between each depth strata. The sediment slices should then either be slid directly into a sterile bag (sterile Whirl-pak® bags are ideal for this purpose). If a smaller volume is required, the slice can be subsampled using a metal spatula (the spatula should be flamed with ethanol immediately prior to subsampling) and the sediment sample placed in sterile 50 mL Falcon tubes. Depending on the molecular procedure to be carried out, the desired fixative should be added prior to sealing the bag/tube, and the sediment stored in a cool, dark place. If there is going to be a significant delay between sampling and laboratory processing, it is recommended that molecular samples are stored frozen at −80°C. If liquid nitrogen is available this is the preferred rapid freezing procedure. Once frozen, samples can then be transferred to a −80°C freezer for longer term storage. If neither liquid nitrogen nor a −80°C freezer is available, the samples can be frozen at −20°C and transferred to −80°C later.

10.4.9 *Sediment parameters*

The sediment of the sampled core can be analysed for a range of environmental variables (e.g. dissolved oxygen in porewater, levels of chemicals such as sulfides, etc.). Most commonly, sub-cores of sediment are taken for granulometric analysis (particle size determination), organic matter determination (CHN), and/or pigment analysis (chlorophyll a etc.). Sediment collected for pigment analysis on board should be stored 'live' in the dark at 4°C (if <24 hours) or in the dark at −20°C until extraction and post-processing can take place (it is the same for later particle and organic matter analysis). If the aim is to target resting stages, such as eggs, sediment should be stored 'live' prior to post-processing using a technique such as sugar flotation (Onbé, 1978) or Percoll separation (Epstein, 1995).

10.4.10 *Sieving*

10.4.10.1 Choice of mesh size

The objectives set at the outset of the research programme, and the fauna that are being targeted, generally dictate what size of mesh should be used to collect the infauna. For the macrofauna, the smallest mesh size currently used is 250 μm (typically 300 μm), whereas for the meiofauna it is 20 μm (typically 32 μm). Sieving the fauna directly on the smallest mesh size may appear from the outset to be advantageous. However, by sieving through stacked sieves, less material is retained on an individual sieve, and it is easier to sort specific size fractions. If time is short, a decision can be made to look at one specific size fraction without the need to re-sieve the whole sample through stacked sieves and potentially damage the individuals. The disadvantages of stacked sieves are the need for a series of sieves with different mesh apertures, and the requirement for more than one set of stacked sieves to allow for more than one operator to sieve the material. Both the latter cases can be expensive.

10.4.10.2 Elutriation and washing techniques

Washing or elutriating samples on board ship must be undertaken using filtered seawater. Filtered seawater prevents the contamination of the sediment sample with planktonic organisms found in the water column. Portable filtration units (generally with a mesh size between 20 and 50 μm) can be inserted between the ship's deck hose and the 'garden-type' hoses used for washing or elutriating the samples.

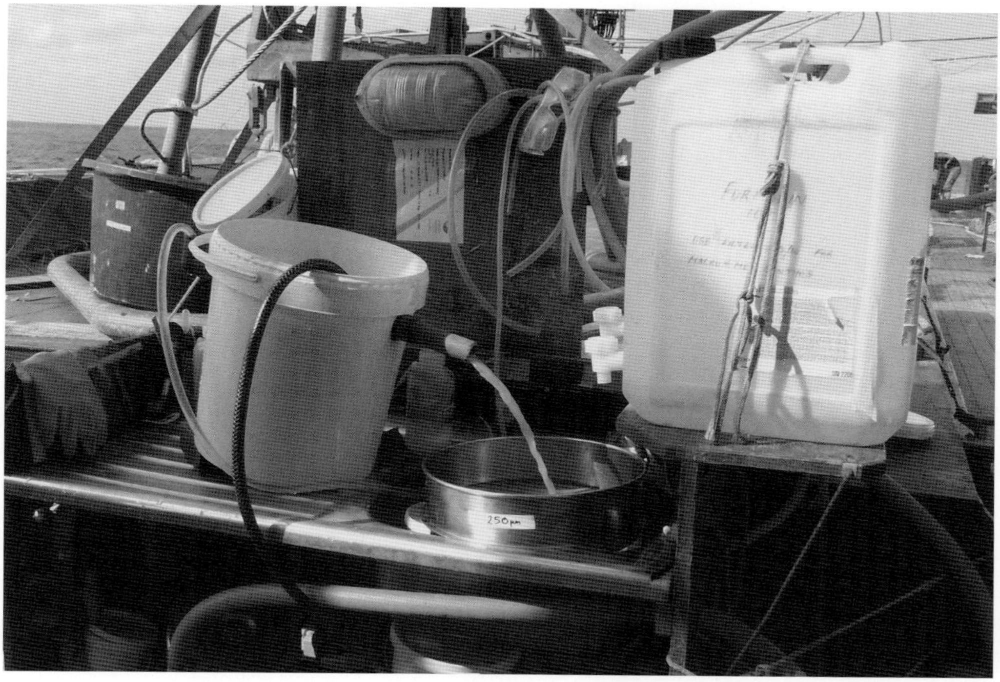

Fig. 10.5 Elutriation of sediment samples at sea. (Photograph courtesy of P Lamont © Scottish Association for Marine Science, UK. Reproduced with permission.)

A gentle method used for elutriating box-core samples, initially described by Sanders *et al.* (1965), is to use a large bucket/dustbin with two hose nozzles fitted near the top. A hose is connected to the slightly higher inflow nozzle allowing a continuous input of filtered seawater to fill the bucket. Water and suspended sediment and fauna flow through the outflow nozzle onto the sieve below (Fig. 10.5). A secondary hose can be used in the bucket to gently break down large sediment aggregates while under water. Care needs to be taken when using this method if the sediment contains glass sponge spicules; these have the ability to block the sieve very rapidly. Material residue on the sieve should be periodically removed and stored in a sample container. Once elutriation is complete, the entire residue can be preserved in the sample container and mixed gently to ensure that all material comes into contact with the fixative.

The method for elutriating multi- or megacorer samples is to place each core sample individually into a round bottom glass distillation flask. As with the larger set-up for box-cores, there is both an inflow of water and an outflow of water and sediment sample. The inflow of water from above gently and continuously washes and agitates the sediment. This process results in the small fauna being gently elutriated out of the flask and on to the sieve below. The material residue on the sieve can be periodically removed from the sieve and stored in a sample container. When the water runs clear, or no more animals are elutriated out of the distillation flask, the entire residue can be preserved in the sample container.

Washing of the sample on a sieve can be undertaken without using elutriation. Commercial or institute-built 'auto-sievers' use upwardly rotating water jets which gently wash the underside of the sieve. The whole system is held in a stainless steel deck table. The continuous pressure of water from underneath prevents the mesh of the sieve from becoming blocked. The sieve is fitted with a lid which seals the sample inside allowing the operator to slightly increase the water pressure. The

Table 10.4 An example of potential headings in a coring log sheet

Record	Comments
Ship name/cruise number	Sufficient to identify the vessel and its particular cruise
Deployment number	Sufficient to identify a particular deployment
Gear	Sufficient to identify a particular piece of equipment in its 'standard form'
Gear modifications	Note any variations from 'standard form', e.g. number and size of core tubes, ancillary equipment attached
Date/time launch	Used to determine total operation time for use in future planning
General pay-out speed	Speed that wire is payed out during the majority of the descent
Final pay-out speed	Speed that wire is payed out during the final approach to the seabed
Date/time bottom contact	This is taken to be the sample time, and so that of the navigation and depth data
Ship position	Ship's position at sample time
Ship sounding	Ship's echosounder depth at sample time
Gear position	When available, absolute or ship-relative position of gear at sample time
Gear depth	When available, telemetered depth of gear at sample time
Initial haul speed	Speed of wire recovery during pull-out phase
Pull-out tension	Maximum wire tension during pull-out phase
Final haul speed	Speed that wire is hauled in during the majority of the ascent
Date/time recovered	Used to determine total operation time for use in future planning
(Degree of) gear triggered	Gear triggered/not triggered; number of core units triggered
(Degree of) sample success	Sample recovered/number of samples recovered
(Degree of) gear damage	Any damage to the deployed equipment

centre of the lid is also fitted with a small hole covered by a mesh which allows for water to overflow, but prevents the loss of fauna in the event of the sieve being overfilled with water. Auto-sievers are particularly useful if a large quantity of sediment needs to be processed. Sieve size is generally 33 cm in diameter, and mesh sizes are typically 250 μm, 0.5 mm or 1 mm. The resulting residue can be stored in a sample container with fixative/preservative and mixed gently.

It is important the sample containers should be labelled on the outside, as well as water- and preservative-proof sample labels being placed on the inside of the container. The labels should contain some of the information highlighted in Table 10.4.

10.4.11 Fixation and preservation

In general, macrofauna and meiofauna samples are initially fixed in a buffered seawater solution of commercial-grade formalin (equivalent to 10%). Borax powder is added as a buffer to prevent decalcification of shelled organisms as the mixture is otherwise quite acidic. Often samples are sieved prior to being fixed in formalin, although there is a suggestion that first fixing the unsieved sediment leads to more intact specimens being recovered and a reduction in the number of animals lost (Degraer *et al.*, 2007). Sieving prior to fixation leads to smaller quantities of formalin being required, thereby reducing the demand for storage space on board ship. However, fixing prior to sieving is extremely useful for the more fragile fauna often found in the top 3 cm of sediment and should be used just for this uppermost fraction whenever possible. Careful on-deck treatment of the sample that results in better specimens can save hours of work in the laboratory dealing with poor quality material. Once the specimens in the sample have been fixed in formalin for several days, the sample can be washed with filtered seawater and transferred to alcohol to which 10% propylene glycol is added. The addition of propylene glycol will prevent irreversible damage to the specimens if for some reason the sample dries out.

Virus, Archaea and bacteria samples should be fixed with glutaraldehyde (Noble & Fuhrman, 1998) to a final concentration of 1% v/v. Ideally EM (electron microscopy)-grade glutaraldehyde should be used if viruses are to be enumerated. Algae and Protozoa can be preserved in a number of ways depending on their end use (see Alongi, 1993; Wickham *et al.*, 2000). Fixing in glutaraldehyde or formalin to a final concentration of 1–2% v/v is common. Lugol's iodine (final concentration 2–5% v/v) is also used regularly by microbiologists investigating, for example, ciliates and dinoflagellates, while foraminifera samples should be preserved in formalin buffered with sodium borate (final concentration 2–8% v/v) and are often stained with Rose Bengal to distinguish protoplasm that was alive at the time of collection (1 g Rose Bengal and 10 g phenol in 1 litre of tap water) (Gooday *et al.*, 2001). Once the fixative of choice has been added the sample container should be gently agitated to mix sediment and fixative. If bags have been used, these can be gently massaged to achieve homogenization.

We recommend reading Chapter 15 for further information concerning the fixing and preservation of samples.

10.5 Data interpretation

Prior to the start of any survey, careful consideration needs to be given to what sampling equipment to use. Some of this is influenced by the research objectives and hypotheses, the environment that is being sampled and the availability of equipment. The first question to ask is whether a quantitative or qualitative sample is required. Non-quantitative or semi-quantitative samples may be sufficient, for example, when the main objective is to provide an inventory of species. In such cases, large grabs may provide suitable samples even from coarse sediments. If quantitative samples are needed, the next issue to consider is the type of sediments being sampled. Sediment type has a major influence on what equipment can be used; for example if the sediment is quite coarse then the use of a multiple corer is not appropriate – the core tubes can be broken and the closing mechanism will not be able to retain a core very easily. However, if the sediment is finer, then a multiple corer is better than a box-corer for undisturbed sampling of the macrofauna and meiofauna.

When both box-corers and multiple corers are suitable, the last consideration is whether accuracy or precision is of prime interest. Accuracy is defined as the closeness of an estimate to the true value of the parameter being estimated (Andrew & Mapstone, 1987). Box-corers, as already discussed, induce a bow-wave effect, whereas multiple corers do not have this bias and thus are more accurate (e.g. Bett *et al.*, 1994). Precision, on the other hand, is defined by the degree of concordance among a number of estimates for the same population (Andrew & Mapstone, 1987). The precision of a density estimate from a sample is computed by dividing standard error by mean density (Elliott, 1977) and standard error is a function of both variance and the number of replicate samples. Consequently, precision increases with densities and with the number of replicate samples. Box-corers, due to their large size, allow achieving a desired level of precision with a lower number of replicate samples than do multiple corers. In environments where macrofauna densities are naturally very low, such as in the abyss, box-corers may be the only option. However, the use of multiple corers may be a more efficient means of obtaining sufficient samples for meiofauna studies. Although individual cores on a single multiple-corer cast are not independent and are not typically considered replicates for macrofauna studies (i.e. they are pseudoreplicates), they are sometimes considered as replicates for meiofauna studies. This is because patchiness in the distribution of, say, nematodes, occurs at a smaller spatial scale than the distance between cores on a multiple-corer.

There are a number of methods that have been proposed to optimize sampling design and achieve the best trade-off between the size of a sampling unit and the number of replicate samples, from the most simple based on precision (e.g. McIntyre *et al.* 1984; Elliott, 1997), to the more complex involving power:cost analyses (e.g. Ferraro *et al.*, 1989).

In conclusion, a sampling scheme will always involve trade-offs between a desired level of accuracy and a desired level of precision, but also practical, logistical or financial constraints. In the deep sea, these later constraints are usually quite large and in many cases unfortunately remain the first-order consideration in the design of a sampling scheme (see Chapter 3).

References

Alongi, D.M. (1993). Extraction of protists in aquatic sediments via density gradient centrifugation. In: Kemp, P.F., Cole, J.J., Sherr B.F., Sherr E.B. (eds), *Handbook of methods in aquatic microbiology*. Boca Raton, FL: CRC Press, pp. 109–116.

Andrew, N.L., Mapstone, B.D. (1987). Sampling and the description of spatial patterns in marine ecology. *Oceanography and Marine Biology: an Annual Review*, 25: 39–90.

Barnett, P.R.O., Watson, J., Connelly, D. (1984). The multiple corer for taking virtually undisturbed samples from shelf, bathyal and abyssal sediments. *Oceanologica Acta*, 7: 399–408.

Bett, B.J., Vanreusel, A., Vincx, M., et al. (1994). Sampler bias in the quantitative study of deep-sea meio-benthos. *Marine Ecology Progress Series*, 104: 197–203.

Billett, D.S.M., Lampitt, R.S., Rice, A.L., Mantoura, R.F.C. (1983). Seasonal sedimentation of phytodetritus to the deep-sea benthos. *Nature*, 302: 520–522.

Boland, G.S., Rowe, G.T. (1991). Deep-sea benthos sampling with the GOMEX box corer. *Limnology and Oceanography*, 36: 1015–1020.

Bowen, V.T., Sachs, P.L. (1964). The free corer. *Oceanus*, 11: 2–6.

Craib, J.S. (1965). A sampler for taking short undisturbed marine cores. *Journal du Conseil Permanent International pour l'Exploration de la Mer*, 30: 34–39.

Danovaro, R. (ed.) (2010). *Methods for the study of deep-sea sediments, their functioning and biodiversity*. Boca Raton, FL: CRC Press.

Degraer, S., Moulaert, I., Van Hoey, G., Vincx, M. (2007). Sieving alive or after fixation: effects of sieving procedure on macrobenthic diversity, density and community structure. *Helgoland Marine Research*, 61: 143–152.

Eleftheriou, A. (ed.) (2013). *Methods for the study of marine benthos* (4th edn). Oxford: Wiley Blackwell.

Eleftheriou, A., McIntyre, A. (eds) (2005). *Methods for the study of marine benthos* (3rd edn). Oxford: Blackwell Science.

Elliott, M. (1977). *Some methods for the statistical analyses of samples of benthic invertebrates*. London: Freshwater Biological Association, Scientific Publications.

Epstein, S.S. (1995). Simultaneous enumeration of protozoa and micrometazoa from marine sandy sediments. *Aquatic Microbial Ecology*, 9: 219–227.

Fabri, M.C., Galeron, J., Larour, M., Maudire, G. (2006). Combining the Biocean database for deep-sea benthic data with the Online Biogeographic Information System. *Marine Ecology Progress Series*, 316: 215–224.

Ferraro, S.P., Cole, F.A., Deben, W.A., Swartz R.C. (1989). Power-cost efficiency of 8 macrobenthic sampling schemes in Puget Sound, Washington, USA. *Canadian Journal of Fisheries and Aquatic Sciences*, 46: 2157–2165.

Gage, J.D., Bett, B.J. (2005). Deep-sea benthic sampling. In: Eleftheriou, A., McIntyre, A. (eds), *Methods for the study of marine benthos* (3rd edn). Oxford: Blackwell Publishing, pp. 273–325.

Gage, J.D., Tyler, P.A. (1991). *Deep-sea biology: a natural history of organisms at the deep-sea floor*. Cambridge, UK: Cambridge University Press.

Gerdes, D. (1990). Antarctic trials of the multi-box corer, a new device for benthic sampling. *Polar Record*, 26: 35–38.

Gooday, A.J., Hughes, J.A., Levin, L.A. (2001). The foraminiferan macrofauna from three North Carolina (USA) slope sites with contrasting carbon flux: a comparison with metazoan macrofauna. *Deep-Sea Research*, 48: 1709–1739.

Hartmann, O. (1955). Quantitative survey of the benthos of San Predo Basin, *Southern California. Part I. Preliminary results. Allan Hancock Pacific Expeditions*, 19(1): 185 pp.

Hessler, R.R., Jumars, P.A. (1974). Abyssal community analysis from replicate box cores in the central North Pacific. *Deep-Sea Research*, 21: 185–209.

Hessler, R.R., Sanders, H.L. (1967). Faunal diversity in the deep sea. *Deep-Sea Research*, 14: 65–78.

Jonasson, A., Olausson, E. (1966). New devices for sediment sampling. *Marine Geology*, 4: 365–372.

Jumars, P.A. (1975). Methods for measurement of community structure in deep-sea macrobenthos. *Marine Biology*, 30: 245–252.

Lisitsin, A.I., Udintsev, G.B. (1955). A new type of grab. (In Russian). *Trudy Vses Gidrobiol. Obsch.*, 6: 217–222.

McIntyre, A.D., Elliott, J.M., Ellis, D.V. (1984). Design of sampling programs. In: Holme, N.A., McIntyre, A.D. (eds), *Methods for the study of marine benthos*. Oxford: Blackwell Scientific, pp. 1–25.

Mortensen, P.B., Roberts, J.M., Sundt, R.C. (2000). Video-assisted grabbing: a minimally destructive method of sampling azooxanthellate coral banks. *Journal of the Marine Biological Association of the UK*, 80: 365–366.

Murton, B. (2010). HY-BIS: hydraulic benthic interactive sampler. www.noc.soton.ac.uk/geochem/Facilities%20 Links/hybis.htm (accessed November 2010).

Narayanaswamy, B.E., Bett, B.J. (2011). Macrobenthic biomass relations in the Faroe-Shetland Channel: An Arctic-Atlantic boundary environment. *PLoS ONE*, 6(4): e18602.

Narayanaswamy, B.E., Bett, B.J., Gage, J.D. (2005). Ecology of bathyal polychaete fauna at an Arctic-Atlantic boundary (Faroe-Shetland Channel, North-east Atlantic). *Marine Biology Research*, 1: 20–32.

Noble, R.T., Fuhrman, J.A. (1998). Use of SYBR Green I for rapid epifluorescence counts of marine viruses and bacteria. *Aquatic Microbial Ecology*, 14: 113–118.

Onbé, T. (1978). Sugar floatation method for sorting the resting eggs of marine cladocerans and copepods from sea-bottom sediment. *Bulletin of the Japanese Society of Scientific Fisheries*, 44: 1411.

Petersen, C.G.J., Boysen Jensen, P. (1911). Valuation of the sea. I. Animal life of the sea bottom, its food and quantity. Report from the Danish Biological Station, 20, 81 pp.

Reineck, H.E. (1963). Der Kastengreifer. *Natur und Museum*, 93: 102–108.

Riddle, M.J. (1989). Bite profiles of some benthic grab samplers. *Estuarine, Coastal and Shelf Science*, 29: 285–292.

Rowe, G.T., Clifford, C.H. (1973). Modifications of the Birge-Ekman box corer for use with SCUBA or deep submergence. *Limnology and Oceanography*, 18: 172–175.

Sanders, H.L., Hessler, R.R., Hampson, G.R. (1965). An introduction to the study of the deep-sea benthic assemblages along the Gay Head – Bermuda transect. *Deep-Sea Research*, 12: 845–867.

Sheremet, V., Efimova, T. (1996). 2D bottom sampling. In: Ivanov, M.K., Limonov, A.F., Cronin, B.T. (eds), *Mud volcanism and fluid venting in the eastern part of the Mediterranean Ridge*. UNESCO Reports in Marine Science No. 68, 126 pp.

Spärck, R. (1956). The density of animals on the ocean floor. In: Bruun, A.F., Greve, S.V., Mielche, H., Spärck, R. (eds), *The Galathea deep sea expedition*. London: George Allen and Unwin, pp. 196–201.

Van Dover, C.L. (2002). Community structure of mussel beds at deep-sea hydrothermal vents. *Marine Ecology Progress Series*, 230: 137–158.

Wickham, S., Gieseke, A., Berninger, U-G. (2000). Benthic ciliate identification and enumeration: an improved methodology and its application. *Aquatic Microbial Ecology*, 22: 79–91.

Chapter 11
Landers: Baited Cameras and Traps

Alan J. Jamieson
Oceanlab, Institute of Biological and Environmental Sciences, University of Aberdeen, Aberdeen, United Kingdom

Abstract

The free-fall lander method is now routinely used to deliver baited cameras and traps to the deep seafloor to observe animals *in situ* or recover specimens to the surface. Baited camera landers operate autonomously under pre-programmed settings, typically with time-lapse still or video cameras to observe deep-sea fauna attracted to the vicinity using bait. The method relies on a basic delivery system comprising ballast weights to sink the lander, and a release mechanism to jettison the ballast and flotation to surface the lander after the experimental period. The scientific payload typically comprises one or more digital still or video cameras with other environmental sensors such as temperature, salinity, pressure or current meters. Baited traps are delivered to the seafloor in the same manner but the payload can comprise a diverse array of traps, designed specifically to target certain faunal groups such as fish or small invertebrates. Free-falling landers enable multiple systems to operate on the seafloor simultaneously and independently of one another. Furthermore, this method provides the opportunity to sample to full ocean depth (11 000 m) from relatively small research vessels. The basic design considerations and equipment configurations of baited camera and trap landers are discussed as well as a review of the types of data typically obtained using this method.

Keywords lander, baited camera, baited trap, hadal depths, autonomous lander

11.1 Introduction

The deep seafloor had long been regarded as lifeless until the HMS *Challenger* expeditions in 1872–1876 (Younge, 1971) found deep-sea communities which were rich and diverse. Since that time, scientists have been pushing back the boundaries of science and technology in order to understand the complex processes that occur in the deep ocean. Until recently, investigation of the ocean floor was limited to ship-board observations or to the depths achievable for divers. A modern revolution in investigative ocean technologies has resulted in diverse and imaginative techniques ranging from simple wire-deployed instruments to satellite imagery to remotely operated vehicles (ROVs) (Priede, 1980).

The recovery of organisms from the deep seafloor can now be achieved by a number of different methods that enable the discovery and cataloguing of new species and facilitate the distributional mapping of species in relation to their environment. However, the recovery of live deep-sea organisms is usually inhibited as a result of the effects of decompression and temperature shifts during their ascent to the surface. Consequently, the direct behavioural or physiological observations that are routinely undertaken during the recovery of live organisms from shallower marine environments

Biological Sampling in the Deep Sea, First Edition. Edited by Malcolm R. Clark, Mireille Consalvey and Ashley A. Rowden.

are not possible with deep-sea fauna. This dilemma prompted the need for *in situ* observations to be undertaken on the seafloor and inspired the development of autonomous free-fall lander vehicles (also known as free-vehicles), a technique that enables deep-sea scientific exploration using tools that were previously restricted to the laboratory.

Most autonomous landers are designed for the study of biogeochemical processes of the seafloor and their use in the deep sea in this respect has already been extensively dealt with in other publications (Tengberg *et al.*, 1995; Pfannkuche & Linke, 2003; Viollier *et al.*, 2003; Tengberg *et al.*, 2004). Landers have also been adapted extensively to enable the delivery of baited cameras and traps to the deep sea to focus on biological processes. In the context of this book, the purpose of the current chapter is to review the methods employed for the delivery of imaging systems and traps to the deep sea (>200 m) using free-fall landers to support biological research and the sampling of deep-sea fauna.

Autonomous landers are platforms upon which scientific equipment is mounted and transported to and from the seafloor. The vehicle, carrying a variety of scientific sensors and instruments, is lowered from a ship into the sea and released. The lander free falls to the seafloor where pre-programmed experiments, observations and monitoring tasks are carried out for a pre-determined amount of time (between 1 hour and up to 1 year). The lander is fitted with positively buoyant flotation and expendable ballast weights. Following the experiment, the lander jettisons its ballast weights either by acoustic command from the ship or via an onboard timing device. Once the ballast has been shed, the lander ascends to the surface by virtue of the positive buoyancy, whereupon it is recovered by the ship.

In the absence of continual ship support, there are obvious cost- and time-related advantages to using landers in long-term applications. Aside from deployment duration, large research vessels can carry in excess of 12 landers at a time, permitting a diverse array of experiments occurring simultaneously across large areas of the seafloor (Priede, 1999). Lander operations can be scheduled to fit in around other activities and preparation for deployment can be undertaken while the ship is changing position. One lander can remain operational on the seafloor for days or weeks yet only a few hours of ship time are required for its deployment, recovery and turnaround. A further benefit is that landers can be deployed from all ships, including relatively small ships such as trawlers (Yau *et al.*, 2001). Even smaller ships that lack the winches capable of reaching the deep seafloor can sample at full ocean depths using landers.

Free-falling landers also work where other methods such as trawling or the use of camera sledges are ineffective or prohibited; these areas include Marine Protected Areas, areas of complex topography (e.g. canyons; King *et al.*, 2008), fragile habitats (e.g. coral reefs; Heger *et al.*, 2007), areas with sub-sea infrastructure (e.g. oil pipelines, sub-sea cables; Walls & Clarke, 2005) or areas of extreme depths (e.g. hadal trenches; Jamieson *et al.*, 2009a, 2009b).

11.2 Experimental design

Using the lander principle to deliver baited cameras and traps involves careful consideration of both the scientific payload (bait, camera, lights, power, traps and other sensors) and the delivery system of the vehicle itself (ballast, buoyancy, ballast release, framework). How and where the equipment is mounted is equally important for the experimental design. The following sections describe the design considerations for lander delivery systems and scientific payloads for baited camera and baited trap applications in the deep sea. An example of a basic baited lander (with traps) and all its primary components is shown in Fig. 11.1.

11.2.1 Scientific payload

11.2.1.1 Bait and odour plumes

Food is supplied to the deep sea in three general forms: (i) as particulate organic matter (POM, or 'marine snow') which is taken up by deposit feeders, (ii) chemosynthetic origins found in areas such

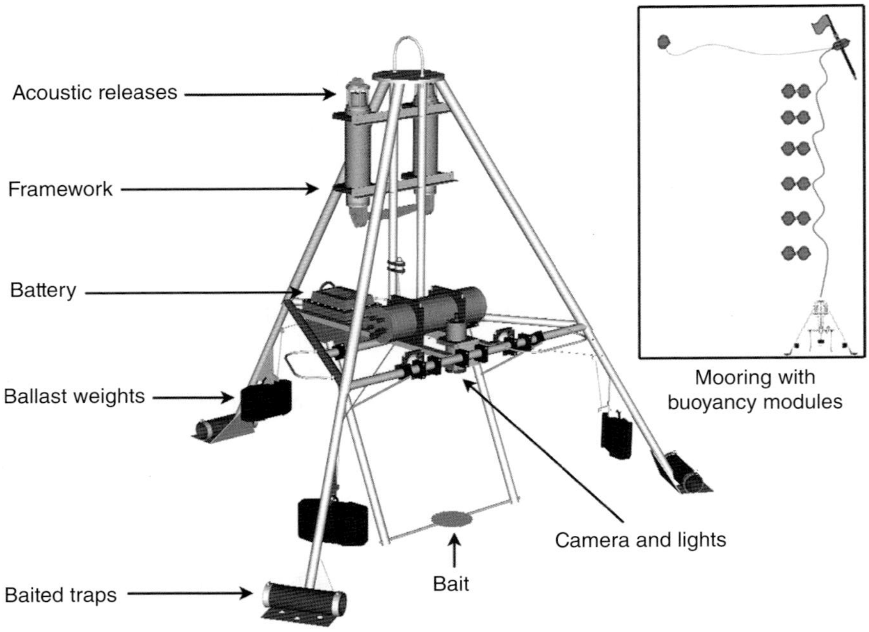

Fig. 11.1 Basic components of a typical baited camera lander for deep-sea applications. The basic delivery system comprises acoustic releases, ballast, buoyancy and framework while the scientific payload comprises a camera, bait, lights, battery and, in this example, small baited traps. (Oceanlab, University of Aberdeen, UK. Reproduced with permission.)

as hydrothermal vents and cold seeps, or (iii) in the form of the sinking carcasses of pelagic organisms, often originating from surface waters (food-falls). Theoretically, food-falls should occur irrespective of water depth as there are few known pelagic processes that impede their descent. Thus, food-falls are considered an extremely important source of highly concentrated organic matter for the deep sea and they are rapidly intercepted and consumed by the mobile benthic and benthopelagic communities (Bailey *et al.*, 2007). Such rapid consumption rates mean that the probability of finding naturally occurring food-falls using manned submersibles or ROVs is extremely low (Klages *et al.*, 2001; Soltwedel *et al.*, 2003; Yamamoto *et al.*, 2009). Therefore, in order to study the significance of food-falls in the deep sea and/or to collect specimens of the scavenging community, scientists use bait to mimic the food source.

Selecting the type of bait to use for baited cameras or traps depends on the experimental design and the scientific questions being addressed. Bait types can be conveniently split into six categories:

- A strong and persistent bait to attract as many species as possible for the greatest length of time. This type is typically used in traps. Many institutes have their own 'recipe' which includes different types and ratios of fish oils, ground fish meat, homogenized mixtures of oily fish (such as mackerel or tuna) and squid, and some use breadcrumbs or sawdust to thicken an oily mixture, or even catfood. The bait mix is placed in a perforated receptacle or in a fine mesh bag to allow odour dispersal without loss of bait.
- Bait used to simulate a natural food-fall (whole carcass). This is normally used in baited camera deployments, first to attract fish and invertebrates into the field of view of the camera and second to investigate the interception, consumption and dispersal of organic matter in the form of surface-derived food-falls (Priede *et al.*, 1991; Priede & Bagley, 2000). Single or multiple fish carcasses

are secured in view of the camera and slashed to release the odour. Depending on the experiment, the bait can be openly secured to permit consumption by scavenging fish and other organisms or can be wrapped in mesh to allow odour dispersal but prevent feeding. Mackerel and tuna are particularly oily fish that produce strong odour plumes.

- Large carcasses to simulate the fate of large animal carcasses. For longer term baited camera experiments, the carcasses of dolphins, porpoises and sharks have been deployed (Jones *et al*., 1998; Witte, 1999; Kemp *et al*., 2006). These carcasses differ from more typical baits (such as mackerel) as they have longer consumption times that often results in the creation of an ephemeral habitat.
- Although not commonplace in deep-sea operations, artificial baits are an option. These are based on natural chemical attractants (amino acids) found in fish and squid. The amino acids tyrosine, cysteine, phenylalanine, glycine and methionine can be mixed to produce a liquid attractant (Løkkeborg *et al*., 1995).
- Whale-falls and wood-falls. Segments of whale bones and packages of wood have been deployed in the deep sea to simulate the deposition of whale carcasses and wood from terrestrial sources or shipwrecks (Palacios *et al*., 2006; Glover *et al*., 2014; Amon *et al*., 2013). Such packages, if left for extended periods of time, are eventually colonized by bone-eating or wood-boring polychaetes, thus simulating a biogenic chemosynthetic habitat.
- Plant debris is another food source found in the deep sea (Menzies *et al*., 1967). Trawls as deep as ~8000 m have revealed that some benthic fauna feed on plant debris originating from the coastal zone (e.g. sea grass; Wolff, 1976). Sea grass has been used in baited traps (Lawson *et al*., 1993) and spinach in baited landers to investigate how it is consumed by the scavenging community (Jeffreys *et al*., 2010).

The high pressure and low temperatures found in the deep sea can be exploited to maintain bait freshness. The freshness of mackerel carcasses can be maintained for at least three months if they are placed in sealed chambers containing sterilized seawater. This technique has been used to introduce new bait at weekly intervals from a single baited lander, thus extending the experimental occupancy (Kemp *et al*., 2008).

Our understanding of bait consumption by individual species is still in its infancy. However, there is evidence to suggest that the use of deep-sea fish as bait causes avoidance in conspecifics (Barry & Drazen, 2007). Though never proven, it is widely considered that the use of surface-dwelling fish that are native to the experimental site is more representative of a natural setting.

When using baited systems the bait creates an odour plume which, due to the currents, laterally disperses in an elliptical shape over time (assuming a flat seabed). Irrespective of the odour plume, some species may detect the lander due to noise, hydrodynamic disturbance (Wilson & Smith, 1984), mechanoreception (Klages *et al*., 2002) or bioluminescent activity (Priede *et al*., 2006a). However, the odour plume is classed as the catchment area for scientific purposes. The effective area of the plume depends on the current velocities and bait characteristics (Bailey & Priede, 2002), but predictions of the spread of an odour plume over time are usually based on assumptions about unknown variables. For example, the first animals to intercept open bait often remove and consume large pieces. This results in a temporary increase in the odour concentration prior to a decrease in concentration with time. In addition, variations in current velocity and direction affect odour concentration at a given distance from the source at any given time. Furthermore, the odour detection threshold for most, if not all, deep-sea organisms is unknown and therefore the point at which the odour plume is no longer an effective attractant is also unknown. Odour plume models are typically used for the first half hour of the deployment to calculate the theoretical population densities of the most abundant scavenging species.

The area of the odour plume can be calculated to provide an indication of the minimum distance at which a second bait could be placed in order to prevent interference between experiments. The area of a simple two-dimensional (2D) plume emanating from a baited camera or bottom deployed trap can be calculated based on the workings of Bailey and Priede (2002), Stanley *et al.* (1985), Sainte-Marie and Hargrave (1987). In 1986, Rowe *et al.* described a three-dimensional odour plume model for midwater baited traps, however the 2D model is usually deemed sufficient for baits placed on the seafloor (see eqns 1–3).

The odour plume length (L_{pl}, m) equals the current velocity (V_w, ms^{-1}) multiplied by time (t, s), therefore:

$$L_{pl} = t \times V_w \tag{1}$$

As the plume spreads, the width (L_{pw}, m) increases:

$$L_{pw} = 2\left(\frac{B_y}{V_w}\right) \cdot L_{pl}^{b}\left(\frac{(b+d)}{b}\right)^{0.5} \cdot e^{0.5} \tag{2}$$

where b and d are assumed to equal 1 and B_y is the diffusivity constant or spreading velocity, estimated at 0.01 ms^{-1} for deep waters. Therefore the area of the odour plume (A_p, m^{-2}) at a given point in time can be calculated using the following equation:

$$A_p = \pi\left(\frac{L_{pl} \cdot L_{pw}}{4}\right) \tag{3}$$

This theoretical model predicts a plume area of 0.01 km^2 after 1 hour, 0.6 km^2 after 6 hours and 2.4 km^2 after 12 hours (assuming an average current speed of 0.05 ms^{-1}). However, in reality the concentration of the plume will decrease with time therefore at some point it will inevitably fall below the detection threshold of a given species.

11.2.1.2 Imaging systems

There are three main types of baited camera systems: digital stills, colour charge-coupled device (CCD) video and ultra low-light video cameras (Fig. 11.2). These cameras are available commercially, with perhaps the exception of ultra low-light video cameras. The selection of a suitable camera is wholly dependent on the scientific questions being addressed.

Underwater digital still cameras currently offer up to 10 megapixel resolution or more as standard, although many institutes develop custom-built designs of higher resolution. Still cameras are frequently used with a time-lapse facility whereby the camera is pre-programmed to capture an image at regular intervals throughout the deployment duration. On short timescales (<24 hours) images are typically captured every 1 or 2 minutes. Over longer timescales (1–6 months) image capture is less frequent; every 3–6 hours. The advantages of time-lapsed still photography include the relatively low power required, and the camera's ability to provide high-resolution images of the seafloor and associated fauna that can be digitally magnified and measured with great accuracy back at the laboratory. In the deep sea where many rare or unknown species exist, high-resolution stills are incredibly useful in aiding species identification.

One disadvantage of still photography is its inability to capture large amounts of behavioural or physiological information, such as the swimming speeds and tail-beat frequencies of fish. High-resolution colour video cameras are, however, capable of observing such parameters. To maximize power and reduce periods of illumination, video cameras also employ time-lapse technology to record for short, regular bursts. For example, video cameras may record for 1 out of every 5 minutes. Deconstructing video sequences into individual frame grabs can result in a loss of resolution, but can effectively slow down fast-moving events such as escape responses. This is advantageous in aiding

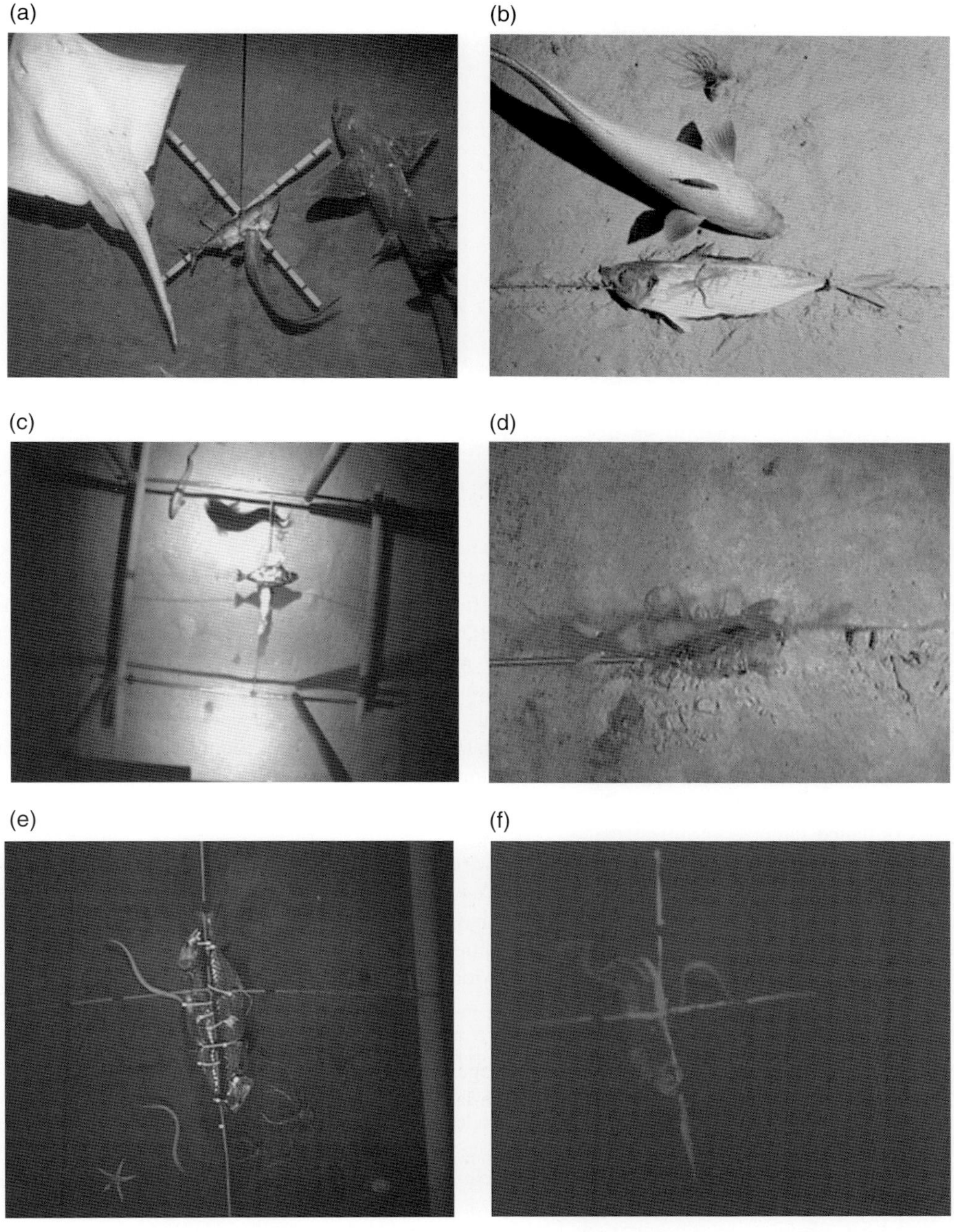

Fig. 11.2 Examples of imaging types used in baited cameras. (a) Digital stills at 2 m above bottom (field of view (FOV) = 2.1 × 1.6 m; 3.4 m^2). (b) Digital stills at 1 m above bottom (FOV = 0.7 × 0.5 m; 0.3 m^2). (c) Frame grab from video at 2.8 m above bottom (FOV = 1.8 × 2.6 m; 4.7 m^2). (d) Frame grab from video at 1 m above bottom (FOV = 0.7 × 0.5; 0.4 m^2). (e) Normal colour CCD video camera with white light and (f) under red light (note the degradation in quality after switching to red light).

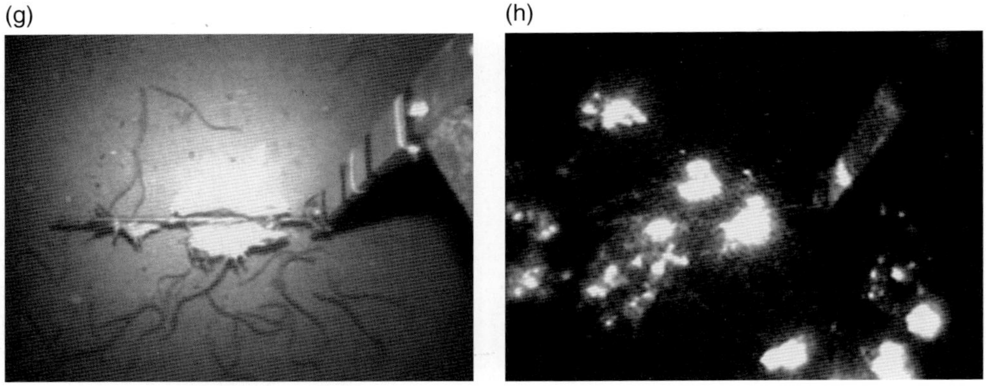

Fig. 11.2 (*Continued*) (g, h) Ultra low-light Intensified Silicon Intensifier Target (ISIT) camera with white light on (g) and no illumination (h). When the lights are off the organisms cannot be seen but bioluminescent activity can. (Oceanlab, University of Aberdeen, UK. Reproduced with permission.) (Please see Plate 18 for colour representation of the figure.)

the interpretation and accuracy of measurements. Video sequences can also be used in frame-by-frame playback to identify a particular physical feature, such as a pectoral fin. Such physical features, used to aid species identification, may only be visible briefly in a video but may be absent or obscured completely in a still image. Using high-definition (HD) camera technology, video frame grabs almost equal the current resolution of digital still photographs, although the usage of HD in autonomous systems is still infrequent. More typically, baited cameras use CCD or 3CCD video cameras with minimum sensitivities and resolutions of 0.5 and 0.9 lux with 470 and 800 television lines (TVL) respectively. Video systems use high power-consuming halogen bulbs for illumination that can severely limit autonomous operations. However, the power consumption of modern camera technology and illumination (e.g. Light-Emitting Diodes (LEDs)) is constantly being reduced, thus extending the *in situ* occupancy of baited cameras.

Low-light cameras can be used as an alternative to colour video systems when artificial illumination may have adverse effects on target species (Widder *et al.*, 2005) or in the study of bioluminescence (e.g. Priede *et al.*, 2006a). These cameras have considerably greater sensitivity and therefore require a far lower intensity of illumination. In addition, they use red or far-red illumination (Fig. 11.2) which most deep-sea fish are unable to detect (Widder *et al.*, 2005). Low-light cameras range from intensified CCD cameras (ICCD; min. sensitivity = 5×10^{-7} lux, resolution = 450 TVL) and silicon intensifier target cameras (SIT: min. sensitivity = 5×10^{-4}, resolution = 600 TVL), to even more sensitive ultra low-light systems such as intensified SIT (ISIT; min. sensitivity = 5×10^{-6}, resolution = 525 TVL) and image-intensified charge-coupled device for deep-sea research (ICDeep) (min. sensitivity = 1×10^{-7}, resolution = 720 TVL). Ultra low-light cameras can also be used without illumination in order to film natural bioluminescent activity on the seafloor, in the vicinity of bait (Fig. 11.2; Gillibrand *et al.*, 2006; Heger *et al.*, 2007). Bioluminescent imaging cannot observe the organism responsible for the light production and, therefore, the dark observations are often interspersed with periods of low-light illumination to identify the source(s) of the light.

Each of the imaging technologies described here has advantages and disadvantages, thus, it is often beneficial to use a combination of systems, either on the same lander or on different landers in the same area. Just as selection of a suitable camera is driven by the scientific question being addressed, the way in which the camera is deployed is equally as important. For example, cameras can be deployed in a baited system for long or short periods, deployed vertically downward-facing or at oblique or horizontal angles at different altitudes above the seabed (Figs 11.2, 11.3 and 11.4). As well as camera resolution, the quality of the image (and subsequent data), whether obtained from digital stills or video, is

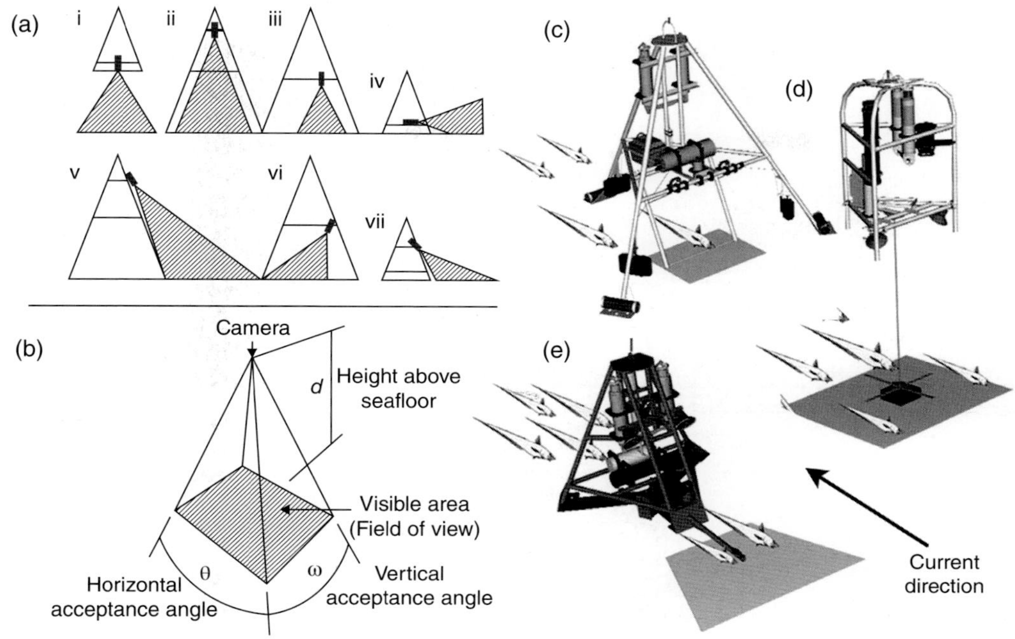

Fig. 11.3 Considerations for baited camera experiments. (a) Examples of vertical photography (i–iii), horizontal (iv) and oblique (v–vii). (b) Camera field of view parameters. (θ, d, ω) Practical consideration of camera mounting position: (c) shows a vertical camera with 360° access with limited obstruction, (d) shows a tethered vertical camera with 360° access and no obstruction, and (e) shows a horizontal (or oblique) camera with only 180° access and major obstruction when the current prevails in the direction shown. (Oceanlab, University of Aberdeen, UK. Reproduced with permission.)

also affected by illumination, angle and altitude considerations. There must, therefore, be a trade-off between these parameters and the desired visible area of seafloor or body size of animal to be imaged.

Vertical photography, where the camera is downward-facing and perpendicular to the seafloor, is probably the most common position for deep-sea baited cameras. The images can be interpreted as point-source quadrats. This method of imaging organisms from above allows 360° unobstructed access to the bait and facilitates the measurement of faunal body lengths. A vertical photography baited system creates a setting similar to an artificial food-fall where fauna such as fish are not required to navigate round or through equipment and neither are they obscured from view by structures. Most scavenging fish follow the bait odour up-current and therefore imaging a 360° area around the bait enables the camera to capture every individual (Fig. 11.3). The altitude at which a camera is held above the seafloor is a compromise between visible area, available illumination and the resulting resolution. The greater the altitude of the camera from the seafloor, the greater the required illumination but the lower the resolution in each image. Most baited cameras image from an altitude of between 1 and 2 m above the seafloor.

Vertical photography requires the camera to be at a greater altitude above the seafloor than horizontal or oblique-angled photography. There are several methods that achieve this without greatly increasing the physical size of the lander. The simplest method involves the construction of a small lander with the camera positioned at the bottom. The frame, when underwater, becomes tethered between the buoyancy above and a ballast weight suspended below. If the length of the ballast tether is, for example, 2 m, then a larger visible area relative to the size of the lander is achieved. Alternatively, the lander can be positioned directly on the seafloor with the camera facing downward inside the frame where the lander structure is designed around the field of view.

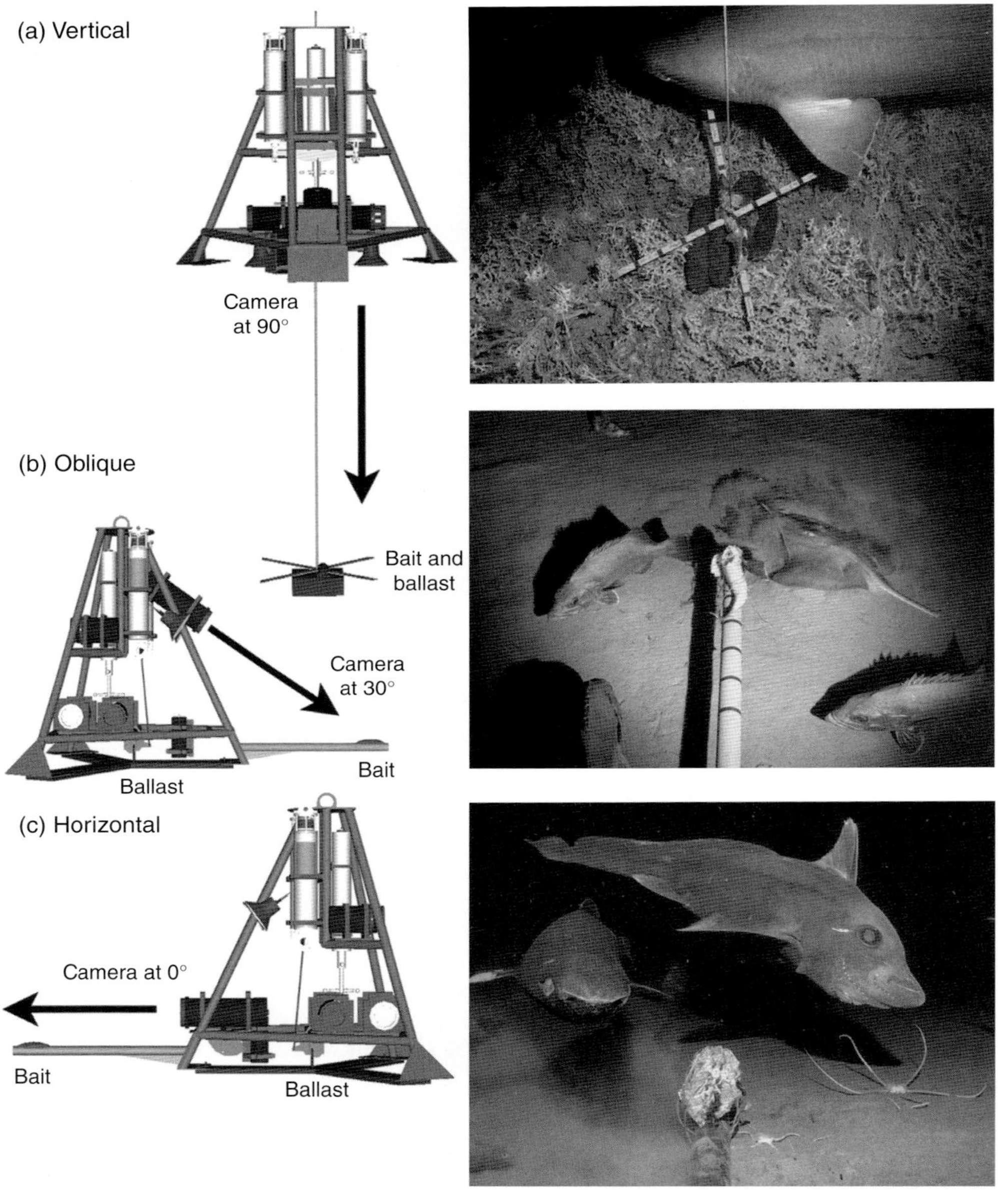

Fig. 11.4 Examples of different camera angles: (a) vertical; 90°, (b) oblique; 30° and (c) horizontal; 0°. The examples shown are from the ROBIO lander at (a) 1000 m in the Northeast Atlantic, (b) 900 m eastern Mediterranean and (c) 2500 m eastern Atlantic. (Oceanlab, University of Aberdeen, UK. Reproduced with permission.)

The field of view (visible Area) in vertical photography can be calculated using eqn 4 (Jones *et al.*, 2009),

$$\text{Area} = 4d^2 \tan\left(\frac{\theta}{2}\right)\tan\left(\frac{\omega}{2}\right) \tag{4}$$

where *d* is the camera altitude (m), θ is the horizontal camera acceptance angle (degrees) and ω is the vertical camera acceptance angle (degrees; schematic shown in Fig. 11.3).

Vertical photography provides an accurate setting for measuring body lengths and observing the time course of interception, consumption and dispersal of bait. However, identification of species is often difficult from above, particularly in areas where baseline data are lacking. This problem can be alleviated using horizontal imaging for ground-truthing identifications. Lateral images of fish or invertebrate species provide more taxonomic features on which to base identifications – such as pectoral and anal fins, mouth characteristics and headshots of fish. These features are obscured if imaged from above.

Oblique camera angles (15–45° from horizontal) are alternative camera positions that allow a compromise between vertical and horizontal photography. Oblique camera angles are not typically used in studies of fish and invertebrates as this method, like horizontal photography, requires the lander structure to be situated behind the camera (out of the field of view). If the lander lands on the seafloor facing up-current, the majority of scavenging fauna will approach from behind the camera. Oblique-angled photography is more commonly employed for non-baited applications to view the activity around specific habitats such as coral (Roberts *et al.*, 2005) or the seafloor and associated epifauna over longer periods of time (Lampitt & Burnham, 1983; Lampitt *et al.*, 1983; Smith *et al.*, 1993). Oblique photography has three advantages: (i) it permits larger areas of visible seafloor relative to the size of the lander, (ii) it images the seafloor outside the lander footprint, thus eliminating any effects of the lander structure, and (iii) it highlights topographical or biologically induced features such as animal tracks more clearly than vertical photography.

To accurately measure objects and organisms from oblique images, perspective grids (such as a 'Canadian grid') can be applied to the image. This procedure is complex and is described in Wakefield and Genin (1987). The angle and altitude of an oblique camera depends on the target species and the scientific question and is a compromise between visible area and resolution. The greater the altitude, the larger the visible area but at the expense of image resolution. In addition, the greater the angle from vertical, the greater the potential for error in measurements.

Some landers, such as ROBIO (Jamieson & Bagley, 2005), can be set up with different camera angles (vertical, horizontal and oblique) that can be changed between deployments at the same site to give a variety of different information (see Fig. 11.4). Other systems, such as the LAFF lander (Jones *et al.*, 1998), have vertical and horizontal cameras operating simultaneously on the same frame.

11.2.1.3 Illumination

The natural optic environment of the deep sea consists of very dim, homochromatic, downwelling light, supplemented by bioluminescence. In the absence of natural light beyond 1000 m from the surface, there is the potential for artefacts caused by artificial illumination that could attract or deter certain species. A time-lapsed still camera only releases a millisecond flash of white light per image, resulting in most of the experimental time being at ambient darkness. Many organisms inhabiting this environment possess highly specialized visual systems that are sensitive to artificial illumination (Douglas *et al.*, 1995; Herring *et al.*, 1999) because they lack irises that would otherwise restrict the amount of light entering the eye. Numerous studies have demonstrated that most retinal photopigments in deep-sea fishes and crustaceans have peak sensitivities between 450 and 500 nm, similar to that of downwelling light and bioluminescent emissions (Frank & Case, 1988; Hiller-Adams *et al.*, 1988; Partridge *et al.*, 1988; Douglas *et al.*, 1995).

Although there are few studies from which to draw information, it is thought that the effects of artificial illumination on deep-sea animals may result in a variety of physiological responses, ranging from permanent and temporary blindness (photobleaching), to having little or no effect. Photobleaching is thought to occur rapidly, however, there are indications that it is reversible with time (Hiller-Adams *et al.*, 1988). There are few data regarding the tolerance limits for temporary or permanent optical damage in most deep-sea species, but visual impairment has the potential to affect behaviour and as a result may lead to uncertainty in baited camera results (as well as potentially causing the death of the effected organism). There are no known reports that indicate an absence of certain species from baited cameras using illumination, although there are indications that the intensity, duration and wavelength of light can alter behaviour of some species. There are five scenarios of behavioural response to artificial illumination:

- No response or behavioural change while exposed to artificial illumination.
- Positive phototactic responses which entices organisms toward the light.
- Negative phototactic response which deters organisms from the light.
- Indirect phototactic responses which entices predators toward positively phototactic prey.
- Temporary blindness (dazzling) effects which may inhibit an organism's behaviour (such as the ability to detect prey or avoid predators).

Bailey *et al.* (2005b) reported that during 15 second periods of illumination with a 25 W lamp, there was a significant decrease in the numbers of the deep-water eel *Synaphobranchus kaupii* over time. Raymond and Widder (2007) observed a flight response of the sablefish *Anoplopoma fimbria* when subjected to white light (390–750 nm) and sometimes red (685 nm), but not with far-red illumination (695 nm), whereas the grenadier *Coryphaenoides acrolepis* showed no behavioural responses to changes in wavelength.

Most reports on damage to visual systems by artificial illumination are derived from exposure to powerful white-light illumination from ROVs or submersibles, in the range of 1250 W (Herring *et al.*, 1999). Furthermore, ROVs and submersibles produce continuous illumination for periods of hours at a time and often directed at a static habitat such as a hydrothermal vent. Baited cameras, however, do not typically image sessile organisms and attract more mobile fauna. Baited camera lights are typically more in the region of 25–75 W and only illuminate for short periods at longer time-lapsed intervals. The intensity of the cone of light from a baited camera falls off logarithmically with distance. The volume of this illuminated cone is minuscule in comparison to the unlit volume surrounding the camera in every other direction. Furthermore, it is assumed that only ultraviolet wavelengths of light are liable to cause irreversible damage to optical pigments.

Understanding the light levels produced by a baited camera are not only important for interpreting biological data, but are also useful in selecting the camera sensitivity, the distance/angle from the seafloor and/or the numbers of lights required. The amount of light reaching the seafloor (the scene illumination) can be estimated using eqn 5:

$$E_{sc} = \frac{It^d}{d^2} \tag{5}$$

where E_{sc} is scene illumination (lux), I is luminous intensity of the light(s) (candela), t is light transmission factor in water (0.9 m^{-1}) and d is distance from the seafloor (m).

Most camera systems quote light sensitivity in terms of faceplate sensitivity (the amount of light reaching the camera faceplate rather than the seafloor), and eqn 6 calculates the camera faceplate sensitivity from the scene illumination:

$$E_{fp} = \frac{E_{sc}\rho\tau t^d}{4f^2} \tag{6}$$

where E_{fp} is faceplate illumination (lux), ρ is scene reflectance factor, τ is lens transmission factor, f is lens aperture (f-stop), t is light transmission factor and d is distance between camera and seafloor.

For simplification, the typical values for the scene reflectance factor (ρ), lens transmission factor (τ) and light transmission factor (t) are 0.2, 0.9 and 0.9 m^{-1}, respectively (note: for video camera the f-stop = 1).

If, however, the camera and light(s) are at the same distance from the seafloor, the approximate faceplate illumination can be calculated using eqn 7:

$$E_{\mathrm{fp}} = \frac{It^{2d} \cdot 0.18}{4f^2d^2} \tag{7}$$

where faceplate sensitivity of the camera (E_{fp}, lux), I is lamp luminous intensity (candela), t is light transmission factor of deep seawater (0.9 m^{-1}), f is lens aperture (f-stop) and d (m) is distance from target. Therefore the value of E_{fp} must be within the maximum and minimum faceplate sensitivity of the camera. If the camera specification is already fixed, this equation can be rearranged to determine the most suitable distance from the seafloor or illumination intensity. These equations should only be used a rough guide as there are many variables present under normal operating conditions, such as water quality (variability in suspended particles), substrata reflectance, or other objects or organisms in the field of view.

Commercially available sources of illumination can specify intensity in different units (e.g. luminous intensity in candelas or luminous flux in lumens). There are no precise methods to convert one to the other as they are theoretically different units. An approximate conversion can be found using a candela–lumen (cd–lm) conversion factor at a known beam angle (see Fig. 11.5) where lumen = candela/cd–lm factor or candela = lumen × cd–lm factor.

The clarity of the image also depends on the position of the lights relative to the camera. If the volume of water within the camera's field of view is reduced, a clearer image is produced since glare from reflected particles in the water (backscatter) is minimized. Using light(s) that have been mounted at an angle and that are further away from the camera can reduce the amount of backscatter in the image (Fig. 11.5). Where only one light is employed, the greater the angle of this light, the

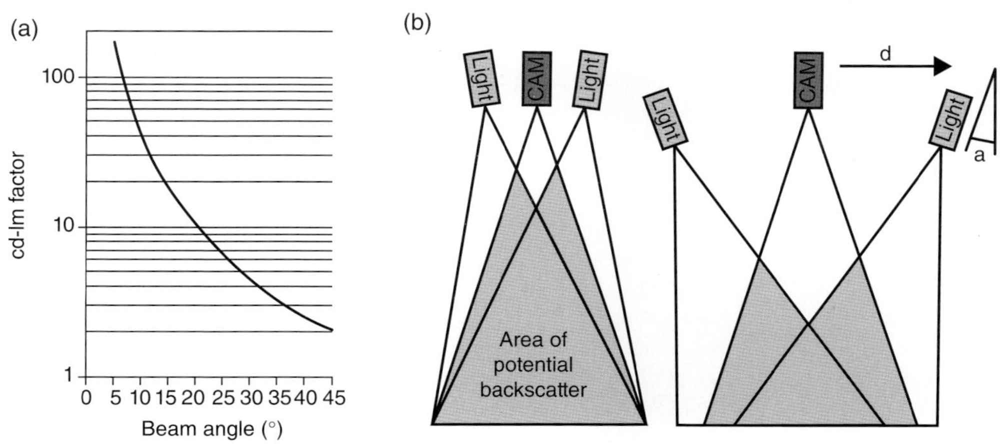

Fig. 11.5 Practical considerations for illumination. (a) Approximate conversion of luminous intensity (candela) to luminous flux (lumens) using a candela–lumen conversion factor (cd–lm) for a specific beam angle (°). (b) Positioning of the light(s) relative to the camera to reduce backscatter in the image; increasing distance (*d*) and angle (*a*) will reduce backscatter. (Oceanlab, University of Aberdeen, UK. Reproduced with permission.)

larger the shadows that will be in the image and this results in a less uniform spread of illumination and a light-to-dark gradient appearing across the image.

11.2.1.4 Baited traps

Imaging scavenging species with baited cameras does not always allow identification of organisms to species level. Simultaneously identifying organisms of drastically dissimilar sizes is also not always possible. For example, fish and crustacean assemblages are often composed of extremely different size classes regardless of diversity. Where a fish can be up to 1 m long there may also be hundreds or thousands of small scavenging amphipods present that are less than 10 mm in size. Therefore, fixing the camera at a distance capable of imaging a 1 m long fish prohibits the resolution required to identify small crustaceans and vice versa. In fact, the accurate identification of fish from photographs can be very difficult or impossible (Barnes, 1955). This is equally the case with even smaller organisms such as amphipods. It is, therefore, common place to include baited traps either as part of the scientific payload of the baited camera or as a separate lander employed in the same vicinity to trap specimens for accurate identification. The specimens recovered from baited traps contribute not only to the accuracy of the baited camera images (through taxonomy), but offer further scope for scientific research using other techniques such as population genetics, physiological measurements, stomach content analysis (diet), chemical composition and length–weight (biomass) relationships.

The use of baited traps is well established (Paul, 1973; Isaacs & Schwartzlose, 1975). Traps have been used extensively to recover mobile scavengers at bathyal, abyssal and even hadal depths (Shulenberger & Hessler, 1974; Hessler *et al.*, 1978; Thurston 1979; Stockton, 1982; Wickins, 1983; Blankenship *et al.*, 2006). Baited trap designs are diverse, but are mostly based on a simple funnel trap design. The size and design varies depending on the target species.

The most common baited trap is a small funnel trap designed for invertebrates, and used to recover small scavenging crustaceans. These traps comprise a tube with a mesh funnel on one or both ends where the bait is secured inside (Fig. 11.6). The scavenger locates the trap by following the odour plume and is funnelled into the trap whereupon they feed on the bait. Cylindrical traps with single or double funnels are usually effective at recovering a sufficient number of deep-sea specimens for further scientific applications. As the funnel exits of the traps are not easily located from the inside, most occupants are recovered, although there may be some loss of samples due to escapees and wash-out during the ascent to the surface. Escapee numbers are reduced by increasing the number of chambers within the traps and more complicated designs can reduce the chance of wash-out. Furthermore, some systems have converted closing water samplers (Niskin bottles) into funnel traps (Blankenship *et al.*, 2006) or use custom-built closing mechanisms (Hargrave *et al.*, 1995) to prevent sample loss.

The simplicity of these traps means they are readily mounted with baited cameras with no significant cost or time implications (Lampitt *et al.*, 1983; Jones *et al.*, 2003; Heger *et al.*, 2007). Placement of traps onto existing baited cameras must be carefully considered, as if the trap is not in view of the camera it will inevitably draw some species away from the field of view temporarily, but placing the traps within the field of view will obscure the resulting images. It is common place to secure baited funnel traps on cameras that are not intended to observe the natural time course of bait consumption. The traps are used therefore on static horizontal cameras or for experiments where information concerning the exact number of fish and/or interception times is not critical. Funnel traps can, however, be integrated into baited camera experiments; for example, when using low-light video to film bioluminescent organisms trapped *in situ* in transparent traps which are then recovered for identification (Oceanlab, unpublished data, Fig. 11.6).

Traps can also be deployed on a single free-fall system, as a cluster on a near-bottom package or on a vertically moored array. Baited trap arrays have been used to investigate the vertical community structure of benthopelagic species (Smith *et al.*, 1979; Ingram & Hessler, 1983; Smith & Baldwin, 1984). These arrays comprise an expendable ballast weight and release mechanism and a long

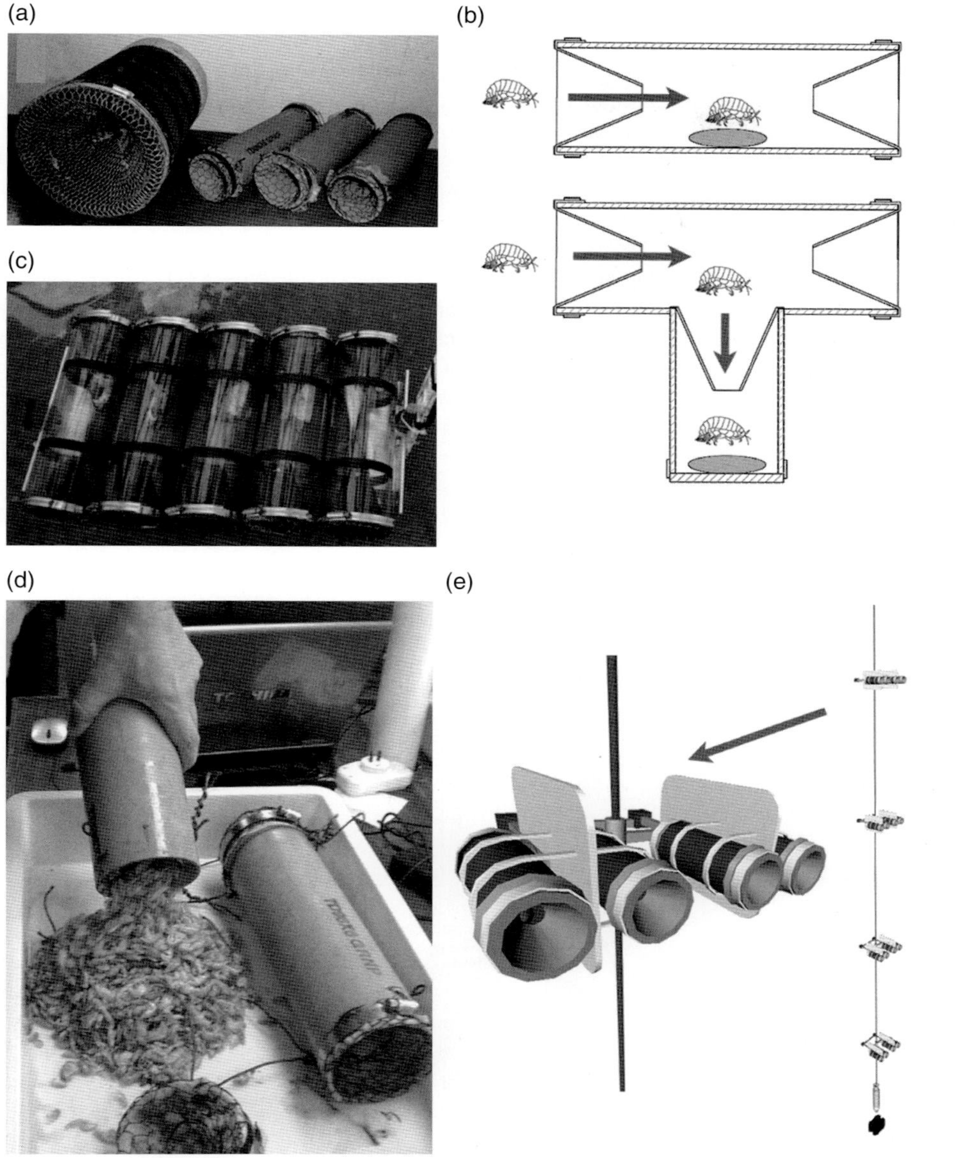

Fig. 11.6 Baited funnels for small invertebrates. (a) Examples of funnel traps where the larger one is 35 cm diameter and the smaller ones are 10 cm diameter. (b) Funnel trap principles: crustacean located trap by following the odour plume into the funnel to the bait (shaded) and is then unable to locate the exit, second chamber design also shown. (c) An array of transparent funnel traps for *in situ* observations. (d) Examples of a large catch of amphipods from just one trap at 9300 m. (d) Array of midwater traps with current vanes deployed on a mooring. (Oceanlab, University of Aberdeen, UK. Reproduced with permission.)

(tens or hundreds of metres) mooring line with flotation. Traps are secured at intervals on the mooring line. Some systems use twin traps or clusters of traps that are swivelled and orientated into the prevailing current using current vanes to increase the catch size (Ingram & Hessler, 1983; Charmasson & Calmett, 1987; Fig. 11.6).

The funnel trap principle can be enlarged to capture larger animals such as fish (Isaacs & Schick, 1960; Rowe *et al.*, 1986). These traps are typically deployed using the free-fall principle and comprise a large cuboid or trapezoidal mesh cage. Bait is placed internally to entice the fish inside.

In contrast to the small invertebrate funnel traps that can recover large numbers of specimens not likely to be caught in trawls, fish traps tend to recover extremely low numbers relative to trawling, however, the samples are in better condition. Low catches of deep-sea fish are probably due to their lack of hiding instinct (Jamieson *et al.*, 2006). This reluctance of fish to enter traps must be compensated by larger entrances. Although most fish traps used in the deep sea are single-chambered, multiple-chamber designs similar to those used in shallow waters are increasing in popularity. The multi-chamber traps work on the principle that the largest exit (the most likely to be found by the fish) from the first chamber is actually the entrance to another chamber and so on. The instinct to escape the trap thus drives the fish deeper into the trap and it becomes less likely to escape with every pass into another chamber. Three basic designs of fish trap are shown in Fig. 11.7. Most single-chamber traps are cuboid or trapezoidal and have two entrances at different heights. Staggering the position of the entrances both horizontally and vertically prevents the chance of a fish swimming

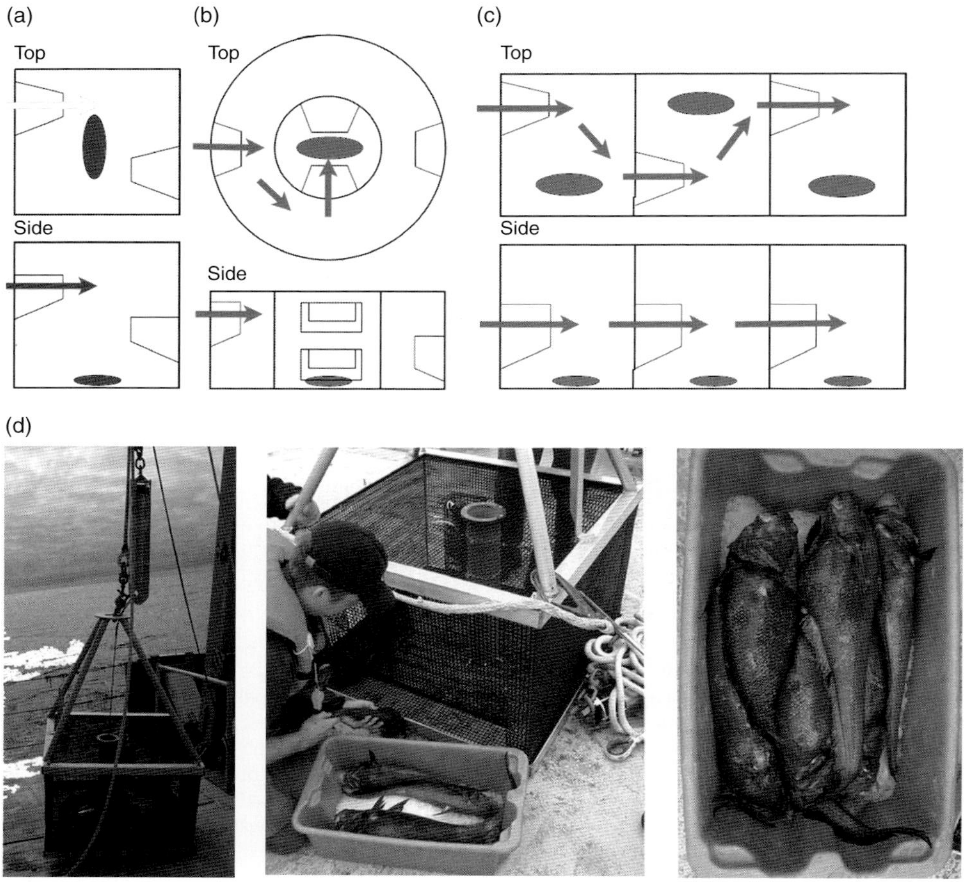

Fig. 11.7 Baited fish trap design. (a) Rectangular single-chamber fish trap. (b) Cylindrical double-chambered fish trap. (c) Rectangular multi-chamber fish trap. (d) A deep-sea fish trap in operation to 4200 m in the Pacific Ocean. (Oceanlab, University of Aberdeen, UK. Reproduced with permission.)

directly into one funnel and straight out of another. Double-chamber traps are similarly square or cylindrical, and the position of the funnels is similarly staggered. Multi-chambered traps are typically rectangular in design with three chambers. Some shallower multi-chambered traps have another funnel that directs the fish vertically into a net in the roof of the trap.

The sinking of whale bones and wood-falls is another baited technique that differs from conventional traps used to collect samples of deep-sea fauna. The deposition of whale bones or wood packages on the seafloor using the lander method simulates the naturally occurring, ephemeral, biogenic, chemosynthetic environments that are quickly colonized by specialized fauna such as the bone-eating *Osedax* and wood-boring worms (Rouse *et al.*, 2004). These localized habitats and the specialized species that are endemic to whale- and wood-falls are not likely to be successfully sampled using any other methodology (Smith & Baco, 2003). Furthermore, the likelihood of discovering a natural whale- or wood-fall is low, thus the lander method is favoured in order to study these phenomena. The method comprises segments of whale vertebrae or bundles of wood that are lashed to the frame of a simple lander delivery system. They are left on the seabed for at least a year before recovery. During the deployment time various species of worm will bore into the wood and *Osedax* will eat into the whale bones with ramifying roots. Both groups of organisms are sufficiently attached to survive the ascent intact (Glover *et al.*, 2013; Amon *et al.*, 2014).

11.2.1.5 Additional equipment

Most baited camera systems used in the deep sea do not carry imaging equipment exclusively. In order to model the odour plume emanating from the lander, current velocity and direction are measured using acoustic current meters. Two types of current meters are employed: the acoustic travel time principle (ATT) which measures a point source within the lander structure, or acoustic Doppler current meters which measure at a distance away from the lander structure. The Doppler method is generally preferred. Measuring current outside of the structure provides a more representative measurement unaffected by hydrodynamic disturbance from the lander itself. The addition of an acoustic Doppler current profiler (ADCP) can measure the current speed and direction at multiple vertical depth intervals above the baited camera. This addition provides a great understanding of the hydrography of the sampling site and can also be used to model the behaviour of the vehicle itself, especially when using vertically tethered trap arrays (discussed further below). The inclusion of such hydrographic data can be used to detect any effects on species abundance, interception times or the catch rates in traps. For example, discrepancies in these factors may be attributed to periods of extremely low current velocities which would result in a small odour plume and not necessarily indicate low population densities, an effect that may be misinterpreted in the absence of hydrographic data.

Other parameters such as pressure, temperature and salinity are measured to provide both a baseline environmental setting and in the case of pressure, an accurate sample depth. These sensors can be either independent or integrated into current meters/profilers. Pressure sensors can also be used to monitor the behaviour of baited cameras and traps. Pressure measurements that are converted to depth profiles during the descent and ascent of the lander will provide an accurate transit speed on which to base future deployments. Placing pressure sensors on each component of a tethered trap array can also provide an exact depth of capture.

11.2.2 Delivery system

11.2.2.1 Ballast and buoyancy

The basic delivery system of a lander vehicle relates to a frame (to mount the scientific payload), the ballast to achieve negative buoyancy on descent, a mechanism to jettison the ballast weight and a source of positive buoyancy to allow ascent to the surface. The following describes the basic design considerations of a lander delivery system.

Lander ballast weights operate by the same principle as deadweight anchors: by providing sufficient negative buoyancy to sink the lander and enough weight to prevent lateral motion in strong currents. There are typically four types of ballast configurations, all of which can be single or multiple:

- *Fixed ballast*: A fixed ballast arrangement is held directly onto the lander frame by clamps, toggles or in cradles. The advantages are that there is no additional coupling hardware required (shackles, swivels, etc.). However, care must be taken to prevent dissimilar metal corrosion while submerged.
- *Ballast footpads*: When using ballast (expendable) footpads, the vehicle is unlikely to suffer any resistance from the seabed when released due to suction. Specifically designed ballast footpads can be expensive to produce as they have to be of a particular dimension.
- *Suspended ballast*: Ballast is often suspended from the frame with releasable hooks. Suspending ballast allows its shape and size to vary depending on the deployment requirements; this enables less costly, irregular shaped, 'scrap' ballast to be used.
- *Tethered ballast*: When ballast is tethered under the frame, the lander itself can be held at the tether height in tension between the ballast and buoyancy. This allows relatively large areas of seafloor to be imaged from small landers without the need for oblique camera angles, wide-angle or macro lenses, thus simplifying image analysis and improving measurement. Landers with tethered ballasts have the advantage of remaining correctly orientated if landed on uneven ground and do not suffer the adverse effects of suction of seafloor snagging on lift off.

Ballast is expendable and therefore the use of inexpensive materials such as scrap iron (e.g. railway line), anchor chain, steel bars or bags of steel shot is commonplace. Generally the ballast is held on the frame by hooks, clamps or cradles secured to a release mechanism. There are three main types of instrument used on landers to release ballast weights:

- *Electromechanical drives*: Electromechanically driven releases are often triggered acoustically from the ship or by an onboard timer. Acoustic releases are used in most landers and have a closed jaw at one end and a hydrophone receiver at the other. Ballast can be held in the jaw and released when a unique acoustic command is sent from the ship. Acoustic releases will also permit simple two-way communication, allowing them to be tracked from the ship when descending or ascending (Bagley & Phillips, 1999).
- *Dissolvable links*: A simpler and cheaper method are dissolvable links (e.g. magnesium) that are designed to corrode in seawater. By changing the cross-sectional diameter of the link, the time it will take to fail can be accurately determined assuming ambient temperature, salinity and oxygen levels are known. Once the links have failed, the ballast is shed. However, dirt on the links and water characteristics (e.g. salinity, temperature and oxygen) can greatly affect timing.
- *Burn wires*: These are sacrificial weak links that drop the ballast when triggered. The corrosion of burn wires is controlled by a timing device.

Timed devices suffer a disadvantage in that the surfacing time for the lander is pre-determined. Unforeseen ship scheduling or adverse weather conditions may lead to the vehicle surfacing at an inopportune time. Most landers therefore use acoustic release systems. Two parallel acoustic releases are often used. They hold the ballast via a yoke, creating a back-up in the event that one should fail. One acoustic release may also be used with a timed release as a back-up.

The disposal of ballast raises ethical and environmental issues of introducing foreign and persistent structures into potentially unimpacted environments, and so use in sensitive habitats should be minimized if possible.

Buoyancy must be provided by static methods and not by inflatable devices since inflating bags or tanks at depth will result in severe loss of buoyancy due to the reduction in volume under pressure. There are two main types of flotation used to provide positive buoyancy:

- *Glass spheres*: Glass spheres provide good positive buoyancy per volume. However implosion at depth can occur due to manufacturing imperfections or if not carefully maintained (regular inspection of the glass to detect early signs of damage or fatigue).
- *Syntactic foam*: Syntactic foam has a greater weight and volume per unit buoyancy relative to glass spheres. However, it is extremely robust and incapable of subsurface implosion. Syntactic foam can also be manufactured in any shape to suit the vehicle, albeit at greater cost. It is, however, much heavier in air than glass and this can cause handling issues.

There are three main buoyancy configurations currently used in lander design: multiple tethered, single tethered and fixed (Fig. 11.8):

- *Multiple tethered buoyancy*: This can comprise either single or paired floats, spaced out on a mooring line to reduce the overall size and weight of the vehicle. Small float modules are easier to handle on deck and can be manually fed off a ship's stern. With multiple tethered moorings, floats can be added or subtracted in response to varying payload. In the event of a glass sphere failure at depth, multiple tethered buoyancy configurations are advantageous in preventing the implosion shock wave initiating a chain reaction implosion of other spheres in close proximity.
- *Single tethered buoyancy*: Configurations comprise one large mooring module suspended above the lander on a mooring line. Although a crane is required to deploy and recover this configuration, it reduces the amount of mooring hardware (shackles etc.), thus reducing the risk of mooring failure.
- *Fixed buoyancy*: This configuration comprises flotation fixed directly onto the lander frame. Although this increases the lander's weight in long-term operations, the risk of rope and shackle erosion and corrosion in seawater is reduced. A fixed buoyancy of glass spheres can be very

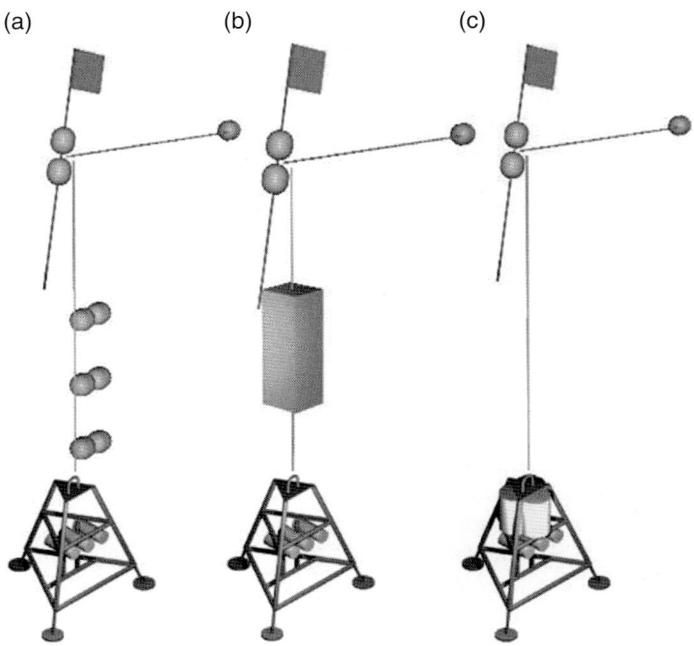

Fig. 11.8 Lander buoyancy configurations. (a) Multiple tethered, (b) single tethered, and (c) fixed buoyancy. (Oceanlab, University of Aberdeen, UK. Reproduced with permission.)

difficult to maintain properly due to the time and effort required to uncouple them from a frame and to disassemble, inspect and reassemble them.

Several devices are available to locate the lander when it surfaces and they are either mounted on the lander frame itself (when buoyancy is fixed) or on a dhan buoy (buoyancy module in which recovery aids are mounted). The devices used are a flag for short-range location, a strobe light for location in low visibility (or at night) and a VHF radio beacon that can be detected by a multi-band receiver on the ship. Both the strobe and the beacon can be pressure-deactivated when submerged to save battery power. Onboard recovery instruments such as directional finders have been used in adverse weather conditions or if there are positional errors. A directional finder can give the ship a bearing and range of the lander's position. Other devices, such as radar reflectors (detectable on a ship's radar) or satellite beacons (e.g. ARGOS, Iridium), have been used. The ARGOS beacons send a position to a satellite wherever and whenever it surfaces which will then notify the lander's registered user of its position. Satellite beacons are an advantage when operating long-term landers in the event of premature surfacing.

To retrieve the lander when surfaced, the recovery aids must be mounted above the water line. The lander also requires a 'lazy' float trailing a floating rope that operates by freely drifting downwind of the lander/dhan buoy. The floating rope becomes fully outstretched, thus providing a suitable target to grapple from the ship.

11.2.2.2 Structural design

Seawater is an aggressive substance that will undoubtedly affect all submerged structures and instruments to some extent (Shifler, 2005). Therefore, when selecting a material for a robust underwater structure such as a lander frame the most crucial factors are primarily mechanical properties and corrosion resistance. The main objective of the frame is to secure the instruments and protect them from damage and in order to do so, it must maintain its original properties for the duration of its intended life. The material selection also depends on the specific application of the component. There are many marine-grade materials that are suitable and often used for frame fabrication in short-term applications (e.g. aluminium 6082 or stainless steel 316) and long-term applications (titanium grade 2). These metals, however, are not strong enough to be used in pressure-resistant housings, where other grades such as aluminium 7075 T-6, stainless steel 255 and titanium grade 5 are used. These grades are far more expensive and therefore are not recommended for the frame fabrication. Using different grades for the frame and instrument housing results in dissimilar metals therefore it is common to insulate the two using plastic clamps or clamps with rubber insulators in order to prevent galvanic corrosion.

11.2.2.3 Marine fouling

The term 'marine fouling' describes the problem created by the attachment and growth of marine plants and animals on surfaces of operational importance (Abarzua & Jakubowski, 1995). Very few known materials are free from fouling when immersed in seawater. This biofouling consists of a thin, slimy secretion of polysaccharides, proteins, lipids and enzymes (Shams El Din *et al.*, 1996). This film of bacteria together with its binding material is referred to as 'biofilm', and is often a precursor to fouling by larger organisms. Structures deployed in the ocean for extended periods of time are vulnerable to biofouling (Guezennac *et al.*, 1998; Shifler, 2005). Biofouling is not easily preventable, nor predictable in the deep sea. Although biofouling can be very problematic in the photic zone and can lead to 'artificial reef' effects, this issue is less in the deep sea. Long-term lander structures are known to foul, and unidentified hydroids have colonized landers as deep as 3000 m over six months (personal observation). However, there is currently no evidence that long-term landers create artificial reefs (Vardaro *et al.*, 2007) that would alter events in the vicinity of bait.

11.2.2.4 Lander hydrodynamics

When in free-fall, ocean currents will affect the lander's descent and can carry it away from the desired position. A faster descent will minimize the distance that the lander could drift from position and in addition can save ship time, if it is being tracked to the seafloor. If the lander is travelling too fast it may cause extensive disruption to the landing site as well as possible damage to the lander. Typically a descent rate of about 0.5–1 ms^{-1} is acceptable. The terminal velocity (V_{limit}) of a falling body in water, such as a lander can be calculated using eqn 8 (from Tengberg *et al.*, 1995):

$$V_{limit} = \sqrt{\frac{2m_{water}g}{\rho_{water}SC_D}} \tag{8}$$

where V_{limit} is terminal velocity (ms^{-1}), m_{water} is lander mass in seawater (kg), g is gravity (ms^{-2}), ρ_{sw} is density of seawater (kgm^{-3}), S is characteristic projected area (m^2) and C_D is drag coefficient.

A lander's descent (or ascent) rate depends on the net buoyancy and the drag force. The drag force can be split into a projected area term (S; m^2) and a drag coefficient (C_D; unit-less), which depends upon the Reynolds number (*Re*) and the lander's structure. The drag coefficient can be assumed to be turbulent as the Reynolds numbers for such structures are estimated to be within a 5×10^5 to 1.5×10^6 range (Tengberg *et al.*, 1997). It is the Reynolds number (*Re*) that defines whether the water flow around the lander is laminar or turbulent (the upper limit between laminar and turbulent is $Re = 4000$). It is therefore considered acceptable to combine the C_D value with the characteristic projected area (S) of the lander, creating a single SC_D value (Hendricks & Rodenbush, 1981) which for an irregular object such as a lander is estimated at 2 (Tengberg *et al.*, 1997).

The lander mass in water (or net buoyancy) is controlled by the ballast–lander buoyancy ratio. On descent, the vehicle must be negatively buoyant (i.e. the weight in water of the ballast, combined with the lander's wet weight, must exceed the total positive buoyancy). After the ballast is released, the positive buoyancy must exceed the remaining weight of the lander. In practical terms, a baited camera only requires a negative buoyancy on descent of between 40 and 60 kg depending on ambient current regimes, particularly if using a tethered lander. The total positive buoyancy on ascent is normally around 60–80 kg. Calculating the terminal velocity of the lander using eqn 8 is sufficiently accurate to schedule operations. It must be noted that due to pressure-induced compression of the lander and changes in seawater density, the velocity of lander actually changes throughout the descent (or ascent). This decrease in velocity does not have a significant effect on typical deep-sea operation from 1000 to 5000 m since the time difference between using terminal velocity estimates and actually velocity is usually a matter of minutes. At full ocean depth, however, these subtle changes in velocity can result in the lander landing, or surfacing, up to 20 minutes late.

11.2.2.5 Non-lander systems

Conventional underwater imaging techniques are widely used for inshore, aquaculture and shelf applications for non-invasively observing, for example, fish, cetaceans and coral reefs (Done, 1981; Klimley & Brown, 1983; Petrell *et al.*, 1997). Known as BRUV (baited remote underwater video) or RUV (if non-baited) systems, these are commonly used to study reef fish assemblages in shallow water (<100 m). BRUV systems comprise a small metal frame which contains a camera, light(s), battery pack and bait (Cappo *et al.*, 2004, 2007; Malcolm *et al.*, 2007). They are deployed on a rope attached to a marker buoy on the surface and left to record autonomously. BRUV systems can be deployed with either vertical or horizontal cameras (Langlois *et al.*, 2006) although horizontal imaging is more common. Although some BRUV systems can be used with a live video feed to the surface vessel and/or stereo video cameras (Costa *et al.*, 2006; Langlois *et al.*, 2006) they are simply shallower water counterparts of a typical deep-sea baited camera lander minus the free-fall delivery system. There are recent instances when the deployment method of a surface line has been used as deep as 1500 m (Zintzen *et al.*, 2012).

The relevance of BRUV systems in a deep-sea context is their suitability for use with ROVs. Without the flotation, ballast and ballast-release system that are required on a lander, the BRUV systems are relatively small and light and can be placed into a specific position using the a ROV or submersible, such as next to a whale-fall, hydrothermal vent or cold seep; areas too precise to be targeted using the free-fall method. The ROV or submersible delivery method has already been used in, for example, the EITS (Eye-in-the-Sea) low-light camera system rated to 6000 m (Raymond & Widder, 2007).

11.3 Interpretation of data

11.3.1 Species composition

It is important to acknowledge when using baited traps and baited cameras that the recovered or observed fauna are only bait-attending fauna and not representative of the entire community. Furthermore, the bait-attending fauna are often referred to as scavenging fauna, which is somewhat misleading as not all bait-attending organisms are necrophagous. In fact, there are many bait-attending species that prey upon other organisms. In reality there are probably no or few obligate scavengers or predators but rather a spectrum between the two. With baited cameras there is also the issue of chance encounters whereby species which have no interest in scavenging or predation happen to simply swim through the field of view of the camera (e.g. members of the Halosauridae, Ipnopidae). Therefore, good background knowledge of each species and their behaviour is required to accurately interpret the results. There has been speculation that there are other detectable sources of attraction such as noise or bioluminescent activity from other species, but these are as yet unproven. Nevertheless, these are factors that would exist at a natural food-fall and are therefore not considered artefacts. Similarly, whether an observed or captured species is a scavenger or predator, it is representative of a natural event; it is simply how these data are interpreted which requires caution and/or clarity in explanation.

There are various measurable parameters that are routinely derived from baited cameras. These range from simple presence/absence, to theoretical abundance and biomass. Typically, each image of video sequence is played back manually and the presence of each organism in the image/sequence is identified, noted and recorded against time (often referred to as 'time after touchdown' (landing on the seafloor)). From these data, arrival times, arrival patterns, maximum numbers, time of maximum numbers and percentage of frames present are recorded for each species.

The simplest use of baited camera and trap data is to examine species composition. To estimate the total number of demersal fish species in the sample area, the rate of discovery of species with increasing number of deployments can be examined. Averaged species accumulation curves are calculated for the camera (and trap) data using a randomization procedure in which the sample order is randomized 100 times and the mean species richness determined for each value of n to produce a smoothed species accumulation curve (Colwell & Coddington, 1994; Gotelli & Colwell, 2001). This procedure permits comparisons of diversity between areas of similar depths (Fig. 11.9). Species accumulation curves can also be used to predict an asymptote to indicate the percentage of species already observed in an area (Fig. 11.9), an effective tool by which to manage baited camera and trap surveys. In areas where extensive trawling campaigns have been undertaken it is therefore possible to estimate the percentage of fish that would be expected to attend a baited camera. For example, Priede *et al.* (2010) found that in the Northeast Atlantic Ocean between 240 and 4865 m there is a predicted asymptote of 120 fish species of which 27 were bait-attending.

From an ecological perspective, baited trap data provide relative abundance and diversity of bait-attending fauna at a given depth and/or altitude above the seafloor (e.g. Ingram & Hessler, 1983; Christiansen *et al.*, 1990; Jamieson *et al.*, 2011). Complementing baited cameras with baited traps

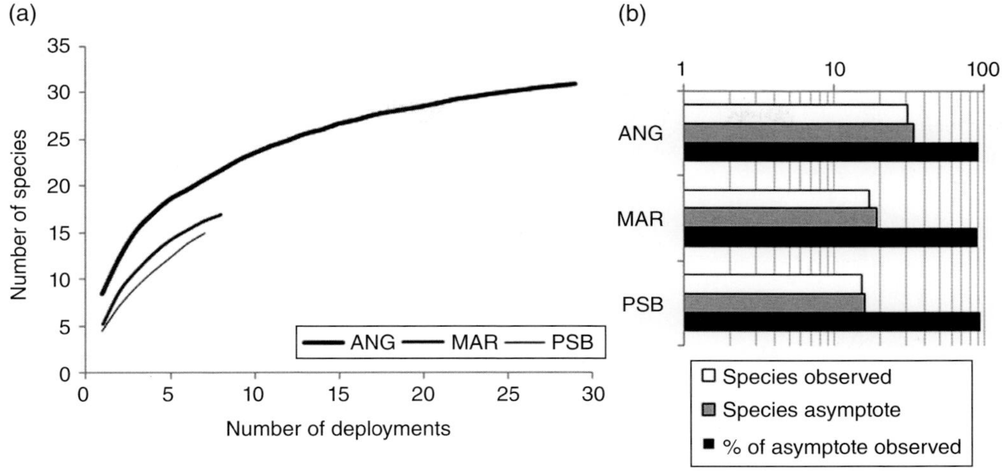

Fig. 11.9 (a) Species accumulation curves from three regions in the Atlantic Ocean between 1250 and 2500 m using a baited camera; off Angola (ANG), the Mid-Atlantic Ridge (MAR) and the Porcupine Seabight (PSB). (b) Actual number of observed species, the predicted asymptote and the percentage of that asymptote that has been observed. (Oceanlab, University of Aberdeen, UK. Reproduced with permission.)

enables the collection of smaller organisms that cannot be identified from camera images and this sampling provides other undetectable parameters such as sex ratio and ontogenetic stratification with depth or altitude (Ingram & Hessler, 1983; Perrone *et al.*, 2002; Thurston *et al.*, 2002).

11.3.2 Behaviour and interactions

The first arrival times, maximum numbers, time of maximum numbers and staying times of species attracted to baited cameras are important parameters in the interpretation of how, and the rate at which, food-falls are intercepted and consumed. From shallow to deep there is a general decrease in fish diversity and abundance (Priede *et al.*, 2006b). This trend is also observed in an increase in arrival time with depth (Fig. 11.10). Interpreting the arrival time and maximum numbers of a species present must be considered on an individual species basis. For example, species that are quintessentially scavengers, such as synaphobranchid eels, tend to arrive quickly (often within the first 5 or 10 minutes) and rapidly increase in numbers (often exhibiting a frenzied feeding bout) before tailing off over time, as the bait is consumed (Fig. 11.10). Predatory species can exhibit different patterns. Natant decapods often arrive in low numbers consistently throughout a deployment to quickly prey on small amphipods and leave. Other species, such as grazing zoarcid fish, arrive several hours into a deployment and gradually accumulate in numbers with time (Fig. 11.10). Thus zoarcid numbers often do not reach a maximum until very late, often days into a deployment (Witte, 1999; Kemp *et al.*, 2008). There are other species such as chimaeras and rays that often do not show any obvious pattern in their presence.

Baited cameras also permit the correlation of behaviour to environmental parameters as well as to time. For example, Thurston *et al.* (1995) reported peak numbers of natant decapods during periods of low current speeds, a relationship also observed by Domanski (1986). They concluded that these decapods utilize weak currents as low-energy mechanisms to search for food. There is also anecdotal evidence to suggest that the camera itself can create artefacts. Ryall and Hargrave (1984) reported that a non-baited camera attracted wreckfish (also visible in Fig. 11.4). The reason for this attraction is unresolved. Jamieson *et al.* (2006) reported that the macrourid *Coryphaenoides armatus*, a common scavenger in the deep sea, are temporarily distracted by the physical structure of a lander which resulted in many individuals exploring the lander structure rather than feeding, a behaviour which is somewhat contradictory to optimal foraging theory.

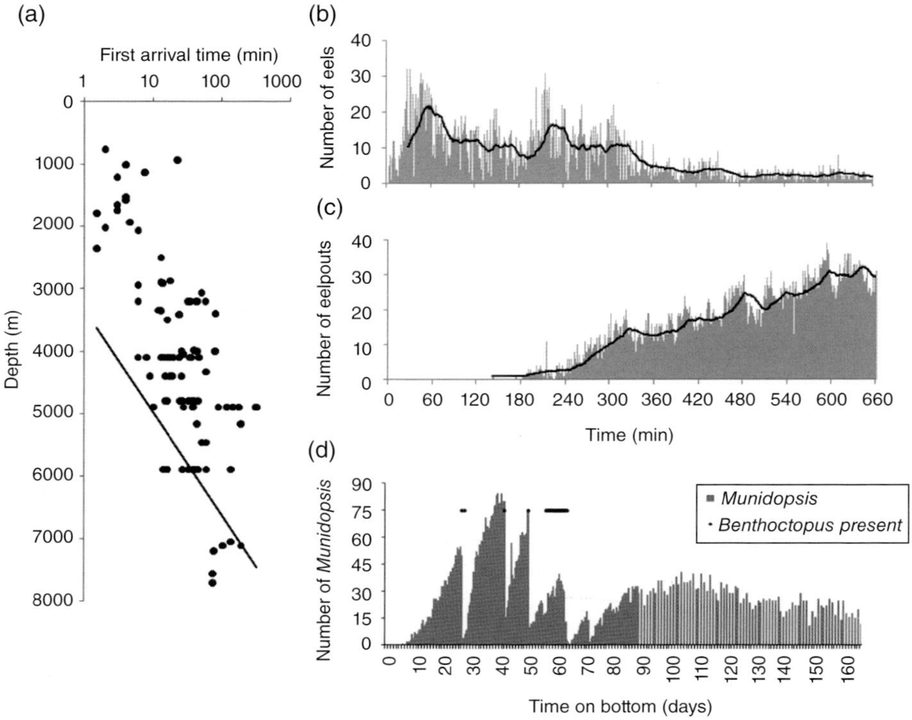

Fig. 11.10 (a) Global patterns of fish arrival times with depth (data from Priede *et al.*, 2006b). (b, c) typical but completely opposite arrival patterns of the scavenging eel *Synaphobranchus kaupii* (b) and the grazing predator *Pachycara crassiceps* (c). (d) Distinct species interactions at simulated food-falls. The number of squat lobsters *Munidopsis crassa* oscillates in response to predation by a benthic octopus. (Oceanlab, University of Aberdeen, UK. Reproduced with permission.)

Illumination may also affect behaviour. Bailey *et al.* (2005b) observed a slight decline in numbers of *Synaphobranchus kaupii* between the start and end of an illuminated sequence, suggesting that these eels have an aversion to light. This was later evident in a more natural setting when Heger *et al.* (2007) observed an interaction between the eels and the bioluminescent ostracod *Vargula norvegica*, and speculated there may be temporary blindness in the eels. Using a combination of a baited time-lapse stills camera, a baited low-light camera and baited traps they concluded that the bioluminescence emitted by the ostracods was a defensive reaction to mechanical stimulation by the eels, thus they were temporarily blinded while the ostracod escaped. Further species interactions have been observed using long-term baited cameras. Kemp *et al.* (2006) deployed a porpoise carcass to 2700 m. Once the bulk of the flesh had been removed by scavengers, the bones became populated by the squat lobster *Munidopsis crassa* after 5 days. Periodically the numbers of squat lobster rapidly decreased, which coincided with the presence of a deep-sea benthic octopus (Fig. 11.10).

These are just some examples of the types of behaviour and interactions that would not be observed using other conventional sampling methodologies.

11.3.3 Measurements from still images

In situ imaging lends itself to the measurement of individual species (Fig. 11.11). With digital photography and accessible image analysis software, morphological measurements can be made quickly with relative ease. Vertical photography is a method that provides the most accuracy as most fish and

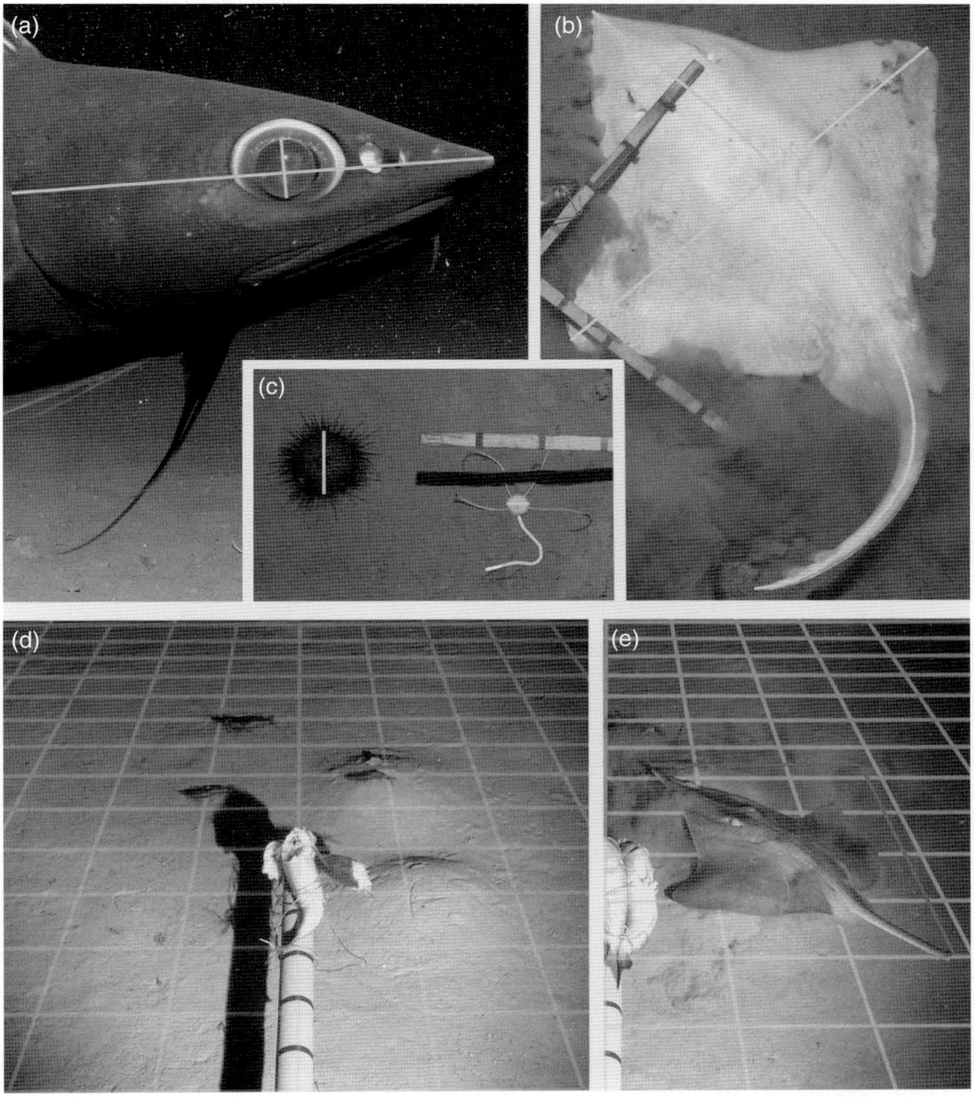

Fig. 11.11 Digital measurements of deep-sea fauna from baited camera images. (a) Relative (%) measurements of eye diameter to head length from horizontal images. (b) Absolute body measurements from vertical photography. (c) Measurements of epifauna such as echinoid body diameter and ophuiroid arm length and disc diameter (the red line indicates the calibration measurement from the *in situ* scale bar). (d, e) Measurements from oblique images using a perspective grid; direct body lengths of natant decapods (d) and body lengths of a deep-water ray using trigonometry (e). (Oceanlab, University of Aberdeen, UK. Reproduced with permission.) (Please see Plate 19 for colour representation of the figure.)

epifauna are observed perpendicular to the camera plane. Placing a scale bar in the field of view provides the calibration measurement. Objects and epifauna in oblique images can be measured using perspective grid techniques, although there is a significant increase in error as the distance from the subject moves away from the bottom of the frame. The perspective grid provides a method which is more applicable to measuring species directly on the seafloor.

Animals such as fish swimming above bottom cannot be accurately measured unless using stereophotography, a technique not commonly used in the deep sea because of the cost of such cameras and associated lighting arrays (and battery requirements). Horizontal images are more accurate for measuring fish than oblique images since fish are often observed perpendicular to the camera plane. However, absolute measurements cannot be made from horizontal images but relative measurements, such as head:body length ratio, can be determined to aid in identification. Digital measurements made from images have the advantage that they can be made non-destructively and include the animal in whole, not always possible from trawl samples due to damage. The disadvantage is in the degree of accuracy. Even in vertical images the subjects are unlikely to be at perfect right angles to the camera or at equal distances from the seafloor. Therefore, to avoid false precision, fish, for example, are typically measured to the nearest centimetre. When compiling length-frequency data there is the issue of duplication of individuals. To overcome this it is good practice to only measure individuals that can be distinguished from one another (by size, colouring, external markings, ectoparasites, etc). In the case of aggregating species, measurements are made from the image with the maximum number of visible individuals. Where there are transient species present in high numbers, for example eels or hagfish, measurements are made at 30 or 60 minute intervals to avoid duplication.

Body length measurements can be converted to mass using either previously published length–weight relationship data (e.g. Wenner & Musick, 1977; Gordon & Duncan, 1987) or voucher specimens from trawl samples or from baited traps.

11.3.4 Measurements from video images

Video sequences are useful for observing movements and behaviour that are not detectable in still images. For example, observations of suction feeding have been noted in liparids from hadal depths (Jamieson *et al.*, 2009b) and liparids exhibiting periods of inexplicable inactivity (Fujii *et al.*, 2010). Within these sequences are events where large amphipods are seen to temporarily grab hold of the fish tails, initiating a tail flip response. It is also possible to take physiological measurements such as tail-beat frequencies and swimming speeds of fish and crustaceans from video sequences (Collins *et al.*, 1999; Priede *et al.*, 2003; Jamieson *et al.*, 2009a, 2009b). Swimming speeds (and locomotion speed of epifauna) can be measured from video using frame-by-frame playback. A point on the organism is measured against a known point on the seafloor. For tail-beat frequencies, video recordings are played back frame-by-frame and the number of frames (i.e. time) per complete tail-beat (contractions of both sides of the fish) are calculated. Analysis of tail-beat frequency permits the physiological comparison of different species (Collins *et al.*, 1999) or the same species at different depths or environmental conditions (Priede *et al.*, 2003). The same method can be used to measure the synchrony between caudal and pectoral fins (Jamieson *et al.*, 2009b).

Asides from routine swimming speeds, organisms are often filmed exhibiting burst swimming (or 'fast-start' escape responses). Burst swimming speeds can be approximately calculated at typical frame rates (25 Hz; Jamieson *et al.*, 2009a). However, this frame rate is often too low to accurately measure such fast events, thus the use of high-speed video cameras for this application is favoured. Since the likelihood of capturing a burst swimming event with a baited camera is very low, baited cameras with integrated electrical stimulator, such as the 'SPRINT' lander have been developed (Bailey *et al.*, 2003, 2005a). The 'SPRINT' lander produces an electrical pulse at regular intervals to incite a burst swimming response that is recorded by the camera.

11.3.5 Theoretical abundance and biomass

Beyond the study of individual organisms lies the challenge of estimating the species' abundance in a given area. Baited cameras are often used to provide estimates of abundance where other quantitative techniques such as trawling are impossible (Priede *et al.*, 1990; Priede & Merrett, 1996, 1998; Priede & Bagley, 2000; Bailey & Priede, 2002; Collins *et al.*, 2002). Population abundance estimates are interpreted from assumptions about foraging behaviour, sensory abilities, scavenger distribution, swimming speeds, hydrographical regimes and odour plume dispersal. The methods assume the following: (i) the odour plume from the bait disperses downstream and either arrives at a stationary animal, or (ii) an animal swims into the odour plume while actively searching for food (Priede & Merrett, 1996; Bailey & Priede, 2002). In reality, however, there is probably a species-specific continuum between these two extremes (Farnsworth *et al.*, 2007). Rheotaxis is triggered on contact with the plume, and the animal follows the plume up-current to the bait and either feeds until all bait is consumed or leaves sooner if better feeding opportunities are perceived to exist elsewhere (optimal foraging theory; Charnov, 1976) or the risk of predation too great.

When estimating the abundance of a species, the challenge lies in there being adequate knowledge of these assumptions and that the animal behaves in accordance with optimal foraging theory. There are, however, many bait-attending species that do not behave in this manner. Some species aggregate in large numbers in order to prey upon smaller crustaceans. Other species arrive at the bait, feed and leave, while many other species exhibit no predictable behaviour. There is also great inter- and intra-species variation in the length of time for which a species remains at the bait, and this in turn can be influenced by interactions with other species.

The models for abundance estimation are therefore only applicable to the most dominant and ubiquitous scavenging species, such as the grenadier *Coryphaenoides armatus*, of which there are also sufficient trawl data for comparison. The absence of extensive and conclusive behavioural data for most other deep-sea scavengers has accurate estimates of abundance using baited cameras are lacking for most species.

The model proposed by Priede *et al.* (1990) for estimating scavenging fish abundance is based on first arrival times at the bait and the dispersal of fish after feeding. The model produced abundance estimates close to those of otter trawls from the same locations (Fig. 11.12; Priede & Merrett, 1996). This model forms the basis of abundance estimates for most deep-sea baited camera surveys (Henriques *et al.*, 2002; King *et al.*, 2006, 2008; Jamieson *et al.*, 2009b).

The relationship between arrival times (t_{arr}) and fish abundance (N; km^{-2}) can be defined as:

$$A = 60\ t_{arr} / \left(1/V_w + 1/V_f\right) \quad (9)$$

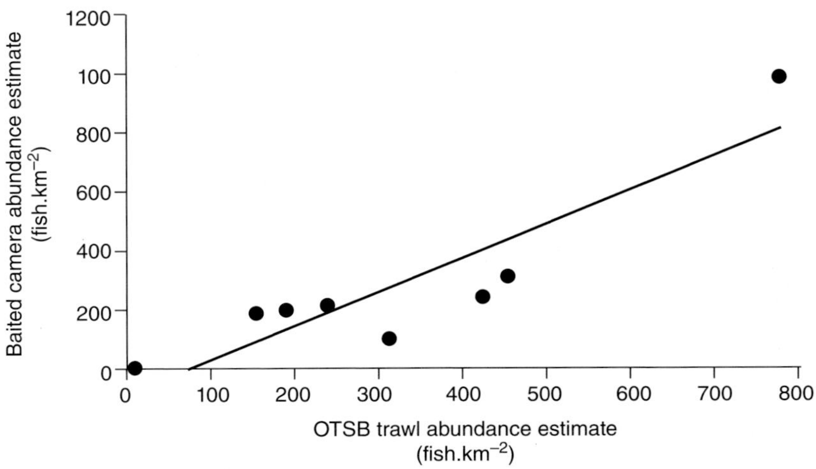

Fig. 11.12 Comparison of theoretical abundance estimation from baited cameras and semi-ballon otter trawl (OTSB) trawls from the same time and location, ($y = 0.97x$, correlation coefficient = 0.86, $p = 0.01$) (from Bailey *et al.*, 2007). (Oceanlab, University of Aberdeen, UK. Reproduced with permission.)

where A is the mean area occupied by one fish (m^{-2}) and v_w is the water velocity of the current dispersing odour over the seafloor (ms^{-1}) and v_f is the velocity of the fish moving upstream toward the bait following odour interception. Then:

$$N = 10^6/3A^2 \tag{10}$$

assuming $v_w = v_f = 0.05$ ms^{-1}, typical of most deep-water localities (Priede *et al.*, 1991).

Farnsworth *et al.* (2007) suggested the use of a stochastic model as an alternative to deterministic models. However, such a model has not yet undergone sufficient field trials or statistical analysis to determine its appropriateness (Bailey *et al.*, 2007).

Once the length frequency of a species is known and theoretical abundance has been calculated, the biomass of the species in a given area can be calculated and a comparison made between different geographical areas (Henriques *et al.*, 2002; King *et al.*, 2006, 2008).

11.4 Future developments

The resolution, physical compactness and storage capacities of digital still and video cameras are ever increasing. The commercial developments inevitably filter into scientific applications which have recently seen the introduction of still photography in excess of 14 megapixel resolution and HD video. The area requiring immediate attention, however, is the issue of replication which baited camera and traps have always lacked. The financial implications of deploying multiple baited cameras simultaneously currently renders this method unrealistic therefore the development of smaller, low-cost versions must be pursued. By miniaturizing the delivery system in conjunction with smaller imaging technology, as described by Hardy *et al.* (2002), multiple systems can be deployed in an area simultaneously and significantly increase sample sizes, replication and the number of images. This technique of deploying multiple, smaller, low-cost systems has revolutionized ocean bottom seismometry with the use of compact ocean bottom systems in an array, and physical oceanography with the use of lagrangian floats with periodic profiling capabilities. In short-term applications, increased replication will provide greater spatial resolution, either across larger areas such as abyssal plains or across a bathymetric range such as the continental slope to hadal trench. In long-term applications, these arrays will achieve the same advantages but in a fraction of the time it would have other taken with single instruments.

Miniaturizing baited traps and cameras will rely on the scaling down of the delivery system as it is the ballast-release system and flotation which account for the relatively large size and often the majority of the cost of the vehicle, especially with traps.

Although both baited cameras and baited traps have their advantages and disadvantages, the complementary capacities of simultaneous operations, coupled with increased replication, will provide synergies that will revolutionize our current interpretation of the diversity, abundance and the role of the scavenging community the deep sea.

Acknowledgements

The following are thanked for their supply of images and helpful discussions; Toyonobu Fujii and Thom Linley (Oceanlab, University of Aberdeen, UK), David Bailey (University of Glasgow, UK), Adrian Glover (Natural History Museum, London, UK), Kirsty Kemp (Institute of Zoology, London, UK), Daniel Jones (National Oceanography Centre, Southampton, UK) and Vincent Zintzen (Museum of New Zealand Te Papa Tongarewa, New Zealand). Many thanks to Rachel Jamieson for reading and greatly improving the text.

References

Abarzua, S., Jakubowski, S. (1995). Biotechnological investigation for the prevention of biofouling. I. Biological and biochemical principles for the prevention of biofouling. *Marine Ecology Progress Series*, 123: 301–312.

Amon, D.J., Copley, J.T., Dahlgren, T.G., *et al.* (2014). Three new species and an expansion of range for the genus Osedax (Annelida: Siboglinidae) in the deep Antarctic. *Zoologica Scripta*, 43(4): 405–417.

Bagley, P.M., Phillips, G. (1999). Acoustic release control and display equipment (ARCADE). In: Yau, C. (ed), *ALIPOR Final Report*. Aberdeen: University of Aberdeen, pp. 23–33.

Bailey, D.M., Priede, I.G. (2002). Predicting fish behaviour in response to abyssal food-falls. *Marine Biology*, 141: 831–840.

Bailey, D.M., Bagley, P.M., Jamieson, A.J., Collins, M.A., Priede, I.G. (2003). *In situ* investigation of burst swimming and muscle performance in the deep-sea fish *Antimora rostrata* (Günther, 1878). *Journal of Experimental Marine Biology and Ecology*, 285–286: 295–311.

Bailey, D.M., Bagley, P.M., Jamieson, A.J., *et al.* (2005a). Life in a warm deep sea: routine activity and burst swimming performance of the shrimp *Acanthephyra eximia* in the abyssal Mediterranean. *Marine Biology*, 146: 1199–1206.

Bailey D.M., Genard, B., Collins, M.A., *et al.* (2005b). High swimming and metabolic activity in the deep-sea eel *Synaphobranchus kaupii* revealed by integrated in situ and in vitro measurements. *Physiological and Biochemical Zoology*, 78: 335–346.

Bailey, D.M., King, N.J., Priede, I.G. (2007). Camera and carcasses: historical and current methods for using artificial food-falls to study deep-water animals. *Marine Ecology Progress Series*, 350: 179–191.

Barnes, H. (1955). Underwater television and research in marine biology, bottom topography and geology. *Ocean Dynamics*, 8: 213–236.

Barry, J.P., Drazen, J.C. (2007). Response of deep-sea scavengers to ocean acidification and odor from a dead grenadier. *Marine Ecology Progress Series*, 350: 193–207.

Blankenship, L.E., Yayanos, A.A., Cadien, D.B., Levin, L.A. (2006). Vertical zonation patterns of scavenging amphipods from the hadal zone of the Tonga and Kermadec Trenches. *Deep-Sea Research I*, 53: 48–61.

Cappo, M., Speare, P., De'ath, G. (2004). Comparison of baited remote underwater video stations (BRUVS) and prawn (shrimp) trawls for assessments of fish biodiversity in inter-reefal areas of the Great Barrier Reef Marine Park. *Journal of Experimental Marine Biology and Ecology*, 302: 123–152.

Cappo, M., De'ath, G., Speare, P. (2007). Inter-reef vertebrate communities of the Great Barrier Reef Marine Park determined by baited remote underwater video stations. *Marine Ecology Progress Series*, 350: 209–221.

Charmasson, S.S., Calmett, D.P. (1987). Distribution of scavenging Lysianassidae amphipods *Eurythenes gyllus* in the northeast Atlantic: comparison with studies held in the Pacific. *Deep-Sea Research*, 34: 1509–1523.

Charnov, E.L. (1976). Optimal foraging, the marginal value theorem. *Theoretical Population Biology*, 9: 129–136.

Christiansen, B., Pfannkuche, O., Thiel, H. (1990). Vertical distribution and population structure of the necrophagous amphipod *Eurythenes gryllus* in the West European Basin. *Marine Ecology Progress Series*, 66: 35–45.

Collins, M.A., Priede, I.G., Bagley, P.M. (1999). In situ comparison of activity of two deep-sea scavenging fish occupying different depths zones. *Proceedings of the Royal Society, Series B*, 266: 2011–2016.

Collins, M.A., Yau, C., Guillfoyle, F., *et al.* (2002). Assessment of stone crab (Lithodidae) density on the South Georgia slope using baited video cameras. *ICES Journal of Marine Science*, 59: 370–379.

Colwell, R.K., Coddington, J.A. (1994). Estimating terrestrial biodiversity through extrapolation. *Philosophical Transactions of the Royal Society B*, 345: 101–118.

Costa, C., Loy, A., Cataudella, S., Davis, D., Scardi, M. (2006). Extracting fish size using dual underwater cameras. *Aquacultural Engineering*, 35: 218–227.

Domanski, P. (1986). The near-bottom shrimp faunas (Decapoda: Natantia) at two abyssal sites in the Northeast Atlantic Ocean. *Marine Biology*, 93: 171–180.

Done, T.J. (1981). Photogrammetry in coral reef ecology: a technique for the study of change in coral reef communities. In: *Proceedings of the Fourth International Coral Reef Symposium*, Manila. pp. 315–320.

Douglas, R.H., Partridge, J.C., Hope, A.J. (1995). Visual and lenticular pigments in the eyes of demersal deep-sea fishes. *Journal of Comparative Physiology A*, 177: 111–122.

Farnsworth, K.D., Thygesen, U.H., Ditlevsen, S., King, N.J. (2007). How to estimate scavenger fish abundance using baited camera data. *Marine Ecology Progress Series*, 350, 223–234.

Frank, T.M., Case, J.F. (1988). Visual spectral sensitivities of bioluminescent deep-sea crustaceans. *Biological Bulletin*, 175: 261–273.

Fujii, T., Jamieson, A.J., Solan, M., Bagley, P.M., Priede, I.G. (2010). A large aggregation of liparids at 7703 m depth and a reappraisal of the abundance and diversity of hadal fish. *BioScience*, 60: 506–515.

Gillibrand, E.J.V., Bagley, P.M., Jamieson, A.J., *et al.* (2006). Deep sea benthic bioluminescence at artificial food falls, 1000 to 4800m depth, in the Porcupine Seabight and Abyssal Plain, North East Atlantic Ocean. *Marine Biology*, 150: 1053–1060.

Glover, A.G., Wiklund, H., Taboada, S., *et al.* (2013). Bone-eating worms from the Antarctic: the contrasting fate of whale and wood remains on the Southern Ocean seafloor. *Proceedings of the Royal Society Part B*, 280, 20131390.

Gordon, J.D.M., Duncan, J.A.R. (1987). Deep-sea bottom living fishes at two repeat stations at 2200m and 2900 m in the Rockall Trough, northeastern Atlantic Ocean. *Marine Biology*, 96: 309–325.

Gotelli, N.J., Colwell, R.K. (2001). Quantifying biodiversity: procedures and pitfalls in the measurement and comparison of species richness. *Ecology Letters*, 4: 379–391.

Guezennac, J., Ortega-Morales, O., Raguenes, G., Geesey, G. (1998). Bacterial colonization of artificial substrate in the vicinity of deep-sea hydrothermal vents. *Microbiology Ecology*, 26: 89–99.

Hardy, K., Olsson, M., Yayanos, A.A., Prsha, J., Hagey, W. (2002). Deep ocean visualisation experimenter (DOVE): low-cost 10km camera and instrument platform. *Oceans 2002 Conference Proceedings*, MTS/IEEE-OES, Washington DC, pp. 2390–2394.

Hargrave, B.T., Phillips, G.A., Prouse, N.J., Cranford, P.J. (1995). Rapid digestion and assimilation of bait by the deep-sea amphipod *Eurythenes gryllus*. *Deep-Sea Research I*, 42: 1905–1921.

Heger, A., King, N.J., Wigham, B.D., *et al.* (2007). Bioluminescent aggregations of *Vargula norvegica* (Ostracoda) at artificial food falls at 1000m depth in carbonate mound provinces of the NE Atlantic. *Marine Biology*, 151: 1471–1478.

Hendricks P.J., Rodenbush, G. (1981). Interpretation of velocity profiles measured by freely sinking probes. *Deep-Sea Research*, 28: 1199–1213.

Henriques, C., Priede, I.G., Bagley, P.M. (2002). Baited camera observations of deep-sea demersal fishes of the Northeast Atlantic Ocean at 15°–28°N off West Africa. *Marine Biology*, 141: 307–314.

Herring, P.J., Gaten, E., Shelton, P.M.J. (1999). Are vent shrimps blinded by science? *Nature*, 398: 116.

Hessler, R.R., Ingram, C.L., Yayanos, A.A., Burnett, B.R. (1978). Scavenging amphipods from the floor of the Philippine Trench. *Deep-Sea Research*, 25: 1029–1047.

Hiller-Adams, P., Widder, E.A., Case, J.F. (1988). The visual pigments of four deep-sea crustacean species. *Journal of Comparative Physiology A*, 163: 63–72.

Ingram, C.L., Hessler, R.R. (1983). Distribution and behavior of scavenging amphipods from the central North Pacific. *Deep-Sea Research*, 30: 683–706.

Isaacs, J.D., Schick, G.B. (1960). Deep-sea free instrument vehicle. *Deep-Sea Research*, 7(1): 61–67.

Isaacs, J.D., Schwartzlose, R.A. (1975). Active animals of the deep sea floor. *Scientific American*, 233(4): 84–91.

Jamieson, A.J., Bagley, P.M. (2005). The ROBIO and DOBO landers: Deep-sea biodiversity surveys in areas of anthropogenic activity. *Sea Technology*, 46: 54–57.

Jamieson, A.J., Bailey, D.M., Wagner, H.-J., Bagley, P.M., Priede, I.G. (2006). Behavioural responses to structures on the seafloor by the deep-sea fish *Coryphaenoides armatus*: Implications for the use of baited landers. *Deep-Sea Research I*, 53: 1157–1166.

Jamieson, A.J., Fujii, T., Solan, M., Matsumoto, A.K., Bagley, P.M., Priede, I.G. (2009a). First findings of decapod crustacea in the hadal-zone. *Deep-Sea Research I*, 56: 641–647.

Jamieson, A.J., Fujii, T., Solan, M., Matsumoto, A.K., Bagley, P.M., Priede, I.G. (2009b). Liparid and Macrourid fishes of the hadal zone: in situ observations of activity and feeding behaviour. *Proceedings of the Royal Society B*, 276: 1037–1045.

Jamieson, A.J., Kilgallen, N.M., Rowden, A.A., *et al.* (2011). Bait-attending fauna of the Kermadec Trench, SW Pacific Ocean: evidence for an ecotone across the abyssal-hadal transition zone. *Deep-Sea Research I*, 58: 49–62.

Jeffreys, R.M., Lavaleye, M.S.S., Bergman, M.J.N., Duinevald, G.C.A., Witbaard, R., Linley, T. (2010). Deep-sea macrourids fishes scavenge on plant material: evidence from in situ observations. *Deep-Sea Research I*, 57: 621–627.

Jones, D.O.B., Bett, B.J., Wynn, R.B., Masson, D.G. (2009). The use of towed camera platforms in deep-water science. *Underwater Technology*, 28: 41–50.

Jones, E.G., Tselepides, A., Bagley, P.M., Collins, M.A., Priede, I.G. (2003). Bathymetric distribution of some benthic and benthopleagic species attracted to baited cameras and traps in the deep eastern Mediterranean. *Marine Ecology Progress Series*, 251: 75–86.

Jones, E.J., Collins, M.A., Bagley, P.M., Addison, S., Priede, I.G. (1998). The fate of cetacean carcasses in the deep sea: observations on consumption rates and succession of scavenging species in the abyssal north-east Atlantic Ocean, *Proceedings of the Royal Society B*, 265: 1119–1127.

Kemp, K.M., Jamieson, A.J., Bagley, P.M., *et al.* (2006). Consumption of a large bathyal food fall, a six month study in the north-east Atlantic. *Marine Ecology Progress Series*, 310: 65–76.

Kemp, K.M., Jamieson, A.J., Bagley, P.M., Collins, M.A., Priede, I.G. (2008). A new technique for sequential periodic bait-release at a camera platform in the deep sea; trial at 3664m depth in the Charlie Gibbs Fracture Zone, Mid-Atlantic Ridge. *Deep Sea Research I*, 55: 218–228.

King, N.J., Bagley, P.M., Priede, I.G. (2006). Depth zonation and latitudinal distribution of deep sea scavenging demersal fishes of the Mid-Atlantic Ridge, 42°–53°N. *Marine Ecology Progress Series*, 319: 263–274.

King, N.J., Jamieson, A.J., Bagley, P.M., Priede, I.G. (2008). Deep-sea scavenging demersal fish fauna of the Nazaré Canyon system, Iberian coast, Northeast Atlantic Ocean. *Journal of Fish Biology*, 72: 1804–1814.

Klages, M., Vopel, K., Bluhm, H., Brey, T., Soltwedel, T., Arntz, W.E. (2001). Deep-sea food falls: first observation of a natural event in the Arctic Ocean. *Polar Biology*, 24: 292–295.

Klages, M., Muyakshin, S., Soltwedel, T., Arntz, W.E. (2002). Mechanoreception, a possible mechanism for food fall detection in deep-sea scavengers. *Deep-Sea Research I*, 49: 143–155.

Klimley, A.P., Brown, S.T., (1983). Stereophotography for the field biologist: measurement of lengths and three-dimensional positions of free swimming sharks. *Marine Biology*, 74: 175–185.

Lampitt, R.S., Burnham, M.P. (1983). A free fall time-lapse camera and current meter system 'Bathysnap' with notes on the foraging behaviour of a bathyal decapod shrimp. *Deep-Sea Research*, 30: 1009–1017.

Lampitt, R.S., Merrett, N.R., Thurston, M.H. (1983). Inter-relations of necrophagous amphipods, a fish predator, and tidal currents in the deep-sea. *Marine Biology*, 74: 73–78.

Langlois, T., Chabanet, P., Pellerier, D., Harvey, E. (2006). Baited underwater video for assessing reef fish populations in marine reserves. *SPS Fisheries Newsletter*, 118: 53–57.

Lawson, G.S., Tyler, P.A., Young, C.M. (1993). Attraction of deep-sea amphipods to macrophyte food falls. *Journal of Experimental Marine Biology and Ecology*, 169: 33–39.

Løkkeborg, S., Olla, B.L., Pearson, W.H., Davis, M.W. (1995). Behavioural responses of sablefish, *Anoplopoma fimbria*, to bait odour. *Journal of Fish Biology*, 46: 142–155.

Malcolm, H.A., Gladstone, W., Lindfield, S., Wraith, J., Lynch, T.P. (2007). Spatial and temporal variation in reef fish assemblages of marine parks in New South Wales, Australia – baited video observations. *Marine Ecology Progress Series*, 350: 277–290.

Menzies, R.J., Zaneveld, J.S., Pratt, R.M. (1967). Transported turtle grass as a source of organic enrichment of abyssal sediments off North Carolina. *Deep-Sea Research*, 14: 111–112.

Palacios, C., Zbinden, M., Baco, A.R., *et al.* (2006). Microbial ecology of deep-sea sunken wood: quantitative measurements of bacterial biomass and cellulolytic activities. *Cahiers de Biologie Marine*, 47: 415–420.

Partridge, J.C., Archer, S.N., Lythgoe, J.N. (1988). Visual pigments in the individual rods of deep-sea fishes. *Journal of Comparative Physiology A*, 162: 543–550.

Paul, A.Z. (1973). Trapping and recovery of living deep-sea amphipods from the Arctic Ocean floor. *Deep-Sea Research*, 20: 289–290.

Perrone, F. M., Dell'Anno, A., Danovaro, R., Della Croce, N., Thurston, M.H. (2002). Population biology of *Hirondellea* sp nov (Amphipoda: Gammaridea: Lysianassoidea) from the Atacama Trench (south-east Pacific Ocean), *Journal of the Marine Biological Association of the UK*, 82: 419–425.

Petrell, R.J., Shi, X., Ward, R.K., Naiberg, A., Savage, C.R. (1997). Determining fish size and swimming speed in cages and tanks using simple video techniques. *Aquacultural Engineering*, 16: 63–84.

Pfannkuche, O., Linke, P. (2003). GEOMAR lander as long-term deep-sea observatories. *Sea Technology*, 49(9): 50–55.

Priede, I.G. (1980). Use of satellites in marine biology. In: MacDonald, A.G., Priede, I.G. (eds), *Experimental biology at sea*. London: Academic Press, pp. 3–50.

Priede, I.G. (1999). *Autonomous lander instrument packages for oceanographic research (ALIPOR): final report*. University of Aberdeen, Aberdeen.

Priede, I.G., Bagley, P.M. (2000). *In situ* studies on deep-sea demersal fishes using autonomous unmanned lander platforms. *Oceanography and Marine Biology: an Annual Review*, 38: 357–392.

Priede, I.G., Merrett, N.R. (1996). Estimation of abundance of abyssal demersal fishes; a comparison of data from trawls and baited cameras. *Journal of Fish Biology*, 49 (suppl. A): 207–216.

Priede, I.G., Merrett, N.R. (1998). The relationship between numbers of fish attracted to baited cameras and population abundance: studies on demersal grenadiers *Coryphaenoides (Nematonurus) armatus* in the abyssal NE Atlantic Ocean. *Fisheries Research*, 36: 133–137.

Priede, I.G., Smith, K.L., Armstrong, J.D. (1990). Foraging behaviour of abyssal grenadier fish: inferences from acoustic tagging and tracking in the North Pacific Ocean. *Deep-Sea Research*, 37: 81–101.

Priede I.G., Bagley P.M., Armstrong J.D., Smith, K.L., Merrett, N.R. (1991). Direct measurement of active dispersal of food-falls by abyssal demersal fishes. *Nature*, 351: 647–649.

Priede, I.G., Deary, A.R., Bailey, D.M., Smith, K.L. (2003). Low activity and seasonal change in population size structure of grenadiers in oligotrophic abyssal North Pacific Ocean. *Journal of Fish Biology*, 63: 187–196.

Priede, I.G., Bagley, P.M., Way S., Herring, P.J., Partridge, J.C. (2006a). Bioluminescence in the deep sea: Free-fall lander observations in the Atlantic Ocean off Cape Verde. *Deep-Sea Research I*, 53: 1272–1283.

Priede, I.G., Froese, R., Bailey, D.M., *et al.* (2006b). The absence of sharks from abyssal regions of the world's oceans. *Proceedings of the Royal Society B*, 273: 1435–1441.

Priede, I.G., Godbold, J.A., King, N., Collins, M.A., Bailey, D.M., Gordon, J.D.M. (2010). Deep-sea demersal fish biodiversity of the Porcupine Seabight, NE Atlantic Ocean: Global trends versus local heterogeneity. *Marine Ecology*, 31: 247–260.

Raymond, E.H., Widder, E.A. (2007). Behavioral responses of two deep-sea fish species to red, far-red, and white light. *Marine Ecology Progress Series*, 350: 291–298.

Roberts, J.M., Peppe, O.C., Dodds, L.A., *et al.* (2005). Monitoring environmental variability around cold-water coral reefs: the use of a benthic photolander and the potential of seafloor observatories. In: Freiwald, A., Roberts, J.M. (eds), *Cold-water corals and ecosystems*. Berlin Heidelberg: Springer-Verlag, pp 483–502.

Rouse, G.W., Goffredi, S.K., Vrijenhoek, R.C. (2004). Osedax: Bone-eating marine worms with dwarf nales. *Science*, 305: 668–671.

Rowe, G.T., Suibuet, M., Vangriesheim, A. (1986). Domains of occupation of abyssal scavengers inferred from baited cameras and traps on the Demerara Abyssal Plain. *Deep-Sea Research I*, 33(4): 501–522.

Ryall, P.J.C., Hargrave, B.T. (1984). Attraction of the Atlantic wreckfish (*Polyprion americanus*) to an unbaited camera on the Mid-Atlantic Ridge. *Deep Sea Research*, 31: 79–83.

Sainte-Marie, B., Hargrave, B.T. (1987). Estimation of scavenger abundance and distance of attraction to bait. *Marine Biology*, 94: 431–443.

Shams El Din, A. M., Saber, T.M.H., Hammoud, A.A. (1996). Biofilm on stainless steels in Arabian Gulf water. *Desalination*, 107: 251–264.

Shifler, D.A. (2005). Understanding material interactions in marine environments to promote extended structural life. *Corrosion Science*, 47: 2335–2352.

Shulenberger, E., Hessler, R.R. (1974). Scavenging abyssal benthic amphipods trapped under oligotrophic central North Pacific Gyre waters. *Marine Biology*, 28: 185–187.

Smith, C.R., Baco, A.R. (2003). Ecology of whale falls at the deep-sea floor. *Oceanography and Marine Biology: an Annual Review*, 41: 311–354.

Smith, K.L., Jr., Baldwin, R.J. (1984). Vertical distribution of the necrophagous amphipods, *Eurythenes gryllus*, in the North Pacific: spatial and temporal variation. *Deep-Sea Research*, 31(10): 1179–1196.

Smith, K.L., Jr., White, G.A., Laver, M.B., McConnaughey, R.R., Meador, J.P. (1979). Free vehicle capture of abyssopelagic animals. *Deep-Sea Research*, 26A: 57–64.

Smith, K.L., Jr., Kaufmann, R.S., Wakefield, W.W. (1993). Mobile megafaunal activity monitored with a time-lapse camera in the abyssal North Pacific. *Deep-Sea Research I*, 40: 2307–2324.

Soltwedel, T., von Juterzenka, K., Premke, K., Klages, M. (2003). What a lucky shot! Photographic evidence for a medium-sized natural food-fall at the deep-seafloor. *Oceanologica Acta*, 26: 623–628.

Stanley, B.H., Hummel, H.E., Ruesink, W.G. (1985). Estimating maximum horizontal area of pheremone plumes. *Journal of Chemical Ecology*, 11: 1129–1146.

Stockton, W.L. (1982). Scavenging amphipods from under the Ross Ice Shelf, Antarctica. *Deep-Sea Research*, 29: 819–835.

Tengberg, A., De Bovee, F., Hall, P.O.J., *et al.* (1995). Benthic chamber and profiling landers in oceanography – a review of design, technical solutions and functioning. *Progress in Oceanography*, 35: 253–294.

Tengberg, A., De Bovee, F., Hall, P.O.J., *et al.* (1997). The Banyuls and Goteborg benthic landers: two modular autonomous instruments for biogeochemical sea floor studies. Paper II: PhD thesis, Analytical and Marine Chemistry, Chalmers University of Technology and Goteborg University.

Tengberg, A., Stahl, H., Gust, G., *et al.* (2004). Intercalibration of benthic flux chambers I. Accuracy of flux measurements and influence of chamber dynamics. *Progress in Oceanography*, 60(1): 1–28.

Thurston, M.H. (1979). Scavenging abyssal amphipods from the northeast Atlantic Ocean. *Marine Biology*, 51: 55–68.

Thurston, M.H., Bett, B.J., Rice, A.L. (1995). Abyssal megafaunal necrofages: Latitudinal differences in the Eastern North Atlantic Ocean. *Internationale Revue der gesamten Hydrobiologie*, 80(2): 267–286.

Thurston, M.H., Petrillo, M., Della Croce, N. (2002). Population structure of the necrophagous amphipod *Eurythenes gryllus* (Amphipoda: Gammaridea) from the Atacama Trench (south-east Pacific Ocean). *Journal of the Marine Biological Association of the UK*, 82: 205–211.

Vardaro, M.F., Parmley, D., Smith, K.L., Jr. (2007). A study of the possible 'reef effects' caused by long-term time lapse camera in the deep-sea. *Deep-Sea Research I*, 54: 1231–1240.

Viollier, E., Rabouille, C., Apitz, S.E., *et al.* (2003). Benthic biogeochemistry: state of the art technologies and guidelines for the future of in situ survey. *Journal of Experimental Marine Biology and Ecology*, 285–286: 5–31.

Wakefield, W.W., Genin, A. (1987). The use of Canadian (perspective) grid in deep-sea photography. *Deep-Sea Research*, 34: 469–478.

Walls, A.H., Clarke, J. (2005). Marine biodiversity – exploring the deeps. Paper SPE 95381 presented at the Society of Petroleum Engineers Asia Pacific Health, Safety and Environment Conference and Exhibition, Kuala Lumpur, Malaysia, 19–21 September.

Wenner, C.A., Musick, J.A. (1977). Biology of the Morid fish *Antimora rostrata* in the western North Atlantic. *Journal of the Fisheries Research Board of Canada*, 34: 2362–2368.

Wickins, J.F. (1983). Catches of large lysianassid amphipods in baited traps at the Nuclear Energy Authority dump site during June 1979. *Deep-Sea Research*, 30: 83–86.

Widder, E.A., Robison, B.H., Reisenbichler, K.R., Haddock, S.H.D. (2005). Using red light for in situ observations of deep-sea fishes. *Deep-Sea Research I*, 52: 2077–2085.

Wilson, R.R., Smith K.L. (1984). Effect of near-bottom currents on detection of bait by the abyssal grenadier fishes, *Coryphaenoides* spp. recorded in situ with a video camera on a free vehicle. *Marine Biology*, 84: 83–91.

Witte, U. (1999). Consumption of large carcasses by scavenger assemblages in the deep Arabian Sea: observations by baited camera. *Marine Ecology Progress Series*, 183: 139–147.

Wolff, T. (1976). Utilisation of seagrass in the deep sea. *Aquatic botany*, 2: 161–174.

Yamamoto, J., Nobetsu, T., Iwamori, T., Sakurai, Y. (2009). Observations of food falls off the Shiretoko Peninsula, Japan, using a remotely operated vehicle. *Fisheries Science*, 75: 513–515.

Yau, C., Collins, M.A., Bagley, P.M., Everson, I., Nolan, C., Priede, I.G. (2001). Estimating the abundance of Patagonian toothfish *Dissostichus eleginiodes* using baited cameras: a preliminary study. *Fisheries Research*, 51: 403–412.

Younge, M. (1971). The inception and significance of the Challenger expedition. *Proceedings of the Royal Society B*. 72: 1–13.

Zintzen, V., Anderson, M.J., Roberts, C.D., Harvey, E.S., Stewart, A.L., Struthers, C.D. (2012). Diversity and composition of demersal fishes along a depth gradient assessed by baited remote underwater stereo-video. *PLoS ONE*, 7 (10): e48522.

Chapter 12
Towed Cameras

David A. Bowden[1] and Daniel O.B. Jones[2]

[1] National Institute of Water and Atmospheric Research, Wellington, New Zealand
[2] National Oceanography Centre, European Way, Southampton, United Kingdom

Abstract

Towed camera systems are used extensively in deep-sea research to acquire photographic samples of the seabed. They are deployed on a cable from a surface vessel, have no internal propulsion, can carry video or still image cameras, and can either be self-contained, recording images to be downloaded on return to the surface, or can transmit data to the surface in real time via a fibre optic cable. In this chapter we describe the principal components of deep-sea towed camera systems in use today, and how these components are configured to achieve optimum image quality. We then review approaches to survey planning and practical considerations for towed camera deployments in the deep sea, and conclude with considerations of photographic sample management and analysis.

Keywords underwater camera, seabed photography, habitat mapping, benthic, visual sampling, video, still image, remote imaging, transect analysis, seabed survey

12.1 Introduction

Towed cameras are technologically the simplest of the methods currently available for collecting transects of moving or still images from the seabed. From their beginnings as still image cameras triggered by contact with the seabed (Owen *et al.*, 1967), they have developed to become versatile, multifunctional tools often carrying multiple cameras and a range of other sensors on a single platform (Chezar & Lee, 1985; Barker *et al.*, 1999; Fornari, 2003; Hill, 2009; Howland *et al.*, 2006; Jones *et al.*, 2009).

Towed camera platforms are deployed on a cable from a surface vessel and have no propulsion mechanism. The platform moves across the seabed either by being towed at slow speed or by passive drifting of the ship, while height above the seabed (altitude) is adjusted by means of the shipboard winch from which the platform is deployed. Thus, these platforms have very low manoeuvrability and tend to follow relatively simple line transect trajectories, during which both direction and altitude can be strongly influenced by surface sea state, wind direction and currents. Either video or still image cameras, or both, can be deployed, and camera orientation and lens focal length are generally, but not always, fixed. A variety of other sensors may be fitted to the platform frame, typically including depth or conductivity temperature and depth (CTD) sensors and an acoustic altimeter to measure height above the seabed. The seabed position of the platform may be monitored by use of acoustic

Biological Sampling in the Deep Sea, First Edition. Edited by Malcolm R. Clark, Mireille Consalvey and Ashley A. Rowden.

positioning systems and additional refinements can include forward-looking sonar to detect obstacles. Although most deep-sea towed camera systems in use today are suspended ('flown') just above the seabed during operation, some systems are designed to be landed on the seabed (Mortensen & Buhl-Mortensen, 2004; Howell *et al.*, 2010), or as sleds to be towed in contact with the seabed (Tuck *et al.*, 1997), while still others can be operated in more than one of these modes (Bellec *et al.*, 2008). Furthermore, high-resolution cameras can now be attached to almost any type of deep-sea sampling gear (e.g. trawls, grabs and multi-corers) allowing them to function as towed cameras (Mortensen *et al.*, 2000; Post *et al.*, 2010; Thurber *et al.*, 2010).

For deep-sea operation, towed camera platforms can be grouped as either surface recording systems, which have real-time image data transfer between the towed camera and the surface vessel (e.g. Barker *et al.*, 1999), or seabed recording systems, which are flown 'blind' with images being recorded in-camera and downloaded after retrieval (e.g. Jones *et al.*, 2009). Seabed recording systems can be deployed on any cable, but surface recording systems usually require fibre optic cables for real-time, full-bandwidth data transmission to the surface (Shortis *et al.*, 2008). At least one hybrid system has been developed, however, in which low frame-rate colour video images are viewed at the surface in real time via a single conductor-CTD cable (Hill, 2009), and others can be used with either conducting or fibre optic cable (Fornari, 2003). Systems may be powered from the surface vessel (e.g. Shortis *et al.*, 2008) or from batteries carried on board the towed camera platform (e.g. Hill, 2009; Jones *et al.*, 2009).

This chapter deals primarily with towed camera systems, but because images captured by autonomous underwater vehicles (AUVs) can have similar characteristics to those from towed platforms, much of the discussion on the fundamentals of imaging systems and the analysis of images are applicable to both (Jones *et al.*, 2005). AUVs are self-propelled and operate without any direct link to the surface. They are not piloted, but they can be programmed to follow a specific survey pattern or to modify their survey in response to water column or seabed characteristics (e.g. constant altitude in relation to topography, or following salinity, temperature or methane gradients) (see Chapter 5). As with towed platforms, AUVs usually carry a range of sensors in addition to cameras with the important difference that the cameras often are not the primary research instruments, priority going to sonar, geomagnetic and physical oceanographic sensors. AUVs rely on considerably more sophisticated technology than towed platforms, but because camera orientation and lens focal length are usually fixed, and both altitude and velocity are usually constant, photographic samples acquired by towed platforms and AUVs share important characteristics. This is in contrast to remotely operated vehicles (ROVs) and submersibles which, while they can be used to run simple linear transects or grids, are often paused at sites of interest and cameras are panned, tilted and zoomed to record small-scale detail (see Chapter 13). Although the technology of cameras, lights, storage media and positioning systems are in constant development, the fundamental characteristics of images from towed cameras and AUVs allow resulting samples to be analysed using the same techniques and for comparisons to be made between survey data collected using the two different platform types (Jones *et al.*, 2005).

In comparison with piloted vehicles (ROVs and submersibles), towed camera platforms have some significant limitations. These are primarily caused by the inability to control direction and speed at fine scales, to stop and zoom-in on features of specific interest, or to manipulate and take samples of seabed features. The direct physical link to the surface vessel also makes the quality of imaging dependent on sea state. They have compelling advantages, however, associated with their greater simplicity of design and construction, their lack of dependence on sophisticated cable technology and their simplicity of operation. These factors result in considerably reduced costs of construction, operation, and maintenance, and generally more robust vehicles which can be used over a wider range of sea states than their more sophisticated and complex counterparts.

There are also more fundamental limitations and advantages that are associated with photographic sampling, regardless of the platform used. In most applications, seabed images can only be used to generate reliable data on the mega- and larger macro-epibiota, and then only those which are emergent and are visible in the 'upper storey' of the assemblage (Owen *et al.*, 1967; Grassle *et al.*, 1975). Thus, much of the benthic macrofaunal diversity associated with, for instance, stands of live coral may not be detected in camera images. Similarly, abiotic substrata in photographic images can only be discriminated between on the basis of their surface appearance: bedrock overlain with a thin layer of fine sediments may be indistinguishable from fine sediment many metres deep. In soft-sediment habitats, however, particularly the fine muds characteristic of continental shelves and ocean basins, towed camera images can yield information on the abundance and diversity of organisms living within the sediments, as well as sparse mobile epifauna not captured in the actual image field, by recording surface features such as burrows, pits and tracks ('lebenspurren', see Solan *et al.*, 2003 and references therein for review). Such features are often distinctive and, where the associated organism has been positively identified, can be used to generate population density estimates at species level (e.g. Tuck *et al.*, 1997). More usually, however, lebenspurren can only be associated with higher taxon levels of identification (Dundas & Przeslawski, 2009).

Photographic surveys in themselves, therefore, may yield only a partial estimate of biological or habitat diversity because of limitations which are inherent in all photographic images. With remote imaging of the deep seabed using towed cameras, however, there can also be considerable variability in the quality and spatial scale of images, both within and between studies, which arises from differences in technical specification, modes of use and operational conditions. Such variability is difficult to control and affects the precision and accuracy of data derived (Mortensen & Buhl-Mortensen, 2005), and ultimately the comparability of data.

These limitations notwithstanding, visual sampling is the most effective method available for quantitative studies of mega-epifauna in the deep sea (Rice *et al.*, 1994), particularly on hard substrata. Despite advances in the interpretation of acoustic data, visual sampling is also the best remote method we have for reliably and accurately detecting transitions between seabed areas with different biological assemblages (e.g. Fossa *et al.*, 2005; Hewitt *et al.*, 2004). A further fundamental advantage of visual sampling methods in contrast with traditional gears is that they are non-destructive. Thus, visual transects, particularly video, enable immediate qualitative assessment of the seabed and its associated biota with all structural and distributional relationships intact. Furthermore, the power of human observation is such that a large amount of useful information is processed, if not always recorded, simply by watching the image in real time (Grassle *et al.*, 1975; Hessler *et al.*, 1985). Without performing any kind of analysis, researchers are generally able to describe the main features of the transect and note where any significant regions of interest occurred simply through having viewed the video footage.

Photographic images generally contain more information than can be extracted within the constraints of the project for which they were originally captured. Therefore, usually only a subset of the potential data is extracted at first pass but the images remain as an archive of information that can be mined for more detailed, or entirely different, data in the future. This archival, or historical, value of images has long been appreciated in terrestrial applications (e.g. aerial surveillance photography) but to date has been less well appreciated in marine benthic studies. It is also relevant in this context that, although silver halide film has been almost entirely superseded by digital imaging technologies, a substantial archive of gelatin–silver images of the deep seabed exists in institutions around the world. This is a resource that remains largely untapped in terms of extraction of ecological data (e.g. Beaulieu *et al.*, 2005), particularly where the original surveys were for geological research. However, the common properties of visual images are such that there is little reason why data from such images should not be incorporated in analyses with data from more recent digital images.

This chapter starts with brief technical summaries of some towed camera systems in use today, then covers aspects of towed camera configuration and deployment procedures, and concludes with considerations of photographic sample management and analysis. A short glossary of definitions used is given in Box 12.1.

Box 12.1 Definitions

- *Still images or stills*: Discrete individual photographs, whether recorded to digital or silver halide media.
- *Video*: Continuous moving images, whether recorded to magnetic tape, computer hard disc drive (HDD), or silver halide film.
- *Visual sampling*: Seabed sampling using any combination of video and stills cameras.
- *Platform*: Any apparatus or vehicle on which cameras are mounted.

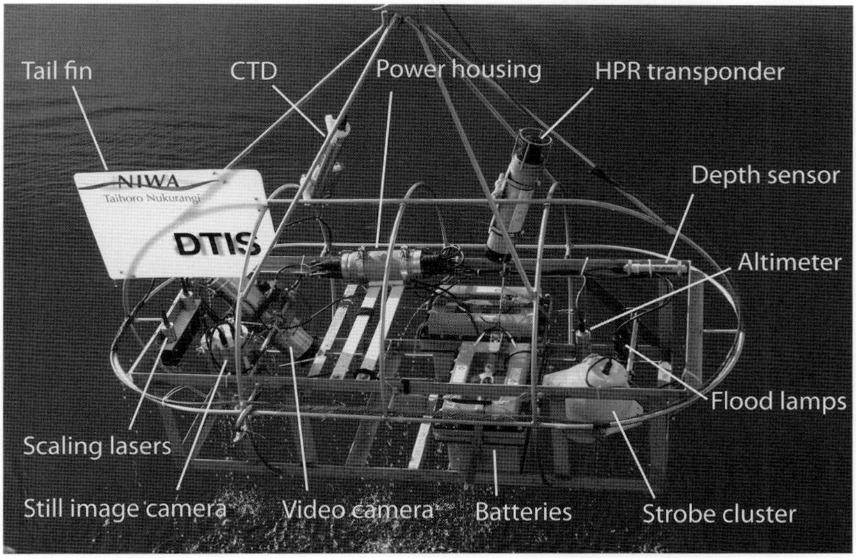

Fig. 12.1 Towed camera platform components: National Institute of Water & Atmospheric Research's DTIS (Deep Towed Imaging System) as an example of a system with internal recording and power supply designed for use with single conductor cables (see Table 12.1 for technical specifications). Note the combination of vertically orientated stills camera and obliquely orientated video (45° forward-facing), and the positioning of cameras and lamps at opposite ends of the chassis to reduce backscatter from suspended particles in the water column. An additional pair of parallel scaling lasers is attached directly to the video camera housing. (Photograph courtesy of NIWA, New Zealand. Reproduced with permission.)

12.2 Towed camera systems

The basic requirements of towed camera systems are simple: one or more cameras, lighting, a means of recording images, a power supply and a cable. The actual specifications of these components and how they are combined differ widely between systems, and these differences can strongly influence image quality and deployment methods. The components of the system are usually attached to a substantial metal chassis, designed to protect the instruments and withstand the inevitable impacts with the seabed, and most have some form of tail fin to maintain constant orientation of the vehicle while towing (Fig. 12.1). The frame itself can take almost any form but they are generally made of marine-grade stainless steel or aluminium to reduce corrosion and enable attachment of, for instance, magnetometers, magnetic compasses or other magnetically sensitive instruments.

Examples of towed camera platforms currently in use are given in Fig. 12.2 and Table 12.1. Most of these systems have been designed specifically as towed camera vehicles but nearly all carry

Fig. 12.2 Towed camera platforms. (a) Deep Towed Imaging System (DTIS), National Institute of Water and Atmospheric Research, New Zealand. (Photograph courtesy of NIWA, New Zealand. Reproduced with permission.) (b) Deep towed camera system, Commonwealth Scientific and Industrial Research Organisation (CSIRO), Australia. (Photograph courtesy of CSIRO, Australia. Reproduced with permission.) (c) TowCam, Woods Hole Oceanographic Institute (WHOI), USA. (Photograph courtesy of WHOI, USA. Reproduced with permission.) (d) CamPod, Institute of Marine Research, Norway. (Photograph courtesy of the Institute of Marine Research (IMR), Norway.) (e) Benthic Optical Acoustic Grab (BOAGS), CSIRO, Australia. (Photograph courtesy of CSIRO, Australia. Reproduced with permission.) (f) Seabed high-resolution imaging platform (SHRIMP), National Oceanography Centre, UK. See Table 12.1 for technical specifications. (Please see Plate 20 for colour representation of the figure.)

Table 12.1 Technical summaries of the six towed camera systems shown in Fig. 12.1

	DTIS	Deep towed camera system	TowCam	CamPod	BOAGS	SHRIMP
Institute	NIWA, New Zealand	CSIRO, Australia	WHOI, USA	IMR, Norway	CSIRO, Australia	NOC, UK
Cable type	Single conductor-CTD	Fibre optic	Single conductor-CTD *or* multi-conductor-CTD *or* fibre optic	Fibre optic	Fibre optic	Fibre optic
Power source	2 × 12 V internal batteries	Surface supply via cable	4 *or* 5 × 12 V internal batteries	Surface supply via cable	Surface supply via cable	Surface supply via cable
Video cameras	1 × Sony HD video (oblique orientation)	2 × HD Gigabit Ethernet video (oblique orientation) as stereo pair 1 × low-light DV video (forward-looking)	n/a	1 × Sony HD video with pan-tilt-zoom (oblique orientation during transects)	1 × Hitachi HD video (seabed observation, oblique orientation) 1 × Hitachi HD video (grab monitoring)	1 x Insite Mini Zeus HD video
Video recording	MiniDV tape in camera	DV tape at surface	n/a		HDD at surface	HDD and DV tape at surface
Video lighting	450 W quartz halogen	2 × 250 W HID	n/a	2 × 400 W HMI	2 × LED (2500 lumens each)	1 × 400 W HMI
Video viewing	Real time at surface via custom telemetry (low frame-rate)	Real time at surface (full resolution)	n/a	Real time at surface (full resolution)	Real time at surface (full resolution)	Real time at surface (full resolution)
Still image cameras	Digital SLR, (vertical orientation)	Digital SLR, (45° orientation)	1 × digital (non-SLR) *or* 1 × digital SLR *or* custom sensor in fibre optic mode (all vertical orientation)	n/a	2 × digital SLR as stereo pair	1 x Insite Scorpio digital (non-SLR)
Still image recording	In-camera (auto at 15 s interval or manual)	In-camera (auto or manual)	in-camera (auto at 10 s interval) *or* HDD at surface in fibre optic mode	n/a	In-camera (auto or manual)	In-camera (manual)

(*continued*)

Table 12.1 (Continued)

	DTIS	Deep towed camera system	TowCam	CamPod	BOAGS	SHRIMP
Still image viewing	Post-deployment	Post-deployment	Post-deployment *or* real-time low resolution at surface via custom telemetry *or* real-time full-resolution for fibre optic	n/a	Post-deployment	Post-deployment
Image scaling	2 × red laser (for each camera)	Multiple red laser array (for stereo image calibration)	2–3 × red *or* green laser	2 × red laser	None for video, stereo imaging for stills	2 × red laser
Tracking	USBL	USBL	USBL *or* LBL *or* layback	USBL	USBL	USBL
Duration	1 h	Indefinite (0.5–1.0 h standard)	5–9 h	Indefinite	Indefinite	Indefinite
Maximum depth	6000 m	2000 m	6000 m		3000 m	6500 m
Additional sensors	Acoustic altimeter Pressure sensor (depth) CTD	Acoustic altimeter Pressure sensor (depth) CTD Fluorometer	Acoustic altimeter Pressure sensor (depth) CTD, Forward-looking sonar Magnetometer	Acoustic altimeter Pressure sensor (depth)	Forward-looking sonar Scientific echosounder (for seabed survey)	Acoustic altimeter Pressure sensor (depth)
Sampling tools			Rock fragment samplers 5 litre Niskin bottles		Smith–McIntyre grab	
Reference	(Hill, 2009)	(Barker *et al*., 2000; Shortis, *et al*., 2008)	www.whoi.edu/page.do?pid=17619 (Fornari, 2003)		(Sherlock *et al*., 2010)	

NIWA, National Institute of Water and Atmospheric Research; CSIRO, Commonwealth Scientific and Industrial Research Organisation; WHOI, Woods Hole Oceanographic Institute; IMR, Institute of Marine Research; NOC, National Oceanography Centre; DTIS, Deep Towed Imaging System; SHRIMP, Seabed High Resolution Imaging Platform; HD, high definition; USBL, ultrashort baseline acoustic tracking system; CTD, conductivity, temperature and depth recorder; SLR, single lens reflex camera; LED, light-emitting diode; HMI, hydrargyrum medium-arc iodide lamp; HID, high-intensity discharge lamp.

additional sensors or sampling tools. Of the systems shown here, CamPod (Institute of Marine Research, Norway) has been designed to be 'landed' on the seabed for detailed inspection of fauna and substrates, TowCam (Woods Hole Oceanographic Institute, USA) can collect water and rock samples (Fornari, 2003), and Commonwealth Scientific and Industrial Research Organisation's benthic optical acoustic grab sampling system (BOAGS) (CSIRO, Australia) is effectively a Smith–McIntyre grab with a camera system built around it (Sherlock *et al.*, 2010). The modular BOAGS system has also been designed such that the grab can be replaced by, for instance, a box-corer, or other sampling tool.

12.3 Fundamentals of towed camera imaging systems

12.3.1 Camera orientation

Camera orientation is a fundamental variable in all seabed imaging. For towed or autonomous systems in which camera orientation is fixed, it is a critical parameter for quantitative interpretation of imagery. The optical axis can be either vertical or oblique. Each of these orientations has its advantages and disadvantages but the methods required for extracting quantitative data from images produced by each may vary considerably, leading to problems with cross-study comparison if methods are not clearly stated.

12.3.1.1 Oblique orientation

Oblique orientation has several advantages, particularly for video when the image is viewed in real-time during the transect:

- The image frame encompasses a greater area of the seabed than with vertical orientation, which increases the probability of detecting specific habitats and larger or rarer taxa.
- The partially lateral viewing angle aids identification of fauna which have upright or laterally compressed body forms.
- Longer transit times through the image frame can aid identification of fauna.
- The forward-facing field of view reduces the risk of collision with seabed obstacles.
- The oblique view is usually more visually appealing and enhances the value of images for illustration of features and for outreach applications.

The main disadvantage of oblique images is that they are more problematic to extract quantitative data from (Wakefield & Genin, 1987). This is because the angled viewpoint introduces inherent gradients of both lens-to-subject distance, and illumination intensity (Fig. 12.3). Clearly, these gradients become steeper as the optical axis is angled further away from vertical.

12.3.1.2 Vertical orientation

Vertical orientation, by contrast, produces images which cover a smaller seabed area, are less easy to make identifications from (for some taxa, such as fish), and increase the risk of collision if used as the only means of viewing. However, the conspicuous advantages are that, in a well-designed system, images from vertically orientated cameras have:

- negligible gradients of lens-to-subject distance, and
- negligible gradients of illumination intensity across the frame.

These properties make vertically orientated images optimal for quantitative analyses of benthic substrata and sessile or sedentary biota.

(a)

(b)

Fig. 12.3 Camera orientation. (a) Oblique orientation encompasses greater seabed area facilitates identification of upright benthic fauna, but introduces strong gradients of illumination and distance from foreground to background. (Photograph courtesy of Julian Gutt, Alfred Wegner Institute for Polar and Marine Research. Reproduced with permission.) (b) Vertical orientation provides even illumination and uniform subject-to-camera distance. Scale bars in each are 20 cm (the two red laser reference points can be seen above the scale bar in the lower image). Images courtesy of Julian Gutt, Alfred Wegner Institute for Polar and Marine Research (Weddell Sea, top), and the New Zealand IPY-CAML programme (Ross Sea, bottom).

12.3.1.3 Combination

Combining oblique and vertical cameras on a single platform makes full use of both the descriptive potential of forward-looking video and the potential for accurate quantitative analyses from vertical images (e.g. Stone, 2006). This potential is further enhanced if the vertical camera takes high-resolution still images, enabling the discrimination of extremely fine detail (millimetre scale) in substrata and fauna.

12.3.2 *Lighting*

The intensity and beam width of lighting determine the maximum altitude at which correctly exposed images can be captured. Typical deep-sea camera systems in use today operate at target altitudes of 2–5 m above the seabed (Fornari, 2003; Hill, 2009), but up to 10 m for those with more powerful lighting (e.g. Lutz *et al.*, 2002) or extremely sensitive cameras (Jones *et al.*, 2005). The evenness of illumination across the image and the degree of backscatter from particles suspended in the water column are critical to the quality of the images obtained and therefore have a major influence on the resolution of data that can be extracted. If the illuminated field is not matched to the field of view of the camera, strong gradients of light intensity are introduced which complicate image analysis and can render significant portions of the image unusable. In addition, if there is insufficient angular separation between the axis (or axes) of illumination and the optical axis of the camera, backscatter from suspended particles can obscure much of the image. Although this is less of a problem in the deep sea, where suspended particulates are less common than in coastal or shelf depths, most systems are designed to maximize angular separation between the axes. Angular separation has another advantage for biological interpretation of photographs in that it causes shadows that enhance the viewer's perception of three-dimensional structures, thus facilitating identification.

As with most other aspects of imaging technology, light sources have developed considerably during recent years, with more compact and robust units that provide increased output for reduced power consumption. This is particularly striking with the use of high-intensity discharge (HDI or HMI) lights, and light-emitting diode (LED) arrays. At present, LEDs are not a substitute for conventional high-capacitance strobe lighting for still image illumination, but because of their low power consumption and robustness they are being used increasingly for video flood lighting (Rinke & Lee, 2009; Sherlock *et al.*, 2010).

The appropriate levels of lighting can be estimated as a function of the camera settings (lens aperture and recording sensitivity), the attenuation of light in water (from backscatter and absorption) and the range to the subject. However, such calculations are complex and, in practice, lighting is usually optimized by consideration of basic parameters during the design phase (Fig. 12.4) and experimentation during initial field trials. Because lights are usually by far the largest power drain on the whole system, the amount of lighting that can be supplied is the main limiting factor in relation to the camera settings used.

12.3.3 *Optical resolution*

For any quantitative comparison between sets of image-derived data it is necessary to know the minimum size of object that can be detected reliably in each study. This is a measure that is rarely reported but is particularly important for comparisons of biological diversity between studies because it affects both the size spectrum of organisms that can be detected and the taxonomic precision with which they can be identified (Bowden, 2005). Image resolution is a property of the whole camera system, not just the grain or number of pixels in the image, and is affected by the focal length, aperture and optical quality of the lens, by the altitude of the camera above the substratum, by the evenness and intensity of illumination, by the sensitivity of the recording medium, and by the clarity of the water column. Thus, all of these factors must be considered during the design and set-up of the system.

12.3.4 *Camera set-up*

Towed camera platforms very rarely operate at constant altitude. Thus camera-to-subject distance varies continuously during deployment, oscillating around the target altitude. It is important, therefore, to optimize lighting and camera settings (aperture, focal length and sensitivity (ISO)) to ensure correctly exposed and focused images across the range of expected altitudes, rather than just the target altitude. The lighting and camera settings are usually fixed for the duration of a deployment. To ensure correctly focused images, the depth of field (DOF) should be maximized to ensure that it

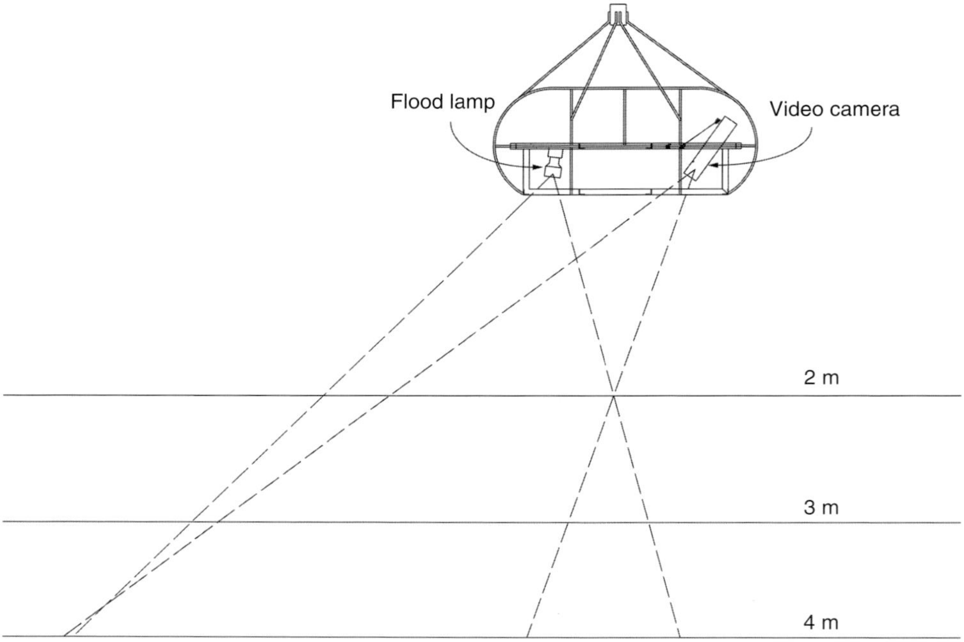

Fig. 12.4 Matching lighting and camera angles, fields of view, and altitude ranges during design and set-up of a towed camera system. The drawing shows a system with a 60° field of illumination and a video camera with 30° vertical angle of acceptance. It is designed to operate at altitudes from 2 to 4 m above the seabed, with an optimum altitude of 2.5 m.

Box 12.2 Depth of field

The depth of field (DOF) of a camera system can be calculated as:

$$DOF = \frac{2 \cdot f^2 \cdot R \cdot c \cdot (f - R)}{(R \cdot N^2 \cdot c - f \cdot c - f^2) \cdot (R \cdot N^2 \cdot c - f \cdot c + f^2)}$$

where f is the lens focal length, N is the lens aperture (*f*-number), c is the acceptable circle of confusion (often taken as the dimension of a pixel in digital photography) and R is the range to subject (i.e. the target operating altitude). Note also, if a flat lens port is used, the imaged field is magnified by approximately 25%. Thus, if the optimal altitude for the system is, say 3 m, camera focus should be set at 2.25 m and this value, rather than the actual distance to the seabed, should be used as R.

exceeds the expected variation in operating altitude (Box 12.2). A trade-off must be made between the DOF and amount of light entering the lens because larger DOF is achieved at smaller lens apertures (high *f*-number) and therefore demands greater lighting intensity for correct exposure. (Note: Lens apertures, *f*-stops, are defined as the ratio of focal length of the lens to the diameter of the iris aperture and are conventionally defined on a scale ($f2.8, f4, f5.6, f8, f11, f16, f22$, etc.) such that changing the aperture by one full *f*-stop either halves or doubles the amount of light reaching the recording medium. Changing from $f8$ to $f11$, for instance, halves the amount of light.)

Optimal lighting across the entire image field is, arguably, the single most important element in achieving high-quality seabed imagery and it is worth investing time and resources to ensure that this is addressed before the system is deployed for research purposes. If sufficient intensity and coverage of lighting is designed into the system at the outset, it allows for greater flexibility with respect to the critical camera parameters (aperture, ISO, field of view) which control basic image quality. For instance, image noise ('grain' in silver halide film) is least at low ISO settings, but if there is insufficient lighting, correct exposure will only be possible at high ISO settings, with concomitant reduction in image resolution. Similarly, a large lens aperture will be needed at lower light levels, resulting in reduced DOF and images being out of focus at the extremes of the altitude range. The intensity of lighting also has an influence on the shutter speed used. In deep-sea applications, ambient light is absent or negligible. Thus, for still image cameras which use electronic strobe lighting, shutter speed is not critical because the short-duration flash freezes motion. For video or still cameras that rely on continuous flood lighting, however, blurring of the image will occur if inadequate lighting necessitates use of slow shutter speeds.

12.3.5 Image scaling

Accurate scaling of the image field is essential if quantitative data are to be extracted. The most commonly used method at present is to project two or more parallel lasers onto the seabed within the framed image area. The known distance between the lasers can then be used to scale the image. Red lasers have been the most readily available type for many years but green lasers are now a practical alternative. Green lasers have considerably greater penetration in sea water because of their shorter wavelength, but backscatter from particles in the water column, particularly phytoplankton, is greater than with red lasers and can detract from the quality of the images.

Clearly, a single pair of lasers can only be used to scale the entire image where the camera orientation is vertical, the imaged field is effectively planar, and lens distortion is negligible. For obliquely orientated cameras, a square array of four parallel lasers enables scaling of the entire image frame, while an additional non-parallel laser in the plane of the two lasers closest to the camera can be used to measure camera-to-substratum distance (Pilgrim *et al.*, 2001). On topographically complex terrain, it is unlikely that any combination of lasers will adequately scale the entire image plane but in many practical applications, the single pair may be adequate for the level of scaling accuracy required.

Despite the wide adoption of lasers, image scaling can be achieved in a number of other ways including: calibration with an object of known dimensions suspended within the image frame (Auster *et al.*, 1995); calculation from acoustic altimeter data (Hessler *et al.*, 1985; Jones *et al.*, 2009); estimation from known size ranges of fauna in the image (Pratt, 1967; Williams *et al.*, 2010), and the use of stereo paired cameras (Abdo *et al.*, 2006; Shortis *et al.*, 2008). Calculation of image area from altitude data is a practical method for vertically orientated cameras. If altitude is recorded continuously to data files (and thus does not have to be transcribed from image overlays) areas of all images from a survey can be calculated rapidly by using simple formulae (Box 12.3) in automated functions (Jones *et al.*, 2009). This contrasts with the laser method, where images are generally scaled individually using image analysis tools (e.g. ImageJ http://rsbweb.nih.gov/ij/), which is more time-consuming. Stereo cameras have considerable advantages for precise measurement throughout the image frame, and developments in camera and data-processing technology are enabling increasing application of this technique for both still and video imaging in the deep sea.

Scaling of entire video transects, as opposed to still images or frame grabs from video, is done using the same principles (laser spacing, altitude, etc.) but is complicated by variations in altitude along the transect. An approximate total area of seabed covered can be calculated as the average video frame width, derived from measurements at a number of points along the transect, multiplied by the length of the transect derived from the navigational data (e.g. Jones *et al.*, 2006). For systems in which camera altitude data are recorded routinely, however, total seabed area covered in the video can be calculated more accurately.

Box 12.3 Scaling calculations

The spatial scale of video and still images can be calculated from the angle of acceptance of the camera system and its altitude above the seabed. The angle of acceptance can be measured horizontally, vertically, or diagonally across the image frame and is calculated using:

$$a = 2 * \arctan \frac{d}{2f}$$

where a is the angle of acceptance in degrees, d is the chosen dimension of the film/sensor (e.g. the width) and f is the focal length of the lens.

As with depth of field calculations, however, if a plane (flat) lens port is used, this value must be modified to allow for refraction. The effective angle of acceptance in water, a', is calculated as:

$$a' = 2 * \arcsin \left[\frac{\sin a/2}{n} \right]$$

where n is the refractive index of seawater (~1.339).

For vertically orientated cameras, areal coverage for individual frames can be calculated from altitude using:

$$\text{Area of photograph} = 4A^2 \tan\left(\frac{\theta}{2}\right) \tan\left(\frac{\omega}{2}\right)$$

where A is camera altitude, θ is horizontal camera angle of acceptance (degrees) and ω is vertical camera angle of acceptance (degrees).

These equations apply only to vertically orientated images; the calculation of area from oblique images is more problematic and is discussed in Wakefield and Genin (1987).

12.3.6 Spatial referencing

Ecological interpretation of visual samples requires that the geographic location of the images is known. In some studies at the scale of biogeographic regions it may be enough to know only the general coordinates of the study, such as the summit of a seamount, but for any smaller scale studies, particularly those concerned with depth distributions of different species, accurate seabed positions are needed.

Acoustic positioning systems enable real-time recording of the seabed position of the camera platform, their accuracy depending on the sophistication of the system used, the distance from the vessel, sea state and the properties of the water column. Horizontal accuracy using contemporary ultrashort baseline (USBL) acoustic tracking systems in 1000 m depth is typically in the region of ±10–20 m but more advanced systems incorporating depth sensors and inertial data, or long-baseline (LBL) systems using multiple seabed transducers, provide greater accuracy. All such systems are a considerable improvement on methods in which camera position is estimated trigonometrically from the ship's surface position and course, the length of winch wire deployed and the depth. For work on steep terrain, depth is likely to be an important positional reference and, where acoustic positioning gear is not available, a depth logger attached to the camera platform can potentially provide sufficient information to relate images to the overall topography if combined with accurate multibeam bathymetric maps and the surface position of the ship (Clark & Rowden, 2009; Williams *et al.*, 2010).

The availability of real-time position data during deployments can be particularly valuable if the seabed position of the camera platform can be displayed against raster maps of multibeam bathymetry, side-scan sonar or other processed acoustic data. Using this capability, specific topographic features or isolated habitats can be targeted precisely and observations from initial transects can be used to refine subsequent deployments (Fig. 12.5). There are several custom and proprietary

(a)

(b)

(c)

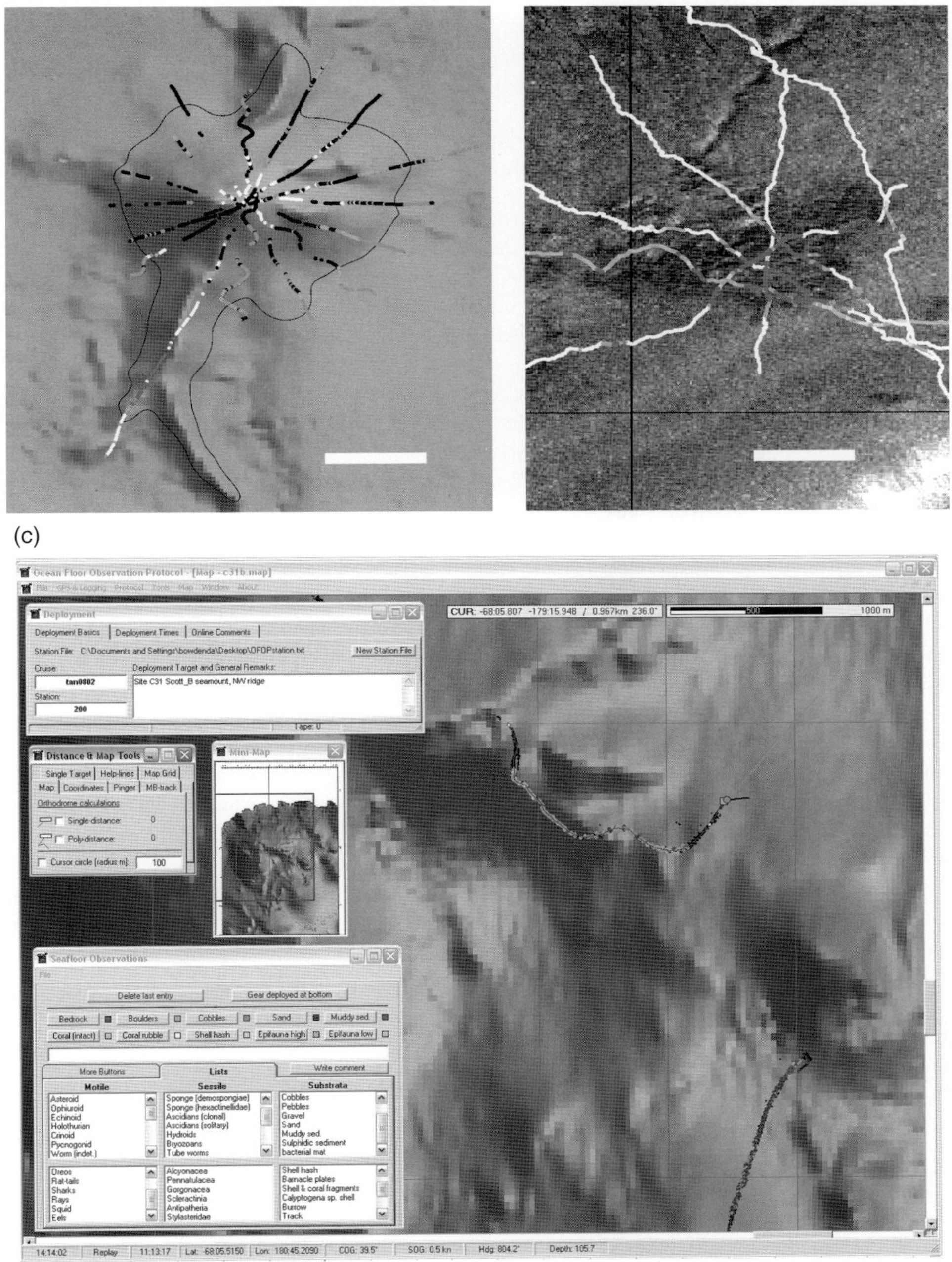

Fig. 12.5 Towed camera survey strategy. (a) Multiple still image transects running downslope from summit to base of Morgue Seamount, New Zealand, planned to map substrates and biota (Williams *et al.*, 2010). Points represent individual still photograph and are colour coded by substrate categories derived from analysis of images (scale bar, 500 m). (b) Intersecting video transects planned to locate and define the extent of North Tower cold seep, New Zealand (Baco *et al.*, 2010; Klaucke *et al.*, 2010). Tracks are displayed against high-resolution side-scan sonar imagery (the seep is the dark area at centre) and are colour-coded by substrate type (scale bar; 100 m). (c) Exploratory video transects on Scott B Seamount, Antarctica (Bowden *et al.*, 2011). Screen shows tracking of the towed camera in real time (transect at upper right) against multibeam sonar bathymetry, while observations of fauna and substrata are recorded using Ocean Floor Observation Protocol software. Observations from an earlier transect are displayed simultaneously at lower right. (Please see Plate 21 for colour representation of the figure.)

software solutions for tracking real-time camera positions against bathymetry (e.g. OFOP, http://ofop.texel.com/; Fledermaus, www.ivs3d.com/) and these are likely to become more sophisticated and widely used in future.

12.3.7 Stills and video: suitability for purpose

The continuous coverage of the seabed afforded by video has major advantages with respect to detection of infrequent taxa and the precise delineation of habitat boundaries. A commonly used procedure to derive quantitative data from video transects, however, is to take a series of still 'frame grabs' from the continuous footage. With standard-definition video this approach results in very low-resolution images and highlights some disadvantages of video for quantitative work:

- It can be difficult to make detailed counts of abundant features from a moving image.
- It can be difficult to make consistent, accurate, identifications from a moving image.
- Image resolution of video is poor by comparison with stills.

The availability of relatively inexpensive, high-resolution digital stills cameras allows a more precise way of working. If still images are taken at regular intervals simultaneously with video, considerable advantages are gained because each still image is effectively a high-resolution fixed quadrat embedded within the continuous video transect. Thus, the still images can be used to partially ground-truth the lower resolution video record. A more immediate advantage is that the set of still images generated during a camera transect can be viewed as a page of thumbnail files (equivalent to a traditional photographic 'contact sheet') giving an immediate overview of the transect and showing clearly where substratum or habitat transitions or other points of interest occur.

The development of high-definition (HD) video opens the possibility of taking frame grabs from video that are closer in resolution to images from dedicated stills cameras (Lutz *et al.*, 2002). However, while developments in video technology are bringing dramatic increases in image resolution, the current standard for HD video cameras equates to an image size of 1920 × 1080 pixels (about 2 megapixels) and thus is still at somewhat lower resolution than equivalent digital still camera images (>10 megapixels). Inexpensive HD video cameras are now widely available and are fitted to many deep-sea camera platforms. In conjunction with rapid developments in image-mosaicing techniques (Pizarro & Singh, 2003; Singh *et al.*, 2004), which join individual video frames to form continuous high-resolution still images of entire transects, they can greatly extend the analysis potential of video records.

Another alternative, enabled this time by developments in digital still camera and strobe technology, is to dispense with the video altogether and take still images at a high enough frequency that each frame overlaps with the next. Multiple images can then be mosaiced together to form extremely high-resolution continuous images of the seabed along the entire length of the transect. Systems now in use can operate at image capture rates of >5 frames per second (fps) and although these are used primarily in shallower shelf environments at present (e.g. Howland *et al.*, 2006; Singh *et al.*, 2007; Rosenkranz *et al.*, 2008) the technology has considerable potential for use in the deep sea. A significant practical advantage of this method is that using strobe lighting (as opposed to flood lighting for video) freezes motion and thus allows increased survey speed without impairing image quality.

12.4 Deployment and survey design

12.4.1 Deployment considerations

Because towed cameras are suspended on a direct cable link from the surface, any motion of the ship will be transmitted to the camera at the seabed. Thus, for a platform designed to operate at an altitude of 2–3 m above the seabed, a 2 m swell is the practical limit of useful operation, but consistently high image quality throughout transects is only achieved in minimal swell conditions. For operational decisions it is also important that the limiting factor is sea state, rather than wind strength. In flat water, stronger winds can actually make it easier to maintain consistent tow speed because the vessel's forward drive can be balanced against wind resistance. Drifting during deployment, as opposed to maintaining headway into the weather, has several potential drawbacks. First, speed when drifting is dictated by wind or current strength and cannot be controlled. Thus, if conditions change, speed may increase, decrease, or even change direction completely during deployment. Second, when drifting the vessel will lie beam-on to the wind and therefore will roll if there is any swell. If the camera platform is deployed from amidships, as is often the case, this will result in exaggerated vertical movement of the gantry which will be imparted directly to the camera platform at the seabed. Third, and most serious, if the camera platform becomes trapped at the seabed while drifting, the vessel will effectively be anchored by the towed camera platform and will place great strain on both the cable and the platform structure while it gets under way and steams back upwind to free the gear. If the vessel is steaming into the weather rather than drifting when the camera becomes trapped, simply disengaging drive and allowing the ship to fall back along its course is likely to free the gear. Some of these considerations are largely overcome if the vessel has dynamic positioning (DP) capability, but the sea state and vessel's orientation to the swell remain fundamental to achieving high-quality results with towed camera gear.

The speed at which the camera travels across the seabed is critical for the quality of video but not still images. Still images are illuminated by near-instantaneous strobe lighting and therefore movement is effectively frozen and good-quality images can be recorded at high tow speeds (>1 $m\,s^{-1}$). For video, maximum acceptable speed varies with altitude and with the frame rate. In practice, at altitudes of 2–3 m and recording at 25–29 fps, optimum detail is retained at tow speeds of ~0.25 $m\,s^{-1}$. Tow speeds in excess of 0.5 $m\,s^{-1}$ (1 knot) will cause blurring of the image. Greater speeds can also have unwanted consequences in terms of colliding with, or becoming trapped under, seabed obstacles because there is insufficient warning time for the winch operator to react.

Maintaining near-constant altitude above the seabed is also critical to all image quality: too high or too low and images will be incorrectly exposed and detail will be lost. All towed camera systems, therefore, require some method for simultaneously monitoring and adjusting altitude in real time. Whether this is achieved by real time video feed, acoustic altimeter, depth sensor, or some combination of these, it is essential that there should be open communication and understanding of deployment aims between scientists, helmsman and winch operator.

Although maintaining constant altitude is relatively straightforward on level seabeds, it is more problematic on the steep, heterogeneous topography often associated with, for instance, seamounts and canyons (Clark & Rowden, 2009; Schlacher *et al.*, 2010). In such environments, it is particularly valuable to have accurate acoustic tracking of the camera platform and for this information to be plotted in real time against high-resolution multibeam echosounder or side-scan sonar maps of the seabed. Arguably the safest approach on steep gradients in many cases is to run towed camera transects downslope, rather than upslope, because this reduces the risk of the camera becoming trapped under overhangs or being hit by rock slides started by accidental contact with the seabed. However, if the principal viewing camera on the towed vehicle is orientated obliquely forward, working downslope results in greater camera-to-seabed distances because the camera will be looking outward, away from the hill. On steep gradients, this can result in poor imaging of the seabed. Working upslope, by contrast, places the camera at a much more effective viewing angle in relation to the seabed, but unless the vehicle is fitted with an additional forward-looking camera or sonar to enable winch-operators to judge impending obstacles accurately, this is a higher risk approach. The other alternative, running transects as traverses across the slope at constant depth, is possible but because the actual track of the vessel, and therefore the camera vehicle, is strongly influenced by surface conditions and is rarely straight, it is likely that the final result will be either an up- or a downslope tow anyway, if not a combination of the two, which can make altitude control more difficult.

12.4.2 Operational modes

Towed camera deployments in the deep sea can be classified into four broad modes of operation associated with different research aims:

- *Searching*: The aim is to find a specific habitat or feature that is known, or suspected, to exist in the study area. Example habitats include hydrothermal vents and cold seeps, the presence of which can be deduced from water column chemistry gradients or acoustic flares but for which camera deployments are needed to pinpoint actual seabed locations and extents (Baco *et al.*, 2010).
- *Exploring*: The aim is to investigate an unsampled area for which there is no *a priori* knowledge (e.g. Bowden *et al.*, 2011).
- *Mapping*: The aim is to produce representative data which can be used to extrapolate, for instance, substrate or faunal assemblage types to the surrounding areas on the basis of broader scale knowledge of the environment (e.g. Kostylev *et al.*, 2001).
- *Monitoring*: This involves revisiting earlier transects in order to record, for instance, recovery of benthic assemblages after disturbance (e.g. Williams *et al.*, 2010).

Within each of these modes and depending on the research priorities, individual camera transects can be arranged in at least four different ways:

- targeted within specific habitats or features, e.g. intersecting transects within a chemosynthetic habitat (Klaucke *et al.*, 2010);
- stratified or orientated by environmental gradient, e.g. with depth on seamounts (McClain *et al.*, 2010; Williams *et al.*, 2010);
- randomly distributed, e.g. to provide statistically robust population estimates within a study area (Tuck *et al.*, 1997; Buckland *et al.*, 2001);
- evenly spaced grid – primarily for mapping and exploration where little is known about environmental gradients or benthic distributions (e.g. Barry *et al.*, 2003).

The actual survey design used in any study will, of course, be decided on the basis of the specific research questions being addressed, the topography of the area, and the sea state at the time of deployment (Fig. 12.5).

12.4.3 Length of transects

For internally recording and internally powered systems, transect length is generally dictated by the capacity of image storage media (whether film, videotape, digital flash memory or computer hard drive) or battery duration. However, while it can simplify subsequent analysis if all deployments are of the same length, there is no theoretical need to standardize the length of camera transects in the way that is done for other seabed sampling gear, such as trawls and sleds. This is because the seabed 'sample' that is brought to the surface from a towed camera transect retains its spatial structure, whereas in sled or trawl samples specimens captured at one point of the transect cannot be distinguished from those captured at any other. Therefore, in practice, transects of video or still images can be of any length and appropriate sections can subsequently be edited out from them for quantitative analysis. Thus, whereas data from conventional sampling gear would be standardized by applying a factor based on a standard distance to all samples (e.g. if one tow was twice the length of the standard distance, the factor applied to it would be * 0.5), the more accurate method with photographic data is to use only data recorded from a standard measured distance on the seabed. For instance, if the shortest transect in a study were, say 900 m in length, all remaining transects would be cropped to this length and only data from these uniform 900 m length portions would be used in analyses.

For systems that are powered from, and save data directly to, the surface vessel, individual deployments can last for many hours and may include turns and re-crossings of earlier paths. For such deployments, post-deployment editing of transects is often essential because the 'turn' segments are often of poor quality because of changes in altitude caused by changes in tow speed, and may be necessary in the first place because the camera has moved away from the seabed area of interest. For quantitative analysis, such tracks also pose greater problems with respect to spatial autocorrelation of data (Fortin & Dale, 2005). The practical decision to be made in these circumstances is whether it would be more efficient to retrieve and re-deploy the camera several times rather than running continuously.

12.5 Management of images and metadata

Visual samples are now generally acquired as digital files, rather than films and prints, and consequently there is enormous potential to make more effective use of them through server-based archives, databases, geographic information systems (GIS) and web visualization tools. If they are to be useful for analyses at any level, however, visual samples must be labelled and catalogued with the same attention to detail as are other scientific samples: much like unsorted fauna samples, they require further processing and considerable investment of time and taxonomic expertise before useful ecological data can be extracted. Furthermore, in the same way that physical fauna samples are archived following full identification, so images should be stored in such a way as to be readily accessible to subsequent studies which may be able to mine more information than was extracted for the original research purpose (see also section on image data in Chapter 16). Perhaps because of the sheer volume of images that can be captured with modern equipment, and the relative novelty of visual sampling as a scientific tool, it is all too easy for image files to be poorly archived. This may also be related to the mode of use, in that cameras are commonly used as a search tool, rather than a sampling tool, and it is often the case that no one 'owns' or has a vested interest in the images as a primary data source for further study.

12.5.1 Metadata

Photographic samples and their associated data are most useful when viewed within the context of all other data collected, including geographic position, bathymetry, water column chemistry and physical structure, current profiles and geology. Most underwater video systems used in seabed research have some facility to write basic data, such as position, depth and time, directly on to the image, from which the footage can be unambiguously identified and linked to the appropriate metadata. This is less commonly the case with still images, however, and there is consequently greater potential for images to become separated from their core reference data. This is of concern both for immediate use of the images and in relation to their future archival value. In commercial photography it is now common practice to write image metadata into EXIF fields (www.exif.org/) embedded in the digital image file using standard image-management software packages. This system of working has the great advantage that the image file then carries with it all data essential for its own interpretation.

The International Council for the Exploration of the Sea (ICES) publishes guidelines for metadata fields for a range of physical and biological oceanographic sample types (see data guidelines on http://ices.dk/publications/library/Pages/default.aspx) but not, as yet, for visual samples. Similar guidelines could usefully be prepared for seabed image samples specifying the basic data required for their cataloguing and interpretation. Minimum data would include:

- date and time,
- cruise identifier,
- station/event number,

- latitude and longitude (start and end of video transects, position of individual stills, ideally with indication of the type of navigation system used and its associated errors),
- depth,
- scale information (e.g. laser spacing),
- sequential number within transect (for stills),
- time code within transect (for video frame grabs) and
- reference details of the towed camera system used.

The most important of these are accurate date and time, appropriately referenced to Coordinated Universal Time (UTC), because with this information images can be referenced to all other data recorded during the transects.

12.5.2 Databases

The most effective way to link visual samples with metadata is by incorporation into a spatially enabled relational database. Most ocean-going research vessels operate some form of database which links navigational information from the ship's systems with scientific survey data, and the major institutions making regular use of camera systems in the deep sea have effective systems to incorporate images into such a database. Having digital image files and their associated metadata automatically uploaded to a server and lodged in a spatially enabled database affords the potential immediately to view new visual information in the context of the full range of oceanographic data collected at the time of image acquisition. There are now many possibilities for using GIS and web mapping tools to interface with spatially enabled databases and to view seabed images directly within these visualization tools (see for example, http://habcam.whoi.edu). Development in this field is rapid and the potential of such visualization methods for synthesizing information from seabed images across studies is immense.

12.6 Data extraction and analysis

12.6.1 Counts and measures

A range of data types can be extracted from towed camera seabed images depending on the quality of the images and the research questions to be addressed. For instance, individual studies might aim to record: all fauna to the finest taxonomic resolution possible (e.g. Kostylev *et al.*, 2001; Teixido *et al.*, 2002); lebenspurren (e.g. Jones *et al.*, 2007); substratum type, grain size and areal coverage (e.g. Wright *et al.*, 2002); population densities of target species (e.g. Mortensen & Buhl-Mortensen, 2004); categorical descriptors of benthic assemblage type (e.g. Greene *et al.*, 1999) or any combination of these.

Identification of fauna from visual samples, particularly video, can be problematic for all but the more distinctive taxa or coarser taxonomic groupings, and the reliability of identifications, can vary not just between studies but within transects and even within individual still image frames. At the same time, accurate counts of individuals or measurements of areal coverage can be both time-consuming and of varying accuracy. For these reasons, many analyses of visual samples employ identification schemes based on high-level taxonomic groupings (e.g. Class, Order), operational taxonomic units (OTUs) or assemblage descriptors (e.g. 'low-relief epifauna', 'sponge community') and use abundance categories rather than counts (e.g. 'low', 'medium', 'high'). This leads to a situation in which there can be large differences between studies in the type of analysis scheme used.

To enable routine comparison between studies, there would need to be some agreement, or means of interpretation, with regard to which objects or qualities are recorded, and which quantitative measures are used. In practice, comparisons between data sets which use different measurement methods or taxonomic resolutions require that all analyses are constrained to the level of 'lowest common denominator' categories, detailed information in finer resolution data sets being aggregated to the level of the coarsest. Thus, if one study reports abundances of named holothurian species but others report only the abundance of holothurians as a Class, the comparison dataset would necessarily group all data at the level of Class (see for example, Williams *et al.*, 2010).

Because benthic taxa are classified in hierarchical taxonomies, arguably the most useful way to analyse seabed visual samples for fauna is to identify all individuals to the finest taxonomic level achievable, regardless of how much this level changes within or between transects. As long as all identification labels are embedded within a standard hierarchy, the final data set can then be aggregated to whichever level of the hierarchy (e.g. Family, Order, Class) is most appropriate. A major advantage of this approach is that it does not apply any prior assumptions as to the level of detail that can be determined, and thus makes maximum use of the information available in the image. In practice, a library of taxon identification images can be generated initially from the camera transects that are to be analysed, either as crops from still images or frame grabs from video during preliminary viewing. This library can then be used by analysts to apply working identification labels as they go through the transects in detail. A useful aspect of this is that accurate formal identifications are not needed before analysis begins: working labels can be assigned (e.g. 'Asteroid_1') and while the analysis proceeds, identification images from the library can be sent to the relevant taxonomic experts for formal identification. As the formal names are confirmed, they are then substituted for the working labels in the analysis database.

A standardized classification and measurement scheme for the analysis of visual samples might be feasible at higher levels of taxon, assemblage or substratum descriptors, and categorical abundance scales, as has been proposed for deep-sea environments in general by Greene *et al.* (1999). However, such a scheme would need to incorporate all schema currently used into a broad hierarchical structure and it is questionable whether standardization at such a high level would either be practicable or would result in appreciable gains over the present situation.

12.6.2 Data extraction methods

There are important differences in the quality of data that can be derived from the two types of photographic sample: video and still images. Because video transects (and extensive mosaics of overlapping still images) provide continuous coverage over relatively large areas of the seabed, they can provide accurate data on the spatial distribution and abundance of features which are sparsely distributed, conspicuous, and readily identified, such as larger, rare epifauna (e.g. Bowden *et al.*, 2011). For the same reasons, video is also well suited to identifying the spatial extent of substratum types and biological habitats. However, for quantification of highly abundant features, or for measurement of areal coverage of substrata and sessile biota, still images are more useful.

The techniques used for estimating abundances of fauna or areal coverage of substrata from both video and still images are closely tied to camera orientation and evenness of illumination across the image. If the optical axis of the camera is perpendicular to the substratum (vertical orientation) and illumination is even across the field of view, the probability of detecting an object of a given size is equal for all regions of the image and direct counts and areal measurements can be made once the image area has been calculated. However, if the camera axis is at an oblique angle to the substratum, two gradients are introduced: a gradient of increasing field width, and hence decreasing spatial resolution, towards the top of the image frame; and a gradient of decreasing illumination towards the top and sides of the image frame. Thus, the probability of detection is highest in the foreground of the image and decreases with decreasing illumination and increasing distance towards the background. For still images or video frame grabs of this sort with strong gradients quantitative data can be derived by treating the image as a point transect in the distance-sampling sense (Buckland *et al.*, 2001; Thomas *et al.*, 2002) with detection probabilities being calculated for different regions of the image based on the known gradients (Smith *et al.*, 2003). For oblique video, the simplest and arguably the best practice is to count only those targets which pass through a horizontal line or 'letter box' superimposed across the foreground of the image in the region of maximum illumination and optimum resolution (Hewitt *et al.*, 2004). If this line is in the plane of two scaling lasers, reliable size estimates are also practicable.

Techniques such as the visual fast count (Kimmel, 1985) are modifications of distance sampling in which it is assumed that the probability of encountering any taxon is equal throughout the transect. The transect is divided into a number of equal time segments and estimates of abundance for each taxon over the entire transect are derived based on the number of individuals observed in the transect segment in which that taxon first occurs (e.g. Magorrian & Service, 1998; Strong *et al.*, 2006). The primary advantages of such methods are that they reduce the influence of variations in field of view caused, for instance, by inconsistent camera altitude, and they reduce the processing time required to derive taxon abundance data. However, in many practical applications, video transects will encompass more than one type of habitat and therefore the assumption that the probability of encountering a given taxon is constant throughout the transect is unlikely to be correct. A useful approach to the analysis of transects which encompass multiple habitat types is to stratify by habitat type (e.g. Williams *et al.*, 2010). This can be done using, for instance, a clustering algorithm operating on substratum data extracted from individual images, or simply by visual assessment of habitat transitions.

12.6.3 Real-time observations

An alternative, or an addition, to exhaustive post-capture data extraction is to make full use of the time spent watching video or still images during camera deployments. This is usually a time when observers are concentrating fully on the images and can be more attuned to changes in substrata and fauna than they are during post-voyage viewings of the recorded tapes. In the past, recording observations in real time entailed writing to a paper log, a procedure which was problematic in terms of consistency of reporting, speed of recording and post-voyage transcription. The development of real-time tracking and logging systems which can be integrated with, for instance, taxonomic databases, maximize the amount of useful data that can be recorded during deployments by providing a simple intuitive interface with standardized observations. Examples include the Video Annotation and Reference System (VARS, www.mbari.org/vars/, see also Chapter 16) developed at the Monterey Bay Aquarium and Research Institute (MBARI), and the Ocean Floor Observation Protocol (OFOP, http://www.ofop-by-sams.eu/). Using these systems, observers record time-stamped observations by clicking on labelled buttons or lists in a software interface (Fig. 12.5). Because the set of available observations (taxon lists, substratum types, etc.) is determined in advance, consistency of reporting between transects is improved and the observations can be stored directly to a spatially enabled database for analysis. The level of detail that can be recorded in real time during deployments is necessarily less than with laboratory-based analyses where slow motion, freeze frame and replay facilities are available. The advantages, however, are that geo-referenced distributional data are immediately available while at sea, which can be valuable for the planning of subsequent sampling, and that the observation files provide a time-coded index for locating points of interest in the video recordings during post-voyage analyses.

12.7 Methods reporting

As with any other research technique, a minimum standard of methods reporting is needed for visual sampling to ensure valid comparisons between studies. For the image acquisition stage, this standard would include details of:

- platform type,
- camera orientation (including angle away from vertical if oblique),
- imaged area and scaling method,

- target altitude above the seabed,
- image resolution,
- spatial referencing method and
- spatial distribution of transects.

In addition, the methods used for image processing, data extraction and analysis must be clearly stated if the resulting data are to be correctly interpreted. For instance, confusion can arise if details are not supplied for any image corrections or enhancements that were used, the criteria used to select video frame grabs or to subsample from sets of still images, decisions to include or exclude faunal groups, and whether counts were based on area or distance sampling protocols. Exemplary models of methodology reporting for seabed image studies can be found in Grassle *et al.* (1975) and Stone (2006).

Image resolution is perhaps the most fundamental property of visual samples to be included in published methods because it dictates the level of data that can be reliably determined from the image. Unfortunately, it is also one of the most difficult factors to quantify consistently, particularly for oblique images in which there are inherent gradients of effective resolution within the image, and for which the actual angle of orientation is rarely documented. Where original images are available and have the relevant scaling information appended, the resolution of the image can be determined empirically. However, where image-derived data are presented alone, as in most journal articles to date, it is important that the effective resolution of the camera system should be clearly stated (e.g. 'organisms >approx. 5 cm were identifiable') or that technical details of the camera system from which resolution can be calculated should be included in the methods. Of these, the former is the simplest and probably the most reliable approach.

12.8 Summary

Photographic imaging of the seabed provides insights into the fine-scale structure of habitats and distributions of organisms that are not captured by conventional sampling methods. In deep-sea research, towed camera systems are an effective and practical means of aquiring high-quality video and still images over relatively large areas. Although they have limitations by comparison with more sophisticated platforms such as ROVs and AUVs, these are balanced by their greater robustness, simplicity of construction and use, and substantially lower build and operational costs. The usefulness of images from towed cameras for quantitative analyses is dependent on careful design of the imaging system, close attention to all aspects of image quality control and metadata recording during deployment, and meticulous archiving of image files and all associated data after deployment. Increasing use of underwater camera systems by science institutes and industry (e.g., oil and gas, and mining sectors) around the world is generating a huge volume of high-quality imagery of the deep seabed which contains a wealth of ecological information. Much of this remains untapped, however, in part because of the variations between studies in both image quality and type, but also because image-derived data have been little used in ecological research to date and thus no consensus has developed about standard analytical procedures. The main challenges are to refine the techniques by which quantitative data are extracted from images, and to work towards greater comparability of data between studies.

Acknowledgements

This chapter is based on a review paper funded by a Census of Marine Life on Seamounts (CenSeam) minigrant and originally prepared for a meeting of the CenSeam Standardisation Working Group at Dartington, UK, in May 2007. Many thanks to: Dan Fornari and Stace Beaulieu, Woods

Hole Oceanographic Institution, USA; Alan Williams, Bruce Barker, Franzis Althaus, Rudy Kloser and Matt Sherlock, Commonwealth Science and Research Organisation, Australia; Roger Coggan, Centre for Environment Fisheries and Aquaculture Research, UK; Pål Buhl-Mortensen, Institute of Marine Research, Norway; Julian Gutt, Alfred Wegner Institute for Polar and Marine Research, Germany; Leigh Marsh, the National Oceanography Centre, UK, and Ashley Rowden and Mireille Consalvey, National Institute of Water and Atmospheric Research (NIWA), New Zealand, for material, discussions, and constructive reviews that helped to structure the document.

References

Abdo, D.A., Seager, J.W., Harvey, E.S., McDonald, J.I., Kendrick, G.A., Shortis, M.R. (2006). Efficiently measuring complex sessile epibenthic organisms using a novel photogrammetric technique. *Journal of Experimental Marine Biology and Ecology*, 339: 120–133.

Auster, P.J., Malatesta, R.J., Larosa, S.C. (1995). Patterns of microhabitat utilization by mobile megafauna on the southern New England (USA) continental-shelf and slope. *Marine Ecology Progress Series*, 127: 77–85.

Baco, A.R., Rowden, A.A., Levin, L.A., Smith, C.R., Bowden, D.A. (2010). Initial characterization of cold seep faunal communities on the New Zealand Hikurangi margin. *Marine Geology*, 272: 251–259.

Barker, B.A.J., Helmond, I., Bax, N.J., Williams, A., Davenport, S., Wadley, V.A. (1999). A vessel-towed camera platform for surveying seafloor habitats of the continental shelf. *Continental Shelf Research*, 19: 1161–1170.

Barry, J.P., Grebmeier, J.M., Smith, J., Dunbar, R.B. (2003). Oceanographic versus seafloor-habitat control of benthic megafaunal communities in the S.W. Ross Sea, Antarctica. *Antarctic Research Series*, 78: 327–354.

Beaulieu, S.E., Singh, H., Smith, K.L. (2005). Analysing time-lapse photographs of the sea floor for changes in benthic community activity. *American Fisheries Society Symposium*, 41: 245.

Bellec, V., Wilson, M., Bøe, R., *et al.* (2008). Bottom currents interpreted from iceberg ploughmarks revealed by multibeam data at Tromsøflaket, Barents Sea. *Marine Geology*, 249: 257–270.

Bowden, D.A. (2005). Quantitative characterisation of shallow marine benthic assemblages at Ryder Bay, Adelaide Island, Antarctica. *Marine Biology*, 146: 1235–1249.

Bowden, D.A., Schiaparelli, S., Clark, M.R., Rickard, G.J. (2011). A lost world? Archaic crinoid-dominated assemblages on an Antarctic seamount. *Deep Sea Research II*, 58: 119–127.

Buckland, S.T., Anderson, D.R., Burnham, K.P., Laake, J.L., Borchers, D.L., Thomas, L. (2001). *Introduction to distance sampling: estimating abundance of biological populations*. Oxford: Oxford University Press.

Chezar, H., Lee, J. (1985). A new look at deep-sea video. *Deep Sea Research A*, 32: 1429–1436.

Clark, M.R., Rowden, A.A. (2009). Effect of deepwater trawling on the macro-invertebrate assemblages of seamounts on the Chatham Rise, New Zealand. *Deep Sea Research I*, 56: 1540–1554.

Dundas, K., Przeslawski, R. (2009). *Deep sea lebenspurren: biological features on the seafloor of the eastern and western Australian margin*. Geoscience Australia Record 2009/26. Canberra: Geoscience Australia, 76 pp.

Fornari, D.J. (2003). A new deep-sea towed digital camera and multi-rock coring system. *EOS, Transactions American Geophysical Union*, 84: 69–76.

Fortin, M.-J., Dale, M.T.R. (2005). *Spatial analysis: a guide for ecologists*. Cambridge, UK: Cambridge University Press, 365 pp.

Fossa, J.H., Lindeberg, B., Christensen, O., Lundalve, T., Svellingen, I., Mortensen, P.B., Alvsvag, J. (2005). Mapping of *Lophelia* reefs in Norway: experiences and survey methods. In: Freiwald, A., Roberts, J.M. (eds), Cold-water corals and ecosystems. New York: Springer, pp. 359–391.

Grassle, J.F., Sanders, H.L., Hessler, R.R., Rowe, G.T., McLellan, T. (1975). Pattern and zonation: a study of the bathyal megafauna using the research submersible Alvin. *Deep Sea Research*, 22: 475–481.

Greene, H.G., Yoklavich, M.M., Starr, R.M., *et al.* (1999). A classification scheme for deep seafloor habitats. *Oceanologica Acta*, 22: 663–678.

Hessler, R.R., Smithey, W.M., Keller, C.H. (1985). Spatial and temporal variation of giant clams, tube worms and mussels at deep-sea hydrothermal vents. *Biological Society of Washington Bulletin*, 6: 411–428.

Hewitt, J.E., Thrush, S.E., Legendre, P., Funnell, G.A., Ellis, J., Morrison, M. (2004). Mapping of marine soft-sediment communities: integrated sampling for ecological interpretation. *Ecological Applications*, 14: 1203–1216.

Hill, P. (2009). Designing a deep-towed camera vehicle using single conductor cable. *Sea Technology:* 49–51.

Howell, K.L., Davies, J.S., Narayanaswamy, B.E. (2010). Identifying deep-sea megafaunal epibenthic assemblages for use in habitat mapping and marine protected area network design. *Journal of the Marine Biological Association of the UK*, 90: 33–68.

Howland, J., Gallager, S., Singh, H., *et al.* (2006). Development of a towed survey system for deployment by the fishing industry, *Oceans 2006*, Vols 1–4: 399–403.

Jones, D.O.B., McPhail, S.D., Bett, B.J., Flewellen, C., Conquer, M. (2005). Seabed photography from an autonomous underwater vehicle. *Journal of Marine Science and Environment*, C3: 29–36.

Jones, D.O.B., Hudson, I.R., Bett, B.J. (2006). Effects of physical disturbance on the cold-water megafaunal communities of the Faroe-Shetland Channel. *Marine Ecology Progress Series*, 319: 43–54.

Jones, D.O.B., Bett, B.J., Tyler, P.A. (2007). Megabenthic ecology of the deep Faroe-Shetland channel: a photographic study. *Deep-Sea Research I*, 54: 1111–1128.

Jones, D.O.B., Bett, B.J., Wynn, R.B., Masson, D.G. (2009). The use of towed camera platforms in deep-water science. *Underwater Technology*, 28: 41–50.

Kimmel, J.J. (1985). A new species-time method for visual assessment of fishes and its comparison with established methods. *Environmental Biology of Fishes*, 12: 23–32.

Klaucke, I., Weinrebe, W., Petersen, C.J., Bowden, D. (2010). Temporal variability of gas seeps offshore New Zealand: multi-frequency geoacoustic imaging of the Wairarapa area, Hikurangi margin. *Marine Geology*, 272: 49–58.

Kostylev, V.E., Todd, B.J., Fader, G.B.J., Courtney, R.C., Cameron, G.D.M., Pickrill, R.A. (2001). Benthic habitat mapping on the Scotian Shelf based on multibeam bathymetry, surficial geology and sea floor photographs. *Marine Ecology-Progress Series*, 219: 121–137.

Lutz, R., Shank, T., Rona, P., *et al.* (2002). Recent advances in imaging deep-sea hydrothermal vents. *Cahiers De Biologie Marine*, 43: 267–269.

Magorrian, B.H., Service, M. (1998). Analysis of underwater visual data to identify the impact of physical disturbance on horse mussel (*Modiolus modiolus*) beds. *Marine Pollution Bulletin*, 36: 354–359.

McClain, C.R., Lundsten, L., Barry, J., DeVogelaere, A. (2010). Assemblage structure, but not diversity or density, change with depth on a northeast Pacific seamount. *Marine Ecology*, 31: 14–25.

Mortensen, P.B., Roberts, J.M., Sundt, R.C. (2000). Video-assisted grabbing: a minimally destructive method of sampling azooxanthellate coral banks. *Journal of the Marine Biological Association of the UK*, 80: 365–366.

Mortensen, P.B., Buhl-Mortensen, L. (2004). Distribution of deep-water gorgonian corals in relation to benthic habitat features in the Northeast Channel (Atlantic Canada). *Marine Biology*, 144: 1223–1238.

Mortensen, P.B., Buhl-Mortensen, L. (2005). Deep-water corals and their habitats in The Gully, a submarine canyon off Atlantic Canada. In: Freiwald, A., Roberts, J.M. (eds), *Cold-water Corals and Ecosystems*. Berlin Heidelberg: Springer-Verlag, pp. 247–277.

Owen, D.M., Sanders, H.L., Hessler, R.R. (1967). Bottom photography as a tool for estimating benthic populations. In: Hersey, J.B. (ed.), *Deep-sea photography*. Baltimore, MD: Johns Hopkins Press, pp. 229–234.

Pilgrim, D.A., Parry, D.M., Jones, M.B., Kendall, M.A. (2001). ROV image scaling with laser spot patterns. *Journal of the Society for Underwater Technology*, 24: 93–103.

Pizarro, O., Singh, H. (2003). Toward large-area mosaicing for underwater scientific applications. *IEEE Journal of Oceanic Engineering*, 28: 651–672.

Post, A.L., O'Brien, P.E., Beaman, R.J., Riddle, M.J., De Santis, L. (2010). Physical controls on deep water coral communities on the George V Land slope, East Antarctica. *Antarctic Science*, 22: 371–378.

Pratt, R.M. (1967). Photography of seamounts. In: Hersey, J.B. (ed.), *Deep sea photography*. Baltimore, MD: Johns Hopkins Press, pp. 145–158.

Rice, A.L., Angel, M.V., Grassle, J.F., *et al.* (1994). Suggested criteria for describing deep-sea benthic communities; the final report of SCOR Working Group 76. *Progress in Oceanography*, 34: 81–100.

Rinke, C., Lee, R.W. (2009). Macro camera temperature logger array for deep-sea hydrothermal vent and benthic studies. *Limnology and Oceanography – Methods*, 7: 527–534.

Rosenkranz, G.E., Gallager, S.M., Shepard, R.W., Blakeslee, M. (2008). Development of a high-speed, megapixel benthic imaging system for coastal fisheries research in Alaska. *Fisheries Research*, 92: 340–344.

Schlacher, T.A., Williams, A., Althaus, F., Schlacher-Hoenlinger, M.A. (2010). High-resolution seabed imagery as a tool for biodiversity conservation planning on continental margins. *Marine Ecology*, 31: 200–221.

Sherlock, M., Kloser, R., Williams, A., Ryan, T. (2010). A combined benthic grab and observation system for rapid assessment of water column and seabed. In: *Oceans 2010*. Sydney: IEEE, pp. 1–5.

Shortis, M.R., Seager, J.W., Williams, A., Barker, B.A., Sherlock, M. (2008). Using stereo-video for deep water benthic habitat surveys. *Marine Technology Society Journal*, 42: 28–37.

Singh, H., Howland, J., Pizarro, O. (2004). Advances in large-area photomosaicking underwater. *IEEE Journal of Oceanic Engineering*, 29: 872–886.

Singh, H., Roman, C., Pizarro, O., Eustice, R., Can, A. (2007). Towards high-resolution imaging from underwater vehicles. *International Journal of Robotics Research*, 26: 55–74.

Smith, C.J., Marrs, S.J., Atkinson, R.J.A., Papadopoulou, K.N., Hills, J.M. (2003). Underwater television for fisheries-independent stock assessment of Nephrops norvegicus from the Aegean (eastern Mediterranean) Sea. *Marine Ecology Progress Series*, 256: 161–170.

Solan, M., Germano, J.D., Rhoads, D.C., *et al.* (2003). Towards a greater understanding of pattern, scale and process in marine benthic ecosystems: a picture is worth a thousand worms. *Journal of Experimental Marine Biology and Ecology*, 285–286: 313–338.

Stone, R.P. (2006). Coral habitat in the Aleutian Islands of Alaska: depth distribution, fine-scale species associations, and fisheries interactions. *Coral Reefs*, 25: 229–238.

Strong, J.A., Service, M., Mitchell, A.J. (2006). Application of the Visual Fast Count for the quantification of temperate epibenthic communities from video footage. *Journal of the Marine Biological Association of the UK*, 86: 939–945.

Teixido, N., Garrabou, J., Arntz, W.E. (2002). Spatial pattern quantification of Antarctic benthic communities using landscape indices. *Marine Ecology Progress Series*, 242: 1–14.

Thomas, L., Buckland, S.T., Burnham, K.P., *et al.* (2002). Distance sampling. In: El-Shaarawi, A.H., Piegorsch, W.W. (eds), *Encyclopedia of environmetrics*. Chichester: John Wiley & Sons, pp. 544–552.

Thurber, A.R., Kroger, K., Neira, C., Wiklund, H., Levin, L.A. (2010). Stable isotope signatures and methane use by New Zealand cold seep benthos. *Marine Geology*, 272: 260–269.

Tuck, I.D., Chapman, C.J., Atkinson, R.J.A., Bailey, N., Smith, R.S.M. (1997). A comparison of methods for stock assessment of the Norway lobster, *Nephrops norvegicus*, in the Firth of Clyde. *Fisheries Research*, 32: 89–100.

Wakefield, W.W., Genin, A. (1987). The use of a Canadian (perspective) grid in deep-sea photography. *Deep Sea Research I*, 34: 469–478.

Williams, A., Schlacher, T.A., Rowden, A.A., *et al.* (2010). Seamount megabenthic assemblages fail to recover from trawling impacts. *Marine Ecology*, 31 (suppl. 1): 183–199.

Wright, I.C., Stoffers, P., Hannington, M., *et al.* (2002). Towed-camera investigations of shallow-intermediate water-depth submarine stratovolcanoes of the southern Kermadec arc, New Zealand. *Marine Geology*, 185: 207–218.

Chapter 13
Submersibles and Remotely Operated Vehicles

Christopher Kelley[1], Terrence Kerby[1], Pierre-Marie Sarradin[2], Jozée Sarrazin[2] and Dhugal J. Lindsay[3]

[1] Hawaii Undersea Research Laboratory, University of Hawaii, Honolulu, United States of America
[2] Institut Francais de Recherche pour l'Exploitation de la Mer, Brest, France
[3] Japan Agency for Marine-Earth Science and Technology, Yokosuka, Japan

Abstract

Submersibles and remotely operated vehicles (ROVs) provide the means for conducting precise and selective sampling of deep-sea organisms both in midwater and in all types of benthic terrain. This chapter introduces the advantages of using these vehicles, their key features and how they differ from each other. Submersible and ROV sampling gear is essentially the same and often involves the use of manipulators. The different types of samplers are described, followed by a brief description of the storage gear used to transport the specimens to the surface. Typical sampling operations are covered, including the collection of ancillary data such as *in situ* images of the specimens as they are being collected as well as their habitat, conductivity, temperature and depth (CTD) data at the collection site, and GPS coordinates. Finally a brief discussion is included about onboard specimen processing.

Keywords human occupied vehicle (HOV), submersible, remotely operated vehicle (ROV), manipulator, manipulator-deployed sampler, pot sampler, net sampler, suction sampler, tube-corer, box-corer, bottle sampler

13.1 Introduction

Submersibles (i.e. human occupied vehicles or HOVs) and ROVs (remotely operated vehicles) provide a level of precision and selectivity for deep-sea biological sampling that cannot be achieved using any other technique. These vehicles place humans and/or cameras directly into the deep sea where they can provide unprecedented guidance during sampling activities. Submersibles are untethered and thus decoupled from the ship, thereby providing considerable flexibility in selecting and sampling specimens. ROVs are coupled to a surface ship via a cable and are somewhat more constrained in their movements. However, ROVs can also achieve a high level of sampling precision under reasonable sea conditions and if operated from a ship with dynamic positioning capabilities. Researchers have used these vehicles to achieve the following sampling goals:

- *Sampling at specific locations and depths*: Highly accurate tracking and geo-positioning of these vehicles has been achieved with the use of ultrashort baseline acoustic positioning systems

Biological Sampling in the Deep Sea, First Edition. Edited by Malcolm R. Clark, Mireille Consalvey and Ashley A. Rowden.

(USBL). USBL systems determine the range and bearing of the vehicle to the support ship on the surface and calculate location by combining these data with the ships' GPS position. The margin of error is dependent on depth and horizontal range from the ship but is generally no greater than 15 m when the systems are properly calibrated. USBL systems allow precise GPS coordinates to be obtained where and when specimens are collected. They also allow submersibles and ROVs to be driven to pre-determined sampling sites or repeatedly return to the same site.

- *Sampling specific species, individuals, and parts of individuals*: Visual examination of potential specimens, whether in person or through video, provides the means to select only specimens of target species, thereby increasing the efficiency and decreasing the environmental impact of sampling. Visual examination of potential specimens also provides the means to sample specific individuals or specific parts of larger individuals (e.g. branches of a coral or particular section of a sponge). For example, the live tips of branching corals are typically collected for genetic analyses whereas the bases of either live or dead corals are most important for age and growth-rate studies (Etnoyer *et al.*, 2006). Repetitive sampling of specific individuals over time is important for certain types of studies, such as those attempting to determine reproductive cyclicity (Wagner *et al.*, 2012).
- *Obtaining micro (<1 m) and macro (1–10 m) habitat and environmental data associated with sampled specimens*: Observers in submersibles as well as the video from both submersibles and ROVs provide a means to collect small-scale habitat data associated with the specimens. The exact type of substrate that specimens were attached to can be documented and included with the collection data. Observations of the other animals at the collection sites can also be recorded for ecological studies examining community structure and potential associations between different species. Submersibles and ROVs are typically deployed with CTD sensors that provide data on depth, temperature, salinity, dissolved oxygen and density during the dives. When needed for specific projects, these vehicles have also been fitted with sensors to measure factors such as pH, hydrogen sulfide and turbidity. All of these data can be precisely correlated to the locations where the specimens were collected via synchronized instrument clocks. CTD sensors, flowmeters and acoustic Doppler current profilers (ADCPs) can also be deployed next to specimens marked for repeat sampling to obtain water-quality profiles over time. This type of precise environmental monitoring associated with biological sampling is invaluable, particularly around hydrothermal vents or in midwater where thermoclines and haloclines result in rapid environmental changes.
- *Live specimen sampling*: Collecting and maintaining live specimens, particularly invertebrates, requires the precision sampling capabilities of submersibles and ROVs. These animals need to be carefully removed from their environments using manipulators or suction samplers and placed into storage containers that can be completely closed for transport to the surface. These storage containers are often designed to control as many environmental variables as possible such as light, temperature and sometimes pressure. For example, a precisely controlled light environment was required for crab retina studies (Frank *et al.*, 2012) whereas pressurized containers have been used for collecting extremophile bacteria (Malahoff *et al.*, 2002) and fish (Takemura *et al.*, 2010).

Submersibles and ROVs achieve these goals through *in situ* visual guidance, which conveys a significant advantage in precision over other sampling techniques. However, the size or volume of the area sampled and specimen holding capacity are general much smaller compared to 'blind' samplers operated from surface vehicles, and the cost of deployment is generally higher. These vehicles are therefore used for projects requiring selectivity and precision over volume and numbers of specimens, or when the target animals are too fragile to be sampled by other means (e.g. those with body water content exceeding 98% such as ctenophores, or composed of semi-independent zooids such as siphonophores).

A feature of some deep-sea ecosystems is their insularity, the importance of chemical and/or physical gradients, and the strength of interactions between organisms and environments at various spatial scales. The study of these ecosystems is impossible without the use of visually guided vehicles due to the small scale of their spatial patterns. Furthermore, the increasing anthropogenic pressure on the deep sea (e.g. offshore fisheries, waste disposal, deep-sea mining) and conservation requirements under the Convention on Biological Diversity, reinforces the importance of observing, sampling, measuring and experimenting in the deep sea using non-invasive approaches and technologies that minimize the impacts on the fragile environment.

13.2 General descriptions of submersibles and ROVs

Visual exploration of the deep sea began in 1934 with the creation of the 'bathysphere' by William Beebe and Otis Barton (Busby, 1976). The bathysphere was not a vehicle per se because it was tethered to the surface, had no propulsion or buoyancy, and was simply dropped over the side of a barge. Power for lights and communications were not independent but supplied from the ship by cables attached to the tether. Nevertheless, this was the first time humans had physically entered the deep sea and saw with their own eyes animals that lived on or near the deep seafloor. Beebe and Barton made more than 30 dives in the bathysphere, reaching a maximum depth of 923 m.

The next generation of manned submersible, called the 'bathyscaphe' was first designed and built by Auguste Piccard in 1948 and was simply called *FNRS-2* after the Belgium source of its funding: Fonds National de la Recherche Scientifique (FNRS, Busby, 1976). This was the first true deep-sea vehicle, being untethered and having its own source of propulsion and power. The *FNRS-2* and all subsequent bathyscaphes used gasoline for buoyancy, which was contained in large tanks that dwarfed the relatively tiny living sphere. The most famous bathyscaphe was the *Trieste*, which in 1960 reached the bottom of the Marianas Trench at a depth of about 10 900 m.

Neither the bathyscaphe nor the bathysphere had the means to collect biological samples and therefore served only as a way to transport humans to the seafloor to make observations. The first submersible that had collecting capabilities was the SP-350 Denise 'diving saucer' (*Soucoupe plungeante*), invented and built by Jacques Cousteau and Jean Mollard in 1959 (Busby, 1976). Like its predecessor, this submersible was untethered and had its own propulsion and power. Unlike bathyscaphes, the diving saucer was much smaller because it was fitted with with hard ballast air tanks for maintaining buoyancy instead of gasoline. This was possible because of its relatively shallow diving range (<400 m). The saucer also had liquid mercury which could be shifted fore and aft in order to adjust trim.

The first 'modern' submersible was the *Alvin* built in 1964 by the US Navy for scientific research. A major innovation was the use of syntactic foam for buoyancy which allowed for a further decrease in the size of the vehicles.

ROVs were invented somewhat in parallel to submersibles, being designed to accomplish different and more dangerous tasks. Historic references are difficult to find but according to some accounts, the Royal Navy's *Cutlet*, built some time in the 1950s, is generally considered to be the first ROV (Matsumoto & Potts, 2011). This vehicle was designed to recover practice torpedoes and was therefore fitted with an early version of a manipulator. The US Navy followed with the development of CURV or 'cable-controlled underwater recovery vehicle', designed to recover objects on the seafloor, including a lost nuclear bomb. More recently, the offshore oil and gas industry has created 'work class' ROVs to support offshore exploitation of deep-water energy resources. Since then, ROVs have become extremely useful vehicles for performing tasks during the building, inspection, maintenance and repair of human structures in the deep sea. While the offshore oil and gas industry continues to lead the use of these vehicles, ROVs are now routinely built and used for scientific purposes, many of which involve biological sampling.

13.2.1 Submersibles

A submersible is occupied, and directly operated, by humans. The core of this type of vehicle is therefore the living sphere, which varies in size and the material it is made of depending on the number of persons it is designed to support and its operating depth. For example, the Canadian one-person

DeepWorker 2000 submersibles were designed for relatively shallow dives to a maximum of 610 m and had a small pressure hull with an acrylic viewing dome. In contrast, the recent and now famous *Deepsea Challenger* was designed and built to transport a single person, looking out of a single view port, to a depth of 11 000 m in the Marianas Trench. Intermediate submersibles include the *Delta* submersible which accommodates two persons, the pilot and observer, has a maximum operating depth of 365 m and has a unique configuration consisting of 19 acrylic view ports. In general, submersibles with large acrylic living spheres such as the *Johnston Sealink* submersibles, operated for many years by Harbor Branch Oceanographic Institute (USA), have only been used at depths above 900 m, whereas submersibles designed for greater depths have metal living spheres with much smaller acrylic port windows.

The *Pisces IV* and *Pisces V* submersibles, currently operated by the Hawaii Undersea Research Laboratory (HURL) at the University of Hawaii, have the general design of a deeper water scientific research submersible (Fig. 13.1a). Their living spheres, rated to 2000 m depth, can hold three persons, usually two observers and the pilot, each being able to view the seafloor from their own front view port. The living sphere fits into a frame outfitted with syntactic foam and both soft and hard ballast spheres for maintaining buoyancy and adjusting trim. The design of the frame includes extended legs and skids that allow these submersibles to work in viscous volcanic mud found on undersea volcanoes without the risk of taking on hundreds of pounds of unwanted mud weight in their bottom pans. Propulsion is provided by two 5 hp electric brushless DC thrusters capable of maintaining an underwater speed of 3 knots in zero current conditions. Power for the thrusters comes from large rechargeable batteries that also are used to run hydraulics, life support, lights and control systems.

The *Pisces* submersibles and the currently inactive *Shinkai 2000* are considered to be intermediate-depth vehicles and their living spheres were constructed of steel. Submersibles designed for greater depths have living spheres made of stronger materials to withstand pressures as great as 8 tons per square inch. The two MIR submersibles (*MIR 1* and *MIR 2*, Fig. 13.1b), operated by the Russian Academy of Sciences (RAS) to depths of 6000 m, have living spheres constructed of maraging steel alloy, a special class of ultra high-strength steel. The *Alvin*, operated by the Woods Hole Oceanographic Institution (WHOI), was originally built with a steel living sphere rated to 2000 m. In 1974, the *Alvin* was upgraded with a titanium alloy living sphere rated to 4500 m depth (Fig. 13.1c). In 2013 the *Alvin* was upgraded with an even stronger titanium alloy living sphere that allows the submersible to reach depths of 6500 m (Fig. 13.1c).

Other 6000+ m research submersibles with titanium alloy living spheres include the *Nautile* (Fig. 13.1d) operated by IFREMER, the French Research Institute for Exploitation of the Sea, the *Shinkai 6500* (Fig. 13.1e), operated by the Japan Agency for Marine-Earth Science and Technology (JAMSTEC), and the new Chinese submersible, *Jiaolong* (Fig. 13.1f), which reached a depth of over 7000 m in 2012.

Submersible dives typically last for 6–10 hours, surface to surface. Most are designed primarily for benthic surveys and consequently were manufactured with downward angled viewports. The acrylic spheres of the *Johnston Sealink I* and *II* were exceptions to this trend and have enabled the vehicles to work efficiently in midwater. Ballast systems were also designed with benthic surveys in mind, where neutral buoyancy would only need to be obtained at the seafloor and where the seafloor depth would not vary by more than 1000 m during the course of a single dive. It is therefore not possible to do midwater transects at multiple depths, such as 1000 m, 2000 m, 3000 m, etc. during a single dive on any of the submersibles currently in operation.

Most submersibles are relatively slow, and move at 1–2 knots; the *MIR* submersibles are the fastest, with a maximum reported speed of 5 knots. All submersibles communicate with the surface via sub-phones, the most common of which are duplexed 27 kHz and UQC (8–10 kHz) underwater communication systems. Communication with the ship is not only important for safety reasons but also for guiding the submersible during the dive since the tracking systems are located on the ship, not the submersible.

Fig. 13.1 Examples of submersibles currently being used for biological sampling. (a) *Pisces V.* (Courtesy of NOAA-HURL Archives.) (b) *MIR 2.* (L. Murphy, Wikipedia Commons.) (c) *Alvin.* (Tom Kleindinst, Woods Hole Oceanographic Institution.) (d) *Nautile.* (IFREMER). (e) *Shinkai 6500.* (JAMSTEC. Reproduced with permission.) (f) *Jiaolong.* (Liu Feng, China National Deep Sea Center. Reproduced with permission.) (Please see Plate 22 for colour representation of the figure.)

The overall weight in air of these vehicles ranges between 13 and 27 tons. As a result, large oceanographic vessels equipped with specially adapted A-frames are generally required for their deployment. Due to recent budget constraints, HURL has resurrected its launch recovery transport (LRT) which it used in the 1980s and early 1990s to deploy its *Pisces V* submersible. The LRT is a 'submergible' platform that the submersible is attached to and towed out to the dive site using relatively small boats of opportunity (presently a 21 m tugboat). The platform is then submerged to a depth of 20 m where the *Pisces V* is released and driven off to begin the dive. At the end of the dive, the submersible is recovered by simply securing it back on to the LRT, which then returns to the surface.

13.2.2 Remotely operated vehicles

ROVs are unoccupied vehicles operated remotely by pilots and technicians aboard a support ship. They are hard-linked to the ship by an umbilical cable containing fibre optics and power transmission wires that transmit video and data up from the vehicle, and electrical power and pilot controls down from the ship. Umbilical cables are mechanically strengthened to support the weight of the ROVs during launch and recovery.

There are presently two ways in which most ROVs are connected to their support ships: (i) a single-cable system or (ii) a combination of a cable and secondary tether system, which when combined with a winch is called a tether management system (TMS). On single-cable systems, such as the one used by the Monterey Bay Aquarium Research Institute (MBARI) with their 1850 m *Ventana* and 4000 m *Doc Ricketts* (Fig. 13.2a), the umbilical cable coming off the ship's winch is directly connected to the ROV, which provides ease of deployment but renders the vehicle sensitive to heaving and other motion of the ship at the surface, which is furthermore required to hold station in a position directly above the ROV. Adequate slack in the cable is one way to avoid this problem but this strategy confers greater vulnerability to snagging in rough, high-slope terrain. To avoid this problem, MBARI has attached floats onto the first 100 m of cable coming off the ROV. This forms a large buoyant loop of slack away from the seafloor. In contrast, combination cable/tether systems use an intermediate vehicle or a depressor weight that is connected to the ship via the main umbilical cable while being connected to the ROV via a slack tether, thereby decoupling the ROV from surface motion. This vehicle then relays data and power to the ROV through a neutrally buoyant, lightweight, 50–250 m long secondary tether. The intermediate vehicle can function as a launcher, a protective cage, hard ballast, TMS or all four at once. The most well-known of these combination systems are WHOI's 6500 m *Jason/Medea* (Fig. 13.2b), the University of Rhode Island/Institute for Exploration (URI/IFE)'s 4000 m *Hercules/Argus* (Fig. 13.2c), IFREMER's 6000 m *Victor 6000* (Fig. 13.2d), JAMSTEC's 7000 m *Kaiko7000 II* and new 11 000 m *ABISMO* (Fig. 13.2e) and the National Oceanic and Atmospheric Administration's (NOAA) new 6000 m *Deep Discoverer* (Fig. 13.2f).

The 5000 m *ROPOS* ROV, owned and operated by the Canadian Scientific Submersible Facility (CSSF), is unique in having both a single-cable system for shallower dives to 3000 m as well as a cable/tether combination system that is used for dives to 5000 m. In 2013 the University of Hawaii acquired a new 6000 m ROV, the *Lu'ukai*. This vehicle is also deployed by a cable/tether combination system.

There are now many different types of scientific and industrial ROVs operating throughout the world's oceans, and a thorough discussion of all is beyond the scope of this chapter. Similar to submersibles, ROVs are equipped with video cameras and lights, manipulators, sonars, and swappable samplers and sensors. ROVs do not require the life support systems characteristic of submersibles. All ROVs are cabled in one way or another to a ship on the surface, and as a result, all have essentially unlimited durations of power for their thrusters, sensors, video cameras and collecting equipment. Furthermore, real-time participation during the dive is not limited to the pilot and observers as with submersibles. Not only can the live video feed from the vehicle be relayed throughout the ship but also to the rest of the world via high-speed networks and high-capacity shipboard transmission systems, a capability referred to as telepresence. Telepresence in deep-sea exploration was first unveiled on NOAA's Ocean Exploration programme's ship R/V *Okeanus Explorer* and is now being used on several ships, including URI/IFE's R/V *Nautilus* and the University of Washington's R/V *Thompson*. Live feeds from these ships and the ROVs they are using can be accessed at: www.explorationnow.org/.

Fig. 13.2 Examples of ROVs currently being used for biological sampling. (a) *Doc Ricketts*. (Todd Walsh © 2009 MBARI. Reproduced with permission.) (b) *Jason II.* (Photograph by Margo Edwards. Reproduced with permission.) (c) *Hercules.* (d) *Victor 6000.* (IFREMER.). (e) *ABISMO.* (JAMSTEC.) (f) *Deep Discoverer.* (Photograph by Art Howard. Reproduced with permission of NOAA Office of Ocean Exploration and Research.) (Please see Plate 23 for colour representation of the figure.)

13.2.3 The Nereus hybrid ROV

WHOI's 11 000 m *Nereus* hybrid ROV (Fig. 13.3) was unfortunately lost during a dive in May 2014. However it had several unique features which will likely be incorporated in similar new vehicles, and hence a brief description of the vehicle is given here. *Nereus* could be operated as either an ROV or an autonomous underwater vehicle (AUV) in the deepest part of the ocean, the Challenger Deep in the Marianas Trench. In order to accomplish that goal, it was built with features that are different from all other types of ROVs. These include the use of ceramic spheres instead of syntactic foam for buoyancy, lithium batteries for power instead of cable-supplied power from the ship, and a micro-thin fibre optics tether for data transfer while in ROV mode (www.whoi.edu/main/nereus). These features were primarily designed to reduce weight in both the vehicle and cable. Standard oceanographic power cables are not strong enough to support even their own weight at depths below 7000 m, which prompted the designers to use batteries for power. The system still used a conventional strong cable to connect the ship to a depressor weight. However, the depressor weight was connected to the vehicle by the ultra-thin, lightweight optical tether that only transmits data. The tether also conferred greater flexibility in positioning the support ship, allowing 4 km or more in horizontal separation from the vehicle (C. Machado, personal communication). While the *Nereus* had relatively few deployments, its batteries provided power for dives lasting up to 20 hours (D. Gomez-Ibanez, personal communication). It had a single 7-function manipulator, cameras, various sensors and could be fitted with various samplers.

Fig. 13.3 *Nereus* hybrid ROV. (Courtesy of Advanced Imaging & Visualization Lab © Woods Hole Oceanographic Institution.)

13.2.4 Submersibles versus ROVs

During the last few years, meetings on deep-sea exploration and sampling have almost always included an animated discussion on which vehicle is best – the submersible or ROV. Gage and Tyler (1991) pointed out that submersibles are useful in areas inaccessible to surface-deployed sampling gear and ROVs, such as steep rocky terrain, conducting manipulative experimentation, and for making detailed observations in areas such as hydrothermal vent fields. They furthermore cite the advantages of 'clearer visualization of the interrelationships of environmental components' mentioned by Heirtzler and Grassle (1976) and reiterate the points made by Grassle (1980) that submersibles are the only means to sample small-scale features, conduct repeat sampling, correctly deploy sediment samplers, survey complex topography and sample specific water column layers.

However, ROV technology has improved during the last 20 years since these arguments were made and many of these activities can now be carried out with unmanned vehicles, without any potential risk to the investigators. As a result, submersibles have clearly lost ground to ROVs in terms of the numbers of each type that are currently active. This fact has led many researchers to believe the manned vehicle is on the way out and will soon be replaced by the unmanned vehicle. However, it is interesting to note that the *Nereus* went back to using battery power similar to submersibles and that some facilities are now limiting their ROV dives to 12 hours, only 4 hours longer than most submersible dives, due to the costs involved in supplying around the clock operators as well as vehicle maintenance constraints (Ferguson, personal communication). The *Jason* ROV is an exception to this trend and recently completed a dive lasting more than 100 hours (C. Offinger, personal communication). The requirements for conducting studies in deep-sea environments are varied and complex. As such, both submersibles and ROVs should be available to researchers since each have their own strengths and weaknesses.

It is not a question of one type of vehicle or the other. Hence new submersibles, as well as ROVs, are being built or upgraded. As a final note to this section, Japan is now considering developing a hybrid submersible. This vehicle would be manned in order to take advantage of the superior optical capabilities of the human eye but would be tethered to the surface via a micro-thin fibre optics cable capable of transmitting real-time video and data. The micro-thin cable technology used by the *Nereus* and being considered for this project was developed by JAMSTEC (D.J. Lindsay, personal communication). This technology may therefore be opening up a new 'hybrid vehicle era' in deep-sea research.

13.3 Submersible and ROV sample collection gear

In general, the various types of biological sampling gear typically found on submersibles and ROVs are the same or very similar. So there will be no distinction made below between submersible gear and ROV gear. Both vehicles rely heavily on robotic arms, called manipulators, to collect biological samples. Manipulators are used directly by simply grabbing the specimen, or indirectly by holding and deploying other types of samplers that perform the actual collection. These include pot and bucket samplers, net samplers, suction samplers, sediment samplers (for infauna) and bottle samplers (for microscopic organisms found in the water column).

13.3.1 Manipulators

There are a variety of manipulators presently in use by submersibles and ROVs that can be distinguished by the number of articulations or functions they can perform, generally 3, 5 or 7, and the manner and sophistication in which they are controlled by the pilot. The more functions a manipulator performs, the more dexterity it has and each function corresponds to an independent 'joint' or means to articulate.

Most scientific submersibles and ROVs are fitted with two manipulators, traditionally one being capable of 7 functions for precision sampling of fauna (Fig. 13.4a) and the other, often called the 'grabber', being capable of either 3 or 5 functions that the pilots use for collections that require brute force (in-place rock sampling) or to assist in collections by the other manipulator (e.g. temporarily holding the specimen,

(a)

(b)

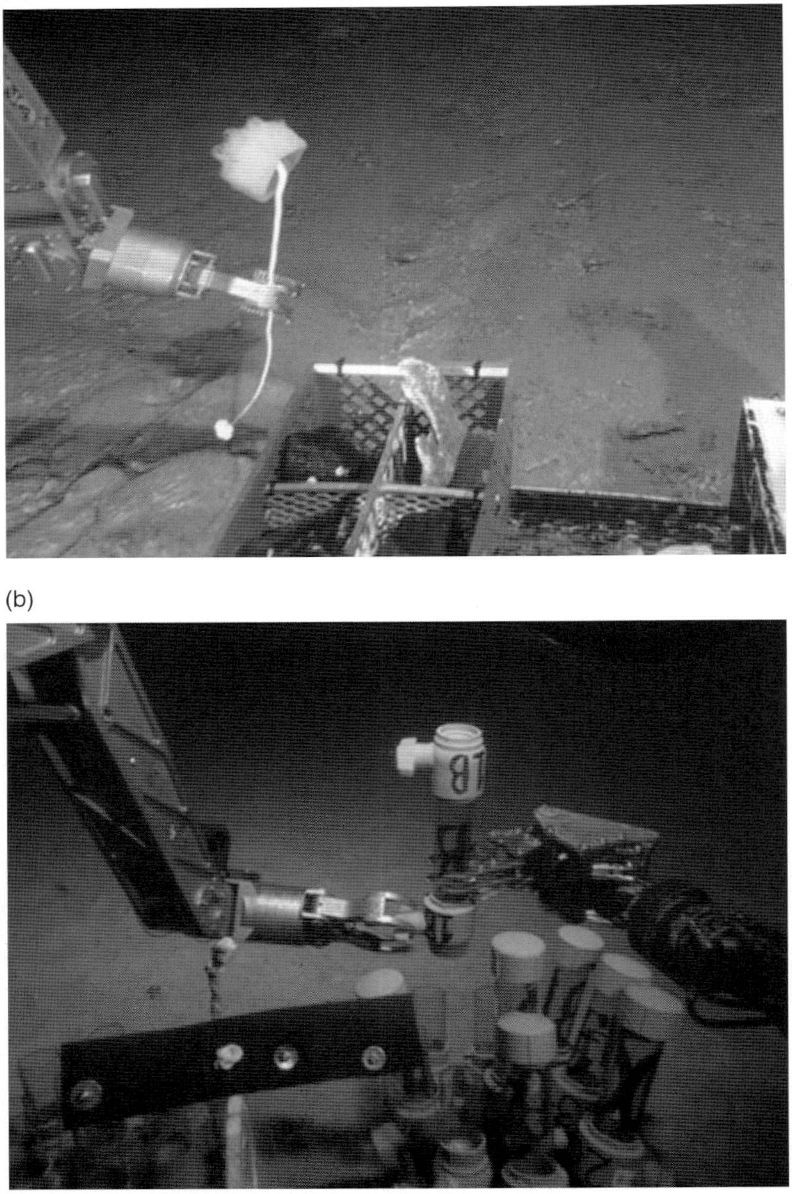

Fig. 13.4 Illustrations of the use of manipulators: (a) Collection of a hexactinellid sponge specimen using a 7-function manipulator. (Courtesy of NOAA-HURL Archives.) (b) Grabber manipulator being used to hold a double-valve plastic tube sampler while a 7-function manipulator is used to close the valve. (Courtesy of NOAA-HURL Archives.)

sampler or holding the storage box lid open while the specimen is stowed, Fig. 13.4b). Newer configurations have two 7-function manipulators, as found on the *Jason 2*, *ROPOS* and *Doc Ricketts* ROVs that are capable of lifting several hundred pounds at their full reach. Seven-function manipulators can perform the following movements: up–down, right–left, in–out, wrist rotate, wrist bend, jaw open and jaw close. The two functions most grabbers are missing are wrist rotate and wrist bend. Seven-function manipulators typically have a sophisticated master controller inside the submersible or in the ROV control room on the ship that resembles a miniature replicate of the manipulator for intuitive, smooth operation by the pilot. Movement capability is the key for the precision sampling these vehicles and their pilots are tasked with performing.

An important consideration for a successful dive is preparing the collecting manipulator for the type of sampling or type of sampler it will be deploying. There are two general categories of interchangeable manipulator claws (also called grippers or end effectors): claws with opposed (or intermeshed) fingers and claws with parallel fingers (Fig. 13.5). Which claw type to use depends on the type of sampling that will be carried out. For example, opposed finger claws are used for collecting whole and/or delicate specimens such as echinoderms, molluscs and whole coral and sponge colonies. These claws are also used for handling most other types of samplers because the fingers have a rounded shape as do many samplers (i.e. sediment, net, suction and bottle samplers). Parallel finger claws are used for collecting pieces of coral and sponges since they can grab small branches or pieces more precisely and hold them more securely for transfer to the storage container. The fingers have been fitted with Nalgene tubing (see http://oceanexplorer.noaa.gov/explorations/04mountains/welcome.html, May 14 log for image) to further minimize damage to specimens during collection. To avoid grip damage, cup extensions can also be fitted to parallel claws to allow direct sampling of sediment or organisms. Both types of claws typically have 'T' handle grooves that allow firm handling and precise orientation of larger sampling instruments.

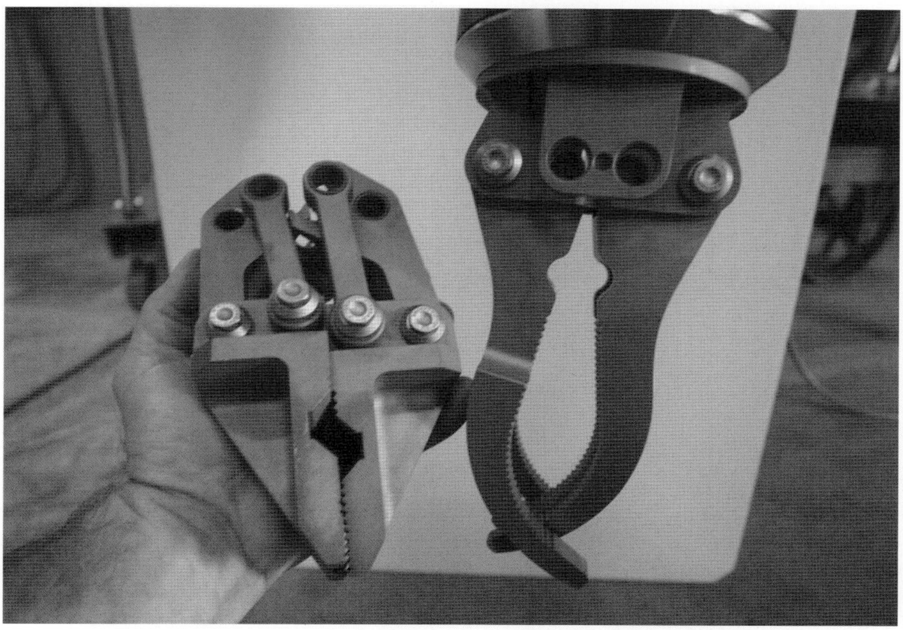

Fig. 13.5 The two basic types of manipulator claws, parallel finger type on left, and opposed or intermeshed finger type on right. (Courtesy of NOAA-HURL Archives.)

Submersible pilots experience a three-dimensional perspective that is very helpful for precision sampling. ROV pilots, however, have only a two-dimensional perspective, which makes judging distance between the claws and the specimen more challenging. One or more additional cameras mounted at a different angle or on a pan and tilt actuator provide more depth perception. In addition, manipulator wrist cameras are now available that can be extremely helpful for sampling from either ROVs or submersibles. One other issue both submersible and ROV pilots face is their inability to feel the force they are applying to the claws, which can sometimes cause damage to delicate specimens (Meng *et al.*, 2006). Force feedback sensors are available on some manipulators, such as the 7-function Kraft Predator II used on the *Jason 2*, *ROPOS*, Deep Discoverer, and *Doc Ricketts* ROVs.

13.3.2 Manipulator-deployed samplers

The pot sampler, originally designed by Van Dover (2002) and recently upgraded by Charles Fisher at Penn State University, is used to sample mussel beds and the associated community in both hard and soft substrata. It consists of a modified 12-quart aluminium-clad stainless pot with a handle and filled with a Kevlar bag. This sampler, similar to most, is deployed by the submersible or ROV manipulator. After enclosing the assemblage, a handle is turned to cinch up the Kevlar bag, firmly securing the sampled mussels and associated fauna. The sampled area is known (531 cm^2), but the sampled volume is variable and depends on the penetration of the pot in the sediments.

Sampling large, dense faunal communities on hard substratum has proven to be a challenging problem. The 'Bushmaster' and 'Chimneymaster' are original instruments created at Penn State University to collect vent communities on and around hydrothermal edifices (Hourdez *et al.*, 2002; Bergquist *et al.*, 2002). These large collection nets are used to sample complete bushes of tubeworms and associated fauna, either on the seafloor (Bushmaster) or on sulfide chimneys (Chimneymaster). While the Bushmaster has proven to be efficient on relatively flat hard bottoms (Bergquist *et al.*, 2002, 2007; Urcuyo *et al.*, 2003), no sampler has been found to be effective on irregular or vertical sulfide edifice surfaces despite numerous attempts. Some researchers have resorted to using suction samplers with a manipulator arm to collect 'grab' samples of the fauna. The video taken during the collection is analysed afterwards to get an estimate of the sampled area (Sarrazin *et al.*, 1997; Tsurumi, 2003; Tsurumi & Tunnicliffe, 2003).

Collection nets deployed with a submersible manipulator have been used to collect specimens of the eel, *Dysommina rugosa*, living in a dense community inhabiting low-temperature hydrothermal vents (www.oceanexplorer.noaa.gov/explorations/05vailuluu/welcome.html). Nets have been used in conjunction with anesthetizing substances to collect small fish specimens that would otherwise have been impossible to sample (R. Moffitt, personal communication). Plankton nets have also been mounted on submersibles to collect benthopelagic and midwater plankton.

Most scientific submersibles and ROVs are equipped with a suction sampler. This sampling tool can be installed or removed from the vehicles as necessary and consists of a centrifugal pump and a long flexible supply tube fitted with a nozzle. Suction samplers can be used with a manipulator to position the nozzle for collecting benthic animals (Fig. 13.6a) and either with or without a manipulator for collecting animals in the water column (Fig. 13.6b). Animals smaller than the diameter of the nozzle are sucked into the tube and deposited into a single bucket (removable jar) or into one of the buckets of a multi-chambered carousel. A mesh of varying size is sometimes placed inside or at the outlet of each bucket to prevent sample loss. The suction sampler is commonly used to collect smaller unattached individual animals off hard surfaces, free swimming animals in the water column or anaesthetized animals before they drift away in the current. Suction samplers have also been used on soft substrates, with care taken to avoid sucking in too much sediment. Although the size of the sampled area can be determined by analysis of the video taken during the collection (Sarrazin *et al.*, 1997), this tool is generally not considered to be quantitative. A flowmeter can be placed in a suction sampler hose to measure the volume of water sampled (e.g. Watanabe & Kado, 2005).

(a) (b)

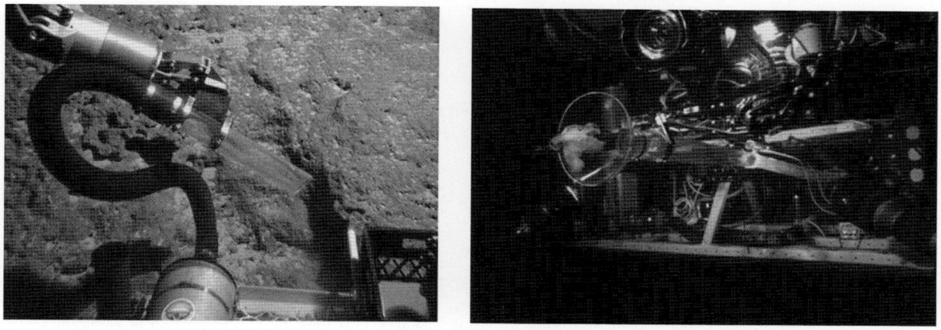

Fig. 13.6 Suction samplers. (a) Manipulator-deployed suction sampler for collecting benthic animals. (Woods Hole Oceanographic Institution.) (b) MBARI suction sampler used to collect animals in the water column. (Ernie Kovacs © 2005 MBARI. Reproduced with permission.)

Tube (or push)-corers, Birge–Ekman box-corers (used by *Alvin*) and blade box-corers (used by the *Nautile*) are the typical submersible and ROV tools used to sample infauna, bacteria and other microscopic organisms (Rowe & Clifford, 1973; Rowe & Sibuet, 1983; Sibuet *et al.*, 1990). Tube-corers are often clear plastic tubes, approximately 30 cm long, with or without a T-handle mounted on their upper ends used to 'push' the corer into the seafloor with the manipulator. A one-way valve secures the sample and the corer is then removed and stored in a box or inserted into a sleeve (i.e. quiver) attached to the submersible or ROV storage container. Box-corers deployed from submersibles and ROVs as opposed to ships, are smaller as well as being modified for deployment by manipulators. These are designed to be carried down in the vehicle baskets, then be removed, inserted into the sediment, and activated by manipulators. These tools are very efficient at sampling a known surface area of sediment but unfortunately no 'generic tools' are available or used by scientists from around the world. Tube cores and blade cores do not necessarily have the same size, and this may pose some comparison concerns. Nevertheless, they are considered to be quantitative and their results can be easily transformed to account for the area or volume sampled. Standardization of sampling tools and survey methods is an area where more coordination would be beneficial, especially where the goal is for environmental impact assessments of undersea mining activities (e.g. Collins *et al.*, 2013).

Bottle samplers, pumped water systems and syringe-based systems have been used to collect water samples containing bacteria and other microscopic organisms. There are two types of bottle samplers: Niskin bottles, sized and designed for use on submersibles and ROVs, and double-valve plastic tube samplers. Niskin bottles have triggered lids and can be rack-mounted and activated by the pilot, therefore not requiring the use of the manipulator. They can also be left unattached and activated by the manipulator when more precise water sampling is required. Precision sampling with the use of a manipulator can also be accomplished with double-valve plastic tube samplers, which, unlike tube-corers, are completely closed off at both ends after the sample is acquired (shown in Fig. 13.4b).

A number of traps have also been designed for deployment and recovery by submersibles and ROVs. These are necessarily small to fit on the vehicle's front rack and be deployable by manipulator. Traps are often used to collect small arthropods such as shrimps, crabs, isopods and copepods. These can be left on the rack or deployed on the seafloor and later recovered.

13.3.3 Midwater bucket sampler

The midwater bucket sampler is the name given to the use of a carousel to collect water column macrofauna without suction or with the assistance of a manipulator. The technique was developed for use on the *Johnston Sealink* submersibles by Harbor Branch Oceanographic Institute and

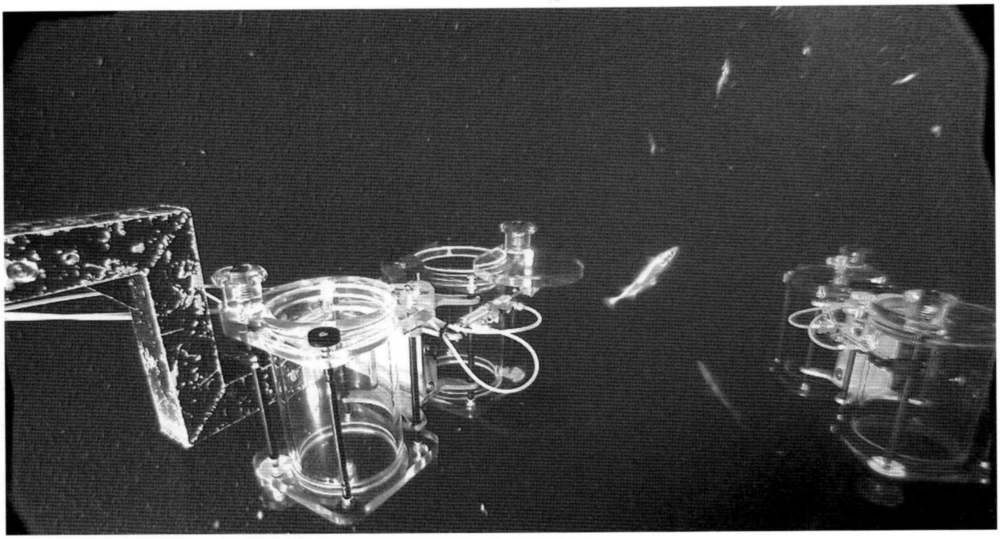

Fig. 13.7 Detritus samplers on MBARI's ROV *Ventana* used to collect midwater animals without damaging them. (© 2000 MBARI. Reproduced with permission.).

basically involves positioning an open bucket of the carousel near the animal, which then enters and the bucket is rotated shut. MBARI developed their own version of this method for use on their *Ventana* and *Tiburon* ROVs which they refer to as a detritus sampler, or simply D-sampler (Fig. 13.7). The containers are either mounted onto a carousel attached to the ROV headbar or mounted in pairs on swing arms. The containers have both top and bottom simultaneously rotating lids. The ROV positions an open container near an animal that then enters and the two lids are shut. This method was often used by both institutions to collect extremely delicate water column animals such as medusae and larvaceans that would otherwise have been damaged during the collection process (Raskoff, 2002; Robison *et al.*, 2005).

A manipulator-operated version of this sampler, called the 'gate sampler' was developed at JAMSTEC for use in conjunction with a suction sampler (Hunt *et al.*, 1997). An animal specimen is sucked into an acrylic box, the ends of which are closed off by a gate originally used in field irrigation systems. Mesh is attached to the outlet gate to prevent the specimen from being sucked right through the box. The flat clear walls of the box allow photographs to be taken while the animal is still alive and without the need to transfer it to a phototank, which could cause further disturbance or damage.

13.4 Submersible and ROV sample storage gear

The storage of biological samples for transport to the surface presents unique problems, particularly for physiological studies. Barotrauma from depths of even 100 m can severely impact the condition of the specimen. Fish collected from 100 m experience exophthalmia and everted stomachs (Demartini *et al.*, 1996). Pressure-sensitive enzymes in deep-sea animals can exhibit reduced functionality when brought to lesser depths (Herring, 2002). Some species of deep-sea corals, such as *Acanthogorgia* sp. and members of the family Isididae (Etnoyer *et al.*, 2006), undergo rapid tissue deterioration after being brought to the surface. The bathypelagic scyphozoans in the genera *Atolla* and *Periphylla* have protoporphyrin-based pigmentation that causes tissue damage when exposed to light (Arai, 1997).

The problems associated with sampling deep-sea organisms are not unique to submersibles and ROVs, and effective solutions have proven elusive.

Cold temperatures and absence of light have been successfully maintained in storage containers mounted on submersibles and ROVs (see Shillito *et al.*, 2001; Ravaux *et al.*, 2003; Frank *et al.*, 2012). In these cases, controlling these factors alone resulted in intact tissue as well as specimens that lived long enough for use in shipboard experiments. Maintaining pressure for very deep collections has proven to be a far more difficult, complex, and expensive problem.

Koyama *et al.* (2002) developed the first and only pressurized storage system, the 'Deep Aquarium', designed for use on a submersible. The system was created to suck animals into a chamber and maintain them at pressures equivalent to 2000 m. A fish specimen in the family Zoarcidae was successfully collected with the system from a depth of 1171 m and a specimen of another fish, *Ebinania brephocephala*, obtained from 500 m was kept alive in the system for 64 days. A recent effort to develop a second 2000 m pressurized specimen storage container for submersibles and ROVs through NOAA's Small Business Innovation Research (SBIR) programme proved unsuccessful.

For the time being, most animals collected by submersibles and ROVs are stored in simple, sealable but uncompensated containers called 'bioboxes' and reach the surface in whatever state the extreme changes in environment have left them in. Larger megafauna collected with the use of the manipulator are generally placed directly into one of these containers mounted on the vehicle front rack. Bioboxes are generally solid sided, which at least keeps the samples wet, and may be divided with partitions or contain racks of smaller numbered containers for keeping smaller animals and/or pieces of animals, such as sponges and corals, separated (Fig. 13.8). Microfauna collected from the water column are stored in their collecting bottles, whereas meiofauna and small megafauna collected from the water column or the substrate surface by a suction sampler are stored in carousel jars. Infauna collected in sediment samples are stored in the tube- or box-corers used to collect them.

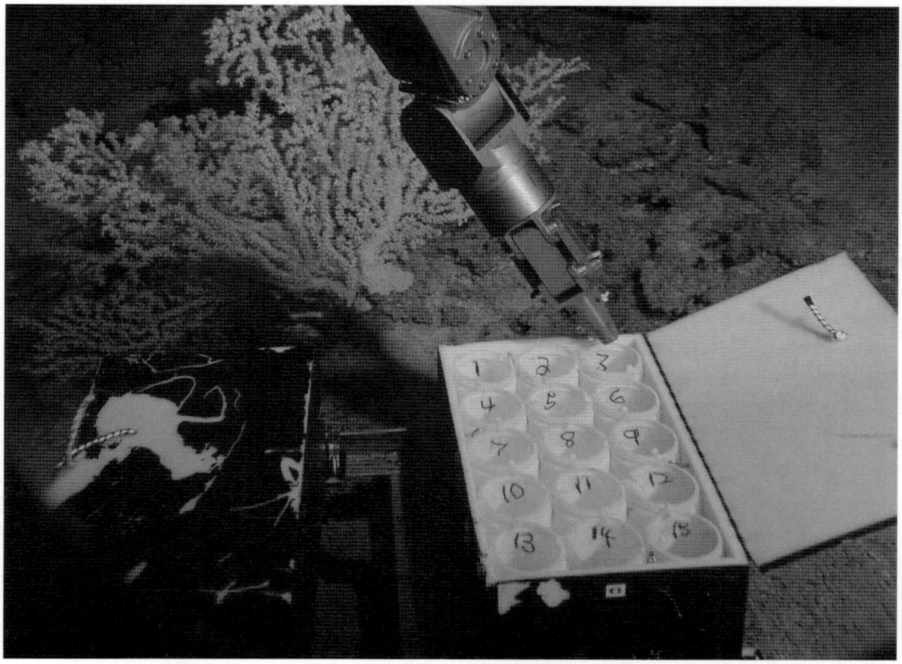

Fig. 13.8 Biobox fitted with partitions for keeping small specimens separated. (Courtesy of NOAA-HURL Archives.)

13.5 Other gear used during submersible and ROV sampling

One of the greatest advantages of collecting biological samples from submersibles and ROVs is the opportunity to collect simultaneous data on each specimen's physical environment and the community of organisms it is a part of. Both types of vehicles can be equipped with up-to-date imagery tools including video and still cameras, lights and laser-based size calibration devices. The main video cameras are often mounted on pan and tilt actuators to follow and record the sampling operations in real time. Camera systems are used to record the substrate type and the topography of the seafloor as well as the other organisms that are in close proximity to each of the specimens. They can furthermore document the *in situ* condition and appearance of the specimens prior to collection. Water-quality data are acquired with the use of sensors (i.e. CTDs and others) that are used to measure temperature, depth, salinity, dissolved oxygen, pH and turbidity.

There are also many other project-specific tools that have been designed and fitted onto these vehicles according to the objectives of the dive (see Gauthier *et al.*, 2010). Markers are commonly used for repeat sampling of the same individuals or at the same location. Among the simplest of these are numbered 5-gallon white bucket lids and plastic superballs wrapped with reflective tape. These are attached to weights and float a metre or so above the bottom, making them highly visible to the pilots. Numbered cement flowerpots, each with an attached piece of bright yellow polypropylene line, have been used to both mark coral colonies as well as protect thermographs for periods of a year or more (Parrish & Roark, 2009).

Highly mobile cryptic animals, such as small planktivorous fishes that shelter in rocks, are generally impossible to collect without using anaesthetics or poisons. To sample these, quinaldine and rotenone are typically squirted directly into holes and crevices where the fish are hiding with the use of dispensers specifically designed for use with manipulators. A 5-gallon collapsible water carboy fitted with a dispensing tube was used by Moffitt (unpublished) to collect small fish from the *Pisces V* submersible. The mouth of the dispensing tube was positioned near the holes by the 7-function manipulator while the grabber was used to apply pressure on the carboy to dispense the chemical. Once the fish or other small animals succumb, they generally drift out of the holes where they can be collected using nets or suction samplers. Suction samplers can also be used to suck or blow out animals that have remained in the holes.

Bait is an olfactory lure used to attract and capture animals that otherwise would not approach the vehicles. These are typically used in conjunction with traps that are deployed on the substrate or left on the front rack during the dive. Baited traps have proven to be a very effective means of collecting crustaceans and fish. It is widely known that the type of bait to use depends on the animals being targeted since they will not be attracted to bait too dissimilar to their natural prey. Often, the bait is placed into small mesh bags to ensure it is retained inside the traps and not consumed too quickly by the attracted animals that enter the trap. Red lights or no lights at all have been used in cases where the vehicle remains in the vicinity of the trap during the sampling period.

13.6 Submersible and ROV sampling operations

13.6.1 Dive preparation

Successful submersible and ROV sampling operations begin with careful and well-thought-out dive preparation. These vehicles have limited storage space and weight constraints for samplers and specimens so the types of samplers and other gear to use and the types and numbers of specimens that will be collected must be planned in advance. The layout of the storage space is an important consideration to ensure easy access of the samplers by a manipulator and securing the samplers after use for safe return to the surface. MBARI's *Doc Ricketts* ROV, like its predecessor, the *Ventana* ROV, has several interchangeable compact toolboxes that can be quickly installed and removed from the vehicle in order to accommodate different types of studies and collecting modes. IFREMER's *Victor 6000* also has a sampling toolbox, however other less modularized operations must make compromises between the amount of gear that is taken down and the amount of storage capacity available for specimens.

Unlike ROVs, submersibles do not have an active tracking display for the pilot and observers to use doing the dive. Coordination between the submersible and tracking room on the ship is therefore critical to a successful dive. To prepare for this, dive site underlays that include multibeam bathymetry or backscatter data, contours and target waypoints are created and loaded onto the ship's tracking computer prior to each dive. Dive maps containing the same data are then made and distributed to the pilot and observers. HURL uses electronic versions of the maps loaded onto iPads and carried down in the *Pisces* submersibles. The iPads are also preloaded with many other materials such as the HURL Deepwater Animal Guide and images of species targeted for collection, thereby serving as a general reference source during the dives.

Submersible observers must also be well prepared before their dives. Observers typically carry down a waterproof notebook and pencils for recording dive and collection notes and, in some cases, a headlamp since the lighting in the living sphere is often quite dim. While the video recorder will document the collection events, including the associated audio, the video is generally not available to the observers for hours after the dive since it must be first downloaded off video cards or other types of media. Therefore handwritten notes by the observers are the only means to properly label and process the specimens immediately following the dive. Finally, one of the most critical steps during dive preparation is synchronizing all the clocks of the sensors and instruments in the submersible with the tracking computers on the ship. These include the CTD, the video overlay generator that creates a text display of the date, time, depth and heading on the video, the video recorders and any still cameras that will be used. Synchronized timecodes are often the only way the various types of data that need to be input into the specimen collection logs can be correlated. Recently, logging systems have been developed that allow synchronous playback of dive video and data, thereby simplifying post dive analysis (Lindsay *et al.*, 2012).

13.6.2 General sampling protocol

Most biological sampling from submersibles and ROVs involves the use of a manipulator. In vehicles that have a 7-function manipulator and a 5-function grabber, the former is the one that is normally used for most megafaunal collections such as sponges and corals. The 7-function manipulators on both *Pisces IV* and *Pisces V* are mounted on the front left side of the submersible, which the pilots keep in mind when approaching the specimens, as do all pilots operating other submersibles and ROVs. Current direction and intensity as well as topography are the other factors pilots consider as they make the approach. Without thrusters, the *Pisces* submersibles have a natural tendency to face into the current and this is generally the preferred orientation for sampling since any disturbed sediment clouds will be quickly carried away from the cameras and view ports.

Once the pilot has manoeuvred the vehicle into a stable position, the storage box lid is opened in preparation for the collection. At this time, the pilot or an observer documents the pre-collection state of the sample with the video and still cameras. A common procedure for sessile invertebrates is to video and photograph the whole animal first with the laser scale. The cameras are then zoomed in for additional close-up images of various parts of the animal, such as the extended polyps on deep-water corals or the arms and scales on sea stars. Lasers are not as effective for obtaining measurements of midwater animals because the pair of beams rarely hit the animal simultaneously and when they do, they do not reflect well off gelatinous, transparent bodies. As an alternative, Lindsay *et al.* (2013) described the use of two cameras and stereogrammetric techniques to obtain size estimates of transparent midwater animals.

After the collection is made and the animal is secured inside the storage box, the observers record the collection data in their notebooks. During ROV operations, the GPS coordinates can simply be read off the display. During submersible operations, the ship's tracking room is contacted for the coordinates and asked to note in the tracking log that a collection was made at this location. The tracking room also calls down the coordinates, which the pilot repeats to confirm the position and to ensure that the audio information has been clearly recorded on the video recorders. If the collection site is to be revisited later in the dive or during another dive, a marker is deployed and the ID number is recorded on video, in the observer notes and provided to the tracking room. In general, the entire collection event is carefully recorded on video, not just during ROV dives but also during submersible dives to ensure specimens are not mixed up after they are transferred from the vehicles to the lab. For the most part, this basic sampling protocol is used during all types of sampling, including sediment sampling with corers, net sampling and midwater sampling with suction samplers.

13.6.3 Sampling with the use of anaesthetics, poisons and bait

Sampling with anaesthetics and poisons is carried out with the use of net and suction samplers. These chemicals are generally applied in small doses and in a localized manner since too many animals anesthetized at the same time are difficult to collect before drifting away. Current direction is a consideration and pilots may have to assume a less than optimal orientation in order to apply the chemicals at the desired location. Bait is an olfactory lure typically used to sample benthic animals in conjunction with traps. Baited traps are typically deployed by a manipulator and left at a desired collection site for a period of time while the submersible continues on with other dive tasks. Markers may be left next to the trap, making it easier for the pilot to relocate.

13.6.4 Live specimen sampling

Live specimens of urchins, cnidarians, crustaceans and seastars have been collected during a number of submersible dives. These dives have required special preparation and methods in order to maximize the chances the animals will survive. On the ship itself there can be many different cold-water aquaria set-ups. Commonly, large coolers were obtained are used as the live animal holding tanks in the ship's laboratories fitted with aerators, thermometers and filled with cold seawater before the dive. The submersible manipulators were fitted with opposed finger claws and the pilot had to use extreme caution not to close them too tightly on the animals. Smaller seastars were collected with an empty tube-corer and cnidarians were collected by grabbing the whole rock that the animal was attached to. Similarly, small commensal crabs were obtained by collecting the coral branches they were perched on.

As soon as the submersible was secured on the deck of the ship, the animals were moved out of the bio boxes and into buckets of chilled seawater for transfer to the lab. Coolers or tanks generally have a small hatch door used for siphoning urchin waste, feeding the animals, and changing ice if used. The animals need to be kept in cold dark conditions for the remaining period of the voyage.

Dissolved oxygen needs to be reduced for vent and seep animals or for animals from the oxygen minimum zone (OMZ) that naturally live in low-oxygen environments. This can be accomplished by using airtight live storage containers and bubbling nitrogen into the water. Midwater animals are also very prone to being damaged when being transported from the sub or ROV and laboratory. To minimize damage, the transfer containers must be completely full of water without any air pockets. Some particularly delicate animals may also need to be kept in a plankton kreisel tank to stop them from colliding with the container walls.

13.7 Submersible and ROV sample processing

Onboard specimen processing is crucial for ensuring each sample is properly documented and preserved for more detailed analysis after the cruise is complete. Accurate collection data is perhaps the most important of these steps since poorly documented specimens are considered to have marginal, if any, value. The responsibility for this step falls squarely on the observers who participated in the dive. The observers must accompany the specimens to the ship's lab and transcribe their dive notes onto the labels that will be placed with each specimen. Chapter 15 covers aspects of these procedures in detail. These labels should include at least the vehicle and dive number, specimen identification number, collection date, time, depth and location in latitude and longitude. The identification number is particularly important for specimens that will be split into subsamples for genetic analysis or other purposes. Specimens for taxonomic identification are often photographed onboard with their labels, along with a small ruler for scale in order to document their coloration and condition before being placed into preservative. Following this, the specimens are placed into containers with an appropriate preservative (see Chapter 15). Furthermore, anaesthetics such as MgCl can be added before the preservative to keep retractible body parts extended, facilitating subsequent examination. Genetic subsamples of specimens are either frozen in an ultra-cold freezer if one is available on the ship, placed in ethanol or buffer solution, or both. Tissue subsamples for histological analysis are preserved in one of various types of fixatives.

Once the specimens have been secured in the lab, then the observers can turn their attention to compiling the ancillary data. A datasheet, spreadsheet or database should be created that contain records with the same information as that put on specimen labels. Sensor data such as temperature, oxygen, salinity, pH and imagery, such as video captures, still camera photos and lab photos, are then added to the records based on matching the instrument, camera and video recorder time codes to the time of the collections noted by the observers. This sort of information is described in greater detail in Chapter 15.

Acknowledgements

We thank Dr Jeffrey Drazen for describing to us his experiences collecting animals from the MBARI *Tiburon* and *Doc Ricketts* ROVs and Dr John R. Smith for his knowledge and helpful comments about various deep-water vehicles. We would also like to thank Casey Machado for comments on the manuscript.

References

Arai, M.N. (1997). *Functional biology of scyphozoans*. London: Chapman and Hall.

Bergquist D.C., Urcuyo I.A., Fisher C.R. (2002). Establishment and persistence of seep vestimentiferan aggregations from the upper Louisiana slope of the Gulf of Mexico. *Marine Ecology Progress Series*, 241: 89–98.

Bergquist, D.C., Eckner, J.T., Urcuyo, I.A., *et al.* (2007). Using stable isotopes and quantitative community characteristics to determine a local hydrothermal vent food web. *Marine Ecology Progress Series*, 330: 49–65.

Busby, R.F. (1976). *Manned submersibles*. Office of the Oceanographer of the Navy, USA, 764 pp.

Collins, P.C., Kennedy, B., Forde, J., *et al.* (2013). VentBase: developing a consensus among stakeholders in the deep-sea regarding environmental impact assessment for deep-sea mining – a workshop report. *Marine Policy*, 42: 334–336. 10.1016/j.marp01.2013.03.002

DeMartini, E.E., Parrish, F.A., Ellis, D.M. (1996). Barotrauma-associated regurgitation of food: implications for diet studies of Hawaiian pink snapper, *Pristipomoides filamentosus* (family Lutjanidae). *Fishery Bulletin*, 94(2): 250–256.

Etnoyer, P., Cairns, S.D., Sanchez, J.A., *et al.* (2006). Deep-sea coral collection protocols. *NOAA Technical Memorandum NMFS-POR-28*, Silver Spring, MD, 53 pp.

Frank, T.M., Johnsen, S., Cronin, T.W. (2012). Light and vision in the deep-sea benthos: II. Vision in deep-sea crustaceans. *Journal of Experimental Biology*, 215: 3344–3353.

Gage, J.D., Tyler, P.A. (1991). *Deep sea biology: a natural history of organisms at the deep-sea floor*. Cambridge, UK: Cambridge University Press.

Gauthier, O.J., Sarrazin J., Desbruyeres, D. (2010). Measure and mismeasure of species diversity in deep-sea chemosynthetic communities. *Marine Ecology Progress Series*, 402: 285–302.

Grassle, J.F. (1980). *In situ* studies of deep-sea communities. In: Diemer, F.P., Vernberg, F.J., Mirkes, D.Z. (eds), *Advanced concepts in ocean measurement*. Columbia, SC: University of South Carolina Press, pp. 321–332.

Heirtzler, J.R., Grassle, J.F. (1976). Deep-sea research by manned submersibles. *Science, Washington*, 194: 294–299.

Herring, P. (2002). *The biology of the deep ocean*. Oxford: Oxford University Press.

Hourdez, S., Weber, R., Green, B., Kenney, J., Fisher, C. (2002). Respiratory adaptations in a deep-sea orbiniid polychaete from Gulf of Mexico brine pool NR-1: metabolic rates and hemoglobin structure/function relationships. *Journal of Experimental Biology*, 205: 1669–1681.

Hunt, J.C., Hashimoto, J., Fujiwara, Y., *et al.* (1997). The development, implementation, and establishment of a Meso-pelagic and Bentho-pelagic biological survey program using submersibles in the seas around Japan. *JAMSTEC Journal of Deep Sea Research*, 13: 675–685.

Koyama, S, Miwa, T., Horii, M., Ishikawa, Y., Horikoshi, K., Aizawa, M. (2002). Pressure-stat aquarium system designed for capturing and maintaining deep-sea organisms. *Deep-Sea Research I*, 49: 2095–2102.

Lindsay, D.J., Yoshida, H., Uemura, K., *et al.* (2012). The untethered remotely-operated vehicle PICASSO-1 and its deployment from chartered dive vessels for deep sea surveys off Okinawa, Japan, and Osprey Reef, Coral Sea, Australia. *Marine Technology Society Journal*, 46(4): 20–32.

Lindsay, D.J., Yoshida, H., Ishibashi, S., *et al.* (2013). The uROV PICASSO, the Visual Plankton Recorder, and other attempts to image plankton. *Proceedings of the International Symposium on Underwater Technology*, 2013: doi:10.1109/UT.2013.6519854

Malahoff, A., Gregory, T., Bossuyt, A., Donachie, S.P., Alam M. (2002). A seamless system for the collection and cultivation of extremophiles from deep-ocean hydrothermal vents. *IEEE Journal of Oceanic Engineering*, 27: 862–868.

Matsumoto, G.I., Potts, T.A. (2011). *Two ways of researching the underwater world*. Boston, MA: Cengage Learning, 31 pp.

Meng, H.W., Li, P., Wang, L., He, Z. (2006). Dexterous underwater robot hand: HEU Hand II. In: *Proceedings of the 2006 IEEE International Conference on Mechatronics and Automation*, June 25–28, 2006, Luoyang, China, pp. 1477–1482.

Parrish F.A., Roark, E.B. (2009). Growth validation of gold coral *Gerardia* sp. in the Hawaiian Archipelago. *Marine Ecology Progress Series*, 397: 163–172.

Raskoff, K.A. (2002). Foraging, prey capture, and gut contents of the mesopelagic narcomedusa *Solmissus spp.* (Cnidaria: Hydrozoa). *Marine Biology*, 141: 1099–1107.

Ravaux, J., Gaill, F., Le Bris, N., Sarradin, P-M., Jollivet, D., Shillito, B. (2003). Heat-shock response and temperature resistance in the deep-sea vent shrimp *Rimicaris exoculata*. *Journal of Experimental Biology*, 206: 2345–2354.

Robison, B.H., Reisenbichler, K.R., Sherlock, R.E. (2005). Giant larvacean houses: rapid carbon transport to the deep sea floor. *Science*, 308: 1609–1611.

Rowe, G.T, Clifford, C.H. (1973). Modifications of the Birge-Ekman box corer for use with scuba or deep submergence research vessels. *Limnology and Oceanography*, 18-1: 172–175.

Rowe, G.T., Sibuet, M. (1983). Recent advances in instrumentation in deep-sea biological research. In: Rowe, G.T. (ed.), *The sea*. New York: John Wiley & Sons, Vol. 8, pp. 81–95.

Sarrazin, J., Robigou, V., Juniper, S.K. Delaney, J.R. (1997). Biological and geological dynamics over four years on a high-temperature sulfide structure at the Juan de Fuca Ridge hydrothermal observatory. *Marine Ecology Progress Series*, 153: 5–24.

Shillito, B., Jollivet, D., Sarradin, P.-M., *et al.* (2001). Temperature resistance of *Hesiolyra bergi*, a polychaetous annelid living on deep-sea vent smoker walls. *Marine Ecology Progress Series*, 216: 141–149.

Sibuet, M., Floury, L., Alayse-Danet, A.-M., Echardour, A., Le Moign, T., Perron, R., (1990). In situ experimentation at the water-sediment in the deep-sea: 1. Submersible experimental instrumentation developed for sampling and incubation. *Progress in Oceanography*, 24: 161–168.

Takemura, A., Tamotsu, S., Miwa T., Yamamoto, H. (2010). Preliminary results on the reproduction of a deep-sea snailfish *Careproctus rhodomelas* around the active hydrothermal vent on the Hatoma Knoll, Okinawa, Japan. *Journal of Fish Biology*, 77(7): 1709–1715.

Tsurumi, M. (2003). Diversity at hydrothermal vents. *Global Ecology and Biogeography*, 12: 181–190.
Tsurumi, M., Tunnicliffe, V. (2003). Tubeworm-associated communities at hydrothermal vents on the Juan de Fuca Ridge, northeast Pacific. *Deep-Sea Research*, 50: 611–629.
Urcuyo, I.A., Massoth, G. J., Julian, D., Fisher, C. (2003). Habitat, growth and physiological ecology of a basaltic community of *Ridgeia piscesae* from the Juan de Fuca Ridge. *Deep-Sea Research I*, 50: 763–780.
Van Dover, C.L. (2002). Community structure in deep-sea hydrothermal vent mussel beds. *Marine Ecology Progress Series*, 230: 137–158.
Wagner, D., Waller, R.G., Montgomery, A.D., Kelley, C.D., Toonen, R.J. (2012). Sexual reproduction of the Hawaiian black coral *Antipathes griggi* (Cnidaria: Antipatharia). *Coral Reefs*. doi:10.1007/s00338-012-0882-3
Watanabe, H., Kado, T. (2005). Planktonic larvae of crustaceans in the chemosynthetic communities. *Bulletin of the Plankton Society of Japan*, 52: 119–122.

Chapter 14

Seafloor Observatories

Marjolaine Matabos[1], Mairi Best[1]**, Jérôme Blandin[2], Maia Hoeberechts[1], S. Kim Juniper[3], Benoît Pirenne[1], Katleen Robert[3‡], Henry A. Ruhl[4], Jozée Sarrazin[2] and Michael Vardaro[5§]*

[1] Ocean Networks Canada, University of Victoria, Victoria, Canada

[2] Institut Français de Recherche pour l'Exploitation de la Mer, Département Recherches et Développements Technologiques, Brest, France

[3] University of Victoria, Victoria, Canada

[4] National Oceanography Centre, Southampton, United Kingdom

[5] Oregon State University, Corvallis, United States of America

Abstract

Seafloor observatories offer earth and ocean scientists unique opportunities to study multiple, interrelated processes over time scales ranging from seconds to decades, and to conduct comparative studies of regional processes. The significant advantage that observatories have over traditional sampling methods lies in the enabling of long-term interdisciplinary monitoring of ocean processes. The two first sections of this chapter provide an introduction and approach to seafloor observatory design. The third section provides details on cabled observatory design through case studies of two cabled networks: VENUS and NEPTUNE Canada (Ocean Networks Canada). The next section highlights the diversity of autonomous observatories using the example of the European setting, and provides an overview of the various observatories in place and the range of research areas offered. The final sections give an overview of data management, archiving and processing, and the importance and potential for outreach and education using online tools. The authors' experience means that examples focus on North American and European observatories, but the chapter provides information that is applicable to any observatory.

Keywords seafloor observatory, cabled observatory, autonomous observatory, time series

*Current affiliation: IFREMER, Département Etude des Ecosystèmes Profonds, Plouzané, France.

**Current affiliation: Consultant on Ocean Observing and Research Infrastructures.

‡Current affiliation: National Oceanography Centre, University of Southampton, Southampton, United Kingdom.

§Current affiliation: Rutgers University, Department of Marine and Coastal Sciences, New Brunswick, NJ, United States of America.

Biological Sampling in the Deep Sea, First Edition. Edited by Malcolm R. Clark, Mireille Consalvey and Ashley A. Rowden.

14.1 Introduction

14.1.1 What is an observatory?

Our knowledge of remote marine environments is primarily the result of a considerable effort made over the last few decades to sample the deep sea using specialized gear deployed from oceanographic vessels. This research and exploration has required a presence at sea, usually for weeks at a time. Such expeditions are expensive in terms of ship and personnel time, and only permit high-resolution sampling and measurements for short periods of time. One general consequence of this limited observational capability has been an underestimation of deep-sea biodiversity because rare species are likely to be missed, especially on abyssal plains which are characterized by low densities of organisms. In addition, a lack of continuous observations limits our understanding of the role of stochastic and even seasonal events in controlling deep-sea ecosystem dynamics. Like most marine environments, the deep sea is a complex, non-linear system subject to disturbances that act at temporal scales that vary from days to geological time, and links between those various processes are not yet well understood. As a result, scientists are missing key elements of a multidisciplinary understanding of marine ecosystem function, including species and community level responses to environmental change.

New technologies and methodologies are needed to carry out experiments and conduct long-term, high-resolution studies in the deep sea. Furthermore, traditional instrumentation and gear used to sample the deep sea (e.g. trawling, coring) are not without impact on bottom communities and there is increasing recognition of a need for 'low-impact' and 'no-take' methods to minimize sampling-related disturbances, particularly for time-series studies. Hence, the concept of 'seafloor observatories' has been developing within the international marine science community (e.g. Thiel *et al.*, 1994). Seafloor observatories offer scientists unique opportunities to study multiple, interrelated processes over time scales ranging from seconds to decades and to conduct comparative studies of regional processes (Juniper *et al.*, 2007). A seafloor observatory is defined as an 'unmanned system, at a fixed site, of instruments, sensors, and a command module connected to land either acoustically or via a seafloor junction box to a surface buoy or a fibre-optic cable' (NRC, 2000).

The idea of deploying instruments on the seafloor for long periods is not new and various types of long-term seafloor sensors and sensor platforms have been developed in the past (e.g. landers, moorings, seafloor cameras). However, these systems have limited power autonomy, local data storage and limited sensor capability.

Two types of real-time or near real-time observatories can be identified:

- *Cabled*: Seafloor observatories linked to land by submarine cables providing power and communications.
- *Autonomous*: Moored buoys that provide power to seafloor instruments and a satellite communication link to land. The landers are discussed in another chapter of this volume (Chapter 11).

The focus in this chapter is on observatories connected to land via a cable or a surface buoy. Aspects of environmental measurements are given in Chapter 4, and baited landers are the subject of Chapter 11.

14.1.2 Time-series observations

The ability to predict variation on longer time scales requires time-series measurements that resolve rapid processes, such as the propagation of internal waves or behavioural responses of marine organisms. Recent oceanographic research has demonstrated that long-term time series are needed to supplement traditional seagoing investigations in order to understand the complex interaction of various ocean systems (NRC, 1999). Seafloor observatories cannot, however, replace traditional seagoing investigations because not all variables of interest to a biologist can be measured. These include, for example, reproductive state, physiological condition and metabolism of the fauna. Current technology does not allow for sampling of living organisms, limiting study of faunal community dynamics in the sediments, population genetics or species colonization history. Seafloor observatories are not intended to replace traditional biological approaches but rather give access to complementary information at temporal scales not previously available.

The significant advantage that observatories have over traditional sampling methods is they enable continuous long-term, interdisciplinary, high-resolution monitoring of ocean processes through which researchers from diverse fields can combine their efforts to understand ecosystem functioning and ecosystem changes in a comprehensive manner. In this context, cabled observatories generally have a greater capacity to support power intensive sensors than do observatories supported by surface mooring. This capability is an important advantage for observing natural processes that are very episodic, statistically unpredictable, or require long time series to detect because they are obscured by higher frequency noise. One of the most scientifically powerful components of the observatory concept is the measurement of multiple oceanographic variables at a single location. The remote control of instruments through communication is another advantage as it increases our ability to adapt observation and sampling schedules to detected natural events and perturbations, and so is the capability to detect and react to events on a very short time scale.

14.1.3 History

Delayed mode observing systems, where data are only retrieved when instruments are recovered, have been in use for decades and have helped provide much of the data that revealed the importance of temporal variation in ocean systems (Lampitt *et al.*, 2001; Smith *et al.*, 2008; Glover *et al.*, 2010). They have been deployed during exploration cruises for periods varying from a few days to several months (Person *et al.*, 2006). Examples of sensors operated in delayed mode include conductivity, temperature probes, current meters, pressure recorders, acoustic listening devices, sediment traps, optodes (i.e. oxygen sensors), time-lapse and video cameras. Delayed mode systems are still in use today and in many cases can provide a high-quality sampling option. Although they are still very useful, the deployment of these arrays has been constrained to relatively short time periods (generally months). Other constraints have included imprecise positioning resulting from surface launch, and limited support for multiple sensors, data storage and power supply (Person *et al.*, 2006). In addition, remote control of instruments is generally not possible after deployment. And when, for whatever reason, these systems malfunction, all recorded data can be lost.

The experience acquired with pioneer ocean observation systems over the last 40 years has led to the development of a new generation of seafloor observatories. The first ‘observatories’ started in the 1950s (reviewed in NRC, 2003). As an example, the Sound Surveillance System (SOSUS) is a formerly classified system developed by the US Navy in the late 1950s to detect, track and classify submarines using hydrophone arrays connected to a shore station by a submarine cable off the coast of Washington State and elsewhere. A small number of long-term ocean measurement sites were later established by civilian research organizations in coastal settings and the open ocean using arrays of moored buoys (e.g. TAO, BTM-Bermuda). The oldest cabled observatory is the Field Research Facility (FRF) off Duck, North Carolina, established in 1977 by the US Army Corps of Engineers. The first cabled observatories were designed to study coastal waves, currents, meteorology, seismology and earth movements (e.g. LEO-15, HUGO and H20 for the US, DONET for Japan and MACHO for Taiwan). A non-exhaustive list of ocean observation systems can be found in *Enabling Ocean Research in the 21st Century* (NRC, 2003).

More recently, integrated cabled observatories equipped with chemical sensors and biological sampling systems such as sediment traps and video camera structures have been deployed. The first ones were VENUS, off Vancouver Island (BC, Canada; Tunnicliffe *et al.*, 2003), MARS in Monterey Bay, California (Massion & Raybould, 2006), and the regional ocean network NEPTUNE Canada,

Table 14.1 Main seafloor cabled observatories/networks for multidisciplinary studies (non-exhaustive list)

Observatory/network	Location	Research areas	Link
DONET (JAMSTEC, Japan)	Sagami Bay (Japan) Nankai Trough (Japan) Kuril Trench (Japan)	Geophysics Geophysics Geophysics	www.jamstec.go.jp/donet/e/
Kilo Nalu Nearshore Ree Observatory	Hawaii	Oceanography, chemistry, biology	www.soest.hawaii.edu/OE/KiloNalu
LEO 15	Off Tuckerton (NJ, USA)	Oceanography, meteorology, biology	http://rucool.marine.rutgers.edu/
Marine Cable Hosted Observatory (MACHO)	Ilan plain (Taiwan)	Geophysics	
Martha's Vineyard Coastal Observatory	Off Martha's Vineyard (MA, USA)	Climatology Oceanography	www.whoi.edu/mvco
Marine e-Data Observatory Network (MEDON)	Northeast Atlantic, Off Britanny (France)	Oceanography, biology	http://www.ifremer.fr/emso-france/EMSO-Molene
Monterey Accelerated Research System (MARS)	Monterey Bay (CA, USA)	Geophysics, oceanography, chemistry, biology	www.mbari.org/mars
Neutrino Telescope Italy, France, Greece (KM3NeT)	Mediterranean sea (Italy)	Astronomy, geophysics	www.km3net.org
North-East Pacific Time-series Undersea Networked Experiments (NEPTUNE Canada)	Northeast Pacific, (BC, Canada)	Geophysics, oceanography, chemistry, biology	www.oceannetworks.ca
Expandable Seafloor Observatory (OBSEA)	Mediterranean Sea (Spain)	Geophysics, oceanography, biology	www.obsea.es
Victoria Experimental Network Under the Sea (VENUS)	Saanich Inlet/Strait of Georgia (BC, Canada)	Geophysics, oceanography, chemistry, biology	www.oceannetworks.ca

off Vancouver Island (BC, Canada). Other smaller coastal observatories devoted to the study of biological and ecological processes included the Bonne Bay Observatory in Newfoundland (Canada) and, more recently, OBSEA off Barcelona, Spain (Manuel-Lazaro *et al.*, 2010). Indeed, the next challenge in observatory development will be to design integrated, multidisciplinary systems that can host instruments to measure physical properties (i.e. waves, tide, currents, pressure) and seismic activity, and also gather chemical and biological data. A list of cabled observatories installed as of January 2013 is presented in Table 14.1. The purpose of this chapter is not to describe all existing observatories but rather provide insights into their implementation and use for deep-sea biology. However, a growing literature is emerging on seafloor observatories and reviews can be found in Favali and Beranzoli (2006), Favali *et al.* (2010) and Ruhl *et al.* (2011).

14.2 Planning an observatory system

The value of a time series is strongly linked to the continuity of the data and the length of the record (Glover *et al.*, 2010). Ocean observatory systems and their instrumentation must be designed for reliable, continuous operation over long periods of time. To ensure the viability of such systems for operation over years and decades, maintenance requirements must be minimized (NRC, 2003). Recommendations for planning seafloor observatories have already been developed and can be found in various reports such as National Research Council (NRC) (2000, 2003) and the European Seafloor Observatory Network (ESONET) final report (Priede & Solan, 2005). In this section, we briefly review the main considerations related to seafloor observatory design based on several successful examples.

14.2.1 Network and organization

Seafloor observatories are costly 'big science' projects that are necessarily greater in scope than research undertaken by individual or small groups of investigators. They are often many years in development and require a considerable level of organization and fundraising at national and/or international level. The first step in establishing a seafloor observatory system should be the development of a detailed and comprehensive programme and project implementation plan, with review by knowledgeable, independent experts. Planning of seafloor observatories should be driven by scientific requirements, which must be clearly defined at the beginning of the process – for example by identifying project leaders and research teams, and exchanging ideas at workshops.

In the case of the large-scale observatories presented below, the implementation of seafloor observatories required coordination among researchers from different agencies and institutions, often at the international level. Examples of the various programmes found around the world can be found in Favali *et al.* (2010).

14.2.1.1 Ocean Networks Canada

The Ocean Networks Canada (ONC) is a not-for-profit corporation created by the University of Victoria (Canada), to oversee the management, development and operation of the VENUS and NEPTUNE Canada cabled observatories. With a 25 year or more expected operating lifespan, the ONC observatories enable ocean research and technology development through a cabled infrastructure that supplies continuous power and internet connectivity to a broad suite of subsea instrumentation. VENUS and NEPTUNE Canada are the coastal and regional components respectively of ONC (Tunnicliffe *et al.*, 2008; Barnes *et al.*, 2012; Best *et al.*, 2012). The two networks provide near real-time chemical, geological, biological and physical oceanographic data covering a wide range of ocean environments from the coastal area of Vancouver Island to the Juan de Fuca Ridge, 300 km off Vancouver Island. In addition to traditional fields of marine science, the technology required to operate the observatory along with the huge amount of data retrieved from the ocean offer considerable potential for research project development in computer science and engineering.

The ONC Data Management and Archiving System (DMAS – also known as 'Oceans 2.0') is the digital infrastructure (DI) component of ONC and supports both VENUS, NEPTUNE Canada as well as smaller local cabled systems (Pirenne & Guillemot, 2012). The mission of DI is to develop software for data acquisition and lifetime storage, and to enable remote interactive control of instruments. All data are freely available to the public through web-based interfaces. The DI is also responsible, in coordination with the science team, for the development of data products to meet user requirements for data format and delivery. Considerable effort is being made to ensure interoperability with other observatories.

14.2.1.2 Ocean Observatories Initiative

The Ocean Observatories Initiative (OOI) is a project funded by the US National Science Foundation (NSF) to create an interactive network of ocean observatories with a projected 25 year lifespan that will provide near real-time chemical, geological, biological and physical oceanographic data on coastal, regional and global scales. The OOI offers great potential for observing ocean processes ranging in magnitude and temporal scale in the fields of climate variation, ecosystem transformation, ocean acidification, carbon uptake by algal blooms, anoxic events and short-term weather patterns. The implementing organizations of the OOI are the Woods Hole Oceanographic Institution (WHOI), Oregon State University (OSU), University of Washington (UW) and Rutgers University (RU). The OOI infrastructure will contain a suite of sensors tailored to record environmental data from the seafloor to the surface at their respective deployment locations.

The cyberinfrastructure (CI) component of the OOI (implemented by Rutgers University) will coordinate the operations of the OOI observatories, store and process the data collected, and communicate these data to the scientific and educational communities. In addition, they will help tailor the data products produced by OOI by responding to the diverse and changing needs of various user groups. The long-term objective is to connect the observation, modelling and mission control systems and to make the resulting OOI data widely available to the public.

14.2.1.3 European Multidisciplinary Seafloor and water column Observatory

The concept of a coordinated chain of deep-sea observatories around Europe was initially known as European Sea-floor Observatory Network (Priede *et al.*, 2003, 2005). However, it was quickly recognized that monitoring throughout the water column is necessary, leading to a slight, but highly significant change in name to the European Seas Observatory Network, Network of Excellence (ESONET NoE) which became a European multidisciplinary research programme led by IFREMER (Institut Français de Recherche pour l'Exploitation de la Mer) and partnering dozens of agencies, including research institutes, universities and companies in Europe. ESONET was funded to integrate researchers for a marine component of GMES (Global Monitoring for Environment and Security), comprising a network of long-term, multidisciplinary seafloor observatories at key locations around the European margin for long-term monitoring advancement in geophysics, chemistry, biology, oceanography, geochemistry and fisheries.

The European Multidisciplinary Seafloor and water column Observatory (EMSO), in collaboration with other projects, is establishing a formal organization to operate ocean observatories in Europe (Favali *et al.*, 2010) building on the experience of ESONET. These efforts aim to create an organization capable of implementing, operating and maintaining a network of ocean observatories in European waters from the Arctic Ocean to the Black Sea. The integration process is based on: building up active sharing groups; acting as one body towards funding institutions, stakeholders, potential users and similar international projects; and advancing the infrastructure policy of observatories in Europe. EMSO comprises the deep-sea observatories themselves and is effectively a large-scale European research infrastructure based on the network of open ocean observatories and platforms (Favali & Beranzoli, 2009).

14.2.2 Observatory location

In the implementation process of seafloor observatories, considerable attention must be given to select a location that is strategic and based on clear scientific priorities, for example, spanning an entire tectonic plate (NEPTUNE Canada + OOI regional-scale nodes (RSN)), existing time-series sites (EMSO) and/or representing potential hazards to the population (Ligurian Sea, Marmara Sea, Juan de Fuca plate). The size and location of an observatory depends on processes and scales of interest:

- Global observatory networks are designed to study large-scale processes and are needed to study the ocean's role in climate change and to understand the structure and dynamics of the Earth's interior (NRC, 2003).
- Regional observatories allow long-term measurement of geological and oceanographic processes at a regional scale using electrical power and real-time network connections (e.g. NEPTUNE Canada).
- Coastal observatories are designed to understand coastal processes and the magnitude and quality of anthropogenic effects on the coastal ocean (e.g. VENUS).

Many location-related issues need to be considered during the planning and construction of an undersea observatory. Some involve interactions with other users of the marine environment. National navies may have concerns about the security of military vessels being compromised by underwater hydrophones and other sensors. Commercial fishing gear can damage underwater infrastructure and instruments, requiring measures to be taken to minimize negative interactions (consultation and agreements with the fishing industry) and mitigate their impacts (e.g. trawl-resistant node frames).

14.2.2.1 Ocean Networks Canada

ONC operates coastal and regional observatories, VENUS and NEPTUNE Canada, respectively. It also operates a small observatory in the Arctic (in Cambridge Bay, Nunavut, 69°N 105°W) where a cabled system provides real-time data from over half a dozen instruments in a coastal Arctic environment (see Juniper *et al.*, 2012). The ONC observatories cover a wide range of ecosystems from coastal waters to a mid-ocean ridge, from Marine Protected Areas (MPA) to zones highly affected by human activities, and includes ecosystems based on both photosynthesis and chemosynthesis.

The NEPTUNE Canada regional observatory lies across the Juan de Fuca tectonic plate (subduction zone to ridge), permitting real-time observation of tectonic processes at the plate scale. Observations from the network will be applicable to other tectonic systems all over the planet. Settings include shallow rocky and soft-bottom coastal habitats, the continental slope, a submarine canyon, gas hydrate fields, the abyssal plain and mid-ocean ridge hydrothermal vents.

VENUS, operational since 2006, is the coastal component of ONC with a node in Saanich Inlet (48°N 123°W), a coastal fjord north of Victoria (Vancouver Island, BC) and two nodes in the Strait of Georgia (49°N 123°W), with both locations displaying interesting oceanographic features (Tunnicliffe *et al.*, 2003). Saanich Inlet is a seasonally hypoxic fjord and one of the best-studied marine basins in the world because of its easy access and unusual properties. The Strait of Georgia, the location of the other two VENUS nodes, is also an interesting oceanographic setting because of complex water mass mixing occurring throughout the year. The two main influences are freshwater from the Fraser River and upwelling on the west coast that generates an inflow of dense, salty nutrient-rich water through the adjacent Juan de Fuca Strait. The resulting high productivity throughout the region supports salmon and herring fisheries, and resident marine mammals. The VENUS setting is thus strategic for studies of changes in water mass properties and their influences on biological communities.

The Cambridge Bay Observatory (CBO), operational since 2012, is located in the Arctic, near the Northwestern Passage. It is nestled in a sheltered bay in Cambridge Bay, Nunavut. The CBO allows for a very cost-effective long-term study of the underwater conditions in an Arctic environment and has therefore attracted a lot of attention from scientists and local governments alike.

14.2.2.2 Ocean Observatories Initiative

There are three components that make up the OOI, on a coastal, regional and global scales.

The *coastal-scale nodes*, implemented by WHOI and OSU, will consist of cabled and autonomous assets deployed on the Atlantic and Pacific coasts to study a variety of long-term processes in shallow (25 m), shelf (80 m) and deep-water (~500 m) environments. The coastal Endurance Array consist of two long-term networks collecting continuous data off Newport, Oregon (44°N 125°W) and Gray's Harbor, Washington State (44°N 127°W) in the Pacific Northwest. A Pioneer Array, designed to be flexible and relocatable, will initially be deployed on the East Coast in the mid-Atlantic Bight, south of Cape Cod. These sites are all areas of complex current dynamics, high productivity and significant seasonal climatic variability; they also have substantial economic importance for fisheries and are impacted by a range of anthropogenic factors.

The aim of the Cabled Array (implemented by the University of Washington) is to provide long-term, near real-time measurements of biogeochemical and oceanographic phenomena in the deep ocean. Electro-optical cables will deliver power and provide two-way communication to the installed sensors, which will observe processes at two locations: Axial Volcano (45°N 130°W) (1000–1500 m), an area of volcanic and hydrothermal activity on the Juan de Fuca plate, and Hydrate Ridge (44°N 125°W) (~800 m), a region characterized by subsurface and exposed gas hydrate formations, authigenic carbonate outcrops and associated chemosynthetic communities. A mid-plate node will permit future expansion of the system midway between the spreading ridge and the subduction zone.

The *global component*, also implemented by WHOI, is made up of a network of full ocean depth (~5000 m) moorings at four high-latitude sites: the South Pacific off the coast of Chile (55°S 90°W), the South Atlantic in the Argentine Basin (42°S 42°W), the Irminger Sea in the subpolar North Atlantic (60°N 39°W) and Ocean Station Papa in the North Pacific (50°N 145°W) (Fig. 14.1). These sites were

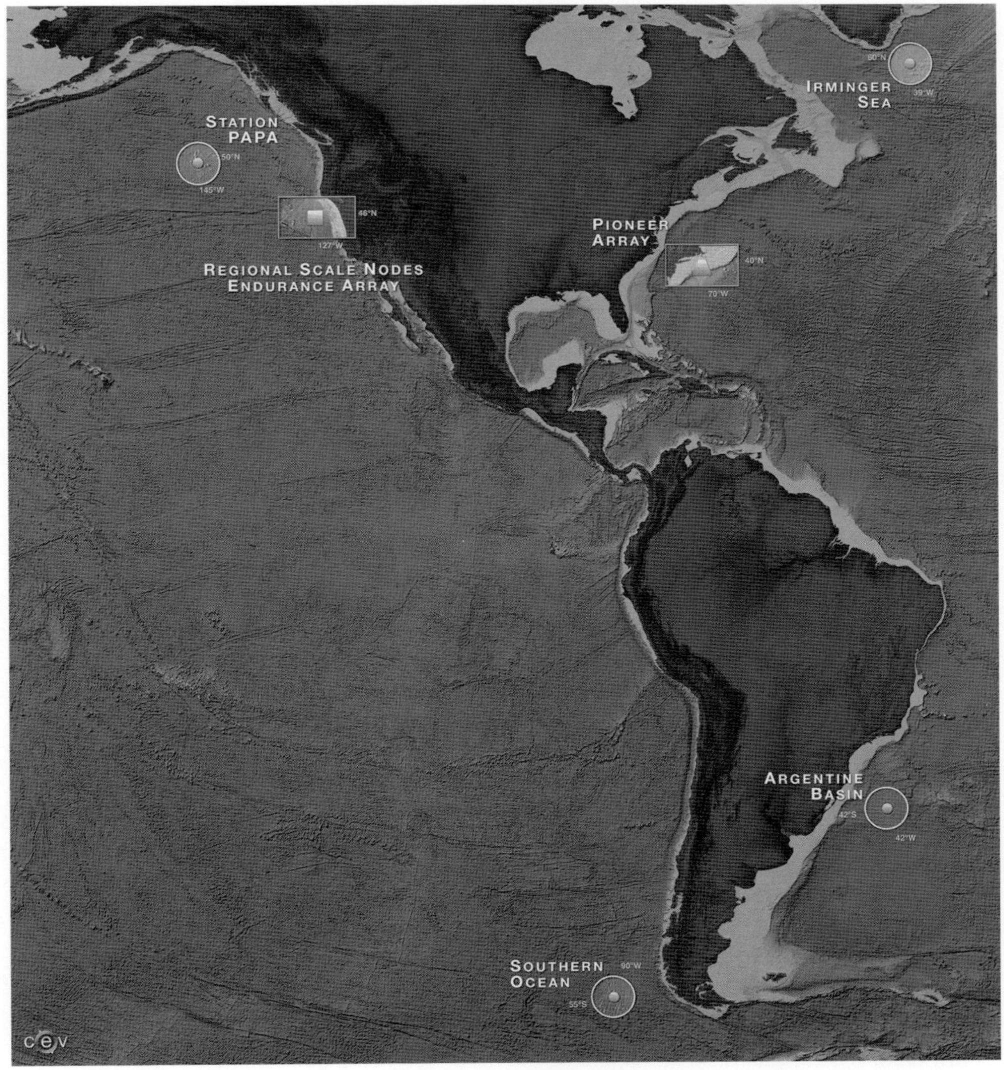

Fig. 14.1 Map of the Ocean Observatories Initiative (OOI) site locations. (OOI Cabled Array program and the Center for Environmental Visualization, University of Washington. Reproduced with permission.)

chosen because high-latitude areas are more vulnerable to, and indicative of, the large-scale effects of climate change, as well as being crucial to ocean–atmosphere exchange and global seawater circulation. Severe weather variability and the remote nature of these locations has also resulted in them being historically undersampled, so the long-term deployment of OOI assets at these sites is particularly important to obtaining a complete picture of the world's oceans.

14.2.2.3 European Multidisciplinary Seafloor and water column Observatory

More than a dozen observatory sites are geographically distributed around European seas, reaching from the Arctic to the Marmara Sea passing through the Mediterranean Sea, and include several coastal and deep seafloor observatories (Favali & Beranzoli, 2009; Ruhl *et al.*, 2011). Each study site has features that align with specific programme goals and research themes. Figure 14.2 illustrates the location of existing or expected EMSO sites around Europe.

Fig. 14.2 Map of European margin 'key-sites'. Candidates as nodes of the EMSO underwater permanent work. (EMSO. Reproduced with permission.)

14.2.3 *Implementation*

14.2.3.1 Planning

The first step in establishing an observatory is planning the location. For cabled observatories this involves mapping and cable route design. The next step is the selection of appropriate technology. Some institutions have already invested a great deal of effort in observatory design, which includes development of new technology and implementation of established technology. Successful prototype cabled observatories such as MARS or the first stage of VENUS incorporated technologies that were available for future development. Another approach to the development of observatory technologies is the example of the test platform operated by the Ocean Technology Lab at the University of Victoria. The Ocean Technology Test Bed (OTTB) is connected to the VENUS network in Saanich Inlet (Sydney, BC, Canada) (Proctor *et al.*, 2007). An instrument platform suspended beneath a surface buoy allows experimental instruments to be lowered to the seafloor without requiring submersible support, permitting field testing of technology at low cost. Cooperation between observatory organizations will permit developing seafloor observatories to take advantage of the knowledge and experience, both good and bad, accumulated by other observatory projects and make use of existing observatory infrastructure for initial testing. This type of exchange will also help with global efforts aimed at improving interoperability of observatory instruments and research methodologies.

14.2.3.2 Environmental impact

The development of seafloor observatories is just one component of an increasing human presence offshore and on the seafloor. The evaluation of environmental disturbance caused by research equipment and observatory infrastructure is a field of research that will be important to regulatory agencies considering sites for future observatories. Subsea cables were shown to have neutral or minor species-specific environmental impacts (Kogan *et al.*, 2006). Effects of other observatory components and activities on the marine environment are not well documented (Burnett & Carter, 2009). Environmental impact assessment requirements vary between jurisdictions, and are usually more rigorous in areas with protected status, such as MPAs. An understanding of the boundaries of the impacted region is essential to determine under which jurisdiction it falls, and therefore what rights and obligations exist for the implementation of an offshore observatory. Laws should be analysed in each country where an observatory or cable system is to be installed. Relevant jurisdictional boundaries include those established by the United Nations Law of the Sea Convention (UNCLOS) in 1982. Five zones are defined: the Territorial Sea, the Contiguous Zone, the Exclusive Economic Zone, the Continental Shelf and the High Seas. Each zone is under different national and international laws. The latter are considered in a review by Burnett and Carter (2009), who conclude that there is no entity competent to broadly regulate structures and cables in the High Seas zone. Their review provides useful guidance on UNCLOS provisions relevant to scientific activities outside of territorial seas.

14.3 Cabled observatories

14.3.1 *What is a cabled observatory?*

Current cabled observatory infrastructure consists of a submarine cable that provides power and real-time communication and control capabilities to scientific monitoring equipment at 'nodes' along the cabled system. Each node can support a range of devices such as instrument platforms, individual sensors, autonomous underwater vehicle (AUV) docks and water column moorings.

The advantage of cabled observatories resides in the real-time, bidirectional, high-bandwidth communication as well as the continuous availability of power to the instruments (up to several kilowatts). They are highly flexible systems that can host many types/categories of instruments with different communication protocols. Cabled observatories can provide or support:

- continuous daily to decadal presence for a variety of sensors;
- sub-second sampling frequencies for most parameters;
- real-time multidisciplinary, interactive experiments;
- event detection and response capabilities;
- an ability to transmit large volumes of data to users in near real time;
- almost complete immunity to weather problems;
- large amounts of continuous electrical power (kW) to the ocean floor/shore connection;
- high reliability of the technology provided by commercial research and development.

14.3.2 Cabled observatory technology

14.3.2.1 Basic infrastructure

A cabled observatory has four essential elements:

- a network of electro-optical cables and seafloor nodes that provide power and two-way, high-speed communications;
- seafloor and water column instrument arrays and their related connectivity (extensions, junction boxes);
- an archive centre with data management and distribution that provides a user interface and long-term access;
- an operations centre to monitor and control all subsea and shore elements.

New or complementary equipment can be attached at any time during the lifetime of the observatory. Sites can be any distance from each other as long as repeaters are deployed along the cable (Barnes *et al.*, 2008). The topology usually includes a shore station providing power and a communication link, and a backbone cable with branches that connect to seafloor nodes and from there to peripheral installations that include secondary cables, junction boxes and instruments (Fig. 14.3).

The following description is based on the NEPTUNE Canada configuration (Barnes *et al.*, 2008).

Power distribution The NEPTUNE Canada power system differs from that of standard submarine telecommunications systems in that it uses parallel rather than serial connections of the loads. Parallel power offers significant advantages in efficiency, reliability and flexibility over serial power, and is also readily compatible with the branching topology of NEPTUNE Canada. The design using a cable loop has the advantage of redundancy, so that if there is a break in the cable, the data can still flow back to the shore station and power is still available to nodes.

System power is provided from the shore station AC utility power, which is converted to 10 kV DC using an off-the-shelf converter. Medium voltage converters at each node convert the 10 kV DC to 400 V. Low voltage DC–DC converters at junction boxes provide outputs at 15, 24 and 48 V (which most instruments need). Junction boxes support up to 10 instruments each, and both nodes and junction boxes can be daisy-chained where necessary.

Communication systems The communication system for the undersea portion of the network consists of three parts: the shore station, the optical fibres and the instrument connection system (through nodes). The baseline design uses Internet Protocol (IP) with packet-switched Gigabit Ethernet (GbE) technology to provide the communication needed for the observatory. The shore station serves as the link from the observatory to the DMAS, and the internet. Onshore software monitors communication system performance and makes necessary adjustments to the undersea communications components. At each node, data from scientific instruments are aggregated and switched onto the

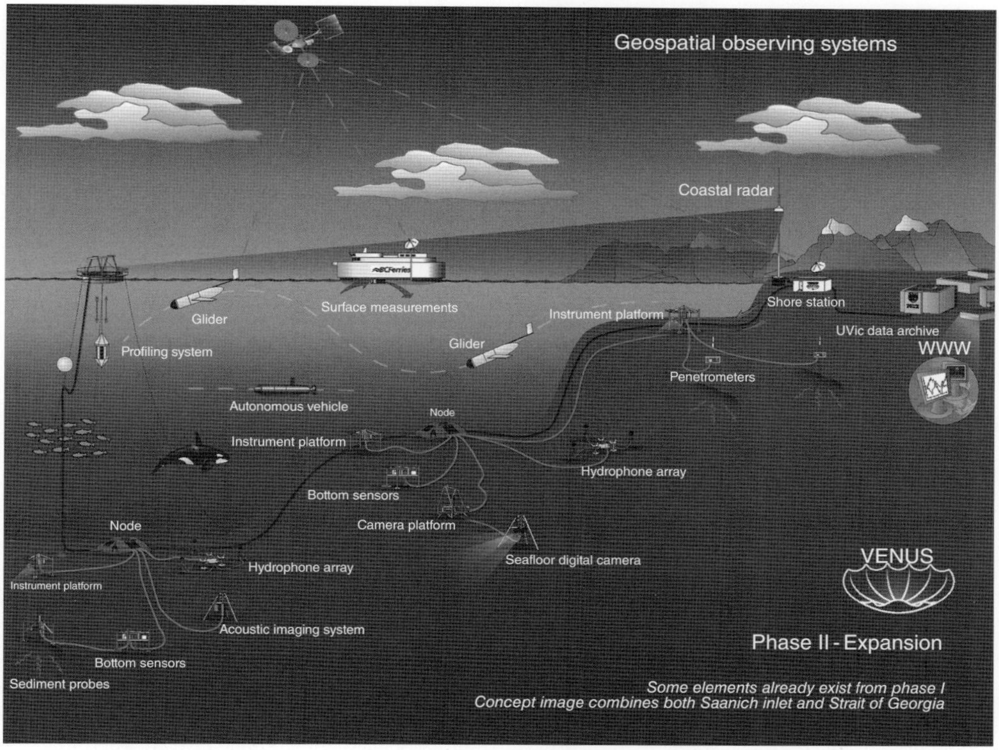

Fig. 14.3 Sketch of a cabled observatory structure based on the example of VENUS (University of Victoria, BC, Canada). (Ocean Networks Canada. Reproduced with permission.) (Please see Plate 24 for colour representation of the figure.)

backbone cable for delivery to the shore station. Command and control information are also delivered to the instruments via switches in the nodes. The ring topology provides redundancy in case of a fault on the main backbone cable; in that case communications can be rerouted around the fault. Each observatory node also contains a pair of routers for redundancy.

Time distribution Accurate and precise time is necessary for some science experiments and for simultaneously recording events detected by different instruments at different locations. This requires a reference standard and the distribution of a time signal across the seafloor network. High-accuracy time can be distributed over the system, allowing high-accuracy time signals to be transferred from the shore station to the nodes and to the instruments, including correction for propagation delays on the backbone cable, and providing synoptic time ticks everywhere on the network. Timing accuracies of 10 μs (worst case scenario) are attainable if the Precision Time Protocol (PTP – IEEE1588) is used. A 10 ms accuracy can be reasonably expected if the Network Time Protocol (NTP) is used.

System construction Cable splices, branching units and cable penetration into housings use existing qualified cable jointing technology from the submarine telecommunications industry. Housings use current designs for seals and pressure vessels, and are built of corrosion-resistant materials (e.g. titanium). Connections to the scientific instruments use remotely operated underwater vehicle (ROV)-adapted wet-mateable electrical connectors. This enables individual instruments or groups of instruments to be recovered for servicing without disruption of neighbouring sensors. Underwater-mateable electrical connectors for this purpose are readily available and have a track record of long-term reliability. Underwater-mateable optical connectors are an emerging technology that has been deployed with success on NEPTUNE Canada.

Observatories should be designed to be installed using conventional cable-laying technology, to take advantage of the worldwide fleet of cable ships and associated technical expertise. Scientific instruments are attached to the infrastructure using standard oceanographic ship and ROV assets. A major design goal is to minimize the cost of maintenance of the seafloor plant. Therefore, wherever possible the infrastructure should be designed to be maintained using the same ships and ROVs that are used by the broader research community. Some specialized shipboard and ROV equipment is required for deployment of large instrument platforms, moorings and secondary cable networks on the seafloor.

14.3.2.2 Instrumentation

Three approaches to instrument selection have emerged as cabled observatory designs have evolved (NRC, 2003):

- *Core or initial sensors and instruments*: These are defined during observatory design for fundamental physical, chemical and/or biological data, depending on the focus of the observatory. Instruments are calibrated according to international standards. Typically, these are instruments with long operational track records that require little development to adapt them to cabled observatories (e.g. seismometers, Conductvity Temperature Depth sensors (CTDs)).
- *Community instruments*: These are specialized scientific instruments critical to the long-term scientific objectives of a particular node. They provide data of interest to a wide range of investigators (e.g. camera systems, borehole instruments).
- *Individually owned/experimental instruments*: These are associated with a single investigator (or laboratory institute). They are often developmental instruments, which may require specialized data analysis.

Researchers invariably identify very specific sensors and instruments (i.e. manufacturers and models) based on their prior experience with autonomous, typically single-instrument, deployments. Procurement regulations and engineering procedures, however, often require that a set of functional instrument requirements (science needs) be developed for a competitive bidding process. In addition, there are many new requirements related to long-term deployments on a cabled observatory that are not as relevant for autonomous deployments. NEPTUNE Canada Observatory instruments therefore need to be assessed with regard to requirement and quality criteria listed in Tables 14.2 and 14.3.

In most instances meeting these criteria require relatively minor changes in the commercially available off-the-shelf (COTS) instruments (e.g. changes in power supplies, removal of grounding through housings to seawater). Substantial issues with COTS instruments are sometimes only discovered during in-house testing. After Bornhold *et al.* (2010) these could include:

- unanticipated noise issues with high-frequency, active acoustic sensors,
- corrosion problems common on most instruments, including the use of dissimilar metals as well as insufficient, or inappropriate, sacrificial anodes, and
- lack of vendor documentation of inrush currents on start-up. While of little consequence in battery-powered deployments, an understanding of these values can be critical in a cabled system.

Although cabled observatories and their instruments can already meet a broad range of research requirements for fields such as seismology and geophysics, marine chemistry and biology are still underdeveloped with respect to cabled technology and methodologies, and a major effort of pilot studies and methodological development is required for these fields (e.g. Pawlowicz & McClure,

Table 14.2 Instrument requirements criteria

Power and grounding	Instrument control and data transmission	Physical interfaces	Documentation
Communications physical media and network layers	Metadata	Reliability	Shore terminal-based supporting equipment
Communications network compatibility	Safety	Maintainability	
Communication data protocol	Engineering	Embedded software	
Timing	Environmental requirements	Instrument testing	

From Bornhold *et al.* (2010).

Table 14.3 Instrument quality criteria

Quality plans	Handling and shipping	Rework, repair and observatory redeployment
Instrument qualification – Design requirements	Testing – Plans, procedure, test instrument calibration, verification of prerequisite tests, final acceptance tests	Service ability – Allowable field service, access to electronics and special tooling availability, training, protection during maintenance and servicing
Manufacturing – Documentation. acceptability of electronic assembles, electrostatic discharge, soldering requirements, wire routing and strain relief, circuit board hardware, electrical isolation., shielding, fibre handing, seals and O-rings, cable moulding	Operation and maintenance – Manual, deployment and recovery instructions, maintenance instructions	

From Bornhold *et al.* (2010).

2010; Matabos *et al.*, 2011, 2014; Robert & Juniper, 2012; Juniper *et al.*, 2013). In particular, chemical and biological sensor technologies are not sufficiently advanced to take full advantage of observatory infrastructure. Many sensors will need considerable improvement before they can operate unattended for long periods of time. As an example, the Monterey Bay Aquarium Research Institute (MBARI) has developed an environmental sampling processor (ESP) that can operate from oceanographic moorings, for *in situ* genetic recognition of bacteria and phytoplankton (Scholin *et al.*, 2006), but a version of the ESP has yet to be deployed on a long-term cabled observatory.

14.3.3 Case studies of observational approaches

Cabled observatories offer a novel means of studying benthic processes. Methodological approaches to conduct studies in this field are currently being developed (Matabos *et al.*, 2011, 2014; Robert & Juniper, 2012; Juniper *et al.*, 2013; Cuvelier *et al.*, 2014) and will help optimize the unique capabilities offered by these systems, particularly high-resolution sampling, long-term monitoring, real-time event detection, and interactive observation. The following examples illustrate some of the work currently underway on the NEPTUNE Canada regional network (Fig. 14.4), the VENUS camera platform array in Saanich Inlet (Fig. 14.5), and the OOI's research plans.

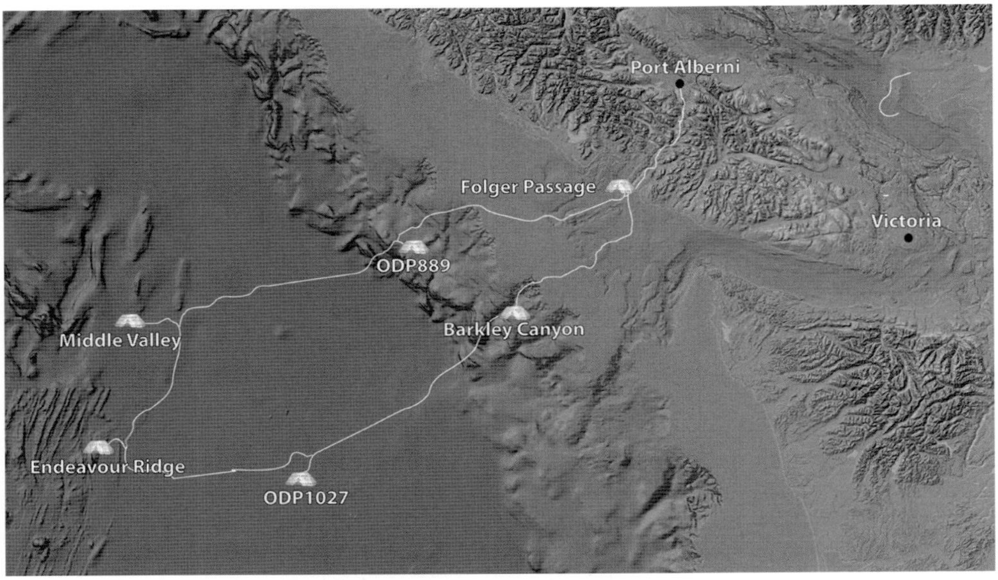

Fig. 14.4 The NEPTUNE Canada regional ocean network deployed on the west coast of Vancouver Island (British Columbia, Canada). The camera system described in this chapter is deployed at the Barkley Canyon node. (Ocean Networks Canada. Reproduced with permission.)

14.3.3.1 Surveys to determine optimal placement of instruments

At Barkley Canyon (48°N 126°W), the locations of four different instrument pods were determined by the requirements of a multidisciplinary project involving ecologists, physical oceanographers and geologists from various Canadian institutions (Dalhousie University (Nova Scotia), University of Victoria (British Columbia), Memorial University (Newfoundland), University of Toronto (Ontario) and Université du Québec à Rimouski (Québec)). The initial deployment includes a sediment trap, plankton pump, acoustic Doppler current profilers (ADCP), CTDs, cameras, hydrophones and high-resolution sonars. This setting provides opportunities to study the responses of benthic communities to variations in organic input from daily to interannual and even decadal scales, and to distinguish between organic matter inputs from the surface and those advected through the canyon. The sonars allow the study of seabed dynamics in relation to benthic faunal activity or transport processes within the canyon (upwelling and downwelling currents, slumps, slides, gravity flows).

Site selection for a camera system in Barkley Canyon An important first step in the deployment of camera systems is site selection, which should be based on prior knowledge of the study area. Site selection can also be constrained by technical considerations. The deployment of mid-canyon instrument pods (i.e. instrument platforms) in Barkley Canyon illustrates the various technical and ecological trade-offs that need to be taken into account when selecting sites for seafloor cameras. The initial experimental design required that instrument pod locations be representative of the major physical habitat types in and around Barkley Canyon, rather than any special localized phenomena such as the gas hydrate field in the mid-canyon area. Logistic requirements included the limited cable length from junction boxes due to data and power transmission constraints. As a result, the mid-canyon pods could be no more than a few hundred metres away from the 870 m deep hydrate field. A pre-deployment ROV video

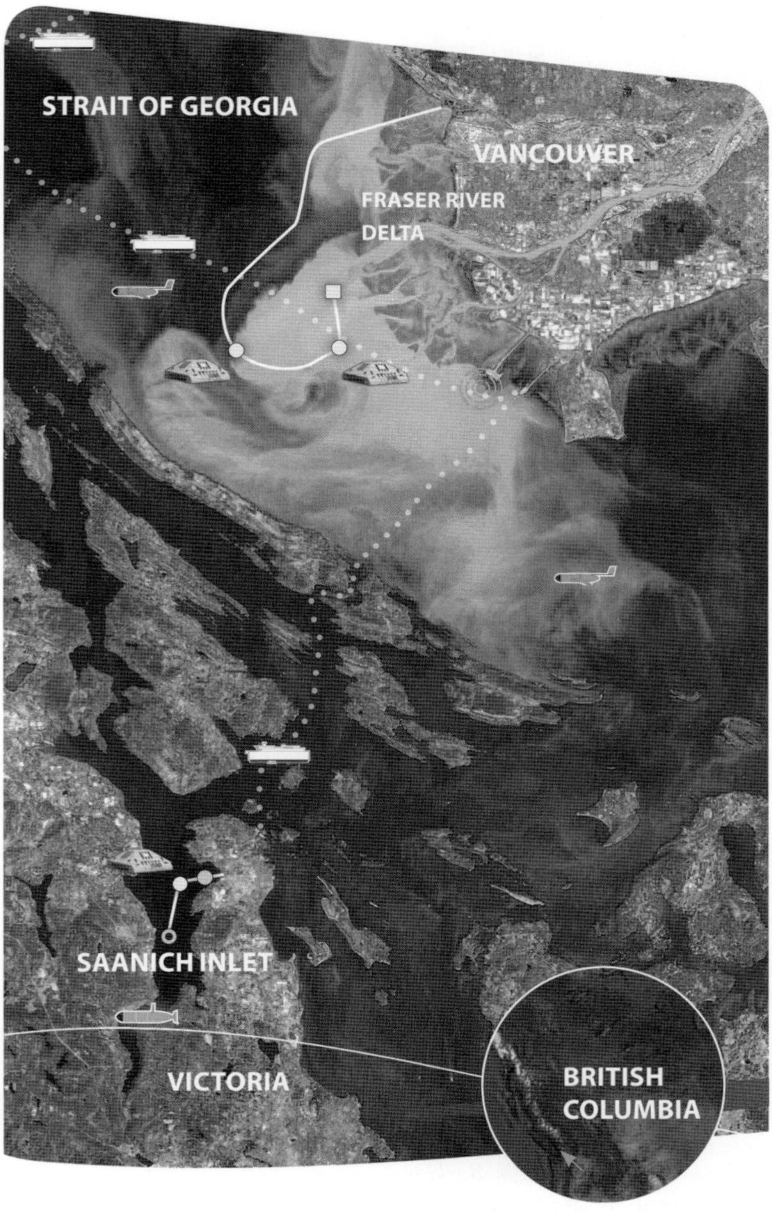

Fig. 14.5 VENUS cabled observatory, node arrays in Saanich Inlet and Strait of Georgia, Vancouver Island (British Columbia, Canada). Map shows the route of instrumented ferries and ocean gliders. The camera system described in this chapter is deployed at the Saanich Inlet node. (Ocean Networks Canada. Reproduced with permission.)

survey was thus conducted to determine the limits of the hydrate field's influence, in terms of the composition of the benthic megafauna. The position on the seafloor of individual megabenthic organisms, identified to the lowest taxonomic unit, was georeferenced and mapped using video data (K. Robert, unpublished data).

14.3.3.2 Deep-sea bioturbation

One of the camera-equipped instrument pods deployed at a 400 m deep location on the outer continental shelf away from the edge of Barkley Canyon was used to develop methods for the study of surface bioturbation by benthic megafauna (Robert & Juniper, 2012). The continuous power supply provided by the cabled observatory offers the potential to record seafloor imagery without interruption, 24 hours a day. However, the manual analysis of such a large quantity of video would be impossibly time-consuming. In addition, the impact of the artificial lighting required for making observations in aphotic environments has not been well documented (Herring *et al.*, 1999) although use of red lights has been suggested as a mitigation measure (Widder *et al.*, 2005). As a precautionary approach, one hour of lighting was allotted per day. The next step was therefore to develop an observational protocol that would make optimal use of the one hour per day of observation, to quantify the activities of the two main groups of bioturbating megafaunal organisms found at this site – sea urchins and flatfish. Below we describe the approach used for the sea urchins.

An iterative method was employed to determine the observation frequency best suited to estimating each of the parameters of interest: bioturbation footprints of individual sea urchins, sea urchin abundance and locomotion rates. In order to take maximum advantage of the increased field of view (FOV) provided by the pan and tilt mechanism, the camera was swept almost 360° at specific tilt angles to cover the seafloor from the horizon to the nadir point. These sweeps took 5 minutes, meaning the observation frequency could be increased to a maximum of 12 times daily and still remain within the lighting time limit. At this observation frequency, the locomotory path of individual sea urchins could be tracked and measured (Robert & Juniper, 2012).

The extraction of quantitative information from imagery involved the use of laser pointers and scaling rulers to build perspective grids (see Wakefield & Genin, 1987) as well as a polar coordinate system, based on the pan and tilt angles of the camera, to track organisms across the FOV (Fig. 14.6). The laser points, set 10 cm apart, provided a horizontal scale for all tilt angles and were used to trace meridian lines for the perspective grid. To draw the horizontal lines, a ruler marked at 10 cm intervals, laid beneath the camera, provided the vertical scale. With this information, we were able to build perspective grids (Fig. 14.7) without knowing the exact height of the camera from the seafloor. This latter information could not be obtained in the current situation because the platform's feet sank to unknown depths in the sediment. In parallel, information from nearby sensors was useful in relating bioturbation rates to environmental fluctuations.

In Barkley Canyon, complementary information on surface chlorophyll levels was obtained through imagery from the SeaWiFS satellite (Fig. 14.8), and assembled using the Ocean Color Radiometry Online Visualization and Analysis tool from NASA's web-based GIOVANNI application developed by GES DISC (http://daac.gsfc.nasa.gov/giovanni/). Monthly averages of chlorophyll-*a* concentration are being compiled for the 600 km- area overlying Barkley Canyon and will eventually be compared to the monthly estimates of bioturbation rates currently being collected.

The arrival of phytodetrital accumulations has previously been observed a posteriori with time-lapse cameras (e.g. Billett *et al.*, 1983). With the current NEPTUNE Canada system, such an occurrence can be detected in real time and updates to observation schedules could be implemented within hours. In a more sophisticated system, co-located sensors such as fluorometers could be programmed to trigger an event response sampling protocol once chlorophyll levels surpass a predetermined threshold (Fig. 14.9).

14.3.3.3 Benthic community dynamics in a seasonally hypoxic fjord

In Saanich Inlet, a seasonally hypoxic fjord, the VENUS node was strategically located in the oxygen fluctuation zone providing a unique laboratory to study ecosystem responses to changes in oxygen concentrations. The VENUS camera platform was located at 103 m depth on the edge of the main basin, slightly upslope from the anoxic waters, within a zone of fluctuating hypoxia. Dissolved

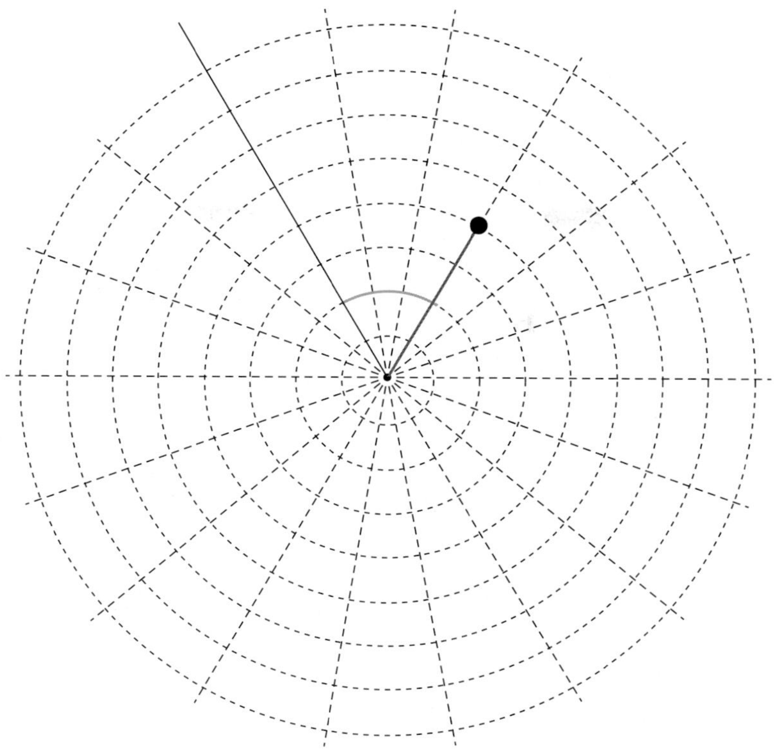

Fig. 14.6 Polar coordinate system representing the field of view of the camera. (Katleen Robert. Reproduced with permission.)

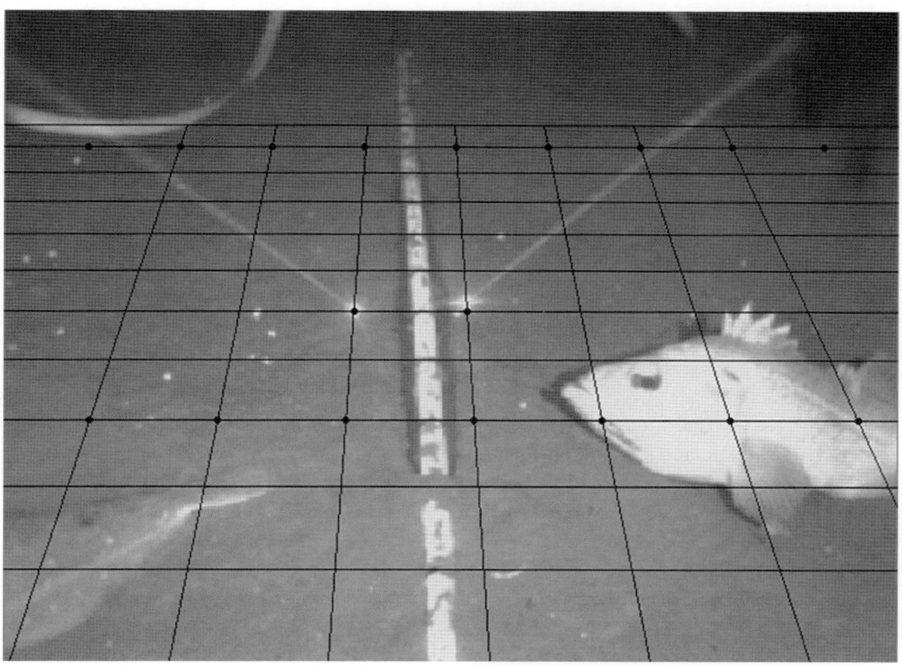

Fig. 14.7 Perspective grid built with the help of the scaling ruler and the two laser beams and overlain over an extracted video frame. (Katleen Robert. Reproduced with permission.)

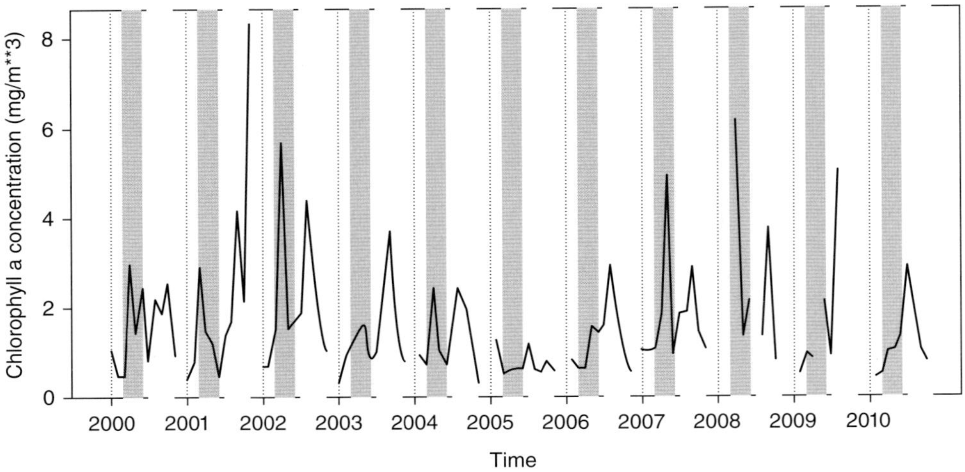

Fig. 14.8 Historical trend in chlorophyll-*a* concentration for the last decade obtained through ocean colour monitoring from the SeaWiF satellite. Grey bars represent April–May. (Katleen Robert. Reproduced with permission.)

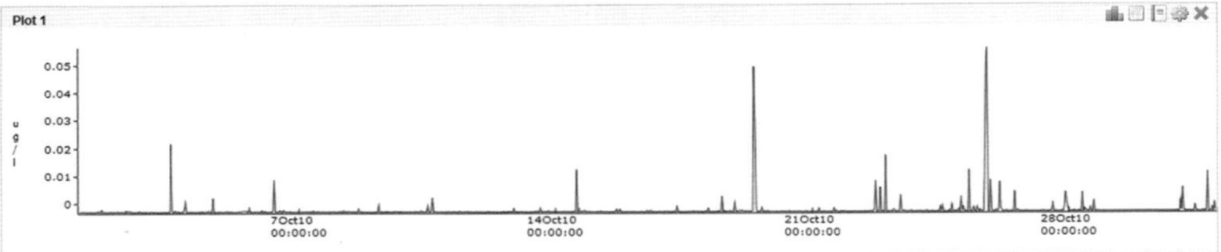

Fig. 14.9 Chlorophyll measurements in Barkley Canyon (871 m), off Vancouver Island (British Columbia, Canada), as measured by the Seapoint Fluorometer 3125. (From Plotting Utility. Ocean Networks Canada. Reproduced with permission.)

oxygen (mL L^{-1}), temperature (°C), conductivity (S m^{-1}) and pressure (dbar) sensors together with the camera platform's remotely controlled digital camera offer a unique suite of instruments to study the response of benthic communities to changes in oxygen concentrations.

During an initial one-year experiment, observations were performed every 2–3 days (Matabos *et al.*, 2012). Imagery data consisted of digital still photos acquired in a 360° panorama using the remotely controlled pan and tilt mechanism of the camera. It is important to maintain consistent camera positions (i.e. consistent pan/tilt values) during the sweeps, in order to facilitate surface area estimation. Using the same methodology as for the Barkley Canyon camera, a perspective grid for each pan position of the camera was created. The total surface area was defined by adding the surface area of each photograph of the sweep. Digital photo acquisition was followed by a reverse-direction video pan along the same sweep.

Image analysis methods A total of 1833 images and 130 videos were analysed manually (Matabos *et al.*, 2012). The biological variables extracted from each image included: density of benthic organisms, percentage of bacterial mat coverage, turbidity of the water and population size frequency distribution. Return-sweep video recordings provided information on inter- and intra-species interactions and feeding strategies.

Often explanations of variations in organism density require an understanding of animal behaviour. With no power limitation and direct connectivity, it is now possible to study organisms' activity rhythms in an ecological context and thus distinguish cyclic signals from non-cyclic ecological responses by adapting sampling frequency to the process of interest. In addition to the above study, hourly acquisition of digital pictures for shorter time periods (one month) allowed estimation of hourly changes in species abundances and densities and areal coverage of bacterial mats, used as proxies of biorhythms for decapods and bacteria (Matabos *et al.*, 2011; Matabos *et al.*, 2015).

14.3.3.4 Oceanic observations in the water column at Folger Passage

Another example of integrated research including biology involves the use of a multifrequency echosounder (Pawlowicz & McClure, 2010). The NEPTUNE Canada network Folger node is located off Bamfield, not far from the Bamfield Marine Sciences Centre (BMSC). This proximity is valuable because time-series ship-based sampling in the area complements data collected by network instruments, providing an opportunity for data cross-calibration. A three-frequency (38, 123 and 210 kHz) upward-looking echosounder is deployed in 95 m of water along with a CTD, an ADCP (300 kHz) and an oxygen sensor. Fish, zooplankton and bubbles are imaged with the echosounder, and the use of different frequencies allows for discrimination between different sizes of particles/organisms. Pawlowicz and McClure (2010) developed a method for visualization of the three frequencies where low, medium and high frequency are mapped in different colours. Intensity of the echo return is indicated with colour intensity, the strongest return from all frequency bands being white. This high-resolution sampling can thus be converted to a map representing 1-minute averaged multifrequency data over months, and eventually over years. This map can then be compared to the tidal signal using pressure data (dbar), wind and current speeds (m s^{-1}). More finely resolved 1-second interval data reveal details such as migration direction and speed of small single scatterers like the zooplankton during their diel migration. This instrument array will help to understand the dynamics of primary and secondary production at temporal scales ranging from hours to years. Interpretation of data from multifrequency echosounders is a highly developed field for ship-borne systems (Simmonds & MacLennan, 2005) and there is great scope for adapting such algorithms to fixed-point active acoustic transducers on seafloor observatories.

14.3.3.5 Ocean Observatories Initiative perspectives

The components of the US OOI have been designed and configured to measure environmental changes in the deep ocean. High-resolution video and still cameras and acoustic sensors will be deployed at all regional-scale and some coastal-scale nodes to observe real-time changes in seafloor sediment characteristics and populations of benthic organisms. Seafloor pressure recorders will measure fluctuations in seafloor sediments due to tectonic forces, and in combination with ADCPs and 3D current profilers they will track the propagation of internal waves and near-bottom currents. Fluid flux sensors will capture changes in sediment pore water chemistry, and pH and pCO_2 instruments will monitor levels of ocean acidification and carbon flux. Dissolved oxygen sensors will track hypoxic water masses from deep water onto the continental shelf, and complementary sensors can capture the impacts of such events on seafloor communities and chemistry. In addition to the core sensors, every OOI site will have a built-in capacity for expansion to accommodate new instruments and experiments and to upgrade already installed sensors with updated technology. The ability to expand and incorporate collaborative projects from independent scientists outside of OOI is central to the design structure of the project. The OOI implementing organizations are currently testing the seafloor cable, assembling moorings and installing seafloor infrastructure. The first production gliders and moorings were deployed at Station Papa in August 2013. Installation will continue in stages through 2013 and 2014, and the full system is scheduled to be commissioned and producing data by the end of 2015.

14.4 Autonomous observatories

14.4.1 What is an autonomous observatory?

Autonomous observatories operate in standalone configuration for power, using battery packs, and either record data internally (the most common case) or have limited connectivity, through release of data capsules or an acoustic link to the surface, which can transfer either system status parameters or a very limited quantity of instrument data. While cabled infrastructures offer fantastically expanded capabilities in terms of connectivity, bandwidth, power supply and time synchronization, autonomous observatories have the advantage of deployment flexibility (see below) and continue to evolve in their specifications and applications. Many contemporary moorings and even benthic landers have either already integrated, or are developing, data telemetry systems. Telemetry technologies typically include a surface buoy that relays data to shore by satellite. The buoys can connect to subsurface networks via power and communication cables and simply provide data telemetry through acoustic modems. Two-way communications are also possible with autonomous systems, albeit with significantly reduced interaction with sensors compared to cabled systems (due to the limited bandwidth).

Autonomous observatories are less expensive and more easily deployed than cabled observatories. They are more convenient when the selected sites are either within the Exclusive Economic Zone (EEZ) of a foreign country or far from land, which strongly limits the deployment of a cable. Furthermore, they are easy to service, recover and relocate. Hence, although there are some examples of cabled observatories, most deep-sea observatory projects within Europe and North America now employ autonomous platforms.

14.4.2 Autonomous observatory technology

Many autonomous observatory systems being designed today include a data transmission infrastructure for at least a portion of the sensors. Several systems that have been in operation for a decade or more are able to periodically transmit data on sea surface conditions and upper water column measurements of temperature and salinity. Data transmission is made possible by a satellite telephone modem transmission from a surface buoy. The buoys are often outfitted with a combination of solar panels, wind or diesel generators, and batteries to supply power for months, and up to a year in some cases.

New designs are also being developed which can deliver data from the deep sea by electro-optical-mechanical (EOM) cable mooring. This mooring type can conduct signals and power to and from the seafloor with induction and slip-ring based mechanisms which allow the mooring line to turn as tides and currents oscillate the mooring's position. On the seafloor, the telecommunication systems can then connect with junction boxes, and local seafloor cables provide the link to benthic observation systems such as seismometers or time-lapse cameras. While these systems can include buoy-generated power and allow transmission of data to and from shore, these capabilities remain limited when compared to cabled systems.

Acoustic modems have also been incorporated into autonomous observatory infrastructures to allow greater flexibility in observatory design. Autonomous benthic platforms, landers, rovers, AUVs and gliders can all communicate to observatory infrastructures via acoustic modem, albeit at substantially reduced bandwidths compared with cabled systems. These modem systems can also reduce requirements for EOM cable (vs. standard mooring rope) by bridging some portions of the water column with acoustic signalling. Acoustic modems have provided a means of transmitting data files of simple scalar variables such as temperature and salinity, and are beginning to be used for transmission of larger files like those from acoustic Doppler current meters and cameras (mainly individual still images). When specifying protocols for data transmission, the frequency and degree to which a digested version of the raw data is transmitted needs to be carefully considered to maintain the key advantages of assurance of operation and recovery of data. Acoustic modems are now capable of transmitting several thousands of

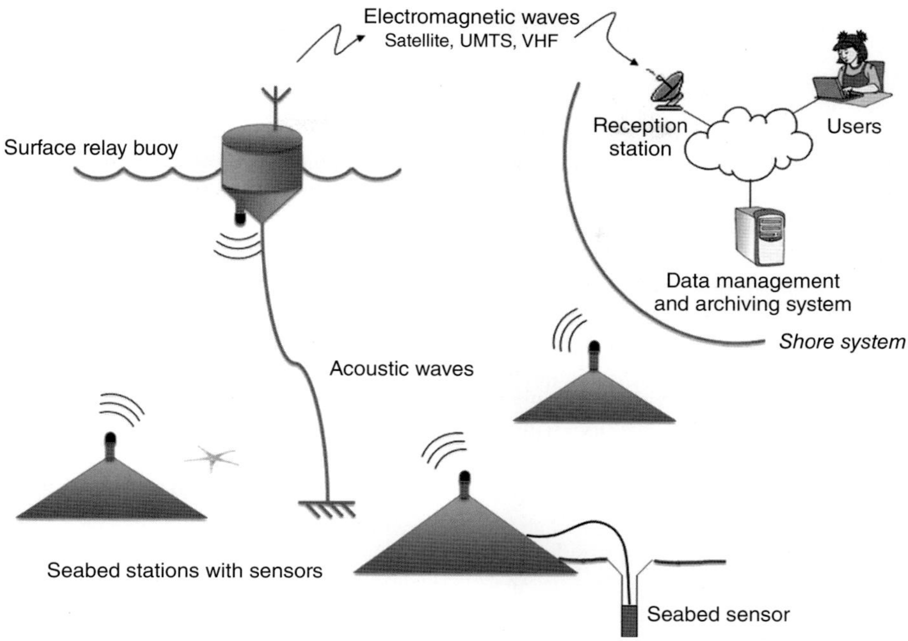

Fig. 14.10 Typical architecture of a non-cabled subsea observing system. In this system, a number of seabed stations are installed on the seabed, with their own energy sources, and periodically acquire measurement data from their local set of sensors. (Jérôme Blandin. Reproduced with permission.)

bits per seconds over thousands of metres through the water, and thus provide a low-cost solution when lower data transmission rates meet scientific design requirements.

Instruments and sensors used on autonomous observatories are the same as those used on a cabled observatory albeit often without providing the same time resolution.

14.4.2.1 Typical system architecture

Figure 14.10 depicts a typical non-cabled observing system, with subsea, surface and onshore components. Several times per day the seabed stations acoustically transmit acquired data to a surface relay buoy moored in the vicinity. The measurement data are then sent to shore from the buoy by radio waves, either directly (coastal buoys) or via a satellite link (offshore buoys). Received data are stored in an archiving system and made available to the users through a web interface.

14.4.2.2 Subsea component level

In contrast to cabled observatories, the design of autonomous observatories is driven by limited power- and data-storage capacities. Strategies for reducing system power requirements include the following:

- The design of very low-power circuit boards, especially for those requiring constant operation. Recent advances in mobile phone technologies are leading to an increasing number of microcontroller boards with more advanced functions. However, the ultra low-power specifications/constraints of autonomous observatories still frequently require the use of specially designed boards. This is also the case for embedded software (an essential factor in the power consumption of microcontroller boards), although the increasing computing capability of very low-power microcontrollers is leading to the creation of entire libraries of low-power functionalities and 'apps'.

- The utilization of efficient sleep modes wherever possible. Much of the time, an autonomous observatory is a sleeping system that can be awoken at pre-programmed intervals for sensor sampling and transmission of data, or by an event detected by a local or remote sensor or by a request from an operator.
- Careful tuning of sensor duty cycles to minimize power consumption. While some sensors cannot be powered off (e.g. sensors with self wake-up or event detection capability) all others should be electrically switched off and controlled by the hosting observing system.
- Careful selection of an energy-efficient means of data transmission, adapted to the specific transmission conditions (geometry, topology, state of the transmission channel) associated with an efficient data coding convention and an intelligent selection of the data to be transmitted.

Power sources Unless a specific site provides suitable conditions for local generation of electrical power, a non-cabled observatory has to rely on a finite quantity of energy between maintenance visits. The diversity of the site-dependent principles that can enable local energy production, known as 'energy harvesting', has until now prevented the design of widely spread and reliable energy production systems to feed installations the size of seafloor observatories. Hence, non-cabled seafloor observatory designers generally use conventional electrochemical devices, seeking products that offer both a high-energy density and a low auto-discharge rate. Although surface buoys have regularly been fitted with solar or wind power, or both, today the most widely used energy sources are assemblies of primary lithium battery cells. As an example of the energy density currently obtained with such electrochemical cells, a 4000 m water depth-rated cylindrical container of external length 890 mm, diameter 174 mm and weight 35 kg can hold around 4000 watt-hours (Wh) at a temperature of +2°C.

Data transmission Electromagnetic waves are attenuated over short distances by seawater. Currently, the only means of transmitting data wirelessly through seawater over distances of more than a few metres is through sound waves.

Observation data from non-cabled observatories are typically transmitted from the seabed to users by means of acoustic modems, the receiver being placed either on a ship for occasional transmission or on a relay buoy towards a permanent data server on shore. These technologies can only transmit between consecutive servicing interventions on the seabed.

14.4.2.3 Surface relay buoy

The surface buoy is the sole gateway to the seabed observation network until servicing. Exposed to potentially harsh surface conditions, it is the weakest link in the transmission chain from seabed to shore. The watch circle of this buoy, within acoustic range of all seabed stations, must be mastered by a precise control of the deadweight position during the installation and an extremely accurate measurement of the mooring line length.

The transmission system typically consists of an acoustic modem with a transducer installed below the sea surface, a satellite or terrestrial cell phone or very high-frequency (VHF) modem, and an interface between the two transmission devices to control the transmissions. Given its importance within the transmission chain, two completely independent and redundant transmission channels should be installed on a buoy.

Damage to buoy systems can come from weather, ice, fishing, vandalism and poor navigation, all of which have caused major damage or loss of observatory buoys in the past. Experience has shown that careful consideration of the effects of intense storm activity should be taken when determining the dynamic loading of the mooring infrastructure. This has been facilitated by finite element modelling, where loads can be evaluated virtually. However, repetitive motion and extreme conditions continue to provide surprises with regard to the anticipated durability and wear of observatory components. Because buoys are susceptible to damage or loss, a new type of buoy development is under way which allows for the buoy to be winched down to a stand-by depth when not transmitting or measuring at the surface. This allows the system to avoid weather, ice and other hazards and still provide a means of conveying subsurface data to satellites. In addition, the winch action of these systems allows for vertical profiling of the water column with each cycle. As winch systems build a track record of operational successes, their use may become more widespread.

14.4.3 *Observational approach*

14.4.3.1 The MoMAR deep-sea observatory

Located on the Mid-Atlantic Ridge (MAR) near the Azores, the 'Monitoring the MAR' (MoMAR) project area contains four known hydrothermal vent fields, located at different depths: Menez Gwen (800 m), Lucky Strike (1650 m), Saldanha (2200 m) and Rainbow (2300 m). This region has been extensively studied for 30 years with more than 20 oceanographic voyages. Its geophysical background is well described, as are the general characteristics of most hydrothermal vents. Each of the vent fields has specific geological, chemical, hydrothermal and biological characteristics (Colaço *et al.*, 2011). (Note: Several research institutes are actively involved in the project: IFREMER and IPGP (France) are leaders of the MoMARSAT cruises with the participation of the University of the Azores, University of Lisbon (Portugal), the National Oceanography Centre of UK (NOC, UK), University of Bremen (Germany) and several other French CNRS laboratories, including the Toulouse Geosciences Environment laboratory (GET), European Institute for Marine Studies (IUEM), the Transfer Mechanisms in Geology Laboratory (LMTG) and the Laboratory for Experimentation and Numerical Methods in Oceanography and the Climate (LOCEAN).)

The MoMAR project was initiated by the international InterRidge programme in 1998 to study the environmental instability resulting from active mid-ocean ridge processes, including changes in flux, composition and temperature of emitted hydrothermal fluids, and its effects on dependent hydrothermal fauna at the slow-spreading MAR (Blandin *et al.*, 2010). The ultimate objective of MoMAR is to monitor the natural dynamics of hydrothermal vent ecosystems in real time, from deep in the ocean crust to the surface of the seafloor, along with its associated fauna. A multiscale approach is being applied, ranging from studying larger regional scale processes (>100 km) such as seismicity, oceanographic circulation and biological dispersion, to processes occurring at local (<10 km) and very local (<1 m) scales that control variations in environmental conditions and faunal community structure.

In 2003, the MoMAR area was chosen as one of the sites for concerted action by the European Network of Excellence ESONET and in 2008, the MoMAR project was selected for a demonstration mission (MoMAR-D, Blandin *et al.*, 2010). The MoMAR-D demonstration mission had objectives to integrate time-series data from a multidisciplinary range of sensors and implement technological solutions required for near real-time transmission of a significant subset of the data during at least one year on the Lucky Strike vent field.

A multidisciplinary observing system, with a satellite link to shore, was deployed during the MoMARSAT oceanographic cruise in October 2010. During MoMARSAT, an ROV deployed a series of autonomous oceanographic instruments. Through this newly installed observatory, the active hydrothermal processes on the Lucky Strike field can now be monitored on a continuous basis. Since 12 October 2010, the data acquired at 1700 m depth have been transmitted to a surface buoy and then by satellite to the land-based IFREMER Research Centre, several thousands of kilometres away. Every year, the system is recovered for a rapid turnaround and redeployed in the same area. This allowed ground-truthing and repairs of some of the problems. Participating scientists can now continuously monitor variations in temperature and physico-chemical conditions (east node) as well as seismic activity and ground deformation (west node) of the area and observe the dynamics of a mussel assemblage located at the base of the Tour Eiffel edifice (east node) (Sarrazin *et al.*, 2014; Matabos *et al.*, 2015). In addition to gathering fundamental knowledge about the functioning of hydrothermal systems, the information gained will help manage the deep-sea resources found in this area, which became a marine protected area in 2006 as part of the OSPAR network.

14.4.3.2 The Porcupine Abyssal Plain – Sustained Observatory

Time-series research into deep-sea biogeochemistry and ecology has been ongoing at the Porcupine Abyssal Plain – Sustained Observatory (PAP-SO) since 1989 (e.g. Lampitt *et al.*, 2010), although research there began more than a decade earlier. Located at 4850 m depth in the subpolar Northeast Atlantic (49°N 16.5°W), this site is one of only two time-series locations with comparable detailed data including particulate organic carbon (POC) fluxes to the abyssal seafloor and variation in the abundance and composition of benthic fauna over a period spanning more than two decades (Smith *et al.*, 2009).

PAP-SO is located in an abyssal plain area between the MAR and the European continental shelf. Research priorities at the site include detecting climate variation and its influence on variability in the Northeast Atlantic, particularly in relation to understanding irregular variations in primary production, export flux of organic matter to the seafloor, the influence of sinking carbon on deep-sea communities and the ultimate fate of carbon in the ocean.

Research at the site has long been funded by the UK's Natural Environment Research Council (NERC), but has also included important contributions from many other partners including the University of Southampton, University of Liverpool, University of Aberdeen, IFM-GEOMAR, IFREMER and the Natural History Museum in London. Most recently the NOC, which operates the observatory, has partnered with the UK Met Office in using common infrastructure. This collaboration builds on efforts started in 2002 and now includes an ocean mooring and meteorological station in conjunction with the subsurface biogeochemical sensors in the mixed layer. The resulting improvements in the breadth and manner of data collection will better enable researchers to examine climate–ocean interactions. The new buoy and mooring system have also been made more robust to withstand both the intense weather of the Northeast Atlantic and damage from fishing activity; both have previously resulted in major equipment and data losses.

The site has been part of the EuroSITES network, which has been the European contribution to the OceanSITES network. EuroSITES focuses on active research and operational data collection using observatories around Europe. Data from PAP-SO are now supplied to the Marine Automatic Weather Stations (MAWS) network, and the Global Climate Observing System (GCOS). PAP-SO contributes to the Global Earth Observation System of Systems (GEOSS) and Global Ocean Observing System (GOOS). PAP-SO has also been included in planning related to the European Seas Observatory Network – Network of Excellence (ESONET NoE) and the EMSO programmes.

One of the recent activities at PAP-SO was the ESONET NoE demonstration Mission called MOdular and mobile Deep Ocean Observatory (MODOO). This project helped generate detailed plans for creating data and sensor management telemetry systems from the surface to the seafloor using an acoustic modem. This system remains to be implemented in an operational sense. The system outlined in MODOO will be capable of two-way transmission allowing for the modification of sensor settings or firmware in response to environmental variation or sensor behaviour. MODOO also included specifications for providing geo-hazard alerts when tsunami-related variations are detected. The autonomous system will therefore have many of the features of a cabled observatory, albeit with near total reliance on locally stored power and the need to tightly manage data transmission rates.

14.4.3.3 Insights into other European deep-sea observatory demonstration projects

LOOME (Long-term Observations on Mud-Volcano Eruptions) The Haakon Mosby Mud Volcano (72°N 14°E) is an active submarine mud volcano located in the Norwegian Sea at 1250 m depth. The 15 m high, 1.4 km wide volcano represents the top of a giant chimney through which methane-rich fluids are pressed upwards towards the seafloor. The objectives of this observatory are to investigate the temporal variability of the active gas-emitting mud volcano and to analyse the effects of mud eruption on gas hydrate stability, seafloor morphology and the distribution and colonization patterns of benthic communities. Sensor systems and biological experiments were integrated around a combination of deeply penetrating, *in situ* temperature and geo-acoustic recording units, which it is hoped will give an early warning of eruptive events. The first deployment of this observatory was accomplished in July 2009, with recovery in September 2010, and an eruption event was recorded by several of the sensors (Feseker *et al.*, 2014). (Note: The partners that are involved in this project are: KDM/IFM-GEOMAR, KDM/UniHB/MARUM, KDM/MPIMM, KDM/AWI (Germany), IFREMER (France), UiT (Norway).)

LIDO (LIstening to the Deep-Ocean environment) The scientists involved in LIDO are studying the effects of noise on marine organisms. LIDO involves teams from two ESONET nodes (the cabled Sicily site and the non-cabled site of the Gulf of Cadiz), NEPTUNE Canada nodes (Folger passage and Barkley Canyon) and DONET nodes (off Kushiro and Hatsushima), thus forming the first nucleus of a homogeneous marine mammal observatory. The main objectives of LIDO are to establish a common seismology monitoring approach for both areas and real-time automatic identification and classification system for non-biological and biological sounds, to monitor marine organisms and population dynamics, to assess and control the long-term effects of anthropogenic sound sources on marine organisms, to educate the public, end-users and regulators, and to approve ethical guidelines and procedures. (Note: The partners involved are: UPC, CSIC, dBScale (Spain), KDM/UniHB/MARUM, BHT-Berlin (Germany), FFCUL (Portugal), INGV, ISMAR, INFN, Tecnomare (Italy) and associated with CIBRA (Italy).)

AOEM (Arctic Ocean ESONET Mission) The Arctic Ocean off the western Svalbard shelf and Fram Strait was a critical site for documenting climate-induced change, as the Arctic Ocean geosphere, hydrosphere and biosphere are among the most sensitive to irreversible perturbations. Obtaining sustained seafloor and water-column measurements and data transmission from this environment with seasonal sea-ice cover represented a significant challenge. AOEM comprised two subprojects: one designed and evaluated data acquisition and real-time transmission methodologies for Fram Strait oceanography, including an acoustic network for ocean tomography and glider navigation. The second project deployed observatory lander technology and associated biogeochemical sensors, sector scanning echosounders, heat flow probes, upward-looking ADCPs and ocean-bottom seismometers to document the temporal variability and the flux of methane release off western Svalbard. (Note: The partners are: NOC (UK), KDM/IFM-GEOMAR, KDM/AWI (Germany), NERSC, UiT (Norway) and FORTH (Greece).)

14.5 Data processing, management and archiving

14.5.1 Data archiving

14.5.1.1 Why data archives?

The establishment of ocean data archives provides a guarantee that data will be preserved and remain accessible for an extended period of time. There are several important reasons for archiving data from environmental observations:

- In many cases, the cost of the initial data collection is so large that there is a clear need to have the data reviewed and analysed by many different researchers for a variety of purposes.
- Scientific questions evolve as our knowledge of the oceans increases, so that future researchers may be able to derive new knowledge from existing data. This is particularly important for time-series data sets.

- Many temporal trends can only become apparent in data from long time-series.
- Observations may contain evidence of phenomena that occur rarely.
- Data may contain information relevant to more than one study and to more than one individual or group of scientists.
- Software systems that perform the data analysis evolve and improve over time. Newer systems might be able to extract further information from the same data set.

14.5.1.2 What does a data archive require?

The resources required to build and maintain a data archive can be significant. In particular, a software system must be set up to streamline the retrieval and recording of the raw data from the instruments, has to bring them to the storage area and finally has to make them available to their various users. Some degree of standardization and a rigorous process within the observatory have to be in place to enable archiving. Moreover, data will only be re-usable if they are fully qualified and described in terms of the conditions under which they were collected. Future users of the data, beyond the individuals in charge of the initial data collection, will not understand the observational conditions (e.g. location, time, instrument configuration details, etc.) unless they are recorded together with the data. Therefore, the proper archiving of metadata is as essential as maintaining the data themselves (and see Chapter 16).

The standardization mentioned above has several features. Some have to do with data formats (which depend on the type of instrument producing the data), and others with the process used to collect data, move them to the archive, record their presence in the archive and associate them with all relevant meta-data. This process should be clear, consistent and predictable. Moving data from instruments to the archive and from the archive to users can also accommodate extra processing steps on the way that will implement automatic event detection, data quality control, calibration, etc. (Owens *et al.*, 2010). The process (or workflow) is illustrated in Fig. 14.11.

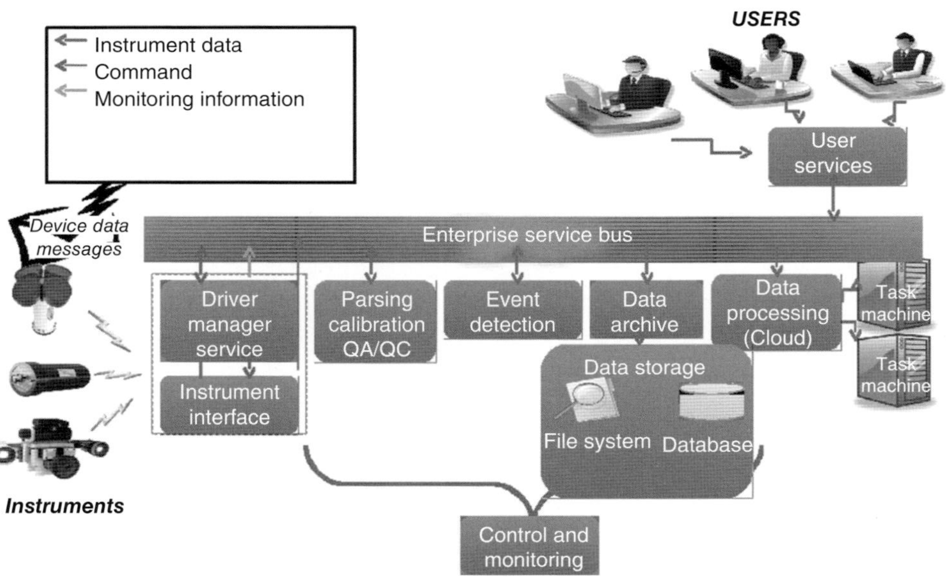

Fig. 14.11 Conceptual diagram of the Data Management and Archiving System (DMAS) in use for the NEPTUNE Canada and VENUS observatory networks. It shows various instrument types making data available to the data management infrastructure using instrument 'drivers'. The rest of the infrastructure will support data calibration, event detection, data archiving, and processing. (Ocean Networks Canada. Reproduced with permission.)

14.5.2 *Data processing*

There are several forms of data processing that are relevant to ocean observatories. They all have different goals and different conditions of execution, yet are all important and should be supported as part of the ancillary services provided by the archive:

- *Data format conversion*: This is important for data distribution (e.g. data to be made available as a CSV file or as a NetCDF or HDF file) in accordance to existing standards.
- *Calibration*: Raw original data are still available in the archive and can be reprocessed to replace initial calibration values should an issue be identified at a later stage.
- *Visualization*: Creation of plots, images and other multidimensional products will allow better understanding of the data.

14.5.3 *Statistical analysis for long time series*

Although cabled observatories bring unprecedented opportunities for time-series observations, they also bring the challenge of analysing and interpreting multivariate time-series data sets. Time-series ecological data are a combination of many patterns and as such are temporally structured at multiple scales. The purpose of time-series analyses is to detect these structures, describe the observed patterns and understand the ecological processes underlying them. This temporal structure has two main origins: the external forcing of environmental variables that are themselves temporally structured (induced dependence) and results from processes internal to the community itself (autocorrelation) (Borcard *et al.*, 2011).

New methods in spatial ecology have been developed to detect existing structures in the data, identify sources of variation, and fit appropriate models (Borcard & Legendre, 2002; Borcard *et al.*, 2011). These include the methods of principal coordinates of neighbour matrices (PCNM; Borcard & Legendre, 2002) later on refined into Moran's eigenvector map (MEM; Dray *et al.*, 2006). They are especially well suited for detecting small-scale structure in the data, and suitable for time-series data sets. More recently, an extension of the MEM, the assymetric eigenvector maps (AEM) was designed to account for directional processes (Blanchet *et al.*, 2008) and appear to be more suitable for assessing dynamics in time-series anallysis (Baho *et al.*, 2015).

Classical time-series methods are also appropriate for observatory data: they include periodograms and spectral analyses in order to detect periodicities/rhythms, and autocorrelations and cross-correlations to describe relationships between different biological and environmental variables. Several different periodogram and correlation methods exist in the literature (e.g. Legendre & Legendre, 1998).

14.6 Outreach for seafloor observatories

As cabled observatories are expensive and long-term endeavours, an active public outreach effort and education programme must be a high-priority component to ensure the success and longevity of the project. Outreach encompasses all communication with individuals outside the observatory organization including policy-makers, the general public, the media, other scientists, educators and students.

Observatories typically engage in some or all of the following types of communication: press releases, websites, workshops, streaming undersea video from cruises and undersea cameras, data dissemination, individual exchanges with stakeholders by phone or e-mail, public lectures, cooperation with navy and fisheries, training of students, contributing to museum displays and other public exhibitions, hosting international visitors and collaborating with other observatories.

It is essential to note that a 'one size fits all' approach to outreach for all interested parties is not sufficient (Kusek *et al.*, 2007). To maximize knowledge transfer to all interested parties in order to make the most use of the observatory investment, communications targeted to each audience should be produced. The HERMES (Hotspot Ecosystem Research on the Margins of European Seas) project 2005–2009 (Gunn & Thomsen, 2009) and its successor HERMIONE (Hotspot Ecosystem Research and Man's Impact On European Seas) are examples of projects that attempt to address the needs of a wide variety of audiences. HERMES staff designed activities and products especially targeted to its various users including educators, researchers, policy-makers, non-governmental environmental organizations and governments, the majority of which were web-accessible. The interactive components of the education materials, such as an interactive ROV simulator 'Mud Taxi' to learn about mud volcanos (Gunn & Thomsen, 2009), were particularly effective in engaging young people. The website also included live feeds of expeditions and cruises (Classroom@Sea). Ocean Networks Canada's SeaTube, which allows viewers to see ROV installation, maintenance and survey dives is another example of a popular outreach communication medium.

Children engaged in an up-to-date curriculum can also become educators for their parents, grandparents, siblings and peers (Evans *et al.*, 1996; Ocean Project, 2009).

Communicating knowledge acquired through observatory studies to inform policy-making is a key responsibility of an observatory organization. To make the installation of a seafloor observatory – a multi-year, major science initiative – relevant to policy-makers, it is imperative that their needs be considered at every stage of the project (Bielek & Schaefer, 2009). Ideally, outreach at all levels would be funded as part of the cabled observatory organization, but in the absence of this, alternative means of dissemination must be found. In addition to preparation of outreach material, partnerships with outside scientists, teachers, writers, educators, website developers, artists and marketing experts can be valuable in communicating scientific results to interested parties outside the organization (Kusek *et al.*, 2007).

14.7 The future

The development of research tools and technology has provided rapid progress in science over recent decades. However, the next giant leap in our understanding of the ocean will require a continuous interactive presence, analogous to the means through which our knowledge and intuition of terrestrial environments have increased. As we enter the twenty-first century, a series of technological advances are enabling a real-time, continuous and interactive presence in the ocean, with sufficient power and bandwidth to operate a broad range of instruments and open the door for development of new sensors. Perhaps even more powerful than the environmental and temporal scope will be the coincident, coordinated sampling of hundreds of real-time data streams across disciplines and the interactive remote control of instruments. This allows responsive real-time sampling of events, and analysis of relationships among disparate data to test relationships and discover process drivers across the earth–ocean system. It is expected that ocean observatories will act as a catalyst for the development of Ocean Informatics that will provide the tools needed to process unprecedented volumes of ocean–earth data and discover relationships between biological, chemical, physical and geological processes.

Acknowledgments

The authors would like to thank Nikolai Korniyuk and Flavia Taggiasco for providing maps and illustrations. We are grateful to the Ocean Networks Canada, Ocean Observatory Initiative and EMSO teams for their support and making these projects possible. ONC received support from the Canada Foundation for Innovation and the British Columbian Knowledge Development Fund. OOI was funded by the US National Science Foundation (NSF). EMSO and its predecessors acknowledge support from successive European Community funding programmes, including FP6, FP7, and H2020. In addition, development and operation of the EMSO nodes is thanks to support from the

EMSO partner countries: Italy (host country), France, Germany, Greece, Ireland, The Netherlands, Norway, Portugal, Romania, Spain, Sweden, Turkey, and The United Kingdom. The Natural Environment Research Council (NERC) supported H. Ruhl.

References

Baho, D.L., Futter, M.N., Johnson, R.K., Angeler, D.G. (2015). Assessing temporal scales and patterns in time series: comparing methods based on redundancy analysis. *Ecological complexity*, 22: 162–168.

Barnes, C.R., Best, M.M.R., Johnson, F.R., Phibbs, P., Pirenne, B. (2008). Building the world's first regional cabled observatory: NEPTUNE Canada. *Journal of Ocean Technology*, 3: 13–18.

Barnes, C.R., Best, M.M.R., Johnson, F.R., Phibbs, P., Pirenne, P. (2012). The NEPTUNE Canada Project: installing the world's first regional cabled ocean observatory. In: Favali, P., de Santis, A., Beranzoli, L. (eds), *Seafloor observatories*. New York: Springer-Praxis Geophysical Science Publishing, Chapter 16.

Best, M.M.R., Barnes, C.R., Bornhold, B., Juniper, S.K. (2012). Integrating continuous observatory data from the coast to the abyss: a multidisciplinary view of the ocean in four dimensions. In: Favali, P., de Santis, A., Beranzoli, L. (eds), *Seafloor observatories*. New York: Springer-Praxis Geophysical Science Publishing, Chapter 1.

Bielek, A.T., Schaefer, K. (2009). A comparison of European and North American approaches to the management and communication of environmental research. *Strengthening Science Policy Links: Study Series no. 2*. Environment, Canada.

Billett, D.S.M., Lampitt, R.S., Rice, A.L., Mantoura, R.F.C. (1983). Seasonal sedimentation of phytoplankton to the deep-sea benthos. *Nature*, 302: 520–522.

Blanchet, F.G., Legendre, P., Borcard, D. (2008). Modelling directional spatial processes in ecological data. *Ecological modelling*, 215: 325–326.

Blandin, J., Colaço, A., Legrand, J., Cannat, M., Sarradin, P.M., Sarrazin, J. (2010). The MoMAR-D project: a challenge to monitor in real time the Lucky Strike hydrothermal vent field. *ICES Journal of Marine Sciences,* advance access.

Burnett, D.R., Carter, L. (2009). Deep-sea observatories and international law. *Proceedings of OCEANS 2009 IEEE Conference*, Bremen, Germany, May 2009.

Borcard, D., Legendre, P. (2002). All-scale spatial analysis of ecological data by means of neighbour matrices. *Ecological Modelling*, 153: 51–68.

Borcard, D., Gillet, F., Legendre, P. (2011). *Numerical Ecology with R*. New York: Springer Science, 306 pp.

Bornhold, B., Best, M.M.R., Hasanen, R., Hofmann, M., Leslie, M. (2010). Testing and integration of a cabled ocean observatory: complexity is greater than the sum of the off-the-shelf parts. *Proceedings of OCEANS 2010 IEEE Conference*, Seattle, USA, September.

Colaço, A., Blandin, J., Cannat, M., *et al.* (2011). MoMAR-D: a technological challenge to monitor the dynamics of the Lucky Strike vent ecosystem. *ICES Journal of Marine Science*, 68: 416–424.

Cuvelier, D., Legendre, P., Laes, A., Sarradin, P.M., Sarrazin, J. (2014). Rhythms and Community Dynamics of a Hydrothermal Tubeworm Assemblage at Main Endeavour Field–A Multidisciplinary Deep-Sea Observatory Approach. *PLoS ONE*, 9(5): e96924.

Dray, S., Legendre, P., Peres-Neto, P.R. (2006). Spatial modelling: a comprehensive framework for principal coordinate analysis of neighbour matrices (PCNM). *Ecological Modelling*, 196: 483–493.

Evans, S.M., Gill, M.E., Marchant, J. (1996). Schoolchildren as educators: the indirect influence of environmental education in schools on parents' attitudes towards the environment. *Journal of Biological Education*, 30: 243–248.

Favali, P., Beranzoli, L. (2006). Seafloor observatory science: a review. *Annals of Geophysics*, 49: 515–567.

Favali, P., Beranzoli, L. (2009). EMSO: European multidisciplinary seafloor observatory. *Nuclear Instruments and Methods in Physics Research A*, 602: 21–27.

Favali, P., Person, R., Barnes, C.R., Kaneda, Y., Delaney, J.R., Shu-Kun, H. (2010). Seafloor Observatory Science. In: Hall, J., Harrison, D.E., Stammer, D. (eds), *Proceedings of the OceanObs'09: Sustained Ocean. Observations and information for society conference 2*, Venice, Italy, 21–25 September, 2009. ESA Publication WPP-306.

Feseker, T., Boetius, A., Wenzhöfer, F., *et al.* (2014). Eruption of a deep-sea mud volcano triggers rapid sediment movement. *Nature communications*, 5: 5385. doi:10.1038/ncomms6385

Glover, A.G., Gooday, A., Bailey, D.M., *et al.* (2010). Temporal change in deep-sea benthic ecosystems: a review of the evidence from recent time-series studies. *Advances in Marine Biology*, 58: 1–95.

Gunn, V., Thomsen, L. (2009). The next generation: providing inspiration and training for future marine scientists. *Oceanography*, 22: 166–176.

Herring, P.J., Gaten, E., Shelton, P.M.J, (1999). Are vent shrimp blinded by science? *Nature*, 398: 116.

Juniper, S.K., Escartin, J., Cannat, M. (2007). Multidisciplinary, time-series observations at Mid-Ocean Ridges. *Oceanography*, 20: 102–111.

Juniper, S.K., McLean, S., Pirenne, B., Moran, K. (2012). Community-based mini-observatories for Arctic ocean science and outreach. Paper presented at ArcticNET Conference, Vancouver, BC, December 2012.

Juniper, S.K., Matabos, M., Mihály, S., Ajayamohan, R.S., Gervais, F., Bui, A.O.V. (2013). A year in Barkley Canyon: a time-series observatory study of mid-slope benthos and habitat dynamics using the NEPTUNE Canada network. *Deep Sea Research II*, 92:114–123.

Kogan, I., Paull, C., Kuhnz, L., *et al.* (2006). ATOC/Pioneer Seamount cable after 8 years on the seafloor: observations, environmental impact. *Continental Shelf Research*, 26: 771–787.

Kusek, K.M., Goehring, L., Robigou, V, (2007). Schoolchildren as educators: the indirect influence of environmental education in schools on parents' attitudes towards the environment. *Oceanography*, 20: 142–151.

Lampitt, R.S., Bett, B.J., Kiriakoulakis, K., *et al.* (2001). Material supply to the abyssal seafloor in the Northeast Atlantic. *Progress in Oceanography*, 50: 27–63.

Lampitt, R.S., Salter, I., de Cuevas, B.A., Hartman, S., Larkin, K.E., Pebody, C.A. (2010). Long-term variability of downward particle flux in the deep northeast Atlantic: Causes and trends. *Deep-Sea Research II*, 57: 1346–1361.

Legendre, P., Legendre, L. (1998). *Numerical ecology* (2nd edn). Amsterdam: Elsevier Science, 853 pp.

Manuel-Lazaro, A., Nogueras, M., del Rio, J. (2010). OBSEA: an expandable seafloor observatory. *Sea Technology*, 51: 37–39.

Massion, G., Raybould, K. (2006). MARS: The Monterey accelerated research system. *Sea Technology*, 47: 39–42.

Matabos, M., Aguzzi, J., Robert, K., *et al.* (2011). Multi-parametric study of behavioural modulation in demersal decapods at the venus cabled observatory in Saanich inlet, British Columbia, Canada. *Journal of Experimental Marine Biology and Ecology*, 401: 89–96.

Matabos, M., Tunnicliffe, V., Juniper, S.K., Dean, C. (2012). A year in hypoxia: epibenthic community responses to severe oxygen deficit at a subsea observatory in a coastal inlet. *PLoS ONE*, 7(9): e45626. doi:10.1371/journal.pone.0045626

Matabos, M., Bui, A.O.V, Mihály, S., Aguzzi, J., Juniper, K.S., Ajayamohan, R.S. (2013). High-frequency study of epibenthic megafaunal community dynamics in Barkley canyon: a multi-disciplinary approach using the NEPTUNE Canada network. *Journal of Marine Systems*, 130: 56–68.

Matabos, M., Cuvelier, D., Brouard, J., *et al.* (2015). Behavioural study of two hydrothermal crustacean decapods: Mirocaris fortunata and Segonzacia mesatlantica, from the Lucky Strike vent field (Mid Atlantic Ridge). *Deep Sea Research II* (in press).

Matabos, M., Piechaud, N., De Montigny, F., Sarradin, P.M., Sarrazin, J. (2015). The VENUS cabled observatory as a method to observe fish behaviour and species assemblages in a hypoxic fjord, Saanich Inlet (British Columbia, Canada). *Canadian Journal of Aquatic and Fisheries Sciences*, 72: 24–36.

NRC (National Research Council) (1999). *Enabling ocean research in the 21st century: implementation of a network of ocean observatories*. Washington DC: National Academy Press, 240 pp.

NRC (National Research Council). (2000). *Illuminating the hidden planet: the future of seafloor observatory science*. Washington DC: National Academy Press, 135 pp.

NRC (National Research Council) (2003). *Enabling ocean research in the 21st century: implementation of a network of ocean observatories*. Washington DC: National Academy Press, 240 pp.

Ocean Project (2009). America, the ocean, and climate change: new research insights for conservation, awareness, and action. http://theoceanproject.org (accessed May 2011).

Owens, D., Guillemot, E., Jenkyns, R., Best, M.M.R., Pirenne, B. (2010). Ocean observatories and social computing: potential and progress. *Proceedings, OCEANS 2010 IEEE Conference,* Seattle, USA, September.

Pawlowicz, R., McClure, B. (2010). Inverted echosounder for continuous high-resolution water column profiling from the NEPTUNE (Canada) ocean observatory. *Proceedings, OCEANS 2010 IEEE Conference*, Seattle, USA, September.

Person, R., Aoustin, Y., Blandin, J., Marvaldi, J., Rolin, J.F. (2006). From bottom landers to observatory networks. *Annals of Geophysics*, 49: 581–593.

Pirenne, B., Guillemot, E. (2012). Oceans 2.0: a data management infrastructure as a platform. EGU General Assembly 2012, 22–27 April, 2012 in Vienna, Austria, p.13346.

Priede, I.G., Solan, M. (eds) (2005). *ESONET (European Seafloor Observatory Network) final report*, Version 3.0.

Priede, I.G., Mienert, J., Person, R., *et al.* (2003). ESONET – European sea floor observatory network. In: Dahlin, H., Flemming, N.C., Nittis, K., Petersson, S.E. (eds), *Building the European capacity in operational oceanography*. Elsevier Oceanography Series 69. Amsterdam: Elsevier, pp. 291–294.

Priede, I.G., Person, R., Favali, P. (2005). European seafloor observatory network. *Sea Technology*, 46: 45–49.

Proctor, A., Bradley, C., Gamroth, E., Kennedy, J. (2007). The Ocean Technology Test Bed – an underwater laboratory. *Proceedings, OCEANS 2007 IEEE Conference*, Vancouver, Canada, October 2007.

Robert, K., Juniper, K. (2012). Quantifying megafaunal surface bioturbation using cameras on the NEPTUNE Canada cabled-observatory: observational protocol development and Bayesian modeling. *Marine Ecology Progress Series*, 453: 137–149.

Ruhl, A.R., André, M., Beranzoli, L., *et al.* (2011). Societal need for improved understanding of climate change, anthoropogenic impacts, and geo-hazard warning drive development of ocean observatories in European Seas. *Progress in Oceanography*, 91: 1–33.

Sarrazin, J., Cuvelier, D., Peton, L., Legendre, P., Sarradin, P.M. (2014). High-resolution dynamics of a deep-sea hydrothermal mussel assemblage monitored by the EMSO-Açores MoMAR observatory. *Deep Sea Research Part I: Oceanographic Research Papers*, 90: 62–75.

Scholin, C.S., Jensen, B., Roman, E., *et al.* (2006). The Environmental Sample Processor (ESP) – an autonomous robotic device for detecting microorganisms remotely using molecular probe technology. *Proceedings, OCEANS 2006* MTS/IEEE Conference, Boston, USA, September.

Simmonds, E.J., MacLennan, D.N. (2005). *Fisheries acoustics: theory and practice* (2nd edn). Fish and Aquatic Resources Series 10. Oxford: Blackwell Science.

Smith, C.R., De Leo, F.C., Bernardino, A.F., Sweetman, A.K., Martinez Arbizu, P. (2008). Abyssal food limitation, ecosystem structure and climate change. *Trends in Ecology and Evolution*, 23: 518–528.

Smith, K.L., Ruhl, H.A., Bett, B.J., Billet, D.S.M., Lampitt, R.S., Kaufmann, R.S. (2009). Climate, carbon cycling, and deep-ocean ecosystems. *Proceedings of the National Academy of Sciences of the USA*, 46: 19211–19218.

Thiel, H., Kirstein, K.O., Luth, C., *et al.* (1994). Scientific requirements for an abyssal benthic laboratory. *Journal of Marine Systems*, 6: 421–439.

Tunnicliffe, V., Dewey, R., Smith, D. (2003). Research plans for a mid-depth cabled seafloor observatory in Western Canada. *Oceanography*, 16: 53–59.

Tunnicliffe, V., Barnes, C.R., Dewey, R. (2008). Major advances in cabled ocean observatories (VENUS and NEPTUNE Canada). in coastal and deep-sea settings. *US/EU-Baltic International Symposium*, 2008 IEEE/OES, 8 pp.

Wakefield, W.W., Genin, A. (1987). The use of a Canadian (perspective) grid in deep-sea photography. *Deep Sea Research A*, 34: 469–478.

Widder, E.A., Robinson, B.H., Reisenbichier, K.R., Haddock, S.H.D. (2005). Using red light for in situ observations of deep-sea fishes. *Deep-Sea Research I*, 52: 2077–2085.

Chapter 15

Sorting, Recording, Preservation and Storage of Biological Samples

Stefano Schiaparelli[1], Kareen E. Schnabel[2], Bertrand Richer de Forges[3] and Tin-Yam Chan[4]*

[1] Department of Earth, Environment and Life Sciences and Italian National Antarctic Museum, Section of Genoa, University of Genoa, Genoa, Italy

[2] National Institute of Water and Atmospheric Research, Wellington, New Zealand

[3] KIWA Consulting, Nouméa, New Caledonia

[4] Institute of Marine Biology, National Taiwan Ocean University, Keelung, Taiwan, Republic of China

Abstract

This chapter describes the most important operations related to the sorting, recording, preservation and storage of biological samples on board a research vessel. The primary aim is to provide information on the sequence of events following collection of biological samples, from the organization of a sorting station on board to the shipping of samples after a voyage. For most activities we provide a number of recommendations that are based on the direct experience of the authors during ship-based expeditions around the world, and give guidance and indicate the 'best practices' for situations and problems that may occur while in the field. Specific attention has been devoted to the sorting and processing of different types of samples, with a focus on larger organisms that are individually sorted, photographed and fixed during a research voyage.

Keywords biological samples, sampling, sorting, recording, fixation, preservation, shipping, photography, storage

15.1 Introduction

The goal of many deep-sea biological research voyages is to document and sample the fauna living in a specific place, and to retrieve enough material for taxonomic, genetic and ecological studies. The flow of activities during and after a voyage can be summarized in five steps (Etnoyer *et al.*, 2006): (i) documenting specimens *in situ*, (ii) collecting specimens to ensure scientific value, (iii) preserving specimens to enhance long-term study, (iv) labelling specimens to ensure accurate transcription, and (v) shipping specimens to accredited scientific institutions. Apart from the first step, which is addressed elsewhere in this book (see Chapters 11, 12 and 13), all the other steps will be treated here. For each step, a system of 'best practices' on board the research vessel is outlined, which should assist a smooth sequence of operations during and after a voyage.

This book is concerned with deep-sea sampling, and so it is implicit here that the activities are carried out from vessels larger than 25 m that are capable of deploying heavy benthic sampling gear

*Current affiliation: MNHN, Paris, France.

Biological Sampling in the Deep Sea, First Edition. Edited by Malcolm R. Clark, Mireille Consalvey and Ashley A. Rowden.

such as trawls (Chapter 7), sledges (Chapter 9), corers and/or grabs (Chapter 10), remotely-operated vehicles (ROVs) and submersibles (Chapter 13). This chapter will focus on providing information for the sorting, recording, preservation and storage of the larger fauna (mega- and macrofauna) sampled by these gear types. The 'best practices' suggested in this chapter provide examples based on the combined and diverse experience of the authors and are intended to provide a starting point for scientists new to this field. However, these recommendations may need to be adapted depending on the specific circumstances of the research being undertaken; each survey is different and the operations on board the research vessel are driven first and foremost by the overall objectives of the voyage, as well as factors such as the type of gear deployed, sampling opportunities (e.g. driven by the substrate type, or by the time available between deployments), spatial constraints due to the size of the vessel, and use of available ship time by different research groups.

15.2 Pre-voyage preparation

In general, two different categories of voyages can be differentiated, based on their research targets, and may affect the choice of sampling gear used. During an exploratory voyage in areas never surveyed before, sampling is usually undertaken in an opportunistic fashion, and is often qualitative. It allows for a first characterization of the organisms and communities present at particular locations. In contrast, if the area to be sampled is sufficiently well known, a survey voyage may be more oriented towards quantitative sampling, such as fishery or environmental impact surveys for ecological characterization and quantification of benthic communities. Irrespective of the type of voyage, much of its success relies on good preparation; such as organization of materials, as well as safe workspaces and sound practices, which need to be established well in advance (and, if necessary, adapted later). It is desirable to know *before* the start of a voyage for what purpose samples will be collected, who will be examining them, and in which laboratory or museum collection they will be deposited. This step is crucial in order to design proper protocols for sample preparation and preservation. If this information is available in advance, it will be possible to manage samples appropriately, avoiding any uncertainty during the voyage and minimizing the handling of the samples during and following sampling activities.

It is also important to consider appropriate permitting regulations before the voyage. This may apply, for example, if the targeted sampling area is protected or if the voyage transits international boundaries or territorial waters that may require permits to be presented to a national biosecurity regulation body. In these cases the necessary permissions need to be acquired for collecting samples, a process that may be time-consuming. The same applies when sampling covers animal groups that are protected by national or international laws, and it must be kept in mind that sampling, custody and shipping of certain organisms are subject to regulations under the Convention on International Trade in Endangered Species (CITES). For example, stony (Scleractinia) and black (Antipatharia) corals, which are routinely collected during deep-sea trawling, fall in this category and may require CITES permits if they are to be shipped (www.cites.org). Custom inspection may cause considerable delays in the delivery of samples back to the laboratory or onwards to taxonomists or geneticists, and it is a good practice to evaluate in advance the documentation needed, both at the international and national level, to ship samples of marine fauna. In all cases, it is crucial to involve representatives from all collaborating institutions to outline the intended taxonomic and regional sample requirements, and engage with the relevant agencies about permit acquisition, well before the beginning of the voyage.

Among the several resources available online that address similar topics to those treated in this chapter, two particularly useful ones are the *Archiving Standards for Census of Marine Life* prepared for the partners of the Census of Marine Life project (Rosenberg, 2005) and the *Marine Invertebrate Sample Processing Procedures* prepared for the Australian Marine Invertebrate Taxonomists (AMIT) group (Wilson, 2005). Other useful resources are the chapter on deep-sea benthic sampling by Gage and Bett (2005), and the book edited by Danovaro (2010) about methods to study deep-sea sediments.

15.3 Sorting

The deck of the vessel is where sampling gear is retrieved and where, in most cases, the samples are initially processed. The deck is also one of the places with the highest inherent safety risks, posed by the gear itself (e.g. triggering mechanisms may suddenly release), the presence of wires to which the gear is linked and which may be under tension, in addition to exposure to the elements. Sorting on deck may take place while gear continues to be deployed, which means that some of the areas that need to be accessed by personnel are close to wires under tension or in motion. These risks may be greater if the weather conditions are harsh (e.g. in presence of ice at high latitudes) or in rough sea conditions.

For these reasons, the paramount consideration for setting up any operations on a vessel (indoor and outdoor) has to be safety. It is therefore essential that all scientists and crew are briefed in safe operational procedures, particularly around heavy equipment, and that personal safety equipment (e.g. helmets, gloves, safety shoes, etc.) is used. The same applies for the use of fixative chemicals that need to be prepared (i.e. diluted or decanted) and brought to the sorting area from their storage location. In this case, safety equipment (which will include gas masks, glasses, gloves, etc.) has to be easily accessible. In addition, all material in use (e.g. jars with samples, tanks with fixatives, etc.) needs to be properly secured so they do not come loose with the ship's movements.

The biological material recovered is usually mixed (i.e. organisms are surrounded by sediment and rocks, entangled in structures or with each other). Although organisms can usually stay alive for some time while they are surrounded by wet substrates in the 'bulk sample', the sample should be sorted as quickly as possible to extract the organisms and transfer them into a fixative, as detailed in the subsections below. This is especially important if weather conditions on deck are warm.

15.3.1 General operations

Different types of gear collect different spectra of organisms (Fig. 15.1a–f) and have advantages and disadvantages for sorting (Table 15.1), which need to be taken into account in the planning as well as the sorting phases. Regardless of the gear used, one of the immediate goals is to protect specimens from further damage by separating them from sediment and/or stones, rocks and even other organisms that may be abrasive, crushing or otherwise further damaging to the specimens. Damaged or incomplete specimens are very often unidentifiable, and large numbers of compromised specimens will typically reduce the value of a study (Wilson, 2005). This separation is not an easy task and, depending on the amount of material retrieved, sorting activities may be organized in different steps starting with a basic sorting on deck (Fig. 15.1g, h, i) followed by more detailed sorting in a better equipped sorting station (Fig. 15.1 j, k), where running seawater, microscopes, directed illumination, jars and vials are available (Boxes 15.1 and 15.2).

In the case of a very large catch it may be necessary to establish 'triage' priorities. Either the catch may be subsampled (only a portion is processed in more detail) or very abundant taxa may be subsampled once the catch is sorted. It is always useful to take notes of these decisions for calculations of abundance or catch weights. For fine residues and those from wood, which are much more time-intensive to sort in detail (see below), the complete sample should be transferred immediately to the desired fixative to prevent further degradation of the samples, and sorted at a later stage.

Before the transfer of the sample to the sorting station occurs, if the sampling gear includes a net, it is also good practice to check for entangled specimens. This is especially true in the case of trawls which have large mesh sizes, since they may trap megafauna (e.g. gorgonians, alcyonarians, stalked crinoids and crustaceans with long legs) in the initial part of the net (wings, belly and throat) and not in the cod-end (Fig. 15.1d). It is also good practice to capture the basic characteristics of the sample by taking photographs of the whole catch (arranged on the deck or in large containers) that includes a station identification card (Fig. 15.1a, b, c).

(a)
(b)
(c)
TAN0802/109
ORH 109
19/02/08
(d)
(e)
056
TRAWL
(f)
(g)
(h)
(i)
(j)
(k)
ORH 106

15.3.2 Sorting large fauna by hand

Large-bodied fauna recovered by trawls and sledges are typically sorted to operational taxonomic units (OTUs), using the approach 'same with same', which does not require detailed taxonomic knowledge but reduces the effort involved in sorting samples after the voyage. However, the accuracy of this process depends on the level of training of sorting personnel, the time available when processing the sample (i.e. sometimes, due to time limitations, samples may need to be clustered into practical groups that contain multiple species, e.g. starfish or polychaetes). The level of identification will vary with taxon. Although large-bodied commercial fish species may be readily identifiable because regional guides are generally well developed and available, deep-sea benthic invertebrates are often poorly known, and reliable identification is impossible except by specialist taxonomists. Nevertheless, science staff without taxonomic experience can still use basic keys, and separate most groups at the order or family level. The types of identification and preservative sheets given in Appendix 15.1 have been used extensively by National Institute of Water and Atmospheric Research (NIWA) science staff with highly diverse backgrounds. It is generally possible to achieve the levels given in the Appendix. Specimens will require more specific identification by specialists once ashore, but the higher-level sorting onboard facilitates post-voyage treatment, and also allows provisional voyage report type analysis. Small-bodied animals that require microscope examination will generally not be sorted at sea, as weather conditions often make such work difficult or impossible.

It is recommended that the largest, most fragile and best condition specimens are immediately isolated from the sampling gear and placed in seawater before the samples is removed from the deck and transported to the sorting area. This reduces damage and stress and keeps specimens in optimal

Fig. 15.1 Overview of the principal operations dedicated to sorting on deck and examples of catches from different gears. (a) Epibenthic sledge sample from a rocky bottom. The catch is naturally clear from fine sediment and the organisms are easily visible among the rocks. (b) Beam trawl catch with a high load of sediment. In this case the specimens sampled are not visible, being mainly within the mass of sediment. These need to be isolated, on a sorting table, after the fine sediment fraction has been washed away by spraying with seawater through a garden hose spray gun. (c) A very clean trawl catch. Here species have not been damaged by rocks and can be immediately processed for photography, tissue sampling, etc. (d) Large gorgonians (*Thouarella* spp.) up to 35 cm in length have been entangled at the mouth of the net. This kind of organism might not reach the cod-end of the net. In this case, the abundance of the octocorals makes their presence very obvious, however, when densities are lower they can be overlooked when entangled at the mouth of a long trawl net. (e) A large catch (about 150 kg) of mainly invertebrates. Large robust invertebrates (e.g. asteroids) are mixed with small, delicate invertebrates (e.g. crinoids) and small fishes. This is a more difficult situation to be handled, as each group of organisms has a different tolerance to air exposure and, due to the amount of material collected, the sorting can take a long time to be finished. Note the white board with the number of the station, which is a 10 × 10 cm square that can be used to give a rough indication of the size of the sampled organisms. (f) Another difficult case of a diverse catch in which there are species and associations which require different treatments. The large amount of wood requires special handling to isolate the associated fauna, which is a time-consuming process. (g) Sometimes a catch may be sorted directly on the deck. In this case, the deployed gear was a dredge with a mesh size of 2 cm: sieving is not required and sorting can take place with the naked eye. (h) During quantitative voyages (e.g. fish surveys), the catch has to be split into bins and weighed before species are isolated from the sediment. (i) In some cases, the large amount of collected biological material may require the involvement of most personnel. Sorting activities can be slow, meaning an accumulation of bins with unsorted biological material. Labelling bins is critical to avoid mixing catches. (j) Sorting of species in a sheltered lab. Note the number of transparent jars (in the white tray on the left corner of the sorting table) filled with seawater to keep specimens alive before photography and fixation. Running seawater is provided through the green tubes equipped with garden hose spray guns. (k) The sorting table after the subdivision of the catch into OTUs. Specimens are now ready to be placed in jars together with a label and fixed. (Please see Plate 25 for colour representation of the figure.)

Table 15.1 Summary of some sampling characteristics of various gear types and suggestions for processing (see text for more detail)

Gear	Targeted fauna	Habitats and depth of deployment	Advantages	Disadvantages	Suggestions
Large towed gears (e.g. Agassiz dredges, trawls, beam-trawls)	Large sessile (especially erect) invertebrates, and slow-moving species. Mainly macro- and megafauna	Mainly on soft sediment seafloor. On rocky substrate (e.g. seamounts) modified gears have to be used, such as rock-dredges and modified rough bottom trawls which are, respectively, more sturdy or have bobbins or rollers on the mouth which allow 'rock hopping' All depths (but usually not deeper than 2000 m)	High numbers of specimens of macro- and megafauna specimens	Species are mixed together Appendages of fragile species easily broken, or deeply entangled into the net meshes of the sampling device Small species (when present) easily overlooked on deck Pre-sorting on deck may require a considerable time according to the quantity of material present causing a delay in fixation of samples	Split the group of people on deck leaving to (at least) one person the specific task of picking up the 'best' specimens and transferring these immediately into water
Epibenthic sledges (EBS)	Small benthic and epibenthic fauna	Soft substrate primarily, but heavy or sturdy types have been designed for use on seamounts with rugged seafloor All depths (but usually not deeper than 3000 m)	Usually only small organisms are caught Good condition of the specimens (i.e. generally complete in all their parts) if tow length is short	Larger mobile organisms not sampled or, if sessile invertebrate taxa are caught, only fragments are present Sorting by hand of small-bodied species is often not possible on board The bulk sample is cleaned from sediment by elutriation which is time-consuming	Elutriate as slowly as possible to avoid damaging organisms; keep water pressure at minimum power Pick up organisms with long appendages immediately after these have been observed in the sieve/water and preserve these in separate tubes, not with the bulk sample-this avoids the limbs being broken If possible, photograph coloured species before fixation

(*continued*)

Table 15.1 (Continued)

Gear	Targeted fauna	Habitats and depth of deployment	Advantages	Disadvantages	Suggestions
Grabs	Macrofauna and meiofauna	Soft sediments only Usually deployed not deeper than 2000 m due to the light weight of the gear: the cable may be heavier than the gear itself causing entanglements	Small and medium sized organisms can be caught Infauna are retained within grab sample Living position of organisms is very often preserved	Gear may be dangerous when being armed/opened and require trained persons for their use Large quantity of sediment may preclude or make difficult the observation of the species present in the sample When sieving is performed the organisms may break (e.g. fragmentation of polychaetes, loss of appendages in crustaceans) Sieving is time-consuming Bulk samples generally fixed in formalin because of the large amount of sediment present and, therefore, are not usable for DNA analysis	If possible, isolate at least some organisms from the sample to be preserved in ethanol for DNA analysis and not place with the bulk sample Photograph coloured species before fixation Sieve gently in seawater, not under high pressure
Corers	Macrofauna and meiofauna	Soft sediments only All depths, down to trenches	Small fauna retained	Usually not sorted on board but 'sliced' for bacterial or meiofaunal studies	Refer to Chapter 10 for further details
ROV/ submersible	Megafauna and macrofauna	All bottom types All depths (although most are restricted to certain depth ranges, and typically less than 6500 m	Perfect specimens can be obtained with the proper tools placed on mechanical arms (e.g. suction pumps) Specimens can be selected for collection. Their *in situ* location is recorded	Only one/few specimens can be obtained on each dive due to limitations of payload. Keeping samples separate can be difficult, extremely expensive, often needing specialized vessels. Highly trained technicians and operating personnel are needed	Each sampling event must be well documented. Samples must be kept separate

Box 15.1 Organization of the sorting station

Sorting samples may be done at different locations, avoiding any overlap of activities where possible. The sorting station should be set up with safety in mind and consider the following:

- Proximity to where the sample arrives on deck, particularly if handling requires heavy lifting.
- Identify possible obstacles that may be present when moving the catch to a sorting station (unused gear, steps, ropes, wire, etc.) and make sure the main route is clear and safe.
- Availability of Material Safety Data Sheets (MSDS) (www.msds.com) for all chemicals. All members of the sorting team should be briefed on the location and use of the personal safety equipment, spill kit(s), and the general health and safety procedures around the use of all chemicals to be used.
- Easy and safe access to preservatives and fixatives, such as tapped carboys of high-strength ethanol and dilute 5–10% formalin. All such containers need to comply with dangerous goods regulations, be clearly labelled, and located in areas with good ventilation.
- Proximity to a sample storage area to facilitate the quick clearance of sorted samples from the station. Ethanol and formalin preserved samples should be stored in separate areas (designated for flammable and corrosive chemicals, respectively, see Section 15.7).
- Samples that are to be frozen (e.g. −20°C or −80°C) will need to be processed quickly to prevent degeneration of specimens, particularly when working with deep-sea samples in regions where surface waters and air temperature are much higher compared to temperatures on the ocean floor. Chilled or frozen seawater should be available to add to the sorting trays for maintaining integrity of specimens.
- Space may need to be allocated for weighing, the relaxation of specimens, histological preparations or tissue subsampling for molecular work prior to fixation.
- Running seawater from a shipboard or standalone submerged pump will need to be available at the sorting station, and ideally should include a filter to avoid contamination of the sorted sample from surface zooplankton. Garden hose spray gun fittings with adjustable water pressure are useful to prevent damage to fragile specimens when sorting the sample.
- Easy access to a good supply and range of types and sizes of sample containers (20 mL–20 L; vials, jars, plastic bags).

condition, and therefore suitable for subsequent photography (see Section 15.6) and preservation for taxonomic study (see Section 15.4).

The remainder of the catch is then placed in buckets or bins (each with a label noting the sampling station) and transported to the sorting station. Here the samples can be sorted in more detail at the sorting table, gently washing it with seawater from hoses if needed. Depending on the nature of the catch, more or less attention has to be paid to separating the fauna from coarse substrates such as gravels or rubble that may have been collected with it. The specimens are first sorted by major taxonomic group into containers filled with seawater, followed by a detailed sorting to taxonomic levels as fine as possible (or practicable) using progressively smaller containers (Fig. 15.1l,m). When all specimens of one OTU have been extracted into a container, they are usually treated together as one sample 'lot'.

Each sample lot is then labelled, weighed (if required), and recorded on the recording sheet to ensure samples are quickly taken away to be fixed and/or frozen to avoid degradation. Sometimes it is necessary to split the sample lot into multiple sub-lots for separate subsequent treatment (e.g. for genetics, taxonomy, etc.). If samples of particular species or OTUs are to be subsampled it is

Box 15.2 Items at the sorting station

- Labels and containers. Note that containers for wet samples (ethanol or formalin) need to be properly sealed (e.g. when using separate jars they should always include lid liners). When using larger buckets or drums for bulk samples, they need to have a tight-fitting lid and, ideally, a rubber seal to prevent leakage or evaporation post sampling.
- Sieves, tweezers, sorting trays and gloves should be ready and clean. Specialized equipment (e.g. small tweezers, pipettes, brushes, etc.) will generally be brought by scientists, however, stocking the sorting station with a full range of standard tools may prove expedient.
- Waterproof paper and pens.
- Logbook for documenting each sample, rating sampling success and noting general sample characteristics such as substrate type, total weight and an initial description of faunal composition. As the logbook usually becomes wet during use, it is recommended that a graphite pencil is used for writing.
- Field guides or other aids for preliminary identification if taxonomic experts are not at hand. Most likely, organisms will only be identified to a higher taxonomic level (such as phylum, order or family). Therefore, having simple illustrations of the main taxonomic groups and preferred names (or placeholder names) used for recording these organisms will aid rapid processing, and make identifications more consistent between different shifts of personnel (see Appendix 15.1).
- Clear instructions on how to fix the different groups of organisms needs to be displayed or within easy reach (e.g. Appendix 15.1). Note that additional treatments may be required, depending on specific requests, and samples may need to be split for different uses (morphology vs. DNA analysis vs. isotope analysis). If a special handling protocol is required, this has to be known in advance and designated persons have to be appointed for treating the subsample in the desired manner. However, quick decisions will have to be made on deck with respect to prioritising treatments (triage), which will depend on the quality/quantity of the sample and the availability of personnel to process samples. It is best if one person is responsible for assigning and delegating sample treatment and preservation.
- Paper-based (Fig. 15.3a) or digital computer-based recording sheets for documenting sample lot details.

important to know in advance the flow of operations in order to ensure that they are handled appropriately and with the necessary tools (e.g. by using gloves and tweezers).

It is important that all staff are familiar with the major taxonomic groups, usually phylum to class level, depending on the group, since this will determine the type of further treatment. If the animals are to be fixed in formalin or ethanol they are generally placed in jars; if they are to be frozen they will be placed in bags. Having a basic guide to the animal groups and their preferred fixation techniques on display by the sorting table (e.g. Appendix 15.1) greatly speeds up this process and will ensure consistency between different shifts of sorting personnel.

The personnel doing the final labelling and recording may also need access to further identification guides to common species. However, the time and skills available and the objectives of the voyage will determine how much effort can be spent on identification of the separate taxa at this point. In most cases, it can be expected that the samples will be sent to experts for detailed taxonomic identification following the voyage. Hence, it is most important that taxa are sufficiently sorted to a level that the relevant groups can be quickly separated and handed to the relevant experts (e.g. splitting Echinodermata into Ophiuroidea, Asteroidea, Echinoidea, etc.).

15.3.3 Sorting small fauna by sieving and elutriation

Although large species of macro- and megafauna sampled by trawls and sledges can be easily sorted by eye on deck or on a sorting table, different methods have to be used to extract the smaller fraction of macrofauna (300 μm to a few millimetres) that is usually embedded in finer sediments. These small organisms are typically recovered in high numbers (e.g. polychaetes) from soft sediments using grabs, corers and fine-meshed epibenthic sledges. In the case samples from these gear types are processed, specimens are generally extracted by 'puddle sieving' or elutriating sample (see Chapter 10 on corers and grabs for detail about these procedures).

15.3.4 Sorting of organisms living in association

Although spatial and biotic interactions can only rarely be reconstructed from a sample on deck, it may still be possible to obtain useful data on small invertebrates that live in close association with larger hosts. Discovering these associations in the field generally requires a keen eye and a good amount of luck; general zoological and ecological knowledge of the different typologies of associations can help guide the search for these cryptic organisms. When previously unsurveyed areas are sampled and there is no indication of what kind of associations can be found, a good practice is to closely scrutinize all common invertebrates in the catch to search for possible parasites, commensals, etc. These species, whose abundance in the catch reflects approximately their abundance in the environment (e.g. Fig. 15.2a), have a high potential to represent a target for other organisms to use as a host. Moreover, the availability of numerous host specimens partially compensates for the unavoidable loss or detachment of associates caused by inherently destructive sampling methods (e.g. a trawl or a dredge). If there is any suspicion that two species could be somehow linked (e.g. one has been found in strict contact with another or there is a colour match between the two (mimicry)), both partners should immediately be placed in water in a container. In this way the associate can re-occupy its position on the host. If this treatment is done quickly and the two partners have not been stressed too much, it is even possible to observe their behaviour for some time in an aquarium.

Deep-sea groups that are most commonly found living in association with larger invertebrates are polychaetes (especially polynoids), crustaceans and molluscs. Their hosts are usually emergent sessile invertebrates, such as octocorals, hexacorals and sponges (Buhl-Mortensen *et al.*, 2010; Shank, 2010), and also echinoderms. As such it is good practice to pay particular attention to the larger organisms that are sampled by trawls and sledges, and proceed with a careful examination of their surfaces (e.g. cavities such as oscula in sponges, or turf-forming epibionts such as hydroids) before these habitat-forming invertebrates are preserved. This search may uncover small parasites, predators or commensals. ROV and submersible technology offers an unprecedented opportunity to sample even very small associates which can be picked by the vehicle's manipulator arm together with their hosts without any damage (see Chapter 13 for further details). See Box 15.3 for some tips on recording animal associations.

15.3.5 Sorting of organisms on/in rock and wood

Samples may contain rocks and stones that provide habitat for organisms on the seafloor. An examination of these substrates may reveal mobile epibionts or other small, cryptic invertebrates (Brachiopoda, Bryozoa, serpulids, etc.). Carefully breaking rocks to reveal hidden organisms is time-consuming but provides valuable information. An alternative to physically breaking the rock is to completely submerge it in saltwater with a squirt of formalin. The formalin usually induces motile specimens to leave their hideaway within about half an hour (if collections on board proceed during that time, it is advisable to attach a label showing the sample station number so as not to cause confusion later).

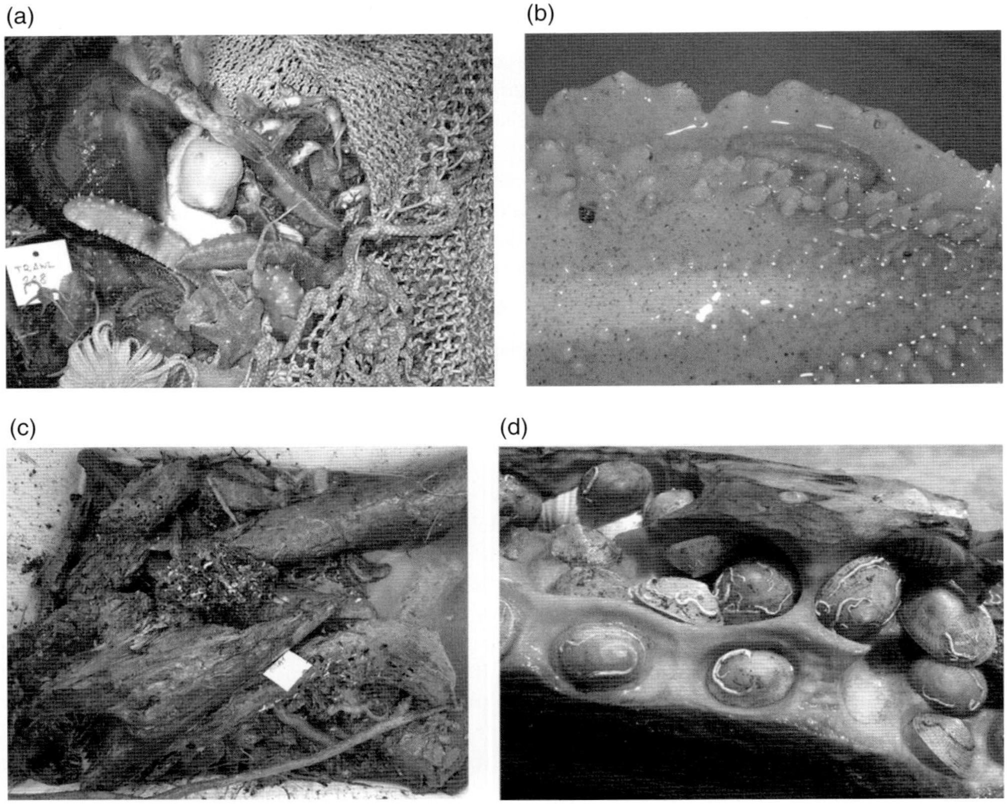

Fig. 15.2 Examples of associations that can be found in samples obtained from towed gears. (a) When large numbers of one species are caught, carefully look for parasites and associates. If associated species are spotted in the catch, these need to be placed in seawater immediately. In this example a parasitic Antarctic polynoid worm (*Eunoe opalina*) has been found on the large holothuroid in the picture (*Bathyplotes bongraini*). (b) The parasitic polynoid *Eunoe opalina* moving along the sole of the holothuroid. If not placed immediately in seawater, the associated polychaete will ultimately detach from its associate, preventing recognition of the association. (c) A catch very rich in wood remains. It is possible to see different degrees of degradation of the wood pieces, due either to the nature of the plant that produced the wood and/or the prolonged period at sea. Each one of these fragments has to be treated individually. (d) An example of a wood-associated community, with limpets, teredinids and chitons.

There is no universal method for the treatment of wood samples: each piece of wood and associated organisms living inside and/or on it has its own peculiarities. Generally, a range of macrofauna can be extracted from a single piece of wood. It is good practice to keep samples of the wood itself for further study to determine its origin and residence time on the bottom (e.g. Pailleret *et al.*, 2007, and see Box 15.4).

Wood that has been submerged for a short time is usually colonized by epifauna such as barnacles and limpets. The large animals can be handpicked, while a knife or chisel can be used to remove attached fauna by shaving off a thin flake of the wood. The piece of wood can then be broken up to find any animals residing within, such as boring bivalves. Bivalves can be extracted by splitting the wood with a wedge or cross-sectioning it with a saw (e.g. a stainless garden saw or a Teflon-coated carpenter's saw). If electric saws are used, cordless ones reduce the risk of electrocution, and stainless blades are better suited in a marine environment.

Box 15.3 Sorting of animals in association

- Consider the use of an additional code or comment for the specimens involved in associations when recording sample lots. This is to avoid any problem arising from the fact that associates usually belong to different phyla and, as such, need to be recorded separately.
- Often the host and the associate need to be preserved in separate fixatives or containers. If possible do not physically separate the two containers, but place the jars in a larger bag or container with a clear label 'association'.
- If it is not possible to keep the associated animals together then the association code can be used to cross-reference the different specimen lots and make it possible to 'reconstruct' the association after the voyage.

Box 15.4 Processing specimens attached to wood and rocks

- Wood should be separated from the rest of the sample as soon as possible, as much of the surface fauna is easily washed off and the association with wood can be lost (Fig. 15.2c).
- The pieces of wood should be gently washed or brushed into a suitable container, preferably with cold seawater. The residue in the container can be gently sieved to separate specimens from the sediment.
- A large-mesh sieve is used to separate material that can be sorted directly on deck; a finer meshed sieve retains animals and particles to be saved for later examination, either alive on the ship (keep in a mixture of seawater and ice) or fix for use in the laboratory onshore.
- Large organisms on the wood surface are handpicked first and then smaller surface organisms removed with tweezers, knifes or chisels.
- Animals residing within the wood are carefully extracted with tweezers after saws, pliers, etc. are used to expose them.
- At least a portion of the remaining wood is retained for later analysis.

When submerged for a long period of time, the wood matrix is progressively broken up by a succession of boring organisms (Fig. 15.2d), and may be so permeated by tunnels from boring animals that only a little of the actual wood remains. This kind of wood usually has the richest animal life, with the boring bivalves being replaced by gastropods, sipunculids, echinoids, brittle stars, polychaetes and crustaceans, which use the wood as shelter or as a hunting ground. Several gastropods, especially limpets, continue the breakdown of the wood, while other gastropods mainly eat the bacterial film that plays an important role in the degradation process (Samadi *et al.*, 2010). Various kinds of pliers, especially wire cutters (which occasionally can be obtained in stainless steel) are useful for breaking up the wood and exposing the animals. A small axe may come in handy for shaving off larger quantities of superficial wood over a container, for later washing.

15.3.6 Sorting of samples for molecular analyses

All steps of molecular analyses can be performed at sea if the research vessel is equipped with the necessary instruments (e.g. Bucklin *et al.*, 2010). But more commonly, specimens destined for DNA sequencing are only fixed, leaving DNA extraction, amplification and sequencing to be done in a well-equipped laboratory on land. Samples for molecular studies can be frozen or fixed in highly concentrated ethanol (>80%). However, specimens release water which will progressively dilute the alcohol concentration. If this happens and the ethanol is not immediately replaced, the sample may not be stable and decay, compromising DNA integrity. It is therefore good practice to use ethanol of the highest concentration available, and substitute it a couple of times over the course of the voyage. However, with this practice, tissue becomes very rigid. Crustaceans, for example, become brittle and specimens may be very difficult to study for taxonomy. Thus, if there are sufficient specimens sampled, a subsample should be preserved separately for molecular studies (e.g. 30 individuals from a sample of 100). An alternative is to remove a portion of the animal's tissue (e.g. a leg), which can be preserved in a small vial with a high volume-to-sample ratio (5–10 times the volume of the sample tissue). This approach allows the rest of the organism to be fixed in another medium (e.g. formalin for morphological vouchers). It is crucial that these subsamples are properly labelled and cross-referenced to avoid any mismatch between samples and sequences (see Box 15.5).

A widely recognized problem in molecular studies is the delay between collection and fixation: the longer the delay, the lower the quality of the DNA and the lower the potential for successful extraction and amplification. At high latitudes, cold temperatures may help, slowing the speed at which an organism dies and begins to decay, but in the tropics, special attention must be paid to ensure rapid sorting and fixation.

Box 15.5 Processing specimens for molecular analysis

- Develop clear protocols for sorting genetic samples from those destined for taxonomy or ecological studies. Specify whether whole or partial specimens are required.
- If specimens are large, it is better to fix a small fraction of tissue separately for molecular analysis.
- Clarify in advance which part of the animal is to be removed for separate fixation. First, it is important to avoid removing taxonomically important body parts. Second, it is important to avoid contamination, for example, species living in association can be contaminated due to the presence of tissue fragments or mucus from the host animal. In this case internal muscle should be subsampled. External debris, parasites and gut contents can also be a source of contamination and have to be avoided.
- Cross-referencing between the original and the tissue subsamples has to be clear and unambiguous in order to avoid any mismatch between sequences and specimens: ensure that all the relevant information is present both on labels of the specimen and of the subsample.
- Consider the use of *ad hoc* codes for the specimens to be used for molecular analysis. For example, if the specimen is being taking for barcoding then it is good practice to pre-print small labels with the DNA sample number which can be placed in vials and jars together with the official labels of a voyage. This supplementary labelling has been proven to save time by diminishing the handling of the samples, avoiding contamination and the inevitable errors in handwritten labels (including errors as a result of national differences in number writing).
- Any pre-printed labels should be produced in pairs, one to be placed with the specimen and one to be placed with the subsample.
- The label information should also be referred to on the recording sheet or in a database where all the other relevant information about station details, etc. is recorded.
- If a subsample has to be taken, ensure that blades and tools are clean (e.g. have not been used to dissect another specimen or species) in order to avoid any contamination.
- Laboratory gloves should be worn when handling specimens in order to avoid contamination of the DNA.
- Pre-cooled ethanol (e.g. placed in the freezer before use) can help make ethanol fixation more effective.

15.3.7 Sorting of samples for biochemical, isotopic and trace element analyses

Specimens destined for biochemical (e.g. fatty acid composition) or stable isotope analyses are usually frozen on board at either –20 or –80°C. For the analysis of fatty acids it may be necessary to dissect and isolate the gut of the animal before freezing the sample. In some cases, such as with holothuroids, it is first necessary to drain the coelomic fluid to avoid contaminating the gut (Ginger *et al.*, 2001). In these specific cases all the necessary tools for dissection and syringes have to be accessible. The working station has to be clean from contamination by other species and free from fixatives. It is therefore necessary to designate a specific area for this type of work, away from the sorting table, and usually at a constant temperature of 4°C. Specific treatment protocols for different animal groups may vary, and these are available for different taxa in the literature. Depending on requirements given a particular situation, clear instructions should be prepared in advance.

Stable isotope analysis, generally involving carbon ($^{13}C/^{12}C$) and nitrogen ($^{15}N/^{14}N$) stable isotope ratios, provides information about food web structure and trophic level. It is preferentially done on frozen samples, although some studies have highlighted an impact of this kind of fixation on the subsequent analysis (for a review see Syväranta *et al.*, 2011). Other isotopes that have been studied in deep-sea animals include silica (^{32}Si) that allows dating or growth measurements of silicate skeletons in, for example sponges or diatoms, and oxygen ($^{16}O/^{18}O$) that is used as a proxy for temperature and/or climate (Ellwood *et al.*, 2007). In these cases preservation in ethanol is adequate or specimens can be simply dried.

Species with a $CaCO_3$ skeleton, especially deep-water corals, can be dated using isotopes (e.g. ^{14}C, $^{234}U/^{230}Th$) or studied for biogeochemical proxies by evaluating trace and minor element ratios (Mg/Ca, Ba/Ca, Sr/Ca, U/Ca, B/Ca, Mn/Ca and P/Ca) which provide coherent high-resolution signals stored in growth bands of their skeletons (Montagna *et al.*, 2005). In these cases dry or frozen samples are preferable to ethanol-fixed ones.

15.4 Preservation

It is important that preservation techniques are decided prior to the voyage, ideally in agreement with the researchers who will study the material afterwards. This will avoid any misunderstanding and will ensure the samples are appropriately preserved for their intended use post voyage. The great majority of specimens collected will be preserved in liquid, which usually is ethanol or formalin.

A first general rule is that the volume of the preservative has to be kept high with respect to that of the sample (a rule of thumb is at least 5 to 1). Species with a large body containing a lot of water and an impermeable integument (e.g. large fishes, crustaceans, holothurians) will have to be injected with preservative to ensure complete penetration into the body cavity. The choice of preservative depends on the later use of the material. A general rule is that samples that will subsequently be used for molecular studies are either ethanol-fixed or frozen. For taxonomic purposes, many organisms should be fixed in formalin, but there are important exceptions listed below. Since the decision about what fixative to use for a particular sample has to be made quickly and consistently on board the research vessel, the provision of a 'guidance sheet' at the sorting station can be invaluable (Appendix 15.1).

The most convenient way of preserving large specimens in bulk is to place the sample in a labelled, closed plastic bag punctured with holes, and then in a drum filled with 80% ethanol or 10% formalin. The holes should be small enough to prevent loss of the smallest specimens or labels, but large enough to allow easy penetration of the fixative. The specimens will be well fixed if the bag with the specimen(s) remains completely submerged. Every 2 days, the bags should be removed and the concentration of preservative checked with a densitometer and increased as necessary. After 6 days the samples can typically be considered to be in a steady state and can be maintained as such in a museum for decades with periodical checks.

All fixatives and preservatives should be assumed to be harmful and all efforts should be undertaken to ensure a safe workplace, with provision of Safety Data Sheets (see www.msds.com) and adoption of approved handling procedures.

15.4.1 Formalin preservation

Formalin is commercialized as an aqueous solution at 37%, which is considered 'full strength' or concentrated formalin, and needs to be diluted before it can be used as fixative. Formalin is also toxic and carcinogenic: it kills specimens and coagulates their protein and will do the same to any careless user (Wilson, 2005).

If no molecular studies are scheduled, then formalin can be used. Formaldehyde cross-links the proteins in the tissue with the result that the specimen tissue remains flexible, which is important for manipulation required for taxonomic identification. However, formalin is acidic, even when buffered with borax, and will dissolve calcium carbonate; consequently, animals with carbonate shells, carapaces or internal ossicles (e.g. holothuroids) should be preserved in ethanol. Zoologists working on animals lacking an exoskeleton (e.g. sea-anemones, antipatharians, hydroids, polychaetes, tunicates or fish), generally prefer a diluted formaldehyde solution at 8–10% (1 volume of concentrated formalin in 9 volumes of seawater). The same concentration is used for fishes and large invertebrates with a high aqueous content, while a dilution of 4% is generally used for small invertebrate specimens. An alternative solution to diluted formalin stock is one part of 37% formaldehyde and one of part propylene glycol; this solution is then mixed with seawater (1 to 9 parts of seawater). The advantages are that less formalin is used and the propylene glycol also acts as a penetrating agent, allowing better fixation. However, this solution is still as toxic as standard formalin and an extra MSDS (Material Safety Data Sheet) for the propylene glycol is needed. After formalin fixation (usually 2–3 days), the samples may be transferred to 75% ethanol for long-term storage, making sure they are soaked in seawater first to clear the formalin. Even in this case samples retain some formalin and when handled special precautions have to be taken.

15.4.2 Ethanol preservation

For ethanol preservation of tissues not intended for molecular analysis, 75–80% is adequate. Pure ethanol dehydrates specimens and they become too brittle for anatomical study. However, because animal bodies contain a high percentage of water, ethanol concentration can decrease quickly, depending on the size and volume of the specimens. The use of an alcohol densitometer permits testing and adjusting the ethanol concentration as necessary to ensure adequate preservation. Below initial ethanol concentrations of 60% there is a high risk that organisms will decay quickly and the samples are rendered useless. Echinoderms, as a consequence of their large size and volume, are among the most sensitive to decay, especially the large sea stars and holothuroids.

15.5 Sample labelling and recording

Every container or bag with specimens, parts of specimens or sediment samples needs to be clearly labelled and recorded on a recording sheet (Fig. 15.3a) or electronically (Fig. 15.3b). It has proven useful to associate a unique identifier (e.g. lot or catalogue number) to each sample in order to link every entry to a specific jar/bag. This approach makes it easier to locate specific samples and to support post-voyage loan requests, particularly when multiple parties are involved.

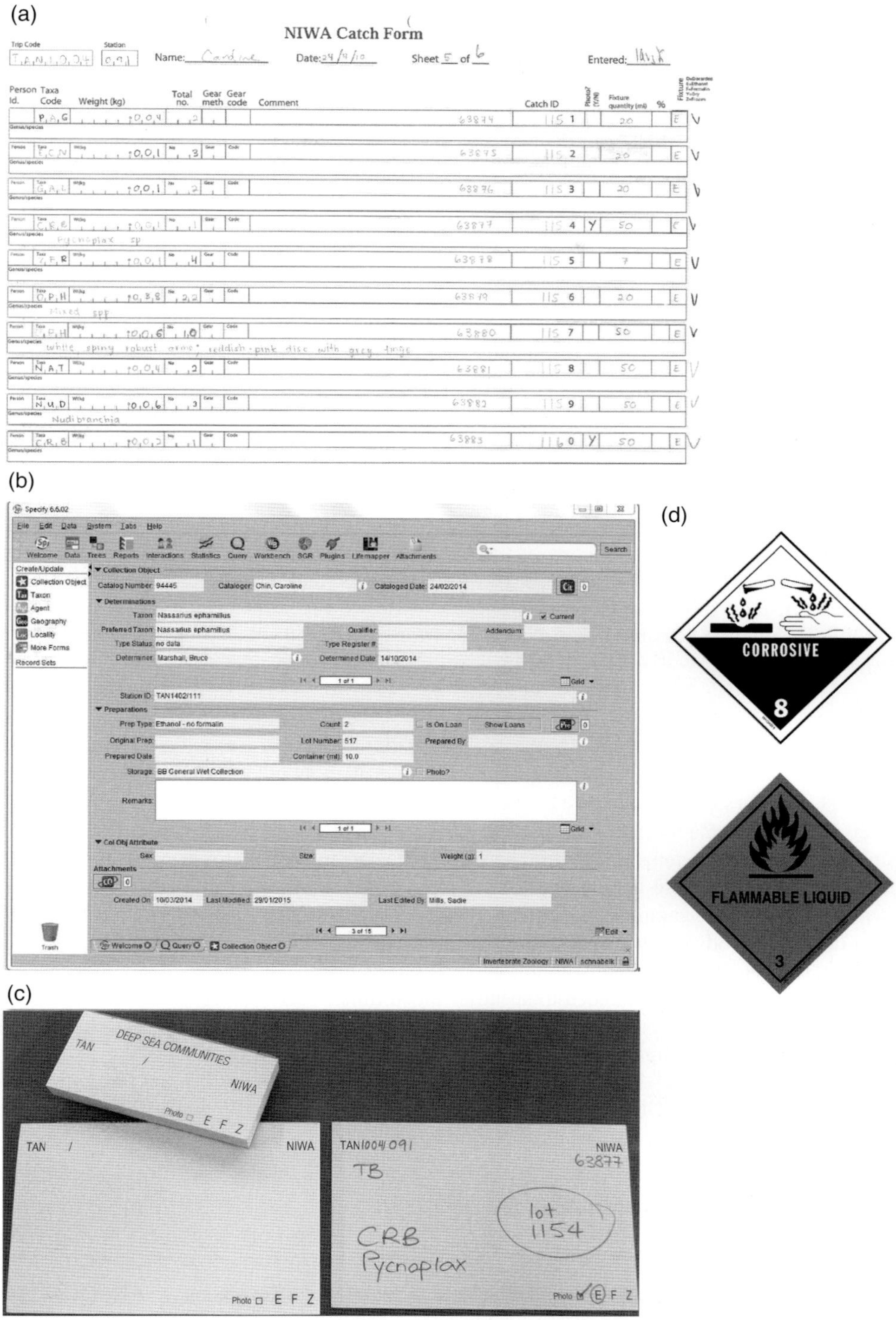

Fig. 15.3 Examples of various forms associated with processing samples. (a) Example of a data recording sheet used by NIWA (b) Example of an electronic data sheet as entered from the catch sheet, transcribing identification, station and preservation information against the assigned catalogue number. To this aim

15.5.1 Labels

The importance of proper labelling cannot be overemphasized as collections can be rendered useless when vital metadata are lost or erroneous. For example, it is crucial that a label is physically inserted into each container unless specifically required otherwise (e.g. to retain sterile conditions for DNA sampling). Usually, solely writing on the outside of jars and of storage containers is not acceptable as it is too prone to dissolving very quickly. Box 15.6 lists recommendations concerning the nature and content of labelling samples.

15.5.2 Recording sheets

Recording sheets (Fig. 15.3a) should be pre-printed with pre-defined fields for the most vital information onto archival-quality paper with high cotton fibre content that can withstand wet and muddy conditions. Details need to include the station name and any associated data that can be cross-checked and cross-referenced (e.g. with the log on the ship's bridge), the identification number (lot number or catalogue number), specimen details (preliminary identification, number of specimens, weight), fixation (this is important for people who will deal with the samples after the voyage for security and safety requirements of toxic and flammable liquids), whether a photograph has been taken and the name of the recorder. Any interesting details (live colouration, texture or if the sample has been requested by someone) should also be recorded here. If samples are split, they should be cross-referenced. While we recommend the use of paper-based recording sheets, it is possible to use digital recording sheets on splashproof/waterproof laptop computers.

15.5.3 Data entry on board

By the end of the voyage all relevant sample and specimen data should be transcribed into electronic format. This action safeguards against the loss of any single-copy data entry sheets and will make the collection data available for querying, and will simplify preparation of any voyage reports and transactions such as loans or donations to third parties. These data include preliminary taxonomic identification, the location where they have been collected and how they have been fixed. Figure 15.3b provides an example of sample data (as seen in Fig. 15.3a) entered into an electronic database used during dedicated biodiversity surveys. The format of data capture may be pre-determined through protocols established by a given institution, ranging from simple spreadsheets to complex relational databases. It is highly recommended to spend as much time as possible on board entering data as there is usually little time for these tasks post voyage and requests for data may begin immediately following the return of samples.

If the data will ultimately be uploaded into existing databases, such as a natural history museum, or data portals such as the Global Biodiversity Information Facility (GBIF, www.gbif.org), it is also recommended that any specific format requirements to facilitate data transfer be considered and adopted beforehand (see Chapter 16).

specific museum collection software can be used (e.g. www.specifysoftware.org). (c) Pre-printed labels identifying the collector (In this case NIWA), the research programme (Deep Sea Communities) and fixative (E: Ethanol, F: Formalin, Z: Frozen). Labels have been produced in two sizes (50 × 80 and 25 × 55 mm) and pre-printed on waterproof paper. Specimen information has been added by using a graphite pencil. (d) Hazard diamonds for concentrated formalin (corrosive substance, class 8) and ethanol (flammable liquid, class 3) to be attached on jars and storage containers.

Box 15.6 Notes on labelling

- Normal bond paper will quickly tear and dissolve in liquid and is unsuitable. The best options are archival-quality paper with high cotton fibre content which is designed for long-term storage in wet and dry conditions and resistant to ethanol and formalin, or plasticized wood fibre paper which may be the more readily available 'waterproof' paper. Labelling should be done exclusively with a graphite pencil. Unlike indelible ink pens, a graphite pencil can write on wet paper and does not dissolve in ethanol. Any writing on the outside of ethanol containers will need to be done using ethanol-proof pens.
- Key survey details should be pre-printed on labels. This saves time, ensures accuracy and that some information (e.g. survey names, collecting agency) is consistently associated with the samples. It is important that this is done using offset printing with indelible ink; the toner used in laser printing will abrade and flake away. Ideally, pads of pre-printed labels are professionally printed, cut and glued to booklets that save time and keep the work station tidy (Fig. 15.3c).
- A diversity of label sizes will allow selecting suitable labels for a given container size (Fig. 15.3c).
- Labels for each sample should always include the following information: date, station number, lot or catalogue number, taxonomic identification (maybe a code) and preservation type. Including as much information as possible provides a degree of redundancy that allows for cross-referencing if details have been incorrectly transcribed. The label should be designed to remain with the sample indefinitely. This may include the gear type with which the sample was caught, and the sieve mesh size if this was used (important if corers, grabs or fine-meshed sledges are used that may be sieved on variable mesh sizes between voyages).
- If samples are fixed in formalin or other toxic solutions (e.g. containing gluteraldehyde), this should be clearly marked on the outside of the container.
- Position the label so it can be read without opening the container. If the sample and label are subsequently placed in a non-transparent container (e.g. a bucket), the primary information (station, lot number, sample identification (ID)) should be written on the lid with ethanol-proof marker pen.
- When placing the label into a container with larger specimens, or with specimens with hard exoskeletons or other hard external parts, it is a good practice to place the label into a small plastic bag first. This will help prevent the label from being abraded.
- For larger specimens that are frozen (usually because they are too big to fit into a preservative container), or placed in a bag inside a drum of preservative, the label should be put inside a small plastic bag with the specimen, and a duplicate tied clearly visible to the outside with string. In some cases attaching the inner label to the actual specimen with string can also be a good idea.
- Beside the final labels, it is always advisable to have a label on specimens being examined or processed (e.g. waiting to be photographed) in order to avoid any confusion when samples from different stations are sorted at the same time.

15.6 Photographing specimens

Specimens set aside to be photographed are usually those that are in very good condition and/or the best representatives of a species. These specimens may consequently provide excellent material for future taxonomic studies, serve as a voucher in museum collections and, sometimes, will become a type specimen. Therefore, it is imperative that they are clearly labelled and traceable to enhance access from collections or archives. Ideally, these specimens should be stored in a separate container, or within the same container but in a separate bag, and labelled as a photographic voucher.

Photographic documentation provides valuable information about the characteristics of a taxon, especially in those groups in which colour patterns may have taxonomic importance. The live colouration usually fades quickly after fixation, especially in ethanol. Colour pattern is frequently used to discriminate between species with otherwise very similar or identical morphology and provides useful characters that may aid in their discrimination. Capturing a range of colour morphs can also be valuable for describing polymorphic species. On the other hand, colour patterns may be extremely variable and hence misleading; the real importance of this character can only be understood by combining morphology with molecular studies. Recent advances in digital technology have greatly improved the quality of the images and allow for real-time verification. Many other technological innovations such as infrared commanders for flashes also give much more freedom to the photographer, avoiding the clutter of connecting cables which greatly limit movements.

For high-quality images, it is essential that an SLR camera is available with a range of lenses from macro to mid zoom. This kind of camera allows, in manual mode, the *f*-stop (and shutter speed) to be varied for a perfect depth of the field. This point is very important in scientific photography where all the morphological characters useful for the identification of a species need to be focused and sharp. Colour-balanced floodlights and off-camera orientable flashes are essential to manage light sources and obtain true colours (Fig. 15.4a). Image-stabilizing lenses, if available, also help to compensate for the effects of shipboard vibration and motion and are preferred to non-stabilized ones. The problem of roll and pitch of the vessel, amplified in adverse weather conditions, can be overcome by using fast shutter speed, with sufficient flashlights.

If SLRs are not available, suitable results have been obtained with compact digital zoom cameras, but these need to have a good-quality fixed lens with a charge-coupled device (CCD) of at least 10 megapixels. These cameras, however, rarely have a manual-mode setting, making it impossible to control *f*-stops and, in turn, the depth of the field. For close-up photography of specimens in air, compact cameras may be equipped with a macro-ring flashlight, which is a cost-effective and a practical solution for illuminating the specimens. However, with a ring flash, it is impossible to avoid mirroring if the subject is immersed in water or if it is photographed through the glass of an aquarium. In general, non-reflex compact cameras with built-in flashes rarely produce publication-quality images.

Irrespective of the camera in use, special care has to be taken to maintain a clean background in order to reduce post-voyage editing (which can be very time-consuming considering the large number of pictures that can be taken).

15.6.1 Organizing a photographic station on board

Photographic set-ups and protocols may differ when documenting vertebrates and invertebrates. Furthermore, the decision to photograph the specimen dry or submerged in water will depend on the size of the specimen and on the taxonomic group; taxa such as crustaceans with a hard exoskeleton can be easily photographed dry, while soft-bodied organisms are better photographed submerged. No matter the technique, all pictures should include a scale placed close to the specimen. A 1 cm wide scale bar cut from archival-quality paper can be used for specimens that are photographed in water and kept in place by inverted forceps (Fig. 15.4e). For specimens photographed in air, a ruler with a non-reflective surface can be used. All images should also contain a relevant archival label for the specimen (Fig. 15.4h). If this is not possible, for example due to size constraints, a picture of the label and/or scale bar can be taken *before* photographing the specimen; this will avoid any uncertainty when pictures are renamed. Considering the high rate of work during a voyage it is often impossible to go back to the fixed specimen to check station or sample identification. It is important that all available information (station, lot number and/or registration number, preliminary identification) is linked immediately to the image.

15.6.2 Photographing small specimens in water

Photography of small soft-bodied invertebrates, or those with a reflective exoskeleton, is best done with the specimen immersed in water (Fig. 15.4d, e, i). A hand-held camera can be used in rough weather conditions and thus photography requires only a small area of bench for the

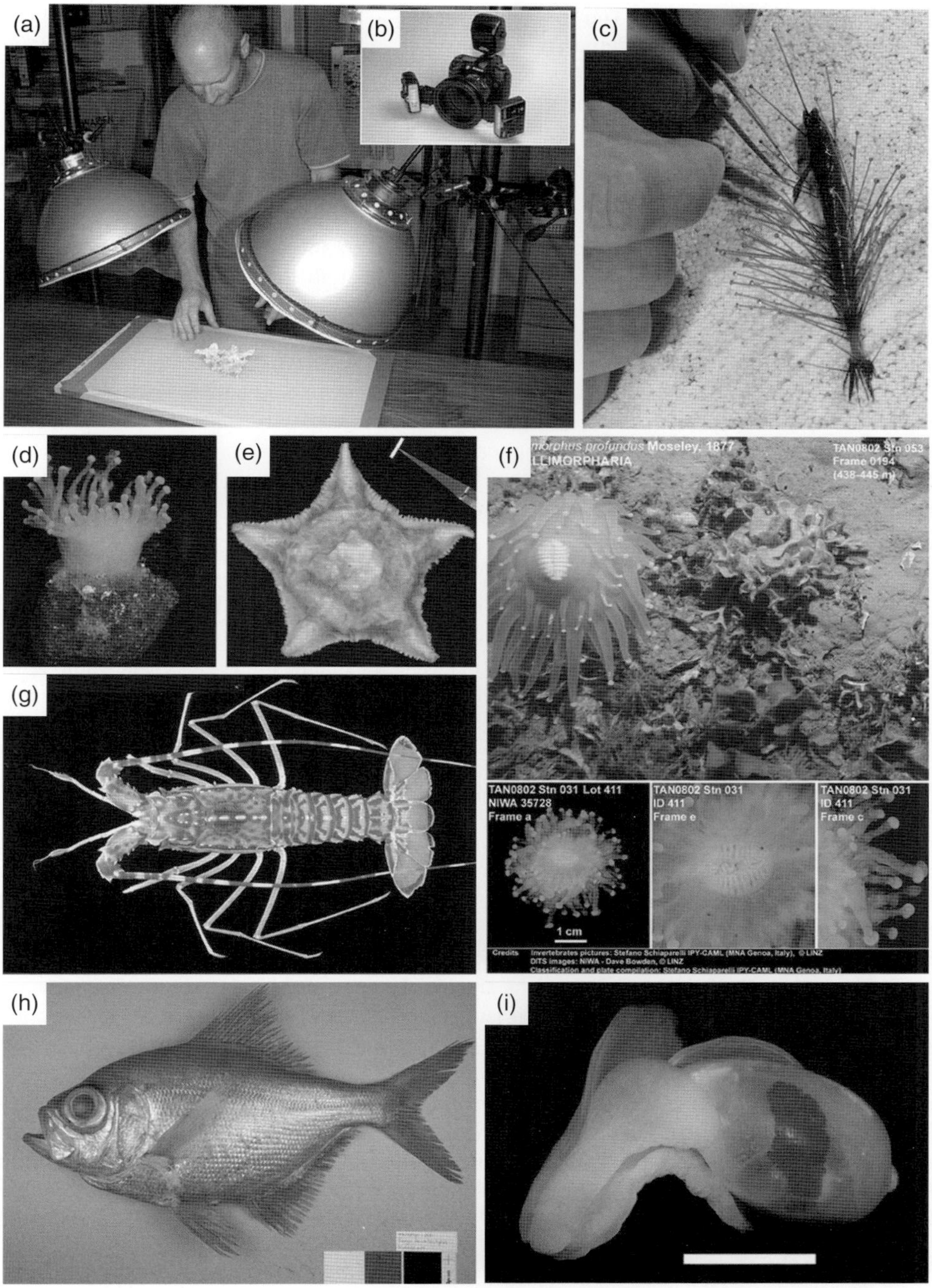

Fig. 15.4 Examples of different photographic set-ups and of images taken on board. (a) Photographic set-up for large species in air. An example of a picture taken with this solution is given in (h). (b) To photograph small species in seawater, placed in containers with a black background, it is possible to use two small orientable flashes triggered by a wireless infrared commander. Examples of pictures taken by using this solution

container holding the specimen. The specimen can be placed in a Petri dish filled with seawater and with a black velvet cloth underneath, or in a small aquarium with the bottom and sides covered by a black adhesive tape. To secure the container to the bench, and to reduce the effects of the ship's vibration, the container can be fitted in a space cut in a ~5 cm thick high-density polystyrene tablet attached to the bench. It is also possible to fix the flashes used to illuminate the container on the same polystyrene tablet. Two to four remote control flashes placed opposite each other are used to avoid the casting of shadows (although some camera models have a flash mounted directly above the lens (Fig. 15.4b)). In calm weather conditions it may be possible to achieve a more homogeneous and clean black background by suspending a glass aquarium above a box whose internal sides are covered with black velvet cloth. In both cases images are taken from directly above the specimen, and flashes need to be oriented at 45° angles in respect to the water surface in order to avoid any mirroring of the light. The same applies if images are taken from one side of the aquarium.

Species photographed in water are usually still alive and may release mucus, sediment or noxious compounds following the stress of capture. It is therefore necessary for high-quality images to frequently change the water to ensure it is optically clear.

15.6.3 Photographing large specimens in air

For large specimens, a larger set-up is used. A well-proven set-up is composed of a photo stand (95 cm high), a clear sheet of glass (80 × 35 cm and 5 mm thick), two cardboard boxes (about 20 × 20 × 30 cm), a large piece of black velvet cloth (about 90 × 150 cm), and two flashes with individual tripods and diffusers. The glass is placed on the two cardboard boxes with the velvet cloth between them to provide a diffuse, solid black background. The specimen is placed in the middle of the glass plate without water. Flashes are positioned on the sides next to the cardboard boxes, at the proper angle to eliminate reflection. The glass must be very clean and free of water marks, which can be difficult to achieve. A sponge soaked in a photographic wetting agent (100 times dilution with freshwater) is used to eliminate water marks and dirt on the glass. However, water marks will reappear quickly in about 1 minute when the glass dries. Further, in many specimens (e.g. in particular shrimps) the integument dries quickly, producing white marks. These can be eliminated using a wet sponge but will reappear quickly. It is thus essential that the focus, lighting, exposure and composition of the image are pre-adjusted as much as possible so that the image is taken quickly. Recommendations for improving photography of specimens are given in Box 15.7.

To obtain a clear image with good depth of field, generally an aperture of *f*/16 or higher is required for specimens less than 10 cm – the smaller the specimen the higher the *f*-stop required. High-output flashes (which tend to be larger and heavier) are thus better, but space can be limited onboard. Balanced lighting from the flashes is also essential. Depending on the shape of the

are in (d), (e) and (i). (c) Photography of fishes requires specific preparation (see Box 15.7 for details). (d) A small coralliomorpharian species still adhering to the pebble on which it has been sampled. It has been photographed in water, in a tray lined with adhesive black background. Note the extended tentacles and its almost 'live appearance'. (e) An asteroid (*Porania* sp.) photographed with the same technique. The scale (1 cm) is a piece of Resistall paper kept in place by inverted forceps. (f) An example of a 'working field guide' page. In the lower part of the plate, there are images taken in the laboratory as shown in (d). (g) A lobster (*Puerulus angulatus*) of about 15 cm body length photographed in air using a black velvet background. Note that the image is a direct result from the camera without any digital editing. (h) A large fish photographed in air. Also included are the photographic colour scale (which will allow a perfect colour balance in post processing) and the specimen label. (i) An example of a small Antarctic mollusc (*Neactaeonina*) photographed in water. In this picture the scale bar, originally captured as in picture (e), was replaced in post production by a digitally added one. (Please see Plate 26 for colour representation of the figure.)

specimens, the flashes are often set to a different output level, and their positions and angles are adjusted. Although each flash is equipped with a diffuser, often reflections are still strong on the body of the specimen (and sometimes on the glass) because of the high output required for an adequate depth of field. To minimize these reflections, use a linear polarizing filter on the lens and polarizer sheets attached to the diffusers. However, both polarizer filter and sheets significantly reduce the amount of light entering the lens and can cause 3–5 *f*-stop differences. Thus, more powerful flashes are needed for taking photographs with minimum reflections from the specimens. The sheet of glass on the cardboard boxes suspends the specimen well above the black velvet and ensures a smooth black background that is out of focus. Specimens larger than 30 mm can be put directly on the black velvet cloth (alternatively cheaper polar-fleece type cloth) (Fig. 15.4g). Because of the longer distance between the lens and the specimen, a smooth black background also can be obtained without using the glass.

A simpler system is to use photographic-quality neutral grey or neutral black plastic background sheets (Fig. 15.4a). These can be wiped clean, rolled up and stored in poster tubes when not in use. They can be purchased in sheets about 1.5 × 1.0 m in size so can be used to photograph large

Box 15.7 Specimen photography

- In order to provide adequate lighting and to eliminate undesirable shadows use at least two orientable flashes or well-diffused fluorescent floodlights for larger specimens. Black velvet cloth can eliminate all shadows but requires protection from water by using a glass or should be changed frequently.
- Flashes should be inclined at a 45° angle to avoid reflecting on glass or water surfaces.
- Prepare a series of white, grey and black backgrounds in advance (including a full set of replacements as they need to be changed at regular intervals).
- Species should be photographed alive or in the best possible condition. Consider relaxation (e.g. with $MgCl_2$ (all taxa) or brief storage at −20 or 4°C (for Crustacea).
- Soft-bodied invertebrates should be photographed in water.
- A scale bar should always be used (in water, this can be kept in place by inverted forceps).
- Photography of fish requires special preparation. This involves pinning the fins to a polystyrene board and then painting them with concentrated formalin (37% formaldehyde) in a chemical hood (Fig. 15.4c). After 5–10 minutes, specimens are rinsed, pins removed and the specimen photographed on a neutral grey background (Fig. 15.4h). With the exception of right-eyed flounder, fish are always imaged laterally with the head to the left, preferably with the mouth slightly open. It is useful to image both sides of flatfish and rays.
- Also image the dorsal aspect of fish as this can be useful for the interpretation of video and stills collected by towed cameras and ROVs (e.g. Stein *et al.*, 2006).
- Photography of some invertebrates may require careful positioning of the specimens, separating appendages or tentacles and a cleaning of the body surface from detritus and mucus with a soft brush in water.
- Crustaceans require specific views and standard positions. Decapods in particular (e.g. shrimps) should be photographed in lateral view with the legs separated as much as possible, the abdomen not flexed tight, displaying eggs under the female abdomen if they are present. Hermit crabs should also be photographed outside their shells. Dorso-ventrally flattened decapods such as crabs and lobsters are photographed dorsally with all appendages spread out as much as possible (Fig. 15.4g).
- Always include at least one image that shows the specimen label for reference.

specimens which are laid directly on the background sheets. These sheets do not produce reflections from the specimen, but are prone to producing unsightly shadows. The background can easily be changed to white, blue, neutral density board or any other colour to enhance the specimen's natural colour, although photographic grey is preferred in scientific images.

15.6.4 Organizing a database of images

The innovation of camera-to-computer acquisition software now allows streamlined image capture and automatic registration with a progressive acronym or number. However, if saved to a flashcard or disk, all images when transferred to a computer can be immediately renamed in batch by using available software (e.g. ACDSee for PC or Bridge® CS3 for Mac), clearly indicating station and all the relevant data. These software packages also allow access to the metadata (International Press Telecommunications Council (IPTC) core) where further information can be stored (e.g. name of the voyage, copyright holder, funding agency, author of the picture, confidence of classification of the species, etc.).

15.7 Sample storage and transport

Most of the chemicals used to fix biological specimens on board research vessels are considered Dangerous Goods (and marine pollutants); storage and shipping of these substances must follow current national and international regulations. Recommendations for the Transport of Dangerous Goods are made by the United Nations Economic and Social Council's Committee of Experts on the Transport of Dangerous Goods Regulations (www.unece.org/trans/danger/publi/unrec/12_e.html), and are implemented through regulations such as those of the International Maritime Dangerous Goods (IMDG) Code (www.maritimenz.govt.nz/Rules/List-of-all-rules/Part24A-maritime-rule.asp), or the International Civil Aviation Organization (www.icao.int/) that publishes the International Air Transport Association (IATA) regulations annually. In addition, specific local health and safety requirements may apply.

15.7.1 Short-term (onboard) storage of samples

As a general rule, hazardous chemicals should always be clearly labelled and stored separately in an area with adequate ventilation and equipped to deal with spills. Ideally, the ship is equipped with chemical safety storage cabinets but alternatively, securing large drums or bins on deck, as much out of the elements as possible, should provide best access and ventilation and can be used to both store the bulk fixative and the sample containers. Storing chemicals inside the vessel should only be considered with caution and only if the space has been specifically designated as a chemical-safe laboratory. Always consult the MSDS for storage, emergency response and transport recommendations and notify relevant staff of any handling requirements.

Sample containers with liquid preservative should be tightly sealed (providing primary containment) to prevent leakage or evaporation. It is recommended to check seals and/or use liners or inserts inside the lids of smaller jars (usually made from compressible polyethylene) in order to provide a good seal and long-term vapour barrier, and prevent sample loss through desiccation. Containers should be stored upright to prevent leakage.

Local health and safety regulations commonly require secondary containment measures that depend on the container size and quantity of liquid stored (total pooling potential or TPP). Consequently, sample containers should be stored in a sealable secondary containment vessel that will collect any liquid that may leak and will typically need to capture between 25–100% of the liquid. The minimum capacity requirement for secondary containment may differ between local authorities, and should be checked before the voyage.

Storage location for frozen samples is most likely pre-determined by available facilities on board. They should be located away from any food freezers to avoid any confusion about their purpose. Ideally, large 'walk-in' freezers designed specifically for scientific purposes should be used to store frozen biological material. These should have secure shelving to enable safe and easy access to samples. When using these types of freezers scientific personnel should wear protective clothing when handling frozen samples to prevent frost bite and adequate footwear as walk-in freezers may have slippery floors. It is important to regularly check power supply to all freezers.

Frozen samples should be stored in containers that can be quickly moved out of the freezer and off the ship and into a freezer unit on shore. Cardboard boxes with plastic bag liners have proven adequate and cost-effective for this purpose. The boxes need to be clearly marked on the outside for easy location post voyage.

15.7.2 Sample transport and shipping

All samples will require transport to their final destination following completion of the voyage. This can be a complex task depending on the location of the voyage. First, samples need to be offloaded and transferred by road to a nearby location for post-voyage processing; in the simplest cases, this may also be their final destination, but they may need to be shipped onwards to other institutions for analysis and/or identification, which involves further shipment by road, sea or air freight. In all cases specific national and international regulations and costs must be considered.

Shipping frozen samples is risky and is usually prohibitively expensive. The most practical and cost-effective way of transporting small quantities long distance is to hand-carry them by air. Most airlines will allow transport in the cabin of 'non-hazardous perishable items' cooled by a set weight of dry ice in a vented cooler box. It is highly recommended to contact the airline directly and request a formal letter of approval to show to ground staff before departure. Also note that frozen samples may be subject to quarantine regulations at their destination and permits may be required.

Shipment of dangerous goods (e.g. ethanol and formalin) requires specific packing and handling (Appendix 15.2) and should be accompanied by declaration forms that clearly state the type of substance (including hazard class, UN number for identifying hazardous substances (assigned by the United Nations Committee of Experts on the Transport of Dangerous Goods) and packing group), weight and volume and full contact details for enquiries. Staff involved in packaging, handling or transporting dangerous goods (DG) may require specialized DG handling and packing training and certification; a special driver's licence may also be required if large volumes of chemicals are transported on the road. Enquire with local authorities about certification and documentation requirements. Shipping DG by air is highly regulated and rules frequently change. Refer to an up-to-date issue of the IATA regulations and contact local international freight agents for details on limits and documentation of all substances (see Appendix 15.2). In all cases, ensure the containers are well sealed and that the specimens remain moist. Draining liquid for the purpose of shipping reduces the weight and cost, but specimens should never be allowed to dry in transit and need to be re-hydrated as quickly as possible after arrival; this is not recommended for transport by sea freight, which can involve lengthy durations. Some specimens may also require further permits. For example, CITES-regulated specimens can be exchanged between CITES-registered institutions and a permit needs to be included in every consignment (see www.cites.org).

Finally, it always helps to carry or attach a letter containing a formal statement from the collector (institution or museum) clearly stating the nature, purpose, permits (if needed) and destination of the samples, on letterhead and with contact details for enquiries (see Appendix 15.3).

15.7.3 Long-term (collection) storage of samples

All samples should ultimately be destined for deposition in an internationally recognized specimen collection such as those maintained by natural history museums. If it is not possible to place specimens in such a collection, then at least a set of voucher or reference specimens should be sent to a permanent collection maintained by a scientific institution. The benefits of depositing specimens in recognized collections are: (i) safe-keeping, as natural history collections are normally required to maintain samples in good condition 'in perpetuity', (ii) visibility and access to a network of specialists for authoritative identification of the samples and for other research projects, which may provide repeat identification and can contribute to other projects (Wilson, 2005). These samples will also support future research with increasingly comprehensive collections. However, curatorial staff of the collections should be consulted well in advance to ensure that resources are available (e.g. staff time, storage space and consumables). If possible, include a budget for consumable supplies that can help offset the often considerable costs of curation, especially if the volume of material is large. In addition, issues of ownership, intellectual property rights, deposition of types and access to samples by third parties need to be formalized well before work begins. Useful resources and access to a network of collections experts is provided by the Society for the Preservation of Natural History Collections (e.g. www.spnhc.org).

It may be necessary to transfer samples to more suitable long-term storage containers before they are deposited in a long-term collection. Plastic bags and containers that have often travelled months on board ship, exposed to the effects of high temperature and/or corrosive effects of the preservative, can become quite brittle and quickly ineffective. Consequently, for long-term storage of samples, the containers will have to be replaced with archival-quality glass jars and closures, which is costly. These final steps, however, usually fall within the responsibility of the museum/institution in charge of the collection and are often outside the remit of the organization responsible for the research voyage.

Acknowledgements

Sadie Mills and Peter (Chazz) Marriott from NIWA have provided valuable discussion and comments on the manuscript. Karen Gowlett-Holmes, J. Galeron and one anonymous referee provided invaluable advice which improved an original draft. K.S. was funded by NIWA through the Ministry for Primary Industries (project C01X0219 Marine Biodiversity and Biosecurity).

Appendix 15.1 Example of forms that help sorting staff with consistent taxonomic identification, recording, and preservation standards

These are forms used by NIWA in New Zealand.

version: Apr 2010

Scientific Name	CODE	Common name	How to fix ben thic specimen?	
PORIFERA	ONG	sponges	**freeze**	
CNIDARIA	COE	unknown cnidarians	F / E [1,2]	**freeze bulk**
	COU	unknown corals	E [2]	
Alcyonacea	SOC	soft corals	E [2]	**freeze bulk**
Gorgonacea	GOC	sea whips, sea fans, bamboo/bubble-gum corals	E [2]	**freeze bulk**
Pennatulacea	PTU	sea-pens	E [2]	**freeze bulk**
Scleractinia	SIA	stony corals	E [2]	
Actinaria	ANT	anemones	F / E [1,2]	**freeze bulk**
Antipatharia	COB	black corals	E [2]	**freeze bulk**
Scyphozoa, etc.	JFI	jellyfish (medusae)	*voucher:* F	**freeze bulk**
Hydrozoa	HDR	hydroids	E [2]	**freeze bulk**
Stylasterida	COR	hydrocorals	E [2]	
CTENOPHORA	ZTE	comb-jellies	F	**freeze bulk**
MOLLUSCA	MOL	unknown mollusca	Z	
Polyplacophora	CHT	chitons	Z	
Bivalvia	BIV	clam-type shells	Z	
Gastropoda	GAS	snail-type shells	Z	
Nudibranchia	NUD	nudibranchs	F / E [1,2]	
Scaphopoda	SPH	tusk shells	E	
Cephalopoda	OCP, SQX	octopods and squid	**Z in salt water (small spm in F)**	
ARTHROPODA				
Pycnogonida	PYC	sea-spiders	E	
Crustacea	CRU	unknown crustaceans	E	
Copepoda	COP	tiny planktonic forms	E	
Cirripedia	BRN	barnacles	E	
Ostracoda	OST	seed shrimps	E	
Amphipoda	APH	hoppers	E	
Isopoda	ISO	sea slaters	E	
Tanaidacea	ZAA	tanaids	E	
Paguroidea	PAG	hermit crabs	E	
Brachyura	CRB	crabs	E	
"Natant decapods"	NAT	shrimps and lobsters	E	
Galatheoidea	GAL	squat lobsters	E	

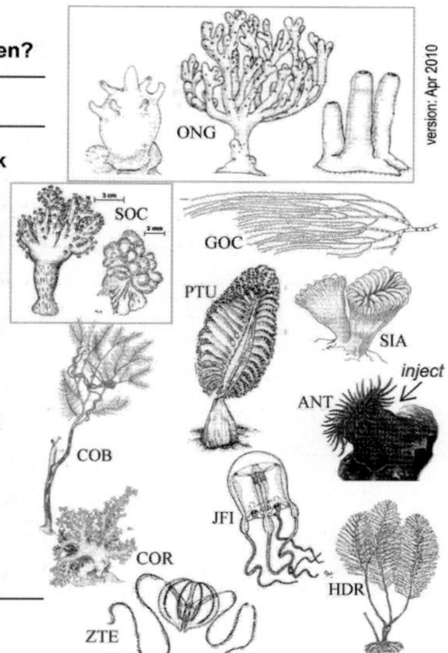

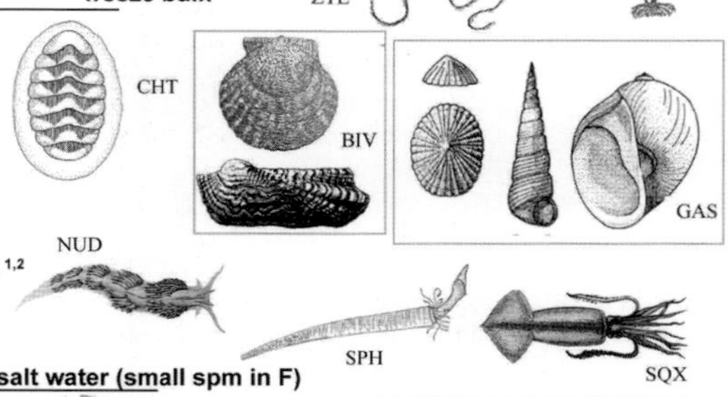

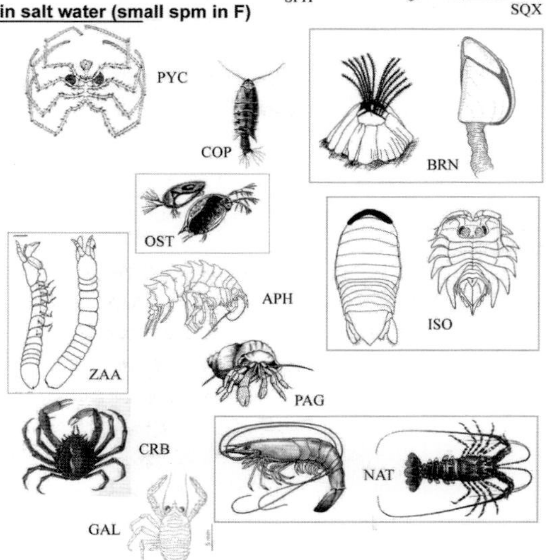

E = 95% ethanol, F = buffered 10% formalin (4% formaldehyde), Z = freeze, Y = Dry, D = discarded

Scientifc name	CODE	Common name	How to fix specimen
"WORMS"	WRM	unknown worm	**F / E** [1,2]
PLATYHELMINTHES	FWA	flat worms	brush onto frozen formalin in flat dish, then fix in ethanol
ANNELIDA		ringed worms	
Polychaeta	POL	bristle worms	**F / E** [1,2] (all in 1 jar)
Hirudinea	LEH	leeches	**E** [2]
ECHIURA	EHI	spoon worms	**F / E** [1,2]
SIPUNCULA	SIP	peanut worms	**F / E** [1,2]
NEMERTEA	NMT	ribbon worms	**F / E** [1,2]
PRIAPULIDA	PDL	penis worms	**F / E** [1,2]
BRYOZOA	COZ	cornflake coral/lace coral	**E or Y**
BRACHIOPODA	BPD	lamp shells	**E**
ECHINODERMATA	ECH	unknown echinoderms	**E** } freeze bulk in plenty sea water
Crinoidea	CRI	sea lilies/feather stars	**E** } freeze bulk in plenty sea water
Asteroidea	ASR	starfish/seastars	**E** } freeze bulk in plenty sea water
Ophiuroidea	OPH	brittle-stars	**E** } freeze bulk in plenty sea water
Echinoidea	ECN	sea urchins	**E** } freeze bulk in plenty sea water
Holothuroidea	HTH	sea cucumbers	**E** } freeze bulk in plenty sea water
TUNICATA	ASC	sea squirts	a) *Colonial*: Relaxed (menthol crystals/MgCl2) **voucher in F, freeze bulk**; b) *Solitary*: **Voucher in E, ethanol-fix bulk**
Thaliacea	SAL	salps	**F** **ethanol-fix bulk**
ALGAE	SEO	unknown seaweed	**F** [1] **freeze bulk**

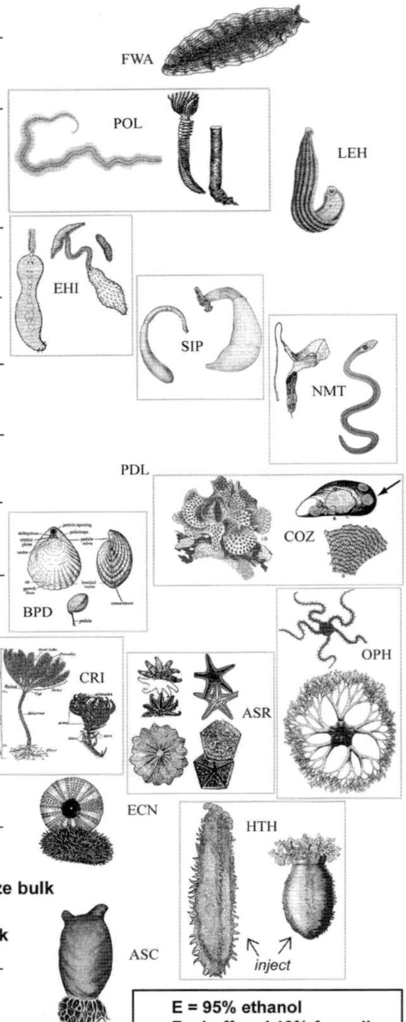

E = 95% ethanol
F = buffered 10% formalin (4% formaldehyde)
Z = freeze
Y = Dry
D = discarded

fixative : specimen, at least 5 : 1

Large specimens: use syringe or knife to get ethanol in the tissue (esp. ANT, HTH, NUD, ASR, ECN)

[1] If several specimens of the same species are caught, place half in 95% ethanol

[2] Relax specimen before fixing when time allows.

version: Apr 2010

Appendix 15.2 Shipping of samples in ethanol or formalin

Ethanol

Ethanol (ethyl alcohol, UN 1170, Class 3) at concentrations above 24% is considered to be a flammable liquid and consequently subject to specific shipping and handling regulations. The recommendations below are based on the 2014 version of the Dangerous Goods Regulations published by IATA. When shipping samples for the first time, it will be important to contact intended shipping companies to check whether further regulations may apply. Note that the person packaging samples may need to have undergone special training.

Temporarily transfer specimens from high-percent ethanol after proper fixation for at least 3 days into an aqueous solution containing ≤24%. If specimens are well fixed, they do not appear to show any adverse effects to the temporary change of concentration (Williams, 2007) but they should be transferred to high-concentration ethanol as quickly as possible on arrival. The consignment needs to be clearly labelled and a declaration for the border authorities need to accompany the samples (see Appendix 15.3 for example wording that has proven useful).

As of 1 January 2011, the IATA Dangerous Goods Regulations include the following Special Provision (A180): non-infectious specimens, such as specimens of mammals, birds, amphibians, reptiles, fish, insects and other invertebrates containing small quantities of UN 1170 [Ethanol], UN 1198 [Formaldehyde solution], UN 1987 [Alcohols], or UN 1219 [Isopropanol] are not subject to these Regulations provided the following packing and marking requirement are met:

(a) The specimens are:
  (1) wrapped in paper towel and/or cheesecloth moistened with alcohol or an alcohol solution and then placed in a plastic bag that is heat-sealed. Any free liquid in the bag must not exceed 30 mL; or
  (2) placed in vials or other rigid containers with no more than 20 mL of alcohol or an alcohol solution;
(b) the prepared specimens are then placed in a plastic bag that is heat-sealed;
(c) the bagged specimens are then placed inside another plastic bag with absorbent material then heat-sealed;
(d) the finished bag is then placed in a strong outer packaging with suitable cushioning material;
(e) the total quantity of flammable liquid per outer packaging must not exceed 1 litre; and
(f) the completed package is marked 'scientific research specimens, not restricted Special Provision A180 applies'. The words 'not restricted' and the special provision number A180 must be included in the description of the substance on the Air Waybill.' (p. 358, IATA Regulations, 55th edn, 2014, section 4, p. 373, www.iata.org/publications/dgr/Pages/index.aspx).

If neither of these two options are possible, samples need to be shipped classified as a Dangerous Good, which requires very specific packaging, is very expensive and requires using freight forwarders.

Formalin

Concentrated formalin is a solution containing 37.5% of formaldehyde, which is classified a DG (UN 2209, Class 8, Corrosive). Dilute formaldehyde solutions containing formalin concentrations between 10 and 25% are also regulated (UN 1198, Class 3 (8), Flammable liquid with subsidiary Corrosive risk). However, formalin at concentrations commonly used as a fixative (<3.75%) does not need to be declared as a DG. Nevertheless, it is an odorous and irritating substance and should be as diligently packaged as indicated above for ethanol.

Appendix 15.3 Recommendations for the completion of a shipping letter (adapted from the Australian Quarantine and Inspection Service, AQIS) for shipping ethanol by air

A 'Manufacturer's Declaration' must be:

- on institution letterhead (including address and country),
- written in English,
- signed by a senior member of staff whose name, title and contact details appear,
- dated and free from erasures and alterations,
- specific to the consignment by referring to a consignment number or tracking number.

The declaration should state:

- exactly what the specimens are (i.e. scientific name and common name),
- the nature of the material (i.e. whether they are 'whole specimens', 'tissue samples', 'DNA samples', etc.),
- their intended use (i.e. scientific research),
- what they are: 'non-viable preserved museum specimens',
- which preservative and concentration the specimens are currently packed in (e.g. 70% ethanol, 24% ethanol, etc.).

If the specimens are being shipped in <24% ethanol, the following statement should also be included: 'This consignment contains museum specimens which are normally preserved in 70% ethanol. They have been temporarily transferred to <24% ethanol for shipment only in order to comply with IATA 'Dangerous Goods' regulations. They will be returned to 70% ethanol immediately upon receipt.'

And it is useful to add the following: 'The sample is of no commercial value but is irreplaceable. If inspected, it is imperative that the specimens remain sealed in their plastic containers and kept in a wet condition at all times. If allowed to dry out, the specimen(s) would become useless for museum scientific research. If damaged, return postage is guaranteed. Please contact me directly if you have any queries.'

References

Bucklin, A., Ortman, B.D., Jennings, R.M., *et al.* (2010). A 'Rosetta Stone' for metazoan zooplankton: DNA barcode analysis of species diversity of the Sargasso Sea (Northwest Atlantic Ocean). *Deep-Sea Research II*, 57: 2234–2247.

Buhl-Mortensen, L., Vanreusel, A., Gooday, A.J., *et al.* (2010). Biological structures as a source of habitat heterogeneity and biodiversity on the deep ocean margins. *Marine Ecology*, 31: 21–50.

Danovaro, R. (ed.) (2010). *Methods for the study of deep-sea sediments, their functioning and biodiversity*. Boca Raton, FL: CRC Press, Taylor & Francis Group.

Ellwood, M.J., Kelly, M., Richer de Forges, B. (2007). Silica banding in the deep-sea lithistid sponge *Corallistes undulatus*: Investigating the influence of diet and environmental conditions on growth. *Limnology and Oceanography*, 52(5): 1865–1873.

Etnoyer, P., Cairns, S.D., Sanchez, J.A., *et al.* (2006). *Deep-sea coral collection protocols*. NOAA Technical Memorandum NMFS-OPR-28. Silver Spring, MD, USA.

Gage, J.D., Bett, B.J. (2005). Deep-sea benthic sampling. In: Eleftheriou, A., McIntyre, A. (eds). *Methods for the study of marine benthos*. Oxford: Blackwell Science, pp. 273–325.

Ginger, M.L., Billett, D.S.M., Mackenzie, K.M., *et al.* (2001).. Organic matter assimilation by holothurians in the deep-sea: some observations and comments. *Progress in Oceanography*, 50: 407–422.

Montagna, P., McCulloch, M., Taviani, M., Remia, A., Rouse, G. (2005). High-resolution trace and minor element compositions in deep-water solitary scleractinian corals (*Desmophyllum dianthus*) from the Mediterranean Sea and the Great Australian Bight. In: Freiwald, A., Roberts, M. (eds), *Deep-water corals and ecosystems*. Berlin, Heidelberg: Springer-Verlag, pp. 1109–1126.

Pailleret, M., Haga, T., Petit, P., *et al.* (2007). Sunken wood from the Vanuatu Islands: identifications of wood substrates and preliminary description of associated fauna. *Marine Ecology*, 28: 233–241.

Rosenberg, G. (2005). Archiving standards for the Census of Marine Life. http://researchdata.museum.vic.gov.au/amit (accessed August 2015).

Samadi, S., Corbari, L., Lorion, J., *et al.* (2010). Biodiversity of deep sea organisms associated with sunken-wood or other organic remains sampled in the tropical Indo-Pacific. *Cahiers de Biologie Marine, Roscoff*, 51: 459–466.

Shank, T.M. (2010). Seamounts: deep-ocean laboratories of faunal connectivity, evolution, and endemism. *Oceanography*, 23(1): 108–122.

Stein, D.L., Drazen, J.C., Schlining, K.L., Barry, J.P., Kuhnz, L. (2006). Snailfishes of the central California coast: video, photographic and morphological observations. *Journal of Fish Biology*, 69: 970–986.

Syväranta, J., Martino, A., Kopp, D., Céréghino, R., Santoul, F. (2011). Freezing and chemical preservatives alter the stable isotope values of carbon and nitrogen of the Asiatic clam (*Corbicula fluminea*). *Hydrobiologia*, 658(1): 383–388.

Williams, S.T. (2007). Safe and legal shipment of tissue samples: does it affect DNA quality? *Journal of Molluscan Studies*, 73(4): 416–418.

Wilson, R. (2005). Marine invertebrate sample processing procedures. http://researchdata.museum.vic.gov.au/amit/Marine_sample_processing_8Sep2005.pdf (accessed August 2015).

Chapter 16
Information Management Strategies for Deep-Sea Biology

Karen I. Stocks[1], Nancy Jacobsen Stout[2] and Timothy M. Shank[3]*
[1] San Diego Supercomputer Center, University of California San Diego, La Jolla, United States of America
[2] Monterey Bay Aquarium Research Institute, Moss Landing, United States of America
[3] Woods Hole Oceanographic Institution, Woods Hole, United States of America

Abstract

Good data management practices are becoming increasingly important in ocean research and allowing more effective integration and reuse of data. This chapter outlines best practices for developing and implementing a data management plan for deep-sea biological research. Topics include determining what data to collect, selecting a metadata standard, organizing and representing data effectively, undertaking quality assessment, selecting a data archive and data repository, and intellectual property concerns. Special considerations for species occurrence data, image data and genetic sequence data are presented.

Keywords ocean data, data management, database, biological data, information system design, metadata, image(s), video, genetic sequences, species occurrence data

16.1 Introduction

The landscape is quickly changing for scientific data management. New instrumentation and satellite connectivity have created unprecedented volumes of data. It is estimated that the volume of scientific data doubles every year (Szalay & Gray, 2006). But advances in data management software, storage devices, and remote internet connectivity now allow data to be collected and managed in adaptable and novel ways.

Funding agencies and scientists are also changing their data management expectations. Increasingly, large programmes and funding agencies require that data be rapidly shared after collection. Researchers now need to store and document their data not only for their own use, but also for future users who might be unfamiliar with – or have different purposes for – the data. Although this demonstrates progress (many large-scale, multidisciplinary research questions can only be addressed by aggregating, integrating and synthesizing data across many studies), it also places new and unfamiliar requirements on scientists.

This chapter gives an overview of some of the data and information management considerations for deep-sea biological research. It begins with a general discussion of the spectrum of information to collect and how to plan an information management process. Special considerations are then outlined for specific data types commonly collected on expeditions: species occurrence data, image data and molecular data. Many of the principles will remain stable through time, but specific technologies change,

*Current affiliation: Scripps Institution of Oceanography, La Jolla, CA, United States of America.

Biological Sampling in the Deep Sea, First Edition. Edited by Malcolm R. Clark, Mireille Consalvey and Ashley A. Rowden.

particularly the accepted data and metadata standards and formats, and available data repositories. Resources for current best practices include national oceanographic data centres, the Global Biodiversity Information Facility (www.gbif.org), Ocean Data Practices (www.oceandatapractices.net), and ESIP (commons.esipfed.org/datamanagementshortcourse).

Throughout this chapter, the terms *data* and *information* are used. The word data is generally defined as facts such as measurements or statistics that have not been interpreted. Information is the knowledge extracted from data. However, the boundary between the terms is blurred and they are often used interchangeably. This chapter does not follow strict definitions, and uses the terms more or less synonymously.

16.2 General information management considerations

It is typically clear what kinds of primary data will be collected on any given research voyage: a study of hydrothermal vent fauna abundances, for example, will collect quantitative data on the abundances of fauna within samples from vent systems. However, in order to interpret the primary data, a great deal of additional information generally needs to be captured. Below are some considerations and recommendations regarding what data to collect in support of the primary data and how to plan and maintain a data management programme.

16.2.1 Data to collect

16.2.1.1 Unique identifiers

Identifiers supply the connections between different pieces of data collected on an expedition. For example, it is important to know the location, date, time, depth and method of collection for each sample, plus the vessel, expedition name and chief scientist. But a taxonomist identifying species within a sample will not want to repeat all of this information on each one of their data sheets – they will want to refer to a sample number. This sample number is an identifier (sometimes called an indexer) – a term or code embedded in one set of data that references information held in another set of data.

The Global Biodiversity Information Facility has recommended using the following identifiers (Kelling *et al.*, 2008). Although these recommendations were developed for species occurrence data, they apply to most expedition data:

- *Project code*. This is a short name for the project being conducted.
- *Sampling event identifier* (also commonly called a sample number). A sample is, for example, one trawl tow, one benthic core or one filtered water sample. In cases where the sampling gear takes replicates or subsamples, such as a vegematic box core, a Niskin bottle rosette, or a frame grab from a video, each sample should have a unique identifier, but they should be constructed so that the relationship of replicates or subsamples to the whole is clear (e.g. Niskin bottle '4-B' might indicate the second bottle on the fourth Niskin rosette; TAN0104_228_3 might indicate the third subsample taken from sample 228 on Tangaroa voyage 104).
- *Survey area identifier* (also commonly called a site name). This indicates a geographic area in which sampling is conducted. It is used to relate multiple samples taken at the same site, such as a benthic core, a trawl and a CTD cast, which have slightly different latitudes and longitudes but are designed to sample the same location. It is also used to relate time-series data taken at the same location over time.
- *Protocol identifier*. This indicates either the sampling or processing method. Each method can be described fully in one part of the data set, and all samples taken or processed in that same way coded with the same protocol identifier(s). A sample may have multiple protocol identifiers – one for the collection of the sample, one for the identification or analysis, one for the quality control procedures applied, etc.

All parties should agree on, and consistently use, the same project code, site names, sample numbers and protocol identifiers. They should be unique within the project. For example, there should only be one sample named '1a' within each project, one protocol 3, etc. Even small differences can generate discrepancies: if chemists number a multi-core sample as 3-a and biologists call it 3A, it may be clear to a person that these represent the same sample, but they cannot be integrated automatically.

It is highly recommended that an event log be used to ensure consistent assignment of identifiers across all sampling activities. A good event log records every sample or measurement taken during an expedition into one master file and includes sample identifiers, the type of sample, time and location, event type (e.g. deployment, recovery, start of an event, etc.), name of the responsible person, and a comments field. An event log also helps ensure the standardization and quality control of metadata. The Rolling Deck to Repository project in the US has developed an open-source event logger for ocean research vessels (www.rvdata.us/about/eventlog) and similar efforts are underway in other countries.

16.2.1.2 Metadata

Metadata are the additional information needed to find, understand, and use a data set – they are the 'data about data'. For expedition data, metadata commonly include the name of the vessel, the voyage the project, the responsible institution(s), the chief scientist and other investigators, the sampling location, the sampling protocol identifier(s), the precision of data presented, references to relevant publications, contact information and any use restrictions placed on the data. Column headers in a spreadsheet, such as 'Depth (m)', can also be thought of as metadata.

Formal metadata standards have been developed to define the information that should be included about a data set and, in some cases, the format in which that information should be expressed. These standards can also be considered a set of rules for creating metadata. They are created by acknowledged authorities and are formally expressed (as opposed to metadata specifications, which may be more informally created within a project or group). A standard can be as simple as specifying that if a water velocity is recorded it be expressed in m/s, or as complicated as the ISO 19115-2 standard that defines hundreds of fields that might be useful in describing geospatial data. Although implementing a metadata standard is time-consuming (and generally unfunded), it is of critical importance. Without metadata, data simply cannot be found, shared, re-used, aggregated or re-analysed. Large-scale or long-term research topics such as how climate change impacts the deep sea, or how biodiversity patterns vary globally, cannot be addressed without integrating data from across studies.

Furthermore, adopting formal metadata standards is often in the best interest of the scientists generating the data. The standards ensure that the details needed to use the information are available years later when critical aspects may have been forgotten. Increasingly, governmental funding agencies require that data meet documentation standards and be contributed to a data repository. Similarly, more scientific journals are explicitly requiring that data supporting a publication be contributed to the journal's data archive or a specified repository. Above all else, good metadata should be considered a critical step in proper scientific methodology, supporting the preservation of scientific records into the future.

Unfortunately, there is no one perfect metadata standard for all purposes. Selecting a metadata standard depends on the nature and intended use of the data, as well as any requirements stipulated by the funding agency, home institution or end repository of the data. The International Standards Organization (ISO) 19115-2/19139 (www.iso.org) standards have been widely adopted internationally, and are becoming the pre-eminent standard. Extensions or profiles that take ISO 19115 as a foundation and extend it or adapt it to particular uses have also been developed, including the Marine Community Profile. The Global Change Master Directory (GCMD, gcmd.nasa.gov) and other organizations have developed transfer methods ('crosswalks') for some of the common standards that help convert metadata from one standard to another.

Such generic standards can apply to any data with a geospatial component. Often, general metadata standards do not capture all the information relevant to the particulars of deep-sea sampling, though they ensure a basic level of documentation and wide interoperability. Ecological Metadata Language (Fegraus *et al.*, 2005, EML, knb.ecoinformatics.org/software/eml) is applicable to a broad array of biological data. Further standards tailored to specific data types are discussed within the sections on species occurrence data, image data, and molecular data. The Biological and Chemical Oceanography Data Management Office (BCO-DMO) Data Management Guidelines Manual (www.bco-dmo.org/), while not a standard, contains lists of suggested metadata to collect for biological and chemical oceanographic studies, many of which are applicable to deep-sea biological collections, and thus offers a good starting point for metadata planning.

Regardless of the standard(s) selected, metadata approaches should be determined in advance of an expedition. However, though advance planning is important, it is common for metadata to evolve in response to additional needs.

Given that standards such as ISO are complex, it is advisable to consult colleagues with data management experience, and to use existing templates (e.g. Morpho for EML, knb.ecoinformatics.org/software/eml) and informative guides and references such as those produced by Ocean Teacher (classroom.oceanteacher.org) and the Marine Metadata Interoperability (MMI) project (www.marinemetadata.org/guides). Although the full standard can seem overwhelming, it is acceptable to omit optional elements that are not relevant.

16.2.1.3 Additional best practices

Here are a few detailed recommendations, drawn from problems the authors have encountered in their experiences as data managers:

- Record the time using the ISO standard format hh:mm:ss (seconds are optional and may have decimals), and include the time zone. For UTC time (also called Greenwich Mean Time), the time zone is indicated with a 'Z' (i.e. '15:46Z'). For other time zones, give the offset from UTC time as +/– hh:mm. For example, Bermuda is 3 hours earlier than UTC, so local times should noted as 12:46–03:00. It is often difficult to determine after the fact what time zone a ship was using: last port of call, UTC and home institution's time are all common. Hence this component is critical.
- Use decimal degrees for latitude and longitude, with a preceding minus sign ('–') to indicate west or south. If degrees and minutes must be used, hold them in separate fields (such as a separate column of data), and do not include symbols for degrees and minutes with the values (i.e. not 30° 2′). Also, record the precision and method (e.g. GPS) of the measurements. If GPS is the method, record the geodetic system, or datum, used (e.g. WGS 1984). Do not rely on the number of digits after the decimal place to indicate precision – researchers frequently convert between decimal degrees and degrees/minutes/seconds, which creates spurious decimal places. Given that data from deep-sea expeditions are of interest to a broad international community and can span large spatial scales, latitude and longitude are preferable to Universal Transverse Mercator and national grid systems.
- Express depth as positive metres below sea surface.

16.2.2 The information management process

A good information management process is critical to ensure proper data collection in the field and accessibility afterwards (Fig. 16.1). Your plan should be laid out and agreed upon prior to the expedition, in written form. It should state exactly what information to collect, the data collection process (where each piece of information will come from, who will collect it, and when), the data formats and structures, and how the data will be organized. As with most large projects, the better the plan, the better the outcome. This is particularly true of open ocean work, where expensive ship time is usually scheduled as intensively as possible, making thoughtful organization particularly important.

Before the voyage

- Develop a data collection plan: what information will be collected, when and by whom? How will files be named, stored, and copied? What metadata standards will be used?
- Create data and metadata templates, and establish controlled vocabularies and critical identifiers.
- Coordinate with other science efforts on the ship and shipboard staff to ensure consistent data naming and collection practices.
- Determine and order hardware and software. Ensure compatibility among components, have redundancy in critical systems. Don't forget the peripherals: cables, batteries, converters, connectors.

⇩

On the voyage

- Collect all data and metadata.
- Use an event logger to record all science party sampling.
- Create regular backups on multiple systems.
- Don't leave the ship without all required information.

⇩

After the voyage

- Quality assess and quality control all data.
- Track the individuals responsible for different components of the data during analysis (e.g. taxonomists identifying particular taxa), and how data will be re-integrated.
- In the short term, copy data onto maintained servers, such as institutional systems, when available.
- In the longer term, submit data to a recognized repository; register data with appropriate registries and catalogues.

Fig. 16.1 The data management workflow.

16.2.2.1 Information management plan

This is the *who, what, where, when* and *how* of the data collection and management process. First, determine what types and sources of information are needed, when it will be recorded and by whom (e.g. bottom temperature recorded by the scientific observer when a sample is collected from the submersible), and how it will be captured (e.g. typed into a spreadsheet on a laptop, written onto a paper data sheet, or captured from the shipboard navigation system).

The information management plan should also cover the movement and processing of data after its initial capture. If the data need to be digitized from hand entry or moved from distributed laptops to a central core, when will that happen and who will do it? When will backups be made, and where will they be stored? What are the quality control steps? As samples are analysed, how will results be stored and maintained, and by whom? It is critical when samples are distributed among many people – such as samples for each major taxon being sent to a different taxonomic expert – to plan how those data will be reunited. Something as simple as each taxonomist using a slightly different spreadsheet format with slightly different ways of representing the sample number will mean that integrating data from across taxonomists will require time-consuming manual adjustment. The plan should also include a data-sharing policy covering the authority and timeframe for data distribution. Developing the plan will usually require some coordination with the scientific, information technology, engineering and operational personnel of the vessel and the other researchers collaborating before, during and after the expedition.

Here are a few specific tips:

- Capture data digitally when possible (e.g. by laptop or 'smart' handheld device with an embedded GPS), as humans are faulty creatures when it comes to transcribing data. For data that are generated digitally by instruments, it is better to access those data files than to transcribe the information.
- Keep all critical information together, now and forever, as metadata are useless if they become disconnected from the data. Embedded headers within text files, relational databases with multiple linked tables, spreadsheets with multiple worksheets, ZIP or TAR file bundles, and GeoTIFF for images are all good approaches, depending on the kind of information.
- Use information-rich filenames. 'Data.xls' is bad, but 'TAN0104_228_20010419_Sled.xls' to indicate data from epibenthic sled sample 228 taken on voyage 0104 of the RV *Tangaroa*, collected 19 April 2001, is good.
- Make backups, but ensure that different versions of the data have different (e.g. sequential) filenames, so there is no confusion later about which version should be used for analyses. It is also advisable to store backups on different computer systems, in different locations, in case of a system-wide crash.
- Give particular consideration for moving data from ship to shore. What information should be in hand when the expedition ends? Are the data organized so they can be copied as a whole disk/folder of data or by individual file? Consider data collected by others (e.g. ship navigation tracks or multibeam files taken by ship operators). Explicitly ask vessel operators what data they record and preserve – they may collect and later archive data from instruments operated by the vessel technicians.
- Be sure that all personnel have a thorough understanding of the information management plan as well as training on all hardware and software they will be using, especially for real-time tasks.
- Document procedures and outcomes in writing, including an inventory of all data collected and an event log of all the instrument operations conducted on the voyage.

16.2.2.2 Hardware and software

Well before the expedition, instrument deployment or other data-generating event, consider the information-related hardware needs and create a packing list. Will laptops, multiple backup hard drives, remote mainframes or other devices be used to capture data? Estimate the volume of data expected, including backup copies, to help decide what storage media are appropriate (internal drives, USB flash drives, memory cards, external hard drives, etc.). Have spares or redundancy for all critical components. Remember peripheral items such as cables, connectors, adapters, batteries and converters.

Similarly, establish a software plan. Determine the appropriate media for each data type. Plan for any additional software needed to visualize, graph, annotate or analyse the data; consider any format conversions that might be required to use those programs, including a file renaming protocol for products derived from raw data. Ensure that all hardware and software systems that need to interact are compatible. All instruments or sensors collecting real-time information should be calibrated and synchronized prior to data acquisition.

16.2.2.3 Data organization

How to organize the data will, in part, depend on the particular data type. The Biocean database provides an example structure for organizing heterogeneous expedition data (Fabri *et al.*, 2006). Here are some general pointers:

- Keep information separated into fields as much as possible. For example, hold genus, species and description authority as three separate fields, instead of one. It is always easier to combine fields than to split them later.
- Do not embed additional information into a field with a specified use; instead, use a separate column for additional information. For example, instead of putting '*Sebastes borealis* – larva' into the species field, put the fact that it was a larva into an additional notes column or one specifically for life history stage. Or instead of having 'Epibenthic Sled, failed to reach bottom' in the field for gear type, put the note about the sled not contacting the bottom in an additional notes field.
- Use controlled vocabularies (also known as taxonomies or code lists) whenever possible, and decide on them in advance of the expedition. A controlled vocabulary is a specified set of terms that forms the allowable content for a field. Allowing people to enter terms in an uncontrolled way can lead to heterogeneous entries and typographic errors (e.g. one person may enter 'Warren Dredge' in the sampling gear field, while another may enter 'Sledge: Warren' and another, just 'sled'). It is preferable to use established controlled vocabularies when they exist, for example those for research vessel names (ocean.ices.dk/Codes/ShipCodes.aspx), ocean place names (www.gebco.net/data_and_products/undersea_feature_names) and species names (e.g. World Register of Marine Species or WoRMS, www.marinespecies.org). The NERC Vocabulary Server (www.bodc.ac.uk/products/web_services/vocab/) and the MMI Ontology Registry and Repository (mmisw.org/orr) are two additional sources of marine community vocabularies. Drop-down menus, also called pick lists, can make entering data from controlled vocabularies simpler and less error-prone. They are supported in most spreadsheet and database applications.
- Leave fields blank when information is missing, instead of entering 'null' or a code like '9999' unless you have a very specific reason for doing so. Most modern data systems can handle missing values appropriately, and adding codes may create confusion for later users.

16.2.2.4 Quality assessment and quality control

After data are collected, it is important to conduct quality assessment (checking for problems) and quality control (fixing problems), all of which should be documented in the metadata as part of the data-processing protocols. Recommended steps include the following:

- Check for null (missing) data in fields that should always be populated.
- Check that all data fall within expected ranges (e.g. latitudes between −180 and 180, months between 1 and 12, etc.). Even parameters that do not have hard boundaries should be checked for outliers (e.g. deep-sea salinities, outside of a few regions, are usually between 34 and 36). Viewing the highest and lowest values for each parameter on a spreadsheet or plot is a simple way to examine outliers that may be erroneous.
- Plot location data on a map to assess potential discrepancies. Errors in latitude and longitude are very common (e.g. transposing latitude and longitude, east vs. west, or north vs. south values), as are errors in conversion from degrees and minutes to decimal degrees.

- Check taxonomic names against a reliable taxonomic resource, such as WoRMS. This will find misspellings (which are very common in taxonomic data sets) and the use of invalid synonyms. WoRMS provides web services that will check a list of names in Excel in one operation, a much more efficient approach than hand entering each name into the WoRMS online search form.

16.2.2.5 Data archival and preservation techniques

Because deep-sea sampling is extremely expensive, deep-sea data are valuable and should be preserved beyond the end of the project. Archiving copies of data onto an institution's network, where regular maintenance and backup are standard, is preferred over saving copies in an individual laboratory.

In the long term, the goal of all deep-sea data management plans should be to preserve the data in a stable, recognized data repository. The unfortunate reality is that most data are stored in a relatively haphazard, under-documented state on a private computer, where the files quickly become lost, forgotten or obsolete. Storage in a national or international data repository ensures that data persist long past the project's end date and the principal investigator's relocation or retirement.

The appropriate repository for a given data set depends on the nature of the data and the country in which the project is funded or the data obtained. When selecting a data repository, consider whether or not the repository:

- is well known and used within the discipline;
- has metadata that is sufficient to appropriately present the data;
- is required by your funding source, project or publication venue;
- places any restrictions or requirements on the use of data in its system (such as requiring that the source be recognized);
- can accommodate new versions of data (if the data set will grow over time or, for example, if species identifications may be refined);
- has long-term archival abilities, including stable funding and the practice of keeping copies of the data in different locations;
- provides Digital Object Identifiers (DOIs) for data sets. DOIs are quickly becoming a data management best practice. They allow for the unique identification and citation of a data set, as well as the ability to move the online location of a data set without having previous URL links to the data (for example in a published paper) break, as long as the DOI registry is kept updated.

The national ocean data centre in the home country, if there is one, is often a good option. The Ocean Biogeographic Information System (OBIS, www.iobis.org), the marine partner of GBIF (Global Biodiversity Information Facility), is the main repository for marine species occurrence data worldwide.

16.2.2.6 Data sharing – why share your data

Another reason for contributing data to a repository is to ensure that they can be found and re-used by other scientists. Many scientists resist the idea of sharing their data, preferring to keep it proprietary, even after the conventional 1–3 year publication window. Box 16.1 gives one example of how a lack of good scientific data management and sharing has hampered progress in deep-sea biogeography. Further arguments for sharing data widely include the following:

- Many governmental funding agencies now require that all data collected from a funded project be shared after a particular span of time. This trend is growing, as is the enforcement.
- It may be required by the scientific journal in which results are published.

Box 16.1 Lessons from SYNDEEP

A consortium of Census of Marine Life projects (see McIntyre, 2010 for project descriptions) covering various deep-sea environments launched an effort in 2008 to pool data and conduct a meta-analysis of global patterns of biodiversity and biogeography in the deep sea (SYNDEEP). The results of this data-gathering effort are illustrative of the current state of information management in deep-sea biology, the importance of moving toward a culture of open access to data, and the need for support, both with respect to funding and expertise, for scientists willing to share their data.

At a meeting in early 2009, 114 existing deep-sea species occurrence data sets were identified as potential components of the meta-analysis. Over the next nine months, SYNDEEP scientists and a supporting data manager made a concerted effort to request access to each of these data sets. Data providers were offered a data agreement showing that the data would be used solely for the proposed meta-analysis, and the data would not be further distributed in any way. All data providers and data sets would be prominently recognized in a table in the planned publication, and any relevant publications describing the data sets would be cited. Scientists providing data were not automatically given co-authorship, but were invited to participate further in the analysis and writing, which would earn co-authorship.

Of the 114 data sets, there were 14 for which a contact could not be made to request the data as the scientist with the data and/or the authority to release it either could not be identified, or could not be found (e.g. a researcher had retired with no known contact information). Seventeen contacts never replied to emailed requests. Five contacts refused to share their data, generally because they requested co-authorship solely for providing data, or because the data were not considered ready. Thirteen contacts agreed to share their data, but never actually sent it.

So, of the 114 requests, 65 data sets were provided (57%). Of those, 8 were missing critical methods information, such as sample locations or gear type used. A further 24 did not provide any sample effort information, such as core diameters or accurate start and end latitudes and longitudes to determine trawl lengths (SYNDEEP was a quantitative study). Overall, only 31 data sets provided sufficient information for analysis *despite the fact that all of this information was known at some time.* This represents 47% of those data sets received, and just 27% of the original 114.

The SYNDEEP analysis effort proceeded with a robust modelling and statistical approach, though not all of the original questions could be addressed, and large spatial and taxonomic gaps existed which the missing data sets would have filled. If all of the data sets, which had been collected at such expense and effort, had been well documented and freely shared, one wonders about the progress that could have been made in understanding the large-scale patterns of life in the deep sea.

- Contrary to fears of being ‘scooped’ or plagiarized, Costello (2009) described benefits of data sharing, including greater citation rate and increased invitations to collaborate. SAHFOS, the organization managing the Continuous Plankton Recorder data set, found a positive correlation between their funding and the amount of data they were sharing over time (Stevens *et al.*, 2006); freely releasing their data led to *more* people pursuing funded collaborations with them.
- It accelerates scientific progress. Disciplines such as astrophysics and genetics that have a culture of data sharing have seen exciting progress directly resulting from the access to data – new kinds of questions can be asked and new kinds of studies done when the majority of a field’s data are widely available. This increased progress can draw new funding to a discipline.
- It increases the efficiency of science. In large part, science is supported through public funds – funds that are often restricted and in competition with other important public needs. When data are shared, there is less unnecessary and costly duplication of sampling. Ethically, scientists have an obligation to use public funds in the most efficient and productive way.

From a legal perspective, in most cases facts do not fall under copyright. Therefore values such as temperature readings or species lists are not considered protected intellectual property; they can be used and re-used by anyone once published or communicated in any form. The particular presentation of the data (e.g. the layout and structure of a specific table) can fall under copyright, as do images and written text, but basic measurements themselves generally do not (though it is always best to check the intellectual property laws of the country in question).

16.2.2.7 Data sharing – how to share your data

The best way to share data sets is through one or more recognized repositories with a public search interface. This will ensure that data can be discovered even by those unfamiliar with a particular study. Serving data through a project or laboratory website is acceptable, but it should not replace a repository, as individual websites typically are difficult to find and are often not maintained over the long term. Distributing data through multiple repositories is also acceptable as long as the metadata includes version information so no confusion arises regarding the different versions that may reside in different repositories. It is also good practice to register data with multiple data registries (also known as catalogues), which hold descriptions and locations of available data sets, but not the data sets themselves.

Ensure the data meet the repository's metadata standards, and exploit the full suite of metadata fields to achieve the richest data description. Distribute data in a non-proprietary format, such as delimited text files, to maximize the likelihood of compatibility and longevity. Verify in advance whether the repository can replace data with new versions. For repositories that do not handle versioning well, contribute data after they are finalized, or break long-term data up into separate data sets and add pieces to the repository as they are ready. Check whether the repository forwards data to other repositories to avoid unintentionally creating multiple versions of data in separate repositories. For example, OBIS forwards all of its data, unless specifically requested otherwise, to GBIF – so submission to both OBIS and GBIF would create redundancy.

The International Ocean Data and Information Exchange (IODE) provides an Ocean Data Publication Cookbook with additional tips (Leadbetter *et al.*, 2013), as well as other data management guidance documents.

16.3 Considerations for specific data types

16.3.1 Species occurrence data

A species occurrence is any kind of data indicating that a particular organism was found in a particular location. These data can come from such sources as sampling events, a visual observation, a photograph or output from a satellite or acoustic tag. Species occurrence records are central to many biological research studies.

16.3.1.1 Data and metadata standards

The Darwin Core (DwC, rs.tdwg.org/dwc) is a widely used data standard for species occurrence data, and is one of the standards adopted by GBIF. DwC provides fields for the taxonomic name of the organism, its full Linnean hierarchy, where and when it was collected, and who identified it. Although DwC provides insufficient coverage of project-level metadata (name of the expedition, chief scientist, etc.) and of the sampling methods for most ecological studies, it is strongly recommended that any species occurrence data set include the relevant DwC fields. Not all the fields are relevant to all projects. For example, information on the accession number and museum is only appropriate for occurrences having a physical specimen. OBIS has created a minor extension of the DwC, called the OBIS schema, to better suit marine observations (www.iobis.org/data/schema-and-metadata).

The GBIF has created a GBIF Metadata Profile that contains DwC and Ecological Metadata Language (EML) components, and is recommended for all GBIF providers (GBIF, 2011).

16.3.1.2 Additional data content considerations

An explicit description of the taxonomic coverage and completeness of the data is critical for any biodiversity or biogeography use, but is frequently missing from species occurrence data. Is every organism collected in a sample represented in the data set or are there gaps? In a benthic core, perhaps only the macrofaunal groups were identified and the meiofauna excluded. Or perhaps certain groups sent out to taxonomic experts are not represented in the data set yet. A complete description of the gaps, as well as the species presence, in the data set is critical for proper interpretation of the species composition of the sample. Lastly, make note of the taxonomy and taxonomic references used.

16.3.1.3 Formats and software

Species occurrence data generally lend themselves to a tabular data format, recording species names and counts per sampling event. Spreadsheets, especially Excel, are likely the most common format used by individual researchers and small projects. They provide a good way of keeping data sets and metadata together by allowing multiple worksheets to be held within a workbook, and they also have sorting and graphing tools that are useful for data browsing and quality control.

Relational databases, such as Microsoft Access or PostgreSQL, have many of the capabilities of spreadsheets and have additional features that make them preferred to spreadsheets in many cases: they can enforce controlled vocabularies and identifiers easily, and they allow more ways to summarize and subset the data. Delimited text has limited capabilities for working with or exploring data, but it is a non-proprietary format that will be readable far into the future, so it is often an excellent choice for long-term archiving or wide distribution. Spreadsheets and relational database files can be easily exported as sets of delimited text files. XML is another non-proprietary format that is often used for machine-to-machine exchange (such as for providing data to GBIF).

Presentation of data in a species-by-station matrix is not recommended as it is difficult to integrate with data from different studies. Instead, use a format in which each row in a table or delimited text file includes the sample identifier, the taxonomic name of the organism and any additional relevant details, such as the number collected and their life history stages.

16.3.2 Image data

Digital imagery provides significant functional advantages, particularly for studies of the deep sea. Imaging devices, now among the core components of sampling gear deployed from most marine research vessels, provide abundant visual samples (see Chapter 12). Accurate and prompt categorization and management of these images are critical for successful scientific analysis, collaboration, publication and documentation of species biodiversity.

16.3.2.1 Creating a visual information system

The first consideration in setting up an imaging system should be determining how the hardware, software and data management elements will be integrated. This integration will require collaboration among the information systems, engineering and operations personnel at the host institution. Compatibility with existing server platforms and databases, for example, will be critical in sharing information.

Hardware The choice of recording medium will depend principally on which video or still-image formats are recorded, and in what quantities. The project budget, science goals and distribution requirements also affect hardware selection.

Recording the entire sampling event (transect or series of transects) as a continuous video stream offers the most flexibility and a higher likelihood of capturing rare species occurrences, behaviours, morphological identifiers or other unexpected events. Still images can also be precisely chosen and acquired directly from the video stream. This is, of course, both technically and monetarily resource intensive. If storage or monetary limitations are of concern, other recording protocols can be devised: recordings can be made at random or timed intervals, or limited to only mission-specific or user-selected highlights.

The selected recording devices will play an important role in devising the image management strategies, and the following should be carefully considered:

- *Primary recording device*: Digital video files are typically captured using direct-to-disk recording (DDR) methods. Data can be recorded to internal drives (single or RAID) or attached storage (e.g., hard disk drive (HDD), flash memory card, or solid-state drive (SSD)). Widely-available data management practices, such as linear tape-open (LTO) data tape archiving, provide ample flexibility.
- *Secondary recording device*: In some cases, a second copy of the video recordings or images may be made during the expedition. This would ideally be a full-resolution duplicate of the original, although it is often a lower resolution proxy format that serves as working copy or as duplicates distributed to the scientists after the expedition.

Once images and video have been captured aboard the platform or vehicle, or over a cabled network, the critical image management issues to consider are the following:

- *Media transfer mechanism*: Tapes will need to be transferred safely to shore and into a temperature- and humidity-controlled facility. Files will need to be moved from portable hard drives, ship's server or memory cards to a permanent archive.
- *Primary archival or backup system*: The type of backup system will vary depending on the type of media. A tape-based system may be backed up by making duplicates onto the same type of tape, whereas backups for a hard-disk drive-based system may be made onto high-density magnetic tape storage (e.g. linear tape-open (LTO)) for data.

Naming and managing files Each media entity should have a unique identifier. A video tape might be named using a scheme that includes voyage or project code, vehicle, dive number and tape number (e.g. Deep Diver ROV, dive number 1272, tape 3 would be DD1272_3). An individual frame of video from that same tape or digital recorder might have the frame-accurate time code appended to the unique identifier for the tape. Similarly, a unique identifier for a single image from a still camera might include the auto-generated number from the camera appended with the date and time (e.g. DateTime field). Be sure that all names are unique, and that extracted frames can be linked to their parent file/media. If the camera or video tape recorder power is cycled off and on, the numbering scheme or tape time code may reset itself, causing duplicate names to be produced and the unique naming convention to be lost. As a safeguard, original media files should be archived in an unaltered state in a secure location.

Metadata No widely accepted metadata standard exists for scientific image data, however, common industry metadata schema can be easily adapted to meet the needs of individual projects. For example, camera data and image-specific properties can be embedded using the Exchangeable Image File Format (EXIF) schema, and ownership and legal rights information using the International Press Telecommunications Council (IPTC) schema. The following fields are strongly encouraged to achieve efficient processing and meaningful data sets:

- Creator (e.g. photographer, principal investigator, responsible organization)
- Provider (normally the institution supplying the image)
- Source (original copyright holder; may be 'photographer/institution')

- Date–Time (of original, or date of modification)
- Image file specifications (e.g. time, make, model, image dimensions, compression, resolution, bits per sample)
- Location of sampled image (e.g. location name, depth, latitude, longitude, altitude, direction)
- Environmental measurements at time of sample (e.g. pressure, salinity, dissolved oxygen, percent light transmission)
- Content information (e.g. description, keyword (including taxonomy), annotator)
- Rights information (e.g. copyright information, credit line, usage terms).

Note that all required metadata fields should be retained within any subsampled image (e.g. frames cropped or extracted from video).

Digital asset management The media content server, or digital asset management (DAM) system, is the heart of the system and is where all aspects of media management occur, such as aggregation of media files, proxy file generation and media distribution. Such a system can be purchased off-the-shelf, built in-house or integrated with an existing system, depending on the needs and resources of the institution or project. Organizations such as the Digital Asset Management News (digitalassetmanagement-news.org) and the Association of Moving Image Archivists (www.amianet.org) provide current information about industry trends, resources and available commercial and open-source systems.

Automated processes Determine which processes can be automated during and after the dive and thoroughly test these prior to deployment. Some processes that are useful to automate include the following:

- Creating new directories for each dive or expedition, based on the adopted sampling plans and naming schemes.
- Embedding metadata items into an image file. Note that some imaging software programs will remove the embedded metadata if the image is opened or re-saved for any reason.
- Creating a copy of each image (the original version should not be modified) with pertinent information overlaid (Fig. 16.2). Information about the file origin, the sampling environment or annotations is particularly useful and can normally be captured from data logs in real time or post dive.

Documentation Since image systems can quickly become complicated, thorough documentation is advisable. Detailed schematics showing how computers, imaging equipment and software components are integrated, and including a data file flowchart, should be completed early and updated regularly. These should be made available onboard for troubleshooting during expeditions when specialized engineers and technicians may not be available.

Data distribution and image rights management With increased availability and portability of images and video on a variety of electronic devices, be sure digital rights management (DRM) issues are clearly understood so that the media are properly protected. A distribution and usage rights plan should encompass the following:

Ownership:

- Are the media files privately owned or in the public domain?
- What is the policy and process for embargoing imagery? Is there an embargo period? Can particular imagery or subject matter be embargoed?
- What are the crediting requirements? Who is the copyright and licensing owner, and who is the author: the institution, the individual, or both?

Usage and licensing:

- How will the images be distributed (e.g. internet, ftp, email)?

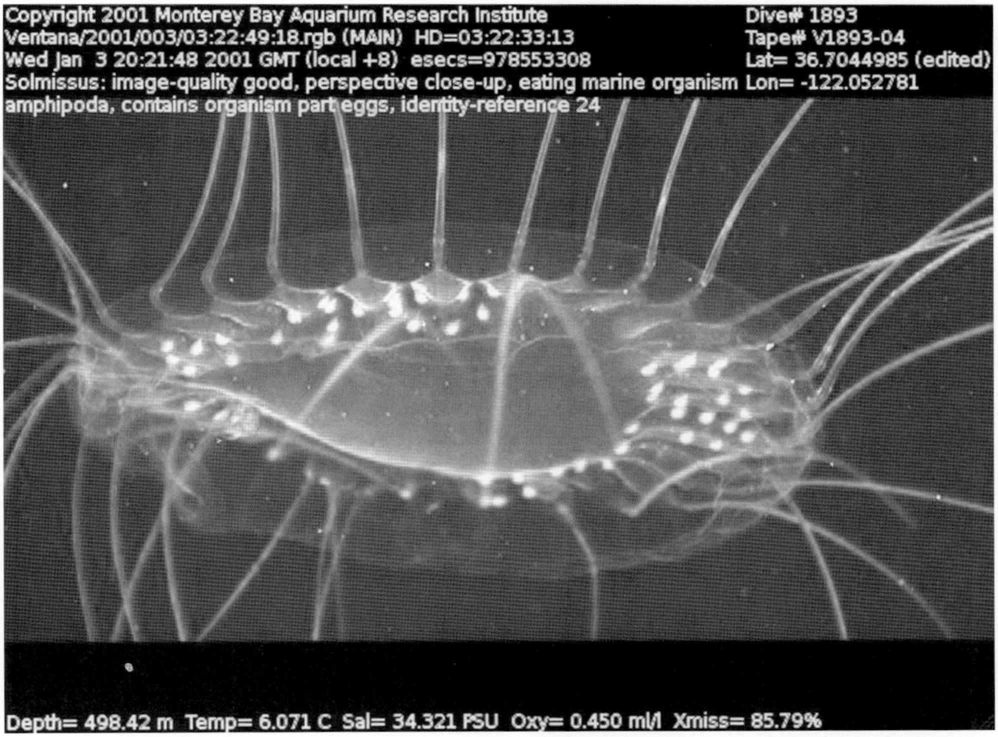

Fig. 16.2 An example of a screen grab from a video overlaid with metadata, including the date, latitude, longitude and environmental conditions. Embedding these metadata within the image ensures they will not be lost as the images are transferred. (© MBARI. Reproduced with permission.) (Please see Plate 27 for colour representation of the figure.)

- What is the scientific data usage policy?
- Will the images or videos be licensed to non-science requestors? How will licences be negotiated and tracked? Will fees be charged?

Preparing media files for distribution:

- Will any special information, such as copyright, be placed on the image overlay at this stage?
- Will the media file be encrypted to prevent unauthorized uses?
- Will the media files be watermarked in order to enable identification in the future?
- Has appropriate metadata to express rights management information been chosen and populated?

16.3.2.2 Annotation and analysis of visual data

Visual data are extremely valuable, but need to be processed to make them accessible and meaningful. Here are some steps to consider:

- Choose an annotation method. Will simple spreadsheets or specialized software be used for viewing, processing and annotating images? See Box 16.2. for information on the Video Annotation and Reference System (VARS), an example of an open-source image management software program.
- Determine how the data will be reviewed during and after the expedition, remembering that logistical factors during the expedition can make detailed analyses difficult. Potential modes of analysis

Box 16.2 MBARI's Video Annotation and Reference System (VARS)

The Monterey Bay Aquarium Research Institute (MBARI) has developed the Video Annotation and Reference System (VARS), a software and hardware system for the creation, storage, and retrieval of video annotation records from ROV dive tapes (Schlining & Jacobsen Stout, 2006). This system is freely available for adaptation by others, and has been used successfully by several institutions or projects. See the MBARI website (www.mbari.org) for further information.

MBARI uses high-definition video equipment to record hundreds of ROV dives per year. In MBARI's first two decades of ROV diving, over 18 000 hours of video tape of biological, chemical, geological and physical interest have been archived, annotated and maintained as a centralized institutional resource. VARS was developed to annotate and manage these video holdings. Using the VARS annotation interface, researchers and technicians take frame grabs and enter annotations, during or after the dive, that describe the content of the image (Fig. 16.3). Annotations are based on the knowledge database, a controlled vocabulary currently containing over 4000 terms. This information then goes into an annotation database, which can be searched according to the query constraints and returns defined by the user (e.g. taxon, physical attributes, locale, depth, scientist, etc.).

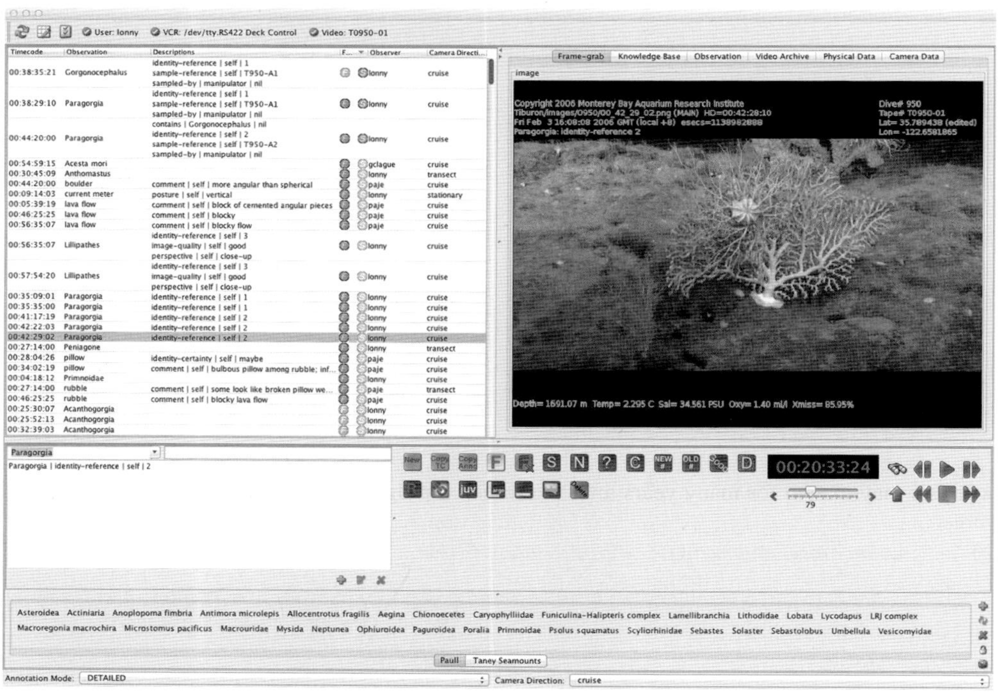

Fig. 16.3 A screen from the Video Annotation Reference System (VARS), showing how an image can be annotated with terms describing the image's content, in this case the taxonomic name of the organism (*Paragorgia*) and its assigned identity reference number. Annotating with controlled terms allows large volumes of images to be efficiently searched for content of interest. (© MBARI. Reproduced with permission.) (Please see Plate 28 for colour representation of the figure.)

include high-level cataloguing or 'outline', detailed description requiring slower playback and multiple passes, or full quantitative analysis (e.g. timed transects or comprehensive species counts). Documenting the mode of analysis is critical so that the level of effort can be accounted for in the data work-up. Take into consideration the time required for each type of analysis. Typically an outline mode will have a 1:1 time requirement (e.g. 1 hour human time per 1 hour of video recorded) whereas a more detailed analysis can take upwards of a 5:1 ratio.
- Establish a controlled vocabulary of annotation terms (e.g. identifiers and descriptors) that can be accessed using pick lists or pull-down menus. This can include species names, life history stages, behaviours and geological features, and will change depending on the particular project and region of study.
- Design the data schema (i.e. the structure of how data will be held and correlated to related data collections). As with other types of data, it is important to record key pieces of metadata related to the annotations in formats that are widely acceptable and compatible.

16.3.2.3 Additional tools

Event detection and classification systems being developed for underwater data collections have already proven effective at detecting an object or a temporal event, and great strides are being made in creating applications that classify these by type (Cline *et al.*, 2009). As imaging systems improve and incoming data streams increase, such systems could become invaluable at various stages of image cataloguing and analysis. Images and video are also increasingly used in combination with bathymetric survey data to enhance visualization, quality assessment and analysis capabilities of seafloor or water column data. Products such as the freely accessible Google Earth (if Google's terms of use are acceptable to your project) and the commercial ArcGIS and Fledermaus support this process.

16.3.3 Genetic sequence data

Molecular genetic data (such as gene sequence, DNA microsatellite, single nucleotide polymorphism and protein expression data) are becoming increasingly important for identifying deep-sea metazoan and microbial species and examining their evolution, dispersal, function and ecological interactions. The state of data management for molecular genetic data is relatively advanced compared to other biological data. Though substantial gaps still exist, there are widely accepted standards, a strong history of data sharing, and well-established data repositories for specific kinds of genetic data.

16.3.3.1 Data collection

Like all data collected at sea, particular attention should be paid to consistent use of identifiers to denote samples and subsequent subsamples, and to processing and handling methods. The genetic data (e.g. DNA sequences) themselves are usually generated and analysed after the voyage, and analysis may extend for many years after the research programme has ended. There are dozens of molecular genetic approaches used on biological samples, from the common direct nucleotide sequencing, to gene expression profiling, population genomics and anti-cancer assays. Each of these approaches has its own method of data collection, data archival and accessibility, which evolves with the technology.

16.3.3.2 Standards

Though there are a growing variety of genetic data types and products, the most common for data on deep-sea metazoans and microbes are genetic (nucleotide or protein) sequence data. These data are generally represented as text strings in a variety of file formats, including Fasta, EMBL, GCG and GenBank (Leonard *et al.*, 2007). Nucleotide (DNA or RNA) sequences are usually stored as one-letter nucleic acid codes in the IUBMB (International Union of Biochemistry and Molecular Biology) format (Leonard, 2003). Similarly, protein sequences are usually stored in the International Union of Pure and Applied Chemistry (IUPAC) one-letter codes (Leonard, 2003).

16.3.3.3 Data repositories

GenBank (www.ncbi.nlm.nih.gov/genbank) is the dominant repository for sequence data. It is an annotated collection of all publicly available DNA sequences (Benson *et al.*, 2015). As of October 2015, there were over 202 billion bases in 188 million sequence records in the traditional GenBank divisions. Each sequence record is identified by an accession number – a unique alphanumeric identifier that is guaranteed to remain with that sequence for the life of the database. If two sequences are merged into one, then the new sequence gets a new accession number and the accession numbers of the merged sequences will be retained as secondary accession numbers.

GenBank, the DNA DataBank of Japan (DDBJ, www.ddbj.nig.ac.jp) and the European Molecular Biology Laboratory (EMBL-Bank, www.ebi.ac.uk), together compose the International Nucleotide Sequence Database Collaboration. All three of these repositories coordinate accession numbers and share data, so that data contributed to one is effectively available to all. But beyond this consortium, there are more than 40 repositories for coding and non-coding DNA (Cochrane & Galperin, 2010). The first month of each new year marks the release of the annual database issue in the publication *Nucleic Acids Research*, which reports on data resources. In 2015, this issue noted descriptions of 56 new data resources and updates to 115 previously published data resources (Galperin *et al.*, 2015). The online Database Collection that accompanies the 2015 issue held 1611 data resources (www.oxfordjournals.org/nar/database/a).

An additional database increasingly used by marine scientists is the Barcode of Life Data System (BOLD, www.boldsystems.org), an informatics workbench to aid in the acquisition, storage, analysis and publication of DNA barcode records. It is a collaborative effort to accelerate acquisition of and access to knowledge of biodiversity using minimal sequences of DNA. These DNA barcodes are essentially 'fingerprints' for species identification through a comparative and comprehensive library of mitochondrial cytochrome oxidase I sequences linked to museum-vouchered specimens. In addition to searchable species identifications and matching/submitting barcodes to a broad spectrum of taxonomic groups barcoded through BOLD, there are other specialized barcoding repositories online, including for zooplankton through the Census of Marine Zooplankton (CMarZ, www.cmarz.org), fish through the Fish Barcode of Life (Fish-BOL) project (www.fishbol.org), benthic and sea ice species through Polar Barcode of Life project (www.polarbarcoding.org) and seamount fauna through the Seamount Barcoding Association (sba.whoi.edu). Whereas GenBank is a repository for a wide variety of genetic data, and stores sequences without express assurance of morphological species designations, barcoding data portals hold data from identified barcode gene regions like COI that are expressly matched with corresponding morphological designations. These barcoding efforts have significantly advanced efforts to discover, understand and manage diversity in marine systems.

16.4 Conclusions

As important as it is now, good information management will become ever more critical in the future. Observatories and other telepresence approaches for collecting and transferring data are creating new volumes and varieties of data at an astounding rate. To effectively manage and use this information, the entire data management process, from collection through archival, must be done carefully and methodically. The collecting scientists have a particularly important role: as the first points of contact for many data sets, they must ensure that all of the critical information is captured with the data in an accessible form. Once lost, it generally cannot be recovered. Although some of the specific standards, websites and technologies promoted in this chapter will inevitably shift over time, the process of planning for the necessary information, and labelling and storing it appropriately, will persist.

Acknowledgements

The authors thank Bob Arko (Lamont-Doherty Earth Observatory), Dave Bowden (NIWA), reviewers and editors for valuable suggestions for improving this chapter. Partial funding was provided by the Census of Marine Life on Seamounts and NSF grant 0606957 for the Marine Metadata Interoperability (MMI) project. We thank MMI for the use of several definitions from their glossary.

References

Benson, D.A., Clark K., Karsch-Mizrachi, I., Lipman, D.J. Ostell, J., Sayers, E.W. (2015). GenBank. *Nucleic Acids Research*, 43 (Database issue): D30–D35. doi:10.1093/nar/gku1216

Cline, D.E., Edgington, D.R., Smith, K.L, Vardaro, M.F., Kuhnz, L. (2009). An automated event detection and classification system for abyssal time-series images of Station M, NE Pacific. *MTS/IEEE Oceans 2009 Conference Proceedings*, October 2009, Biloxi, Mississippi.

Cochrane, G.R., Galperin, M.Y. (2010). The 2010 Nucleic Acids Research database issue and online database collection: a community of data resources. *Nucleic Acids Research*, 38 (Database issue): D1–D4. doi:10.1093/nar/gkp1077

Costello, M.J. (2009). Motivating online publication of data. *BioScience*, 59: 418–427. doi:10.1525/bio.2009.59.5.9

Fabri, M-C., Galérron, J., Larour, M., Maudire, G. (2006). Combining the Biocean database for deep-sea benthic data with the online Ocean Biogeographic Information System. *Marine Ecology Progress Series*, 36: 215–224.

Fegraus, E.H., Andelman, S., Jones, M.B., Schildhauer, M. (2005). Maximizing the value of ecological data with structured metadata: an introduction to ecological metadata language (EML) and principles for metadata creation. *Bulletin of the Ecological Society of America*, 86: 158–168.

Galperin, M.Y., Rigden, D.J., Fernández-Suárez, X.M. (2015). The 2015 *Nucleic Acids Research* Database Issue and Molecular Biology Database Collection. *Nucleic Acids Research*, 43 (Database issue): D1–D5. doi:10.1093/nar/gku1241

GBIF (Global Biodiversity Information Facility) (2011). *GBIF Metadata Profile, how-to guide* (contributed by O'Tuama, E., Braak, K., Remsen, D.). Copenhagen: Global Biodiversity Information Facility, 19 pp. http://links.gbif.org/gbif_metadata_profile_how-to_en_v1

Kelling, S., Ingole, B., Daly, B., *et al.* (2008). *Recommendations of the GBIF Observational Data Task Group*. Copenhagen: Global Biodiversity Information Facility, September 2008, 21 pp. www.gbif.org/communications/resources/print-and-online-resources/download-publications/reports/

Leadbetter, A., Raymond, L., Chandler, C., Pikula, L., Pissierssens, P., Urban, E. (2013). *Ocean data publication cookbook*. Manuals and Guides. Intergovernmental Oceanographic Commission, 64, IOC/MG/64. Paris: UNESCO, 41 pp. & annexes.

Leonard, S.A. (2003). IUPAC/IUB single-letter codes within nucleic acid and amino acid sequences. *Current Protocols in Bioinformatics,* Appendix 1A.

Leonard, S.A, Littlejohn, T.G., Baxevanis, A.D. (2007). Common file formats. *Current Protocols in Bioinformatics, Appendix* 1, pp. B1–B9.

McIntyre, A.D. (2010). *Life in the world's oceans: diversity, distribution, and abundance*. Oxford: Blackwell Publishing, 384 pp.

Schlining, B., Jacobsen Stout, N. (2006). MBARI's Video Annotation and Reference System. In: *Proceedings of the Marine Technology Society/Institute of Electrical and Electronics Engineers Oceans Conference*, Boston, Massachusetts.

Stevens, D., Richardson, A.J., Reid, P.C. (2006). Continuous plankton recorder database: evolution, current uses and future directions. *Marine Ecology Progress Series*, 316: 247–255.

Szalay, A., Gray, J. (2006). 2020 Computing: science in an exponential world. *Nature*, 440: 413–414. doi:10.1038/440413a

Chapter 17
Data Analysis Considerations

Timothy D. O'Hara[1], *Thomas A. Schlacher*[2], *Ashley A. Rowden*[3] *and Derek P. Tittensor*[4]

[1] Museum Victoria, Melbourne, Australia
[2] The University of the Sunshine Coast, Maroochydore, Australia
[3] National Institute of Water and Atmospheric Research, Wellington, New Zealand
[4] UN Environment Programme – World Conservation Monitoring Centre/Computational Ecology and Environmental Science group, Microsoft Research, Cambridge, United Kingdom

Abstract

Investigating patterns of deep-sea biodiversity over large spatial scales is often problematic; existing data are frequently incomplete or spatially biased, and there is little scope for obtaining new data. In particular, the confounding effects of depth can severely limit research findings, because faunal assemblages change much more rapidly along a bathymetric gradient than horizontally across the oceans. The challenge is to extract scientific secrets from this messy data through appropriate analytical techniques. This requires careful construction of the hypotheses, assessment of the types of biological and environmental data available, the accounting of sampling biases by the analytical methods, and restrained interpretation of results.

Keywords data analysis, faunal data, environmental data, sampling bias, stratification, covariance

17.1 Introduction

Ideally, analysis of ecological field samples is the final step in a seamless process beginning with a question, progressing through adoption of an appropriate design, and full execution of concise sampling according to that design. Quantitative ecological data are often difficult to collect, but sampling in deep-sea habitats is especially challenging. Many factors, often outside the control of the researcher, such as access to remote locations and expensive ship time, impose limits on sampling. Operational difficulties at sea can derail the best-laid plans. Many samples are still collected remotely, for example by dredge and trawl gear, often making it difficult for the researcher to judge their performance (e.g. see Chapters 7 and 9).

This chapter does not deal with the analysis of data obtained from samples taken in a controlled and/or experimental manner, such as push-cores taken with submersibles or incubation chambers mounted on landers, nor human-impact assessments. These types of data have analogues from other habitats in terms of their conceptual and methodological approach to analysis that are covered in standard textbooks on experimental design and statistics (e.g. Sokal & Rohlf, 1995; Legendre & Legendre, 1998; Zar, 1999; Quinn & Keough, 2002).

Instead, this chapter focuses on what is characteristic of deep-sea samples, dealing with the typically messy and incomplete data that you have available – often gathered with substantial

Biological Sampling in the Deep Sea, First Edition. Edited by Malcolm R. Clark, Mireille Consalvey and Ashley A. Rowden.

effort – and how these data are best used to detect, describe and predict large-scale patterns of biodiversity in the deep sea. Even if you received the funding for the perfect study to unravel biodiversity in the deep ocean, there is a good chance that some of your survey time will be cancelled due to inclement weather or crew sickness, gear will fail or be lost, the target habitat will not be found at all sites, samples will go missing between the ship and laboratory, the catch will be badly preserved, or you will find that one of your formerly reliable taxonomists is no longer available to identify samples. It would be uncommon to finish up with data necessary to test a balanced design. These problems are compounded once you try to combine samples from more than one expedition to examine patterns and processes at regional to global scales. The surveys will probably have been conducted at different times of the year, collected with slightly different gear, cover a different depth range and have variable location accuracy if collected before the widespread use of Global Positioning System (GPS). Yet, good data have been and continue to be collected that can tell fascinating stories about the origin, evolution, dispersal, diversity and distribution – and potentially the future – of life in the deep sea. The challenge is then to overcome the methodological and statistical problems in the data analysis to get at the scientific secrets hidden within your data, rather than have the story be obscured by them.

This chapter discusses a number of steps in the analysis of deep-sea data intended to clear some statistical hurdles and to overcome the problems encountered – or at least make you fully aware of them. This should enable you to focus more strongly on the question at hand. Using examples, many from our own research, our intent is to foster analyses that contribute to community ecology and biogeography, but our suggestions are broadly applicable to some other fields of research. Rather than expecting to 'clean up' your data to remove errors and biases – which may not be obvious – we focus on the selection of appropriate models and acknowledge issues rather than submerging them beneath an inappropriate analysis. We refrain from discussing individual statistical approaches in detail, but rather discuss general principles of how to approach an analysis of this sort.

17.2 Hypotheses – what is your question?

It is important to carefully spell out specific, logically coherent hypotheses, and to clearly identify questions, before you begin to assemble or analyse data (see Chapter 3). Arguably, it matters whether the traditional approach of formulating null hypotheses is used (e.g. 'Mean diversity will not vary between seamounts and slopes'), or whether predictive hypotheses are formulated (e.g. 'Diversity of corals is predicted to be significantly higher on seamounts than slopes due to the hypothesized greater abundance of hard seafloor on seamounts'). Predictive hypotheses have the advantage that they require a theoretical underpinning and conceptual understanding that may draw upon ecology, evolution, biogeography, geology and species biology for their formulation. It is preferable to take into consideration mechanisms and theory in building your hypotheses, rather than just constructing them and then seeking explanations after the analysis.

One alternative to hypothesis testing is 'data dredging', where you incorporate as much data as you can into the analysis to find what is 'significant'. While the chance discovery of new relationships between variables does happen occasionally, there are many risks with this approach. Most notably you will run the risk of 'discovering' spurious relationships or you will end up with an overfitted model (i.e. one that matches your data well but cannot be generalized to other situations). There is a place for exploratory analysis on pilot data at early stages of the investigation, or field of study. However, to prevent the circularity of generating hypotheses from your data and then confirming them using the very same data, any such preliminary findings should be confirmed with additional data sets (Burnham & Anderson, 2002).

We recommend that you use the supporting framework of your hypothesis-driven questions and work through the following steps.

17.3 Faunal data – what type of data do you have?

One of the first steps is to assess what type of data you have. The fundamental question is: how quantitative were your samples? Any measure of biodiversity, such as the number of species, abundance or biomass, depends on sample volume, sample area and the efficiency of the gear. It is common that investigators do not know exactly how much of the seafloor a trawl traversed or the exact volume of sediment that was collected by that old, rusty Van Veen grab. Historical samples held in museums do not tend to have this information. On the other hand, stereo cameras coupled with multiple, calibrated lasers can accurately determine the area of seafloor an image had captured, and hence yield satisfactory data on animal abundance in the deep sea (Williams *et al.*, 2010).

For archival museum data it can be difficult to reconstruct the original sample. Subsampling large catches on deck retains a few specimens, while discarding 'superfluous' individuals over the side of the ship. This is common practice when biodiversity discovery ('assembling species lists') and taxonomy, rather than ecology, are the main scientific objectives of a voyage. There are also storage issues to consider if the species is large, and hence the taxonomy of relatively large animals, such as the commercially important holothuroids (bêche-de-mer), is surprisingly inadequate; this is partly because of the bias against collecting really large specimens (Uthicke *et al.*, 2004). Museums also frequently donate material to other museums to facilitate morphological comparisons between specimens and species from different regions. All of the above may make your abundance and biomass data less reliable and affect whether you can examine community composition as well as structure.

A common practice is to transform data to binary records. In practice, your data are reduced to a species being either present or absent in a sample. Binary transformations reduce the scope from comparing the structure of assemblages (i.e. species composition and the abundance of species) to comparing composition (i.e. species composition) only. Binary transformation reduces variability in abundance data but it comes at the cost of losing information on how individuals are distributed among the species in a sample – arguably an important measure in most community ecology. Moreover, binary transformation does not eliminate errors arising from variation in the number and size of samples between habitats, sites, times or any other category. Accounting for sampling biases is discussed in Section 17.5.

A further consideration is whether your absence data are accurate records of species being truly absent from a locality. True absence data are frequently not available, because the investigator seldom knows with confidence whether target species X did not occur at the site sampled (Dorazio *et al.*, 2006). For example, the area may not have been thoroughly searched across space and time, the collecting gear was inefficient in capturing the animals, or the species uses the area only seasonally or during certain stages of its life history. The lack of absence data leads to different approaches in species–habitat modelling. The first is to assume that the absence of a species from a sample (a 'pseudo-absence') is an adequate proxy for a real absence given the lack of anything better (Ferrier & Guisan, 2006). The second is to model your occurrence ('presence-only') data against a stratified-random collection of pseudo-absences drawn from across the study area (e.g. Hirzel *et al.*, 2002; Phillips *et al.*, 2006). However, it is important to realize that presence-only models do not estimate the probability of presences directly, but only the probability of presence given the sampling regime and capacity of the study to detect the presence of species. New approaches to modelling presence-only data are emerging (Wharton & Shepherd, 2010; Vierod *et al.*, 2014).

An alternative approach is to use modelled, rather than raw sample data, in a community or biogeographic analysis (Ferrier & Guisan, 2006; O'Hara *et al.*, 2011). Models of habitat suitability use environmental data to predict species occurrences across a landscape. In other words, they predict the potential distribution of a species given the environmental conditions in the samples in which it is found (although the actual distribution may represent only a subset of this wider distribution predicted by the models, due to other factors such as biotic interactions influencing its distribution). These models can be run for multiple species in the same area. Individual grid cells (or pixels) from the predictive maps are then aggregated into pseudo-samples and used as input into a subsequent multivariate analysis, such as an ordination or cluster analysis.

Biotic data should also be assessed for taxonomic consistency and resolution. The same taxonomic groups need to be consistently identified across all the samples for most analyses in community ecology. If the decapods from some samples were 'lost', because they were mislabelled on the ship or mislaid somewhere in the identification or documentation process, then either the sample, or decapods as a group, have to be removed from the data prior to statistical analysis. The species-sample matrix needs to be consistent. A time-consuming taxonomic reconciliation process will probably be required when bringing samples together from several surveys. Scientific names change over time as the taxonomy is revised. Almost certainly there will be undescribed species (those without a given scientific name); these species are often given a number or some other identifier (e.g. *Ophiura* species 'A'). If more than one taxonomist has contributed identifications for your project, then you will have to make sure that the 'species A' of one taxonomist is the same as the 'species A' of the other. Thus, the ongoing involvement of taxonomists, who can check for synonyms and correct any taxonomic inconsistencies, is fruitful and highly recommended.

It is important to consider the taxonomic resolution of your data set: have the taxa been identified to phylum, class, order, family, genus, species or sub-species? Can your hypothesis be tested with the available level of taxonomic resolution? There are many papers in the literature showing that genus-level or family-level identifications are sufficient for detecting some human impacts, for example oil pollution (Olsgard *et al.*, 1997). On the other hand, studies that seek to examine levels of endemism or biogeographical differences between neighbouring regions are generally best done at species-level, because many marine genera and families are widespread, thus obscuring regional differences. For example, a Russian study did not find any genera that were restricted to the north compared to the middle Atlantic across a number of benthic and planktonic groups (Zezina, 1997). This means that it is more difficult to use photographic or video data in these types of studies as the identification of species may require you to examine small anatomical characters that cannot be observed in many images.

If your resources are limited, it is also worthwhile to consider which would be the best taxonomic group(s) to use for your study. The candidate taxa should occur in most of your samples, be diverse enough to facilitate statistical analysis (i.e. more than just a few occurrences in your samples) but not too diverse to become a taxonomic impediment, and have a variety of natural history strategies (reproduction, dispersal, feeding) that might facilitate testing of some explanatory hypotheses. On this basis, O'Hara and Tittensor (2010) selected brittle stars and Rowden *et al.* (2010) selected squat lobsters to test hypotheses about the distribution of seamount animals throughout the southwest Pacific. However, squat lobsters are largely absent from polar regions and consequently would not be suitable for an investigation of faunal patterns from the equator to the pole (cf O'Hara *et al.*, 2011).

Finally, it is important to check the accuracy of the sample location data, particularly for historical information. Published lists can have errors, for example the latitude 38°N can become 30°N (or even worse, 38°S) on data entry. Plotting the sample coordinates on a Geographic Information System (GIS) bathymetric layer is a quick way of assessing the accuracy of the latitude, longitude and depth recorded for your samples.

17.4 Environmental data – what should you use?

Many analyses in community ecology or biogeography attempt to determine the relationship between observed biological patterns and potential environmental drivers, or they seek to model the biological data using environmental parameters. Typically, the relationship with environmental variables is tested using categorical (factorial) analyses and/or examined using gradient analyses (see Box 17.1). For example, McClain *et al.* (2010) examined how the structure of benthic assemblages changes with

Box 17.1 Design influences analysis options

Conventional teaching in statistics traditionally places a heavy emphasis on designs where one or several processes or factors that are thought to influence the response of some ecological or biological attributes are controlled by the investigator. A classic example (and probably oddly familiar to many) is to hold goldfish in solitary confinement, place them on a magnetic stirrer, vary the temperature by heating the jar and measuring oxygen consumption of the fish (making sure to set an upper temperature limit on the hotplate before going to lunch!): here oxygen consumption is the *response* variable and temperature is the experimental *treatment*. Given a set number of fish and aquaria (say 20) you have two very basic choices on how to vary the treatment and use your fish: either allocate a number of fish to each of a limited number of temperature *categories* (e.g. 5 fish at 5, 10, 15 or 20°C), or use fewer fish over a wider *gradient* of temperatures (e.g. 2 fish at 2, 4, 6, 8, 10, 12, 14, 16, 18 and 20°C). In essence, these two approaches amount to a 'categorical design' versus a 'gradient design'.

Field investigation in marine ecology make use of many design types, but perhaps the most familiar type are studies that are based on categorical designs where some ecological or biological response is compared between a usually limited number of treatment categories (e.g. shorebird mortality between oiled and non-oiled beaches; crab recruitment among small, medium and large seagrass patches; species richness of fish between turbid and clear estuaries). Such categorical designs are conventionally analysed by derivates of General Linear Models, such as ANOVAs in case of single response variables, or ANOSIM/PERMANOVA for multiple variables (Clarke & Warwick, 2001; Anderson *et al.*, 2008). Tight control of treatment conditions is often difficult in deep-sea settings, but categories can be clearly defined such as fished versus un-fished seamounts (Althaus *et al.*, 2009; Williams *et al.*, 2010) or differences in benthic biomass between distinct habitat types such as continental slopes and seamounts (Rowden *et al.*, 2010).

Distinct categories may, however, not be an appropriate design choice in deep-sea ecological studies for several reasons:

- in unknown, unmapped and remotely sampled environments (hundreds to thousands of metres below the ship's hull) it can be difficult to accurately and consistently define the boundaries of categories (e.g. sediment types grade into each other, new habitat types are only encountered as the voyage progresses);
- there is no a priori rationale for defining distinct boundaries of categories, either on conceptual or theoretical grounds (e.g. fishing pressure can vary over a wide range with variable response curves of the impacted biota); or
- limited ship time and resources are juxtaposed against the need to obtain data over a large geographic area, such as in biological seabed mapping for conservation planning where spatial coverage becomes the overriding design criterion (Schlacher *et al.*, 2010); or
- the primary interest is to document biological and ecological responses along the gradient itself (O'Hara *et al.*, 2010).

In all of these 'gradient-type' designs, the primary purpose of the analysis is not to compare means or other parameters between categories, but to:

- assess the strength of the relationship between 'predictors' (e.g. depth, grain size, organic content, salinity) and response variables (e.g. diversity, density, biomass, etc.),
- model the nature of the relationship, and
- identify which predictors are most important in explaining the observed responses along the gradient.

The most commonly encountered analytical techniques are various regression approaches, distance-based redundancy analysis (McArdle & Anderson, 2001) and matrix correlation.

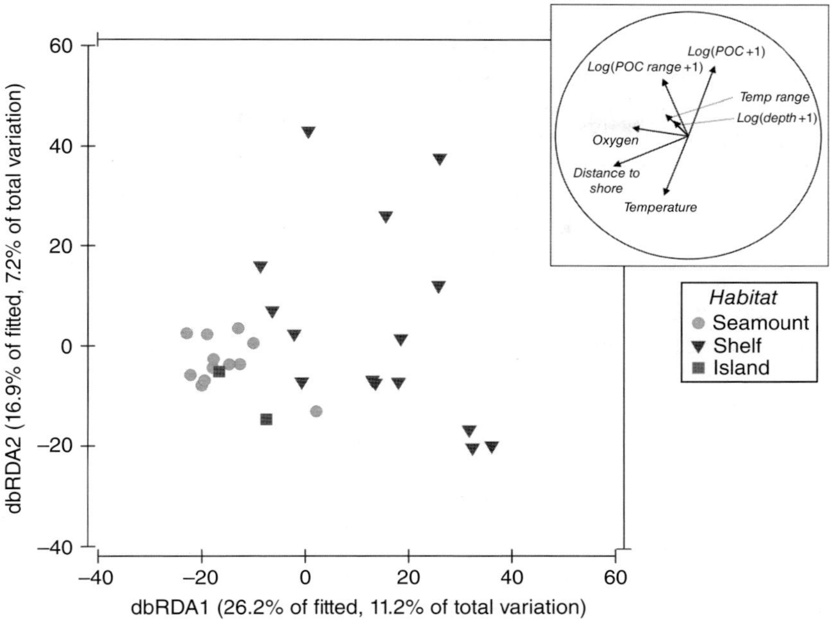

Fig. 17.1 Distance-based Redundancy Analysis (dbRDA) ordinations of invertebrate samples from the REVIZEE 5 survey along the Vitória-Trindade seamount chain and adjacent Brazilian continental shelf. The ordination shows that the samples from the shelf habitat group separately from the samples from the seamount and islands. The vectors show the strength and direction of the relationship between the environmental variables and the assemblage pattern (Anderson *et al*., 2008). POC, particulate organic carbon. (Adapted from O'Hara *et al*., 2010. Reproduced with permission of John Wiley & Sons Ltd.)

depth down the flanks of Davison Seamount. O'Hara *et al.* (2010) tested for differences between several habitat types (shelf, seamount, island slopes) and identified the environmental factors that were most closely correlated with changes in the composition of benthic assemblages among these habitats off the coast of Brazil (Fig. 17.1). Other examples include Schlacher *et al.* (2007) who assessed how the diversity and community composition of sponge communities in submarine canyons responded to variations in depth, seafloor characteristics, and habitat heterogeneity; and Tittensor *et al.* (2011) applied a species-energy framework to examine the relative importance of environmental variables (as proxies for energetic factors) in structuring deep-sea molluscan communities.

Studies in the deep sea that relate biological attributes of the benthos to environmental conditions tend to use three main classes of environmental predictors:

- physical (e.g. temperature, pressure, currents, seafloor characteristics, etc.),
- biological (e.g. suspended or deposited particulate organic matter available as food, bioturbation, overlaying water-column productivity, etc), and
- chemical variables (e.g. calcite and aragonite saturation, salinity, oxygen, etc).

It is important to remember that the spatial and temporal variability of many factors may be as biologically relevant as their mean, minimum or maximum value. For example, seasonal variation in the delivery of organic matter to the seafloor can measurably influence the composition of assemblages in polar regions, and it may even drive long-term evolutionary histories of polar species (Krug *et al.*, 2009). Similarly, frontal areas, where water masses meet, are characterised by gradients in temperature and productivity that may limit species ranges (McClain & Hardy, 2010).

Whatever environmental predictors you choose for your analysis, the issue of scale must be considered (Storch *et al.*, 2007). Most important are the spatial and temporal scales at which environmental variables influence the fauna most. Some predictors, such as seafloor sediment grain size, can change markedly over small distances of metres to hundreds of metres. Others, such as temperature or aragonite saturation, might drive patterns at larger scales of kilometres to hundreds of kilometres (Tittensor *et al.*, 2009). In many studies, *in situ* measurements are only available for a few variables, while others have to be interpolated from global or regional oceanographic models. In this situation the data and associated interpolation errors must be carefully assessed. Using, as an example, modelled current velocity to interpret the distribution of corals on seamounts – which are known to be sensitive to water movement – the first question is whether corals in the real world are truly experiencing the average value of current velocity predicted for this area and over this scale? In fact, abrupt geomorphological features are known to substantially modify currents and seafloor characteristics, leading to corresponding changes in the fauna at scales of metres to hundreds of metres (Genin *et al.*, 1986). Even in situations where the correct scaling between environmental drivers and biological responses is known, it may not be possible to implement this in practice.

One of the main reasons for the inability to match scales is the rather coarse resolution of many environmental data, typically 0.1–1.5 degrees latitude/longitude. This means that values can be applied to unrealistically large areas: for example at the equator, a degree of latitude/longitude covers approximately 12 370 km^2. Coarse-resolution estimates of carbon availability may be problematic, because organic carbon can vary at small scales (e.g. large food falls of whale carcasses; Smith & Baco, 2003) or over short time intervals (e.g. seasonal phytoplankton settlement in the deep Atlantic; Lampitt, 1985). In addition, many interpolated environmental data layers are likely to have lower errors in areas with greater sampling intensity (such as the North Atlantic shelf), and higher errors in other regions (the abyssal equatorial Pacific, for example). Notwithstanding some shortcomings and possible mismatches in the scale and resolution between oceanographic and biological data, the increased availability of this information can measurably benefit our understanding of how biota are distributed throughout the oceans (Tittensor *et al.*, 2009, 2010). Often all you can do is try to match the scale, resolution and spatial placement of your environmental predictors as closely as feasible to the organisms' known, sampled or modelled habitat range – while remaining aware of, and documenting, the compromises you have made.

Oceanographic variables such as temperature, pressure, salinity and calcite saturation, which can change rapidly with depth (especially in shallow water), generally change more gradually over latitudinal and longitudinal clines. We therefore recommend that you investigate how data sets were constructed that, purportedly, provide data on conditions near or on the seafloor (e.g. Lutz *et al.*, 2007). For example, was the variable extracted for the mean depth of each grid cell? This can be misleading if the seafloor varies considerably in depth over these scales: conditions on the summit of a tall seamount (e.g. at 200 m depth) may differ considerably from that at the mean depth (e.g. 800 m) of the cell in which it is placed. In these cases, it is preferable to extract the variables from a three-dimensional model, matching the depths of model predictions to the depths of your samples (e.g. O'Hara *et al.*, 2011).

Finally, many analytic techniques require that you prune your list of environmental variables to a smaller subset. For example, when identifying a model that best approximates your biological system from a set of possible candidate models, it is recommended that the number of variables should not exceed the number of samples (Lukacs *et al.*, 2009) to avoid over-fitting (see Section 17.2). The most parsimonious solution would use the smallest number of variables to achieve an adequate model of the data (Box & Jenkins, 1970). The most important criterion for selection is to include only those variables that are likely to influence the distribution of an organism (i.e. those for which there is a plausible mechanism of operation). For example, it may be questionable to include sea-surface temperature as an environmental predictor when the species whose distribution is modelled never rises to the surface. This selection should be done prior to the analysis to avoid 'data dredging'. Secondly, it can be important to avoid including variables that are highly correlated with each other, as many oceanographic variables are. This 'collinearity' is a methodological challenge for some statistical techniques (Quinn & Keough, 2002), but more generally, it also makes it very difficult to identify the actual environmental drivers of changes in assemblage composition (see Section 17.7). One solution has been to use a principal component analysis to combine a set of correlated variables into a smaller number of uncorrelated variables, although these composite variables may be harder to interpret and assume that there was a linear relationship between the combined variables (Quinn & Keough, 2002).

17.5 Sampling biases – how can you account for them?

When analysing biodiversity metrics such as species richness that are strongly influenced by the number of samples collected, great care must be taken to avoid confounding any analysis with differences in sampling effort, size, depth and area. Because virtually all estimates of species richness and diversity depend on sampling properties, it is not appropriate to compare biodiversity between sites, times or treatments, when these categories were not sampled equally (i.e. they do not have the same number of samples, cover the same area or depth range, were collected with different equipment or have similar numbers of collected individuals).

Since data sets often do differ in terms of sampling effort, a number of methods have been developed to account for such differences. Rarefaction is widely used and essentially makes samples comparable by giving the expected number of species for a given number of individuals (species richness) or samples (species density) (Fig. 17.2). Initially, rarefaction curves were generated

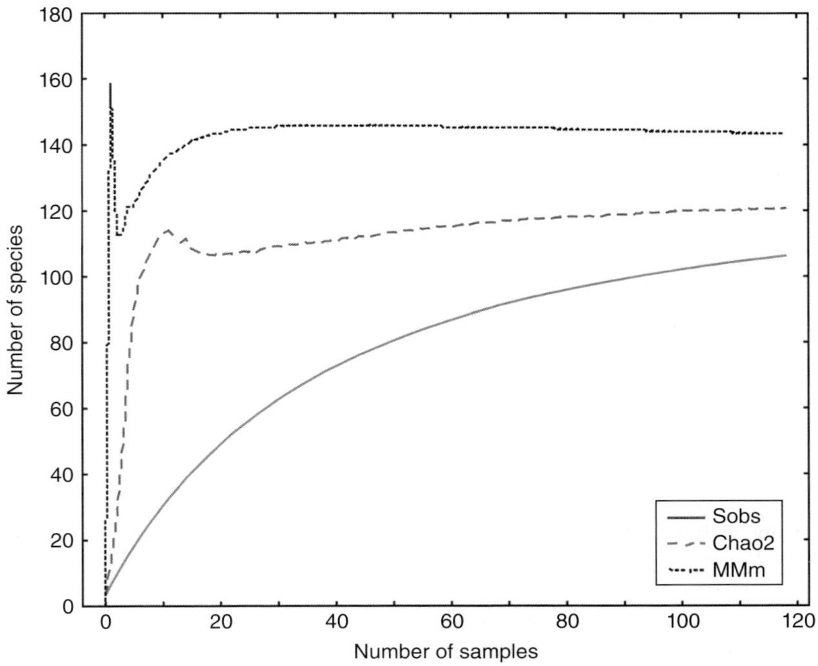

Fig. 17.2 Species sampling curves calculated from samples of deep-sea invertebrates found within a 500 km radius of the Rumble V Seamount, northeast of New Zealand. The 'Sobs' curve is a randomized sample rarefaction curve that rises non-linearly to an asymptote where all available species have been collected. The Chao2 and Michaelis–Menten (MMm) curves are two parametric methods of extrapolating to the asymptote (Rosenzweig *et al.*, 2003). Not only do these estimates differ in this example, but they are also particularly unstable when calculated using a relatively low number of samples. (Data from O'Hara & Tittensor, 2010.)

repeatedly from randomly ordered subsets of the samples or individuals and averaged, creating a curve with the mean expected number of species from n samples or individuals (Gotelli & Colwell, 2001). An algebraic method with confidence limits is available for sample-based rarefaction (Colwell *et al.*, 2004) removing the need for re-sampling, although this method assumes that each species has the same probability of being present in every sample which may be a problem if the study area contains habitat heterogeneity. If the number of individuals per species is suspected to vary markedly between samples (e.g. before and after a recruitment event), it is better to re-scale species accumulation against the number of individuals rather than samples in the final plot (Gotelli & Colwell, 2001). Sample-based rarefaction curves allow you to interpolate species richness for a standard number of samples.

An important property of sample-based rarefaction curves is that they are non-linear beyond the first few samples or individuals. Consequently, it is rarely valid to account for sampling biases using a simple ratio of diversity over effort (e.g. dividing the number of species by the number of samples) (Gotelli & Colwell, 2001). Secondly, the curves rise until they reach an upper maximum (or asymptote) when all species in the area have been recorded (although, of course, species numbers are continually changing over both ecological and evolutionary time scales). It can be useful to plot an accumulation curve of species against effort, and check how close you are to this asymptote to gauge the adequacy of the collection effort – how many of the species making up the assemblage were sampled. It will be rare that deep-sea survey samples approach an asymptote (Ramirez-Llodra *et al.*, 2010), especially in huge habitats that stretch for thousands of kilometres such as the abyssal plains.

While rarefaction reduces samples to a common low level, abundifaction attempts to extrapolate all samples to a common high level. There are a number of abundifaction methods that estimate the asymptote of the species rarefaction curve by extrapolating from a known set of samples (e.g. Rosenzweig *et al.*, 2003; Colwell *et al.*, 2004). These include parametric approaches which fit a curve that is unrelated to the underlying mechanism, and non-parametric methods based on the proportions of rare species in the samples. However, all these methods can produce unpredictable results, particularly when only a few samples are available (Colwell & Coddington, 1994) – a situation not uncommon in deep-sea surveys. In such circumstances, the predicted richness estimates can change markedly as new data are added. The required number of samples for such estimators to settle to a relatively stable value depends on both the species richness and evenness of the assemblage. Ideally, simulations should be conducted to ascertain how much data are required to get a stable extrapolation of richness; plotting estimated (extrapolated) richness against the number of samples should show whether the estimator is 'settling' to a relatively steady value (Fig. 17.2). This generally only occurs when species richness in the samples is beginning to approach an asymptote. This may seem like an awful lot of work to compare samples, and indeed the utility of these approaches in the deep sea may be relatively limited given the sampling difficulties (Ramirez-Llodra *et al.*, 2010). However, generating predicted total species richness can lead to analyses that would not be possible with other approaches (e.g. Mora *et al.*, 2008; Tittensor *et al.*, 2010). In addition, they can provide information on how much extra sampling effort may be needed to properly census the community (Colwell & Coddington, 1994).

Sometimes the total number of species in a region is estimated by constructing lists of species from a study area by consulting the taxonomic literature (e.g. Witman *et al.*, 2004). However, museum collections, on which the taxonomic literature are based, can also be affected by sampling effort and are likely to underestimate the total number of species present. Museum collections are often based on qualitative samples from easily accessible locations, not from systematically sampling across the seascape. Deep-sea expeditions still regularly discover new undescribed species. Discovery curves can be used to predict the unknown number of species for large areas such as ocean basins (Bebber *et al.*, 2007; Costello & Wilson, 2011). Conversely, not all species can exist at every habitat across the region, so a regional inventory may overestimate the number of species available at your site (Hillebrand & Blenckner, 2002).

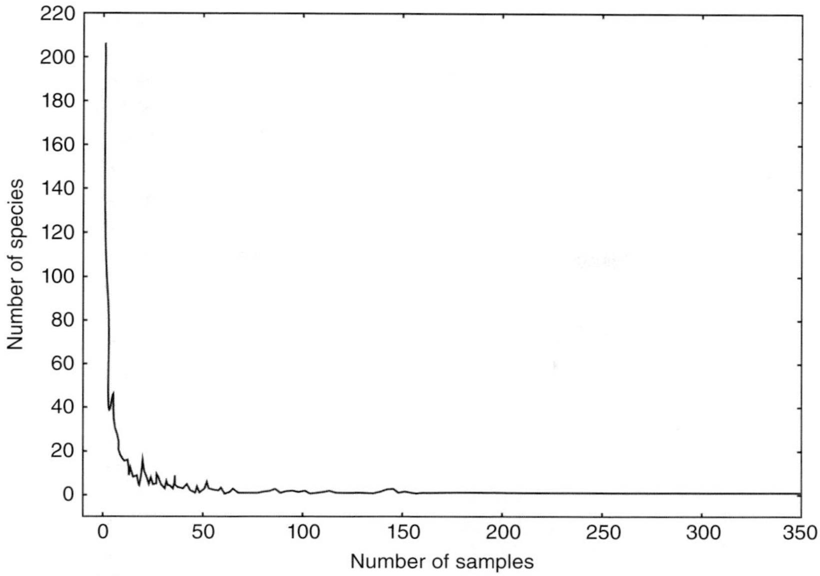

Fig. 17.3 Frequency of species collection. This curve is based on data for 6950 samples of ophiuroids collected from throughout the southern Indo-Pacific Oceans from 0 to 2000 m depth. This graph shows that most collected species were rare. Of the 923 species, 206 were collected once and a further 98 twice; 267 species were present in more than 20 samples and only 14 species were present in more than 150. (Data from O'Hara *et al.*, 2011.)

Sampling problems also arise when attempting to compare levels of endemism between regions. For most data sets and taxa it is very difficult to distinguish narrow-range endemics (species only occurring in a limited area or habitat, e.g. on a single seamount) from those that are rare in the sense that they occur over a larger area but at a low level of abundance (Rex, 2002) and thus are missed by many samples. 'Spot'-endemics, those species known from one sample or site, can be common in benthic communities in the deep sea. For example, O'Hara *et al.* (2011) show that in a large data set of ophiuroid distributional records, 204 of 923 (22%) species were recorded from only one 0.04° cell within the study area (0–70°S, 100–190°E, 0–2000 m depth; Fig. 17.3). Thus, if such high levels of spot endemism are common for marine species, it is not a straightforward task to make meaningful comparisons of endemism among groups of sites or regions. Importantly, whether the level of endemism in the deep sea is indeed high or whether it is inflated due to undersampling remains an open question.

Rarity is another measure which can be heavily influenced by the scale and volume of sampling. Larger or more samples will probably contain more rare species because of the greater number of individuals collected (hence the more likely chance discovery of uncommon species) and because of the greater heterogeneity of habitats encompassed. Conversely, as sampling effort expands, more individuals of previously 'rare' species may be found or species previously considered rare may be encountered in more localities, both depressing the number of what are considered 'rare' species in the final sample.

One approach to overcome potential errors due to sampling effort and volume is to compare your sample diversity metric against that expected from a null model. Null models are produced by randomizing data from a known or expected distribution and are designed to produce patterns that would be expected in the absence of an ecological mechanism (Gotelli, 2001). Comparison with a null model allows for biological meaningful patterns to be distinguished from stochastic and sampling effects (Gotelli & Graves, 1996). For example, O'Hara (2007) tested whether seamounts had greater

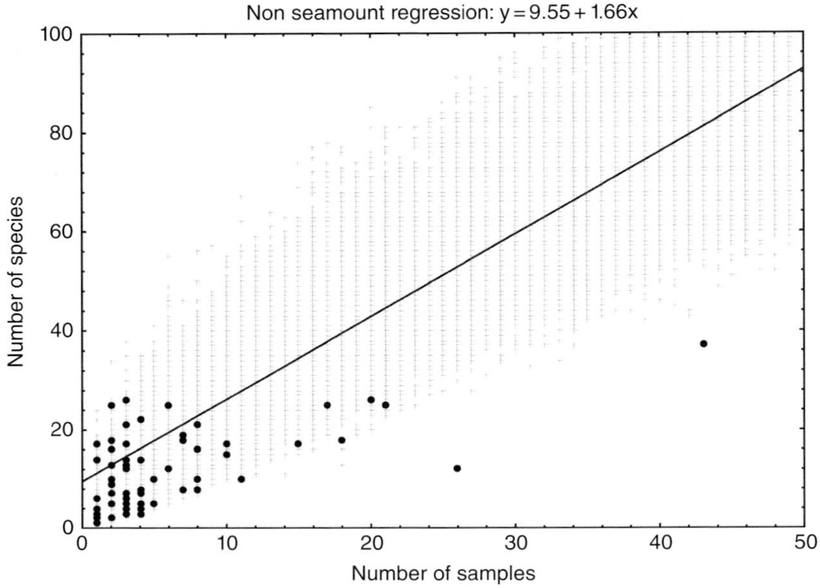

Fig. 17.4 The relationship between the number of species and number of samples for seamounts and non-seamount areas. Dots represent each of the 70 seamounts (many overlap). Crosses represent 25 000 randomly selected populations of between 1 and 50 samples from areas of the continental slope of 50 km radius and 500 m depth range, and the solid and dotted lines show the linear regression and 95% prediction limits. The majority (67) of seamounts have species richness that falls within that expected from the continental slope. (Adapted from O'Hara, 2007. Reproduced with permission of John Wiley & Sons Ltd.)

endemism or species richness than comparable areas on the continental slope (Fig. 17.4). In this study, a null model was generated by randomly selecting samples from kernels of 50 km radius and 500 m bathymetric height that were randomly placed on the slope across the study area. He found that with a few exceptions, seamount endemicity and species richness fell within the range expected from the continental slope.

Spatial aggregation of samples may be a problem for regression analyses, which are sensitive to outliers, such as if data points are skewed or 'bunched' at either extreme. The regression residuals should be checked to ensure that they are randomly distributed along the gradient of interest. A more balanced design may be achieved by transforming the gradient variable or removing outliers from the available data. If such situations cannot be resolved, however, you may have to rely on more 'robust' techniques such as m-estimation or rank-based (non-parametric) regression (Quinn & Keough, 2002).

Multivariate techniques of community analysis are not immune to sampling errors. The most common measure used to assess similarity between samples, the Bray–Curtis coefficient, is sensitive to variations in species richness (Lennon *et al.*, 2001). For example, outlying samples on an ordination or cluster diagram are frequently those with low species richness because these samples are less likely to share many (or any) species with those samples that are more species-rich (Clarke *et al.*, 2006). These results should be examined with care. They can be ecologically meaningful when low species richness is expected from anthropogenic disturbance or samples from extreme environmental conditions. On the other hand, low species richness can also indicate that the collection gear did not work properly, or that the process of identification is incomplete, and hence the sample needs to be excluded from the analysis.

Sometimes the samples are aggregated into larger units to assess diversity or community relationships at larger scales or for ease of interpretation. However, aggregating samples can be problematic when sample sizes differ between units. An obvious solution is to standardize sampling effort during the field stage or to randomly eliminate samples from units with disproportionally greater effort, but here much valuable information is lost. This step would only become necessary if variation in sampling effort is demonstrably and significantly linked to the observed pattern between sites. For example, both O'Hara (2007) and Rowden *et al.* (2010) tested whether the outputs of multivariate analyses with unbalanced numbers of samples in each region (all available data) were correlated with the outputs of analyses with standardized (but fewer) numbers of samples in each region.

Another method to explore whether sampling effort is a problem is to correlate a similarity matrix based on the number of samples against one generated from the biological data of the aggregated units (e.g. the Global BEST test in Clarke *et al.*, 2008). If the chief objective of the comparison is to contrast geographic regions per se, or treatments (e.g. fished vs. protected regions), care must be taken that aggregated samples cover the same spread of bathymetry and habitat in each of the regional data sets. As a minimum, investigators should examine box plots of depth and run inferential statistics on depth metrics such as means, median and quartiles. Because biological assemblages change so dramatically with depth, introducing biases into the analysis through unbalanced depth coverage will almost certainly undermine the validity of the analysis, making it impossible to distinguish between the effects of depth and the effects caused by the factor of interest (see also Section 17.6). An example is the study by Koslow *et al.* (2001), where all the samples from seamounts affected by benthic trawling were at shallower depths than those from unfished seamounts. Although the damage from trawling was clear from *in situ* photographs, it could not be demonstrated from an analysis of the seafloor fauna. In a similar manner, the finding of Brey *et al.* (1994) that faunal diversity was similar between tropical and polar deep-sea sampling was confounded by differences in collection gear and sampled area between the two regions (Rex & Etter, 2010).

Finally, species–habitat models are also sensitive to variation in sampling effort. Samples need to have been taken across the range of environmental space inhabited by the species to accurately predict the extent of a species' distribution (Barry & Elith, 2006). The distribution of pseudo-absences (see above) should be carefully stratified to match sampling effort (Phillips *et al.*, 2009).

17.6 Stratification and covariance – how can you partition out main effects?

Marine benthic assemblages vary considerably in composition and structure with depth, latitude, habitat and collection methodology. Depth is probably the most important environmental factor (Rex & Etter, 2010). Species turnover with depth has been estimated to be at least 1000 times greater than latitudinal gradients (Longhurst, 1998). Different depth strata may not be connected ecologically and have different evolutionary histories, and thus should possibly be considered as distinct biogeographical provinces. The upper bathyal (used here to cover 250–2000 m) fauna off southeastern Australia at temperate latitudes shares more species with the upper bathyal fauna in the equatorial Pacific Ocean than it does with the shallow-water fauna in the same geographic region of Australia (O'Hara *et al.*, 2011). Thus, it is vital that any study of large-scale biodiversity patterns, both ecological and biogeographical, accounts statistically for depth. It may be also necessary to account for bathymetric range; for example, a regional fauna list accumulated from samples along a 1500 m depth gradient will have a different composition from one generated from a 300 m transect because it covers more depth strata (see Section 17.5).

Although latitudinal gradients in assemblage composition are less pronounced than bathymetric ones, they should also be factored into regional ecological studies. Distinct tropical, temperate and polar faunas can be recognized at both shelf and slope depths for ophiuroids (O'Hara *et al.*, 2011; Fig. 17.5) and probably for many other groups as well (Briggs & Bowen, 2012). Differences

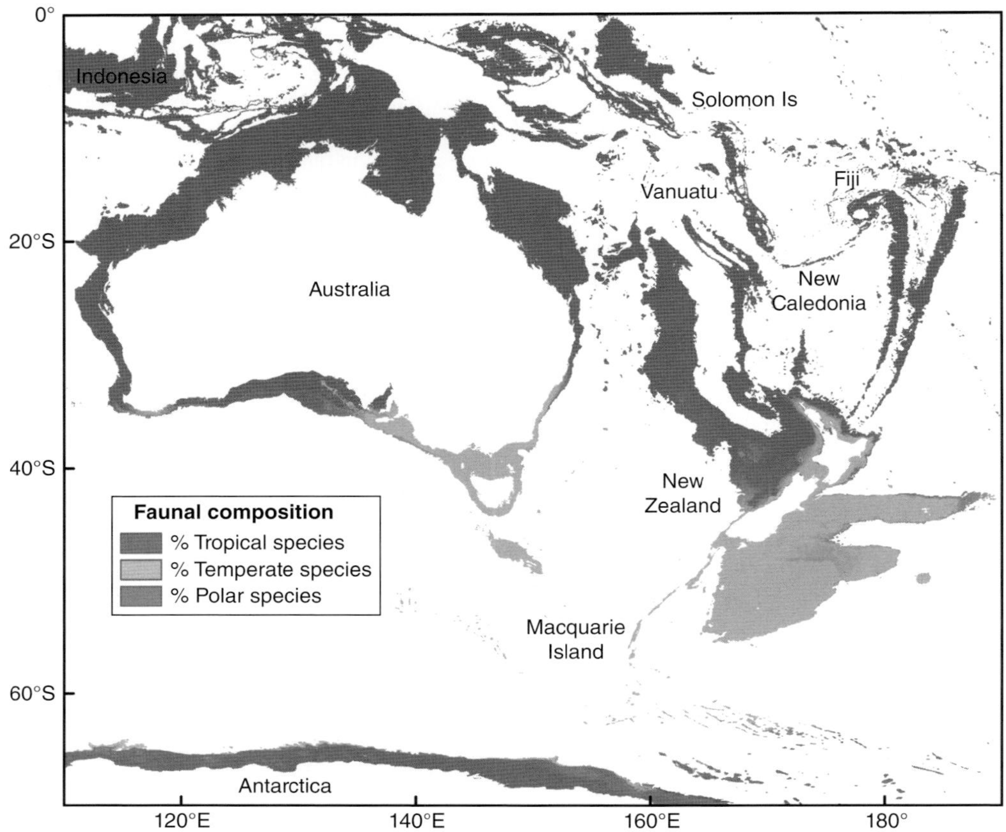

Fig. 17.5 Map showing the overlapping distribution of tropical, temperate and polar brittle stars from the equator to Antarctica. Both shallow water (less than 250 m deep) and deeper (250–2000 m) faunas exhibit a strong latitudinal pattern. (Adapted from O'Hara *et al.*, 2011. Reproduced with permission of Elsevier.) (Please see Plate 29 for colour representation of the figure.)

between assemblages may be due to historical as well as current ecological factors; this may be especially pronounced in studies that encompass a significant latitudinal extent, particularly when samples between 30° and 60° latitude are included, where assemblage turnover can be considerable (O'Hara *et al.*, 2011).

Finally, it is also important to account for any variation in collection methodology. Collection gear will differ in the area it samples, mesh size and sampling efficiency (see earlier chapters). Variation in collection technique can explain much of the assemblage variation in some data sets (Rex & Etter, 2010).

All these factors must be accounted for if they are not of primary interest in the study. The best post-survey solution is to investigate whether the explanatory variables (e.g. the levels of available organic matter) are related to the response variables (e.g. the assemblage composition or species richness) after mathematically accounting for the effects of depth, latitude, habitat or collection gear (termed 'co-variates' in these situations) using statistical methods such as analysis of covariance (ANCOVA), generalized linear modelling or partial correlations. The objective is to determine if a significant response in the variable of interest remains after the effects of the co-variates have been taken into account.

However, sometimes it can be preferable to stratify the data set into two or more groups which are analysed independently. For example, it may not be a good idea to combine samples collected by a grab and trawl in one regression analysis; the resulting measurements, such as species richness, are likely to be separated into two groups based on gear-type and not spread evenly along the gradient of interest (see Section 17.5). Determining how the data set is divided into groups of samples based on depth, latitude or other continuous variables requires some thought to avoid adding additional artefacts. The cut-offs between groups should be objectively determined based on existing knowledge (e.g. biogeographic regions or environmental clines; see Schlacher *et al.*, 2014). Some outlying samples may have to be excluded from the analysis. One of the main disadvantages of stratification is that it can result in groups with low or uneven sample sizes for the factors of interest.

17.7 Interpretation – how can you make the best sense of your results?

Getting the results from statistical tests or exploratory graphs is only one in a series of steps in producing good science (Fig. 17.6). Here we provide some broad guidelines on what to consider when interpreting the outputs of your numerical analysis.

First, be careful not to imply causality if your sampling design was inappropriate. Finding that oxygen was the best predictor of variations in assemblage composition does not automatically mean that differences in oxygen concentrations have caused the observed biological differences, because other unmeasured variables (such as seafloor composition) may have influenced the result. It just happens to be the best predictor from the range of variables you included in your analysis. Is there a mechanism by which oxygen may have caused the observed patterns? One needs to know something about the physiology and natural history of the organisms to judge which factors are relevant. To unequivocally demonstrate causality in ecology requires an experiment. Without applying a process such as cross-validation, it is also more difficult to identify which of the correlated variables are important.

As noted in Section 17.2, it is better not to 'data-dredge' by combining all the available global environmental parameters into a single massive analysis to see what emerges as 'significant'. Overparamaterization (including too many variables) will tend to lead to models that more 'accurately' fit the data in hand, but without an approach such as cross-validation they will be less general, as they fit the idiosyncrasies of the specific data set that you gather.

It is also important to consider the size of the effect you have found as well as its statistical significance. For example, a difference in oxygen concentrations of 0.2 mL L^{-1} may be statistically significant ($p < 0.05$) if all replicates in each site are tightly clustered around the mean (technically, this is called high 'precision'). However, such a small difference in mean values may not be biologically relevant and thus provides a spurious explanation for biological patterns. Similar issues arise with correlation coefficients. In situations where sample size is large ($n > 50$) small values of the correlation statistics may be significant but indicate only a weak association between the two variables. Finally, the term significant carries with it an error trade-off. Conventionally significance is set at $p < 0.05$, meaning there is a 1 in 20 chance of falsely concluding a positive test result (Type I error). But if the 0.05 threshold is lowered there is often a greater chance of making a false negative (Type II) error, when there really is a significant relationship or difference but it is not detected (Quinn & Keough, 2002). As an alternative to the, arguably, arbitrary significance levels of traditional methods, other approaches provide information on the relative importance of variables (Stephens *et al.*, 2006). These approaches include effect size statistics, Bayesian statistics and information theoretic (IT) model comparisons (Stephens *et al.*, 2006). The latter average the contribution of variables across a candidate set of models selected using an information criterion, such as Akaike's Information Criterion (AIC) (Burnham & Anderson, 2002).

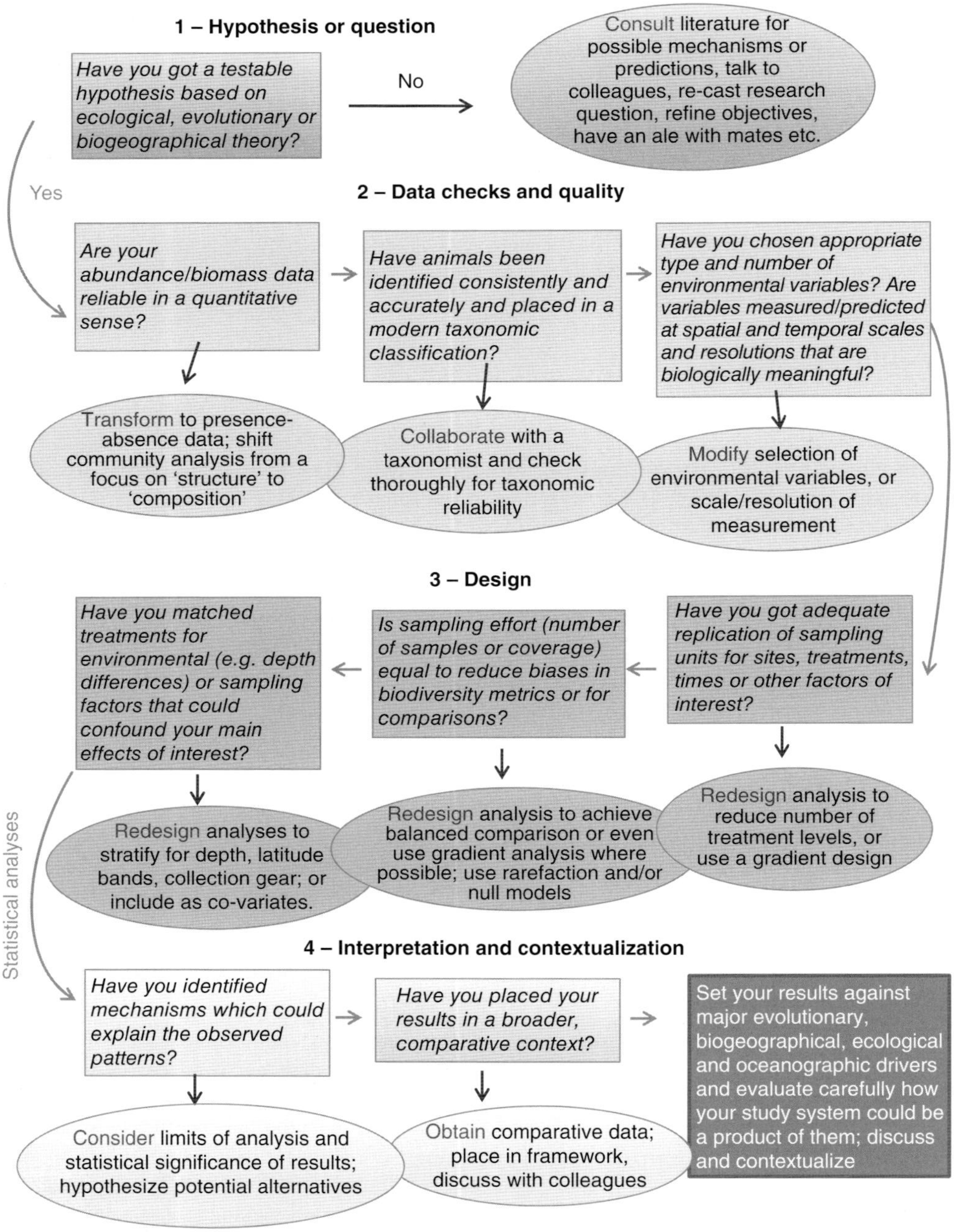

Fig. 17.6 Flow chart of data analysis decisions discussed in the text.

Many tests and analyses gain from being put into a wider ecological, biological, evolutionary, oceanographic and historical context. For example, although it can be demonstrated that assemblages on seamounts differ with latitude, it is more informative to compare this result against similar analyses for other habitat types. If the latitudinal gradient is greater for a chain of seamounts then for continental slopes, then it may indicate that population connectivity differs between these two environments. This is particularly important when measuring scale-dependent measures such as levels of endemism or species richness. Reporting a level of endemism of 2% for a seamount is, arguably, not particularly informative if data for other areas are not presented.

There are a multitude of environmental, biological and historical factors that can influence assemblage composition. Many are interrelated and it can be difficult to discuss these factors in a structured way. Vellend (2010) has proposed a framework that integrates ecology with longer term evolutionary processes, organizing factors into four headings analogous to those that influence population genetics: selection, drift, speciation and dispersal. Selection encompasses how individuals and species respond to their environment (including going extinct), drift includes stochastic changes, speciation is the creation of new species, and dispersal is the spatial movement of organisms. Using a framework such as this will assist in the comparison of your results against the extensive literature.

Finally, you must also be aware that few ecological or evolutionary findings are truly general in nature: variations to the rule and exceptions are common in the natural world (Lawton, 1999; Gaston, 2000). Thus, beware of over-generalizing your results. Evolution has resulted in diverse ways of ensuring survival and there will be spatial and temporal variation in many environmental processes. Consequently, few researchers attempt to claim they have discovered general laws of ecology. A more modest aim is to describe statistical generalizations that assist us in understanding patterns and processes and lead to better management of our oceanic environment.

References

Althaus, F., Williams, A., Schlacher, T.A., *et al.* (2009). Impacts of bottom trawling on deep-coral ecosystems of seamounts are long-lasting. *Marine Ecology Progress Series*, 397: 279–294.

Anderson, M.J., Gorley, R.N., Clarke, K.R. (2008) *PERMANOVA+ for PRIMER: guide to software and statistical methods*. Plymouth, UK: PRIMER-E.

Barry, S., Elith, J. (2006). Error and uncertainty in habitat models. *Journal of Applied Ecology*, 43: 413–423.

Bebber, D.P., Marriott, F.H.C., Gaston K., Harris, S.A., Scotland, R.W. (2007). Predicting unknown species numbers using discovery curves. *Proceedings of the Royal Society B: Biological Sciences*, 274: 1651–1658.

Box, G.E.P., Jenkins, G.M. (1970). Time series analysis: forecasting and control. London: Holden-Day.

Brey, T., Klages, M., Dahm, C., *et al.* (1994). Antarctic benthic diversity. *Nature*, 368: 297.

Briggs, J.C., Bowen, B.W. (2012). A realignment of marine biogeographic provinces with particular reference to fish distributions. *Journal of Biogeography*, 39, 12–30.

Burnham, K.P., Anderson, D.R. (2002). *Model selection and multimodel inference: a practical information-theoretic approach* (2nd edn). New York: Springer.

Clarke, K.R., Warwick, R.M. (2001). *Change in marine communities: an approach to statistical analysis and interpretation* (2nd edn). Plymouth, UK: Natural Environment Research Council and Plymouth Marine Laboratory.

Clarke, K.R., Somerfield, P.J., Chapman, M.G. (2006). On resemblance measures for ecological studies, including taxonomic dissimilarities and a zero-adjusted Bray-Curtis coefficient for denuded assemblages. *Journal of Experimental Marine Biology and Ecology*, 330: 55–80.

Clarke, K.R., Somerfield, P.J., Gorley, R.N. (2008). Testing of null hypotheses in exploratory community analyses: similarity profiles and biota-environment linkage. *Journal of Experimental Marine Biology and Ecology*, 366: 56–69.

Colwell, R.K., Coddington, J.A. (1994). Estimating terrestrial biodiversity through extrapolation. *Philosophical Transactions of the Royal Society B*, 345: 101–118.

Colwell, R.K., Mao, C.X., Chang, J. (2004). Interpolating, extrapolating, and comparing incidence-based species accumulation curves. *Ecology*, 85: 1717–1727.

Costello, M.J., Wilson S.P. (2011). Predicting the number of known and unknown species in European seas using rates of description. *Global Ecology and Biogeography*, 20: 319–330.

Dorazio, R.M., Royle, J.A., Söderström, B., Glimskär, M. (2006). Estimating species richness and accumulation by modeling species occurrence and detectability. *Ecology*, 87: 842–854.

Ferrier, S., Guisan, A. (2006). Spatial modelling of biodiversity at the community level. *Journal of Applied Ecology*, 43: 393–404.

Gaston, K.J. (2000). Global patterns in biodiversity. *Nature*, 405: 220–227.

Genin, A., Dayton, P.K., Lonsdale, P.F., Spiess, F.N. (1986). Corals on seamount peaks provide evidence of current acceleration over deep-sea topography. *Nature*, 322: 59–61.

Gotelli, N.J. (2001). Research frontiers in null model analysis. *Global Ecology and Biogeography*, 10 337–343.

Gotelli, N.J., Colwell, R.K. (2001). Quantifying biodiversity: procedures and pitfalls in the measurement and comparison of species richness. *Ecology Letters*, 4: 379–391.

Gotelli, N.J., Graves, G.R. (1996). *Null models in ecology*. Washington, DC: Smithsonian Institution Press.

Hillebrand, H., Blenckner, T. (2002). Regional and local impact on species diversity – from pattern to processes. *Oecologia*, 132: 479–491.

Hirzel, A.H., Hausser, J., Chessel, D., Perrin, N. (2002). Ecological-niche factor analysis: how to compute habitat-suitability maps without absence data? *Ecology*, 83: 2027–2036.

Koslow, J.A., Gowlett-Holmes, K., Lowry, J.K., O'Hara, T.D., Poore, G.C.B., Williams, A. (2001). The seamount benthic macrofauna off southern Tasmania: community structure and impacts of trawling. *Marine Ecology Progress Series*, 212: 111–125.

Krug, A.Z., Jablonski, D., Valentine, J.W., Roy, K. (2009). Generation of Earth's first-order biodiversity pattern. *Astrobiology*, 9: 113–124.

Lampitt, R.S. (1985). Evidence for the seasonal deposition of detritus to the deep-sea floor and its subsequent resuspension. *Deep Sea Research Part A*, 32: 885–897.

Lawton, J.H. (1999). Are there general laws in ecology? *Oikos*, 84: 177–192.

Legendre, P., Legendre, L. (1998). *Numerical ecology* (2nd edn). Amsterdam: Elsevier.

Lennon, J.J., Koleff, P., Greenwood, J.J.D., Gaston, K.J. (2001). The geographical structure of British bird distributions: diversity, spatial turnover and scale. *Journal of Animal Ecology*, 70: 966–979.

Longhurst, A.R. (1998). *Ecological biogeography of the sea*. San Diego, CA: Academic Press.

Lukacs, P.M., Burnham, K.P., Anderson, D.R. (2009). Model selection bias and Freedman's paradox. *Annals of the Institute of Statistical Mathematics*, 62: 117–125.

Lutz, M.J., Caldeira, K., Dunbar, R.B., Behrenfeld, M.J. (2007). Seasonal rhythms of net primary production and particulate organic carbon flux describe biological pump efficiency in the global ocean. *Journal of Geophysical Research*, 112, C10011, doi:101029/2006JC003706

McArdle B.H., Anderson M.J. (2001). Fitting multivariate models to community data: a comment on distance-based redundancy analysis. *Ecology*, 82: 290–297.

McClain, C.R., Hardy, S.M. (2010). The dynamics of biogeographic ranges in the deep sea. *Proceedings of the Royal Society B*, 277: 3533–3546.

McClain, C.R., Lundsten, L., Barry, J., DeVogelaere, A. (2010). Assemblage structure, but not diversity or density, change with depth on a northeast Pacific seamount. *Marine Ecology*, 31 (suppl. 1), 14–25.

Mora, C., Tittensor, D.P., Myers, R.A. (2008). The completeness of taxonomic inventories for describing the global diversity and distribution of marine fishes. *Proceedings of the Royal Society B*, 275: 149–150.

O'Hara, T.D. (2007). Seamounts: centres of endemism or species-richness for ophiuroids? *Global Ecology and Biogeography*, 16: 720–732.

O'Hara, T.D., Tittensor, D.P. (2010). Which environmental factors drive variation in species richness on seamounts? *Marine Ecology*, 31 (suppl. 1), 26–38.

O'Hara, T.D., Consalvey, M., Lavrado, H.P., Stocks, K.I. (2010). Environmental predictors and turnover of biota on a seamount chain', *Marine Ecology*, 31 (suppl. 1), 84–94.

O'Hara, T.D., Rowden, A.A., Bax, N.J. (2011). A southern hemisphere bathyal fauna is distributed in latitudinal bands. *Current Biology*, 21: 226–230.

Olsgard, F., Somerfield, P.J., Carr, M.R. (1997). Relationships between taxonomic resolution and data transformations in analyses of a macrobenthic community along an established pollution gradient. *Marine Ecology Progress Series*, 149: 173–181.

Phillips, S.J., Anderson, R.P., Schapire, R.E. (2006). Maximum entropy modelling of species geographical distributions. *Ecological Modelling*, 190: 231–259.

Phillips, S.J., Dudík, M., Elith, J., *et al.* (2009). Sample selection bias and presence-only distribution models: implications for background and pseudo-absence data. *Ecological Applications*, 19: 181–197.

Quinn, G.P., Keough, M.J. (2002). Experimental design and data analysis for biologists. Cambridge UK: Cambridge University Press.

Ramirez-Llodra, E., Brandt, A., Danovaro, R., *et al.* (2010). Deep, diverse and definitely different: unique attributes of the world's largest ecosystem. *Biogeosciences*, 7: 2851–2899.

Rex, M.A. (2002). Biogeography of the deep-sea gastropod *Palazzia planorbis* (Dall, 1972): An uncommon form of rarity. *The Nautilus*, 116: 36–38.

Rex, M.A., Etter, R.J. (2010). *Deep-sea biodiversity: pattern and scale*. Cambridge, MA: Harvard University Press.

Rosenzweig, M.L., Turner, W.R., Cox, J.G., Ricketts, T.H. (2003). Estimating diversity in unsampled habitats of a biogeographical province. *Conservation Biology*, 17: 864–874.

Rowden, A.A., Schnabel, K.E., Schlacher, T.A., Macpherson, E., Ahyong, S.T., Richer de Forges, B. (2010). Squat lobster assemblages on seamounts differ from some, but not all, deep-sea habitats of comparable depth. *Marine Ecology*, 31 (suppl. 1), 63–83.

Schlacher, T.A., Schlacher-Hoenlinger, M.A., Williams, A., Althaus, F., Hooper, J.N.A., Kloser, R. (2007). Richness and distribution of sponge megabenthos in continental margin canyons off South-Eastern Australia. *Marine Ecology Progress Series*, 340: 73–88.

Schlacher, T.A., Williams, A., Althaus F., Schlacher-Hoenlinger, M.A. (2010). High-resolution seabed imagery quantifies habitat attributes and megabenthos diversity in submarine canyons to underpin spatial conservation planning on continental margins. *Marine Ecology*, 31 (suppl. 1), 200–221.

Schlacher, T.A., Baco, A.R., Rowden, A.A., *et al.* (2014). Seamount benthos in a cobalt-rich crust region of the central Pacific: conservation challenges for future seabed mining. *Diversity and Distributions*, 20: 491–502.

Smith, C.R., Baco, A.R. (2003). Ecology of whale falls at the deep-sea floor. *Oceanography and Marine Biology: an Annual Review*, 41: 311–354.

Sokal, R.R., Rohlf, F.J. (1995). *Biometry: the principles and practice of statistics in biological research* (3rd edn). New York: WH Freeman.

Stephens, P.A., Buskirk, S.W., Martínez del Rio, C. (2006). Inference in ecology and evolution. *TRENDS in Ecology and Evolution*, 22: 192–197.

Storch, D., Marquet, P., Brown, J. (eds) (2007). *Scaling biodiversity*. Cambridge UK: Cambridge University Press.

Tittensor, D.P., Baco, A.R., Brewin, P.E., *et al.* (2009). Predicting global habitat suitability for stony corals on seamounts. *Journal of Biogeography*, 36: 1111–1128.

Tittensor, D.P., Mora, C., Jetz, W., *et al.* (2010). Global patterns and predictors of marine biodiversity across taxa. *Nature*, 466: 1098–1103.

Tittensor, D.P., Rex, M.A., Stuart, C.T., McClain C.R., Smith, C.R. (2011). Species-energy relationships in deep-sea molluscs. *Biology Letters*, 7: 718–722.

Uthicke, S., O'Hara, T.D., Byrne, M. (2004). Species composition and molecular phylogeny of the Indo-Pacific teatfish (Echinodermata: Holothuroidea) bêche-de-mer fishery'. *Marine and Freshwater Research*, 55: 837–848.

Vellend, M. (2010). Conceptual synthesis in community ecology. *The Quarterly Review of Biology*, 85: 183–206.

Vierod, A.D.T., Guinotte, J.M., Davies, A.J. (2014). Predicting the distribution of vulnerable marine ecosystems in the deep sea using presence-background models. *Deep-Sea Research II*, 99: 6–18.

Wharton, D.I., Shepherd, L.C. (2010). Poisson point process models solve the 'pseudo-absence problem' for presence-only data in ecology. *Annals of Applied Statistics*, 4: 1383–1402.

Williams, A., Schlacher, T.A., Rowden, A.A., *et al.* (2010). Seamount megabenthic assemblages fail to recover from trawling impacts. *Marine Ecology*, 31 (suppl. 1): 183–199.

Witman, J.D., Etter, R.J., Franz Smith (2004). The relationship between regional and local species diversity in marine benthic communities: A global perspective. *Proceedings of the National Academy of Sciences of the USA*, 101: 15664–15669.

Zar, J.H. (1999). *Biostatistical analysis* (4th edn). Upper Saddle River, NJ: Prentice Hall.

Zezina, O.N. (1997). Biogeography of the bathyal zone. *Advances in Marine Biology*, 32: 389–426.

Chapter 18
Application of Biological Studies to Governance and Management of the Deep Sea

Cindy Lee Van Dover[1], *Sophie Arnaud-Haond*[2], *Malcolm R. Clark*[3], *Samantha Smith*[4]*, *Andrew D. Thaler*[1] *and Sybille van den Hove*[5]

[1] Duke University, Beaufort, United States of America
[2] Institut Francais de Recherche pour l'Exploitation de la Mer, Brest, France
[3] National Institute of Water and Atmospheric Research, Wellington, New Zealand
[4] Nautilus Minerals, Queensland, Australia
[5] Autonomous University of Barcelona, Barcelona, Spaina

Abstract

The deep sea remains a frontier of biological exploration and characterization through mapping and sampling. Increasing access to and exploitation of deep-sea resources, combined with developing layers of governance and management, place us at a crossroads where dialogue and collaboration among stakeholders (scientists, industry, policy-makers) can lead to responsible and effective practices. We consider management approaches and outline a process for developing biological sampling programmes in the deep sea that delivers what a manager (or policy-maker) needs and a scientist desires. Using three case studies, we show (i) how scientists and industry can work together at a baseline project level and (ii) with managers at a national policy level, and (iii) we illustrate the interface process between scientists and policy-makers at an international level. We review key enabling functions of effective stakeholder interfaces and conclude with a reminder that biological studies underpin effective governance and management of the deep seas. Governance and management must operate on an evolving knowledge basis, embrace risk, uncertainty, indeterminacy, ambiguity and ignorance. There must also be cooperation, cross-fertilization, joint learning and sharing of best knowledge and practices across issues, areas, scales and sectors for effective management and policy.

Keywords Areas Beyond National Jurisdiction, deep-sea resource management, information and knowledge transfer, expert consultation, science–policy interfaces

18.1 Introduction

Until recent years, most biological studies in the deep sea have been curiosity-driven (e.g. Gage & Tyler, 1991; Sibuet & Olu, 1998; Van Dover, 2000; Koslow, 2007). Scientists continually challenge paradigms and dogmas about the deep-sea environment, with each successive decade yielding fundamental new insights about diversity, dynamics, services and fragility of deep-sea ecosystems.

*Current affiliation: Blue Globe Solutions, Toronto, Canada.

Biological Sampling in the Deep Sea, First Edition. Edited by Malcolm R. Clark, Mireille Consalvey and Ashley A. Rowden.

Not surprisingly, given the vast extent of seafloor still to be explored, those of us who study the deeper waters and underlying seabed are accustomed to discovery; we still can point to blank areas of the map where biological novelties are to be found with as much confidence as might have been expressed by explorers two centuries ago.

Not all deep-sea science, however, is curiosity-driven and exploratory. Where human activities in the deep sea are proposed, there are often (but regrettably not always) studies designed to assess the impact of those activities on the benthic ecosystem and to provide recommendations for responsible management policies. An early example of deep-sea scientific studies designed to inform environmental management resulted from multinational interest in exploitation of polymetallic (manganese) nodules from the Pacific seabed. Talk of mining the seabed led to the call by many nations for the seabed in Areas Beyond National Jurisdiction (ABNJ) (including the 'Area' (seabed and ocean floor and subsoil thereof) and High Seas) to be treated as a *common heritage of mankind*. This 'common heritage of mankind' concept was incorporated with regard to the Area and its mineral resources into the 1982 United Nations Convention on the Law of the Sea (UNCLOS), which came into force in 1994 and includes in its Part XI, what can be considered the first ever Underwater Mining Code. Interest in mining manganese nodules led to studies of the environmental impact assessment (e.g. Burns, 1980; Ozturgut *et al.*, 1981) and baseline characterization of the communities that would be impacted by mining activities, including organisms colonizing the nodules themselves (e.g. Mullineaux, 1987). Large-scale experiments and impact assessments were conducted in the 1970s, including the US-sponsored Deep-Ocean Mining Environmental Study (DOMES) (Bischoff & Piper, 1997), the German-sponsored Disturbance and Recolonization Experiment in a Manganese Nodule Area of the Deep South Pacific (DISCOL, Thiel *et al.*, 2001) and the multinational-sponsored (USA, Soviet Union, Japan, India, Interoceanmetal Joint Organization (IOM)) Benthic Impact Experiment (BIE) (Trueblood & Ozturgut, 1997).

Forty years later, interest in mining of manganese nodules continues and mining opportunities related to cobalt-rich ferromanganese crusts and polymetallic sulfides are explored. The International Seabed Authority (ISA), an autonomous international organization that administers seabed mineral resources in the Area, is engaged in assessment of environmental baseline studies related to nodule exploration activities in licensed areas in the equatorial Pacific (e.g. ISA, 1999), as well as polymetallic sulfides and cobalt-rich ferromanganese crusts (e.g. ISA, 2007). The work of the ISA includes, for example, consideration of environmental management plans. These plans may consider designation of management areas of environmental interest on the seabed within the manganese nodule province of the Clipperton–Clarion Fracture Zone (Anon, 2008; ISA, 2008, 2012; Fig. 18.1), a concept that provoked considerable discussion among stakeholders, and evolved over several years (ISA, 2011; Wedding *et al.*, 2013).

Mineral resources are not the only target for exploitation in the deep sea within or beyond areas of national jurisdiction. Oil and gas extraction and fisheries have been active deep-sea industries for decades (Davies *et al.*, 2007; Ramirez-Llodra *et al.*, 2011). Bio-prospecting and the exploitation of marine biodiversity in the deep sea are steadily increasing, with interest from health, cosmetics and other industries (Arrieta *et al.*, 2010; Thornburg *et al.*, 2010), and there is emergent interest in carbon capture and storage in deep waters (Davies *et al.*, 2007).

Perhaps because it is out of sight and too often out of mind, human activities in the deep sea often take place in advance of any attempt to determine the impact of the activities on seabed ecosystems. This was the case, for instance, in the ocean dumping of sewage sludge off the New Jersey coast in the late 1980s (Fig. 18.2). Only after dumping was initiated were deep-sea biologists engaged to determine the effect of sewage sludge on the seabed environment (EPA/NOAA Report to Congress, 1995). The destruction of fragile cold-water coral reefs and other benthic communities by deep-water fishing (bottom trawling) is another situation where assessment of impact has largely been *ex post* (see Roberts, 2002; Fössa *et al.*, 2002; Clark & Koslow, 2007), after damage to extensive areas of centuries-old growths of corals had already occurred (Fig. 18.3). In many existing deep-sea fishing grounds, scientific assessment of trawling impacts has been as much a requiem for biodiversity loss as it has been a tool for management. Recent actions to protect some hard-substratum features

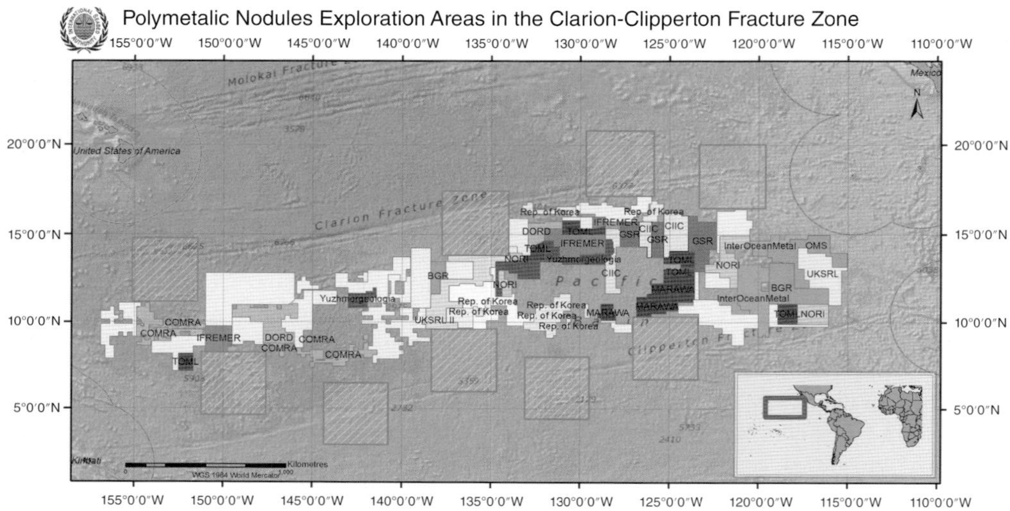

Fig. 18.1 Areas of Potential Environmental Interest in the manganese nodule province of the Clarion–Clipperton Fracture Zone. (Courtesy of the International Seabed Authority.) (Please see Plate 30 for colour representation of the figure.)

such as seamounts, knolls and pinnacles from trawling (UNGA, 2006) are expected to reverse this trend (e.g. CBD, 2009; Morato *et al.*, 2010).

In addition to these extractive seabed industries, there are many other human activities or pressures that may impact deep-sea wildlife and ecosystems, including telecommunication cables, pipelines, dumping, military activities, scientific research, shipping, pollution from land-based activities, atmospheric deposition of elements and contaminants, climate change and ocean acidification (e.g. Fujita *et al.*, 2004; Halfar & Fujita, 2007; van den Hove & Moreau, 2007; Ramirez-Llodra *et al.*, 2011). There is now a clarion call for deep-sea science to inform management, policy and governance decisions, and to this end we can use the improved molecular, geospatial and undersea mapping and sampling technologies that are described in this volume.

18.2 What is managed and who are the managers?

If we are to seek a balance between resource use (be it oil, gas, minerals, fish, marine genetic resources, etc.) and maintenance of a deep-ocean ecosystem that includes its full complement of biodiversity, then the sphere of management should include functional linkages of the ecosystem (where they occur) with the subsurface biosphere, the water column, the atmosphere and the coasts, as well as the full range of goods and services that the ecosystem provides (Armstrong *et al.*, 2010; Thurber *et al.*, 2014). This agenda captures the goals of ecosystem-based (rather than sectoral or species-based) management with marine spatial planning as an important (but not exclusive) tool (e.g. Garcia *et al.*, 2003). As Botsford *et al.* (1997) emphasize in their assessment of the management of fisheries, such an ecosystem-based approach requires integration of information from a wide range of disciplines, across different levels of ecological and socio-economic organization, and on a range of temporal and spatial scales. In the deep sea, as in other ecosystems, we must also manage for and under uncertainty, using current knowledge to prioritize and design scientific research programmes that most effectively reduce critical uncertainties where possible, in support of

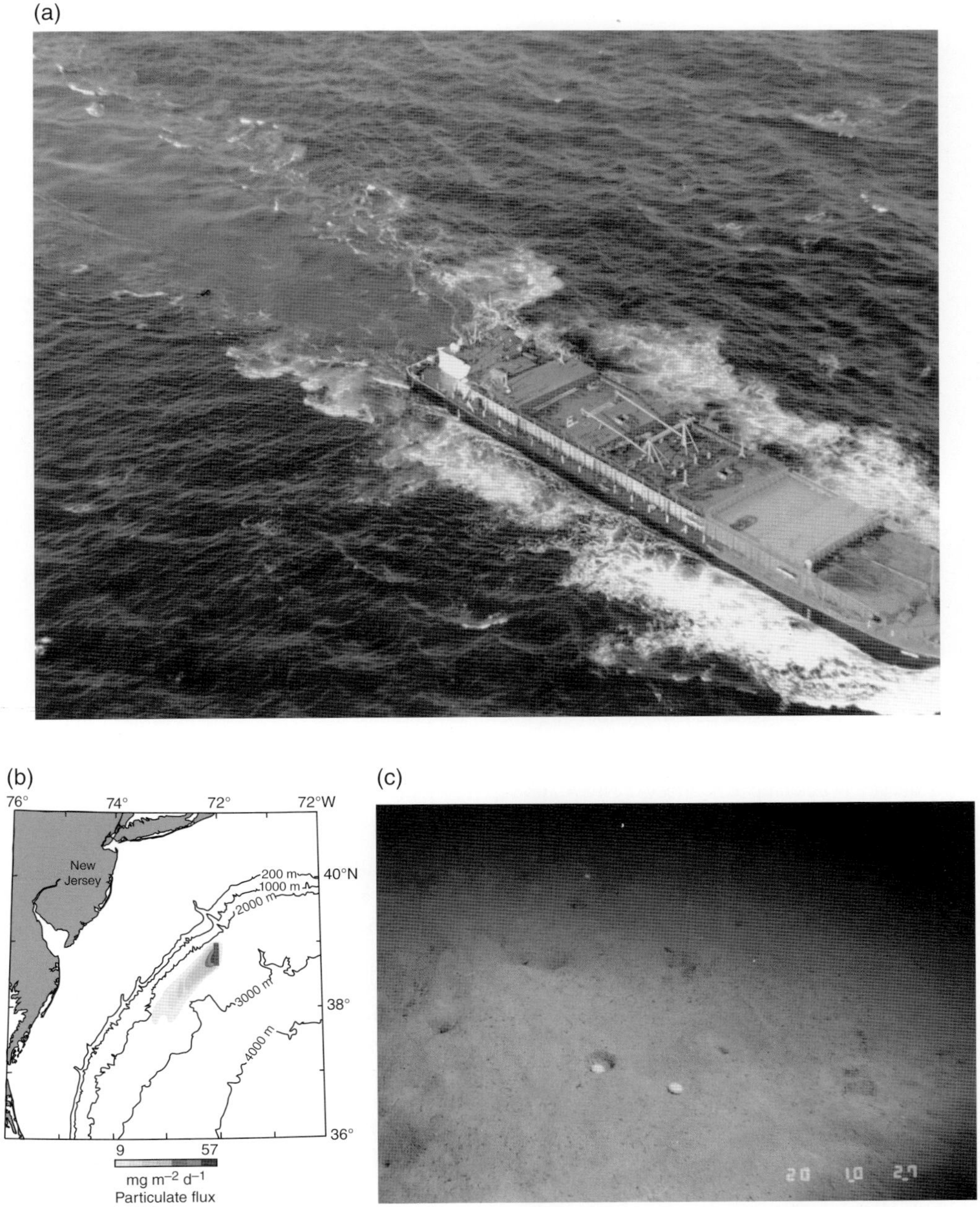

Fig. 18.2 Sewage sludge from metropolitan New York and New Jersey in the late 1980s was barged to *Deepwater Dumpsite 106* and dumped into the water column (a), where sewage-derived organic carbon subsequently dispersed (b) and sank to the seabed where it was incorporated into tissues of sea urchins (c). Environmental impact assessments were initiated only after dumping was sanctioned and had commenced. ((a) Courtesy of the US Environmental Protection Agency. (b) Adapted from Fry & Butman, 1991.) (Please see Plates 31 and 32 for colour representation of parts a and b.)

(a) (b)

Fig. 18.3 Impact of trawling activity by deep-water fisheries. (a) Reef-like formations of stony coral on a small, unfished seamount off New Zealand (depth 950 m). (b) Coral rubble and gouges in the seafloor from trawl doors on a heavily fished seamount off New Zealand (depth 900 m). The speed at which environmental impact can occur from human activities emphasizes the need to integrate science and management from the outset of the activity. (Photographs courtesy of NIWA.) (Please see Plate 33 for colour representation of the figure.)

more informed management decisions, and implementing adaptive management schemes where uncertainties remain. We are barely beyond a frontier, 'wild west' approach to the High Seas and Area; there remain critical knowledge gaps (see discussion below), yet nations and industries are eager to exploit the resources of ABNJ.

As certain terrestrial mineral and coastal fisheries resources have become increasingly exploited in the latter half of the twenieth century, minerals and fisheries exploration have expanded into deep waters (e.g. Clark *et al.*, 2007; Hoagland *et al.*, 2010). Initially there was no governance of these deep-sea activities; now legal instruments and management regimes for the deep ocean are becoming complex and multi-layered (Kimball, 2005; Gjerde, 2006). UNCLOS provides the international legal framework for management of the mineral resources of the Area, administered by the ISA and including protection of the environment from harmful effects that might arise from mining activities. Additional legal instruments include the 1992 Convention on Biological Diversity (CBD), the 1995 United Nations Fish Stocks Agreement, the 1993 Food and Agriculture Organization (FAO) Compliance Agreement and the 1979 Convention on Migratory Species (CMS); these are 'soft laws', with no mechanisms to prevent anyone from doing anything. Complementary to the legal framework of UNCLOS for ABNJ are regional seas conventions (e.g. the Oslo–Paris (OSPAR) Convention), regional fisheries management organizations (RFMOs) and species-specific conventions (e.g. Convention on International Trade in Endangered Species (CITES)).

There remain, however, critical gaps in the governance of ABNJ regarding protection of biodiversity in general and regulation of exploitation of marine genetic resources in particular (Arico & Salpin,

2005). The diversity of life forms in deep habitats is viewed as a virtually endless source of new species and molecules exhibiting useful properties with potential for pharmaceutical and industrial applications (van den Burg, 2003; Leary, 2009; Arrieta *et al.*, 2010; Thornburg *et al.*, 2010). Biochemical, physiological, morphological and behavioural adaptations are required in organisms living in extremes of high or low temperatures (hyperthermophiles in hydrothermal vents; psychrophiles in deep sediment cold seeps or polar regions), high pressure (piezophiles) and peculiar chemical conditions considered toxic for most. Lack of a clear governance authority regarding biological resources of the deep sea currently prevents agreement on measures for protection of and equitable access to these biological resources (Blaustein, 2010).

The United Nations General Assembly (UNGA) has established the Ad-Hoc Open-Ended Informal Working Group (BBNJ) to study issues relating to the conservation and sustainable use of marine biological diversity in ABNJ. In parallel, the CBD has prepared a scientific framework in which Ecologically or Biologically Significance Areas (EBSAs) can be identified (CBD 2009) and Marine Protected Area (MPA) networks could be designed with a view to their management by competent intergovernmental organizations. The UN Working Group is discussing the need for new governance structures and policy development to address management challenges in the deep sea (Arrieta *et al.*, 2010; Blaustein, 2010).

Within areas of *national* jurisdiction, governance and management systems may be complex, responding to historical, political, cultural and socio-economic characteristics of coastal nations. Most countries have some form of fisheries and biodiversity strategies and management systems in place, but each is typically tailored to the specific national situations. Often there is little consistency except where groups of countries have an overarching system (e.g. European Union). This is not necessarily a problem, except where there is limited coordination between agencies responsible for different sectors. In New Zealand, for example, fisheries management (including closed areas) is the role of the Ministry for Primary Industries (Formerly the Ministry of Fisheries), seabed mining (oil and gas, minerals) is under the Ministry for the Environment (formerly the Ministry for Economic Development), and environmental protection is largely the responsibility of the Department of Conservation inshore and the Ministry for the Environment offshore. Complex governance linkages can lead to confused conservation objectives, with areas closed to trawling but available for minerals exploration.

In addition to international and national policy-makers and managers, some industries employ environmental managers and require environmental impact assessments and establishment of environmental baselines before undertaking any activity that may have an impact on the environment. These industries are often proactive in setting environmental standards and protocols (e.g. International Marine Minerals Society Code of Conduct), and individual companies can also take a leading role establishing benchmark surveys and environmental impact assessments (e.g. Nautilus Minerals, 2008).

For High Seas fisheries, the FAO has been proactive in protecting and making fisheries sustainable (see Box 18.1 for some of the main organizational acronyms). Implementation of FAO International Guidelines for the Management of Deep-Sea Fisheries on the High Seas (FAO, 2009b) and FAO calls for ecosystem-based fisheries management are steps toward improved management of deep-sea fisheries resources (Rogers *et al.*, 2008). Fisheries management on the high seas is the responsibility of RFMOs of varying competencies and results. Most RFMOs or agencies focusing on species types (e.g. Tuna Management Commissions) have sustainable fisheries and environmental conservation as objectives, and the scientific requirements that enable managers to achieve a balance between these objectives are, or should be, no different from those for seabed mining. Establishment of MPAs that include deep-sea ecosystems has occurred, but so far these areas are driven at regional or national scales (e.g. Gjerde, 2007). For example, the 2010 Ministerial meeting of the OSPAR Convention in Bergen adopted a resolution to establish six deep-sea MPAs covering more than 36 000 km^2 in the Northeast Atlantic region (Decisions 2010/01-06). Such decisions,

Box 18.1 Acronyms of major international organizations and terms referred to in this chapter

- ABNJ Areas Beyond National Jurisdiction
- CBD Convention on Biological Diversity
- FAO Food and Agricultural Organization
- ISA International Seabed Authority
- OSPAR Oslo–Paris Convention
- VME Vulnerable Marine Ecosystem
- EBSA Ecologically or Biologically Significant Areas
- UNCLOS United Nations Convention on the Law of the Sea

however, require a higher level of review and endorsement by the UN General Assembly in order for those protected areas to be effective.

A more global approach toward spatial conservation practices is underway since the adoption in 2008 by the Conference of the Parties of the CBD of criteria for identifying EBSAs in need of protection, with guidance for design of representative networks of MPAs. At the scientific level, the EBSA effort is supported by a recently established international partnership called the Global Ocean Biodiversity Initiative (GOBI). The GOBI mission is to help countries and other organizations use and develop data, tools and methodologies to identify ecologically significant areas in the oceans, with an initial focus on ABNJ (www.gobi.org). Identification of a site as an EBSA affords no protection per se; it is still not clear what body, if already existing, may be competent to establish and manage MPAs or other spatial conservation practices in ABNJ.

18.3 The role of science

At all levels – international, national and private industry – governance and management need a strong scientific underpinning to address complex socio-ecological issues (e.g. Cash *et al.*, 2003). Biological sampling (or knowledge gained from biological sampling shared through expert panels) is a critical component of the knowledge needed to support assessments and ecosystem-based management.

The scientific community uses biological sampling and analysis to inform policy relating to management of deep-sea resources, biodiversity, and ecosystem structure and services. The expectation is that the scientific approach will follow effective practices such as those outlined in this volume, including work within the context of ecosystem-based studies wherever possible, with appropriate sampling design and analysis, and least impact on the natural environment. From a bottom-up perspective, a first layer of quality control on this process is imposed by the scientific community itself, through the process of peer-review of proposals and results. A further layer of self-governance and management may exist within scientific disciplines, such as that provided by the InterRidge code of conduct for scientific research at hydrothermal vents, which offers six guidelines for responsible research activity (see www.interridge.org/IRstatement) and its derivative, the OSPAR code of conduct for responsible marine research in the deep seas and High Seas of the OSPAR maritime area (OSPAR 08/24/1, Agreement 2008-01; see www.ospar.org; search under Programs and Measures).

From a top-down perspective, regulatory bodies that approve the scope and scale of any extraction, conservation or other activity in the deep sea represent the first layer of management. These bodies may have sectoral or holistic authority and approaches, with management infrastructures designed as appropriate for the authority. The process of soliciting scientific expertise used by these bodies can be sophisticated, with specialized analysis and reviews often undertaken in a consultative manner with outside specialists. The ISA for example, publishes regulatory guidelines that are drawn up with input from the scientific community. The environmental studies collected in response to these guidelines are reviewed by the Legal and Technical Commission (LTC), an organ of the Council of the ISA, whose members are elected based on personal qualifications relevant to scientific aspects as well as technological, economic and legal matters relating to the exploration, exploitation and processing of mineral resources. Similarly, FAO holds regular expert workshops and publishes reports and guidance on a wide range of fisheries management issues that are taken up by Regional Fisheries Management Organizations. Most countries also have some form of scientific working group structure that links with management and policy agencies. However, the nature of these is highly variable, and beyond the scope of discussion here.

Superimposed on the regulatory management are entities responsible for developing environmental baselines, assessing potential impacts of human activities, designing conservation approaches and other tasks that benefit from biological sampling. These responsibilities may be undertaken within university laboratories or through governmental or corporate environmental offices, which may in turn subcontract work for field and laboratory analyses that comply with best practices for environmental studies. This is where science and management can interact to great effect. Given a particular issue, academic or applied research agencies then undertake the appropriate research to address the scientific objectives in order to improve management. There are countless examples of this, but a highly relevant one in the context of the current discussion within this chapter is that of DISCOL. Faced with little knowledge about the long-term effects of manganese nodule mining, a multidisciplinary research programme was initiated by several agencies to examine the nature and extent of a dredging operation at abyssal depths. This was subsequently able to provide important information to other researchers as well as managers on potential long-term changes in benthic communities (see Thiel *et al.*, 2001). Further surveys in 2015 are extending the time series.

18.4 Management approaches and needs

An array of management approaches and concepts for marine systems has developed in recent years that trend toward a shift in focus from practices applied to single sectors (e.g. fishing, shipping, oil, gas) to practices integrated across sectors (e.g. ecosystem-based management, large ocean management areas, regional management areas). Spatial planning that engages the multiple users of marine systems in making coordinated and informed decisions about potentially conflicting uses of space and resources is often an important tool in regulating, managing and protecting the marine environment (e.g. Pomeroy & Douvere, 2008; Ehler, 2008). To be effective, spatial planning and integrated approaches must be informed by comprehensive understanding of where natural resources exist, of the spatial and dynamic characteristics of ecosystems associated with those resources and their exploitation (e.g. Ehler & Douvere, 2009). Quantitative information is also needed about where, how, and when human activities are taking place (Halpern *et al.*, 2008; Benn *et al.*, 2010) as well as quantitative information about potential or actual indirect anthropogenic impacts in the area. Acquisition of spatial and dynamic information is especially challenging in the deep sea. Adaptive management, which allows for iterative cycles of predictive modelling, monitoring, assessment and feedback (Fig. 18.4), is a tool that managers may use to integrate new knowledge into plans to improve future management (e.g. Walters & Hilborn, 1978). Adaptive management can be an especially valuable approach when knowledge is limited (especially in deep-sea environments). Adaptive management approaches, however, beg the questions of how scientific findings may be matched to managers' needs and how results may best be communicated. Practical advice for overcoming these gaps is addressed below.

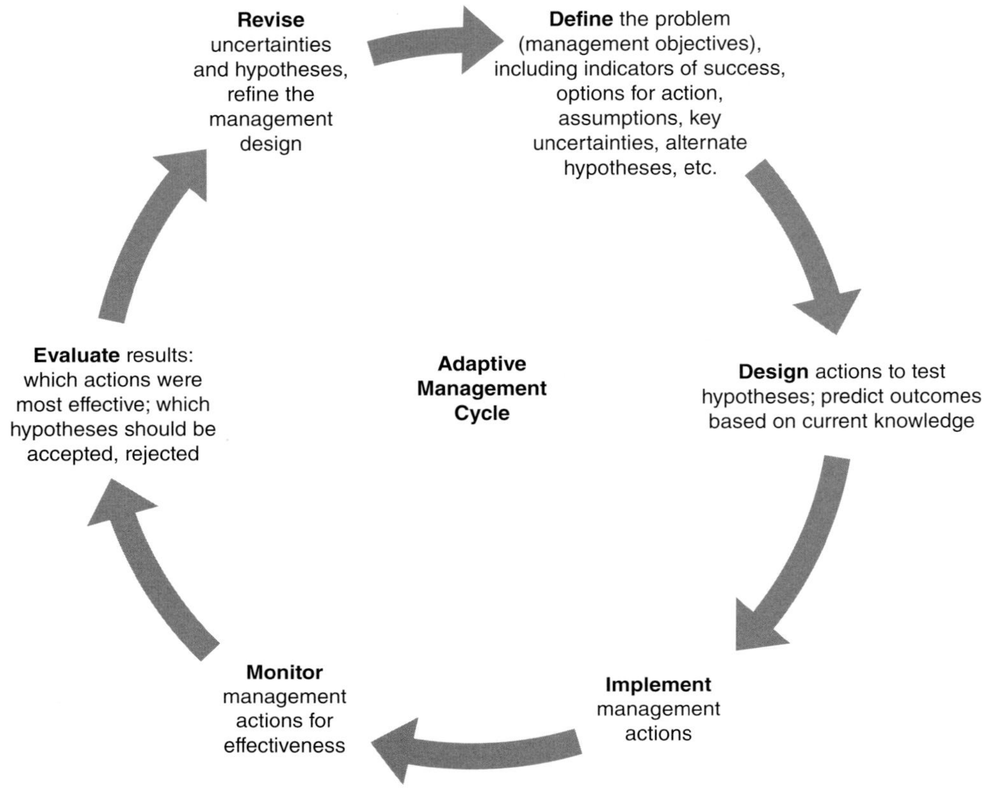

Fig. 18.4 The adaptive management cycle as a systematic approach for exploring means to meet management objectives based on the current state of knowledge. (Redrawn from van den Hove & Moreau, 2007).

18.4.1 Practical advice

It is well enough to discuss management needs and approaches in the abstract, but how does abstract become concrete under conditions where biological sampling and analysis – a costly enterprise under any circumstance, but even more so in the deep sea – is essential? There are multiple steps in the process of developing a biological sampling programme in the deep sea that delivers what a manager (or policy-maker) needs and a scientist desires.

18.4.1.1 Define the problem

Undertaking deep-sea mapping and sampling programmes can be very expensive in a technologically challenging field environment. It is therefore essential that attention be paid to carefully defining the problem and associated questions in order to deliver the necessary data to inform management and to maximize the likelihood of increasing our understanding of deep-sea ecosystems. Both the manager and the scientist see the questions as a means to achieve their goals.

18.4.1.2 Develop collaborative and cooperative approaches

Collaborative and cooperative approaches can bring together individuals, disciplines and sectors with the common interest of responsible management, logistical expertise and research infrastructure that can make best use of expensive facilities and limited human resources.

Expectations of all collaborators should be agreed upon prior to the start of a collaborative effort. Collaboration should involve as many parties as possible, and in most cases include industry relationships as well as science–management. This can help ensure better understanding of scientific results, improving uptake in a practical sense rather than having to drive compliance through regulation.

18.4.1.3 Design sampling and analytical methods

All biological sampling should have clear and achievable objectives that contribute to addressing the problem(s) and questions. The sampling design needs to take into account the priority of objectives, to ensure that primary tasks are completed well, rather than spreading survey resources too widely and not gathering enough samples or data to achieve an adequate or statistically robust result. Any field campaign should strive to achieve as many objectives as possible, without compromising quality (see Chapter 3). For example, there is often a conflict between biodiversity objectives that favour a widespread sampling plan with multiple gear types, and site-specific assessment objectives that involve concentrated sampling in fewer locations. A small shift in survey design may accomplish both objectives, but an *a priori* understanding of the implications of changing the type or numbers of samples to be collected should be evaluated in terms of analyses required for gaining a robust result.

18.4.1.4 Assess assumptions, costs, and limitations

Every sampling design makes assumptions about distributions of organisms, and about what analyses will reveal about community structure and biodiversity, ecosystem function, the appropriateness of the statistical design, etc. An assessment of underlying assumptions helps scientists and managers (and other end-users as well) engage in a transparent process of establishing priorities. Similarly, both scientists and managers need a shared understanding of costs associated with sampling and analysis and of the limitations that may arise when sampling under field conditions often encountered while working in the deep sea. Science as a process constantly adapts to new knowledge and improved methodologies, resulting in the possibility that outcomes of a project may advance from what was initially proposed. Weather, equipment failures, difficult access to remote locations and other unforeseen circumstances at sea or in the laboratory often result in setbacks. Any programme should have back-up plans, in case the optimal design cannot be achieved. A risk assessment process may be used to highlight potential (and most likely) failures during the planning process.

18.4.1.5 Implement sampling and analysis

Although primarily a scientific endeavour, sampling campaigns benefit from involvement of management and end-users through increased understanding of sampling issues and shared perspectives of the ecosystem under study (what organisms/habitats occur where, what they look like, etc.) that are difficult to obtain from numbers in a report. A partnership in the field strengthens the collaborative nature of the venture and allows for consideration of both scientific and management needs as the sampling programme and field campaign evolve. Once samples have been collected, the analytical process begins. Analysis can be a slow process, and in the academic world (e.g. universities) may involve training of students and exploration of new approaches to a problem to advance knowledge beyond the original scope of the project. These considerations should be understood *before* a science–management partnership is established. An important and often overlooked benefit of such partnerships is a better understanding of each other's world, which may contribute to the success of the overall programme, and continuity of interactions.

18.4.1.6 Report and recommend

Scientists typically report research results in peer-reviewed literature and thus are adept at gathering together background material, methods, results and discussions with appropriate tables, figures and illustrations. Peer-reviewed papers should be published as Open Access (free internet access) where possible to allow all stakeholders access to the primary literature. In addition to these peer-reviewed publications, a series of science–management interfacing products and activities should be planned at the start of the project and delivered as the project evolves and at its conclusion. These include reports, executive summaries, policy briefs, recommendations, management tools (e.g. publicly accessible (on-line) baseline data sets, photo-identification catalogues, maps, and other products) and participatory workshops that engage stakeholders and respond directly to managers' needs and allow for discussion of the research strategy and results.

18.4.1.7 Review

At the conclusion of a project, scientists, managers and other stakeholders should review project outcomes and identify best practices, options to address weaknesses and, where appropriate, next steps.

18.5 Case studies

In this section, we provide three case studies drawn from the authors' experiences to illustrate collaborative partnerships between science and industry, and science and management. The first example is a survey and sample-based collaboration between industry and academic partners that illustrates application of scientific investigations contributing to environmental baseline data relevant to deep-sea mineral extraction at hydrothermal vents. The second, survey-based example, represents an association between management and scientists and illustrates the role of sampling surveys in testing hypotheses about the efficacy of conservation measures related to fisheries and seamount biodiversity. The third example has no direct scientific sampling or survey effort; instead, it illustrates the application of expert knowledge gathered from prior biological sampling to formulate deep-sea fisheries policy. Key factors in the success of each of these case studies were effective communications, realistic expectations, timely delivery of resources and products, and mutual trust and respect.

18.5.1 Case study I: Connectivity among invertebrate populations at hydrothermal vents in Manus Basin, Papua New Guinea

18.5.1.1 Context

Seafloor massive sulfide (SMS) deposits associated with hydrothermal vents in the deep sea may contain high-grade ore, enriched in copper, gold, zinc and silver. There is emergent industry interest in deep seafloor mineral extraction of these massive sulfides (SPC 2013) and Nautilus Minerals (Nautilus), a Canadian company with a project office in Australia, is engaged in the commercial exploration of SMS deposits in Manus Basin, Papua New Guinea (PNG). The first candidate for mineral extraction is a site known as Solwara 1.

Potential environmental impacts from the Solwara 1 Project were identified with the help of a number of consultants with experience in marine science and marine environmental impacts. A workshop was convened that engaged Nautilus staff, consultants, marine scientists and other stakeholders including government and non-governmental organizations (NGOs) in a review of the impacts and assessment of studies required to describe the baseline environment. Nautilus then engaged consultants and academic institutions to conduct recommended studies for the Solwara 1 environmental impact assessment (EIA). The EIA culminated in the development of an environmental impact statement (EIS) that described the baseline environment, summarized anticipated impacts from the project, and recommended strategies to minimize and, where possible, avoid impacts to the environment.

In the Solwara 1 Project, impact mitigation strategies were also developed in consultation with international science experts. One of the proposed impact mitigation strategies involves setting aside a reference area (South Su, a site ~2 km from Solwara 1), with the intent that South Su could serve a source of recruits for colonization of vent habitats post mineral extraction and as a reference site, where variations in the natural environmental conditions could be observed away from the impact of direct mineral extraction. Biodiversity and connectivity questions relating to the suitability of South Su as a reference site and set-aside area were assessed through a collaborative effort between Nautilus and Duke University (USA) that investigated connectivity among invertebrate populations at hydrothermal vents in Manus Basin.

18.5.1.2 The problem and questions

The suitability of a reference area, or reserve, depends on both the reference area sharing the same species as the proposed impacted site and on the ability of the reference site populations to serve as a source of recruits to populate a site after the impacting activity has ceased. Little is known of how far offspring of vent species disperse although estimates for some species infer distances on the order of <100 km (e.g. Marsh *et al.*, 2001). No studies of dispersal potential existed for species occurring at Manus Basin vents, and planktonic larval durations and dispersal distances are challenging to measure in the deep sea. Genetic structure, however, can identify natural conservation units defined by significant differences in genetic diversity among populations. It was recommended that Nautilus use genetic characterization to answer the following questions important for developing their extraction strategy:

- What are the spatial scales at which different levels of genetic differentiation can be detected within Manus Basin?
- Is there evidence for directional gene flow in Manus Basin?
- Does the spatial scale of genetic structure differ among species in Manus Basin?

In addition, Nautilus wished to establish baseline genetic data for key species in the system; and both Nautilus and scientists wished to place Manus populations in a broader geographic context (i.e. genetic structure across the entire species ranges).

18.5.1.3 Collaborative and cooperative approaches

Together, Nautilus and scientists at Duke University developed a nested sampling strategy to assess genetic structure at scales ranging from a few metres to tens of kilometres within Manus Basin. Nautilus funded the field sampling programme, including remotely operated vehicle (ROV) and vessel charter and subsea sampling equipment, and provided seed money for genetic analyses. Nautilus and Duke University personnel staffed the field campaign. Duke University scientists contributed to the collaboration by providing molecular instrumentation and expertise to undertake the genetic analyses; science training and capacity building for PNG nationals were included in this effort. The Monterey Bay Aquarium Research Institute collaborated by providing conspecific samples from vents in Lau and Fiji Basin to allow a broader geographic context for the study of genetic structure. An important feature of the collaboration between Nautilus and Duke University was freedom for Duke University to publish results of the research, subject to a 30-day review period by Nautilus prior to submission. The intent of the review period was not to alter results or material published, but to ensure the company would be forewarned if an environmental issue had been identified and to confirm accuracy of vessel and ROV specifications.

18.5.1.4 Sampling design and analytical methods

To assess the suitability of South Su as a reserve site, Nautilus Minerals wished to test whether community structure and diversity at the proposed reserve site matched that of the proposed mine site. An intensive 2 × 2 × 3 × 3 × 3 nested sampling design was used: two habitat types (active and inactive vents) × 2 sites (Solwara 1, South Su) × 3 mounds per site × 3 assemblages per site (*Ifremeria, Alviniconcha, Eochionelasmus*) × 3 replicate samples per assemblage.

For genetic sampling, six species representing diverse dispersal strategies (from partial brooding to dispersal by larvae and swimming juveniles and adults) were identified as targets. Several of these target species are patchily distributed in the vent environment and occupy discrete thermal and chemical regimes; the patch (typically <2 m diameter) was the basic sampling unit in a 3 × 3 × 3 × 30 sampling design for Manus Basin (3 sites: Solwara 8 (45 km to the northwest of Solwara 1); Solwara 1; South Su) × 3 mounds per site × 3 patches per mound × 30 individuals per patch) for each species (Fig. 18.5). Mitochondrial CO1 (mtCO1) gene sequences (partial) were used to assess genetic diversity. In addition, five or more microsatellite markers were to be developed for key species and deployed on the same individuals for which CO1 data were obtained. These different genetic markers are genotypes that are used to determine if and where related genotypes co-occur. The objective was to determine the scale at which exchanges occur among populations and to identify pathways that might allow a 'rescue effect' from an outside source in case of local extinction. Where possible, Manus samples were supplemented by, and compared with, material collected outside of Manus Basin to assess connectivity at different spatial scales. A variety of analytical tools and software were selected to assess genetic diversity and population structure, directional gene flow and time since divergence of populations.

18.5.1.5 Assumptions, costs and limitations

The sampling design and spatial scales of the genetic sampling effort were unprecedented in the deep sea; the project was recognized as ambitious from the start. While there exists a scientific rationale for the choice of each target species, an underlying assumption was that some preliminary screening would allow the team to reprioritize targets for genetic studies. An earnest effort was made to estimate realistic analytical costs and timelines for deliverables, but these estimates proved optimistic, resulting in the need to reduce the analytical effort, extend the project timeline and raise additional funds through university opportunities for faculty and students, bursary support from Nautilus for traineeships, and third-party fellowship support for capacity building.

18.5.1.6 Sampling and analysis

The NorSky Campaign (July/August 2008) accomplished the nested sampling goals outlined above using a tethered ROV and resulted in 8000 tissues from more than 20 species for DNA analyses. A series of other studies was also initiated during this campaign to maximize deliverables, including sampling for food-web analyses, and behavioural and recruitment experiments. Genetic analyses began immediately upon return of samples. The results comprise partial requirements for a number of undergraduate and graduate student degrees and analysis is ongoing.

18.5.1.7 Report and recommend

The project broke new ground in several directions that can inform the management process. With respect to the suitability of South Su as a reference site, the results showed that biodiversity at South Su vents was equal to or greater than that at Solwara 1, and that vent communities at both sites shared dominant species (Collins *et al.*, 2012). From this perspective of community structure and diversity,

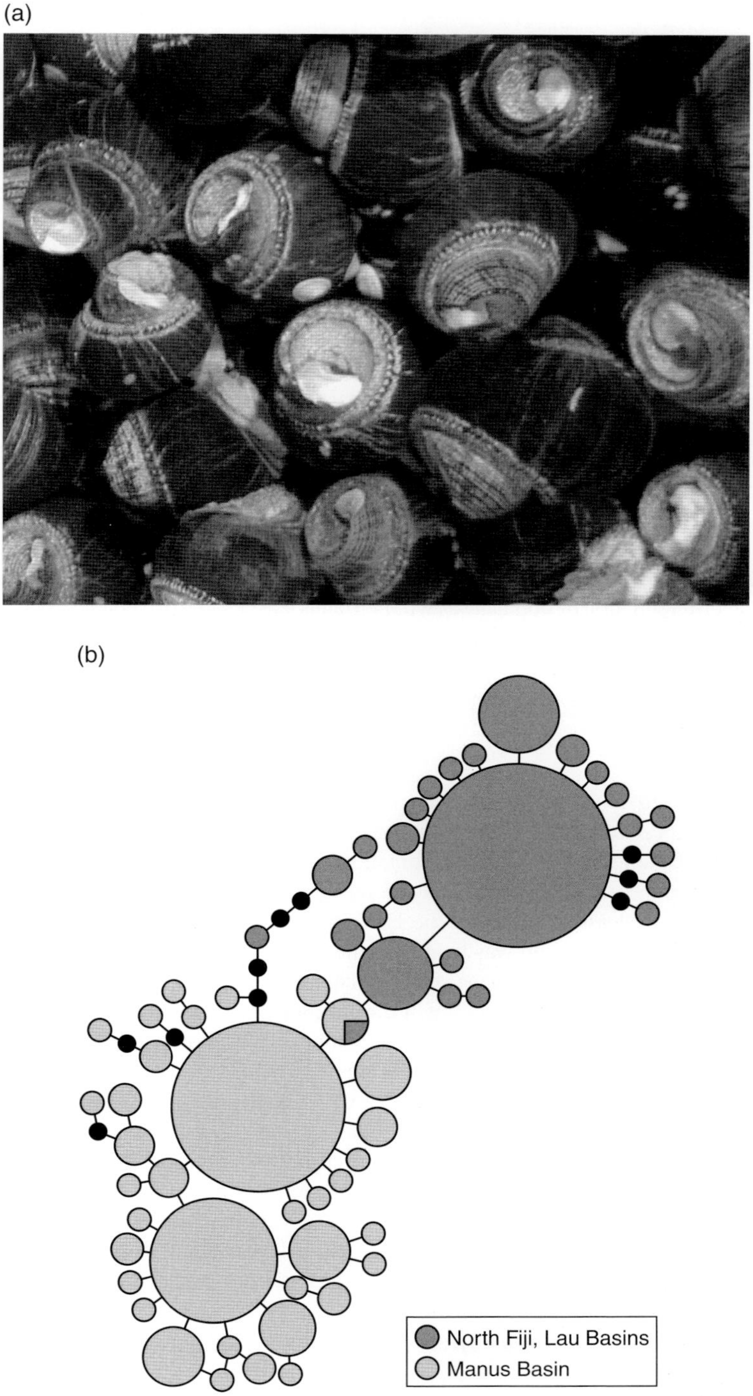

Fig. 18.5 Case study I: Connectivity of hydrothermal vent fauna across basins in the southwestern Pacific. (a) *Ifremeria nautilei*, a hydrothermal vent snail that host chemoautotrophic symbionts. (Image courtesy of Nautilus Minerals.) (b) Haplotype (phenotype) map based on cytochrome c oxidase gene sequences (404 base pairs). Relative sizes of circles represent haplotype frequency. Black circles represent 'missing' haplotypes. (Data from Thaler *et al.*, 2011.)

South Su was deemed a reasonable candidate reference site that may serve as a source of recruits for colonization of vent habitats post mineral extraction.

In addition, microsatellite development has been completed for three species (a provannid snail *Ifremeria nautilei* (Thaler *et al.*, 2010); an alvinocarid shrimp *Chorocaris* sp. (Zelnio *et al.*, 2010); a bathymodiolin mussel *Bathymodiolus manusensis* (Schultz *et al.*, 2010)). The microsatellite markers represent a significant investment of both funds and time; they are now readily available for post-impact assessment of genetic diversity or other applications. Microsatellite markers and mtCO1 sequences were used to determine the spatial scale of genetic differentiation in the brooding *Ifremeria nautilei* (Thaler *et al.*, 2011). This snail displays little genetic structure (populations appear to be panmictic) within the Manus Basin sites sampled (i.e. 40 km maximum separation) or between Lau and Fiji populations (~1000 km separation), but Manus Basin populations are distinct from those found in the Lau and Fiji Basins (minimum of ~2500 km separation; Fig. 18.5) based on pairwise comparisons of genetic markers. Coalescence-based techniques provide genetic evidence for directional gene flow from Lau to Fiji Basin and between vent sites within Manus Basin. Scientific recommendations to date include identification of best practices during the analytical phase (e.g. use of next-generation sequencing methods to develop microsatellite markers (Schultz *et al.*, 2010), establishment of laboratory protocols and standards for sequence editing and analyses, etc.), the need to add gene markers on a case-by-case basis, and the need to explore genetic structure in additional taxa and additional geographic areas to understand the full scope of spatial differentiation of populations within Manus Basin and to model the impact of mining on genetic diversity. From a management perspective, the data indicated that populations of *I. nautilei* were well mixed in Manus Basin and that a single extraction event at Solwara I would be unlikely to result in loss of genetic diversity on a basin scale.

18.5.1.8 Review

Results were reviewed with the programme manager through interim reports and shared with colleagues through peer-reviewed publications and presentations at international conferences, with goals that included establishment of effective practices and providing information useful for environmental assessment, mineral extraction design and mitigation strategies. The Nautilus–Duke collaboration was one of several associated with the EIA for the Solwara 1 Project. Once all of the EIA studies were complete, a workshop was held with the scientists and consultants involved as well as other experts in different fields of marine science. There were four objectives of this workshop: (i) review of results to develop an understanding of the baseline environment; (ii) identification of strategies to minimize impacts of seafloor resource extraction; (iii) development of a series of hypotheses related to system recovery; (iv) draft of a monitoring programme that would allow tests of proposed mitigation strategies and recovery hypotheses. The EIS synthesized the EIA studies (included as appendices) within the context of the environmental legislation of PNG and was reviewed by the contributing scientists. To facilitate transparency, the specialist studies were placed, unedited, on the Nautilus website (www.cares.nautilusminerals.com), together with the complete EIS (refer Nautilus Minerals, 2008). After the EIS was submitted to the PNG government, the Department of Environment and Conservation engaged an independent and international consultant to review the EIS. The PNG government approved the EIS in August 2009. Mining is scheduled to begin in 2017–18.

18.5.2 Case Study II: Protection of biodiversity of deep-sea seamounts around New Zealand

18.5.2.1 Context

Seamounts (including knolls and hills) are often sites of high biological diversity and can host high densities of fragile yet long-lived benthic fauna (e.g. cold-water corals). They are targeted in New Zealand waters and elsewhere by commercial fishers seeking species such as orange roughy (*Hoplostethus atlanticus*) that form aggregations in association with seamounts. The New Zealand government (at the time the Ministry of Fisheries (MFish)) has a legal responsibility to avoid adverse effects of fishing on the aquatic environment where possible. Hence fisheries managers were faced with how to protect vulnerable benthic fauna on seamounts from damaging impacts of bottom trawling. A draft Seamount Management Strategy prepared in 1999 had as its main objective the implementation of measures to manage the impacts of fishing on a representative sample of seamounts in New Zealand waters (Brodie & Clark, 2003). In 2001, 19 seamounts in the New Zealand EEZ were closed to all bottom fishing; this was followed in 2007 by designation of 'Benthic Protection Areas' (BPAs), which included additional seamounts (Helson *et al.*, 2010). These seamount closures included several located on the continental slope of the Chatham Rise (east of New Zealand) in a cluster of features termed the 'Graveyard seamounts', which became the subject of research to determine the efficacy of this management.

18.5.2.2 The problem and questions

Seamounts that were closed under the 2001 decision, as well as those subsequently protected within BPAs, were largely unfished. Little was known about their benthic biodiversity. Fisheries managers, therefore, wanted to know three key things:

(1) What was the composition of benthic invertebrate communities on the closed seamounts (i.e. exactly what was being protected)?
(2) How representative are these communities of wider seamount habitat (i.e. are all the seamounts the same, or does their biodiversity differ)?
(3) Is closing seamounts to fisheries an effective management approach (i.e. can closed seamounts recover from the impacts of trawling)?

The research programme to date has focused on question 1, which is described in greater detail below.

18.5.2.3 Collaborative and cooperative approaches

A discussion process was initiated between scientists involved in seamount research at the National Institute of Water and Atmospheric Research, New Zealand (NIWA), and the science, management, and policy staff at MFish. Discussions centred on identifying research required to assess the impacts of fishing through comparison of fished and unfished seamounts in the Graveyard cluster, as well as the likelihood of longer term monitoring of the effects of the seamount closures. It was decided that in addition to the closing of unfished seamounts, one previously fished seamount would be closed to evaluate recovery of benthic habitat after intensive trawling activity.

18.5.2.4 Sampling design and analytical methods

Unfished/lightly fished and fished seamounts of similar sizes and in close proximity to one another were used in a compare-and-contrast sampling design or natural experiment (Clark & Rowden, 2009). Replication was important for statistical tests, so four seamounts of each 'treatment' were included in the sampling design, and each seamount was to be sampled using multiple photographic transects and small epibenthic sled deployments. Such sampling would determine the distribution and abundance of benthic fauna across a range of body sizes, and data could be analysed to evaluate the difference in community structure between unfished and fished seamounts while accounting for the differences in communities within seamounts. The photographic surveys would also recover information on the seabed character which could be correlated with the observed patterns of community structure.

For question 2 on representativity, the sampling was designed as part of a larger study on seamount communities throughout New Zealand waters. This work is still in progress, although provisional classifications of both New Zealand, and global, seamounts have been undertaken, using several environmental proxies to group seamounts which are expected to have similar benthic communities (Rowden *et al.*, 2005; Clark *et al.*, 2011).

18.5.2.5 Assumptions, costs and limitations

The first survey in 2001 was largely exploratory, and was jointly funded by the Public Good Science Fund (PGSF) of New Zealand and MFish. Each of these agencies was interested in different management objectives (metrics relating to biodiversity (PGSF) vs. fish stocks (MFish)). The combined objectives and funding strategy allowed for more ship days, with minimal compromise in survey design because designs to achieve both the biodiversity and fisheries management objectives were similar. This co-funding situation has continued with subsequent surveys.

18.5.2.6 Sampling and analysis

Surveys were carried out in 2001, 2009 and 2015. The seamounts sampled during each survey differed slightly, with the 2001 survey comprising sled tows on eight seamounts, camera tows on four seamounts. The four seamounts that were surveyed using photographic gear have been included in each subsequent survey, with the other four seamounts added to the photographic survey in 2006), and two of the four seamounts in 2009 and 2015). Sled tows were carried out on all the seamounts in 2006, but not in 2009 or 2015, where only camera gear was deployed to avoid unnecessary impact from scientific tows where the fauna was already reasonably well described. Processing of photographic images (video and stills) is continuing and a complete analysis of data from all four years is underway.

18.5.2.7 Report and recommend

The results of the 2001 survey have been published in the scientific literature (Clark & Rowden, 2009), while the photographic data from 2006 has been published in a report to MFish (Clark *et al.*, 2010). Comparative analyses between 2001 and 2006 have been carried out, with some results presented in Clark *et al.* (2010), and more extensive analyses in Williams *et al.* (2010). The 2001 survey confirmed the existence on several closed seamounts of extensive areas of cold-water coral, which had not previously been documented on New Zealand seamounts. The coral species had been recorded, but it had not been appreciated that they could form reef-like structures over large areas of the summit and upper flanks of the small seamounts. The survey also indicated that >15% of the benthic invertebrates caught in the sleds were new species or new records for the New Zealand region. There was also very strong contrast between the coral cover on fished versus unfished seamounts, almost certainly due to damage by bottom trawling which was evident from gouges and grooves in the sediment caused by trawl door and bobbin rig gear. Variability in coral cover was considered to be responsible for the differences in macrofauna community structure observed between the two types of seamounts. Similar results were found during the 2006 survey, with high levels of intact stony coral cover on unfished seamounts compared to fished seamounts. Analysis of the 2009 and 2015 survey data is still ongoing.

Overall, the research confirmed the validity of closing the two unfished seamounts on account of their coral cover and associated invertebrate communities. In combination with similar surveys of seamounts in other parts of New Zealand, the Graveyard seamounts were identified as being different from those in other regions of the EEZ (e.g. Rowden *et al.*, 2003). Hence the value of their closures to protect representative cold-water coral communities was realized. Repeat surveys of the previously fished seamount showed that some small hydrocorals were either recolonizing the seamount, or were resistant to the impacts of trawling through their small size. There were no indications of recovery of the cold-water coral fauna between the 2001, 2006 and 2009 surveys. These results were presented at several seminars to fisheries scientists, managers, conservation groups and deepwater fishing industry representatives. Recommendations from the study included ongoing monitoring in order to determine the nature and rate of recovery of the benthic communities from fishing. The potential for recolonization of seamounts after human disturbance remains a key unknown in seamount ecology (Clark *et al.*, 2012).

18.5.2.8 Review

Initial discussions between fishery managers, scientists, NGO and fishing industry representatives failed to reach agreement on the necessity or validity of closing seamounts. The study showed that Graveyard closures protect rich deepwater coral reef-based communities on unfished seamounts, but also that recovery from fishing impact is slow (not evident after 5 years), and an established monitoring programme will enable long-term changes in faunal composition to be recorded. Data available at the time of recommending candidate seamounts for closure was poor, but subsequent research has confirmed assumptions underlying the original management decisions around the selection of representative seamounts. The closures are now accepted and the original 19 seamounts closed to bottom fishing in 1991 are included in the industry-driven BPA designations.

18.5.3 Case study III: Improving management of bottom trawling on the high seas

18.5.3.1 Context

Deep-sea fisheries have generally not proven sustainable in the long term (e.g. Clark *et al.*, 2007; Sissenwine & Mace, 2008; Clark, 2009; Norse *et al.*, 2012). In addition to placing fish stocks in jeopardy, deep-sea fishing operations disturb and damage benthic invertebrate communities (e.g. Clark & Koslow, 2007; Clark *et al.*, 2015). Thoughtful management strategies can reduce risks of overexploitation of fish stocks and minimize the impacts of trawling, but fishing is often poorly regulated in ABNJ. In recent years, there have been calls to protect vulnerable deep-sea habitats such as seamounts and canyons that host fragile and long-lived faunal communities, including cold-water corals. The UNGA passed resolutions in 2006 and 2009 calling for states and RFMOs to take specific time-limited measures to protect vulnerable marine ecosystems on the high seas from bottom fishing activities, and to ensure the sustainable use of deep-sea fish stocks. The UNGA has reviewed progress in implementing these resolutions in 2011.

18.5.3.2 The problem and questions

Unregulated fishing on the high seas can rapidly deplete fish stocks and damage benthic faunal communities. Fisheries managers must be able to develop and implement policies and actions that will reduce these risks, but there is typically a paucity of published data to inform and support the management process. Key questions were to determine:

(1) What type of fishing was occurring and where was it taking place?
(2) What are the impacts of the fishing operations on the habitats and biological communities?
(3) How can these impacts be managed?

The challenge was to gather expert knowledge, putting to use extant biological sampling and analysis rather than to gather new data.

18.5.3.3 Collaborative and cooperative approaches

The Committee on Fisheries (COFI) of the FAO met in 2005 to discuss management challenges of deep-sea fisheries. Following from that meeting, FAO initiated a consultative process among managers, policy analysts, maritime lawyers, conservation groups, industry representatives and scientists involved in deep-sea fisheries research and biodiversity to develop guidelines for deep-sea fisheries.

18.5.3.4 Sampling design and analytical methods

A series of four meetings and workshops were convened to collate scientific information and ideas toward formulation of guidelines for management of deep-sea fisheries:

- Expert Consultation on Deep-Sea Fisheries in the High Seas, November 2006
- Workshop on Vulnerable Ecosystems and Destructive Fishing in Deep-Sea Fisheries, June 2007
- Expert Consultation on International Guidelines for the Management of Deep-Sea Fisheries in the High Seas, September 2007
- Workshop on Knowledge and Data on Deep-Sea Fisheries in the High Seas, November 2007.

The two expert consultations were the main focus of this series, with the other two providing more specific input and review to the process. Each of the workshops were attended by invited experts who had either a wide range of experience, or had specific expertise in an aspect of the workshop. Numbers were restricted to ensure full engagement and active discussion. During the workshops, sub-groups were formed to focus discussion and formulate recommendations on a number of topics and issues, ranging from basic fisheries data requirements, through definition of terms such as 'significant adverse impact' and 'vulnerable marine ecosystem', to specific aspects of regulating fishing activity. The sub-groups reported back to a plenary session, where recommendations were further discussed and revised. The preparation of several background documents in advance of the first workshop was an important element to focus discussion during the meeting.

18.5.3.5 Assumptions, costs and limitations

The process depended upon getting a diverse and appropriate group of people together. Several countries contributed funds to assist the FAO in assembling individuals with expertise relevant to key topics, especially at the two expert consultation meetings.

18.5.3.6 Sampling and analysis

No surveys were carried out as part of this process. Several review papers were commissioned by FAO prior to the meetings, but the primary analyses were based on these reviews and the previous research and experience of the participants.

This differs from the research-driven case studies presented earlier. Nevertheless, workshop format followed the structure of defining the problem, discussing ways to address the problem and then providing potential solutions. The two expert consultations demonstrate this. The first (FAO, 2007) involved opening presentations about deepwater fishery resources and fisheries, fishery management examples, legal aspects of fisheries on the high seas, and protected area considerations. The focus of discussions at, and resolutions from, the workshop was on fisheries, with more general conclusions about habitat conservation. The second expert consultation (FAO, 2008a) emphasized the latter, with a sub-group considering the definition of vulnerable marine ecosystems (VME), and how to manage activities to avoid significant adverse impacts (SAI) – both VME and SAI terms have become central to international fisheries regulations.

18.5.3.7 Report and recommend

The two expert consultation meetings (FAO, 2007, 2008a) and complementary workshops (FAO, 2008b, 2008c) produced recommendations concerning the conduct of deep-sea trawling. These recommendations were circulated to all of the consultation meeting participants to ensure agreement among scientists, managers, conservation groups and deepwater fishing industry representatives. These were further discussed by FAO at a meeting with several fishing industry skippers and fleet managers in 2008 to assess their practicality to actual fishing operations. The final recommended guidelines (based on FAO, 2008a) covered a range of issues designed to give guidance on management factors, including an appropriate regulatory framework, components of a good data collection programme, identification of key management considerations, and measures necessary to ensure conservation of target and non-target species and benthic habitat. These guidelines were then circulated for high-level negotiation (see Review section). In addition, a supporting report was prepared that outlined the background and scientific rationale (Rogers *et al.*, 2008). The process was neither simple, nor straightforward, but the multiple meetings proved an effective way of integrating a number of different perspectives, and achieving agreement based on an understanding of all the issues and different viewpoints.

18.5.3.8 Review

Subsequently, two sessions of a high-level and more formal Technical Consultation on international guidelines for the management of deep-sea fisheries in the High Seas were held in 2008 (FAO, 2009a). These sessions involved national fishery managers, scientists, representatives from NGOs and fishing industry representatives. This broad representation was important to engage all parties and stakeholders with varying interests in deep-sea fishing and to improve the level of communication and understanding between different sectors. Voluntary guidelines were reviewed at these meetings and adopted in August 2008 (FAO, 2009b). The participatory nature of the process – involving fishery managers and the fishing industry – led to general agreement at the meetings that the guidelines represented compromise options and actions available to the industry and environmental managers that could lead to an effective balance between fishery exploitation and habitat conservation. The litmus test will be the level of compliance with these voluntary guidelines. At a UNGA review in 2011 there was concern that greater compliance was needed (Rogers & Gianni, 2011), but an overall consensus that progress was being made (UNGA, 2011). UN reviews are ongoing.

18.6 Biological studies at the science–policy interface

Deep-sea ecosystems and the interplay of human activities taking place and/or having an impact on deep-sea systems constitute complex socio-ecological systems (Young *et al.*, 2006). This complexity has implications both in terms of the knowledge needed to underpin the governance and management of these systems and in terms of the governance itself.

As exemplified by the case studies above, knowledge is needed about deep-sea biodiversity and ecosystems, and their abiotic environment, ecosystem functions, and the goods and services they provide, as well as information about human activities and potential or actual impacts and synergetic effects. It is clear that to satisfy these knowledge needs for deep-sea governance and management (Box 18.2) biological studies are required such as those that can establish baselines and trends using long-term data sets (e.g. Bailey *et al.*, 2009), as well as those that determine connectivity patterns and scale. Although biological studies are often at the heart of this knowledge, such research is by nature interdisciplinary, and broader multidisciplinary endeavours spanning geology and oceanography to socio-economics, law, and political sciences are necessary.

Box 18.2 Knowledge needs for deep-sea governance and management

- Biodiversity and ecosystems
- Ecosystem function, structure
- Role of biodiversity in ecosystem structure and function
- Status and trends of biodiversity and ecosystems
- What constitutes a 'significant adverse impact' of an activity
- What is the potential and time scale for recovery following an activity
- Parameters for a sustainable deep-sea fishery
- Natural drivers and evolution of biodiversity and ecosystems
- Geographical occurrence and abundance of genotypes, species and ecosystems
- Direct human interactions with ecosystems (anthropogenic pressures)
- Indirect human influence on biodiversity and ecosystems (anthropogenic drivers)
- How various direct and indirect impacts may interact and combine: synergetic effects
- Effects of impacts on the provision of ecosystem goods and services
- Scenarios of future socio-economic trends
- Scenarios of global environmental change (climate change, ocean acidification, etc.)
- Existing institutional framework and its potential for evolution
- Actors and power distribution
- Uncertainties and scientific disagreements
- Individual and social values and value conflicts
- *Ex ante* and *ex post* policy assessments

The complexity of the socio-ecological system at hand has three key consequences for governance and management. First, they need to operate on an evolving knowledge basis, subject to regular reviews and open to a redefinition of issues (and options) as knowledge and societal priorities develop. Second, the models and paradigms underpinning governance and management must embrace risk, uncertainty, indeterminacy, ambiguity and ignorance. And third, they need cooperation, cross-fertilization, joint learning and sharing of best practices across issues, areas, scales and sectors for the production and sharing of knowledge and for policy-making. Hence, as illustrated in this chapter, deep-sea governance and management need to build on concepts such as the ecosystem approach, ecosystem services, the precautionary principle, prior impact assessment and adaptive management.

An essential requirement for ecosystem governance and management is the existence of well-functioning stakeholder interfaces (van den Hove, 2007; van den Hove & Chabason, 2009). This is as much the case for the deep sea as for any other type of ecosystem. Such interfaces have multiple functions, from exchange and co-production of knowledge to raising awareness (Box 18.3). They may take multiple forms and include processes that operate at different levels from the local to the international, and operate at different stages of the policy or management process (early warning, issue identification, design, implementation, assessment, review). The case studies above illustrate some of this variety: the first two show how interfaces between scientists and industry/managers can take place at a project level or sectoral national policy level, whereas the third case depicts an interface process at the international and intergovernmental level.

In the first section of this chapter, we briefly touched upon the complicated governance framework for the deep sea in general and ABNJ and the high seas in particular. We have also presented some of the science–policy interface processes that take place in the framework of UNGA, CBD,

Box 18.3 Stakeholder interfaces: some key functions

- Allow for exchange and co-production of scientific, policy and management knowledge
- Contribute new thinking to address complex problems
- Facilitate timely translation of research into policy or management option or advice, and early use of results in practice
- Facilitate or produce integrated assessments, demand-driven targeted assessments or advice, environmental impact assessments and scientific-driven advice
- Alert decision-makers about emerging issues
- Contribute to the scientific quality process by allowing critical assessment of scientific outputs in light of user's needs and of other types of knowledge
- Ensure strategic orientation of research and appropriate funding of research in support of societal issues, policies, and management
- Raise public and stakeholder awareness
- Raise willingness to act and to support policy among the public and stakeholders

(Adapted from van den Hove, 2007 and van den Hove & Chabason, 2009.)

UNCLOS, FAO, ISA and other regulatory bodies. Today, there is an increasing recognition of the need to strengthen science–policy interfaces to support international oceans governance, as stressed in particular by the group of experts on a Regular Process for Global Reporting and Assessment of the State of the Marine Environment (UNEP/IOC-UNESCO, 2009). In the case of biodiversity, the need to strengthen the international science–policy interface is recognized and addressed by the setting up an Intergovernmental Science–Policy Platform on Biodiversity and Ecosystem Services (IPBES, www.ipbes.net), which was approved in December 2010 by the United Nations 65th General Assembly. At all management levels, from local to international, there is a similar move toward more effective interfaces between knowledge holders and decision-makers or managers.

There are many different needs for science–policy and science–stakeholders processes and many different ways of designing and implementing such processes. In addition, many of these interfaces are intertwined or embedded in one another. Because they are the holders of unique knowledge about our environment, scientists have a responsibility to contribute to and improve these interfaces. This is particularly true for those doing biological studies, as their knowledge is a cornerstone of the broad interdisciplinary body of knowledge required to address the issues at hand. As illustrated in this chapter, there are many ways and places in which interactions between deep-sea biologists and managers or decision-makers can take place, ranging from face-to-face meetings, participation in expert groups, circulation of studies, reports or targeted briefs, targeted meetings and workshops with relevant stakeholders, contribution to international environmental policy-making processes, and many more. It is up to each scientist to reflect on where and how they feel they want and could make a difference, and to get involved accordingly. This engagement may require some changes in the way we as scientists see ourselves and in the way we inform and train young scientists about the role of science as a support to solve societal challenges. We stress the need for collaboration among stakeholders, including scientists, decision-makers, managers, industry, NGOs, etc., from definition of the problem to the final stages of the research and recommendations that inform appropriate management actions.

Acknowledgements

The authors gratefully acknowledge the editorial guidance and advice from participants at the 2010 Brest Workshop on Biological Sampling in the Deep Sea, organized and funded through CenSeam. We are also appreciative of the very thoughtful comments of several anonymous reviewers. Contributions to this chapter by Sybille van den Hove and Sophie Arnaud-Haond benefited by funding from the European Community's Seventh Framework Programme (FP7/2007-2013) HERMIONE project, grant agreement no. 226354, and Malcolm Clark from 'The Ecology of Seamounts' and 'Vulnerable deep-sea communities' programmes in New Zealand (FRST C01X0204, C01X0906).

References

Anon (2008). First international deep-sea marine protected area proposed. *Oceanography*, 21: 10.

Arico, S., Salpin, C. (2005). *Bioprospecting of genetic resources in the deep seabed: scientific, legal and policy aspects*. Yokohama, Japan: United Nations University – Institute of Advanced Studies, 72 pp.

Armstrong, C., Foley, N., Tinch, R., van den Hove, S. (2010). Ecosystem goods and services of the Deep Sea. HERMIONE report. http://www.eu-hermione.net/policy

Arrieta, J., Arnaud-Haond, S., Duarte, C.M. (2010). What lies underneath? Conserving the ocean's genetic resources. *Proceedings of the National Academy of Sciences of the USA*, 107: 18318–18324.

Bailey, D.M., Collins, M.A., Gordon, J.D.M., Zuur, A.F., Priede, I.G. (2009). Long-term changes in deep-water fish populations in the northeast Atlantic: a deeper reaching effect of fisheries? *Proceedings of the Royal Society B*, 276: 1965–1969.

Benn, A.R., Weaver, P.P., Billet, D.S.M., *et al.* (2010). Human activities on the deep seafloor in the North East Atlantic: an assessment of spatial extent. *PLoS ONE*, 5(9): e12730. doi:10.1371/journal.pone.0012730

Bischoff, J.L., Piper, D.Z. (eds) (1997). *Marine geology and oceanography of the central Pacific manganese nodule province*. Plenium Press, New York, USA.

Blaustein, R.J. (2010). High-seas biodiversity and genetic resources: science and policy questions. *Bioscience*, 60: 408–413.

Botsford, L.W., Castilla, J.C., Peterson, C.H. (1997). The management of fisheries and marine ecosystems. *Science*, 277: 509–515.

Brodie, S., Clark, M.R. (2003). The New Zealand Seamount Management Strategy – steps towards conserving offshore marine habitat. In: Beumer, J.P., Grant, A., Smith, D.C. (eds), *Aquatic protected areas: what works best and how do we know?* Proceedings of the World Congress on Aquatic Protected Areas, Cairns, Australia, August 2002. Cairns: Australian Society of Fish Biology, pp. 664–673.

Burns, R.E. (1980). Assessment of environmental effects of deep-ocean mining of manganese nodules. *Helgoländer Meeresuntersuchungen*, 33: 433–442.

Cash, D., Clark, W.C., Alcock. F., *et al.* (2003). Knowledge systems for sustainable development. *Proceedings of the National Academy of Sciences of the USA*, 100: 8086–8091.

CBD (Convention on Biological Diversity) (2009). *Azores scientific criteria and guidance for identifying ecologically or biologically significant marine areas and designing representative networks of Marine Protected Areas in open ocean waters and deep sea habitats*. Secretariat of the Convention on Biological Diversity, Montreal, Canada, 12 pp.

Clark, M.R. (2009). Deep-sea seamount fisheries: a review of global status and future prospects. *Latin American Journal of Aquatic Research*, 37: 501–512.

Clark, M.R., Koslow, J.A. (2007). Impacts of fisheries on seamounts. In: Pitcher, T.J., Morato, T., Hart, P.J.B., Clark, M.R., Haggan, N., Santos, R.S. (eds), *Seamounts: ecology, conservation and management*. Fish and Aquatic Resources Series, Oxford: Blackwell Publishing, chapter 19, pp. 413–441.

Clark, M.R., Rowden, A.A. (2009). Effect of deepwater trawling on the macro-invertebrate assemblages of seamounts on the Chatham Rise, New Zealand. *Deep Sea Research I*, 56: 1540–1554.

Clark, M.R., Althaus, F., Schlacher, T., Williams, A., Bowden, D., Rowden, A.A. (2015). The impacts of deep-sea fisheries on benthic communities: a review. *ICES Journal of Marine Science*. doi: 10.1093/icesjms/fsv123

Clark, M.R., Vinnichenko, V.I., Gordon, J.D.M., Beck-Bulat, G.Z., Kukharev, N.N., Kakora, A.F. (2007). Large scale distant water trawl fisheries on seamounts. In: Pitcher, T.J., Morato, T., Hart, P.J.B., Clark, M.R., Haggan, N., Santos, R.S. (eds), *Seamounts: ecology, fisheries, and conservation*. Blackwell Fisheries and Aquatic Resources Series 12. Oxford: Blackwell Publishing, chapter 17, pp. 361–399.

Clark, M.R., Bowden, D.A., Baird, S.J., Stewart, R. (2010). Effects of fishing on the benthic biodiversity of seamounts of the 'Graveyard' complex, northern Chatham Rise. *New Zealand Aquatic Environment and Biodiversity Report*, No. 46. 40 pp.

Clark, M.R., Watling, L., Rowden, A.A., Guinotte, J.M., Smith, C.R. (2011). A global seamount classification to aid the scientific design of marine protected area networks. *Journal of Ocean and Coastal Management*, 54: 19–36.

Clark, M.R., Schlacher, T.A., Rowden, A.A., Stocks, K.I., Consalvey, M. (2012). Science priorities for seamounts: research links to conservation and management. *PLoS ONE*, 7(1): e29232. doi:10.1371/journal.pone.0029232

Collins, P.C., Kennedy, R., Van Dover, C.L. (2012). A biological survey method applied to seafloor massive sulphides (SMS) with contagiously distributed hydrothermal-vent fauna. *Marine Ecology Progress Series*, 452: 89–107.

Davies, A.J., Roberts, J.M., Hall-Spencer, J. (2007). Preserving deep-sea natural heritage: Emerging issues in offshore conservation and management. *Biological Conservation*, 138: 299–312.

Ehler, C. (2008). Conclusions: benefits, lessons learned, and future challenges of marine spatial planning. *Marine Policy*, 32: 840–843.

Ehler, C., Douvere, F. (2009). Marine Spatial Planning: a step-by-step approach toward ecosystem-based management. *Intergovernmental Oceanographic Commission Manual and Guides*, No. 53, 99 pp.

EPA/NOAA Report to Congress (1995). Monitoring, research, and surveillance of the 106-Mile Deepwater Municipal Dumpsite and environs. Office of Water, EPA/NOAA EPA-B42-R-95-001. http://www.epa.gov

FAO (Food and Agriculture Organization) (2007). Report and documentation of the Expert Consultation on Deep-sea Fisheries in the High Seas. FAO Fisheries Report No. 838, 203 pp.

FAO (Food and Agriculture Organization) (2008a). Report of the Expert Consultation on International Guidelines for the Management of Deep-sea Fisheries in the High Seas. FAO Fisheries Report No. 855, 39 pp.

FAO (Food and Agriculture Organization) (2008b). Report of the FAO Workshop on Vulnerable Ecosystems and Destructive Fishing in Deep-sea Fisheries. FAO Fisheries Report No. 829, 18 pp.

FAO (Food and Agriculture Organization) (2008c). Report of the Workshop on Data and Knowledge in Deep-sea Fisheries in the High Seas. FAO Fisheries Report No. 860, 15 pp.

FAO (Food and Agriculture Organization) (2009a). Report of the Technical Consultation on International Guidelines for the Management of Deep-sea Fisheries in the High Seas. FAO Fisheries and Aquaculture Report No. 881, 86 pp.

FAO (Food and Agriculture Organization) (2009b). *International guidelines for the management of deep-sea fisheries in the High Seas*. Rome: FAO, 73 pp.

Fössa, J.H., Mortensen, P.B., Furevik, D.M. (2002). The deepwater coral *Lophelia pertusa* in Norwegian waters: distribution and fishery impacts. *Hydrobiologia*, 471: 1–12.

Fry, V.S., Butman, B. (1991). Estimates of the sea floor area impacted by sewage sludge dumped at the 106-mile site in the mid-Atlantic Bight. *Marine Environmental Research*, 31: 145–160.

Fujita, R., Bonzon, K., Wilen, J., *et al.* (2004). Rationality or chaos? Global fisheries at a crossroads. In: Glover, L.K., Earle, S. (eds), *Defying ocean's end: an agenda for action*. Washington DC: Island Press, pp. 139–150.

Gage, J.D., Tyler, P.A. (1991). *Deep-sea biology: a natural history of organisms at the deep-sea floor*. Cambridge UK: Cambridge University Press, 504 pp.

Garcia, S.M., Zerbi, A., Aliaume, C., Do Chi, T., Lasserre, G. (2003). The ecosystem approach to fisheries. Issues, terminology, principles, institutional foundations, implementation and outlook. FAO Fisheries Technical Paper No. 443, 71 pp.

Gjerde, K.M. (2006). *Critical ocean issues: ecosystems and biodiversity in deep waters and High Seas*. Switzerland: UNEP/World Conservation Union (IUCN). http://www.iucn.org/dbtw-wpd/edocs/2006-007.pdf

Gjerde, K.M. (2007). High Seas marine protected areas and deep-sea fishing. FAO Fisheries Report No. 838: 141–180.

Halfar, J., Fujita, R. (2007). Danger of deep-sea mining. *Science*, 316: 987.

Halpern, B.S., Walbridge, S., Selkoe, K.A., *et al.* 2008. A global map of human impact on marine ecosystems. *Science*, 319: 948–952.

Helson, J., Leslie, S., Clement, G., Wells, R., Wood, R. (2010). Private rights, public benefits: Industry-driven seabed protection. *Marine Policy*, 34: 557–566.

Hoagland, P., Beaulieu, S., Tivey, M.A., *et al.* (2010). Deep-sea mining of seafloor massive sulfides. *Marine Policy*, 34: 728–732.

ISA (International Seabed Authority) (1999). *Deep-seabed polymetallic nodule exploration: development of environmental guidelines*, Kingston, Jamaica: International Seabed Authority, 289 pp.

ISA (International Seabed Authority) (2007). Polymetallic sulphides and cobalt-rich ferromanganese crusts deposits: establishment of environmental baselines and an associated monitoring programme during exploration. *Proceedings of the 6–10 September 2004 Workshop* held in Kingston, Jamaica, 481 pp.

ISA (International Seabed Authority) (2008). Rationale and recommendations for the establishment of preservation reference areas for nodule mining in the Clarion–Clipperton Zone. Summary outcomes of a Workshop to Design Marine Protected Areas for Seamounts and the Abyssal Nodule Province in Pacific High Seas, held at the University of Hawaii at Manoa, Hawaii, United States of America, from 23 to 26 October 2007. ISBA/14/LTC/2*.

ISA (International Seabed Authority) (2011). *Environmental management plan for the Clarion-Clipperton Zone*. ISBA/17/LTC/7.

ISA (International Seabed Authority) (2012). *ISA Decision of the Council relating to an environmental management plan for the Clarion-Clipperton Zone.* ISBA/18/C/22.

Kimball, L. (2005). The international legal regime of the High Seas and seabed beyond the limits of national jurisdiction and options for cooperation for the establishment of Marine Protected Areas in marine areas beyond the limits of national jurisdiction. CBD Technical Series No. 19. Secretariat of the Convention on Biological Diversity, Montreal, Canada.

Koslow, J.A. (2007). *The silent deep: the discovery, ecology and conservation of the deep sea.* Sydney, Australia: University of New South Wales Press, 312 pp.

Leary, D., Vierros, M., Hamon, G., Arico, S., Monagle, C. (2009). Marine genetic resources: a review of scientific and commercial interest. *Marine Policy*, 33: 183–194.

Marsh, A.G., Mullineaux, L.S., Young, C.M., Manahan, D.T. (2001). Larval dispersal potential of the tubeworm *Riftia pachyptila* at deep-sea hydrothermal vents. *Nature*, 411: 77–80.

Morato, T., Pitcher, T.J., Clark, M.R., *et al.* (2010). Can we protect seamounts for research? A call for conservation. *Oceanography*, 23(1): 190–199.

Mullineaux, L. (1987). Organisms encrusting manganese nodules and crusts: distribution and abundance at three North Pacific sites. *Deep-Sea Research*, 34: 165–184.

Nautilus Minerals (2008). Environmental impact statement: Solwara 1 project. 226 pp. http://www.nautilusminerals.com/irm/content/environment-reports2.aspx?RID=413

Norse, E.A., Brooke, S., Cheung, W.W.L., *et al.* (2011). Sustainability of deep-sea fisheries. *Marine Policy*, 36: 307–320.

Ozturgut, E., Lavelle, J.W., Burns, R.E. (1981). Impacts of manganese nodule mining on the environment: Results from pilot-scale mining tests in the north equatorial Pacific. In: Geyer, R.A. (ed.), Marine environmental pollution, 2, Dumping and mining. Amsterdam: Elsevier Oceanography Series, pp. 437–474.

Pomeroy, R., Douvere, F. (2008). The engagement of stakeholders in the marine spatial planning process. *Marine Policy*, 32: 816–822.

Ramirez-Llodra, E., Tyler, P.A., Baker, M.C., *et al.* (2011). Man and the last great wilderness: human impact on the deep sea. *PLoS ONE*, 6(7): e22588. doi:10.1371/journal.pone.0022588

Roberts, C. (2002). Deep impact: the rising toll of fishing in the deep sea. *Trends in Ecology and Evolution*, 17: 242–245.

Rogers, A.D., Gianni, M. (2010). *The implementation of UNGA resolutions 61/105 and 64/72 in the management of deep-sea fisheries on the High Seas*. Report prepared for the Deep-Sea Conservation Coalition. International Programme on the State of the Ocean, London, UK, 97 pp.

Rogers, A.D., Clark, M.R., Hall-Spencer, J.M., Gjerde, K.M. (2008). *The science behind the guidelines: a scientific guide to the* FAO *Draft International Guidelines (December 2007) for the Management of Deep-sea Fisheries in the High Seas and examples of how the guidelines may be practically implemented.* Switzerland: IUCN, 39 pp.

Rowden, A.A., Clark, M.R., O'Shea, S., McKnight, D.G. (2003). Benthic biodiversity of seamounts on the southern Kermadec volcanic arc. Marine Biodiversity Biosecurity Report No. 3.

Rowden, A.A., Clark, M.R., Wright, I.C. (2005). Physical characterisation and a biologically focused classification of 'seamounts' in the New Zealand region. *New Zealand Journal of Marine and Freshwater Research*, 39: 1039–1059.

Schultz, T.F., Hsing, P.Y., Eng, A., *et al.* (2010). Characterization of 18 polymorphic microsatellite loci from the deep-sea hydrothermal vent mussel *Bathymodiolus manusensis* (Mollusca, Bivalvia, Mytilidae). *Conservation Genetic Resources*, 3: 25–27. doi: 10.1007/s12686-010-9272-8

Sibuet, M., Olu, K, (1998). Biogeography, biodiversity and fluid dependence of deep-sea cold-seep communities at active and passive margins. *Deep-Sea Research II*, 45: 517–567.

Sissenwine, M.P., Mace, P.M. (2007). Can deep water fisheries be managed sustainably? *FAO Fisheries Report No.* 838: 61–111.

Smith, C.R., Gaines, S., Friedlander, A., *et al.* (2008). Preservation Reference Areas for nodule mining in the Clarion-Clipperton Zone: rationale and recommendations to the International Seabed Authority. Unpublished report, available from the ISA, Jamaica.

SPC (Secretariat of the Pacific Community) (2013). Deep Sea Minerals: Seafloor massive sulphides, a physical, biological, environmental, and technical review. In: Baker, E., Beaudoin, Y. (eds). Vol. 1A. Secretariat of the Pacific Community.

Thaler, A.D., Zelnio, K., Jones, R., Carlsson, J., Van Dover, C.L,, Schultz, T.F. (2010). Characterization of 12 polymorphic microsatellite loci in *Ifremeria nautilei*, a chemoautotrophic gastropod from deep-sea hydrothermal vents. *Conservation Genetics Resources*, 2: 101–103. doi 10.1007/s12686-010-9174-9

Thaler, A.D., Zelnio, K., Saleu, W., *et al.* (2011). The spatial scale of genetic subdivision in populations of *Ifremeria nautilei*, a hydrothermal-vent gastropod from the southwest Pacific. *BMC Evolutionary Biology*, 11: 372.

Thiel, H., Schriever, G., Ahnert, A., Bluhm, H., Borowski, C., Vopel, K. (2001). The large-scale environmental impact experiment DISCOL – reflection and foresight. *DeepSea Research II*, 48: 3869–3882.

Thornburg, C., Zabriskie, T.M., McPhail, K.L. (2010). Deep-sea hydrothermal vents: potential hot spots for natural products discovery? *Journal of Natural Products*, 73: 489–499.

Thurber, A.R., Sweetman, A.K., Narayanaswamy, B.E., Jones, D.O.B., Ingels, J., Hansman, R.L. (2014). Ecosystem function and services provided by the deep sea. *Biogeosciences*, 11: 3941–3963.

Trueblood, D.D., Ozturgut, E. (1997). The Benthic Impact Experiment: a study of the ecological impacts of deep seabed mining on abyssal benthic communities. *Proceedings of the Seventh (1997) International Offshore and Polar Engineering Conference, Honolulu*, USA, May 25–30, pp. 481–487.

UNGA (United Nations General Assembly) (2006). Sustainable fisheries, including through the 1995 Agreement for the Implementation of the Provisions of the United Nations Convention on the Law of the Sea of 10 December 1982 relating to the Conservation and Management of Straddling Fish Stocks and Highly Migratory Fish Stocks, and related instruments. General Assembly Resolution 61/105, Section 83.

UNGA (United Nations General Assembly) (2011). Sustainable fisheries, including through the 1995 Agreement for the Implementation of the Provisions of the United Nations Convention on the Law of the Sea of 10 December 1982 relating to the Conservation and Management of Straddling Fish Stocks and Highly Migratory Fish Stocks, and related instruments. General Assembly Resolution 66/68. Section X.

UNEP/IOC-UNESCO (2009). *An assessment of assessments: findings of the Group of Experts.* Start-up phase of a regular process for global reporting and assessment of the state of the marine environment including socio-economic aspects. UNEP.

van den Burg, B. (2003). Extremophiles as a source for novel enzymes. *Current Opinion in Microbiology*, 6: 213–218.

van den Hove, S. (2007). A rationale for science–policy interfaces. *Futures*, 39: 807–826.

van den Hove, S., Chabason, L. (2009). The debate on an Intergovernmental Science–Policy Platform on Biodiversity and Ecosystem Services (IPBES): exploring gaps and needs. Idées pour le débat No. 01/2009. Iddri, Paris. www.iddri.org.

van den Hove, S., Moreau, V. (2007). *Deep-sea biodiversity and ecosystems.* Cambridge, UK: UNEP World Conservation Monitoring Centre, 58 pp.

Van Dover, C.L. (2000). *The ecology of deep-sea hydrothermal vents.* Princeton, NJ: Princeton University Press, 424 pp.

Walters, C.J., Hilborn, R. (1978). Ecological optimization and adaptive management. *Annual Reviews in Ecological Systematics*, 9: 157–188.

Wedding, L.M., Friedlander, A.M., Kittinger, J.M., *et al.* (2013). From principles to practice: a spatial approach to systematic conservation planning in the deep sea. *Proceedings of the Royal Society of London, Series B*, 280: 1684.

Williams, A., Schlacher, T.A., Rowden, A.A., *et al.* (2010). Seamount megabenthic assemblages fail to recover from trawling impacts. *Marine Ecology*, 31 (suppl.): 183–199.

Young, O., Berkhout, F., Gallopin, G., Janssen, M., Ostrom, E., van der Leeuw, S. (2006). The globalization of socio-ecological systems: an agenda for scientific research. *Global Environmental Change*, 16: 304–316.

Zelnio, K.A., Thaler, A.D., Jones, R.E., *et al.* (2010). Characterization of nine polymorphic microsatellite loci in *Chorocaris* sp. (Crustacea, Caridea, Alvinocarididae) from deep-sea hydrothermal vents. *Conservation Genetics Resources*, 2(1): 223–226. doi:10.1007/s12686-010-9243-0

Chapter 19
The Future of Biological Sampling in the Deep Sea

Mireille Consalvey, Malcolm R. Clark and Ashley A. Rowden*
National Institute of Water Atmospheric Research, Wellington, New Zealand

Abstract

We do not know how the deep sea will be visited and sampled 50 years from now, but current pace and developments in data collection, management and analysis will undoubtedly improve our ability to investigate the deep sea. In imagining the future landscape of deep-sea research we might dream of exciting technological advances, but stark reality tells us that financial limitations will continue to impact on our research endeavours. Researchers need to deliver high-quality science in the most cost-effective way, and leave a legacy of archived data for future generations. We may see greater data collection from ships of opportunity, and certainly from autonomous vehicles, and seabed observatories. The need for baseline data remains a challenge, and visual data will become more important. Data collection and analyses need to be multidisciplinary and collaborative, using geospatial methods to identify spatial and temporal change. The need for trained taxonomists to identify deep-sea organisms will remain, although genetic techniques will greatly assist in this important task. Information exchange underpins the world in which we live, and an increasingly connected and more informed public will place even greater demands on scientists to anticipate the likely human impacts in the deep sea, and inform deep-sea management, walking a tightrope between conservation and exploitation.

Keywords biological sampling, future prospects, data collection, data management, data analysis

19.1 Introduction

It took four days for the news of Edmund Hillary and Tenzing Norgay's 1953 summiting of Mount Everest to break around the world. Fast forward to 2012, and James Cameron shared news of his successful descent to the Challenger Deep in the Marianas Trench, the deepest known place on earth, via Twitter (Fig. 19.1).

In a specially created submarine that literally turned conventional engineering on its head, James Cameron descended to depths only once before visited by Don Walsh and Jacques Piccard in 1960. Marketed as a new 'space race', Cameron hit bottom ahead of three other teams and although the race element was largely media driven, the romance of this human endeavour highlights that the era of deep-sea exploration is very much alive. It is impossible to foresee exactly how we will conduct deep-sea research in the future, however, the pages of this book are testament to the strong foundation from which technological breakthroughs can and will be made.

*Current affiliation: Victoria University of Wellington, Wellington, New Zealand.

Biological Sampling in the Deep Sea, First Edition. Edited by Malcolm R. Clark, Mireille Consalvey and Ashley A. Rowden.

Fig. 19.1 James Cameron's tweet announcing the successful Challenger Deep descent, March 2012.

Researchers have moved from conducting a census of deep-sea fauna to examining the structure and function of deep-sea communities, and in doing so have initiated cross-habitat and multidisciplinary research programmes such as those nested within the Census of Marine Life. The chapters of this book attest to the technologies that have been developed to push the envelope of what we can sample, and where. Yet in spite of advances in deep-sea research, our limited sampling still prevents broad-scale statements to be made about deep-sea ecosystems. Put simply, we need to do more deep-sea research with even greater integration of methods and disciplines. Many of the technological advances outlined in the volume, and to come, will afford us increased opportunity to conduct better designed multidisciplinary studies.

In this closing chapter we present a personal view of where the future of biological sampling in the deep sea might be heading, which can hopefully stimulate and encourage progress.

19.2 Data collection

For the future it is vital that we have a complete record of what data have been collected to date so that we can identify the geographical and bathymetric confines of our research, and hopefully direct sampling to help establish a baseline against which we might assess future change. In stepping up to this challenge we must be mindful of collecting, where possible, data which are comparable to historical data, but also moving on from the pitfalls of the past and the collection of non-robust data sets.

The financial reality of deep-sea sampling has undoubtedly limited the ability of many nations to extensively sample the deep sea, and thus impacted upon the distribution of the sampling effort throughout the world's oceans. The shallower parts of the deep sea are more intensively sampled than areas of greater depths, especially the huge expanse of the abyssal plains. In part this bias has been due to management needs requiring data from areas of the continental shelf and upper slope where direct human impact is greatest; and in part, technical limitations have constrained our abilities to sample the deepest parts of our oceans. Although single-wire towed gears can be used to considerable depths, detailed and precise sampling in the deep sea generally requires specialist equipment. The concept of sending humans to the deepest reaches of our oceans captures the imagination but the large costs put manned dives outside the realms of most nations, and instead remote or autonomous systems are more frequently deployed. Such sampling methods can often collect more data than traditional approaches for a fraction of the operational cost. Technological developments will continue to increase the opportunity for cheaper, smaller, remote or autonomous sampling systems which can embark on longer range, more accurate and deeper deployments.

The images now returned from the deep are at a resolution and quality once unimaginable, and it follows that what we see today will also one day be eclipsed. Such images will be the product of technological advances and increased data storage capacities. Consequently we anticipate that visual data will become an increasingly important and utilized source of information. Furthermore, the increasing miniaturization and durability of camera systems further opens deployment opportunities, for example, attaching such camera systems to nets, sleds and terrain-following autonomous underwater vehicles (AUVs). As scientists extend the depth and duration of deployments there will also need to be improvements in battery power capabilities, not only of cameras but of all electronic equipment, in order to cope with extended transit times to the seafloor.

The resolution at which we can sample has increased both spatially and temporally. We have moved from metre and sub-metre scales to microscopic and beyond (e.g. *in situ* genetic recognition of bacteria and phytoplankton). Advances in sensor technology are also enabling us to sample over increasingly short time periods to gain an increased understanding of biophysical and small-scale biological processes, and to conduct such sampling autonomously. The hope is that future developments will follow a similar trajectory with increasing sensory capabilities, as well as reduced costs.

A stark reality of deep-sea sampling is the inherent risk of deploying equipment into 'unknown' and rugged territory. Improvements in mapping the deep sea by multibeam technology will continue to increase our confidence with gear deployments, and help to minimize the risks of gear loss and damage. However, the fact remains that some areas of the deep sea will always be impossible to sample using more traditional gear types (e.g. sleds, grabs, corers), and researchers will need to rise to the challenge of modifying submersibles and remotely operated vehicles (ROVs) to conduct quantitative sampling in the most extreme of situations, such as canyons and seamounts with rough terrain and steep sides, overhangs, and fast currents.

To improve information used in management of the deep sea we need to not only understand spatial variation but also temporal changes. However, collecting time-series data remains a huge challenge. Along with technological developments in lander and AUV capabilities (e.g. improved acoustic communications, increasing miniaturization) and a decrease in cost we hope to see not only deployments across a greater number of sites but also serial deployments at the same sites. However, it remains that we are reliant upon the retrieval of undamaged equipment, with considerable time and finances resting on their performance in recording data for the duration of their deployment. Seabed observatories go a step further, and enable a continuous interactive presence in the deep sea where there is sufficient power and bandwidth to communicate with, operate and troubleshoot a broad range of instruments. As we look to the future we envision coordinated and interactive sampling of hundreds of real-time data streams to enable specific event sampling, and analysis of relationships among disparate data to test relationships and discover the drivers of biological, chemical, physical and geological processes across the oceans (Fig. 19.2).

When we consider how deep-sea data collection might evolve in the future our thoughts turn obviously to bold technical advances, but a stark reality is the financial challenge of deep-sea research. With research budgets ever tighter, researchers of all disciplines need to capitalize on so-called 'ships of opportunity' as well as facilitate equipment sharing initiatives. The advent of increasingly affordable and more portable research instruments will expand the range of vessels that might be deployed in deep-sea research such as smaller, less specialized, vessels. In December 1981 Bob Ballard published his dream in *National Geographic Magazine* to one day have telepresence technology for voyage participation; this is now a reality at the Inner Space Center (ISC) based in Rhode Island (http://innerspacecenter.org/). Many scientists will argue that you cannot replace the experience of being at sea for optimum data collection, but it is likely that such telepresence initiatives will increase. For beyond the initial outlay costs, such voyages are considerably more economical than expeditions staffed by a large number of highly experienced (and thus expensive) scientists.

19.3 Data management

As the rate of data collection increases, so does the volume and variety of data that needs to be handled. Digitization of data sets has enabled the development of increasingly sophisticated automated extractions, enabling researchers to keep pace with the rapidity with which huge volumes of data can now be collected. However, good data management practices are often overlooked, yet they are vital links in the research chain. Comprehensive metadata are essential, as they alert an increasingly global research community to the existence of the survey or study result. There are a number of

Fig. 19.2 Artist's impression of a future coordinated and integrated deep-sea sampling programme. (Erika Mackay, NIWA. Reproduced with permission.) (Please see Plate 34 for colour representation of the figure.)

metadata formats, and the hope is to standardize these as much as possible so the data can readily be located via the internet.

The volume of data is ever increasing, especially as remote-sensing techniques become more widely used, and satellites will become more sophisticated and able to take more frequent and precise measurements. Technological advances and greater global connectivity should enable access to more, and more varied data. Similarly the ongoing development of image systems involve much more and higher resolution data, with data storage issues that need to be addressed ahead of data collection. The amount of information that can be collected by seabed observatories further emphasizes the critical role of planning for large amounts of data collection. This situation of data volume often dictates that only a sub-set of the potential data is extracted for analysis in the first instance. However, researchers have a responsibility to not only store and document all data for their own use, but also for future users who might be unfamiliar with the nature of the data collection or have different research questions.

The goal of all deep-sea data management plans should be to preserve the data in a stable, recognized data repository. Storage in a national or international data repository ensures regular back-up and maintenance, and ensures that data persist long past the project's end date and the principal investigator's tenure.

There is still a long way to go before there is a consistent international approach to handling, processing and storing the wide array of biological data that is collected by national projects with their own objectives and own methods of doing things. However, the state of data management for molecular genetic data is relatively advanced, and offers guidance to other disciplines. There are widely accepted standards, a strong history of data sharing, and well-established data repositories for specific kinds of genetic data.

19.4 Data analysis

In the early days of deep-sea research to merely deploy and retrieve gear was considered success enough, and much of our early understanding was based on fortuitous and qualitative data collection. In recent years, great efforts have been made to analyse these older deep-sea data sets to maximize what has been accomplished to date. However, the reality is that a great deal of data, due to inadequate recording, are unusable for comparisons between different sample sets. A common problem with merging data sets is sampling gear comparability and although it remains a challenge to retrofit 'older' data, future research voyages will hopefully return more comparable data. One of the central aims of this book was to encourage international consistency in sampling approaches; this is vital if we are to understand the deep-sea ecosystem at a global scale.

Just as sampling techniques and capabilities should develop towards greater consistency we must also strive to conduct well-designed studies that recover data suitable for analysis using the most appropriate statistical techniques available. However, in reality even the best laid plans for sampling in the deep sea can fail and a continuing challenge is to meet the accompanying methodological and statistical problems in subsequent data analysis. International collaborative programmes, such as those established during the decade-long Census of Marine Life, have had to tackle these issues and have established protocols and methods to specifically address the challenges associated with the analysis of deep-sea data. We anticipate more widespread use of geospatial methods and models, greater accessibility of analytical tools (such as growth of freeware), and the development of more diverse and sophisticated techniques to further tackle the idiosyncrasies of deep-sea data sets.

The consistent identification of species among different data collections is another issue that needs to be addressed in order to maximize the rewards of merging data sets for global analyses. The general lack of trained taxonomists to describe and identify deep-sea species (the so-called 'taxonomic impediment') is hindering the achievement of consistency. It is vital that taxonomy networks, such as those developed under the Census of Marine Life and other international research programmes, be maintained and developed to continue consistent identifications at a global scale. Ongoing support and investment in databases such as the World Register of Marine Species (WoRMS; http://www.marinespecies.org) will strengthen this. Although traditional taxonomy will always be needed, 'new' methods such as barcoding and metagenomics (including recent developments in environmental DNA analyses) can help make the identification of species in a sample quicker, as well as ensure accuracy. Timely identification of species to the lowest possible taxonomic level is becoming more critical for managers needing advice for pressing resource use and conservation issues. Scientific understanding of the composition and function of communities needs to keep pace with the political demand for management of the deep sea.

With that in mind, a challenge for science in the future is to embrace the concept of the 'ecosystem approach'. A broad multidisciplinary and multiple sampling design is common in many deep-sea research programmes, but the science is typically still more descriptive (species composition, community structure) rather than analytical (community function). Improved data collection is one part of the story, but a clear process involving a suite of ecosystem-scale methods and models is needed to give results that can help managers evaluate different resource-conservation options that consider the biological complexity of the deep sea.

19.5 Future motivations for sampling

Economics has frequently been a driver and guide of deep-sea research, especially given the expensive nature of scientific exploration offshore. Just as deep-sea fishing in the 1970s produced samples, and a management need, on which our understanding of seamount communities is based, current interest in deep-sea petrochemical and mineral resources will likely motivate and support new research. However, it is a very different societal landscape in which we find ourselves today, and a more informed public places greater demands on scientists to anticipate the likely human impacts in the deep sea, and inform deep-sea management, rather than demonstrate the effects after the fact.

Human impacts in the deep sea are unequivocal and will likely increase with climate change, and as technological advances extend the depth range over which marine industries can operate. The potential impact of human activities in the deep sea was demonstrated very vividly in 2010 with the release of large volumes of oil in the Gulf of Mexico from the drilling rig *Deepwater Horizon*. This unfortunate incident demonstrated that despite the relatively large amount of sampling that had taken place in the Gulf over many decades, there was still much that we did not know about the area in the vicinity of the rig. The need to document and understand the effect of the accident subsequently facilitated a great deal of deep-sea research, including the extensive use of ROVs. Further funding, levied from oil companies and operators, has been directed to studies examining the impact of the oil spill to help better understand the biological responses and to inform our wider understanding of deep-sea ecosystems.

Many governments (and under increasing pressure from NGOs) are moving towards comprehensive and integrated regulation of the activities of deep-sea industries within Exclusive Economic Zones. Furthermore, the fishing, oil and mining industries themselves are also increasingly becoming proactive in efforts to improve environmental responsibility and their public perception, as well as instigating deep-sea research programmes not previously possible. Initiatives such as SERPENT (Scientific and Environmental ROV Partnership using Existent Industrial Technology) enable scientists to increase sample and data collections using commercial equipment when it is not required for the company's use – or dovetailing in with their operations. In the high seas the International Seabed Authority (ISA) is the international organization that administers mineral resources, and this agency also funds independent research. To be efficient and effective, scientists must in future engage and collaborate more and more with stakeholders, and encourage multidisciplinary sampling and research programmes to evaluate the latest threats posed to the deep-sea ecosystem by human activities.

The concept of "privately" funded deep-sea exploration entered the public conscious in 2011 with the National Geographic and Rolex-funded Deepsea Challenge expedition. The increasing prospect of 'private' exploration of the deep sea raises issues of the extent to which scientists should be included in their projects, and how their findings can be shared via traditional channels such as peer-reviewed publication. It is to be hoped that such expeditions will complement more traditionally funded research programmes, and make a very positive contribution to understanding the deep sea.

The people who masterminded the Census of Marine Life in 2000 could not have foretold of the success that was to follow. Funded by the Alfred P. Sloan Foundation, the ambitious 10 year programme aimed to galvanize national research into a worldwide effort. From the deep-sea perspective, the Census of Marine Life was successful not only in facilitating a globally focused deep-sea research effort, but also in uniting previously separate fields of research and serving as an example of how globally focused research might work. We hope that this book in some way can further facilitate rigorous scientific study to inform complex management and societal needs.

Acknowledgements

This chapter was written thanks to support from the Census of Marine Life field programme CenSeam (a global census of marine life on seamounts). The authors would like to thank all those present at several planning meetings for this book (Alan Jamieson, Lenaick Menot, Piere-Marie Sarradin, Stefano Schiaparelli, Cindy Lee Van Dover, Martin White) and the brainstorming session that formed the starting point for this chapter.

Glossary

Abyssal plain An extensive, flat region of the ocean bottom from 3500 to 6000 m comprising fine sediments or oozes.

ADCP (acoustic Doppler current profiler) An acoustic current meter that estimates water currents from the backscatter of an acoustic beam from passive particles in the water, such as plankton.

Annotation Within data management, the inclusion of additional information, such as keywords, labels, or comments, with data. For example, an image might include an annotation describing the content of the image, or its location.

Antipatharia Black corals. Corals with a hard axial skeleton (typically black or brown) that can form tall whip-like or tree-like colonies.

Aphotic zone Where light does not penetrate.

Asexual reproduction Reproduction without the production of gametes. Can be used to produce new individuals or for growth of a colony. Offspring are usually identical.

Assemblage A neutral substitute for 'community', but implying no necessary interrelationship.

AUV (autonomous underwater vehicle) A robot that is powered and can be programmed to undertake a certain path independent of the vessel, and can contain a range of sensors.

Bank A large, gentle elevated topographic feature.

Baroclinic The type of water motion that is driven by internal pressure gradients within the ocean caused by horizontal changes in density. Baroclinic flows have a vertical change in the velocity structure (i.e. a vertical current shear).

Bathymetry Measurement of the depth of the ocean floor; the ocean equivalent of topography.

Bathypelagic zone Depths below the euphotic zone, from 200 m to the abyssal zone at 3500 m.

Benthic Living on or pertaining to the seafloor.

Benthopelagic fish Living and feeding near the bottom up to 1000 m as well as in midwaters.

Benthos Organisms living on or in the seafloor.

Biodiversity The variety of living things in an area, measurable through habitat diversity, species diversity and genetic diversity.

Biogeography The scientific study of the geographic distribution of organisms.

Bioluminescence The production and emission of light by an organism.

Biomass The total weight of animals in an area at a particular time.

Bioregion A spatial representation of the boundary of areas represented by each level in a hierarchical classification of habitat types (provinces, biomes, habitats, etc.).

Biotope An area of uniform environmental (physical) conditions providing habitat for a specific assemblage of plants and animals.

Bottom longline A type of fishing gear comprising thin nylon lines with hooks attached, deployed close to the seafloor by weights.

Bridles Wires that connect the sweeps to a trawl net.

Biological Sampling in the Deep Sea, First Edition. Edited by Malcolm R. Clark, Mireille Consalvey and Ashley A. Rowden.

Broadcast spawning Shedding of eggs and sperm into the water column. Fertilization often takes place externally to the body.

Brooding Retaining developing eggs and larvae internally or externally to the body for a period of time prior to releasing them.

Bycatch The catch of non-target species in fishing.

Canyon A relatively narrow, deep, depression with steep sides, the bottom of which generally has a continuous slope.

Chemosynthesis The use of chemicals as an energy source rather than light to produce organic matter.

Controlled vocabulary A managed list of terms. Management typically includes careful selection of terms, maintenance of terms over time (i.e. addition, deprecation, modification), and presentation of the vocabulary in an accessible format. (Definition courtesy of MMI, marinemetadata.org)

Cnidaria Phylum of animals that include corals, sea anemones, jellyfish, box jellyfish, sea firs and siphonophores.

Cold seeps Areas where hydrogen sulfide, methane and hydrocarbons seep onto the seafloor, and can support chemosynthetic communities.

Cold-water coral reef Reef formed by corals lacking symbiotic algae, generally in deep, cold water. Found on the continental margins, in fjords and on seamounts.

Community An assemblage of organisms that occurs together at a location at the same time.

Continental margin Edge of the continental plate. A submerged elongation of continental landmass comprising continental shelf, slope and rise, but not the deep ocean floor.

Continental shelf All waters between the inter-tidal zone to the shelf break. Depths are generally down to 200 m, although markedly deeper in the Antarctic (600 m).

Continental slope The edge of the continental plate where the seabed descends from the shelf-break to the abyssal plains.

Corers Gear designed to take a sample of sediment containing infauna from the seafloor.

CTD An oceanographic instrument that measures conductivity, temperature and depth.

Cryptic species Animals that are difficult to observe through camouflage, subterranean lifestyle, transparency or mimicry.

Data archiving The long-term preservation of data. Data archives are designed to store data over years, decades, or even centuries. In contrast to data repositories, the focus is on data preservation instead of access.

Data repository A collection of data sets available through a centralized access method (such as an online search page). Data warehouse and data clearinghouse are closely related terms.

Data registry A centralized location where metadata about multiple data sets are available for searching, though the data sets themselves are not necessarily held. Also known as a data catalogue.

Deep scattering layer (DSL) Zooplankton and small swimming animals such as shrimp and fish that migrate towards the sea surface at night and return to deep water at daybreak. These animals are visible as reflective layers using acoustic equipment (echosounders).

Deep sea Areas of ocean beyond the edge of the continental shelf, often defined as deeper than 200 m.

Delimited text A format in which information is stored in a text file structured so that each row of text is a separate record, and the fields for each record are separated by a known character.

Demersal species An animal (typically a fish) that lives on or near the bottom, feeding on benthic organisms.

Deposit feeder Bottom-dwelling organism that feeds on particles of decomposed organic matter.

Detritivores Organisms that feed on dead or decaying organic matter.

Digital asset management (DAM) A system comprised of software and hardware that provides a means of managing and distributing media files over a network; can be purchased off-the-shelf, or, with the proper expertise, customized using individual components.

Diurnal migration The movement of animals through the water column on a daily basis.

Diversity The number of species found in a specific habitat or community (see biodiversity).

DNA barcoding Use of genetic information, specifically DNA nucleotide sequences, to help with the identification of species.

Dredge A fishing and research tool used to collect animals from the seafloor – usually a rigid box-like structure.

Ecosystem engineer A species that, by its presence, alters conditions so as to provide habitat for a community of other species.

Ecosystem A functional unit resulting from the interactions of abiotic and biotic components; a combination of interacting, interrelated parts that form a unitary whole.

Ekman dynamics Dynamics associated with the flow near surface or bottom boundaries whereby frictional forces break the geostrophic balance between pressure and Coriolis forces.

Endemic Native and restricted to a particular area (e.g. a single seamount or seamount chain).

Endemism The proportion of species in a community found only in a locality or geographic area.

Epibenthic living on the surface of the seabed.

Epifauna Organisms living on or just above a surface, such as the seabed. May be attached or mobile.

Filter feeder An organism that obtains food by straining seawater using a specialized feeding structure. Food may include small planktonic organisms and organic particles.

GenBank A database of publicly available DNA sequences provided by the National Center for Biotechnology Information. GenBank shares data with DDBL and EMBL through the International Nucleotide Sequence Database Collaboration. (www.ncbi.nlm.nih.gov/genbank).

GPS (Global Positioning System) A satellite-based navigation system often providing location data for research vessels.

Grab A type of sampling gear designed to collect sediment and infauna, using a scoop mechanism.

Groundgear The bottom-contact parts of a bottom trawl, comprising bobbins or rockhopper discs that travel on the seafloor and are connected to the bellow of the trawl.

Habitat The place where an organism lives and which provides it with requirements to live, grow and reproduce.

Habitat mapping The representation of habitats delineated by distinct combinations of physical, chemical, oceanographic and biological conditions.

Headline The top part of the trawl net, usually held open by floats.

High Seas/Areas beyond national jurisdiction Area of the ocean that lies outside Exclusive Economic Zones.

Hydrocoral Erect or encrusting corals with a calcareous skeleton.

Hydrodynamic modelling A model able to describe or represent the motion of a water mass.

Hydrography The measurement and description of water.

Infauna A diverse group of aquatic animals that live within marine and freshwater sediments.

Isopycnal A layer of constant density, or density surface in the ocean.

Lander An autonomous frame that is usually not tethered, and can be left on the seafloor for a period completely separate from the movements of the ship.

Layback The amount the headline extends back from the groundrope, helping to open the mouth of the net.

Life history The entire life cycle of a species including reproduction, larval development, growth and mortality.

Longevity Life span; oldest animal recorded for a species or stock.

Macrofauna Small benthic animals that live in or associated with sediments and are retained by 0.3 mm mesh sieves.

Megafauna Large animals visible to the naked eye and discernible on photographs or video imagery (generally large than 2 cm).

Meiofauna That part of the microfauna which inhabits superficial layers of the muddy sea bottom and passes through a 0.3 mm sieve.

Mesopelagic The pelagic zone below the euphotic zone, in the mid-depths of ocean waters, generally 200–1000 m.

Metadata Data about data. Metadata provides a context for research findings, ideally in a machine-readable format.

Metadata standard A set of documented rules which define the creation of metadata by providing a combination of terminology (vocabularies), syntactical rules, format rules and other requirements.

Mid-ocean ridges Volcanic mountain chains formed where tectonic plates are being formed and pulled apart.

Mixed layer The upper layer of the ocean that is well mixed and reasonably homogeneous in its temperature, salinity and nutrients.

Model calibration When a model is fit to available appropriate data (e.g. biomass or catch time series or maps). Also termed 'model tuning'.

Monte-Carlo simulation When a solution set is derived by repeatedly running a model, drawing parameters or stochastic parameters randomly from a specified distribution.

Multibeam echosounder an echosounder that transmits many beams and rapidly enables a three-dimensional picture of the seafloor to be determined over a wide area under the ship or towed instrument.

Natural mortality The component of total mortality caused by natural causes such as predation, diseases, senility, pollution, etc.

Nekton Aquatic animals that can swim against prevailing currents, such as fish, squid and whales.

Netsonde Term for an electronic sensor that is attached to the headline of the net, and beams down to detect fish swimming into the net, as well as depth.

OBIS (Ocean Biogeographic Information System) A resource for marine species occurrence data (www.iobis.org).

Oceanic front The density boundary between two water masses with differing temperature–salinity characteristics.

Oceanography The study of the ocean, embracing and integrating all knowledge pertaining to the ocean's physical boundaries, the chemistry and physics of sea water, and marine biology.

Octocorals Diverse group of corals. Name is derived from the fact that the polyps have eight tentacles. Include soft corals and gorgonians.

Oligotrophic waters Waters relatively low in nutrients leading to low levels of primary productivity and unable to support much plant life, such as the open oceans and some lakes.

Oxygen minimum zone Zone of low dissolved oxygen content, usually between 500 and 1000 m water depth.

Parameterization Entering the values for the model parameters (e.g. growth and mortality rates) so that the resulting model represents the system of interest.

Pelagic Referring to organisms that live in the water column and not in association with the seafloor or coastlines. Can include the larval stages of benthic species.

Phytoplankton Small plants that drift in the ocean.

Plankton Small, free-floating aquatic organisms that cannot swim against the prevailing currents.

Population A group of individuals of the same species living in a particular area or ecosystem.

Primary production The conversion of carbon dioxide to carbohydrates within an ecosystem by photosynthesis or chemosynthesis.

Pycnocline The region of rapid vertical change in water density. It is often associated with the thermocline. A seasonal pycnocline will develop associated with the summer heating regime and development of a seasonal thermocline.

Rarefaction curve A curve plotting the expected number of species for a given number of individuals or samples.

Relative vorticity The tendency for a parcel of fluid to rotate (in addition to the Coriolis forces).

Replication Refers to sampling multiple sites with similar characteristics.

ROV (remotely operated vehicle) A tethered underwater robot operated from a vessel.

Seafloor observatory An unmanned system of instruments deployed on or above the seafloor collecting data autonomously and serving them to stations ashore.

Seamount An underwater mountain. In ecological terms they have an elevation above the seabed of 100 m or greater. Geologically, they are divided into seamounts (1000 m or greater elevation), knolls (500–1000 m) and hills (100–500 m).

Secondary circulation Minor circulation patterns that are set up in response to the dominant flow pattern.

Sediment Small mineral particles such as sand, mud or carbonate ooze that cover most of the seafloor.

Sessile organism A species that, as an adult, lives attached to the seabed and is unable to move away from the point of attachment.

Shelf break Where the seabed shows an increase in gradient as it plunges towards in to the deep sea, marking the seaward edge of the continental shelf.

Spatial referencing Determination of the location of asmaple, whether a trawl, sled or photographic image.

Species distribution The geographical distribution of species usually dictated by latitude, depth, temperature or proximity to required habitats.

Species evenness A component of species diversity that measures how evenly individuals are distributed among species. An area dominated by many individuals of one or a few species and having few individuals of most other species would have low evenness.

Species habitat models Modelling techniques that use knowledge of the distribution of a species and associated environmental information to predict the distribution beyond areas sampled using only environmental data.

Species richness A count of the number of species in an area.

Stratification A means of subdividing an area for sampling, or for analysis, to reduce the variability in results.

Surface longline A fishing method using a line with hooks that is kept close to the surface with floats (as opposed to a bottom longline).

Stock A local animal population, generally independent of neighbouring populations.

Submersible An underwater free-moving vehicle for exploring the deep sea, piloted by humans onboard (also termed HOV or human operated vehicle).

Substrate The surface on which an animal or plant lives.

Sweeps Wires that connect the trawl doors to the bridles, and then to the net, helping to spread the sweeping distance of the net.

Taxon A grouping used by taxonomists based on evolutionary similarities.

Taxonomy The classification of life in to categories according to evolutionary relationships.

Taylor column/cone Features associated with the changes in vorticity that occur when a steady flow encounters an isolated obstacle. Under certain conditions, water will divert around the obstacle and an anti-cyclonic flow pattern may develop over the seamount.

Thermocline A steep gradient in temperature through the water column.

Towed camera frame A camera system contained within a frame that is towed behind a ship.

Trawl A net that is towed by a vessel across or through the water column.

Trawl doors Heavy doors that shear outwards with water and ground pressure, and spread the net.

Visual survey Sampling over an area using video or still cameras, deployed from ROVs, AUVs, HOVs or towed camera frames.

Warp Wire or wires that connect between the winches on the vessel, and the sampling gear that is being deployed (a trawl, sledge, dredge, CTD, etc).

Zooplankton Floating and drifting small animals that have little power of independent horizontal movement. Some spend only some time as plankton as they pass through a larval or reproductive stage (Meroplankton), and others spend their entire lives in the plankton (Holoplankton).

Index

Page numbers in **bold** refer to glossary

Biological Sampling in the Deep Sea, First Edition. Edited by Malcolm R. Clark, Mireille Consalvey and Ashley A. Rowden.